CHEMISTRY
in the Marketplace

5th Edition

Ben Selinger

Department of Chemistry
Faculty of Science
Australian National University, Canberra

First edition published in 1975 by Harcourt Brace
Fifth edition published in 1998
Reprinted by Allen & Unwin in 2002

Allen & Unwin
83 Alexander Street
Crows Nest NSW 2065
Australia
Phone: (61 2) 8425 0100
Fax: (61 2) 9906 2218
Email: info@allenandunwin.com
Web: www.allenandunwin.com

National Library of Australia
Cataloguing-in-Publication entry:

Selinger, Ben. 1939- .
Chemistry in the marketplace.

5th ed.
Includes index.

ISBN 978 1 86508 255 4

1. Chemistry. 2. Chemistry—Popular works. I. Title.

540

Editor: Ken Tate
Internal design: Tania Edwards
Cover: Mark Hand
Cover photograph: Norman Nicholls
Publishing editor and
Production manager: Penny Martin
Index: Keyword Editorial Services

Set in 10/12 Palatino by Peter Davix and Priti Chakraborty

The paper this book is printed on is certified by the © 1996 Forest Stewardship Council A.C. (FSC). SOS holds FSC chain of custody SGS-COC-3047. The FSC promotes environmentally responsible, socially beneficial and economically viable management of the world's forests.

Printed and bound in Australia by The SOS Print + Media Group.

10 9 8 7 6 5

Contents

Preface

'It was over three glasses of cold, artificially coloured, artificially foam stabilised, enzyme clarified, preserved, gassed amber fluid that Mal Rasmussen, Derry Scott and the author came to the realisation (in 1973) that we ought to be teaching consumers some *real* chemistry relevant to our lives, so that they could hope to make some sense of the arguments that rage in the media.' Thus spake the first edition.

The course ran under the auspices of the Australian National University Centre for Continuing Education with the alliterative title 'Chemical Consciousness for Concerned Consumers'. I need to re-emphasise the original acknowledgment to Mal and Derry. The amber fluid is now probably fermented with a genetically engineered organism.

Since 1973, science teaching internationally has caught up with the need for relevance and social setting. In the UK, in 1988, the national guidelines that came into effect for chemistry teaching took into account the approach used in *Chemistry in the marketplace*, an approach later adopted by the Salter project. Across the Atlantic in the same year, the American Chemical Society produced *ChemCom* (Chemistry in the Community) for schools, under the project management of Sylvia A. Ware. A tertiary version, 'Chemistry in Context: Applying Chemistry to Society', followed in 1994.

The first commercial edition of *Chemistry in the marketplace* appeared in 1975. It has been widely used as a reference text to which teachers and lecturers have added their own 'spin'. The book seems to have also fulfilled its major aim: to provide a reference book for the general reader. Unexpectedly, it has also found strong favour among innovators in small chemical-related enterprises. It seems that most people seeking information want to consult a chemical general practitioner first, rather than an academic specialist, because they find a huge communication gap with the latter.

The problem faced by the author in the first few editions was that of obtaining information 'off the beaten track'. In the intervening decades, both industry and government have become more open and user-friendly. There has been an explosion in popular publications, and increased access to the world wide web via the internet. In both cases it takes time and experience to assess the glut professionally, and separate the wheat from the chaff.

I have been surprised how little 'added value' chemical educators give to their teaching examples. Richard Feynman's dictum for physics: 'the same equations have the same solutions' is also true for chemistry. My favourite in this edition is all the ideas that arise from experiments that can be done with a metal pencil sharpener.

Publicly, I've often regretted the paucity of a beast called 'popular chemical critiques'. In the print and electronic media we have access to a surfeit of reviews by critics of films and theatre, food and wine, books and TV programs, economic policy and investment offerings, and the like. But public science tends to expose us either to the serious text and PR release, or the whiz bang and quirky, as well as the flip side from predictors of gloom and doom. This edition contains critiques of popular (and textbook) explanations of the greenhouse effect, the ozone layer, viability of alcohol as a motor fuel, chemical risk, and many others.

During a sabbatical in Europe in 1991 I toured with a lecture called 'Chemistry for Tourists'. It struck me that what I wanted as a tourist was for someone else to have done all the hard work, used their experience to sift through the totality, and extract what was really of interest. They had to provide a succinct readable guide that appealed to me, as well as to many other people of diverse interests and backgrounds. What the fifth edition has become is a 'Tourist Guide to Chemistry', somewhere between a Michelin and a Lonely Planet. I'm taking you through a foreign country pointing out and explaining landmarks I'm sure will interest you; giving the cultural setting, telling a few funny stories, helping you with the shopping, and explaining that the natives (the chemical industry) are reasonably friendly.

It's Chemistry on $49.95 (about every five years) — about ten bucks a year.

Chapter 1 is an introduction to the language of chemistry. A number of basic concepts are explained in appendixes and the glossary.

We start in Chapter 2 with an approach to toxicology, occupational health and safety, LD_{50}, ADI, MRL, epidemiology, and risk assessment. Earlier editions lacked a focus in this area.

Chapter 3 enters the house through the laundry.

Chapter 4 moves us to the kitchen, and starts us cooking and evaluating food technology. It ends with fat chemistry (in us and then our foods), the biochemistry of sport, cholesterol, chocolate, ice-cream, fake fats, and so on.

Chapter 5, in the boudoir, delves into fragrances and perfume, deodorants, sunscreens, and other cosmetics. This is where subtle chemistry mixes with less subtle marketing.

Chapter 6 is a new cross-linking chapter. It takes a major move into surface chemistry and provides ideas and concepts needed for a surprising number of the other topics dealt with elsewhere in the book.

With Chapter 7, we move into the garden and are confronted with soils and pesticides.

Chapter 8 is another cross-linking topic based on Q&A on swimming pool chemistry. However, this link is now with 'straight' topics in senior high school and junior tertiary curricula.

Chapters 9 and 10 take us into the hardware store, with a look first at the science of materials such as plastics and glass, and then metals.

Chapter 11 caters for fibres, fabrics and other yarns, including the important non-woven fabrics.

Chapter 12 considers plasters, paints, adhesives, enamels, glues, concrete, and fun materials.

Chapter 13 unlocks the medicine cabinet.

Chapter 14, the chemistry of energy, has grown again. This major chapter looks at thermodynamics in a user-friendly mode, takes us for a chemical ride in cars (fuel, oil, air bags), and looks at solar energy and new batteries. (Appendix VIII gives us the 'The entropy game' to provide both enjoyment and deeper insight.)

Chapter 15, in the dining room, looks at food, both chemically and culturally.

Chapter 16 introduces ionising radiation, and the controversy associated with food irradiation.

The appendixes have been expanded to include some of the more technical matter that was previously dispersed in the book.

Simple experiments and simple lecture demonstrations (for professional teachers) are placed where relevant throughout the book, and interesting titbits are placed in the second column. To assist the reader, icon ✲ denotes an interesting anecdote or titbit of information, ♉ denotes an experiment, and ◎ denotes an exercise.

Some unfamiliar terms and abbreviations are explained at first mention in the text, but an expanded glossary is intended to give further assistance to readers.

I hope you enjoy reading this book as much as I have enjoyed writing it.

WARNING

While all care has been taken to check the experiments and provide safe instructions, readers are warned that chemistry experiments can be dangerous, and should be carried out with safety always in mind. The author does not warrant that any instruction, recipe, or formula in the book is free of possible danger to the user.

See Plate XXVII, 'Doing a chemical experiment safely'

Note

The author welcomes constructive comment on this edition and can be contacted on selinger@grapevine.com.au

Acknowledgments

As an author I am conscious of the debt one owes to other writers and hope that I have dealt with this issue fairly. In over 25 years of activity in consumer chemistry, I have become a 'gatekeeper' for interesting snippets and extensions, not the least of which came from a regular radio talk-back show, 'Dial-a-Scientist', that I ran for several years.

This edition has been substantially rewritten and the database (with over two thousand entries) is based on much oral as well as written material. A full 'academic' format has therefore been abandoned; direct reference to specific material is given in the text or at the end of the chapter. Only a very short bibliography is provided. Earlier editions have more complete references to earlier work. As in the past, I welcome correspondence on errors of omission and commission, and updates where things have changed.

I hope all that have helped me will accept a global 'thank you'.

Chapter 1. May and Baker Ltd (Essex, UK) provided Figure 1.1, the periodic table of the elements (one of the few that gives each element its full name). For help with the section on nomenclature, and much else, I thank Dr Tom Bellas and Dr Malcolm Rasmussen. *Song of the elements* is reproduced with the permission of Tom Lehrer.

Chapter 2. Mr Michael Tracey (then Director, CSIRO Division of Food Processing) gave permission for the extract from his paper 'The price of making our foods safe and suitable'.

Chapter 3. Mr Peter Strasser is thanked, both as author and editor, for the use of his timeless published lecture on detergents, as well as for his continuing advice on updating this material, while New Science Publications allowed Ariadne's photograph of the royal statue attacked by detergent to appear (Fig. 3.11). The *Canberra Times* provided the photograph of the foaming Molonglo River (Fig. 3.15). The poem 'Foam' by A.P. Herbert, is reproduced by kind permission of A.P. Watt Ltd on behalf of Crystal Hale and Jocelyn Herbert.

Chapter 4. I am indebted to the CSIRO Division of Food Processing for permission to use material from *Food Research Quarterly* and a large section of its Information Sheet No. 17–1, *Nutritional value of processed food,* compiled by CSIRO editor, Gordon Walker. The quintessential mashed potato comes from ChemTech and Al Schwartz. Help with microwave cooking came from Victor Burgess (CSIRO Radiophysics) and permission to use the experiments on saliva and starch came from Dr David Hill, RACI.

Chapter 5. Particular thanks are due to Mr John Lambeth of Dragoco, and Michael Edwards for his help with perfumes. Use was made of several excellent articles from *New Scientist* on cosmetics, including 'A hair piece for Christmas', by J. Chorfas, *New Scientist* 19(26), December 1985; 'Only Skin Deep', by E. Geake, 25 December, 1993, 52–54, with permission. Table 5.2 came from E.M Kirschner, C&EN, 1 July, 1996, 13–19. While reviewing 'Antiperspirants and Deodorants', Volume 7 in a series 'Cosmetic Science and Technology', edited by K. Laden and C.B. Felger, Marcel Dekker, New York, 1988, I made use of its wisdom. The photograph of *in vivo* testing of sunscreens (Fig. 5.9) is by courtesy of Greiter, A.G. The two diagrams on the ultraviolet penetration of skin (Fig. 5.8) and eye (Fig. 5.10) were provided by Dr Keith Lokan, Director of the Australian Radiation Laboratory, Melbourne.

Chapter 6. Figure 6.2 is supplied courtesy of 'On Food And Cooking', H. McGee, Unwin Hyman, (their figure p. 357).

Chapter 7. The editor of *Chemistry in Australia* has allowed me to reuse material I published in that journal. Figures 7.1 and 7.2 are reproduced from *Soils: An outline of their properties and management* (CSIRO, Melbourne, 1977), with permission of the CSIRO publications department. Data on pesticide usage came from the National Registration Authority for Agricultural and Veterinary Chemicals and acting CEO Greg Hooper is thanked for checking material. The photograph of the use of chemical weapons in Iraq, Plate VII, came from John Dunn and *Chemistry in Australia.* The section 'The Perfume', in 'Six Songs for Chloë' by A.D. Hope is reproduced by courtesy of Harper Collins. The interview with Mrs Willison on the discovery in Australia of fruit fly pheromones was done by Tom Bellas. The photograph on p. 193 came from the Mitchell Library.

Chapter 9. The Plastics Institute of Australia is thanked for providing material on plastic films, and for material from the seminar, 'Plastics and Food Packaging in Perspective'.

Chapter 10. The information on the Solvay process came from Penrice Pty Ltd. Mr Werner Beek of Staedtler (Pacific) Pty Ltd supplied details of the composition of their most versatile magnesium pencil sharpeners.

Chapter 12. Mr Timothy Crick at Canberra CAE made me aware of the importance of the design of consumer products, and their chemical basis. The American Chemical Society granted permission to reprint Figure 12.6 on the chalking of paints. Selleys provided the information for Table 12.2.

Chapter 13. I am grateful to the editor of Chemistry in Britain (Royal Society of Chemistry, London) for permission to use its material on the chemistry of drugs. The late Professor Adrien Albert provided much help and support on the topic of drugs. Professor Bryan Reuben gave me access to his data on international drug usage and allowed me to quote some of it, particularly for Tables 13.3 (USA 1992), 13.4 (worldwide) and 13.5 (Germany), from 'The Consumption and Production of pharmaceuticals', Bryan Reuben, Chapter 42, pp. 904–938, in 'The Practice of Medicinal Medicine', ed. G.C. Wermuth, Academic Press, 1996. Mt Peter McManus of the Drug Utilisation Sub-committee is thanked for the data on total prescriptions (not just the Pharmaceutical Benefits Scheme) through community pharmacists. Dr Sam Wong (Commonwealth Department of Community Services and Health) helped update aspects of the pharmaceutical sections. John P. Gregan at the NHMRC, Canberra, allowed me to use his history of scheduling. Mr Vashin Demurian of the National Pharmacy Guild gave comments on the section on pharmacies. Mr Brian Richings, Australian Bureau of Statistics, supplied the data on drug usage in Australia.

Chapter 14. The Australian Consumers' Association allowed reproduction of material from its magazine *Choice.* CSIRO encouraged the use of material from its publications, such as *Rural Research* (courtesy B.J. Woodruff), for the section 'Energy down on the farm'. Richard Manuell, Industrial Hygienist at Esso Australia gave comments on engine oil and refinery fractions. Figure 14.10 showing a comparison of the capacity and power of storage batteries with those of a cyclist is included by courtesy of MIT Press. Prof. A. Götzberger, Fraunhofer Institute for Solar Energy Systems, Freiburg, Germany provided details of the solar house. Steve Downing of the RSC at ANU enthusiastically updated all things to do with batteries.

Chapter 15. Mr Bill McCray (Director of Biochemistry, Animal Research Institute, Queensland) allowed me to use a section of his talk on 'The myth of pure natural foods'. Emeritus Professor F. H. Reuter is thanked for allowing the generous use of many articles from the journal *Food Technology in Australia.* Dr Gordon Burch (ANZFA) updated the Food Additives list. Permission to reproduce the following figures is also acknowledged: Fig. 15.1 , United Features Syndicate Inc.; Fig. 15.12, *New Scientist* and Monsanto; Fig. 15.13, *New Scientist*; Figs. 15.18, 15.19 and 15.20, Bread Research Institute of Australia. John Casey is thanked for many discussions on bubbles in sparkling wine and for use of his articles.

Chapter 16. Dr Keith H. Lokan (Australian Radiation Laboratory) helped check the chapter and supplied the radon dosimeter artwork (Figure 16.4). Figures 16.1 and 16.2 are from *New Scientist*, 11 February, 1988.

Appendixes. I thank Terry Sedgwick for the ceramic goanna (Fig. A1.10). Standards Australia is thanked for granting permission to reproduce sections from Australian standards (Appendix II). The National Health and Medical Research Council gave permission for reproduction from their standards and booklet, 'Lead glazes in pottery' (Appendix IV). The Department of Pharmacy, Royal Adelaide Women and Childrens' Hospital provided information on poisoning (Appendix V); The

Australian Bureau of Statistics provided the data on drug use in Australia (Appendix VI). The Information Section, Pharmaceutical Benefits Branch, Australian Department of Health, provided the statistics on the prescription frequency of pharmaceutical benefits (Appendix VII).

Mr Mike Vernon AM, founder of Canberra Consumers Inc., and first chairman of the ACT Consumer Affairs Council, has sadly passed away (2 April 1932–6 November 1993). There is no individual in Australia who can fill the gap on consumer standards and so the section on sources of consumer information has been omitted.

General. Dr Tom Bellas (Division of Entomology, CSIRO) applied his considerable editorial skills to checking details in some of the manuscript. Dr Jesse Shore and the Powerhouse Museum supplied colour plates from their exhibit 'Chemical Attractions' (which I had the honour to launch on 14 February, 1996), as well as help with texts. ICI (Australia) contributed to the cost of colour production for this edition.

Patricia Croft, an experienced editor with ANU Press, encouraged me to write the book in 1974, then edited it from a set of scrappy course notes. She became a very close friend and sadly passed away on 1 August, 1995. ANU Press was closed down in 1984 and publication of the third edition was taken over by Harcourt Brace Jovanovich. I would like to thank Dallas Cox for a major reworking of that edition. Although the fourth edition was meant to be a minor revision, this was not to be. HBJ senior editor, Carol Natsis, has my gratitude for ensuring that considerable rewriting was incorporated smoothly. The fifth edition was crystallised by Penny Martin, senior publishing consultant at Harcourt Brace. She pulled down the barriers and opened the book to a really creative composition, both in content and in presentation. An author's reputation for care and quality is heavily dependent on the editing, and Ken Tate's broad experience and tenacity for detail was a very welcome reassurance for the author of what has, by now, become a heavily cross-linked (hyper-) text.

Community and other activities of the author

One has to admit that community service is not entirely altruistic as it provides an insight into areas not generally accessible to an author. The material thus gleaned from the listed articles can provide vitality and immediacy, and an occasional touch of scandal.

- Consumer representative on the Food Standards Committee, NHMRC, Commonwealth Department of Health member, (1973–75)
- Consumer standards committees of Standards Australia. The author has chaired CS/11 (soaps), CS/2 (detergents), CS/42 (sunscreen agents), and served on CS/53 (sunglasses), CS/64 (solaria), CH/3 (paints) and FT/8 (plastics for food contact) (1973–1986).
- Executive of Canberra Consumers Inc. (1974–79).
- Council of the Australian Consumers' Association (1980–86).
- Editorial panel, Australian Academy of Science School Chemistry Project, *Elements of chemistry: earth, air, fire and water* (1980–81).

- Partner; Versel Scientific Consultants, POB 71, Garran, ACT, 2605, fax 61-2-62852832, (1982–)
- Advisory Council of CSIRO, observer (1983–84).
- ACT Asbestos Advisory Committee, Chair (1983–86).
- National Health and Safety Commission, now Worksafe Australia, Commissioner for the ACT (1984–85).
- Advisory Committee to the Australian Government Analytical Laboratories (1986–1994).
- Chair, ANZECC Independent Panel on Intractable Waste (1991–92).
- Chair, ANZECC Scheduled Waste Working Group (1992–93).
- Report to the Commonwealth Department of Health (consultancy) *Illicit Drug Manufacture in Australia,* 1993.
- Report to ANZECC (consultancy) *Hexachlorobenzene Waste: Background & Issues paper,* 1995.
- Deputy Chair of ANZAAS, 1995–6.
- 'Dial-a-Scientist' radio talk-back on ABC radio stations countrywide.
- Foundation Chair, National Registration Authority for Agricultural and Veterinary Chemicals (1993–1997).
- Emeritus Professor of Chemistry at the Australian National University (1998–).

Thus alongside traditional academia, I have spent decades as a consumer activist, and more recently for four years as the foundation Chair of the Board of Australia's premier chemical regulatory authority; life as both poacher and gamekeeper. Common to all three lifestyles is the art of being critical, an attribute rapidly fading in the trend towards fundamentalism of diverse kinds; economic, religious, environmental, and techno-fix.

Below is a selection of some public lectures that always provided an opportunity to test new ways of presenting ideas.

Compulsory mass chest X-rays do more harm than good, JCSMR (ANU), Canberra, 1974.

Food standards—for industry or consumers?, ANZAAS, Canberra, 1975, publ. Canberra Consumers, **49**, 1976.

Consumer chemistry, Aust. Chemical Specialities Manufacturing Assoc. (N.S.W.) Sydney, 1975.

How consumers view food additives, CSIRO Food Research Institute, North Ryde, NSW, 1977.

Chemistry in the marketplace, Nyholm Memorial Lecture, NSW, 1978, publ. *Chem. Aust.,* **45**(1): 11, 1978.

Chemistry — the best buy in the scientific supermarket, Bayliss Youth Lecture, Perth, 1980, publ. *Chem. Aust.* **47** (12) 491, 1980.

Chemistry in the tropical marketplace, publ. R.A.C.I. in 'Chemistry in the Wet–Dry Tropics', Darwin, 1984.

Science in the witness box, (the Chamberlain case) ANU Convocation Luncheon, 1984, publ. *Chemistry in Australia,* **51**(8), 201, 1984; *A Scientist looks at the Chamberlain Case, Legal Services Bulletin ,* **9**(3): 108, 1984.

Chemistry spoken here, Medal Award Oration, R.A.C.I. Chem. Ed. Div., Melbourne, 1984, publ. *Chemistry in Australia,* **51**(12): 321–7, 1984.

Statistical expert evidence, to Australian Law Reform Commission, Canberra, 1985.

Expert evidence and the ultimate question, Aust. Institute of Criminology, Canberra, 1986, in 'The Jury', ed. D. Challinger, AIC, Canberra.

Introducing chemistry to the citizen, 9th Int. Conf. on Chemical Education, Sao Paulo, Brazil, 1987.

Statistical significance and the public perception of evidence, in 'Evatt Revisited', ANU, Canberra, publ. *CHAST*, 1989, pp. 61–68; *Evatt revisited: Interpretation of scientific evidence, Search* **21**(2) 52–54, 1989.

How to pick an expert, 2nd International Criminal Law Congress, Queensland, 1988.

The meaning of likely, South Australian Legal Services Commission, Tanunda, SA, 1988.

Chemistry in the marketplace, CEPUP Program, Lawrence Hall of Science, University of California, Berkeley, 1989.

Chemistry for tourists, University of Bonn, and Technical University Braunschweig, Royal Dutch Chemical Society, Royal Society of Chemistry (UK), 1992.

Beyond reasonable scientific doubt, Aust. Acad. Forensic Science, Canberra, 1992, publ. *A forensic standards proposal, Criminal Law Journal*, **10**(4), 246–54.

Committing chemistry in a public place, ANZAAS Medal Address 1993, *Search*, **24**(9), 240–41, 1993, *Chem in Aust*, 605–10, Nov. 1993.

Social and preventive chemistry, The Ralph Basden Lecture, RACI, Newcastle, 1993.

How do you regulate waste disposal?, Centre for Environmental Law, ANU, 1993.

New recipes for toxic waste, Science and Technology fora, UTS, Sydney, publ. *Australasian Science*, 31–34, Summer 1993.

Intractable Waste, in 'Waste Management and the Planning System', Annual Court Conference, NSW Land and Environment Court, Macquarie University, Sydney, 1993.

The National Registration Authority for Agricultural and Veterinary Chemicals, AVCARE Conference, Launceston, Tas., 1993.

World series debates, ABC TV: *That science is a health hazard*, Australian Science Festival, 1994.

The NRA's role in clean and green, National Farmers' Federation Annual Conference, Canberra, 1994.

A Matisse for science, Liversidge Lecture, ANZAAS, 1995 , published as *Changing our perception of risk, Search* **26** (100): 313–15, 1995.

SNACA, sensitive new age chemical analysts, Keynote lecture, AK1, 13th Australian Symposium on Analytical Chemistry, RACI, Darwin, 1995.

RACI: Making chemicals cuddly, plenary session, 10th National Convention, Adelaide, 28 Sept. 1995, published in *Chemistry in Australia*, **26** (10), Oct. 1995.

Chemical attractions at the Powerhouse Museum, Launching exhibition, *Chem Aust.* pp. 192–3; 1995, *Australasian Science, Spring 1995*, **17**(3), pp. 11–12.

Communicating risk — two ways, in Science Communication: Possibilities for the Future, Seminar, Communications Research Institute of Australia, Canberra, 11 May, 1995.

Risky business: The communication of risk and hazard, Keynote address; 4th International Conference on the Public Communication of Science and Technology, Melbourne, 1996.

Risky business, Friday Night Discourse, the Royal Institution, London, UK, 24 Oct., 1997.

Dedication

This edition is dedicated to the memory of our mothers, Hilde Selinger and Helen Hollander-Kery, who both passed away since the fourth edition was published.

It is to my late father, Herbert Selinger, that I am particularly indebted. His long involvement with the consumer movement (he was also on the council of the Australian Consumers' Association), as well as his firm belief that scientists should do useful things (at least some of the time) has had a profound influence on me.

My 24-year-old son Michael is thanked for providing many new cartoons for this edition, and 27-year-old Adam for becoming a colleague by following a kindred career at the Edinburgh Science Festival.

Finally, I thank my wife Veronique for her patience, understanding, and sense of fun.

CHAPTER

Some basic ideas

The periodic tablets

This chapter gives an historical flavour to chemistry. Some basic ideas are covered, in particular the language and naming of chemicals, and further depth is provided in Appendix I.

> *'The antidote (against the powerful, the wealthy and the unscrupulous in undermining the ideal of government) is vigorous support of the expression of unpopular views, widespread (scientific) literacy, substantive debate, a common familiarity of critical thinking and scepticism of the pronouncements of those in authority — which are all also central to the scientific method.'*
>
> (Carl Sagan, prize-winning author and astrophysicist, shortly before he died, 20 December, 1996)

Introduction

The meaning of the word chemistry[1]

The meaning of the word *chemistry* has changed many times over the centuries. In the third century, it referred to the fraudulent practice of

Belief systems
Most scriptures provide examples of episodes which just beg for a chemical explanation. This edition tackles some such stories in the Bible. A belief system should not feel threatened by attempted scientific deduction, and certainly should not teach rejection of the scientific method because it disagrees with scientific conclusions. Let's face it, for miracles, **timing** is everything! Science deals with our ever-changing understanding of the physical world, while religion deals with those issues of human existence and behaviour that really never change. Men still 'steal, murder, and covet their neighbour's wife'.

imitating precious metals and stones (Greek *khemeia)*. By the Middle Ages, this practice had evolved into the quest for a substance that would actually transmute base metals into gold, and into the study of matter associated with this process (Arabic *alkimia)* to give *alchemy.* In the sixteenth century, following the redirection of alchemy towards medicine by Paracelsus, the word came to mean the making of medicines. A dealer in medicinal drugs became a *chemist.* With Boyle's critique of alchemical theory in his *The Sceptical Chymist* of 1661, *chemistry* came to mean also the study of matter in a scientific manner.

Whereas sciences such as astronomy, biology, geology and physics continued to *study* the natural world, chemistry quickly went out to *change* it. We say there is a chemical industry, but not a physical or biological or geological industry. While there are industrial offshoots from the other sciences, that is not their dominant theme. Most of the excitement in chemistry comes from creation rather than observation. Chemistry is perhaps less a *natural* science and rather more a *synthetic* science. Because of its dependence on raw materials and its provision of materials that are essential to other industries, the chemical industry plays a major part in world trade *and a significant part in world politics* (for example, Middle East oil, South African minerals). The various environment debates have made chemistry even more central to political science.

Chemistry is applied to a very wide range of activities, comprising the chemical industry (that is, manufacture of materials) and the service industry (purification of water supplies, toxicology, forensic chemistry, etc.). The materials used by the chemical industry, and the equipment that is used to produce them, vary very widely from pig iron made in blast furnaces and petrol in refineries, to pharmaceuticals made in glassware, and precious metal salts made at the bench.

The nomenclature of chemistry is very difficult and is made even more confusing when the purest of purist chemists say *ethyne* but welders use *acetylene.* Linnaeus introduced the two-part nomenclature into botany, such as *Bellis perennis* for 'daisy', and this was in turn adopted into systematic inorganic chemistry, for example, magnesium sulfate for Epsom salts. We'll try and talk as far as possible about daisies!

Social and chemical interactions

Equivalent
As an example, chemistry teachers might like to consider the word **equivalent** (of all words!). In how many different ways is **equivalent** used in chemistry alone?

One of the great difficulties of translating from one language into another is that each language groups concepts in a slightly different way. There is no neutral abstract way of assigning words to meaning. A word means what most people using it want it to mean. This consensus is not static within a language, and even less so between languages. Even within chemistry there are many different meanings given to a word.

If we were to define the social structure of a community, we would start by categorising the social interactions. Starting with the strong interactions we find that they are fewer in number than the weak ones. A (nuclear) family unit has strong interactions, in the sense that the members of a family interact over long periods of time and to a large degree the individuals influence each other mutually. Families are relatively small units (2–8 members) and, initially at least, are highly localised. The next step up gives a choice of categories, for example, a circle of close family friends, or the people at work. Then we have the club or the church group. The larger the group, the weaker are the interactions.

The categories are arbitrary but some are more obvious and useful than others. They are *never* clear cut and there are always questions at the edges. Is Uncle Joe, three times removed, family or close friend? Anyone making out a wedding invitation list knows the problem. With chemical bonding the same problem arises. With extreme examples (that is, the ones always chosen to illustrate the point) everything is clear cut. An isolated gaseous molecule of hydrogen chloride is like a couple on their honeymoon; a unique, unperturbed, bond between hydrogen and chlorine, oblivious to all other interaction. But when the molecule dissolves in water, the honeymoon is over and all the interactions change.

Systematic chemistry deals with the family life of the 90-plus individual elements. The most basic categorisation in chemistry is the periodic table of the chemical elements (Fig. 1.1, overleaf). It is chemistry's icon. The chemical elements, in order of increasing relative atomic mass (much later corrected to atomic number), were found to exhibit periodic behaviour. This was one of early chemistry's major achievements. In the periodic table these systematic variations in the properties of the elements change slowly down a column and also along a row.

We are no longer dealing with 90-plus individuals but with a system that allows extrapolation of the properties of one element from those of others. Missing elements were defined according to their predicted properties, which were later confirmed when they were found. Additional elements at the end of the table had their properties extrapolated from within the table before their artificial discovery and production. It was a change as dramatic as switching from doing arithmetic with Roman numerals to doing it with Arabic numbers. Only much later was a rational basis for the periodic table provided by an understanding of the substructure of the atom in terms of protons, neutrons, and electrons.

Matter under investigation

Matter is traditionally described as occurring in the solid, liquid, and gaseous, phases, but these three categories are not exclusive. Liquid crystals are materials that have ordered structure in one or two dimensions, in contrast to solids which are structured in three dimensions, and liquids, in none. However, this last statement is not strictly true either. Liquids do show some ordering, and solids invariably have some disorder. Solids resist change in shape and volume, liquids resist change in volume but not shape, and gases resist change neither in shape nor in volume.

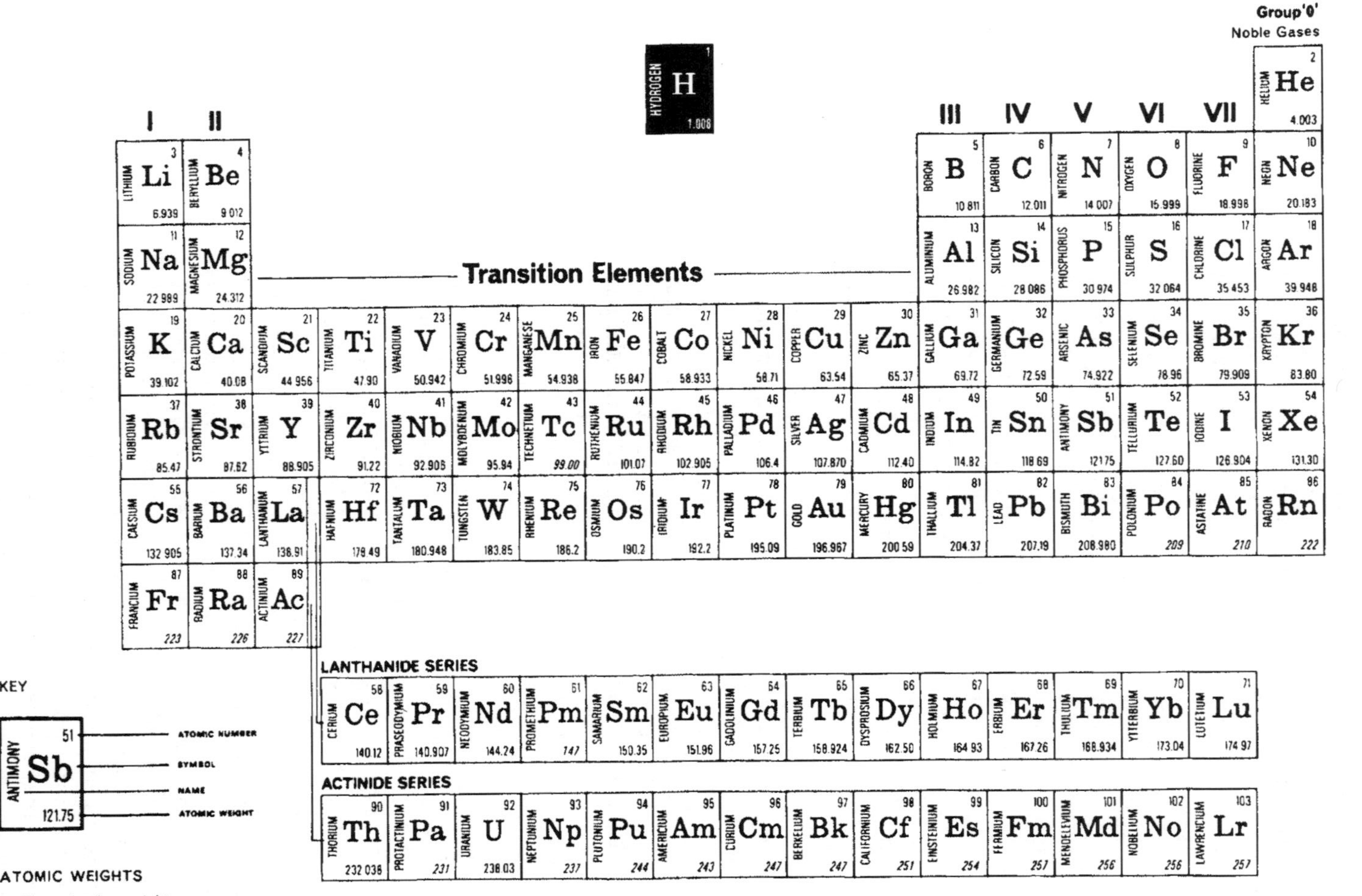

ATOMIC WEIGHTS

The atomic weights shown, except those italicized, are those listed by the International Commission on Atomic weights (1963). Values given in italics are those of the isotope with the longest half life. All are based on $^{12}C = 12$.

Fig. 1.1 Periodic table of elements (courtesy of May and Baker Ltd, Essex, UK)

When two phases of matter are in contact, the thing that separates them is a surface. Two liquid phases, oil and water, will be separated by such a surface. This surface can be disrupted by a detergent, which allows a form of mixing of the two liquids, the formation of an emulsion, with droplets of one liquid suspended in the other. The total surface separating the two liquids is now enormous. The shape of a solid is defined by the solid–air surface.

A fundamental question is: What is matter and what holds it together? There are things we call atoms and molecules, elements and compounds. The atoms and elements are like letters. Just as letters form words, so the atoms form molecules. Only certain orderings of letters form sensible words, and only certain orderings of atoms form sensible molecules. What are the rules of sensibleness? Words are those combinations of letters that stay in the language long enough to be given a defined meaning. Molecules are combinations of atoms that persist long enough to be worth giving a defined meaning. Most molecules, like most words, exist for a long time. But aren't molecules real? Are they just categories for sensibly dividing the world of matter into chunks suitable for human discourse, like words?

If you have a molecule isolated in space, say as a gas molecule, then it is a very real category. But when it is part of a liquid or a solid, its isolated existence is more a convenience than a reality.

We are generally interested in the bulk properties of chemical substances, not their intimate microstructure and relationships, but a real understanding requires this as well. But that is for traditional textbooks to deal with. Before we leave, there is one subject we must deal with, and that is solubility.

Solubility

Ordinarily, when we say a substance is soluble, we mean that it dissolves to an appreciable extent in water. However, we do make everyday use of other solvents; dry-cleaning spirit to dissolve grease stains, turpentine to dissolve paints, and so on. The question of solubility in oils and fats is of great importance to the physiological action of, for example, pesticides and drugs, and the cleaning action of detergents.

There is a simple rule that works well in predicting solubilities of substances in various solvents; *like dissolves like.* This means that ionic and highly polar substances are usually soluble in polar solvents such as water, whereas covalent non-polar substances are soluble in non-polar solvents such as benzene, petrol, and carbon tetrachloride. The stain removal guide (Table 3.4) in Chapter 3 depends on this.

Although many substances do not dissolve in water or other solvents, it is often possible to produce a stable to semi-stable mixture or *dispersion* of solute and solvent. This process is called *solubilisation.* The *solute* is what is dissolved; the *solvent* is what does the dissolving. The resulting dispersion may be stabilised by the addition of another substance, such as a detergent. *Emulsions* are an important group of these dispersions.

We finish with a story on not-so-soluble sugar.

SUGAR NOT SO SOLUBLE

Dear Sir:

It is common folklore that sugar added to an automobile gasoline tank will cause major damage to the engine. However apocryphal this notion is, it is widely believed, and has prompted at least some grudge holders to pour table sugar into the gas tanks of automobiles belonging to their enemies.

The premise that sugar added to a gasoline tank will foul an engine seems to imply that sucrose is soluble in gasoline and that the sugar will therefore be carried to the engine by the gasoline. This premise, which the present work tends to dispel, seems to be embraced by vandals and police investigators alike. Police investigators, in pursuit of 'sugared' gasoline, are more likely to submit to the laboratory samples of suspected gasoline siphoned from the tank or delivered by the fuel pump to a disconnected fuel line. Rarely will an investigator cause the entire contents of a fuel tank to be submitted to the laboratory.

Sugar can be detected in fluid gasoline only to the extent that it is soluble. Chemical principles of solubility would predict marginal if any solubility of sugar in gasoline. We demonstrate here that this is the case. From the standpoint of criminal responsibility it would make no difference whether the sugar is soluble or not, but from the standpoint of sampling and testing, it makes a great deal of difference.

Two experiments were conducted. In the first experiment, a carefully weighed amount of oven-dried sucrose was added to gasoline at room temperature and stirred for 30 min. The mixture was filtered through tared filter paper; the filter paper containing undissolved sucrose was oven dried, allowed to cool, and weighed. The recovery of sucrose in replicate runs was, respectively, 97.97% and 99.58%. While this experiment suggests that sucrose is virtually insoluble in gasoline, the limits of sensitivity of the method do not put the issue at rest.

A second experiment was conducted in which a known amount of ^{14}C- labelled sucrose was added to a known amount of gasoline. The mixture was equilibrated by stirring, and an aliquot taken for scintillation counting. The detected concentration of ^{14}C- labelled sucrose in gasoline ranged in replicate experiments between 1.26 mg/L and 1.44 mg/L. If the upper limit is rounded off to 1.5 mg/L, then the total amount of sucrose that would go into solution in a 15 gallon tank of gasoline would be on the order of 90 mg.

The implications of the solubility of sucrose in gasoline to sampling and analytical considerations are patent. If sugar is added to gasoline, virtually all of it will be found, undissolved, on the bottom of the tank. Even if the gasoline is saturated with sucrose, the concentration of sucrose is too low to be detected by simple means. A 100 mL sample of gasoline, for example, would contain only 150 μg of sucrose. Accordingly, investigation of cases of motor fouling caused by the suspected addition of sugar to the gasoline must include a sampling of any solid residues in the fuel tank.

Submitted to J. Forensic Science **38**(4):753, 1993, by the following 'heavies': Keith Inman, M.Crim.; George F Sensabaugh, D.Crim., Senior Criminalist; John I.Thornton, D.Crim., California Dept. of Justice; Professors of Forensic Science. DNA Laboratory, School of Public Health, University of California Berkeley, Glenn Hardin, M.P.H. Criminalist, Minnesota Bureau of Criminal Apprehension.

SONG OF THE ELEMENTS

by Tom Lehrer

There's antimony[1], arsenic[2], aluminum[3], selenium[4]
And hydrogen[5] and oxygen[6] and nitrogen[7] and rhenium[8]
And nickel[9], neodymium[10], neptunium[11], germanium[12]
And iron[13], americium[14], ruthenium[15], uranium[16]
Europium[17], zirconium[18], lutetium[19], vanadium[20]
And lanthanum[21] and osmium[22] and astatine[23] and radium[24]
And gold[25] and protactinium[26] and indium and gallium[27]
And iodine[28] and thorium[29] and thulium[30] and thallium.
There's yttrium[31], ytterbium[32], actinium[33], rubidium[34]
And boron[35], gadolinium[36], niobium[37], iridium[38]
And strontium[39] and silicon[40] and silver[41] and samarium[42]
And bismuth[43], bromine[44], lithium[45], beryllium[46] and barium[47].
There's holmium[48], and helium[49] and hafnium[50] and erbium[51]
And phosphorus[52] and francium[53] and fluorine[54] and terbium[55]
And manganese[56] and mercury[57], molybdenum[58], magnesium[59]
Dysprosium[60] and scandium[61] and cerium[62] and caesium[63]
And lead[64], praseodymium[65] and platinum[66], plutonium[67]
Palladium[68], promethium[69], potassium[70], polonium[71]
And tantalum[72], technetium[73], titanium[74], tellurium[75]
And cadmium[76] and calcium[77] and chromium[78] and curium[79]
There's sulfur[80], californium[81], and fermium[82], berkelium[83]
And also mendelevium[84], einsteinium[85], nobelium[86]
And argon[87], krypton[88], neon[89], radon[90], xenon[91], zinc[92] and rhodium[93]
And chlorine[94], carbon[95], cobalt[96], copper[97], tungsten[98], tin[99] and sodium[100].
These are the only ones of which the news has come to Harvard
And there may be many others, but they haven't been discarvard.

Reproduced by permission of Tom Lehrer, entertainer and professor of mathematics (© 1959, 1987 Tom Lehrer). This song is sung to the well-known tune, 'I am the very model of a modern major-general', from Gilbert and Sullivan's **Pirates of Penzance**.

Notes

The origins of the elements are taken from diverse and contradictory sources, including The Handbook of Chemistry and Physics (51st edn, 1970–1971). Abbreviations: Arab., Arabic; Ger., German; Gk, Greek; Icel., Icelandic; It., Italian; L, Latin; OE, Old English; ON, Old Norse; Sp., Spanish; Swed., Swedish.

1. L **stibium,** mark. It marks paper like lead.
2. Gk **arsenikon**, orpiment — artist's pigment, As_2S_3.
3. L **alumen**, light, because of the brightening effect of alum (**alumen**), used for dyeing
4. Gk **selene**, the moon.
5. Gk **hydros** + **genes**, water forming.

6. Gk **oxys** + **genes**, acid forming. The theory that oxygen was essential to acids is wrong.
7. L **nitrium**; Gk **nitron**, native soda.
8. L **Rhenus**, the Rhine.
9. Ger. **kupfernickel**, copper demon.
10. Gk **neo** + **didymos**, new twin.
11. The planet Neptune.
12. Germany.
13. OE **iren**, **isen**; Ger. **eisen**. Symbol Fe from L **ferrum.**
14. The Americas.
15. L **Ruthenia**, Russia.
16. The planet Uranus.
17. Europe.
18. Pers. **zargun**, gold-coloured (mineral) zircon (other varieties of which are called argon, hyacinth, jacinth, or ligure).
19. Lutetia, an ancient name for Paris.
20. Icel. **Vanad(is)**, epithet of Norse goddess Freya, discovered in Sweden.
21. Gk **lanthanein**, to escape notice.
22. Gk **osme**, smell or odour. The most dense element forms pongy compounds.
23. Gk **astatos**, unstable.
24. L **radius**, ray.
25. ON **gull**, Gothic **gulth** (ghel, yellow; Ger. gelb), Sanskrit **Jvual**. Symbol from L **aurum**.
26. Gk **protos**, first (of a series), + **actinium**.
27. L **Gallia**, France.
28. Gk **iodes**, rust-coloured (assigned by a colour-blind chemist?).
29. Scandinavian god of thunder.
30. Thule, an early name for Scandinavia.
31. From Ytterby, a town in Sweden.
32. From Ytterby, a town in Sweden.
33. Gk **aktis**, beam or ray.
34. L **rubidus**, red (in allusion to the two red lines in its flame spectral emission).
35. Pers. **borah**; Arab. **buraq** for borax.
36. Finnish chemist J. Gadolin (1760–1852)
37. Gk **Niobe**, mythological daughter of Tantalus (niobium found with tantalum), originally called Columbium.
38. L **iris**, rainbow iridescence in solution.
39. Strontian, a town in Scotland.
40. L **silex**, flint.
41. Anglo-Saxon **seolfor**, **siolfur**. Symbol from L **argentum**.
42. Samarskite (a mineral named after a Russian mine official).
43. Ger. **Wismut**, from Ger. **Weisse Masse**, white mass.
44. Gk **bromos**, stench.
45. Gk **lithos**, stone.
46. Gk **beryl**, beryl (a gem), originally (1798) called **glucinium** from Gk **glykos**, sweet.
47. Gk **barys**, heavy.
48. L **Holmia**, Stockholm.
49. Gk **helios**, the sun.
50. L **Hafnia**, Copenhagen.
51. Ytterby, a town in Sweden.
52. Gk **phosphorus**, light-bearing; a name applied to the planet Venus when appearing as a morning star.
53. France.
54. L **fluere**, to flow; the mineral fluorspar, CaF_2, used as a flux for metalwork.
55. Ytterby, a town in Sweden.
56. L **magnes**, magnet, from the magnetic properties of pyrolusite, It. **manganese**, corrupt form of magnesia.
57. The planet Mercury.
58. Gk **molybdaina**, galena; lead sulfide.
59. Magnesia, a district in Thessaly.
60. Gk **dysprositos**, hard to get at.
61. L **Scandis**, Scandinavia.
62. The asteroid Ceres.
63. L **caesium**, bluish-grey.
64. Icel. **leidha**. Symbol from L **plumbum**.
65. L **prasius**, a leek-green stone, + Gk **didymos** double, twin.
66. Sp. **plata**, silver; **platina**, diminutive.
67. The planet Pluto.
68. Named after the asteroid Pallas, then recently discovered (1803).
69. Gk **Prometheus**, one of the gods.
70. Pot-ashes from early Dutch **potasschena**, L **kalium** from Arab. **gali**, alkali, gives symbol K.
71. Poland, birthplace of Marie Curie, who discovered polonium in 1898.
72. Gk Tantalus, Greek god.
73. Gk **technetos**, artificial, the first element to be made that does not occur in nature; once called masurium (Masuria region in N.E. Poland).
74. Gk Titan, gods of great strength and size.
75. L **tellus**, the earth.
76. L **cadmia**, calamine (a zinc ore containing small amounts of cadmium); calamine lotion, suspension of zinc oxide + 0.5% iron oxide.
77. L **calx**, lime (calcium hydroxide).
78. Gk **chroma**, colour.
79. Marie Curie, Polish chemist who worked in France.
80. Sanskrit **sulvere**; L **sulphurium**.
81. California.

82. Enrico Fermi, Italian nuclear physicist who fled from the Fascists to the US.
83. Berkeley, California.
84. Dimitri Mendeleev (1834–1907), Russian discoverer of the periodic table of the elements.
85. Albert Einstein, German physicist who fled from the Fascists to the US.
86. Alfred Nobel (of prize fame).
87. Gk **argos**, idle.
88. Gk **kryptos**, hidden.
89. Gk **neos**, new.
90. From radium.
91. Gk **xenon**, stranger.
92. Ger. **Zink**, origin uncertain.
93. Gk **rhodios**, rose-like.
94. Gk **chloritis**, kind of greenstone.
95. L **carbo**, coal or charcoal.
96. Ger., **Kobalt**, evil spirit or goblin.
97. L **cuprum**, from the island of Cyprus.
98. Swed. **tung** + **sten**, heavy stone.
99. Germanic **tinam**, of unknown origin.
100. Arab. **suwwad** or **suda** headache; glasswort plant (L **sodanum**) used as a headache remedy. L **natrium** gives symbol Na.

Naming chemicals

Many chemical names are 'trivial', in the sense that they are not part of a systematic approach. These trivial names are often a great entry into the history of discovery and the mythology of the chemicals. Sometimes they have a very prosaic origin.

One of the most common forms of asbestos once found as insulation in Canberra homes is amosite, which was imported from South Africa. Its name? Some exotic river, or Zulu myth? No. The name comes from Asbestos Mines of South Africa condensed with the customary *ite* ending for minerals. You may have tried to test for asbestos yourself at one stage, by heating a sample of insulation to a high temperature. Perhaps you had vermiculite? On heating it forms flakes and opens up into worm-like threads. The name? From the Latin *vermicular* to be worm-ridden, from *vermiculus*, a diminutive of vermis, a worm.

Another mineral, diasporite, an aluminium hydroxide occurring in clay as orthorhombic crystals, also got its name from the effect of heating. When heated it crackles and disintegrates. Hence (1805) Greek *diaspora* a scattering, a dispersion.

Well, if you are not into minerals, at least you drive a car. No, a carburettor is not a laboratory burette stuck in the engine of a car! The term *carburet* was originally (about 1795) used for a combination of carbon with another element, carb(on) + -uret. Marsh gas (methane) was therefore called carburetted hydrogen. Similarly sulphuretted hydrogen was hydrogen sulfide. The ending *-uret* was derived originally from Latin *-ura* (via the French *-ure*) indicating an action or process, or result of the action, (for example, enclosure in English is the result of enclosing). Thus the process of combining fuel with air in an automobile uses a carburettor.

Cars need coatings. Lacquers and ducos have a long history. For example, dammar is the name given to various resins obtained from 'dammar' trees growing in the East Indies, New Guinea and New Zealand, *Dammara orientalis*. This was the basis of the Dammard Lacquer Company that started life in Edwardian times. It is alleged that they made three grades of lacquer, 'dammard', 'dammarder', and 'dammardest'!

In 1906 Dr Baekeland discovered phenol–formaldehyde resin and so this company became the 'Bakelite Company', producing the first synthetic plastic and lacquer.

The precious stone, amethyst, consists of quartz with an iron impurity which turns blue to violet on irradiation. In ancient times the stone was added to wine glasses as a protection against intoxication. Hence the derivation from the Greek *a-*, not, *methustos*, to be drunken. It's enough to drive you to drink!

In 1834, Liebig discovered an efficient condenser needed for distilling wine into brandy; (brandy from the German Brandt-wein, or burnt wine). He once obtained a dirty white amorphous substance by heating ammonium thiocyanate. He called this *melam*. In his original paper he said that he refrained from giving an explanation for the names he was using: 'Names are, if you like, grasped from the air (i.e. pure inventions), and serve the purpose just as well as if they were derived from the colour of one of the properties'. By boiling melam with a solution of potash, he obtained melamine, and set out the chemistry which was to usher in female emancipation! This occurred later by the development of melamine polymer table tops which freed women from the endless task of scrubbing porous wooden, or marble kitchen tables, and allowed them time to think of higher things.

While these products have short etymological histories, the chlorine added to purify our water supply or our pool has had a long and confusing history. Discovering the real nature of chlorine and distinguishing it from oxygen, and understanding the general 'burning' process, covered a few decades of chemical history. In 1774 Scheele studied the action of muriatic acid (from the Latin *muria*, brine, from which it is made) on pyrolusite (manganese dioxide). He produced the choking, greenish gas that we now call chlorine. In accordance with the theories of the day, he regarded the gas as marine (i.e. brine) acid air which had been deprived of its phlogiston. Phlogiston was an eighteenth century chemical antimatter used to explain burning processes. The theory was completely wrong but has excited philosophers of science ever since. Anyway, according to this theory, the new gas had to be dephlogisticated marine acid air. Lavoisier, who later lost his head in the French revolution, had just discovered oxygen and its role in combustion, so what was to be chlorine became oxy-muriatic acid. In 1810 Sir Humphrey Davy proved the gas was an element and called it chlorine, from the Greek *chloros*, green-yellow, plus the ending *ine* used for the elements fluorine, bromine and iodine as well, a custom the French and Germans refused to follow, (i.e. Fr *chlore*, Ger. *Chlor*).

It was from Sweden in 1811 that the chemist Berzelius proposed a logical system of naming based on Latin. Some metallic elements were already in this form, such as uranium, chromium and barium. On the continent, the names of sodium and potassium were Latinised to *natrium* and *kalium*, but not in England. However, the symbols Na and K prevailed, as did Fe, Sn, and Sb for ferrum (iron), stannum (tin), and stibium (antimony). After that date, all metallic elements were given names ending in -ium (occasionally -um). Examples are cadmium, lanthanum, lithium, thallium, and radium. Helium is an exception. It was first discovered on the sun using a spectroscope (Greek *helios*, sun) and assumed to be a form of iron with its electrons stripped off. When it was later discovered on earth, the mistake was realised, but it was too late to change the name.

In the ancient world there were seven known celestial bodies, to which seven metals were assigned. The Sun (gold), Moon (silver), Venus (copper), Jupiter (tin), Saturn (lead), Mars (iron) and Mercury (mercury). Mercury took its actual name from the planet. So when your local apothecary recommends *lunar caustic* for removing warts, what you will get is a solution of silver nitrate that releases nitric acid, which burns away the growth on the skin, leaving a black stain of silver.

Many centuries later the heavens were again inspiration for names. While helium actually had a legitimate connection with the Sun, this is not so for Uranus (uranium), Neptune (neptunium) and Pluto (plutonium). There is one planet we have forgotten, Earth (Latin *tellus, telluris*), from which we have tellurium.

As the metals were linked to the planets, and these were linked to the ancient gods, why not have a direct link? Tantalus was a mythical king of Phrygia, son of Zeus, who, for revealing the secrets of the gods, was condemned to stand in water up to his chin with fruit hanging above him. The water receded when he tried to drink, and the fruit evaded his grasp; hence to *tantalise.* In Finland in 1802, A.G. Ekeberg had a hard time dissolving the oxide of what turned out to be the metal tantalum. He chose the name 'partly in allusion to its incapacity when immersed in acid to absorb any and be saturated'. In other words it couldn't drink. Tantalus had a daughter Niobe, who gave us niobium, found as a minor constituent in ores of tantalum. Actually this pair, when first discovered as a mixture, was called columbium (from a reference to America). This name is still found in some more chauvinistic U.S. texts.

The Nordic god Thor gave us thorium, while Vanadia (goddess of beauty) inspired vanadium from its beautifully coloured compounds.

What do gods eat? The food of the gods, Greek *theo-*, + *broma* gave us theobromine (which incidentally contains none of the element bromine), the major stimulant in chocolate (see Chocolate in Chapter 4).

The gods were not always kind. German miners in the 16th century were a suspicious lot. They insisted that the mines were inhabited by sprites or goblins who placed poisonous ores in their path. These were the kobolds, (probably from the Greek *kobalos*, a knave). The term gradually focused on certain ores regarded as useless for metal extraction (for example, cobalt arsenide) but which were found to be able to impart a blue colour to glass, hence cobalt blue.

Even more frustrating was a mineral that resembled copper ores to the extent that it also gave glass a green colour. Repeated efforts to extract copper (kupfer) were futile and the devil was blamed for deliberately colouring the material to fool the poor miners. The mineral was called kupfer nickel, the devil's copper. The old German *nickel,* is found in the English expression, old Nick.

A miner's lot was not a happy one and they had problems with the air in the mines as well as the minerals. The word *damp,* corresponding to similar words in German, Dutch and Danish, means steam, vapour, smoke. In 15th century English it referred to noxious gases, as in *choke damp,* carbon dioxide. In 1677, *fire damp* (in coal mines) was used to describe methane. After an explosion, the remaining gases were called *after damp.*

The English, as we mentioned earlier, refused to give up the names sodium and potassium in preference to natrium and kalium, because of long tradition. Soda, or sodium carbonate had been extracted from wood ashes, and used to make soap and glass since antiquity. We still use soda today to describe soda water, that is, water supersaturated with carbon dioxide, because this was once made from sodium carbonate. Some believe the name soda originates from the Arabic *suda,* a splitting headache, for which soda was an early alleged remedy. An alternative derivation is from the Arabic *suwwad,* a plant whose ashes yield sodium carbonate.

The metals sodium, potassium and calcium, unlike, say, copper, were much more difficult to isolate from their salts. So while we saw the early development of systematic names based on the metal such as sulfate of copper or copper sulfate, ignorance meant staying with sulfate of soda and carbonate of lime. Only about one hundred and fifty years ago did these names change to today's sodium sulfate and calcium carbonate.

NITRE AND NITRATE

What do Na^+ and NO_3^- have in common? Nothing at first glance; one is the symbol for the sodium ion and the other for the nitrate ion. Sodium and nitrogen both gained their symbols (Na and N) from the Latin **nitrium**. Together they form the salt, sodium nitrate, known as nitre. In German it is called **natrium nitrat** and in Russian, approx. **neetraht nahtrya**. Why? Enter the prophet Jeremiah (Jer. 2:22).

> **For though thou wash thee with nitre**
> **And take thee much soap,**
> **Yet thine iniquity is marked before me,**
> **Saith the Lord God.**

There can be no doubt that when the prophet was speaking of washing with **nitre** he meant washing soda (Na_2CO_3), found as trona in salt lakes in the area (see Chapter 9, glass, trona, Bertholet in Egypt). While some nitre was probably formed in the mixture of dung and salt (see introduction to food additives Chapter 15) the nearest substantial nitrate was in Chile (as the potassium salt, saltpetre). For the Greek version of the Old Testament

(Septuagint, 3rd to 2nd century BC), the translators used the Greek **nitron**, and when St Jerome did the Latin translation (Vulgate, 340–420 AD), he used **nitrum**. In both cases the words meant Na_2CO_3 to their readers. The same can be said of the translators producing the King James version, although by then the changed meaning of **nitre** (sodium nitrate) was already under way. To modern readers, using nitre as a cleansing agent is rather odd, and the chemically literate baulk at Proverbs 25:20.

> **And as vinegar upon nitre, so is he that singeth songs to a heavy heart.**

This obviously refers to the reaction of vinegar (acetic acid) on sodium carbonate, which releases bubbles of carbon dioxide, the evolution of gas symbolising relief of heavy emotion. That makes one wonder about another aspect of the earlier Jeremiah verse, the use of the word **soap**. Soap would not appear until about seven centuries later, but then Jeremiah was a prophet! The Latin **sapon**, for soap, is mentioned first in Pliny's book of natural history (27–79 AD), (it is probably of Teutonic origin). It became the root for the word for soap in many languages (for example, see Soaps in Chapter 3). However, the word Jeremiah used was not **sabon** (this Hebrew word for soap was introduced much later), but **boriit**, which was vaguely described as an alkali used for smelting metals (**barar** to purify, polish), (**bor**, cleanliness, pureness). Perhaps we are dealing here with borax (Arabic **baraqa**, to glisten), a respectable cleaning agent as well as a flux in metallurgy. In any event the word he used later indeed came to be applied to borax.

(See also From compost to gunpowder in Chapter 7.)

Tradition dies hard in the commercial world, and the older names for common compounds are still found on labels in hardware stores. The spelling of sulphate with *ph* is maintained only in medical circles.

Place names have also been an inspiration for chemists. While europium, francium, germanium, polonium and scandium are pretty obvious, to what places do holmium, ruthenium and strontium refer? Which little town has the honour of having four metals named after it? Ask Tom Lehrer in the notes to his poem.

It is interesting to note that the word amalgam (as in amalgamations), originally meant to soften. It arose from the softening action of mercury on gold or silver, and hence to any mercurial alloy, for example as used in dentistry. The word can be traced back to the Latin and Greek *malagma*, an emollient, and Greek *malakos*, soft.

Some sense in the system

In order to communicate information about chemistry and chemicals, there has to be an agreed way of identifying and naming them. A trade or brand name can be given to a chemical or a formulation of chemicals, and registered by a firm for its exclusive use. Such a trade name gives no information, and the composition of the product to which it refers can vary. In many cases a trade name can be so popular that it becomes, by common usage, a common name (for example aspirin, biro, cellophane, bakelite). Manufacturers must keep on insisting that their trade names be spelt with a capital letter because once a trade name has become a common name it may be legally defined as the *generic* or non-proprietary term for a particular material. Today 'aspirin' is not the Bayer trade mark but the generic name for a particular chemical substance. Whereas there can be an enormous number of trade names for a chemical, there are only one or two generic names. The generic name can then be qualified to provide further generic names for related compounds (for example penicillin and penicillin G).

In order to have an unambiguous precise name for a chemical substance, a system has been devised to provide a *systematic name* for any substance. These names are built up according to strict rules. Each compound has only one correct systematic name, and that name conveys the complete information about the detailed structure of the compound: it is a stylised written description of the chemical structure. For example, Aspro is a trade name; aspirin, or acetylsalicylic acid, the generic term; and 2-acetyloxybenzoic acid, the systematic name. To ensure that the chemical can be tracked down when required, each one is also given a sort of 'tax-file, or social security, number'. This is particularly important because chemicals appear under a variety of aliases, some more logical than others.The correct naming of chemicals is very complex and difficult, even for professionals. More details on naming are given in Appendix I.

Chemical formulae and diagrams

Chemical formulae and diagrams are symbolic representations of the composition and structure of molecules and compounds. There is a variety of conventions for them, depending on the sort of information to be conveyed.

At the lowest level of information is the *empirical formula,* which lists only the numbers and types of atoms present in the molecule, and tells us nothing about the structure. For example, for aspirin it is $C_9H_8O_4$. Many different chemicals can have the same empirical formula.

The *group formula* places atoms together in groups that correspond to the grouping in the actual molecule, and gives some indication (using prefix symbols) of how the groups fit together. For example, the group formula for aspirin is $2CH_3COO\text{–}C_6H_4COOH$. As you become more familiar with them, these group formulae can give a fairly complete (marginally ambiguous) description of the structure.

The *condensed structural diagram,* the most common form of line diagram, gives a two-dimensional representation of a three-dimensional structure. It leaves out a lot of the atoms, but their presence is implied by the shorthand conventions. For aspirin, the condensed structural diagram is as shown in Figure 1.2.

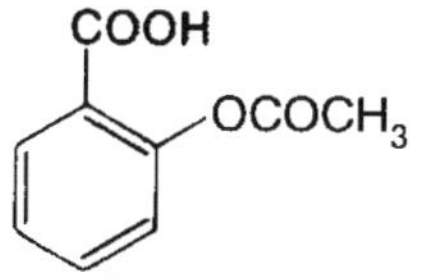

The hexagon is a ring of six carbon atoms joined by alternating double and single bonds. This ring is called the benzene ring. Carbon has a valency of four, so any missing bonds are taken up by a bond to a hydrogen atom. There are other apparent ambiguities (although convention clarifies them). In fact this diagram is a condensation of the *full structural* diagram, which is rarely used, in which the three-dimensional structure of the compound and the bond angles are indicated. The full structural diagram for aspirin is shown in Figure 1.3.

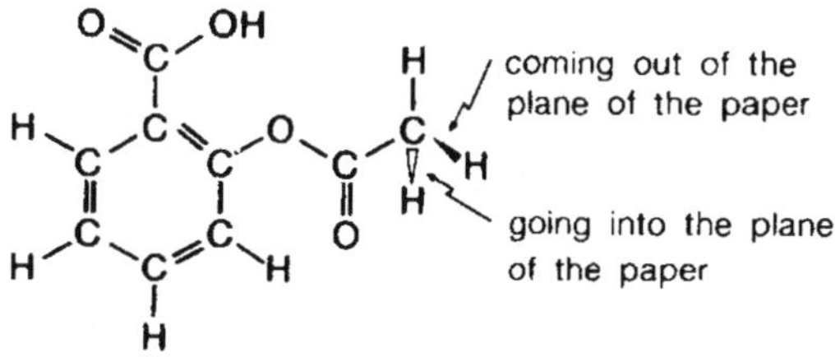

COOH
$OCOCH_3$

Fig. 1.2
Condensed structural diagram for aspirin

coming out of the plane of the paper
going into the plane of the paper

Fig. 1.3
Full structural diagram for aspirin

Note that each carbon atom C, and hydrogen atom H, is depicted specifically, whereas in the more usual condensed version the carbon atoms are implied to be at the intersections of the bond lines, and the hydrogen atoms are implied to fill positions so that the valence of four for carbon is satisfied. Throughout the text of this book, condensed structural diagrams will be used, but extra information will be included when required to emphasise a particular feature.

The stick and space-filling model is useful to indicate the actual geometry of a molecule. This may be essential for explaining its biological activity (see also the section on DDT in Chapter 7, under 'Pesticides and alternatives — Organochlorine compounds', and Plate IV). The stick and space-filling model for aspirin is shown in Plate I.

If you want to know more, go to Appendix I.

Bibliography

The elements, C.R. Hammond, *CRC Handbook of Chemistry and Physics*, Ed. R.C. West, pp. B-4 to B-38, 51st edn, The Chemical Rubber Company, Cleveland, Ohio, 1970–71.

Chemistry of health and risk

Alfred was never any good at communicating risk to others

This is a chapter that deals with the fine print. It contains more detail than some readers may want. The focus is on some aspects of the health and safety of chemicals tackled by regulatory agencies on behalf of the community. The devil is in the detail, and so there is considerable discussion on the nature of measurement and its interpretation.

TEETH DRIPPING WITH FOUL SANIES

'Chemists boast that they have mastered the art of subduing every kind of mineral, yet they themselves do not come off scot-free from their pernicious influence. They very often bring on themselves the same ailments as do other workers who deal in minerals, and in spite of their persistent denials, the colour of their faces reveals the fact. . .

. . . I used to know Carlo Lancilloti, my compatriot, a well-known chemist; he was palsied, blear-eyed, toothless, short of breath, and disgusting: the mere sight of him was enough to ruin the reputation of the medicaments, the cosmetics especially, that he used to sell. But far be it for me to condemn such researches as mischievous. Chemists generally deserve praise, for so devoted are they to investigations of abstruse matters and to the enrichment of natural science that they do not hesitate to risk their lives for the good of the public. . .'

Source: Tony Finlay in Chemistry in Australia, **49**(11), pp. 418–19, 1982, quoting De morbis Artificum Bernardini Ramazzini Diatriba, Ramazzini's Diseases of

Workers, the Latin text of 1713 revised with translation and notes by Wilmer C. Wright, University of Chicago Press, Ill, 1940.

Animal experiments
The use of healthy animals to test the toxicity of chemicals to humans of various ages and states of health (liver and kidney function, etc.) is not always reliable. This explains why, with three exceptions (4-aminobiphenyl, nitrogen mustards, and vinyl chloride monomer) the carcinogenic action of various industrial compounds was detected primarily by exact medical observations of humans, and not in **animal** experiments. Confirmation by animal experiments has often lagged several decades behind the medical observation — over 40 years in the case of 2-naphthylamine (humans, 1895; animals, 1938). Other examples are tar (1775, 1918), asbestos (1930, 1941), chromates (1912, 1958) and benzidine (1940, 1946).

Introduction

Occupational health and safety have improved for chemists and other workers since 1713. No more 'teeth dripping with foul sanies', brought about by gross overexposure to mercury vapour; no more the 'cramping pain in the stomach, difficulty in breathing, bloody urine, painful colic and convulsions in every limb' in those who worked at the sublimation of arsenic. Today, occupational disease is slow to develop and difficult to distinguish from non-occupational sources.

It took a long time for people to become aware of long-term risks. Some lead compounds taste sweet (e.g. sugar of lead) and a modest intake has no short-term effect, so the ancient Romans poisoned their brains and bodies (either from lead water pipes or from using lead as a wine sweetener). Later generations took some time to realise the work-place dangers of hat felting, glassblowing, woodworking, radiation, asbestos and many other hazards.

Routes by which chemicals enter the body

By saying that a chemical is 'in' the body we mean that it has to pass through a barrier of some sort. One can argue that the human animal is like an elongated bead on a string, where the gut, ranging from the mouth through the stomach, small and large intestines and colon to anus, is the string. The gut (string) is topologically (i.e. spatially) outside the body. It is only when substances pass from the gut passages *through* a membrane (in the lining of the mouth or stomach or intestines) that it is 'in' the body. Another route is *through* a lung lining into the blood. Or it can pass *through* the skin.

Once across the membrane and inside the body, most substances enter the bloodstream. The liver is a chemical factory. Its enzymes metabolise foreign substances to change them into a form that allows excretion through faeces or urine. These changes try to make the chemical less harmful, but sometimes the reverse happens. The product from the liver (metabolite) can be more toxic than the original. Different animals have different enzymes and so what is dangerous for one species may not be for another. The notorious dioxin is 5000 times more toxic for female guinea pigs than for hamsters (see p. 20).

The kidneys filter the blood, remove what is not needed and store the resultant urine in the bladder. (The kidneys and bladder may be dam-aged in the process.) Other exit routes are the respiratory tract and lungs, sweat glands, nails and hair (as with mercury and arsenic).

The damage that a chemical does can be immediate (acute toxicity) or can initiate long-term damage (chronic toxicity) — see below. The damage is often specific to certain organs, called the *targets* of the toxicity. Chemicals causing cancer or birth deformities are special forms of toxicity.

Acute and chronic toxicity: the concept of LD_{50}

Toxicity

A *poison* is a substance in quantity or dose which exceeds the body's ability to deal with it without harm. Even though people vary in their sensitivity to poison, it is generally assumed that in most cases there will be a dose below which no one will be harmed. Examples usually quoted are 'natural' poisons like cyanide in bitter almonds. There is a great deal of argument on this issue, but let us keep it simple at this stage. Poisons are then divided into two groups:

(a) *Cumulative poisons:* The poison is absorbed easily but is excreted slowly, and therefore builds up in the body. Because of this, the material is easily detected.

(b) *Non-cumulative poisons:* The poison is absorbed quickly, and also is excreted quickly. The damage it does is cumulative. One of the best examples is alcohol. Alcoholics with central nervous system (CNS) damage, cirrhosis of the liver, and kidney damage may be completely 'dried out' but the damage is irreversible.

Next the effect of poisons is divided into two groups, acute and chronic toxicity.

Acute toxicity

An *acute effect* is an effect that occurs shortly after contact with a *single dose* of poison. The effect depends on how poisonous the substance is, and on where it is applied. A drop of sulfuric acid is more dangerous on the eye than on the skin, and arsenic compounds are more toxic than those of sodium. Because the way in which an acute poison acts is generally understood, acute toxic responses can usually be given in a quantified form and for these, at least, it can be properly argued that below a certain dose, a substance is not toxic.

Measurement of acute toxicity followed this path.

LD_{50} — A HISTORICAL PERSPECTIVE

Until about 50 years ago, the toxicity of a substance was usually expressed as the lowest dose that had been observed to kill an animal, even though it was realised that another animal might survive after a much larger dose.

In 1926, however, it was shown that the individual minimal lethal doses of digitalis for 573 cats followed a log-normal distribution, and in 1927 Trevan, who obtained similar results using an enormous number of frogs, plotted per cent mortality at each dose and interpolated a value for the dose that would kill 50% of the animals. This dose he designated as the LD_{50}. After another quarter century, in the early 1950s, the value of this measure of toxicity had been completely accepted and the methods necessary for obtaining it had been well worked out.

The practical value of this approach is now being questioned, again on grounds of biological variability. It has been shown, for example, that the LD_{50} for a drug used on a single species of laboratory animal varies with strain, sex, age, diet, litter, season, social factors, and temperature. It seems hardly worth standardising all these variables when LD_{50} variation between species is so great, sometimes more than a hundredfold between rats and mice.

Humans differ very greatly from the rat and other experimental animals in their response to some toxins. Some species of animals used in routine laboratory toxicity testing can live happily all their natural days on the seeds of a vetch that, when consumed by humans in amounts of a few hundred grams a day, produces irreversible paralysis of the legs. Much concern was recently aroused by the demonstration that lysinoalanine, a compound formed in foodstuffs treated with alkali, as in the preparation of protein isolates, or more traditionally in the primitive use of maize, is severely toxic to rats, producing kidney damage when fed at levels of 100 ppm in their diet. It has now been shown that 10 times this level in the diet of quail, mice, hamsters, rabbits or monkeys has no discernible effect. Thus it appears that the rat is quite unusual in its susceptibility, and that in retrospect, the primitive Middle Americans who developed maize as a staple of the diet using alkali treatments in its preparation were not foolhardy.

Unfortunately there is no convenient escape from the truism that the proof of the pudding is in the eating, and that must be by humans, for it is clear that lack of effect of a new food on experimental animals does not guarantee that it is harmless to us, nor does toxicity in an animal species mean that we must at all costs avoid it.

The variability of response between animals of the same species occurs in humans too, and ignorance of this principle has led to many unsuccessful murder and suicide attempts in which the presumed lethal dose of a toxic substance was not enough for the purpose envisaged. With very toxic substances, the lethal dose resulting in death in 99% of a population may be several times the LD_{50}, but it is still a very small amount, and the distinction is perhaps not very important. It is interesting to speculate on the application of this biological phenomenon to the long-term deleterious effects of particular diets on individuals.

Let us suppose, for example, that the consumption of a total of 2.5 tonnes of saturated fatty acid could be shown to result in the death from heart disease of 50% of those consuming it. It would take about 100 years to consume that much on a normal Australian diet, and we would perhaps be entitled to say that therefore it could not be regarded as toxic, as no one (or **almost** no one) would have an opportunity to consume that much. We might, however, be wrong to make this judgment, for in a highly variable population such as ours, one could expect death to result in **some** individuals in half or a third of the time.

What I am suggesting is that, with the articles of diet present in major amounts, it is not possible to test for indications of long-term human toxicity using animals, simply because to increase the dosage of the suspect material in the diet enough to give reasonably clear-cut experimental results in animals is impossible. To do this the component would have to amount to the total weight of the diet or more. Still less is it possible to make such tests on humans, for the institution of slavery was abolished in advanced societies half a century before the emergence of human nutrition as a science, which is one reason why we know so much less about human nutrition than about the nutrition of domestic animals.

Source: M.V.Tracey, The Price of Making Our Foods Safe and Suitable, Food quality in Australia (Academy Report No. 22), Australian Academy of Science, Canberra, pp. 69–71, 1977.

The LD_{50} is expressed in terms of mg of poison per kg of body mass of the species. Such a method of representation eliminates the variation of body mass between species, so that when rats and people are compared, at least the mass ratio is taken into account, even though many other characteristics may differ. When, for example, adult male rats are used, it should be remembered that female rats are on average more sensitive than males, as are also young and sick rats. Not all people exposed to chemicals are adult male and healthy!

It should be noted that the smaller the LD_{50}, the *more acutely toxic* is the substance. The units of mg/kg are equivalent to μg/g (or ppm) body mass. An approximate description of toxicity is:

$LD_{50} < 1$ mg/kg	**Extremely** toxic;
$1 < LD_{50} < 50$ mg/kg	**Highly** toxic;
$50 < LD_{50} < 500$ mg/kg	**Moderately** toxic;
500 mg/kg $< LD_{50} <$ 5000 mg/kg	**Slightly** toxic

In common with many other variable factors of a biological nature, the susceptibility of individual members of a group within a species varies. This variation follows an approximately normal (bell-shaped) distribution if the variable, such as dose of poison, is *plotted on a logarithmic scale* — the so-called log-normal distribution — see Figure 2.1, overleaf.

A plot of the number of individuals responding (on average) against the logarithm of the concentration (increasing) is shown in Figure 2.2 (overleaf) and follows a sigmoidal curve.

As the logarithm of the dose is increased, more animals die and they do so at an increasing rate until the dose where 50% die. At this point the rate falls off again as the increasing number dying flattens out to 100%.

An unexpected reaction

The infamous Russian monk Grigori Rasputin, the power behind Tsar Nicholas II, was assassinated during the First World War when Russia was doing very badly. He proved difficult to kill and was finally shot, but some attempts involved botched chemistry. At the palace of Prince Yusupov a Dr Lazovert mixed into some cakes generous quantities of white powder from a box allegedly containing potassium cyanide. Rasputin appeared to suffer no ill effects. What had presumably occurred was a reaction between carbon dioxide from the air and the potassium cyanide. Carbonic acid ($pK_a = 6.35$) is a stronger acid than hydrocyanic acid ($pK_a = 9.22$) and so readily displaces it. The potassium cyanide had probably been converted to harmless potassium carbonate.

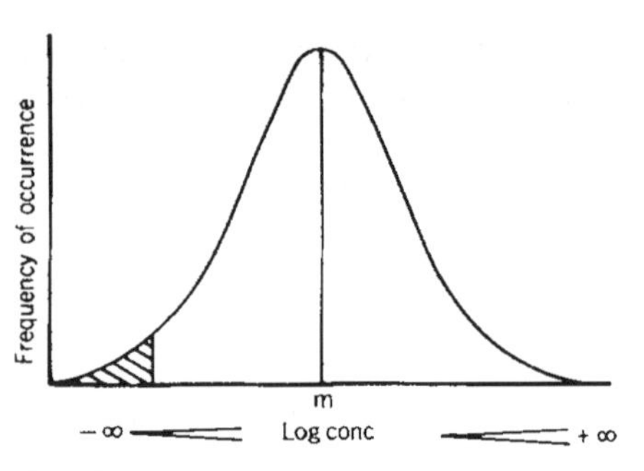

Fig. 2.1
The log-normal distribution

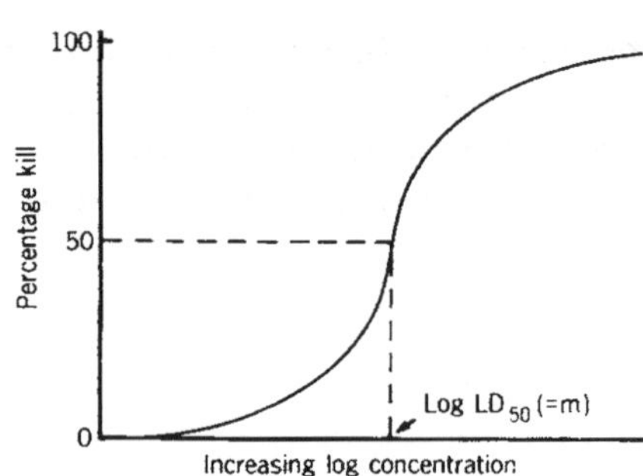

Fig. 2.2
The cumulative log-normal distribution

The dose for 50% dead, LD_{50}, is chosen as the most sensitive point on the graph to the change of (log) dose. Other points, such as LD_5 can be chosen when cost or humanitarian considerations apply. Note that the sigmoidal curve is just the graph of the cumulative area under the normal curve as we go from left to right. Remember that the concentration scale is logarithmic and not linear, so high concentrations are compressed.

For a discussion, explanation, and some further theory on this issue of logarithmic response, see Appendix II.

Chronic toxicity

We now move from acute effects to the long-term chronic effects. A *chronic effect* is the result of exposure to *repeated, small non-lethal doses* of a potentially harmful substance that causes cumulative damage over a long period of time. Asbestosis is a classic example.

Many materials act as *sensitisers*. These are not strictly dose dependent. Examples are turpentine, which cause dermatitis and asthma; epoxy resins cause dermatitis in about 50% of industrial workers. Toluene di-isocyanate (TDI) used as one component in the manufacture of polyurethane is a very potent sensitiser.

An *allergy* is an unnecessary response from the immune system — allergies take time to develop. The longer a person works with a sensitiser, the more likely an allergy is to develop. Once an allergy to one chemical develops, a person can be automatically sensitised to other chemicals (cross-sensitisation), and can become more and more sensitive to smaller and smaller amounts of material. For further detail see Allergies in Chapter 15.

Mixtures of chemicals

Interaction of materials

Materials can interact additively, for example when working with, say, turpentine it is absorbed through the lungs during the day. It reaches the brain and acts as a neural toxin, causing dizziness. At the same time it is detoxified in the liver. If you then go home and have a few glasses of red wine, the alcohol behaves in a similar manner — acting as a neurotoxin until detoxified. Additivity occurs when two substances attack the same target in the same way, and share a common path for their destruction.

Synergism

This occurs when one substance reinforces the effect of a second. Smoking reduces the ability of the lungs to clear dust, so lung cancer from smoking (say × 10 normal) and lung cancer from asbestos (say × 5 normal) increases in an asbestos worker who smoked by a factor not of 10 × 5 = 50 but × 90. The use of some chlorinated hydrocarbon solvents (carbon tetrachloride) followed by the consumption of alcohol has caused many deaths, because the enzymes used in detoxification of alcohol are inhibited by CCl_4. If a medication counterindicates the consumption of alcohol, then this should be extended to include exclusion of similar-acting solvents.

The acceptable daily intake (ADI)

The ADI is the amount of a food additive, pesticide, or veterinary drug, calculated on the basis of body mass, that can be eaten daily for a lifetime without adverse effect. If a particular ADI is x mg/kg body weight, an average 70 kg man can afford to eat $70x$ mg/day for a lifetime. A small child weighing 10 kg could thus afford to eat $10x$ mg/day if it were assumed to have the same metabolism as an adult (differences in metabolism are allowed for separately). The method of establishing the ADI is to do experiments on animals, increasing the quantity of the additive to establish at which level acute and chronic toxicity occurs in any animal (that is, the level at which an immediate poisonous effect is observed, and the one at which long-term effects occur). At least two species of animals are used, and the most sensitive species is taken for determining the level. The number obtained is then divided by a factor — generally 100 — to take into account possible variation both among and within species, sometimes ten when the data are very good, or 1000 when they are inadequate. These factors are called 'uncertainty factors'.

The toxicological arguments for these factors are not as strongly based as we would like, and contain a multitude of uncertainties. Interestingly, the uncertainty factor for essential metals is set much lower at 3–10, because the range between helpful and harmful is often not very large. We evolved over aeons in a chemical soup with large 'natural' variations in mineral levels in soils.

As well, infants and children cannot be considered as just miniature adults. They eat proportionately more food per weight. With less-developed detoxifying mechanisms, they are often more sensitive to chemicals.

We don't need synthetic pesticide residues in our food, so of course there are no positive (vitamin-like) aspects for pesticides in food plants, and so there is no need for a minimum intake! However, there is the same need to set a maximum, to protect human health from potential toxic effects.

Veterinary remedies can also leave residues in the animal flesh, especially at the point of injection. For food-producing animals, the same issues of residues arise as for plant-protection chemicals. The regulation of medicines for humans has been spared these problems ever since the demise of cannibalism!

Modern pesticides are designed to degrade after use. The levels of non-systemic pesticides are further reduced by peeling and cooking foods before eating.

Some metals such as lead, cadmium and mercury, are treated like pesticides — a maximum but no minimum. Others such as copper, iron and zinc are essential minerals and are treated like vitamins; both minimum and maximum levels need to be set. Absence of essential minerals means, for example, that consumption of pure distilled water in large quantities is a potential health hazard.

The ADI approach is based essentially on animal studies and checks for an important, yet limited, set of effects. It is a link in the regulatory chain that always needs strengthening. The ADI are derived from the results of long-term feeding studies with laboratory animals and from observations in humans. The studies use available safety data from a wide range of animal studies. ADIs are set by public health authorities and internationally by UN agencies such as the FAO/WHO (JMPR).

Having set an ADI for a chemical, it is now necessary to look at the produce, what it might contain, how long it should be left after spraying before harvesting, and how it will be processed, cooked and eaten. The upper level of what is acceptable in the produce is called the MRL (maximum residue limit), or in the USA, a 'tolerance'.

(See also Chapter 7, Pesticides and alternatives, and Chapter 15.)

How maximum residue limits (tolerances) for pesticides are set

To meet the requirement for MRLs, farmers using pesticides must adhere to good quality control. This is called good agricultural practice (GAP). GAP is a quality assurance approach determined by the registration authority, and set as a condition of use; MRLs are the corresponding quality assurance parameters to ensure that GAP has been used. GAP is a complex business encompassing factors such as

1. application techniques,
2. crop cultivation and weather conditions, and
3. biodegradation properties of the chemicals.

The GAP rules include:

- Use of only authorised pesticides, under safe conditions, and at the levels recommended for effective and reliable pest control under actual conditions.
- Setting withholding periods and other measures to ensure that a residue is the smallest practical amount.
- Ensuring that occupational health and safety, as well as environmental consequences, are accommodated.

No compromise
Once an MRL is established, however, it is morally wrong to argue that health is not compromised by an occasional small amount of produce that has breached the MRL. This produce is stopped for sensitive export markets, but has occasionally been allowed onto the domestic market. While health is most unlikely to be compromised, such an act creates an enormous breach of trust with the community. It is just like arguing that it is alright to 'crash' a red light on the odd occasion when it seems obviously safe. This is a slippery slope and is just not acceptable.

The MRL is thus a measure of quality control, not a measure of public (i.e. consumer) health risk. Residue surveys on produce 'at the farm gate' are used as an indicator that best practices are in fact in place. (One weakness is that no attempt is made to consider the possible residue effects of mixtures of pesticides either used together, or occurring on the produce together because of consecutive use.) Before the sale of a new agricultural or veterinary chemical is approved the manufacturer is obliged to determine the nature and levels of residues that can be expected to occur in the raw agricultural commodities when the product is used as recommended — that is, with the principles of GAP.

Although considerable safety factors are also built into setting MRLs, the system is only as good as the ability and willingness of the user to follow label directions. In Australia the *monitoring* of chemical use remains the function of the State jurisdictions.

A variety of pesticides are now used on a range of foods. Some pesticides are used more widely than others. Some foods are eaten in greater quantities than others. Diets vary widely in the community. There may be additional crops in which that pesticide may be used in the future. When setting an MRL, 'room' must be left in the ADI for the potential residue from that pesticide's use in new applications.

The total of pesticide residues in the total diet is the ultimate consideration. There is some flexibility, limited by practicalities, in setting the MRL for each pesticide/crop combination (e.g. increasing withholding times). If the residue intake comes close to the ADI (as has been suggested for infants in the USA), then some registered uses of the product may have to be deleted from the label.

Allocation of MRLs between produce classes means you push down the residue levels for each as far as 'reasonable', to keep the weighted total as low as 'possible'. There are value judgments involved. While the worst case assumptions are made, these are in real life mitigated by factors such as:

- When growers apply pesticides, they do not always apply them at the maximum rate or at the closest allowable time to harvest.
- Only some of each commodity reaching markets (and eaten) will have been treated with a particular pesticide.
- Harvesting of some crops, such as cucumbers and tomatoes, takes place over a long time and the residues will decline further.
- Some crops, such as potatoes and cereals, are stored before release.

These factors are likely to generally reduce the residue in food. Conversely, non-adherence to GAP, by not observing withholding periods sufficiently, overdosing, and other misapplications (deliberately, or from not understanding the instructions) are likely to increase pesticide residues. Pesticides used closer to consumer purchase, such as grain protectants, are more likely to run into trouble with the ADI.

The bottom line is that the sum of the mass of residues eaten, calculated from actual dietary surveys over all the foods, in a variety of diets, for a spread in population ages, must be kept well below the health-determined ADI for that chemical. (See also Pesticides and alternatives in Chapter 7.)

Mass movement of chemicals: some physical chemistry

CASE STUDY

Hexachlorobenzene (HCB) and other chlorinated waste in contaminated soil in an industrial site has sunk into the subsoil (medium 1). The material is of concern because it is fat soluble and can bio-accumulate. HCB is not soluble in water but it can be carried physically on soil particles into the ground water beneath the plant (medium 2). This water can move into the wider environment. HCB can move in the river and ocean sediments (medium 3). Organisms living in sediments take it up with their food (medium 4). These organisms in turn are eaten, and the material moves up the food chain (medium 5). Chemical and biological degradation processes are not considered important paths for removal of HCB from water and sediments. HCB is slowly biodegradable; 3 to 6 years in soil and 30–300 days in water.

HCB is non-volatile, which means it has a very low but not negligible volatility (in contrast to non-volatile materials such as granite which have no vapour at all). HCB is virtually non-miscible with water, but once carried into water, it co-evaporates from water into the air (medium 6) and is carried far afield. The process is akin to the more familiar steam distillation, but occurs at ambient temperature. This co-evaporation is the basis for the international concern with POPs (persistent organic pollutants) such as polychlorinated biphenyls (PCBs), DDT and HCB which can travel globally. For a more technical discussion, see Henry's Law in Appendix III.

Reporting amount of material

Different scientific cultures use diverse ways of displaying their results. This has arisen historically and often reflects the specific needs of the discipline. It can be very confusing. The following discussion attempts to clarify some of the ambiguities.

Mass/volume basis

If beer contains say 2 milligrams of vitamin C in each litre, then the concentration of vitamin is said to be 2 milligrams per litre. As a litre of beer (essentially water) weighs one kilogram, the vitamin concentration can also be described as 2 milligrams per kilogram. This concentration is also described as two parts per million (2 ppm). Use of ppm assumes (sometimes incorrectly) that the measure is on a weight to weight (w/w) basis. It is more correct to say mass/mass rather than weight/weight, but w/w is the symbol used.

Take, instead of a level of 2 milligrams per litre of beer, a level of only two *micro*grams per litre. A microgram is a millionth of a gram, and one kilogram is a thousand grams, and a million times a thousand is a (US) billion. The concentration is therefore also described as one part per billion (ppb) — mass to mass assumed.

The term ppt (parts per trillion) should not be confused with the marine scientists' use of this abbreviation (for salinity) as parts per thousand.

A concentration is also known as a *level.*

US health warnings
Have you ever wondered why the health warnings on cholesterol from the USA seem to suggest quite unrealistic targets for blood tests? The following explanation might help. In the USA, the target level of cholesterol in the blood is 200 milli**grams** per 100 millilitres of blood. In Europe (and Australia) the magic number is 5.2 milli**moles** per litre of blood. A mole of cholesterol is the molecular mass of cholesterol in grams, 386 grams. A millimole is 386 milligrams. Thus, $5.2 \times 386 \div 10$ = 200 mg/100 mL.

Mole/volume basis

Chemists also measure the amount of material in terms of moles. A mole of a substance is M grams of that substance where M is the molecular mass (MM) of the substance. By using moles, molecules of one substance are counted in relation to molecules of another substance, irrespective of the mass (weight) of each substance. It is a one-to-one comparison on the basis of molecular entity.

Volume/volume basis

Gases and vapours in air are also measured in terms of volume per volume; that is, millilitres per cubic metre (m^3) or kilolitres of air, giving a ratio of a molecule of pollutant to a molecule of air. This result does not depend on the air temperature or pressure of the day, nor on the molecular mass (MM) of the compound of interest. Those setting threshold limit values (TLV) use this approach and call it ppm (parts per million — v/v).

THRESHOLD LIMIT VALUES (TLV) FOR POLLUTANTS IN AIR

- TLV are advisory limits on the concentration of fumes/vapours to which average, healthy workers may be exposed.
- TLV are determined by an unofficial body — the American Conference of Governmental Industrial Hygienists. TLV is defined as 'airborne concentration of substances to which it is believed that nearly all workers may be repeatedly or continuously exposed without adverse effect'.

- The TLV is set from data based on human experience (epidemiological studies) or testing of animals for short-term effects of exposure to chemicals. They can be expressed in distinct ways. They are usually expressed as mass/mass, that is in ppm (mg/kg). Alternatively, an m/v measure is used such as mg/m^3. These can be related (see below).
- There are three categories of TLV:
  1. TLV–Time Weight Average —'average' permissible exposure over an 8-hour day (40 hour week);
  2. TLV–Short-Term Exposure Limit — maximum concentration to which a worker should be exposed for not more than 15 minutes, no more than four times per day;
  3. TLV–Ceiling — maximum concentration that should not be exceeded even instantaneously.

Worksafe Australia publishes Exposure Standards for the Atmospheric Contaminants in the Occupational Workplace — National Exposure Standards (NOHSC: 1003 (1995)). International values are available on the Internet.

The smaller the number, the more toxic is the substance.

Substance	TLV ppm (mg/kg)
Ethanol	1000
Acetone	750
Xylene	100
Glycol ether	5

Mass/mass basis

In liquid and solid mixtures, concentrations are generally given as mass per mass, mg/kg, but this is also called ppm (w/w). If applied to gases, this measure is independent of the molecular mass of the compound of interest, but depends on the temperature and pressure of air on the day. The average MM of air is 29 and the volume occupied by one mole of air at NTP (normal temperature and pressure) is 24.5 L. Thus the mass of air in one litre is 29/24.5 = 1.18 g, that is 1.18 kg/m^3, and so the following conversion factors apply:

$$1 \text{ ppm (w/w)} = 1 \text{ mg/kg} = 1.18 \text{ mg/m}^3.$$
$$\text{Therefore } 1 \text{ mg/m}^3 = 0.85 \text{ ppm (w/w)}.$$

For this reason, when reporting pollutants in air, ppm (w/w) and mg/m^3 tend to be used interchangeably when high precision is not warranted (i.e. most of the time).

Conversion of units

Data are normally required in terms of mass of pollutant per volume of sampled air, reported as mg/m^3. To convert data in ppm (v/v) to mg/m^3, multiply by the MM of the compound, and divide by the molar volume of air.

Thus for HCB (MM = 285) the conversion factor is 285/24.5 = 11.6.

$$\text{Thus } 1 \text{ ppm (v/v)} = 11.6 \text{ mg/m}^3 \text{ and, conversely,}$$
$$1 \text{ mg/m}^3 = 0.09 \text{ ppm (v/v).}$$

Body chemical warfare

Chemicals are not divided neatly into toxic and non-toxic compounds. The dose can generally convert them in either direction. The body produces for its own purposes such nasties as hydrogen cyanide, nitric oxide, and bleach, albeit in small amounts and at very specific locations. These chemicals are used to kill invading organisms through internal chemical warfare.

On the basis of molecules

- Is the report in terms of the mass of active ingredient(s), or the formulation, or the principal component of a residue?
- Are residues reported in terms of ash weight, dry weight (how dried?), flesh weight (with or without body fluids?), wet weight, or lipid (fat) weight? This is critical when, say, comparing the level of HCB in nursing mothers' milk.
- For air contamination, is the result expressed as the concentration in the particulate collected, or in the air volume from which the particulate was collected?

The data may relate to a preselected part of the sample (residue in the fat of milk), or the results may be normalised (adjusted) afterwards to refer to a wider situation (the whole milk). Extraction of soils, sediments, and sludges is necessary before analysis, and the efficiency of the extraction procedure(s) can vary widely and may not be comparable. For example, the US Environmental Protection Agency sets a limit for nitrate (NO_3^-) in ground water used for drinking of 10 mg nitrate as nitrogen per litre. The Europeans define their level on the basis of NO_3^- itself, and so the equivalent amount would be about four times the US level (44 mg/L). On top of that, the US level is set only on the basis of causing blue baby syndrome — not on the basis of cancer-causing potential. The level of cadmium in Australian fertilisers was once specified as mg/kg of fertiliser but is now specified as mg/kg phosphorus, (i.e. on the amount of phosphorus in the fertiliser). The numbers are not directly comparable.

The dose — a pharmaceutical example

A single tablet of regular-strength aspirin contains 325 milligrams (mg) of drug. An adult takes four tablets per day, by mouth. The total *weight* of aspirin ingested on that day is 1300 mg or 1.3 g. *But the weight is not the dose.* Aspirin, like most chemicals, works on the body (to relieve pain, etc.) by dissolution in body fluids, so it is the concentration, not the absolute weight, that it is important. (For example, it is blood alcohol concentration, 0.05 mg/100 mL, rather than the absolute amount of alcohol,

Saccharin consumption
In the 1976 saccharin cancer scare, surveys showed that a tiny portion of the US population consume more than 36 cans of Diet Coke per day. With more and more eating taking place in restaurants, accurate data are lacking, but scientists are still under increasing pressure to supply firm answers.

that is the legal limit.) The amount of aspirin in the body, divided by the weight of the person, is the critical measure. For a 65 kg person, the aspirin dose is 1300 mg/65 kg = 20 mg/kg body weight (b.w.). The dose period is usually taken as one day, so this person's dose would be 20 mg/kg b.w./day.

For a 20 kg child, the same *amount* of aspirin would become a dose of 1300 mg aspirin / 20 kg b.w. = 65 mg/kg b.w./day. For the same intake, the lighter person receives the higher dose.

Calculation of doses for environmental chemicals follows the same logic.

Environmental example

If the ground water becomes contaminated with say 2 micrograms of X in each litre of water, and adults drink two litres of this water, children one litre each day, then:

Each adult takes in 4.0 μg/day. Each child takes in 2.0 μg/day.

If adults weigh 80 kg and children weigh 10 kg, their respective daily doses are:

Adults 0.05 μg/kg b.w./day.
Child 0.20 μg/kg b.w./day.

A second important factor is the number of days the dose continues.

Duration of exposure, as well as *dose,* is important. Finally a decision needs to be made as to whether this constitutes a health hazard.

In air, it is the weight of material in a given volume of air, usually mg per cubic metre (m^3) and the volume of air a person breathes each exposure period, usually m^3 per day. Suppose the air in a petrol station contains 2 mg of carbon monoxide per cubic metre and a worker breathes that air for an eight-hour work day. An adult working at a moderate level of activity would breathe in about 10 m^3 of air in eight hours. In each shift the worker thus inhales 20 mg of carbon monoxide. He probably inhales lesser amounts in the non-working part of the day as well from street levels.

For food, the intakes are estimated in the same way, although with less confidence. Toxicologists are concerned to ensure that risks to individuals with highly eccentric eating habits exposed to higher than average amounts of chemicals are still not at risk. See the above discussion of the 'acceptable daily intake'.

Faith in a result of a measurement

The most *precise* answers come from direct chemical analysis. Very low levels of chemicals can now be detected, often so low as to be irrelevant. Even with all this precision, the *accuracy* of the answer and its ultimate usefulness depend on how representative the sampling was. Many samples from all the different circumstances must be included to give a *reliable* result.

Sampling is critical to analysis.

Example: When sampling large-scale soil contamination at an industrial plant, it is impossible to test 'all' the soil. Samples must be 'representative', which means individual samples must provide a reliable indication of the range and distribution of contamination of the whole site. Then statistical analysis will give a 'confidence range' within which the 'real' result can be expected to be found.

Example: If a pesticide is used on a tomato crop, the amount that ends up on the tomato you eat is not so easily worked out. It depends on the amount that was actually sprayed, how much actually landed on the tomato, to what extent wind and rain removes it, how quickly the chemical is broken down biologically, and the length of time between spraying and picking, selling, and eating the tomato. If tomato juice is made, you need to know whether the chemical concentrates in the juice, and so on.

Sometimes computer 'modelling' is used to predict the behaviour of chemicals in the environment. For movement of chemicals in the air (and water) a mathematical model is often devised and tested against real data. If it works well, it can be used to try and predict chemical movement in 'what if' situations, where measurements are not available or cannot be easily made. Models vary in their quality, and are very dependent on the initial assumptions made in setting them up.

Why should we trust the experts?

In the majority of cases there is no shortage of available toxicological data, either in the form of reference works or as the original literature. The difficulty arises in selecting the valid data from the mass of sometimes conflicting information. The international agencies differ in the extent to which they use the 'grey' literature — the large body of unpublished industrial research. The OECD stands out in making use of this material.

Whereas errors in computer data banks can be corrected, once a conventional publication is distributed, that error survives as long as the document survives. The International Agency for Research on Cancer (IARC) publication on HCB residues in agricultural products confused and misspelt chemical names. *Ortho*-Toluidine is not the same as the carcinogenic chemical once used in chlorine test kits, *ortho*-tolidine. The Chemical Abstract Service (CAS) number must be checked; it is the only secure reference point. (See Appendix I, Nomenclature).

Older works are superseded by newer ones, although the older material can be useful if its limitations are noted. N-Phthalylglutamic acid imide was reported in the literature of the 1960s as an extremely safe sedative. It was not addictive, and so non-toxic that it could not be used for suicide attempts. Today, (better known by one of its trade names — Thalidomide™), a different story is told.

Where important decisions are going to be based on data, the data must be validated. The published report is not enough; the experimental

Peer review

details on which it is based must be examined by a competent third party. Are the data for the appropriate chemical substance? Cadmium sulfide has a very different toxicity to cadmium sulfate although both contain cadmium. Methylmercury is very different to mercury vapour or inorganic mercury; arsenic trioxide differs from arsenic pentoxide, and both differ from the situation of eating extremely high concentrations of arsenobetaine (arsenic-protein) complex in shellfish, which passes through the human gut without being broken down.

The work quoted in the reports of national and international agencies is subjected to an assessment process using criteria discussed below, so a lot of the sifting for their reports will have been done.

Peer review of material is much touted as a quality measure, but it is not infallible; it does ensure that a substantial quantity of incorrect information does not see the light of day. Journals vary enormously in the quality of their refereeing. Conversely, 'unpopular research topics' out of favour with the establishment do have problems in being published in the mainstream journals. (See also Caffeine in Chapter 15.)

Questions that should be asked are:

- 'Who did the work?'. 'Trust me, I'm a . . . fellow of a learned society, a chartered chemist, a professor in a famous department'. These are not guarantees, but reputations are important to these people, and they are much easier to lose than they were to gain. Multi-author papers, even with 'a famous man' at the front are more likely to have poorer overall quality control — the chances are that the lead author had little direct involvement in the day-to-day activities.
- 'Who sponsored the research?' While industry submitting data on its own behalf inherently arouses suspicions, much toxicology is done by third parties on behalf of industry, anyway. This work can be done in research institutes, universities, or commercial laboratories. In each case the researcher is being paid to do the work unless it is on an unfettered grant. These are rapidly disappearing.
- 'Are the data presented in the publication an accurate reflection of the work that was actually carried out?' The 1970s revealed wholesale fabrication in contract toxicology testing in the USA. This resulted in

a code of good laboratory practice (GLP) which is enforceable for any organisation wanting to submit test data to regulatory bodies. The GLP encompasses, *inter alia*, a traceable pathway through the research, giving a written scheme of work with the responsibility for all aspects clearly defined. All the original data, including any corrected errors, must be dated and signed; it must be securely indexed and archived to allow complete reconstruction of all the experiments from the written record alone.

Good laboratory practice
GLP ensures that the data are a true reflection of the experiments; it does not ensure that the experiments are relevant, or have any scientific merit. GLP is about 'how', not 'what', work was undertaken.

- 'What test animals were chosen and why?' In ecotoxicology, all species will have some relevance since they all live in the natural environment. However, the environment is highly variable, and some species will obviously be more relevant than others. Where the object is to extrapolate animal data to humans, there is a real problem. Where all species give a similar result, we can expect humans will behave likewise. Where there is a wide discrepancy among species, the mechanism for toxic action needs to be addressed. The chemical 2,3,7,8-tetrachlorodibenzo-*p*-dioxin (TCDD) has an LD_{50} of 1 μg/kg for the guinea pig, which is three orders of magnitude more sensitive than the hamster at 5000 μg/kg. Penicillin is also very toxic to guinea pigs. It was just as well that for its introduction during World War Two there was no time to test penicillin before using it on people!

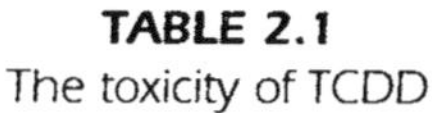

TABLE 2.1
The toxicity of TCDD

Animal	LD_{50} (μg/kg body weight)
Guinea pig	1
Monkey	70
Rat	200
Dog	3000
Hamster	5000

The duration of the test must be long enough for effects to appear. The number of animals must be large enough to detect low levels of effect. One random death in five animals represents a 20% response; one in 20 only a 5% response.

Testing with a number of levels of dose allows a dose–response curve to be established and its nature can be very revealing; it may not be linear, it may be discontinuous, or it may have upper and lower thresholds. Experiments at single levels of dosing are far less useful.

- 'How were the results interpreted?' Results showing high statistical significance may be of little biological importance; conversely, important biological trends should be examined further, even in the absence of statistical significance. A wide range of mutagenicity tests is now available. These are rapid and inexpensive, in contrast to the time consuming and expensive animal carcinogenic experiments. However, they ignore the important DNA repair mechanisms which

Statistics
Statistical significance is an arbitrary ruler devised for mathematically illiterate experimenters.

continuously protect us against assault of natural as well as synthetic chemicals and agents (such as UV and cosmic radiation). There is a great deal of biological variability between species, within species, and even within a single strain of a species. Allowance for this variability must be made in quoting results. LD_{50} and no observable effect level (NOEL) values with two or three significant figures are suspect.

EXPERIMENT

Use of the invertebrate Daphnia to demonstrate that common substances of all types, food or drugs, can have a toxic effect on Daphnia.

The effect is due to the concentration (amount per unit volume) used, as well as the potency (extent of biological effect per same amount).

Visible hearts, eyespots and 'babies' endear Daphnia to children. They can be collected in spring from ponds, or purchased from supply companies.

Method

Substances used:

- coffee — use undiluted drip coffee.
- aspirin — dissolve a 500 mg tablet in 10 mL of water.
- allergy capsule — dissolve one Benadryl caplet in 10 mL of water
- cooking sherry — use undiluted
- chewing tobacco — mash a pinch ($\approx$ 1 g) in 10 mL of warm water with a strirring rod until the liquid is amber (or add a few mL of saliva as well).

Transfer a Daphnia to each of five test tubes filled with pond water or growth medium to a volume of 10 mL. With a medicine dropper full of test solution, deliver one drop of the test solution into the first test tube, two drops into the second, three drops into the third, and four drops into the fourth tube (the fifth remains as a control — if the Daphnia die in this tube, the experiment must be aborted). Set 20 drops equals 1 mL, 1 cup = 5 oz, and 1 oz = 30 g. Assume that 80% of the nicotine is extracted.

Note the test tube that required the least number of drops to kill the Daphnia. Study the five test tubes every few minutes. Note the macroscopic changes in movement, as well as changes in heart rate, gill movement, etc. under a microscope.

Typical results

Chewing tobacco causes quick death at the lowest doses. Cooking sherry first slows the heartbeat rate, and then is lethal at higher doses. Aspirin and allergy capsules are lethal at the highest doses; and coffee causes a racing of the heart, but is not lethal. (One could also compare decaffeinated with normal coffee to see whether it is the caffeine that is important, low nicotine with normal tobacco, or low or no alcohol with normal beer.)

Common substance	Active ingredient	Amount of active ingredient	Conc. in first test tube	Rodent LD_{50}
Aspirin	acetylsalicylic acid	500 mg/ tablet	0.25 mg/mL	1.5 g/kg
Allergy capsule	diphenhydramine	50 mg/ tablet	0.03 mg/mL	0.5 g/kg
Chewing tobacco	nicotine	3% w/w	0.01 mg/mL	0.23 g/kg
Coffee	caffeine	150 mg/cup	0.005 mg/mL	0.13 g/kg
Sherry	alcohol	17% (v/v)	0.7 mg/mL	10.6 g/kg

The lowest active ingredient concentration in mg/mL is the concentration given in the table for the first tube, times the number of drops needed to kill the Daphnia. Thus for aspirin it was 0.25 x 4 = 1 mg/mL; for the allergy capsule it was 0.03 x 4 = 0.12 mg/mL; for the chewing tobacco it was 0.01 x 1 = 0.01 mg/mL; for the sherry it was 0.7 x 3 = 0.21 mg/mL.

This can be plotted as a horizontal bar (range 0–2.0 mg/mL) for the four materials, one above the other.

Such a figure shows that the different substances have different potencies. Nicotine is very potent, but alcohol is not. However, alcohol is toxic in this experiment because of its high concentration in a common product like sherry. Caffeine was not toxic at the highest concentration tried, but this was only 4 x 0.005 = 0.02 mg/mL, so we know only that it is less toxic than nicotine, nothing more.

Can we assume that what is toxic to Daphnia is toxic to humans? Well there are differences. Daphnia swim around in the chemicals, whereas we absorb them through mouth, or lungs, or skin.

We have tested only short-term immediate effects, while long-term chronic effects (e.g. cancer) are also of great importance to humans. Perhaps the studies on rats are likely to be closer to the human case than studies on Daphnia. The last column of the table gives the LD_{50} doses for rodents. LD_{50} values are reported as grams of compound needed to kill, on average, 50% of a sample of rodents whose actual weights are scaled to 1 kg weight. These values approximately parallel those found for Daphnia, except for caffeine which is the most toxic (least LD_{50} value to kill) of all the active ingredients for rodents.

Watching Daphnia struggle under the influence of substances such as tobacco and sherry can send a powerful deterrent message about the abuse of substances.

Source: Toxicology for Middle School, S. Parrish, Ruth N. Russo, J.Chem. Ed. **72**(1) pp. 49–50, 1995 .

We now move from the laboratory to the committee room where the decisions are made.

Regulatory toxicology

How does regulatory toxicology really work? Ronald E. Gots is with the National Medical Advisory Service, Bethesda, Maryland USA, and is a regulator with a wider view. Policy relies on admixture of the known, the possible, and the prudent.

The purpose of scientific toxicology is to uncover truths. It uses rigorous methodology, testing, and retesting, according to scientifically agreed protocols. These protocols change only as new scientific knowledge makes this appropriate.

The purpose of regulatory toxicology is to make decisions. Its aim is to protect in the presence of uncertainties, and to develop systematic approaches to those decisions. It must interpret scientific toxicology results, assuage public perceptions, and respond to political pressures.

Scientific toxicology	Regulatory toxicology
Generates and interprets scientific results	Develops policies
Studies responses of biological systems to toxic agents	Weighs risks versus benefits
Based upon experimentation	Makes safety judgments
Characterised by reproducibility	Responds to public and political demands. Varies from time to time
Apolitical	Heavily reliant upon assumptions: • animal to human; • high to low dose; and so on

Mathematical risk assessment (in the biological domain) is a technique which attempts to convert biological data into regulatory action. It is a method which allows judgments to be made about low-dose human exposure by extrapolation from the high-dose animal experimental evidence. Mathematical modelling can be used to describe known biological behaviour, but it cannot be used to predict biological behaviour that is scientifically unknown.

A procedure which guarantees quality assurance does not necessarily reflect scientific truths, but is designed to bring a semblance of order to chaos. It allows agencies to codify their approaches, make decisions less arbitrary, and explain in some relatively consistent fashion the basis for those decisions.

(Source: Ronald E. Gots)

Gots lists many modern 'syndromes' of today's medical clinical ecologists, and compares them to the Middle Age practitioners' belief in 'humours' as the causes of disease; 'Modern antiscience sentiment is linked to anecdotal illness'. Gots would enjoy the following:

CARROTS KILL!

- Nearly all sick people have eaten carrots at some time.
- 99.9% of people who die from cancer have eaten carrots.
- 97% of Fascists have eaten carrots.
- 90% of juvenile delinquents come from homes where carrots are eaten.

Among all people born in 1839 who later ate carrots, there has been 100% mortality.

All carrot eaters born between 1900 and 1915 have wrinkled skin, have lost most of their teeth, have brittle bones and failing eyesight, if the ills of eating carrots have not already been a cause of death.

Animal studies

Rats force fed 10 kg of carrots a day for 30 days developed bulging abdomens. Their appetites for wholesome foods were destroyed.

Carrots can be linked to AIDS; eating them breeds wars and economic rationalism.

What to do?

Eat orchid petal soup. An exhaustive literature survey including the Internet shows that there are no statistics linking orchid petal soup to any problems.

(Modified from (US) Rural Business — July 1990)

Science defines itself by the attempt to eliminate ambiguity. However, one should never be under the illusion that in areas such as toxicology, science is anything more than temporarily successful. Like everyone else, scientists behave according to the culture in which they were brought up.

Thomas Kuhn recognised that each discipline has its paradigms, basic hard-wired concepts that form the essential belief system of that discipline. Like the courts of law, regulatory authorities are presented with evidence; good, bad, and indifferent. Some of it is by scientific experts (working for the applicant); some from ordinary witnesses and lay activist researchers.

Regulatory authorities have to make decisions, and these must be on the basis of the current evidence. Unlike the courts, however, these decisions can be reviewed regularly.

While the 'regulatory court' places the 'onus of proof' on the applicant to prove safety against a finite number of (current) criteria, once in the system, an almost impossible 'onus of proof' has often shifted *de facto* onto the community to prove danger before deregistration.

The anthropology of the regulatory culture — risk assessment

The Office of Regulatory Review of the Australian Industries Commission surveyed the practices of national and Commonwealth regulatory agencies to discover how they actually went about their tasks of analysing safety risk and then regulated the acceptable risk. The following are extracts from that report, but it is often not clear whether the odds refer to lifetime risk, or risk during operations — quantitative comparisons are therefore difficult.

> In terms of setting regulatory objectives, two distinct groups of agencies emerge . . . The first group have as their target an 'arbitrary' level of risk. This target level of risk might be implicitly set in legislation or in an agency's charter. These 'risk-targeting' agencies do not use cost–benefit analysis, or any other formal analytical techniques when setting objectives. Once the arbitrary level of risk has been set, these agencies attempt to formulate regulations and controls that keep safety risk at or below this predetermined level. In the Department of Health Chemical Safety Unit (CSU), the National Food Authority (NFA) and the Therapeutic Goods Authority (TGA), all appear to fall into this category. So does the Civil Aviation Authority (CAA), which will accept an accident rate of one in 10^7 in many circumstances . . . It accepts an accident rate of one in 10^4 hours [flying] for light aircraft because, in reality, that has long been the accepted rate.
>
> The other agencies, Australian Radiation Laboratories (ARL), Federal Bureau of Consumer Affairs (FBCA), Federal Office of Road Safety (FORS), National Occupational Health and Safety Commission (NOHSC), and National Road Transport Authority (NRTC) do not specifically set out to achieve a target level of risk. Rather, the level of risk that these agencies are prepared to tolerate arises from other considerations . . .
>
> ARL advises regulatory control if the individual risk of contracting fatal cancer at work is likely to exceed about one in 25 000, although the average occupational exposures in the radiation industry lead to a risk level of around one in 10 000, which is surprisingly high.
>
> The agencies in this second group adopt varying degrees of formal economic analysis of regulations . . . However, they also have some methodological differences, particularly in the way non-monetary costs and benefits are valued. Other agencies in this group undertake 'partial valuation' cost-benefit analyses which omit to place a value on human life . . . Regardless of the method they use to determine the level of safety risk, the agencies' actions show a wide range of safety risk levels, some much higher than others . . . There are substantial difficulties in attempting to precisely rank the agencies in order of the level of risk that they tolerate. For example, difficulties arise because few of the agencies actually quantify risk levels and, among those that do, there are differences in the way risk is measured and denominated . . .

There are at least four possible explanations for the spread of agencies along this spectrum. Some of these explanations are clearly borne out by the survey data; others remain speculative. There may also be some overlap in these explanations.

- First, there is a strong link between the processes used to set target risk levels and the size of those risk levels. Agencies clustered at the low end of the risk spectrum are those that set arbitrary risk targets. Agencies that primarily derive the appropriate level of risk from other considerations, such as cost–benefit analyses and community consultations tend to tolerate higher levels of risk.
- Second, the level of risk tolerated by the agencies appears to be related to the number of people who may be involved in a particular safety incident. For example, one air accident can imperil several hundred people, and a safety problem associated with a particular chemical, food, or therapeutic drug is likely to affect a large number of people. Agencies dealing with these safety risks (CAA, CSU, NFA, and TGA) are clustered at the low end of the risk spectrum. On the other hand, accidents associated with particular consumer products are rarely likely to affect more than a few people, as is also the case with individual motor vehicle accidents. The agencies dealing with these risks (FBCA and FORS/NRTC) lie in the middle, or at the upper end, of the risk spectrum.
- Third, there is some evidence of a link between the level of risk tolerated by the agencies and the extent to which the related safety risks can be influenced by the person who bears the risk. At the low-risk end of the spectrum, risks associated with chemicals (CSU), food (NFA), or therapeutic drugs (TGA), are arguably difficult for individuals to understand, and therefore influence, because detailed technical issues are involved. In the case of air travel (CAA), beyond choosing airlines with safer aviation records, travellers have little personal influence over the safety of a flight they take.
- Fourth, the level of risk tolerated by the agencies may be related to the cost or disutility of reducing the relevant safety risks. For example, for FORS/NRTC to achieve the extremely low risk levels pursued by some agencies, they might need to mandate very low speed limiters in vehicles or, at the theoretical extreme, ban vehicles altogether.

Thus each day bureaucrats in the regulatory agencies try to bridge the gap between logic and perception.

Exercise
Using the concepts of a logarithmic scale of risk perception suggested in Appendix II — risk in sels (dB) (a new term to replace bels), place the acceptable risk odds of the different agencies on such a scale (consult the full report), along with data you find on everyday risks, say in the BMA guide 'Living with Risk' (see bibliography). Use equivalent data in your own jurisdiction.

Epilogue on epidemiology

The neighbours are sick, With the same thing? And it's something really rare? That's absolutely terrific!!

Someone finds a possible link between an environmental variable — be it mobile phones, proximity to a power plant, or certain types of work — and cancer, miscarriages, or some other malady. The correlation is

Inconclusive proof?
After a long interview, the epidemiologist still insisted that the results of his comprehensive study on the cluster of leukemia cases near the nuclear power station were inconclusive.
As I left his office, I noticed that the clock said three o'clock, so I began to set my watch. 'Wait', he cautioned, 'just because my clock says it's three doesn't mean it really is. The clock could be wrong'. The epidemiologist also had a clock on his desk. It said three. That still doesn't prove anything. Both clocks run off the mains, the power could have failed'. The epidemiologist also wore a watch. It also said three. 'An interesting association, but it still doesn't prove anything'.
As I left, I walked past his (long-time) secretary. 'Honey', she said and rolled her eyes, 'it **is** three o'clock'.

tenuous, statistically significant for some groups but not others. Nonetheless, headlines appear. Critics point to the flaws in the study (correlation is not necessarily cause and effect). Defenders shout 'cover-up'. An angry debate begins, backed by precious little data on either side.

The health effects involved are small and so there is virtually no research funding unless the political imperative is strong.

How did it begin? Epidemiology could be said to have started with the British doctor John Snow who wondered about the cholera epidemic in London in 1854. If cholera was spread through the air as believed, then all neighbourhoods should be equally affected. However, his study showed that all the clusters related to a single water supply. On the strength of the association he postulated a cause and effect. Because he couldn't convince anyone of this, he pulled the handle off the pump and kept it off.

Epidemiology considers four main factors:

1. the cluster size;
2. its specificity (who is affected);
3. the background rate (who is affected in the general population);
4. the politics of the researcher.

Cluster size and specificity

Small clusters can be significant if the natural background is rare. In 1971, a rare cancer of the vagina turned up in six young women. This type of cancer was only ever reported once or twice before in the literature. In a remarkable piece of detective work, researchers found that the mothers of these women had taken the oral contraceptive, diethylstilbestrol (DES) while pregnant. Cause and effect was established. A remarkable US court verdict assigned liabilities based on market share because the actual company whose drug the mothers had taken could not be established.

On the other hand, suspicions that computer terminals (VDUs) are a minor cause of miscarriages are impossible to confirm because miscarriages are so common; the normal background rate is 15%. Thus a company with 1000 pregnant VDU workers could normally expect about 150 miscarriages. (See also Appendix VIII section Epidemiology and Figs. A8.5 and A8.6 for a more detailed discussion on clusters.)

What about politics?

1. Well, in the US they are still arguing whether excess cases of encephaly (babies born without brains) in towns near the Rio Grande on the Mexican border are due to pollution (activists) or lack of folic acid in the diet of poorer people who happen to live in the cheaper areas on the dirty river (politicians). Maybe it is both.
2. How meaningful is the statement that the rate of cancer has doubled, if the increase is from one in 500 000 to two in 500 000? The statistics are highly significant but the effects are very weak.

Can we afford to close down microwave communication links if a weak correlation to exposure is indeed found? No more TV, perhaps. If we find a subpopulation that is particularly at risk from an environmental effect, it may be easier to protect say 1% of the population (specific genetic make-up, gender, race, etc.) than to shut down every microwave tower. On the other hand, selective exposure to risk of those who are 'safe' has in general been rejected industrially with the argument that the workplace should be safe for all. The exception has been women who are likely to become pregnant who are not allowed to be exposed to certain hazards (e.g. low levels of lead or radiation). Genetic screening has implications for the community rating of life and medical insurance, once such information is available.

3. Death rates from lung cancer in Britain were 166 per 100 000 for heavy smokers and seven per 100 000 for non-smokers, an almost 24-fold increase in risk from death from lung cancer. Yet that can be put differently: a smoker has 99.8% of the chance of a non-smoker of escaping death from lung cancer. We die from so many other causes.

Another perspective
Of 1000 **young** people who smoke cigarettes regularly, on average one will be murdered, six will die in traffic accidents and 250 will die prematurely from smoking cigarettes.

(Source: Sir Richard Doll, Oxford epidemiologist)

Are we still sure it was three o'clock?

That wasn't the sort of profits statement we were looking for

References

Watchdog for biocidal chemicals: Mixing chemicals with people and trade, B. Selinger, *Chemistry in Australia*, March : 124–27 RACI, Melbourne, Australia.

Risk management of chemicals, Ed. M. L. Richardson, Roy Soc. Chemistry, UK, 1992.

The analysis and regulation of safety risk: A survey of the practices of national and Commonwealth regulatory agencies, Office of Regulation Review, Industries Commission, Canberra, pp. 80–83, [2] p. 35, [3] p. 36,[4] p. 28, Feb. 1995).

Bibliography

Casarett and Doull's toxicology: The basic science of poisons, Eds M.O. Amdur, J. Doull, C. D. Klaassen, 4th. edn, McGraw Hill, 1991. Classic reference.

Principles and methods of toxicology, Ed. A.H. Hayes, Raven Press, NY, 2nd edn, 1989. Classic reference.

CRC handbook of chemistry and physics, D. R. Lide, 77th edn, 1996/1997. New material added and older material (but often just what you are looking for) dropped with each edition.

The Merck index: an encycopedia of chemicals, drugs, and biologicals, 12th edn, Merck and Co, NJ, 1996. Over 10 000 entries and much else. First edition 1899.

Dangerous properties of industrial materials, N.I. Sax, 6th edn, 1984, Van Nostrand Reinhold, NY, 1984. Huge, and revised regularly.

The structure of scientific revolutions, T.S. Kuhn, Uni. Chicago Press, 1962. The beginning of paradigm shifts.

Epidemiology for the uninitiated, G. Rose and D.J.P. Barker, 2nd edn, *British Medical Journal,* London, 1986. Simple paperback.

Living with risk, British Medical Association Guide, Wiley, 1987. Won some awards.

Calculated risks: The toxicity and human health risks of chemicals in our environment', J.V. Rodricks, Cambridge U.P., 1992. One of many in this area.

CHAPTER 3

Chemistry in the laundry

Did you use that extra strength biodegradable detergent again?

In this chapter we start our saga in the laundry and discuss soaps and detergents. The chapter is based on a timeless lecture given many years ago by Peter Strasser to whom much gratitude is due for many things.

Introduction

In primitive societies, even today, clothes are cleaned by beating them on rocks near a stream. Certain plants, such as soapworts, have leaves that produce *saponins,* chemical compounds that give a soapy lather.

Today, the word *detergent* implies synthetic detergent, rather than the older *soap.* In fact, commercial formulations consist of a number of components, and the *surface-active agent,* or its abbreviation *surfactant,* is used to describe the special active ingredients that give the formulated detergents their unusual properties. Soap is the oldest surfactant and has been in use for more than 4500 years.

Some soap manufacture took place in Venice and Savona in the fifteenth century, and in Marseilles in the seventeenth century. By the eighteenth century, manufacture was widespread throughout Europe and North America, and by the nineteenth century the making of soap had become a major industry. The first formulated detergent came on the market in 1907 when the German firm Henkel & Co added sodium perborate, sodium silicate, and sodium carbonate to soap to produce

Persil (*per*borate + *sil*icate) in Düsseldorf. The trade mark actually belonged to a Frenchman called Ronchetti.

Soaps

Ordinary soaps are the sodium salts of long-chain fatty acids. They have the general formula $RCOO^-Na^+$, where R is a long hydrocarbon chain, $CH_3(CH_2)_{10-16}$. These salts are made by neutralising the fatty acids with alkali.

Fig. 3.1 Neutralisation of a fatty acid with sodium hydroxide to give soap plus water

$$\underset{\text{acid}}{R{-}\overset{\overset{\displaystyle O}{\|}}{C}{-}OH} + \underset{\text{base}}{NaOH} \longrightarrow \underset{\text{salt}}{R{-}\overset{\overset{\displaystyle O}{\|}}{C}{-}O^-Na^+} + \underset{\text{water}}{H_2O}$$

However the cheapest sources of the fatty acids are animal fats and certain vegetable oils, which are largely *esters*. In practice, therefore, soaps are made by the *saponification* reaction:

Fig. 3.2 Reaction of a fat with sodium hydroxide to give soap plus glycerol

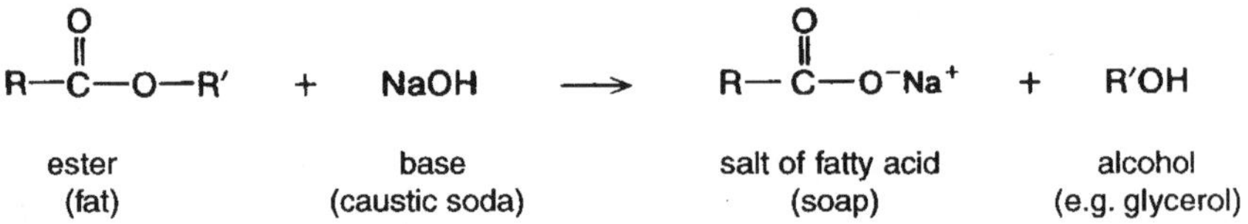

$$\underset{\text{ester (fat)}}{R{-}\overset{\overset{\displaystyle O}{\|}}{C}{-}O{-}R'} + \underset{\text{base (caustic soda)}}{NaOH} \longrightarrow \underset{\text{salt of fatty acid (soap)}}{R{-}\overset{\overset{\displaystyle O}{\|}}{C}{-}O^-Na^+} + \underset{\text{alcohol (e.g. glycerol)}}{R'OH}$$

Toilet soap usage
The combination of frequent showering and soft water means a high use of toilet soap in Australia. Consumption per person (kg/inhabitant) rose from 1.9 kg in 1977 to 2.25 kg in 1982, compared to a rise in West Germany from 0.90 kg to 0.97 kg. In the same period, consumption fell in the UK from 1.48 kg/person to 1.25 kg; and in France, from 0.67 kg to 0.63 kg. They probably use more deodorants.

which is essentially the reverse of the *esterification* reaction. Beef tallow gives principally sodium stearate, $CH_3(CH_2)_{16}COO^-Na^+$, the most common soap. Palm oil gives sodium palmitate, $CH_3(CH_2)_{14}COO^-Na^+$, a component in more expensive soaps. The standard for personal soap in Australia provides for not less than 70% of fatty matter; actual soap plus so-called *superfatting* agents, which can be fats, fatty acids, wool wax, etc., but these agents are limited to a maximum of 10%. The total amount of water allowed is 17%. In addition, there will be some sodium chloride and glycerol left from the production process. Also added are preservatives, antioxidants, perfume and colouring matter (titanium dioxide in the case of white soaps). In laundry soap the amount of fatty material allowed is lower (60%) and the amount of water is higher (34%).

If the sodium ion of ordinary soap is replaced by other metal ions, soaps with different properties are produced. When potassium hydroxide is used instead of sodium hydroxide in the manufacturing process, *soft soaps* are formed. These are semi-solid soaps, once used in shampoos and special-purpose soaps. However they are more expensive than the ordinary soaps.

Most other metals give soaps that are insoluble in water. Clearly these are not much use for washing, but they do find applications as additives for greases and heavy lubricating oils, where their principal function is still as detergents. Copper stearate has been used as a waterproofing colour agent; not only is it water repellent, but the copper ion is poisonous to mildew. The heavy metal stearates are also used as stabilisers or release agents in plastics such as poly(vinyl chloride) (PVC) and polythene (See Vinyl polymers, in Chapter 9).

Experiment
This problem with soap can be demonstrated by a simple experiment in which a concentrated solution of hard-water salts is added to a 0.1% solution of soap, and also to a 0.1% solution of synthetic surfactant. The soap precipitates but the synthetic surfactant remains clear because its salts are water soluble.

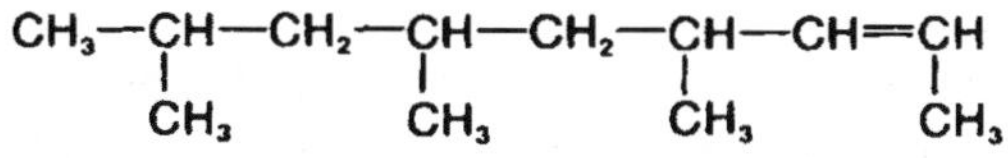

Synthetic surfactant or soap?

The most important reason for the displacement of soap is the fact that, when a carboxylic acid soap is used in hard water, precipitation occurs. The calcium and magnesium ions, which give hardness to the water, form insoluble salts with the fatty acid in the soap and a curd-like precipitate occurs and settles on whatever is being washed. By using a large excess of soap it is possible to redisperse the precipitate, but it is extremely sticky and difficult to move. The pre-sewerage septic tanks and grease traps became clogged and it was no joy to have the task of cleaning them out!

You may live in an area (such as in Melbourne, Australia) where the water is extremely soft. But calcium and magnesium ions are still present in the dirt that you wash out of your clothes, so some precipitation still occurs if soap is used, and gradually deposits are built up in the fabric.

There are other disadvantages with soap; it deteriorates on storage, and it lacks cleaning power when compared with the modern synthetic surfactants, which can be designed to perform specialised cleaning tasks. Finally, and very importantly from a domestic laundry point of view, soap does not rinse out completely; it tends to leave residues in the fabric that is being washed. These gradually build up and cause bad odour, deterioration of the fabric, and other problems.

The development of the first detergents in an effort to overcome the reaction of soaps with hard water provides a good illustration of one of the standard chemical approaches. If a useful substance has some undesirable property, an attempt is made to prepare an analogue, a near relation chemical, which will prove more satisfactory. The petroleum industry had, as a waste product, the compound propylene $CH_3–CH=CH_2$, which used to be burnt off. By joining four of these propylene molecules together, propylene tetramer is obtained (Fig 3.3).

$$CH_3{-}CH(CH_3){-}CH_2{-}CH(CH_3){-}CH_2{-}CH(CH_3){-}CH{=}CH(CH_3)$$

Fig. 3.3
Propylene tetramer

If benzene is attached at the double bond, the resulting compound reacts with sulfuric acid (H_2SO_4). Then sodium hydroxide is added to neutralise the sulfonic acid, and the sodium salt shown in Figure 3.4 is obtained, which is a branched-chain alkylbenzene sulfonate (ABS).

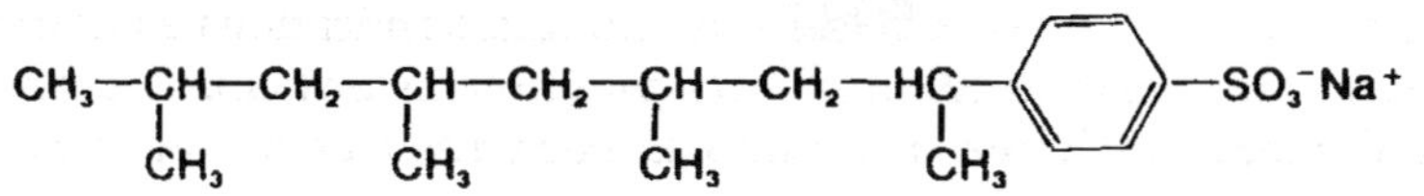

$$CH_3{-}CH(CH_3){-}CH_2{-}CH(CH_3){-}CH_2{-}CH(CH_3){-}CH_2{-}HC(CH_3){-}C_6H_4{-}SO_3^-\,Na^+$$

Fig. 3.4
Alkylbenzene sulfonate

Clearly, the new substance is closely related to an ordinary soap (Fig. 3.5).

Fig. 3.5
Soap and detergent

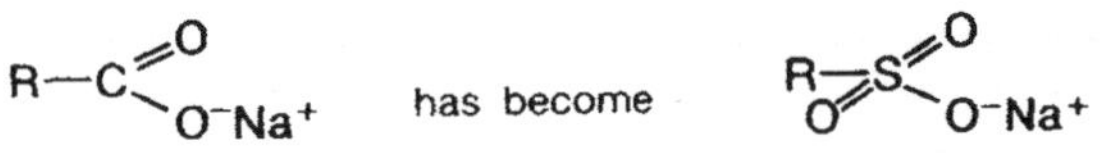

$R{-}C(=O){-}O^{-}Na^{+}$ has become $R{-}S(=O)_2{-}O^{-}Na^{+}$

The detergents produced in this way are much more soluble than soap and their calcium and magnesium salts are also soluble, so a scum is not formed with hard water. However, they are more stable than soaps and persist in the waste water long after use. The consequence was the fouling of sewage works and rivers with tremendous masses of froth. (See poem 'Foam'.) The increased stability of the detergents resulted from both the greater stability of the sulfonate grouping and the fact that the raw material hydrocarbon chain molecules contained large proportions of carbon chains that were *branched,* in contrast to the straight-chain hydrocarbons from animal fats. As bacteria break down the branched chains more slowly, detergents were once considered not to be biodegradable at all (see Biodegradability later in this chapter).

How do surfactants work?

The surfactant molecule is often described as tadpole-like because it has a fairly long fatty tail, which is water insoluble or *hydrophobic,* and a small, often electrically charged head, which is water soluble or *hydrophilic.* There are four possible combinations (Fig. 3.6):

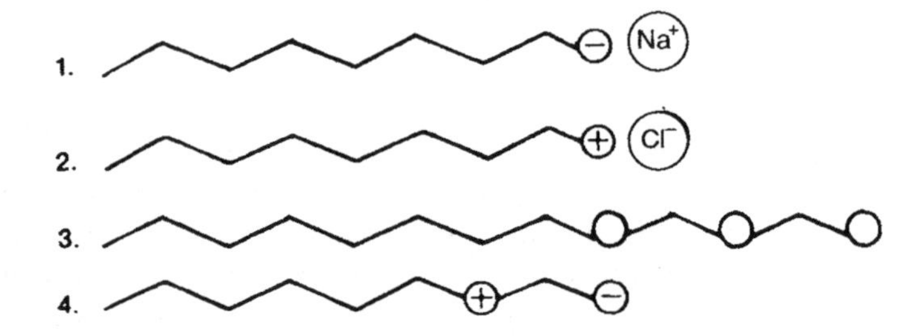

Fig. 3.6
Diagrammatic representation of the shapes and electric charges of surfactant molecules

1. The *anionic* surface-active agents, in which the surfactant is an anion (that is, it carries a negative charge) and the charge is concentrated in the hydrophilic or water-soluble head.
2. The *cationic* products (the opposite of anionic) in which the head carries a positive charge.
3. The so-called *non-ionic* detergents. These do not have a specific charge but the hydrophilic or water-soluble portion of the molecule is usually achieved by incorporating a polyethylene oxide group into the molecule (see below in the section on surfactants in domestic laundry detergents). You can see that, because it is less polar than an ion, the hydrophilic portion of these molecules is usually rather bigger than in the case of the ionic surfactants.

4. Finally, there are some specialised products that carry both a positive and a negative charge in the same molecule. These are called *amphoteric,* and they are particularly useful for very specialised applications (such as hair conditioners and fabric softeners). Because they carry both an anionic and a cationic centre, they behave as either an anion or a cation, depending on the pH of the solution in which they are used.

In domestic detergents, anionic surfactants are predominant. Nonionics are increasingly used, but cationic surfactants are not. Cationic surfactants have two interesting and useful properties. First, they are mildly antiseptic and may be used (in combination with non-ionic surfactants) in nappy washes, hair conditioners, and throat lozenges, and as algacides in swimming pools. Second, the positive charge on the chain makes them useful for washing plastic articles, but not glass. Glass normally acquires a surface negative charge, which to a certain extent attracts dirt. Anionic detergents can remove this dirt, but cationic surfactants are attracted to the glass so strongly that a thin layer adheres to the glass, with the long hydrophobic (fatty) chain outwards, thus making the glass non-wettable and apparently greasy. The reverse is true for plastic articles, which normally have a positive surface charge. Have you ever noticed how dirt clings to plastic and is hard to remove by ordinary washing? The positive charge also makes cationic surfactants useful as fabric softeners. These are liquids and are added to the rinse cycle. The cationic charge has a strong affinity for wet, negatively charged fabric and forms a uniform layer on the surface of the fibres, thus lubricating them and reducing friction and static.

The cleaning action of surfactants

The molecules of surfactants tend to concentrate in the surface layers of the water (because the water-insoluble portion wants to get out of the water), lowering the *surface tension* of the water and allowing the water to wet non-wettable surfaces. Some of the cleaning power of surfactants thus results from the enhanced ability of the water to wet the normally hydrophobic surface and lift off the dirt.

The long hydrocarbon tails of the detergent molecules are soluble in non-polar substances such as oil, whereas the polar carboxyl or sulfonate groups are soluble in water. Thus the molecules promote solubilisation of oil in water by lying across the oil–water interface. When the concentration of the detergent molecules in the water reaches a certain value, called the *critical micellar concentration,* the molecules aggregate into communes called *micelles,* which contain roughly 40 to 100 molecules. In these aggregates, which at high enough concentrations give soap solutions their cloudy appearance (because they scatter light just as dust does in the air), the hydrocarbon tails lie towards the centre, while the surface of the micelle contains the water-soluble polar ends (see Fig. 3.7).

How does a surfactant remove the oily soil and, with that, frequently a lot of the particulate soil? This is a complex process. The insides of the

Experiment
If you place a strip of cotton fabric, weighted at one end to pull it into the liquid, in a 0.1% solution of surfactant, and another in pure water, you will see a difference in behaviour. Because the pure water alone does not wet the cotton, the fabric strip remains 'upright', whereas the fabric in the surfactant solution wets out and sinks immediately.

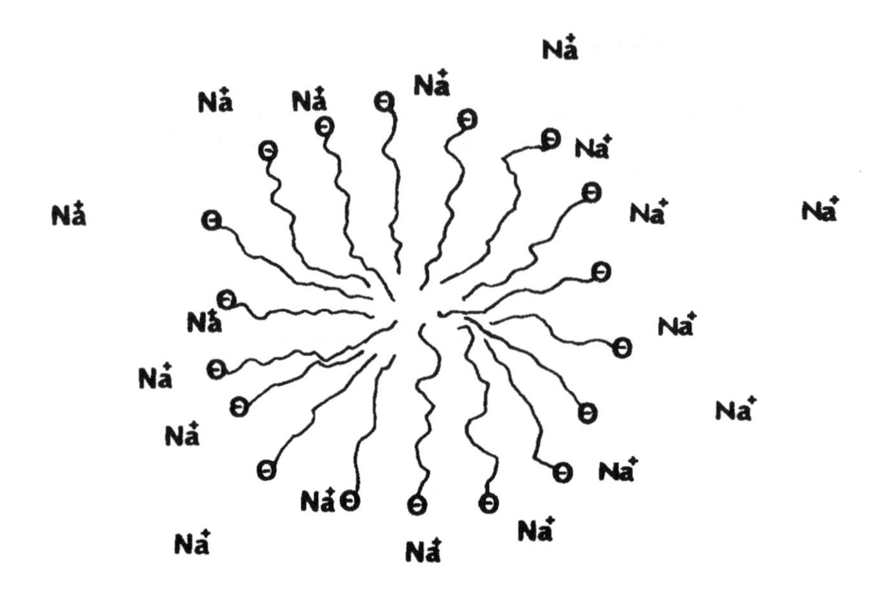

Fig. 3.7
A surfactant micelle

Experiment
If some olive oil is carefully poured into water, and also into a solution of surfactant, and each vessel is vigorously shaken and then put down again, the oil immediately rises to the surface of the water, but remains emulsified and dispersed in the detergent solution and can therefore be rinsed out.

micelles are virtually small oil droplets and so can dissolve oily materials, but the main action of the surfactant is to stimulate emulsification.

The process of emulsification (e.g. the removal of fat from a plate) is illustrated in Figure 3.8.

The fabrics

At a conference in 1955, Professor Speakman, doyen of wool chemistry from Leeds, said in a lecture 'in twenty years' time, sheep will only be grown for their meat'. In the 1960s many were convinced that cotton would be replaced by the new synthetics. Now they expect cotton to dominate until at least the year 2000, holding over 50% of the market. Of the synthetics, nylon has diminished in importance and polyester holds 50%. The change is towards physical modification of current fibres and cotton blends.

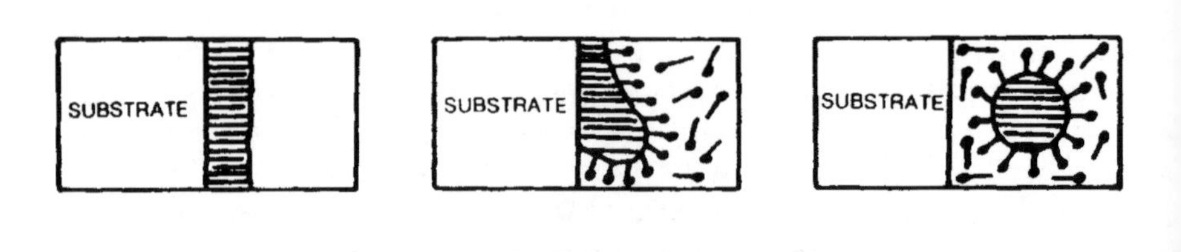

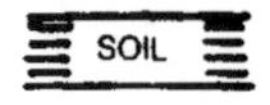

Fig. 3.8
Emulsification

The detergents

The ingredients in any commercially produced detergent fall into six groups: the surfactant system; the builders (inorganic and organic); a filler; bleaches; fluorescers; and enzymes.

The surfactant system

The starting material for the synthetic surfactant that forms the major active ingredient is a material known as alkylbenzene. In appearance it is a kerosene-like liquid with a slightly oily odour. It is a product of the petroleum industry and is made by the condensation of an α-olefin with benzene. As such it is completely insoluble in water. By treating this material in a process called *sulfonation* it is converted to the corresponding sulfonic acid, which is a viscous, dark-brown material (see Fig. 3.4). Chemically, sulfonation is achieved by treating alkylbenzene with an excess of sulfuric acid, giving the sulfonic acid plus water.

Inorganic builders

Builders are included in modern domestic laundry detergents to assist the surfactant system in its action. Builders are both organic and inorganic. Among the inorganic builders, *sodium tripolyphosphate* (STPP) is the major one used. It is a polyphosphate, equivalent to having been formed from two molecules of disodium monohydrogen phosphate and one molecule of monosodium dihydrogen phosphate to give what is correctly called pentasodium triphosphate, by the elimination of water, as shown in Figure 3.9.

$$2Na_2HPO_4 + NaH_2PO_4 \longrightarrow Na_5P_3O_{10} + 2H_2O$$

dipolyphosphate
(tetrasodium)
[pyrophosphate]

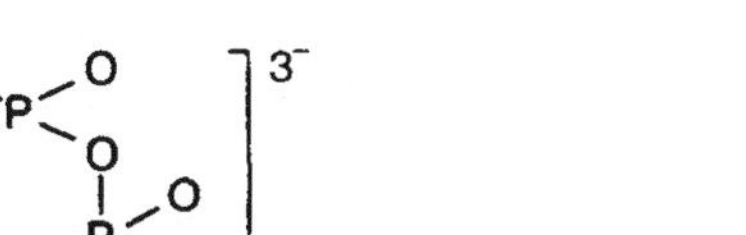

tripolyphosphate
(pentasodium)

linear polyphosphates

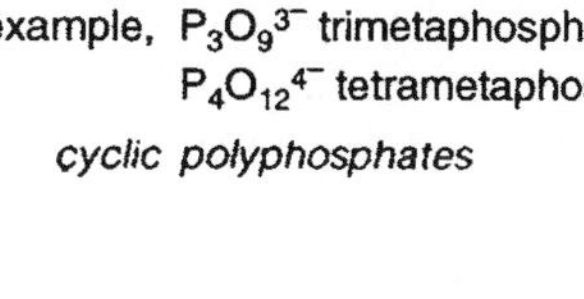

For example, $P_3O_9^{3-}$ trimetaphosphate
$P_4O_{12}^{4-}$ tetrametaphosphate

cyclic polyphosphates

Fig. 3.9
Polyphosphates

Why should tripolyphosphate be put into a detergent? It buffers the washing water to a milder pH than would otherwise be obtained. One of the drawbacks of the early versions of Persil was high alkalinity (sodium carbonate and sodium silicate) and the damage that this could do to fabrics. On the other hand, a certain amount of alkalinity is needed to wash fabrics, particularly cotton, successfully. Sodium tripolyphosphate sequesters hard-water ions. Although synthetic surfactants do not precipitate with the ions (such as calcium and magnesium), in hard water the presence of these ions does tend to decrease detergency to some degree. The full cleaning power of the surfactant is preserved if the ions are sequestered by tripolyphosphate. Finally, tripolyphosphate is important in its deflocculating action; that is to say, it helps to keep a clay-type dirt in suspension.

The zeolite/sodium carbonate/polycarboxylate builders have increasingly replaced polyphosphates. Zeolite Na Al (the most common commercial product) is made by reacting one mole each of soda with two moles of silica and 4.5 moles of water. The basic building blocks of the zeolite structure are tetrahedra. Each silicon and aluminium atom is the centre of a tetrahedron of oxygen atoms, and these oxygens also serve as bridges between the metals. A two-dimensional representation is given in Figure 3.10 and a model of the cubic crystal structure of zeolite A is shown in Plate II (courtesy of Dr J. Thompson, RCS at ANU).

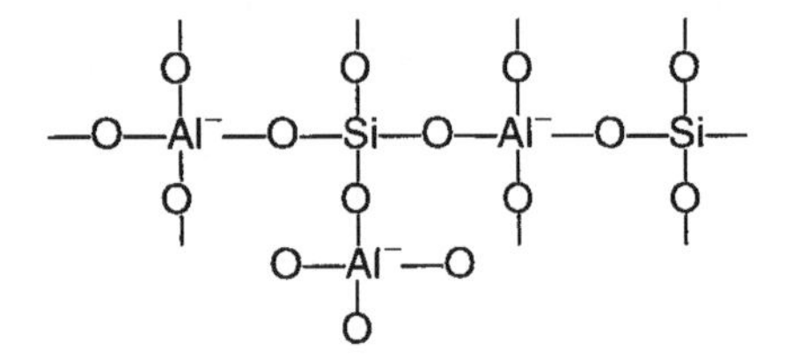

Fig 3.10 Two-dimensional representation of aluminosilicate framework

The silicon atom with a 4+ charge is electrically balanced by four shared oxygen atoms. However the aluminium atom only has a 3+ charge and so is left with a 1– charge. This negative charge in the framework is balanced by a positive charge from a nearby mobile sodium ion. In three dimensions, the framework has the structure of a cube where each of the six faces has a well-defined pore or window, leading to a central cavity.

The mobile sodium counter-ions can be rapidly exchanged for calcium ($\approx$ 60 seconds) and to a lesser extent magnesium, and this is how zeolites soften hard water. The size of particles used is determined by the speed of exchange needed and the flowability of the product. Zeolites allow much higher levels of liquid non-ionic surfactants to be used in powder products. Because zeolites are insoluble, they can only be used in liquid formulations if special attention is given to preventing particles settling by using organic polymers. Sodium silicate must also be avoided because it creates two layers which require shaking before use. A whole new variety of zeolites with better properties are coming into play.

At recommended dosages they work as well as polyphosphates, but are not nearly as tolerant of underdosing. On the other hand, repeated underdosing with phosphate builders causes a deposition of calcium phosphate in the textile fibres, and consequent accelerated wear of both fabric and machine. The zeolite, being a clay, bulks up sewage.

Current zeolite production capacity is USA, 453 000 tonnes; Europe, 1 077 000 tonnes; and Japan, 193 000 tonnes. Zeolite is not used in domestic laundry powders manufactured in Australia and New Zealand. It is only used in imported powder.

The contribution of hardness salts from mains water is readily calculated from data supplied by a water supply authority. The distribution for Australia is shown in Table 3.1.

To this base level of hardness must be added the hardness contributed by the soiled fabrics to be washed. The equivalent of 20–30 mg/kg as calcium carbonate is assumed as typical. Water hardness can be measured by a dipstick test, and the amount of builder a detergent suitable for local conditions should contain can be calculated.

The height of controversy over phosphates in detergents came in the early 1970s in the USA and Australia, and later in Europe. Many countries, including Austria, Germany, Italy, Japan and Canada control phosphate levels by legislation. Canada set a limit on the use of phosphates in household detergents of 2.2% phosphorus (equivalent to 9% of polyphosphate). Switzerland and Norway have outright bans. Sweden, the Netherlands and France have voluntary controls. In the USA, Desoto went to sodium carbonate, Lever Bros reformulated Wisk liquid detergent with sodium citrate and Henkel began to capitalise on its nearly worldwide patents based on sodium aluminosilicate–zeolites. In Japan, for example, zeolites are the sole builder in more than 90% of all detergents sold. In August 1983, zeolite production capacity in Europe was at least 210 000 tonnes per year.

The problem appears to be related to the total amount of detergent used, and the level of pollution, not the phosphate. Unpolluted lakes can absorb a surprising amount of phosphate because zooplankton eat the algae that grow and are themselves eaten by fish. Adding phosphate to polluted water, however, causes unchecked algal growth. The slimy floating masses cut off the light supply, the weeds die, and decompose, using up the oxygen, causing sulfurous smells and killing the fish.

In Australia there is a voluntary labelling scheme whereby NP denotes less than 0.5%, and P indicates less than 5% phosphorus in a product.

The second inorganic builder is a *sodium silicate*, a material that is better known by the name 'water-glass' or soluble silicates. It is produced from sand and sodium carbonate. Sodium silicate can tie up magnesium (and to a lesser extent calcium) and also acts as a corrosion inhibitor, especially protecting diecast washing-machine parts. It has an important role in strengthening the physical part of the detergent powder. (see also Chemical gardens in Chapter 9).

The organic builder

The organic builder used in detergents is a product called *sodium carboxymethylcellulose*, which is produced by treating pure cellulose with

TABLE 3.1
Distribution of water hardness in Australia

Water hardness (mg/kg as calcium carbonate)	Population (%)
10–20	26
21–40	11
41–60	37
61–120	8
121–180	14
181–240	1
Over 240	3

Death of Norfolk pines at Sydney beaches
Undegraded surfactants from sewer outfalls, not phosphates, caused the death of trees lining Sydney beaches. The surfactants lowered surface tension in sea spray blown onto the trees. This led to excessive penetration of salt into normally resistant pine needles, ultimately killing the trees. However, some were planted in 1820 so, with an average lifespan of 150 years, old age was creeping up on them anyway. Secondary treatment of sewage was considered too expensive, so the outfalls were extended out to sea, to dilute the spray contamination. Now, new trees are surviving and some old ones are recovering.

caustic soda and chloracetic acid. It is used at a concentration of less than 1%, and its major function is to act as an anti-redeposition agent: it *increases the negative charge* in fabrics, which then repel the dirt particles because they are themselves negatively charged. Imagine that you have a white handkerchief with a black sooty spot in one corner. If you washed it in a detergent that did not contain an anti-redeposition agent, the soot would be dislodged, but would tend to redeposit all over the handkerchief, so it would emerge a uniform grey.

This material is active only on cellulose fabrics (cotton, rayon, etc.) and on fabric blends with a cellulose component. Polyacrylic acids and polyacrylates (1–6%) in a formulation can handle synthetics and synthetic blends.

Concentrates

Concentrates (called compacts in the US) have made smaller inroads in the Australian market, both in liquids and solids, because the density difference compared to normal products is less than in Europe and the US, and the pack sizes tend to be smaller.

Fluorescers, 'whiter than white'

In the 'good old days', we added washing blue so that cotton aging naturally to yellow would look white. Today, very small amounts of *fluorescers* are added to detergent powders. They are in fact already in the new, gleaming business shirt when you buy it, but wear and washing remove them. These compounds absorb ultraviolet light (which is invisible) and re-emit blue light (which yellow fabrics do not reflect fully from sunlight) and so restore the mixture of colours reflected to what a white fabric would reflect. The exact nature of the brighteners differs with geographic location. We are conditioned to blue–white as the accepted hue for cleanliness, whereas in South America a red-white coloration is culturally accepted for 'clean'. Fluorescers do not clean, but they do whiten the fabric. Different fabrics carry different charges. Nylon carries a positive charge, and cotton a negative one; thus oppositely charged fluorescers are needed.

Foam

The relationship between foaming power and detergency has always been of interest, and foaming power has become associated in many consumers' minds with high detergent power. The first liquid detergent on the Australian market was Trix. It was non-foaming and was soon replaced because of consumer resistance. However, it is generally conceded by detergent technologists that foam has no direct relationship to detergency in ordinary fabric-washing systems.

In systems where the amount of washing fluid is low, foam may have an important role. The individual foam films tend to take up and hold particles of soil that have been removed from the item, preventing them from being redeposited, and allowing them to be washed or scraped away. This effect is very important in the on-location shampooing of carpets and to a certain extent in the cosmetic shampooing of hair. Front-loading washing machines work by bashing clothes against the side of the tub, the 'high-tech' version of beating clothes on rocks. Front-loaders clean clothes better than top-loaders, but only if a low-suds detergent is used, because the suds cushion the impact and reduce the cleaning action. The suds can also cause electrical short circuits in time switches, etc.

Designer wash balls

A French company has manufactured 'designer wash balls' to throw into your washing machine. A dozen 30 g, 4 cm diameter, tough, high-temperature and detergent-resistant balls will allegedly halve the amount of detergent you need. Made from multiple components, these high-friction pounders and scrubbers take washing back to fundamentals.

FOAM

A. P. Herbert

How difficult the days in which we live!
Build how you will, some dyke decides to give.
And here's the Housewife, worn with cleanly care,
But spreading dirt and danger everywhere.
Six thousand years ago—it makes you smile—
They washed with common water on the Nile.
There was no soap, historians have reckoned
Until the time of Charles I — or II.
Water and soap sufficed the human race
Till Hitler happened, and a war took place.
Our soap was rationed; fats and oils were few;
And chemists madly sought for something new.
Now no one 'cleans' the saucepan or the serge;
No woman 'scrubs' or 'washes'. They 'deterge.'
The sink resounds with scientific chatter,
And even dirt gives way to 'soiling matter.'
I could of course from A to Z explain
The technicalities, but I refrain:
For I imagine you are not in touch
With non-ionic sulphates very much.
But briefly, soap plays second fiddle now
To new synthetic things like Buz and Wow.
With other chemicals they all combine
A little laurylalkanomaline[1]
(I may have missed a syllable or two,
I may have added some: but that must do)
This is an 'additive'; it causes foam,
And so does much to make a happy home:
For ladies — and they're not alone, I gather —
Believe there's virtue in a lot of lather,
Though foam and froth are only bubbles still,
And bubbles are, notoriously, nil.

[1] laurylalkylolamide

But there it is. Alas, they do not think
What happens when the foam goes down the sink.
Into the sewer it descends, and can
Become a menace to the sewer-man;
Then at the sewage-works, in monstrous banks,
Makes brave men tremble who must tend the tanks —
Though things are better when the high winds beat
The poisoned bubbles over field and street;
Foam fills the culverts, foam obscures the streams,
Foam rides the river where the pale moon gleams,
And, what is more important, you and I
Expect our water in a pure supply.
The barge is buried in the filthy flecks
And sailors fear to venture on their decks.
Foam round the boats pollutes the virgin paint;
Foam is the fisherman's supreme complaint,
And, though the evidence admits of doubt,
Can not be beneficial to the trout.
Great ills have often claimed a grand excuse:
But foam, in fact, is not the slightest use.
'Why, then, if that be certain,' you exclaim,
'Why not dispense with Lauryl-what's-its-name?'
The manufacturers will tell you why:
'If we omitted it they wouldn't buy.'
O Homo Sapiens! Preserve me from
A world of women! Burst, the atom bomb!

Source: Permission of A.P. Watt Ltd on behalf of Crystal Hale and Jocelyn Herbert. This poem was written in 1953 and so pre-dates EEO considerations. Included with apologies.

In some detergent formulations a small amount of soap is included to serve a number of purposes. Depending on the hardness of the wash water and the balance of the formulation, any soap added will act either as a water softener or as a surface active agent. However, the primary aim for including soap is to bring about a rapid collapse of foam during rinsing after the wash, although silicones are increasingly being used for this purpose — for more on foam see Chapter 6. Because washing conditions and habits vary around the world, detergent formulations also vary greatly (see Table 3.2).

Sodium perborate bleach

In water, sodium perborate (5–15%) releases hydrogen peroxide, which is a powerful *oxidising* agent. The oxidation removes much of the stains, while generally not affecting fast colours. It is particularly effective when

TABLE 3.2
Laundry detergent in Asian and Pacific countries

	Japan	Korea	China	Taiwan	Hong Kong	Philippines	Thailand	Malaysia	Singapore	Indonesia	Australia
Main type of detergent	Compact powder	Powder	Powder	Compact powder	Compact powder	Detergent bar	Powder	Compact powder	Powder	Paste	Powder
Washing method	Machine	Machine and hand	Hand	Machine and hand	Machine	Hand	Hand	Hand	Machine and hand	Hand	Machine
Dose (g/L)	25 g/30 L	30–45 g/ 30 L	25 g/30 L	25 g/30 L	—	—	40 g/ 5–10 L	40 g/30 L	40 g/30 L	—	100 g/64 L
Typical (%) formulation											
Anionics	30–35	18–22	15–20	30–35	0–35	30	20–25	20–25	25	25	15
Non-ionics	3–5	0–2	—	2–4	2–15	—	—	—	—	—	1–2
Total surfactant	40	18–25	15–20	30–40	10–40	30	20–25	20–25	25	25	15–20
Phosphate (% STPP)	none	none	15–20	none	0–50	10	20–30	20	20	5	15–20
Zeolite (%)	20	15–20	15–20	20–25	15–30	20–25	15–20	15–20	20	10	20
Alk. builder	25	15–20	15–20	20–25	15–30	20–25	15–20	15–20	20	10	20
$CaCO_3$	—	—	—	—	—	30–40	—	—	—	—	—
Sulfate	1–3	20–30	35–45	3–5	3–20	—	20–30	25–40	20–30	5	—
Enzyme	0.5–1.5	0.2–0.6	—	0.5–1.5	0.5–1.5	—	—	+	+	—	+
Bleach activator	Various	Various		Perborate TAED							Perborate
Water	3–6	3–6	6	5–10	5–10	2–5	5–10	5–10	5–10	45–55	5–15

Notes
1. The non-biodegradable branched alkylbenzenesulfonates (ABS) are still used in China, the Philippines and Indonesia, but the move to the more degradable linear alkylbenzenesulfonates (LAS) is under way.
2. Use of phosphates is not restricted by law in any of the countries listed except Korea, (4.6% as STPP), but industry has self-imposed limits in Japan, and Australia voluntarily labels with P those with a maximum of 20% STPP.
3. There is a steady move from phosphates to zeolites.
4. Sodium sulfate can be a by-product of the sulfonation process, but is also used as a filler.
5. Calcium carbonate is a filler.

Source: Akira Susuki, Kao Corporation, Japan, at Fourth World Detergent Conference, Montreux, Switzerland, 1993.

Warning
Perborate and chlorine should not be used together.

Is it blood?
Presumptive blood tests are generally catalytic tests involving hydrogen peroxide and an indicator that changes colour (or luminesces) when oxidised. Release of oxygen by the peroxidase activity of the haemoglobin causes the indicator to change. Common indicators used include o-tolidine (once used for testing for chlorine in swimming pools) and luminol. Unfortunately other enzymes (such as vegetable peroxidases) also give false positive results.

Fig. 3.11
A royal statue attacked by detergent (Courtesy of New Scientist Publications)

the material is left to soak, but requires a fairly high temperature to be effective during a wash. Unlike chlorine, this bleach has virtually no adverse effect on textile fibres or on most dyes.

The most striking current change worldwide that is affecting the soap and detergent industry is a move towards lower washing temperatures. Cottons and linens can be washed at the boil at 95° C, at which temperature perborate releases 90% of its oxygen and bleaches heavy soil and wine stains. At 55° C, perborate is only about 60% effective. The rapid rise in synthetic fibres, which are adversely affected by chlorine, has boosted the use of perborate.

Bleach activators such as pentaacetyl glucose and tetraacetylethylenediamine (TAED), decompose to the unstable peracetic acid, and this acts as a bleach at low temperature. This is used in 50% of heavy duty detergents in western Europe (at 1.5–5%). Other activators are also used in smaller quantities. Sodium percarbonate and nonoyloxybenzene sulfonate (NOBS) are used in North America.

Another problem at the lower temperatures is that an enzyme found in many biological strains, called catalase, decomposes hydrogen peroxide rapidly (at a rate of a million molecules a minute) and so can destroy the perborate bleach at room temperature in a few minutes (the enzyme is deactivated at higher temperature). Conversely, the high concentration of a related enzyme in blood stains is used in a presumptive forensic test for dried blood.

The only disadvantage is that boron is toxic to citrus crops and detergent run-off must then be avoided in irrigation.

Washing machines

Europeans have traditionally used front-loading drum machines while the US, Australia and Japan have favoured top-loaders. In the US they are now changing to what they call 'horizontal axis' or 'tumble action' washers. A recent entry can be found on the internet (http://www.frigidaire.com./kitch11.html). A strong feature of front loaders is that they allow addition at the appropriate point in the cycle of extra detergent charges, and in particular, softeners (which are often incompatible with the surfactant). They tend to be more water and energy efficient as well.

A significant technical change in the market has been 'fabric care positioning'. Several manufacturers offer laundry detergents with built-in fabric care, such as enzymes that deal with cotton lint (see Enzymes below), and clays that give cotton a soft feel.

The Asia-Pacific region contains one-third of the world's population, with nearly 1.7 billion people. The consumption of detergents is below the level of Western countries but is expected to grow substantially. Because much washing is still done by hand, detergent bars and pastes are common. The temperatures used are low, and the water is generally soft. This in turn affects the composition of detergents used as shown in Table 3.2.

The second surfactant used in detergents is frequently a non-ionic one, and it may be either a *coconut diethanolamide* (an alkylolamide) or a *synthetic fatty alcohol ethoxylate.* Both are rather waxy products, substantially all active material. The alkylolamide is prepared by making the fatty acids obtained from coconut oil react with an ethylene oxide derivative called monoethanolamine. A condensation reaction takes place with the elimination of water and a waxy product is formed (Fig. 3.12).

Exercise
Use the Periodic Table (Fig. 1.1) to convert the percentage of phosphate given between quoting it as elemental P, P_2O_5, and STPP (formula given in Fig. 3.9).

Answer
STPP:P is 4; P_2O_5: P is 2.3 (care, keep the number of P atoms per formula constant!). The same confusion in impression/comparison by using different units, occurs with pollutants, see Nitrates in Chapter 2, and Fertilisers, see Chapter 7.

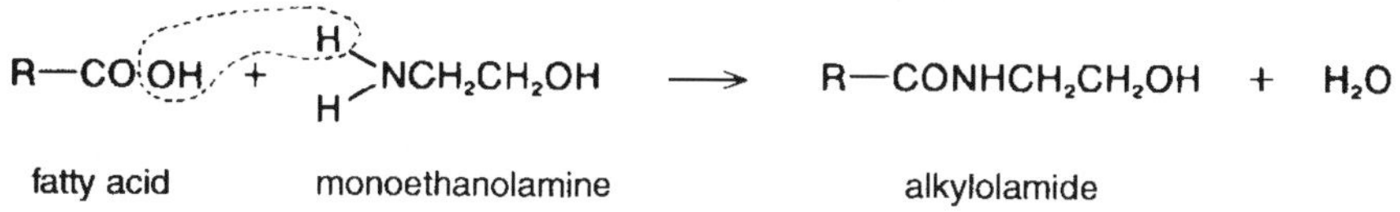

$$R{-}CO{\cdot}OH + H_2NCH_2CH_2OH \longrightarrow R{-}CONHCH_2CH_2OH + H_2O$$

fatty acid — monoethanolamine — alkylolamide

Fig. 3.12
Condensation reaction

On the other hand, the formation of a fatty alcohol ethoxylate illustrates the method of manufacture of an ever-increasing class of surfactants, the ethylene oxide condensates. They are now made in very large quantities world wide (including Australia). Ethylene oxide is a toxic gas, flammable, and forms explosive mixtures with air in any proportion from 3% upwards. Whereas in the sulfonation reaction the chemical nature of the product is relatively definite (as is also the case for the coconut monoethanolamide) this is not so for the ethylene oxide condensation products, because we are now dealing with a type of *polymerisation* (see Condensation polymerisation, in Chapter 9). It is usual for these condensation reactions to take place by starting with a fatty alcohol, which is also a waxy material, and chemically is the long fatty chain forming the hydrophobic part of our surfactant. It has a terminal hydroxyl group to which a molecule of ethylene oxide can be added thus:

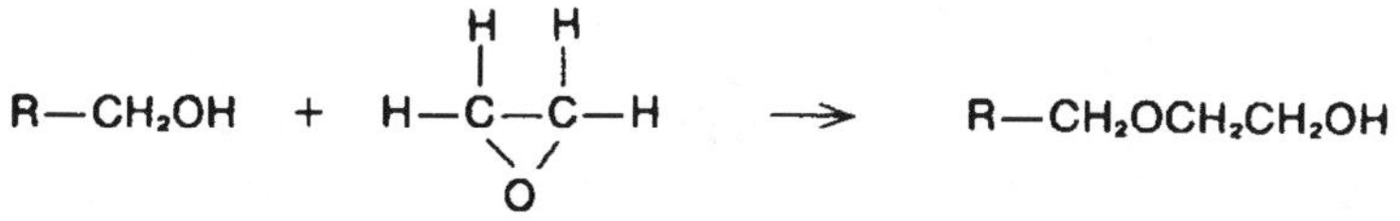

$$R{-}CH_2OH + H{-}\underset{\diagdown\;O\;\diagup}{\overset{H}{C}{-}\overset{H}{C}}{-}H \longrightarrow R{-}CH_2OCH_2CH_2OH$$

Fig. 3.13
Addition of ethylene oxide to a fatty alcohol

Further molecules of ethylene oxide can then add on to the terminal hydroxyl group which is formed at each step of the addition:

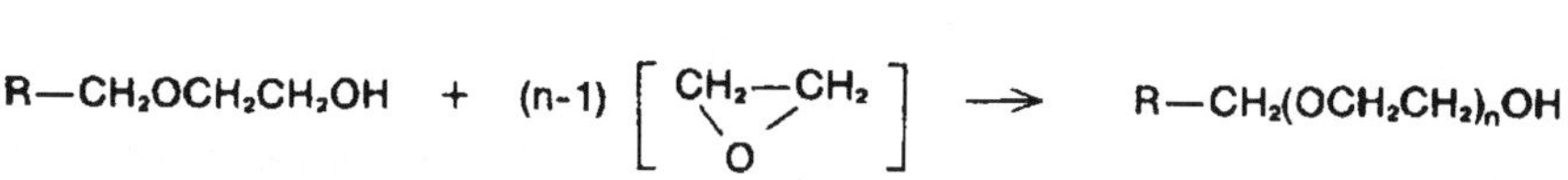

$$R{-}CH_2OCH_2CH_2OH + (n-1)\left[\underset{\diagdown\;O\;\diagup}{CH_2{-}CH_2}\right] \longrightarrow R{-}CH_2(OCH_2CH_2)_nOH$$

Fig. 3.14
Addition of further ethylene oxide to the end of the chain

If the ethylene oxide and tallow fatty alcohol are reacted together in the proportion of ten molecules of ethylene oxide to one molecule of tallow fatty alcohol, not all of the ethylene oxide molecules will combine with the tallow alcohol molecules in the proportion of ten to one. We will in fact get some molecules with fewer ethylene oxide groups, and some molecules with considerably more.

Biodegradability

A sight such as that shown in Figure 3.15 was once relatively common in the 1960s but not today.

They did not degrade quickly enough for the surfactants to be destroyed in the conventional sewage treatment plant, or to be decomposed reasonably rapidly in flowing rivers.

Biodegradability is a term that requires careful definition. In essence, it means the process of decomposition of an organic material by naturally occurring micro-organisms. Note that it applies only to organic materials. Such a process is one that obviously depends on time, concentration and temperature. Speed of degradation is not the only criterion. The avoidance of intermediates which may have undesirable effects, such as nonylphenol (suspected to be an endrocrine mimic) is much more crucial.

Fig. 3.15 Detergent foam in the Molonglo River below the Scrivener Dam, which holds back Lake Burley Griffin in Canberra ACT, during one of the overflows of the (upstream) Queanbeyan sewerage system.

(Courtesy of Canberra Times, 25 June 1975)

Enzymes

The use of enzymes for washing has a long history, starting with a patent in 1913 for soda plus a small amount of impure proteolytic (protein digesting) trypsin enzyme marketed for pre-wash as *Burnus*. There was some sporadic activity in Switzerland around 1935, but it was the fat shortage during World War Two, and lack of surfactants, that led to the

Bio 38 launch in 1945. This contained purified pancreatic trypsin. However, it had limited commercial success and was replaced in 1957 by Bio 40 containing a bacterial protease, which still suffered from needing a neutral pH. In 1960 Novo was introduced in Denmark to clean the white coats worn by workers in the meat and fish industries. The most common enzyme type is alkaline protease, which digests protein in alkaline conditions. It is produced from the bacterium *Bacillus licheniformis* or *B. subtilis* in large fermenters. By working at pH 7, the enzyme produced is inactivated and prevented from digesting the bacteria producing it! At pH 9–10 it becomes active again and will work in the presence of polyphosphate and perborate.

Although enzymes can digest proteins in stains, they can also cause severe allergic reactions in people, just like the proteins in bee stings. They are now used in a granulated form, coated with polyethylene glycol, which melts and releases them in the wash.

Other enzymes are amylases used in detergents to degrade starches to water-soluble sugars. They are useful as washing temperatures are decreased because they hydrolyse the 'starch glue' that binds the soil to the fabric.

Cellulases are another group of enzymes, which remove cellulose fine fluff (microfibrils) released in cotton after repeated washing and which cause stiffness and greying, particularly noticeable in towels.

Lipases are enzymes that hydrolyse fats, and have been developed for dealing with fatty soil in clothes. Lipolase™ was the first genetically engineered commercial enzyme and it efficiently converts fats in food, cosmetics, and sweat into fatty acids and glycerol (without the use of caustic). Interestingly, its effect is low during washing but when the water content of the clothes is reduced on spin drying, it is active and the waste products are removed in the next cycle. This makes it particularly effective as a pre-spotting agent a few minutes before the normal wash. There is also a great future for biotechnology in this area.

Other household cleaning agents

Machine dishwashing detergents

Generally, machine dishwashing detergents contain only about 2% of low-foaming, non-ionic surfactant (usually the block co-polymers, or propylene oxide and ethylene oxide). Their efficiency depends more on their physical characteristics. A typical formulation of dishwashing detergent is shown in Table 3.3 (overleaf).

Sodium metasilicate, in particular, is quite caustic and very dangerous if swallowed. Some commercial powders contain 65% hydrated metasilicate. Preparations with greater than 5% sodium carbonate can cause sheet erosion of glassware, which gradually thins the glass.

TABLE 3.3
Typical formulation of a machine dishwashing detergent

Ingredient	Percentage of total formulation
Anhydrous sodium tripolyphosphate	30
Anhydrous sodium metasilicate	30
Anhydrous sodium carbonate	37.5
Low-foam surfactant	0.5
Sodium dichloroisocyanurate (56-64% available chlorine)	2
Corrosion inhibitors	9.5

GOLD FROM SEWAGE

Why would a mining exploration company take out a mining lease on the land of a sewage disposal farm? Well, everything goes into the sewer, including heavy metals from industry such as chromium, cadmium and mercury. But the kitchen sink, and particularly the dishwashing machine, contributes gold.

When that gilded crockery that shouldn't, does in fact go into the dishwasher, the caustic detergent etches the glass ever so slightly. But this is enough to allow some of the gold to be lifted off the plate and washed down the drain. Wear on gold jewellery on those housepersons' hands dumps in a little more.

Melbourne Water found to its surprise that Echidna Exploration had taken a lease out on its Werribee sewage treatment plant sludge farm. As gold can be economically extracted from granite at a level of 5 parts per million, and that includes the cost of crushing the rock, the sludge is probably worth treating at about half that level. With so many tonnes of sewage sludge per year, there is a living to be made from recycling the effluent from the affluent burghers on Melbourne's Sunset Boulevard. In fact most of the metals in the sludge can probably be recycled economically .

Chlorine-containing powders can cause deterioration of plastic kitchenware, but without chlorine the tannin stains from tea are not removed. Aluminiumware is also attacked quite strongly; therefore your saucepans are probably losing weight at an appreciable rate. Corrosion inhibitors, consisting of aluminium salts, are added to protect aluminium components — a case of shifting the equilibrium back again!

Liquid formulations have been introduced for machine dishwashing as well. In order to keep the liquid from seeping from a standard powder dispenser, something like bentonite clay is added to give it thixotropic properties (see Fun with fluid flow in Chapter 12).

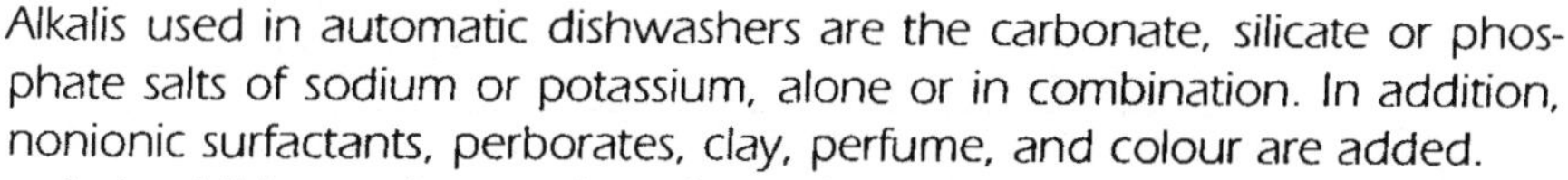

DISHWASHER DETERGENTS

Alkalis used in automatic dishwashers are the carbonate, silicate or phosphate salts of sodium or potassium, alone or in combination. In addition, nonionic surfactants, perborates, clay, perfume, and colour are added.

It is children who tend to be poisoned by automatic dishwashing detergents following exposure to highly concentrated solutions or neat powder-in-door receptacles of the washing machine, or directly from the packet.

Alkalis interact with fatty throat tissues, changing the fat into soap, a process called 'liquefaction necrosis'. The nature and degree of injury depends on the concentration, pH, quantity, physical form, and duration of exposure.

The standard control on these materials based on a limit of pH (11.5) for a 1% solution of the product, is inadequate. How can dangerous formulations be produced to avoid this scheduling?

What would be the advantages and disadvantages of using the pH of a 50% solution as the control parameter? What variation in pH would you expect between a 1% and a 50% solution?

Caustic soda is traditionally used for pulping fibre to produce cellulose. Boiling caustic is an occupational hazard, particularly for the eyes. It is also hygroscopic and absorbs carbon dioxide from the air to produce the much weaker soda, and attacks amphoteric metals and their alloys. All these problems were faced by the makers of the early dishwashers. They found that alkaline silicates worked sufficiently well. Could you pulp vegetable fibre for paper with dishwashing powder instead of caustic?

The standard rinse agent consists of 60% low-foaming wetting agent plus 20% propanol and 20% water.

The koala's killer diet

There is a contradiction between the concept of using bleaches to 'destroy' odours and then for manufacturers to turn around and use perfume in bleach! Antiperspirants with aluminium hydroxychloride will chemically destroy most delicate perfumes. An exception is eucalyptus oil (1,8 cineol — see Steam distillation in Appendix III), which is stable in 4% bleach. Incidentally, 5 g of the 'natural' eucalyptus oil will kill an adult — for example eating 2 kg of eucalyptus-flavoured lollies. See also deodorants and perfumes in Chapter 5.

Scouring powders

Scouring powders consist mainly of abrasive powder (≈ 80%), which can be screened silica, felspar, calcite, or limestone, with the size of most of the material 44 micrometres or smaller. The rest of the powder is sodium carbonate or similar alkaline salts, with about 2% surfactant, and in some cases chlorine bleach. The blue dye which appears on wetting and which is for appearance only can be finely divided copper phthalocyanine, and this sometimes bleaches while in use. (See AS 1962, 1976 Hard Surface (Scouring) Powder.) General purpose liquid cleaners are simply a suspension of scouring powder.

Drain cleaners

Drain cleaners are a mixture of caustic soda ($NaOH$) and aluminium filings. They react in water to provide heat to melt and saponify the fat. The gas released is hydrogen.

Dangers
Mixing household bleach with preparations containing ammonia leads to the formation of chloramines (NH_2Cl, $NHCl_2$), which form acrid fumes and cause respiratory distress, but have no permanent consequences (see also Chapter 8). If swallowed, household bleach causes immediate vomiting but with no serious consequences, except those associated with vomiting in general. The treatment for the ingestion of bleach is to swallow immediately one or more of milk, egg white, starch paste, or milk of magnesia. Avoid sodium bicarbonate because it releases carbon dioxide. **Do not** use acidic antidotes.

Bleach

Hypochlorite salts (e.g. sodium, potassium, calcium, magnesium) serve as disinfectants, bleaches, and deodorisers. Specifically, they are used to disinfect contaminated utensils, nappies, and water for drinking or swimming. Liquid household bleach is usually 5% sodium hypochlorite (NaOCl). Commercial solutions are made by adding gaseous chlorine to 12–16% sodium hydroxide solution (NaOH) until alkalinity is just neutralised, and the resulting solution is diluted to 5%. Even a little too much free chlorine results in an acidic solution that is unstable.

Mould removers

The domestic moulds found in kitchens and bathrooms in Australia belong to a small range of common fungal types. The moulds are normally found growing between the tiles. Some, especially *Phoma*, can cause allergies, but the main household problem is their unsightliness. A survey by the Australian Consumers' Association (ACA) found the common species around the country to be: Adelaide (*Phoma, Penicillium*), Brisbane (*Phoma, Rhizopus*), Canberra (*Phoma, Penicillium*), Darwin (*Phoma, Phialophora, Rhizopus*), Hobart (*Phoma, Phialophora*); Melbourne (*Phoma, Fusarium*); Perth (*Phoma*); Sydney (*Phoma, Fusarium*). Mould removers are generally only thickened, stabilised bleach solutions in bottles with spray caps. The same safety instructions as for bleaches apply.

Cloudy ammonia

When ammonia (NH_3) was first made from coal tar, the solutions were very murky. Later the Haber process for 'fixing' the nitrogen of the air gave a very pure product, but by this time people were used to 'cloudy ammonia'. For this reason soap is added to keep pure, clear ammonia cloudy. Fresh household ammonia ranges in concentration up to 10% actual NH_3. Ammonia vapour is extremely irritating, the solution is very alkaline, and it acts as a caustic. A little ammonia in a dish of hot water placed in a cold oven overnight will saponify the fat to a certain extent and make oven cleaning easier.

Stain removal

To a large extent, stain-removal procedures are based on solubility patterns (like dissolves like) or on chemical reactions. Stains caused by fatty substances such as chocolate, butter, or grease can be removed by treatment with typical dry-cleaning solvents such as tetrachloroethylene, $CCl_2{=}CCl_2$. Removal of iron stains (e.g. rust and some inks) involves a chemical reaction: treatment with oxalic acid (poisonous!) forms a soluble complex with the iron.

Much stain removal is carried out by oxidation using oxidising bleaches such as hydrogen peroxide, sodium perborate (which forms hydrogen peroxide in water), or laundry bleach (sodium hypochlorite).

These bleaches work on mildews and blood, for example, but bleach should not be used on wool because it attacks the linkages that hold the wool together.

The working of some methods is not completely understood. For example, the use of copious amounts of salt on a red wine stain on a table cloth probably operates by *osmosis;* that is, the salt draws out the water from the fibres of the cloth and takes the red stain along with it. This method only works on a fresh stain before the red dye becomes firmly attached to the cloth. Table 3.4 is a guide to stain removal.

TABLE 3.4

A guide to stain removal

Type of stain	Treatment 1	Treatment 2	Treatment 3
Beer	1		
Beetroot	1		
Bleach	1		
Blood	10	6	
Burn or scorch	5		
Candlewax	3		
Chewing gum	7		
Chocolate	1	2	
Cocoa	2	10	1
Coffee/tea	2	10	1
Cooking oils	2	1	
Crayon/colour marker	2	1	
Egg	1		
Faeces/urine/vomit	1		
Fruit juice	9	2	
Furniture polish	2	1	
Grass	4		
Gravy/sauce	9	1	
Grease	2	1	
Ice cream	1		
Ink–ballpoint	4	1	
Ink–fountain	9	1	6
Lipstick	2	1	
Mildew	1	5	
Milk	9	2	1
Nail polish	8	2	
Oil/salad dressing	2	1	
Paint–emulsion	2	10	1
Paint–oil-base	3	2	1
Rust	2	1	11
Shoe polish	2	1	
Soft drinks	9	1	5
Tar	3	2	1
Wine–red	12	9	6
Wine–white	1		

Code to treatments

1. Use a mixture of wool detergent with one teaspoon of clear vinegar in one litre of warm water.
2. Use an organic solvent such as dry-cleaning fluid, mineral turps, light petroleum. (**CARE!** ensure good ventilation and away from flame.)
3. Use a mixture of mineral turps with dry-cleaning fluid (**CARE!** ensure good ventilation and away from flame.)
4. Methylated spirits. (**CARE!** ensure good ventilation and away from flame.)
5. Hydrogen peroxide (20 vol.). Dilute one part to 10 with **cold** water. Do not use on dark or patterned material.
6. Dye stripper. Dilute one part to 50 with **cold** water. Do not use on dark or patterned material.
7. Use a freezing agent to solidify the gum, and then scrape it off.
8. Acetone (nail-polish remover).
9. Clean, warm water (do not use hot water).
10. Cold water only.
11. Weak acidic solution, vinegar, or lemon juice diluted with cold water.
12. Absorbent powder such as salt or talc. Sprinkle on spillage, leave overnight, and vacuum off.

- Clean stains immediately to prevent further chemical bonding to the material; scrape off excess solid and soak up excess liquid.
- Try the treatment first on an inconspicuous part of the material.
- Work inward from the edge to prevent outward diffusion and spreading of the stain.

- Do not rub.
- Allow material to dry before consecutive treatments.
  Note: PVA chamois must not be used with solvent-based agents.

Dry cleaning

The term *dry cleaning* was introduced generally to cover the cleaning of textiles with organic solvents rather than water. When cleaning in water, the water-soluble components of the soil are not taken into account because they spontaneously dissolve in the water and so their removal is not a problem. In dry cleaning, the situation is quite similar with regard to oily and greasy dirt. The necessity for also removing water-soluble substances such as salt and sugar is one of the reasons why water is usually added to the dry-cleaning bath.

The organic solvents most frequently used in dry cleaning can dissolve only extremely low quantities of water. Generally, chlorinated hydrocarbons such as chlorinated ethylene are used. However, the surfactants form 'inverse' micelles, where the polar groups are *inside* and the hydrocarbon tails *outside* in the solvent phase (compare Fig. 3.7). The interior of these micelles can dissolve additional water (about 1.5 molecules of water per surfactant molecule). The removal of solid particles is promoted by surfactants in dry cleaning, as well as in normal washing. Sufficient water is solubilised to maintain a reasonably high relative humidity of water vapour above the solvent; this is a convenient measure of the 'activity' of water *in* the solvent. In practical dry cleaning, however, an upper limit to the relative humidity is determined by the fact that textile material shrinks, wrinkles, or felts if the relative humidity is too high.

Organic pollution

Organic pollution is the most widespread type of river pollution. 'Organic' materials are those complex carbon-based compounds that are such important parts of living matter — proteins, carbohydrates, and fats. But many other substances, including detergents and pesticides, are chemically very similar. Untreated sewage, the discharges from sewage works, and waste from paper factories and food factories, all add organic matter to our rivers.

If the discharge of this organic waste is small compared to the amount of water in the river, it is broken down to simple inorganic substances by bacteria and fungi in the river water. It is a natural process, and is known as self-purification. The main chemical elements in organic matter are carbon (C), hydrogen (H), some oxygen (O), nitrogen (N), sulfur (S), and phosphorus (P). The large organic molecules are broken down and the elements form carbon dioxide, water, nitrates, sulfates, and phosphates, which are harmless in small amounts.

Figure 3.16 illustrates that oxygen is needed for self-purification.

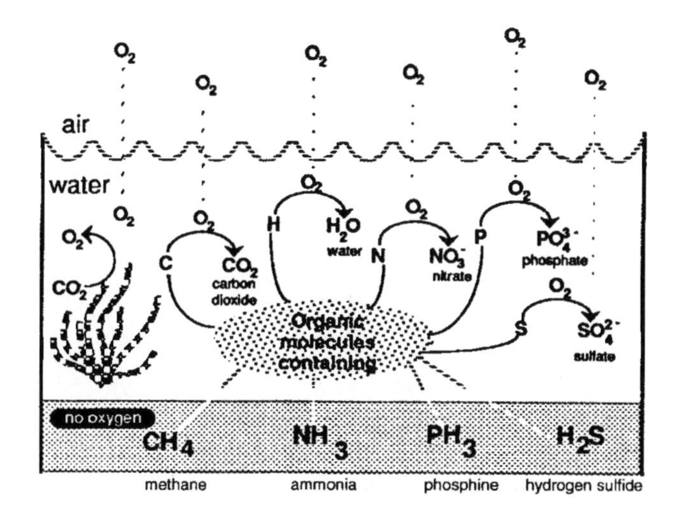

Fig. 3.16 Movement of molecules between air and water

This oxygen must come from the oxygen dissolved in the river water. The dissolved oxygen removed in this way by the bacteria during self-purification is replaced by oxygen seeping through the surface of the water and, during daylight hours, by oxygen given off by the submerged green plants. However, if the pollution is heavy and if there is a lot of organic matter, oxygen will be removed from the water faster than it can be replaced, and self-purification stops. When there is no oxygen present at all, some types of bacteria can still break down organic matter, but marsh gas (methane) and offensive and poisonous substances such as ammonia and hydrogen sulfide (rotten egg gas), are formed, instead of the mild products of self-purification. The smells and colour of the bottom of a pond or stagnant water come from this airless breakdown of organic matter.

EXERCISE: SOLUBILITY AT THE SEA

When you see a rock pool at the sea drying out, you probably think the white stuff crystallising out at the edge is salt. As the solids in sea water are about 96.5% NaCl, that is not an unreasonable assumption. It just happens to be wrong. While magnesium is more abundant than calcium, it is also more soluble. Calcium is the only cation near saturation. As water evaporates from the seawater, the order of precipitation is:

1. Carbonate: mainly calcium carbonate (calcite) but also calcium magnesium carbonate (dolomite).
2. Sulfate: calcium sulfate dihydrate (gypsum).
3. Only when the volume of the seawater is down to about 10% of the original will sodium chloride crystallise out.
4. This is followed by potassium and magnesium salts.

This process is called fractional crystallisation. If all the seawater is allowed to evaporate, the final solid will be about 0.5% carbonates, 3% gypsum, and the rest mostly sodium chloride.

As an exercise, try a little of the vinegar (that you have for treating jellyfish stings) on a sample of the white precipitate at the edge of the rock pool to see if it releases carbon dioxide gas bubbles.

What would happen if you dissolved more carbon dioxide in the seawater; would you precipitate more calcium carbonate? No, intuition is wrong again. The pH of seawater is 7.8 and carbon dioxide is present mainly as bicarbonate (HCO_3^-); about 90%. Dissolving more carbon dioxide lowers the pH; this is what happens in deep water where the pH drops to 7.5 and the equilibrium shifts to converting more carbonate to bicarbonate. Thus calcium carbonate shells tend to redissolve in dropping deeper into the ocean and this means we don't have as many fossils as we might have expected. On the surface, warming reduces the solubility of carbon dioxide and also photosynthesis removes some of it, and so the pH can rise to as high as 8.4. There is a shift in equilibrium to carbonate. Even though there is less total carbon dioxide present, carbonate has increased and calcium carbonate is close to saturation if the pH is too high — see the carbonate titration curve (Fig. 8.3).

Reference

Surface activity in the service of women, the Hartung Youth Lecture 1973, P.H.A. Strasser, published as a supplement in *Proceedings of the Royal Australian Chemical Institute* , **41**, (April), 1974. This lecture was the original basis for this chapter. I am further grateful to Peter Strasser for continuing to keep me up to date on this topic for each new edition.

Bibliography

Surfactants in consumer products, J. Falbe, Springer, Berlin, 1987. Industrial reference.

Detergents, a series of articles, *Chemistry and Industry,* 19 March 1990, pp. 160–191.

Synthetic detergents, A.S. Davidson and B. Milwidsky, 7th edn., Longman, 1992. UK industrial reference.

A guide to the surfactant world, X. Domingo, Edicions Proa, S.A., Barcelona, Spain, 1995. Most recent available.

How to clean practically anything, S. Pemberton, Australian Consumers' Association, Sydney, 1997. Popular.

CHAPTER 4

Chemistry in the kitchen

Ever since she bought that chemistry set, I'm afraid to eat her cooking

In this chapter we move into the kitchen and deal with nutrition, food preparation, microwave ovens and commercial processing. We take a good hard look at fats in food, including ice cream and chocolate. Other aspects, such as food additives, are left until we go into the dining room in Chapter 15.

Nutrition

Have you ever lost a friend after serving chilli con carne? Red kidney beans have caused several outbreaks of serious food poisoning. The beans contain a poison that is normally destroyed during cooking. Soaked raw beans in a salad, or beans that have been slowly cooking in a casserole, can cause vomiting and diarrhoea. Slow cooking at 80°C can increase the amount of poison fivefold, whereas boiling for ten minutes renders the beans harmless.

Red kidney beans contain traces of a protein poison belonging to a class called lectins that coagulate red blood cells and attach themselves to the carbohydrate in the cell walls of the intestine. The Bulgarian emigré George Markov was murdered in London by such a poison injected from an umbrella. That poison has a close relative (ricin) that comes from the

The chemistry of cooking

Our knowledge of the chemistry of cooking has come from our forebears' experiments with various foods and methods of preparation. Those whose experiments worked, survived — and led to us. No wonder food habits tend to be quite conservative. Yet food habits vary enormously from one culture to another.

castor oil bean. Other beans (runner, broad and French) also contain small quantities of lectins, which form during the ripening process.

Chemistry has shown that all staple diets supply basic nutrients of proteins, fats and carbohydrates. Foods do not have 'special' attributes; expressions such as 'fish is good for brain' are without foundation. The food is broken down, and we build up the components 'in our image' again. The plants and animals we eat didn't design their body composition to suit us as food. Much of what we eat, we break down and remake, but some substances we ingest are active individual chemicals, and not just 'generic' food (see box, The myth of pure natural foods).

Our level of understanding is not impressive. Until recently we believed that protein deficiency (kwashiorkor) was the major nutritional disease of the Third World, whereas the problem is simply starvation. Fact and fiction are difficult to separate. An apple a day was supposed to keep the doctor away. But apples are a very poor source of vitamin C, their roughage is less effective than cereals and bran, and they make a poor toothbrush. So there!

Vitamins

VITAMINS AND MINERALS

Vitamin A — is the general name for the group of substances that includes retinol and retinal. Beta-carotene is their precursor. It has an important role in eyesight. When light strikes the retinal/opsin complex in the retina, a double bond in retinal is converted from the cis- to the trans-configuration, sending a signal to the optic nerve. Vitamin A is also important in maintaining epithelial tissue. Normal blood contains 15 to 60 µg of retinol per 100 mL of serum.

Vitamin B_1 (also called thiamine) — is a water-soluble factor that is a co-factor for many enzymes, including some that release CO_2 from beta–keto acids. Thiamine was the first vitamin to be isolated. It is found in the brown coating of unpolished rice and other cereal grains.

Vitamin D_6 — is a family of closely related compounds that includes pyridoxine, pyridoxal and pyridoxamine. The latter two substances are co-factors for many metabolic enzymes, including those catalysing the biosynthesis and degradation of amino acids.

Vitamin C — is ascorbic acid. It is the water-soluble antioxidant and has many roles in the body, including a major one in the formation of collagen, a substance that holds muscle cells together and is also found in bones, teeth, tendons, etc. A lack of ascorbic acid results in scurvy. It also has a role in amino acid metabolism. Recommended daily allowance (RDA) is 60 µg for young adult males in the USA, but this level is set to prevent scurvy (see below).

Vitamin E — is alpha-tocopherol (see Fig. 4.1), an oil soluble anti-oxidant found in polyunsaturated oils in amounts necessary to protect them against oxidation.

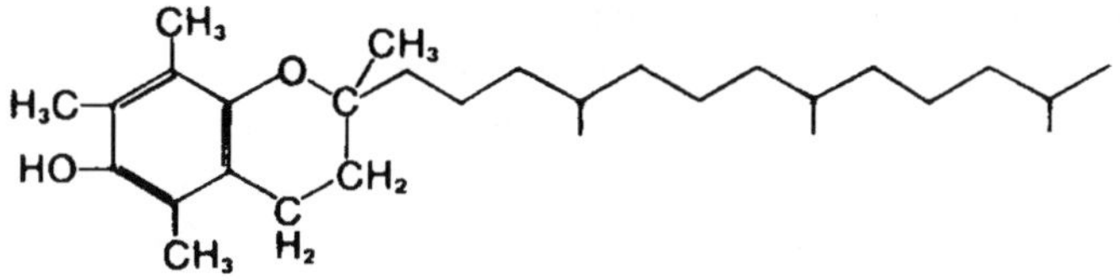

Fig. 4.1
Vitamin E

Calcium — is the most abundant mineral, and the fifth most abundant element, in the body. Bone tissue contains about 99% of the 1.2 kg of calcium in the average body. The remaining 1% is used in nerve transmission, muscle contraction and numerous other functions.

Chloride — accounts for about 0.15% of body weight. There is about 450 to 600 mg of Cl^- per 100 mL of blood. It is a constituent of gastric juice (which is approximately 0.03 M HCl). It also has a role in controlling the transport of O_2 and CO_2 by haemoglobin in red blood cells. Adults require a minimum of 750 mg of chloride daily.

Copper — The human body contains 75 to 100 mg of this trace mineral. Copper competes with zinc for entry from the intestines, so an increase in dietary zinc may result in copper deficiency. Copper is a co-factor for many enzymes and proteins, and it is therefore important to the normal development of nerve, bone, blood and connective tissue. The RDA is 1.5 to 3 mg.

Fluoride — occurs in the blood at 0.28 mg per 100 mL of blood. It is essential for teeth and bones. (See Toothpaste in Chapter 5).

Iodine — The human body contains 20 to 50 mg of iodine. Half of this is in the muscles, and much of the rest is in the thyroid gland where it is part of the chemicals thyroxine and triiodothyronine. People with iodine deficiency develop hyperthyroidism and an enlarged thyroid gland, or goitre, and in more severe cases brain damage. The RDA is 150 μg.

Iron — The human body contains 3 to 5 grams of iron. Half of this is found in haemoglobin, the protein that carries O_2 and CO_2 in the bloodstream. Oxygen is carried by binding directly to the iron, which is in its 2+ oxidation state. Women are close to the deficiency borderline. Monthly blood loss from menstruation carries with it an average loss of 28 mg of iron. The RDAs are 10 mg for adult males and 15 mg for adult females. (See Chemical changes, later in this chapter.)

Lecithin — is the common name for diacylphosphatidylcholine, a component of cell membranes. The lecithins are mixed esters of glycerol and choline with long chain fatty acids and phosphoric acid.

Magnesium — although it is a crucial cation, it accounts for less than 0.1% of body weight. The serum concentration is 1 to 3 mg per 100 mL. The bones contain about 60% of the total magnesium as a reservoir to guarantee adequate supplies in times of shortage. Among other functions, magnesium forms a complex with adenosine triphosphate (ATP) (see Sports chemistry and biochemistry in this chapter). Without Mg^{2+}, ATP could not participate in the vast array of reactions in which it is involved. The RDA is 350 mg for adult males, and 280 mg for adult females.

Manganese — Normal levels are 0.005 to 0.02 mg per 100 mL of blood. Manganese is a co-factor for many enzymes, but other metals, including magnesium, can substitute for it in many cases.

Molybdenum — is essential in the function of two enzyme systems, xanthine oxidase and aldehyde oxidase. Hard water can provide up to 40% of the daily intake of molybdenum.

Phosphorus — Twelve grams of phosphorus are found in each kilogram of fat-free tissue. About 85% of it occurs as inorganic calcium phosphate in bones and teeth. Phosphorus has a level of 30 to 45 mg per 100 mL of blood. The ATP molecule contains three atoms of phosphorus. One of these is regularly transferred (as a phosphate ion) to other molecules in phosphorylation reactions. The RDA is 1200 mg.

Potassium — Potassium is the primary cation in intracellular fluids, where its concentration is 30 times higher than in extracellular fluids. The normal blood level is about 5 mmol per litre of serum. More than 8 mmol per litre is dangerous. It is often forgotten that potassium is lost in sweat (10 mmol per litre of potassium, compared to 25 to 30 mmol of sodium ion per litre of sweat).

Selenium — is a trace mineral that functions either alone, or as an enzyme co-factor. It acts to destroy hydrogen peroxide, and is believed to have a role in preventing cancer. RDAs are 70 μg for males, and 55 μg for females.

Sodium — constitutes about 0.15% of total body weight. It has greater concentrations in extracellular fluids than in intracellular fluids. In this respect it is the opposite of potassium. The level of Na^+ in blood is maintained between 310 and 333 mg per 100 mL of serum.

Zinc — The human body contains 2 to 3 grams of zinc. It is a co-factor for some 20 enzymes, including alcohol dehydrogenase (which breaks down ethanol) and carboxypeptidase (which catalyses the hydrolysis of proteins in the small intestine). RDAs are 15 mg for males and 12 mg for females.

Source: The nutrition desk reference, by R.H. Garrison, Jr and E. S. Keats Publishing New Canaan, CT, 1990.

Our bodies are metabolic factories. All efficient manufacturers must make decisions about which chemicals should be 'outsourced' and which should be synthesised. The bulk materials (fats, carbohydrates, proteins and minerals) are imported, degraded into simpler units, and then reconstituted. Minor essential chemicals that occur with the major ones are obtained at no extra cost, so there is no need to make them. We call such a group of unrelated substances, vitamins. Their only common characteristic is that they are *not* produced in the body.

Different 'bodies' have made different decisions. We share with the fruit bat, guinea pig, anthropoid ape, and red-vented bulbul (a bird) the decision not to produce vitamin C (ascorbic acid) but to obtain it from fruit. This evolutionary decision can also be seen as a genetic deficiency disease that, unlike other such diseases, happens to be common to the whole human species (and four other species). Because of the wide availability of vitamin C in the tropical climates where these five species evolved, such a deficiency disease had no evolutionary disadvantage, and released extra production capacity of the body to produce some other vital material. That decision must have been correct because it gave a metabolic edge that ensured the survival of that decision, at least for those five species.

How much vitamin C do we need? Can excess do us good? Can excess do us harm? Chemistry can suggest some answers.

EXPERIMENT 4.1

Take a dipstick, like the sort your doctor gives you to test your urine, only this one is for ascorbic acid (vitamin C) not glucose. (Dipsticks are available from chemical supply houses.) In fruit juices we would expect to find vitamin C, and in fact the law requires that you should find minimum amounts. You can use the dipsticks to measure vitamin C content in fruit juices. You normally find more than in the original fruit because the processor adds the legal minimum, anyway, to replace any vitamin naturally present that has not survived processing and storage (particularly for apple juice because apples have very little, but everyone expects some in the juice).

Just how stable is the vitamin? There is considerable information available, but why not do your own experiments? Change the following variables: (a) the effect of boiling in water, (b) cooking at different pH (e.g. by adding bicarbonate of soda), (c) the effect of boiling in the absence of oxygen. If you blend vegetables and measure vitamin C content before and after boiling, you will find a large increase, because the boiling extracts the water-soluble vitamin from the food. (See later in this chapter.)

Legally, a chemical is defined on the basis of what it is used for. When vitamin C is added to beer, it is there as an antioxidant (not as a vitamin), so it cannot be mentioned as a vitamin on the label. If you are keen on vitamin C you may be better off eating ham, because vitamin C is added to the nitrate preservative to prevent formation of nitrosamines.

Kalokerinos quotes that ascorbic acid appears in the urine when the concentration in the plasma reaches 1 mg/100 mL and suggests that a value of 25 μg/mL in the urine should be established. Why not test your own urine and establish how you excrete after taking a dose (1–2 g) over a period of one day? You will need to measure both the volume of urine and the concentration of vitamin C. You can also plot the amount of the original vitamin remaining, and the rate of excretion during the day (which peaks rapidly). Every person appears to have a different profile.

Dipstick for vitamin C

The dipstick was developed in the 1960s for Dr Archie Kalokerinos, who asked the Ames company of Indiana, USA, for a quick, cheap diagnostic test that would enable him to test his theory that a great deal of ill health in the Australian Aborigines was caused by a lack of sufficient vitamin C in their diet.

Toxicology testing with animals
Most animals used for testing the response of the liver to toxins protect their bodies by producing extra ascorbic acid. Humans require increased dietary intake to provide similar protection, of the order of several grams at the time of exposure. Thus the results of such animal testing, applied to humans, may not be valid.

(**Source**: The Changing Field of Toxicology, Mark Donohoe, reproduced in Toxic Network News, Oct 1996).

The intake of excess vitamin C has been criticised by those who have learnt enough biochemistry to know of its metabolism to oxalic acid (calcium oxalate is the main component of kidney stones), but not enough to realise how slow this is in comparison to the rate of excretion. (There are reports of withdrawal symptoms (like scurvy) from those who suddenly stop taking a long-term high dose, but that is a dumb thing to do for the intake of any material.)

This high proportion excreted may encourage the conservationists among readers to evaporate their urine and recycle the vitamin C! But then, perhaps not.

The amount of vitamin C *intake* needed to reach the maximum rate of *use* of the vitamin in the body varies. Since the 1940s it has been known that smokers, for example, need a much higher level — perhaps over twice as much. Megadose levels, on the other hand, have been suggested as palliatives for anything from the common cold to cancer. However, Linus Pauling, Nobel laureate for chemistry and for peace, and Albert Szent-Györgyi, Nobel laureate (for the isolation and characterisation of the vitamin in paprika in 1928), took about 1 g per day. They both lived into their nineties — a somewhat unreliable sample of size 2. Linus presented his arguments in Sydney in 1973 and he made a good case.

Unlike the other vitamins, which seem to have very specialised roles, the range of activities of vitamin C is very wide. It also has an important role in food technology. There are sound chemical reasons for adding it to diverse products such as beer and bacon. On the other hand it decolorises the legal synthetic coal-tar food colours (both azo- and triphenylmethane types), particularly in sunlight, and so is not added to food when these are used. There are also patents specifying the use of ascorbic acid for the removal of chlorine from drinking water.

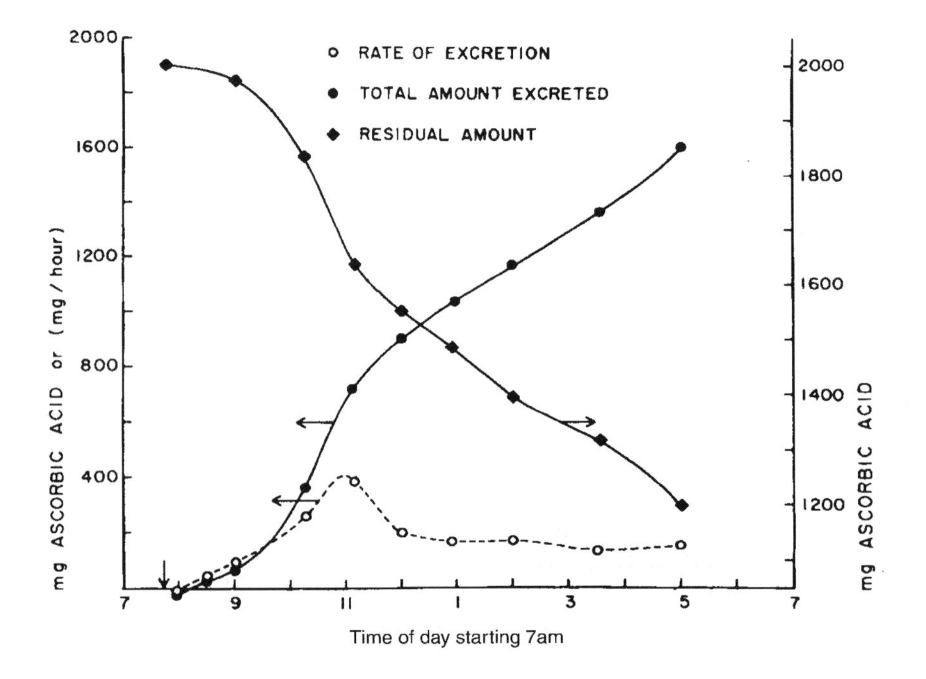

Food processing

Prehistoric methods of food preservation had a common mechanism for making food inedible for micro-organisms. This involved either removing the water component by drying, or rendering the water 'unavailable' to the microbe by increasing its salt content. This meant the use not only of common salt (NaCl) but of other 'salts' (such as in curing with sodium nitrate and nitrite). Such concentrated brines have a high osmotic pressure and microbes in them dehydrate as the water is extracted from their cells by osmosis. Bacteria, moulds and yeasts have different sensitivity and thus need different levels of dissolved material to prevent their growth (see Fig. 4.3). The nature of the dissolved ions or molecules is not important, only the amount dissolved. Thus sugar can be used instead of salt, and this accounts for the longevity of honey and jam.

Precisely the same chemical argument is used in selecting antifreeze materials, except that there are good reasons for not using salt or sugar. Likewise antifreeze is not a recommended food preservative (see Artificial sweeteners in Chapter 15).

The percentage of water can be determined by using the Dean and Starke distillation (Appendix III) for most of the products shown on the figure. The lowering of activity needed to stop various micro-organisms from growing is also shown. (As we saw earlier, the pH is also very important.)

The next level of sophistication involves adjustment of the acidity of the food because many dangerous microbes are sensitive to acid (e.g.

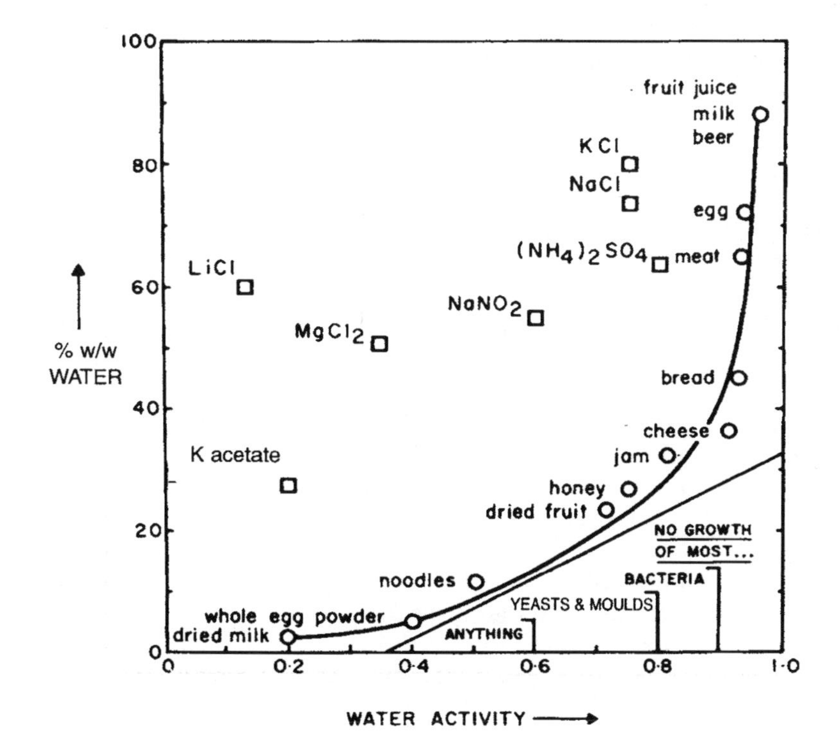

Fig. 4.3
Water activity versus water concentration for selected foods and selected standard saturated salt solutions

Clostridium botulinum, from L. *botulus*, sausage). This is the source of the most deadly form of food poisoning in the absence of oxygen.

Fermentation of food to produce acid is common to all cultures and cuisines, and the list of fermented foods is enormous. Some examples are dill pickles, sauerkraut, coffee beans, vanilla, kimchi, tarhana, sajur asin, kishk, salami, yoghurt, cheese, sour dough bread, and soy sauce. Bacteria, yeasts and moulds are used to produce mainly lactic acid or acetic acid (or both).

Alcohol is only a preservative at high concentrations. While curry and spices are alleged to have minor preservation action, their original use was to cover the bad taste of already decomposing food.

Modern food technology has expanded the types of preservation techniques and reduced the hazard of traditional methods. Fat-soluble antioxidants mimic the action of vitamin E, while vitamin C is added to cured meats to reduce the production of cancer-causing nitrosamines. Canning, followed by (commercial) sterilisation, has given us products with long shelf-lives and has made modern urban living possible. On the other hand we are able to feed ourselves bizarre diets equally well from the supermarket or the health-food store.

We are subjected to bewildering propaganda from the processed-food industry, nutritionists, the medical profession, and consumer organisations. Ignore the junk research.

In summary, the reductionist view of nutrition is wrong. Food is not just the sum of its component parts, and while research showing that vitamin X does Y is important, it is only a small part of the whole story. Concentrating on the linear relationships (one-to-one, cause and effect) can send the wrong message. Food as a whole is the important input (plenty of fruit and vegetables). Specific deficiencies are almost unknown in a balanced diet. Good nutrition is easy. It is not hard to toss a salad, or stir-fry some vegetables.

Natural versus artificial foods

Perhaps the best place to start is with the concepts of *natural* and *artificial*. There is no doubt that natural is not always best. Our introductory discussion on beans should have dispelled that misconception. Moulds produce both penicillin and aflatoxin (probably one of the most carcinogenic chemicals ever, both natural and synthetic). Natural sugar contains, in concentrated form, a great variety of material of unknown composition or effect, and is certainly less desirable than purified sugar, which would seem to be a very natural food. Taken unknowingly (in processed food) and in excess, it causes problems. Is vanillin produced from softwood less natural than vanillin from a vanilla bean? Is a nature-identical synthetic vanillin produced in the laboratory from a coal-tar precursor unnatural? Ascorbic acid (vitamin C) is almost entirely produced synthetically from glucose, although a crucial step in the process requires a microbiological reaction performed by courtesy of *Acetobacter suboxydans*.

THE MYTH OF PURE NATURAL FOODS

In the introduction to his talk 'The myth of pure natural foods', which was given at a public seminar on Nutrition, Health and the Consumer, organised by the Queensland Consumer Affairs Council in July 1977, Bill McCray, then director of the Biochemistry Animal Research Institute of the Queensland Department of Primary Industry, said:

'To the nutritionist, and I quote, food is "nutritive material taken into an organism for growth, work, repair, or the maintenance of vital processes". An objective view? — Agree? — It is not. It is the highly subjective view of the nutritionist. The truth is that food, for all holozoic creatures like ourselves, consists of almost any readily available, independently living fellow creature, and in our case, often the carefully preserved remains of such fellow creatures. Further, these unfortunate victims of the gratification of their fellow creatures are complex mixtures of chemicals, each victim containing a host of chemicals that are inimical to the "growth, work, repair", etc. quoted in the so-called objective definition.

George Bernard Shaw, the Irish wit and playwright of my youth, said it for us — he refused to make his stomach a graveyard for dead animals. He was a confirmed vegetarian. Hence his stated wish was that his funeral cortege include little lambs, calves and chickens to represent those dead animals for whom he had not made his stomach a graveyard. Though he did not admit it, he was forced to make his stomach a compost heap for murdered plants.

It is easy to accept the predator–prey concept for the tiger, or the eagle, but even the gentle lamb devours the living grass. For ourselves, we can readily accept this thesis for meat and fish, but, consider the staff of life — bread. Here we take the grain into which a plant, our fellow creature, has poured all its reproductive energy; its hopes and aspirations for the future generation lie in the germ, the energy for whose early struggle is in the starch, and we crush both to flour, snuffing out the life in the grain. Then into this bleached and whitened corpse we place our fellow creature, yeast, and when it has grown, reproduced, and prospered on the energy of the grain, we put it, living, into an oven at 200°C to die to make our daily bread.

In the multimillennium that man existed as a food gatherer, he existed as a fringe species under constant threat of the extinction that overtook many hominoid species, not the least threat being that posed by his food. Then, as now, eating provided the greatest exposure to exotic chemicals that most of us ever receive.

Man evolved in an environment that provided many poisons produced by micro-organisms, plants and other animals, and is clearly not without biochemical defence mechanisms to protect himself. In his role as a sort of metabolic crematorium for his fellow creatures, he manages to burn up on most occasions the harmful and the helpful with equal facility. Usually it is only when his defence mechanisms are overwhelmed by the sheer numbers of toxic molecules — the total amount of the toxin rather than its toxicity — that man is adversely affected: the dose makes the poison.

That man's fellow creatures are often far from wholesome is a matter for history or prehistory; primitive men developed not only an extraordinary lore about the use and hazard of foods, particularly plants, but also an extraordinary technology to render useful the more recalcitrant of their plant brothers. There is no need to go to prehistory to study food gatherers — just go to the museum to learn how our own Aborigines worked the highly toxic seeds of the Moreton Bay Chestnut to make food of it in the absence of more prolific, or less seasonal, other foods . . .

The list can be extended greatly. Bananas contain serotonin (one of the biologically active amines); chick peas, while highly nutritious, are also toxic. So fad diets are to be avoided. One nutmeg is nice; two can cause an abortion, while three at one sitting could be fatal. Avocados will poison animals, while onions cause anaemia in animals. Half a kilo of horseradish (containing isocyanates) can kill a pig in three hours. Broad beans cause a disease (favism) in people with a particular genetic inclination. Aflatoxin is the most potent known carcinogen and is formed by a fungus that attacks peanuts and other nuts, and cereals such as corn, rice and wheat, when the humidity and temperature are right for it. It is believed that it causes liver cancer which is prevalent in humans in parts of Africa and Asia.

Along with other pesticides, the fungicides used to control fungi are becoming increasingly less effective, so problems will arise in protecting our 'natural' food. Dairy herds in Tasmania in the mid-1950s were being fed on kale which transferred an oxazolidine to the milk. This increased benign goitre among children in spite of the use of iodised salt. Then came the problem of excess iodine in milk from iodophor disinfectants used in the dairies. I have quietly slipped back to an "unnatural" additive. The difference in definition is marginal, and the only sensible approach appears to be to spread any risk by spreading the diet to cover a large selection of foods.'

Chemistry of cooking

Experiment
A very simple experiment is to place some peas in a measuring cylinder and measure the volume of water needed to fill the cylinder to a mark. Do this before and after 15 minutes boiling. Is water gained or lost? Calculate the percentage.

For an experimental chemist, cooking can be divided into four basic methods:

- dry heating (baking, roasting, etc.) < 250°C;
- wet heating (boiling, steaming, etc.) 100–120°C;
- hot-oil frying (frying, etc.) < 300°C;
- microwave.

These methods bring about changes in three classes: physical, chemical and microbiological changes.

Physical changes

When starch grains are suspended in a drop of water and viewed under a low-power microscope (× 100) the discrete grains have a characteristic appearance, depending on the source. Compare potato, rice, wheat, corn, etc. When starch grains are boiled, the starch swells, and the cellulose envelope bursts, allowing the starch to form a gelatinous mass, leaving

TABLE 4.1
Gelating action of some starches

Starch	Barley	Tapioca	Pea	Potato	Wheat	Corn	Rice
Gel temp. (°C)	51–61	52–64	57–70	58–66	59–64	62–70	68–70

Source: Chemistry of cooking, D. J. Smith, School Science Review, **58** (202), 1976.

empty cellulose envelopes behind as 'ghosts'. There is a characteristic temperature range at which this occurs for different starches (Table 4.1).

Chemical changes

Dextrins form from starches during toasting, and samples of bread, before and after toasting, can be compared. An electric coffee grinder can be used to grind up the bread (toast), and the grind is shaken with water and filtered with a Buchner funnel. Dextrins are detected with the iodine solution. Saturation of the solution with crystals of ammonium sulfate will precipitate starch and other large molecules, leaving smaller dextrins in solution. Confirm that thin slices of toast produce more dextrins than thick slices. The effect of the salivary enzyme amylase on raw and boiled starch to produce sugar can be tasted in the mouth and can be confirmed using dipsticks specific for glucose.

Much of cooking has to do with dealing with starches, as the following illustrates:

Cooking and digestion
Although we are told that cooking aids digestion, it would be nice to see some (simple) experimental evidence for this. A test based on the colouring of starch by iodine works well. When starches are heated, the long polymer chain is broken into shorter lengths, called dextrins, which react differently with a solution of iodine (iodine, 1 g; potassium iodide, 2 g; water to 100 mL). The large polymers give the characteristic blue colour, the intermediate ones (of size equivalent to glycogen) give a brown to red colour, and smaller molecules give no colour.

THE QUINTESSENTIAL MASHED POTATO

In a home where eating meat and milk together was seen to be against the laws of the divine being, my mother deftly skirted the culinary minefields by using 'instant' mashed potatoes. She, who had to be obeyed, induced jaw-cramping lumps in that powdery substance for us to digest as best we could.

Since married, she who must now be obeyed insists that I make mashed potatoes from real ones. As an organic chemist, trained in the most vigorous of gustatory experiences — industrial petrochemistry — I needed to fight the beige lumps extruding into starchy dreadlocks. It was, after all, a side-dish conquered by the Irish.

Mashed potatoes consist mainly of insoluble disk-shaped starch particles which interact strongly with themselves and water caught in between, just like library paste, and this connection warns of terrible disaster ahead. I thought of phase inversion back at the plant where lubricating cream gets beaten too hard and turns into grease. Mix oil and water and shake, and it quickly settles out; add a surfactant and the oil is emulsified in the water. If the surfactant is egg yolk (i.e. lecithin) you have made salad dressing. If you beat harder to give smaller oil droplets, you have mayonnaise.

Back at the bowl, the mashing was producing a glutinous mass that more and more closely resembled spackling compound. My mind flooded with surface energies and hydrophilic/phobic ratios, and I was held back at the last moment from squirting in the dishwashing liquid. Suddenly I thought of — **protein**. Molecules with bits that love water, and other bits that love oil (see HLB in Chapter 6). A foot in each camp and keep the bastards apart. Protein from where? Milk (like my mother would never use). Nay, too watery. Butter or margarine? Too greasy. Phase inversion has put the wrong component (oil) in the continuous phase. Something in between? Hurry. Sour cream! In with a dollop of sour cream, and before my widening eyes, the crumbly dreadlocks of mortified starch disintegrated into satiny velvet swirls. The last little starch particles had their Machiavellian machinations thwarted by molecular layers of protein encapsulation, imprisoning them in minute emulsified ball bearings.

Twenty years of schooling, followed by fifteen years of R&D, had marshalled their dreadful power to bring forth the quintessential pot of mashed potatoes. The patent search will start on Monday.

Source: Lightly edited from Mashed potatoes and chemistry, Chemtech 1993, by Al Schwartz who lives in California with a lovely lady and a psychotic cat.

In our Western cuisine we carefully boil superb quality vegetables, and often throw away the water containing most of the vitamins and minerals, in order to eat the almost indigestible remains. When vegetables such as spinach, brussels sprouts, and spring greens have been cooked and filtered, you can see the colouring of beta-carotene (vitamin A precursor) in the water. You can also test for vitamin C with dipsticks. The presence of iron in the water can be demonstrated by reaction with potassium ferrocyanide (1% potassium ferrocyanide plus 1% hydrochloric acid, mixed just before use). A deep blue precipitate indicates iron. Iron pots will release iron and mask the effect.

It should be noted that, in Third World countries today, and in Australia until recently, much of the human dietary need for iron came from iron cooking pots or from iron salts in clay vessels. Stainless steel (and aluminium) pots and quality enamelware have possibly contributed to anaemia in women.

Iron in the flakes
Demonstrate that there is iron in breakfast cereal. Pour some water in a Petri dish on an overhead projector. Float a few flakes of breakfast cereal claiming lots of iron. Chances are that it is there as the metal. So use a strong magnet to pull the flakes around.

A pineapple garnish for ham (gammon) makes digestive sense if it is fresh (it softens the meat) but not when it is canned. On the other hand, fresh pineapple plays havoc with setting gelatine and whipped egg whites because the enzyme is active.

Blanching inactivates enzymes. Potatoes cut for chips, boiled for two minutes, and then cooled in a refrigerator for a further two minutes, do not brown anywhere near as fast as potatoes that are only cut and left.

How are nutrients lost in processing food?

The nutritional value of a processed food is rarely better than that of the raw ingredients from which it is produced, unless nutrients are added in purified or concentrated form. Nevertheless, there are some beneficial effects of processing (e.g. the destruction of trypsin inhibitor in legumes, and the liberation of bound niacin in cereals).

While loss of nutritional value cannot be judged directly, a reduction in nutritional value is usually accompanied by a reduction in other food qualities, such as colour, flavour and texture, which can be judged by the consumer.

During processing, nutrients are lost because they react with other constituents of the food, oxygen, light or heat, or because they are leached by water at some stage of the process. Trace elements and enzymes may also catalyse the destruction of the nutrients. Loss of nutrients may occur at any or all of the following steps between garden and gullet: harvesting, handling and transport; preparation and processing; storage; distribution; and handling and preparation in the home, restaurant or institution. Of these the greatest loss of nutrients in many processed foods occurs during handling and preparation in the home, restaurant or institution. The reasons for this are described below.

Experiment
Inactivation of enzymes can be demonstrated in many ways. A very simple but effective experiment consists in dropping small pieces of raw and cooked liver into test tubes containing 10 vol. (3%) hydrogen peroxide and comparing the rate of effervescence. The enzyme peroxidase catalase is responsible for the reaction. Fresh pineapple contains a powerful proteolytic enzyme called bromelase, but in tinned pineapple it is inactivated.

How stable are nutrients?

Along with the major food constituents such as carbohydrates, fats and proteins, about 50 other nutrients (amino acids, minerals and vitamins) are essential for adequate nutrition by humans.

Foods from plants and animals contain nutrients in varying amounts, and they are unequally sensitive to temperature, light, oxygen and leaching. Consequently the nutrient loss resulting from processing varies according to the type of food, the time that processing takes, and the particular nutrient. Table 4.2 (overleaf) indicates the stability of some vitamins, amino acids and minerals under various conditions.

Loss of nutrients during processing

If we ate a balanced diet, there would be no cause for concern about nutrient losses during processing. Inevitably, great differences exist between the diets of people of different ages and social groups. For example, the diet of babies and of the aged living alone often lacks variety. The diets of some social groups are quite unbalanced. Over-indulgence in alcohol, or dedication to macrobiotic diets bring other problems. In these groups, where one or more essential dietary components may be in limited supply, any loss of nutrients may be critically important.

Foods of animal origin

The successful canning of meats for human consumption depends on the application of sufficient heat to a sealed container to render the contents 'commercially sterile'. Some thiamine is destroyed by the

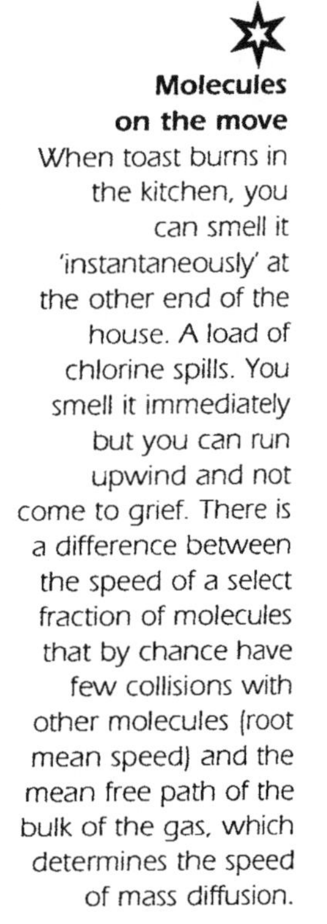

Molecules on the move

When toast burns in the kitchen, you can smell it 'instantaneously' at the other end of the house. A load of chlorine spills. You smell it immediately but you can run upwind and not come to grief. There is a difference between the speed of a select fraction of molecules that by chance have few collisions with other molecules (root mean speed) and the mean free path of the bulk of the gas, which determines the speed of mass diffusion.

TABLE 4.2
Stability of nutrients

	Effect of solutions that are:			Effect of exposure to:			Cooking losses (%)
Nutrient	acid	neutral	alkaline	air	light	heat	
Vitamins							
Vitamin A	U	S	S	U	U	U	0–40
Vitamin C	S	U	U	U	U	U	0–100
Biotin	S	S	S	S	S	U	0–60
Vitamin D	n.a.	S	U	U	U	U	0–40
Folic acid	U	U	S	U	U	U	0–100
Vitamin K	U	S	U	S	U	S	0–5
Niacin	S	S	S	S	S	S	0–75
Riboflavin	S	S	U	S	U	U	0–75
Thiamine	S	U	U	U	S	U	0–80
Essential amino acids							
Isoleucine	S	S	S	S	S	S	0–10
Leucine	S	S	S	S	S	S	0–10
Lysine	S	S	S	S	S	U	0–40
Methionine	S	S	S	S	S	S	0–10
Phenylalanine	S	S	S	S	S	S	0–5
Threonine	U	S	U	S	S	U	0–20
Tryptophan	U	S	S	S	U	S	0–15
Valine	S	S	S	S	S	S	0–10
Mineral salts	S	S	S	S	S	S	0–3

n.a. indicates not available
S: stable (No important destruction)
U: unstable (Significant destruction)
(Source: Nutritional value of processed food, G. J. Walker, CSIRO Information Service, 1979).

extended heat treatment used to sterilise them, in comparison with fresh meats cooked by roasting, braising or boiling.

Foods of plant origin

In their preservation by canning, freezing or dehydration, most vegetables are blanched by immersion in steam or hot water. This inactivates enzymes that would otherwise cause deterioration of the food during storage or thawing. Values for the loss of vitamins resulting from blanching are given in Table 4.3.

Pasteurisation of milk

The introduction of compulsory pasteurisation of milk caused more controversy than has attended the introduction of any other food-processing procedure. What effect does pasteurisation have on the nutrients in milk?

TABLE 4.3
Vitamin losses resulting from blanching

Food	Loss of vitamin C (%)	Loss of thiamine (%)
Asparagus	10	n.a.
Green beans	20–25	10
Lima beans	20–25	35
Sprouts	20–25	n.a.
Cauliflower	20–25	n.a.
Peas	20–25	10
Broccoli	35	n.a.
Spinach	50	60

n.a. indicates not available.

(Source: Nutritional value of processed food, G. J. Walker, CSIRO Information Service, 1979).

- Minerals, vitamins A and D, pyridoxine, niacin, pantothenic acid, and biotin levels are unaffected.
- Riboflavin and vitamins E and K are practically unaffected.
- Thiamine is reduced by 3% to 20%.
- There may be a coagulation of up to 10% of the proteins, albumin and globulin. Since milk is not a rich source of vitamin B_1 (thiamine) and vitamin C (ascorbic acid), the moderate loss from pasteurisation is not significant. On the other hand, pasteurisation destroys all pathogenic organisms in milk, and most other bacteria so the great advance in reducing the risk of milk-borne infection gained by processing far outweighs the insignificant losses of nutrients.

Loss of nutrients during commercial storage

The amounts of nutrients lost from a processed food depend on the temperature experienced, the time that elapses between processing and consumption, the nature of the food, and sometimes the nature of the container. Processed food may be held in a warehouse at the factory; in railway vans, trucks or ships during transport; in distribution warehouses; and in retail stores. The lengths of time that foods remain in any part of this distribution chain, and the temperatures that they experience, are extremely variable. Table 4.4 (overleaf)presents figures that show how vitamins are retained in canned foods stored at 10°C, 18°C and 27°C.

TABLE 4.4
Percentage retention of vitamins in canned foods

	Temp. (°C)	Peas Vit. C	Peas Thiamine	Peas Carotene	Orange juice Vit. C	Orange juice Thiamine	Tomatoes Vit. C	Tomatoes Thiamine
12	10	93	92	98	97	100	95	94
months	18	91	87	94	92	98	94	93
storage	27	86	74	91	77	89	82	82
24	10	91	90	94	95	101	89	91
months	18	89	85	91	80	89	87	87
storage	27	81	70	90	50	83	70	70

Source: Nutritional value of processed food, G. J. Walker, CSIRO Information Service, 1979.

Loss of nutrients during preparation in the home

Foods of animal origin

The major foods of animal origin consumed by Australians are beef, pork, veal, lamb, poultry, fish, eggs, milk and milk products. The nutrients in these foods that cooking and processing affect to an important extent are the B vitamins, proteins and fat.

Freezing, thawing and cold storage: In frozen storage of a few months duration, muscle meats and liver lose 20% to 40% of the thiamine, 0% to 30% of the riboflavin, and up to 18% of the niacin and pantothenic acid. In refrigerator storage for two weeks, losses are 8% to 10% for thiamine and pantothenic acid, 10% to 15% for riboflavin, and less than 10% for niacin. The method of thawing usually has little effect on the nutrient levels, except that thawing in running water or by warming results in a loss of thiamine and pantothenic acid. Up to 33% of pantothenic acid is lost in the 'drip' from thawed beef, which is much more than for other B vitamins. To control bacteria, temperatures should be either very high or very low (see Fig. 4.4).

TAKEAWAY AND FAST FOODS

In 1995, a population of 18 million Australians purchased 1.1 billion takeaway and fast food meals (TAFF). We spend 27 cents per dollar of total household food expenditure on TAFF. Their convenience also causes problems. Each year around 1.5 million Australians are afflicted by some type of illness from eating foods, that is 1 in 12. Microbes that cause illness are called pathogens, for example Salmonella. These microbes can be in the food to begin with, or come in through wrong storage or transport conditions, or bad food handling. Meat, chicken, seafood, dairy products, eggs and even cooked rice are susceptible. Proper cooking, handling and storage can eliminate these risks almost entirely.

Check out the premises where you are going to buy or eat food. Are they clean? Do the staff present a hygienic image (hair, nails and dress)? Can you watch the food being prepared?

One of the golden rules is to keep hot food hot and cold food cold. This includes food served direct to you to eat on the premises, in hot display cabinets, and as delivered at home. Cold food should be refrigerated or on ice. Food in freezers should be stored below the 'load line', a Plimsoll line marked at the ends of the freezer.

Steaks and whole roasts are safer eaten when not very rare. If you are presented with an undercooked product, send it back and demand fresh accompaniments such as vegetables, because these can have become contaminated from the undercooked products.

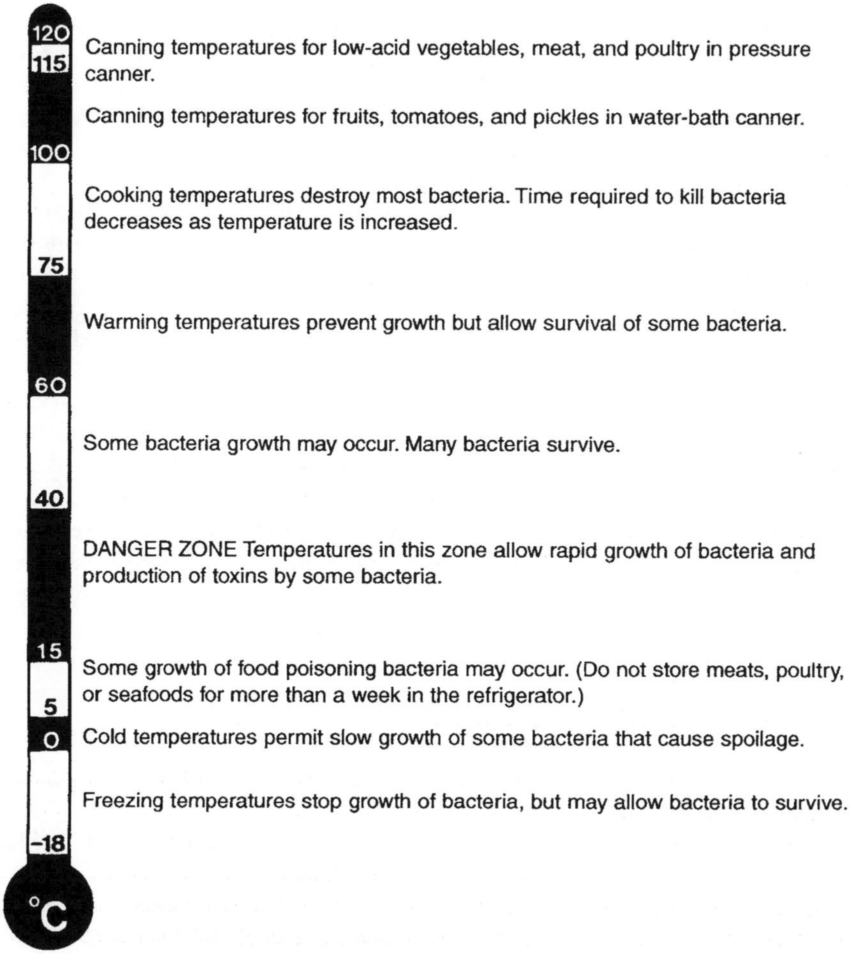

°C	
120 – 115	Canning temperatures for low-acid vegetables, meat, and poultry in pressure canner.
115 – 100	Canning temperatures for fruits, tomatoes, and pickles in water-bath canner.
100 – 75	Cooking temperatures destroy most bacteria. Time required to kill bacteria decreases as temperature is increased.
75 – 60	Warming temperatures prevent growth but allow survival of some bacteria.
60 – 40	Some bacteria growth may occur. Many bacteria survive.
40 – 15	DANGER ZONE Temperatures in this zone allow rapid growth of bacteria and production of toxins by some bacteria.
15 – 5	Some growth of food poisoning bacteria may occur. (Do not store meats, poultry, or seafoods for more than a week in the refrigerator.)
5 – 0	Cold temperatures permit slow growth of some bacteria that cause spoilage.
0 – –18	Freezing temperatures stop growth of bacteria, but may allow bacteria to survive.

Fig. 4.4
Temperature of food for control of bacteria

How to retain nutrients in food

1. Take frozen food home from the supermarket in an insulated container and place it in the home freezer immediately.
2. Store foods at the lowest appropriate temperature and for the shortest possible time.
3. If possible, minimise the amount of chopping and cutting of vegetables for cooking.
4. Do not use sodium bicarbonate in cooking as it increases the rate of loss of thiamine and vitamin C.
5. Try to avoid storing cooked foods for lengthy periods in the freezer or refrigerator.

Thermal processing, cooking and canning: Pyridoxine (vitamin B_6), folic acid and pantothenic acid are the most vulnerable of the B vitamins in foods of animal origin. From 50% to 90% of folic acid and pyridoxine, and 10% to 50% of pantothenic acid, may be lost during cooking. Choline is stable, and biotin relatively so.

Foods of plant origin

A few years ago, a survey of the vitamin content of some 2000 foods consumed by a selected group of Australians showed that, particularly in institutions, cooked fruits and vegetables contained virtually no vitamin C. The most disturbing finding of the survey showed low concentrations of vitamin C in the plasma of many of these people. These low concentrations probably resulted from diets deficient in citrus fruits or juices, and other rich sources of vitamin C. The destruction of the vitamin resulting from institutional cooking practices, and poor handling of the food in some of the domestic kitchens, also led to low concentrations of vitamin C in the foods consumed by some people.

Boiling: Vegetables with large surface-to-mass ratios (e.g. spinach) are especially sensitive to loss of vitamins. Vitamin C is a most unstable nutrient in neutral and alkaline conditions, and in the presence of oxygen; folic acid may be even more unstable; and thiamine, riboflavin, carotene and niacin are sensitive under certain conditions. Nutrient losses increase if large amounts of cooking water are used. Cooking for longer times also results in a greater loss of nutrients.

Baking: Retention of nutrients may be low during baking; for example, the thiamine retention in potatoes has been reported to be 41% to 48%. Apples retained 40% of their vitamin C when baked for 60 to 90 minutes at 200°C in an open dish. When baked in a covered dish, retention was 25%, and only 20% when baked as a pie. Baking powders destroy thiamine; slightly acid conditions are desirable to keep thiamine stable during baking. Bread loses little of its nutritional value during baking, but toasting results in a greater loss of protein and vitamins, depending on the degree of drying out and browning. Manufactured breakfast cereals are prepared in a variety of ways. The heating, rolling and flaking processes do not affect the protein but the more severe process of explosion puffing has been shown to reduce the nutritive value. However, it is necessary to consider these foods in perspective as they are almost invariably eaten with milk.

Retention of nutrients

Table 4.5 gives approximate figures for losses of nutrients from some foods during cooking.

TABLE 4.5
Percentage of original vitamin content lost from foods after commercial cooking, then followed by home cooking

	Thiamine	Riboflavin	Niacin	Pantothenate
Bacon				
Crisped at 175°C	5	—	—	—
Fried at 290°C	5	—	—	—
Beef				
Boiled 55 min,	30	0	10	25
then freeze dried,				
then reheated	90	0	50	
Canned and fried	80	0	10	20
Chicken				
Simmered 50 min,	50	30	40	50
then freeze dried,				
rehydrated, boiled 30 min	55	70	60	—
Boiled 35 min, canned,	98	15	20	30
then reheated 30 min,				
contents drained	98	—	30	

Vitamin C

Carrots		**Green beans**	
Boiled 5 min, dried,	30	Steamed 4 min, dried	70
then cooked 12 min	70	Simmered 10 min	70
Steamed 15 min, dried,	20	Steamed 9 min, freeze-dried,	70
then boiled 20 min	70	then reconstituted	80
Steamed 5 min, canned,	50	Steamed 4 min, canned,	75
then reheated	80	then reheated	75

Source: Nutritional value of processed food, G. J. Walker, CSIRO Information Service, 1979.

Microwave cooking

Microwaves, like radio waves and light waves, are part of the electromagnetic spectrum. From 300 MHz (megahertz) to 300 000 MHz (corresponding to wavelengths of 1 m to 1 mm) is the microwave region, and microwave ovens operate at the fixed frequency of 2450 MHz (12 cm) and powers of 500 to 850 W (about half that supplied in the electrical input). The microwave radiation is generated in an electronic tube called a magnetron, and passes along a waveguide into the metal oven cavity (For more on conversion of energy units, see Solar cells in Chapter 14.)

Other uses of microwaves include radar and TV microwave links. Medical diathermy operates at around 27 MHz (10 m); the much longer wavelength penetrates more deeply. Mobile phones operate at intermediate frequencies, 835 MHz (36 cm) for analogue and 915 MHz (33 cm) for digital, but at relatively low powers (0.6 W for analogue and 2 W for digital).

Why are microwave ovens so efficient for cooking? After all, in a conventional oven, the electricity is converted with 100% efficiency into heat,

Microwave ovens
There was intense research into microwaves during World War Two for the development of radar. The first patent for its use for heating food went to Percy Spender (Raytheon Co.) in 1945, who allegedly had his chocolate bar melt in front of him while working with a magnetron. The first ovens went onto the market in 1947 but were used mainly by the military (submarines, aircraft and field kitchens). By 1961, however, they were in the restaurant cars of the Japan National Railway, and soon on domestic flights in the US. Australia imported its first ovens in 1969 and by 1989, 50% of all households had one. By 1993, 5 million had been imported (just about one for every three people) at an unadjusted import bill of over one billion dollars — big bucks from one bar of chocolate!

The conventional oven

In a conventional oven the heat generated consists of radiation spread over an enormous range of frequencies (Planck or black body distribution), from radio waves to infra-red, with a little bit of visible light if the oven is hot enough. Only limited sections of this radiation are absorbed by the food and the rest is absorbed by the oven walls and conducted to the room. The longer wavelength radiation in the microwave oven is reflected by the oven walls and is later absorbed preferentially by the food.

whereas, in a microwave oven, electricity is converted with only about 50% efficiency into microwaves. The reason is that microwave energy is totally absorbed by some food components, particularly *mobile polar molecules* such as water, but also sugar and fat. The electric field of the passing radio wave can force rotations and vibrations in the polar molecule, and this is a form of heat. Collisions pass this heat on to other molecules in the food by the usual (slow) conduction process.

Water is a very strong absorber of microwaves, see Infra-red radiation and the greenhouse effect in Chapter 14. Cooking occurs through boiling of the water inside the food.

Because the cooking of water inside could occur in our body, it is important to be properly shielded from the microwaves (particularly the eyes). For this reason, microwave ovens are designed with at least two safety interlock switches that cut the power when the door is opened.

Even when electromagnetic radiation is reflected from a material, it can penetrate the surface up to a distance of about one wavelength. The infra-red heat radiation from a conventional oven is less than a tenth of a millimetre in wavelength and so heats only the outside of the food, from where slow conduction takes it into the interior over the period of extended cooking. The 12 cm wavelength microwaves penetrate 2 to 4 centimetres into the food, but because microwave cooking tends to be short, conduction deeper into large food chunks often does not occur, and this can create problems and dangers from undercooking in the centre. (See Chapter 14 box, Approximate conversions . . .) However, there is evidence that microwave cooking causes less destruction of vitamins than conventional cooking.

The standard for microwave ovens requires that the power flux density of microwave radiation shall not exceed 5 mW/cm^2 at any point 5 cm or more from the external surface of the oven (any closer than that and the simple testing devices that measure the electric field power are not accurate). The microwave oven has a clear glass door which is transparent to microwaves but is covered with a wire mesh to hold back the radiation.

TESTING FOR MICROWAVES

To stop radiation, the size of the mesh in the door needs to be smaller than the wavelength of 12 cm microwaves. Thus the few millimetre diameter mesh is much more than sufficient. (A similar effect is the loss of radio reception in tunnels, where the tunnel opening is the 'mesh', and as you go in past a wavelength or so, the sound disappears. The shorter wavelength FM radio reception tends to be less affected than AM.)

If a microwave oven door or door frame becomes distorted through mechanical abuse, or the mesh in the window is damaged, there is a danger of radiation leakage. The cheaper monitoring detectors consist of a piece of wire of length just under half a wavelength (5.7 cm) with a diode in the centre across which a light emitting diode (LED) or meter is placed to indicate a power reading. These devices should be brought in from a distance so that the diode does not saturate and give a zero reading when in fact the level is high.

You can check that the test device is working by placing it next to the antenna of a mobile phone when switched on, and when calling (higher reading). The device should respond, but the result is not a quantitative test for many reasons. Biological effects from mobiles are being questioned and future phones will probably be designed with the antenna transmitting away from the head. Questions of passive exposure to microwaves will then arise!

SUPERHEATING IN A MICROWAVE

When tap water is heated on a conventional stove top, the maximum superheating is to 100.75°C. In a microwave oven, however, superheating temperatures of 105°C to 106°C can be sustained, with periods at 110°C. Superheating can lead to a sudden boiling of almost the entire liquid and it then 'bumps' out of the container.

The microwave oven heats the water directly and the beaker indirectly, the opposite of what happens when heating by conduction with a burner. The water is not heated near the beaker surface where nucleation is most likely to occur. Evaporation occurs only at the water surface and is not fast enough to cool the bulk water. There is much less superheating in plastic cups because water does not wet plastic and many more nucleation sites are available for boiling to start (see Chapter 6).

Fat also absorbs microwave energy strongly and boils at temperatures much higher than for water. Reheating a casserole or soup from the fridge with a layer of fat on the surface, can cause very high temperatures, and if the container is a low melting plastic like polythene, it can melt at the oil level.

Sugar raises the boiling point of water (and lowers the freezing point) so foods with high levels of sugar can heat to well above 100°C and cause similar problems for unsuitable containers.

Materials for microwave cooking

Materials suitable for use as containers or covers in a microwave cooker should not absorb the radiation strongly. Therefore they should be made from material with low polarity or, in other words, low dielectric constant and low dissipation factor (at 2450 MHz), for example, some plastics and glass. Metals reflect microwaves and are generally unsuitable. (The first counter to microwave radar developed during World War Two was the dropping of metallic strips called 'chaff' or 'window'.)

Heat susceptors used for microwave cooking are thin, grey strips or disks of metallised plastic that become hot in a very short time and act like a little frying pan in the microwave oven. An example is the large silver surface (metallised PET) on which your microwave pizza cooks. In microwave popcorn you might have to hold up the bag to see it. The temperatures reached by the susceptors can be as high as 260°C and they can release additives or breakdown products from the plastic.

Data for plastics at 2450 MHz are not available for this book, but Table 4.6 gives data for lower frequencies. The table can be used to suggest which materials might be suitable for use in a microwave oven. The plastic used must also be stable at high temperature and have good resistance to oil vapour.

TABLE 4.6

Dielectric properties of plastics and glass

	Dielectric constant[a] ϵ at 100 MHz (1 MHz)	Dissipation factor[a] 100 MHz (1 MHz) [1 kHz]	1 kHz
Corning	6.5	0.015–0.06	
Pyrex	4–5	0.005–0.02	
ABS	2.4–4.5	0.004–0.007	
PVC/PVA	5.2 (1 MHz)	0.006–0.02 (1 MHz)	
PVC/PVDC	2.82	0.015	
Melamine–formaldehyde	5.5	0.03 (1 MHz)	
Urea formaldehyde	5.2		
Phenol formaldehyde	4.5–6	0.03–0.08	
Polyacrylate (Perspex)	2.6	0.03–0.05	
Polyamide (nylon)	3.1	0.04 (1 MHz)	0.02–0.04
Polycarbonate	2.96 (1 MHz)	0.01 (1 MHz)	0.002
Polyester	2.94–2.98	0.002–0.03	
Polyethylene	2.26	<0.0005 (1 MHz)	<0.0005
Polypropylene	3.2–2.6 (1 MHz)	0.0005–0.002 (1 MHz)	
Polystyrene	2.55	0.0001–0.0004 (1 MHz)	0.0001–0.0003
Polysulfone	—	3.13 [1 kHz]	0.001
PVC	3.3 (1 MHz)		0.009–0.017
PVA	5.2 (1 MHz)		
Polytetrafluoro-ethylene (Teflon)	2.04 (1 MHz)	0.0002 (1 MHz)	<0.0002

[a] Note some data at 1 kHz, 1 MHz and 100 MHz.

Source: Various

EXPERIMENT 4.2
Hydrolysis of starch with saliva

Equipment

Two small glass containers, cooking pot, and a source of starch (such as noodles, or potato).

Procedure

Prepare a solution of starch by boiling a few noodles (or a piece of potato) in a small quantity of water. Pour about 10–20 mL of this aqueous solution into each of two small glass containers. Add about 5 mL of saliva to one of the containers of starch solution.

After about 10 minutes, add a small drop of iodine solution to each container to test for the presence of starch (see Experiment 4.3).

Starch and cellulose are both polymers made up of glucose units, but the units are linked in a slightly different way in the two polymers. In cellulose the polymer chains are essentially linear, whereas in starch the chains have a helical structure. These differences give the two polymers vastly different properties; for example, cellulose is the water-insoluble, fibrous building material of plants, but starch has no fibrous structure and has different solubility properties. (See Recipe 3 in Chapter 12, dilatant and cornflour.)

Reaction

Saliva contains an enzyme (ptyalin) that can break the links between the glucose units in starch, so the solution treated with saliva no longer gives a positive reaction for starch following the action of the enzyme.

$$\text{starch} + \text{ptyalin} \longrightarrow \text{maltose (a sugar)}$$

Maltose is a compound made up of two units of glucose joined together. The enzyme is very selective. It breaks the bonds in starch, but not those in cellulose.

Note

You can test the effect of temperature on the rate of your hydrolysis reaction by bringing the solution to a higher or lower temperature before adding the enzyme and then keeping it at that temperature.

EXPERIMENT 4.3 Test for starch

Equipment

Tincture of iodine, potato, saucepan, cotton wool, funnel.

Procedure

Cut up the potato and boil it with water in a saucepan for a few minutes. Allow the water to cool. Decant or filter the solution to separate the soluble amylose from the insoluble amylopectin of the starch. (A cotton wool pad in the neck of a funnel will provide an ideal filter.)

Add a drop of tincture of iodine to the filtered starch solution. An intense blue colour will appear if any starch is present. If your test doesn't work, try adding some iodised salt to your potatoes when you boil them.

Reactions

The soluble part of the starch contains the substance β-amylose, which has the empirical formula $C_6H_{10}O_5$. The blue starch colour is believed to be caused by the formation of a complex between iodine and β-amylose of the general composition:

$$(\beta\text{-amylose})p(I^-)(I_2)_r\ (H_2O)_s,$$

in which r is very much smaller than p, and s is very much greater than p. (The small value of r means that this reaction can also be used as a very sensitive test for the presence of iodine.)

Notes

1. Try your test on other plants and foods (e.g. beans, peas, bread, spaghetti).
2. Tincture of iodine is used to treat wounds and may be in your home medicine cabinet. If not, it is available from pharmacies.

Our body fat

Statistics

Current social values equate high status with low energy expenditure. Who has the closest parking spot to the workplace — who catches cabs everywhere — who has others run their messages?

Before we start on fats in the kitchen, we will look at body fat, dieting, exercise and sport. On average, Australian men have put on 3.5 kg in the past decade. The average person in the UK in the 1990s expends 3350 kJ less energy a day than the average person in the 1970s. We drive more, and park closer to our destinations. We do less in our everyday life. We also eat less, 3150 kJ/day, but this does not compensate. The effect of the net gain of 200 kJ depends on whether it is fat, which will add to the body weight, or carbohydrate, which is not so easily stored. The fat content of the British diet has, however, increased 50% over the last 50 years. If these official figures are correct, that means a fraction under an extra gram per day.

In Singapore, all 18-year-olds have to do 2 years national service. However, the clock doesn't start until the conscript is down to that army's weight requirements. A quarter of all conscripts are overweight and spend an extra four months on a program; quite an incentive to get right before call-up.

People are obsessed with their body weight. At the extremes are the anorexics who starve themselves to skin and bone, and the bulimics who have an unhealthy hunger. Between these extremes, there are many who expend a lot of energy thinking about food and struggling to avoid eating it!

Although weight loss is easy for most people, maintaining the lower weight is achieved by only a fraction of all dieters. Once overweight we are better insulated and also take less exercise, so our energy needs decrease, and indeed we may actually eat less. This cycle is natural, considering that lack of sufficient food was the norm for the whole human race in earlier times, and so the efficient storage and use of fat was valuable for survival. There is also a theory that suggests that our bodies defend against weight change. This set-point or 'natural weight' depends on heredity factors.

Attractive fat

We are genetically programmed to find fat so attractive. No-one drives through a blizzard to get some broccoli, but for chocolate there are some precedents.

There is another key biochemical element, called the metabolic rate, which is the fuel the body burns when it is 'idling'. Just as cars differ in the petrol they use when idling, so do people when they are not active. In contrast to cars, this idling energy is the major part of the total energy we use each day and so is critical in determining our weight balance. Even with strenuous exercise, it still contributes a significant amount.

Research suggests that taking regular exercise *increases* the idling metabolic rate for up to 15 hours after the activity. You can apparently make your body less biochemically efficient by being in better trim! On the other hand, dieting can be counterproductive. The metabolic rate slows down and we become more energy efficient when the energy intake is reduced. The body doesn't know we have decided to diet, and protects itself against what it believes is famine. When we give up the diet, the body becomes more energy efficient and a return to the 'normal' rate of eating results in weight being putting on at a faster rate than before.

Sports chemistry and biochemistry

The sports area has stimulated much research in energy use. You are an average 70 kg male sitting quietly in a comfortable armchair reading this wonderful book. Ever wondered how the body engine works?

While just idling over, to burn up fuel you are using up 5 mL of oxygen per second, (300 mL per minute, 18 L per hour). You have the output of a 100 watt light globe (or 8.7 MJ/day). This resting metabolic rate goes down, the more fat you have (better insulation), and is increased by stress and through stimulation by some hormones. About half the oxygen goes to running your muscles (heart, lungs and skeletal), your brain uses 20% (more if you have to think hard — no, just kidding), 20% goes to the liver, the chemical factory that makes all this possible, and 7% to the kidneys, that clean up afterwards.

So where does the fuel come from? Food; carbohydrate, fat, and protein.

Carbohydrate is taken in from food as sugars and starches, converted to glycogen for storage in the liver and muscle and released as glucose from the liver when needed. The liver stores around 110 g of glycogen (it can supply 1870 kJ), muscle stores 250 g of glycogen (it can supply about 1050 kJ), and about 15 g of glucose is stored in other parts of the body (they can supply 60 kJ).

Fat (triglycerides) is less accessible for energy because it must be broken into glycerol and free fatty acids (FFA). The FFA are used for energy, and then the energy per mass is over twice what is available from carbohydrate (38 kJ/g, compared to 17 kJ/g). Most fat is stored. The leanest male has at least 3 kg of fat. On average, fat stored under the skin amounts to 7.8 kg (can supply 30 000 kJ) while that stored in muscles amounts to 160 g (can supply 670 kJ). There is almost 50 times as much stored fat energy as there is stored carbohydrate, but it is harder to utilise.

Protein must be broken down into constituent amino acids, converted to glucose before it can be used for energy.

The machinery takes these basic fuels and works them over as follows. Every 24 hours, your 'body machine' in idling mode turns 160 g of fat

The robust ATP molecule

Just think what all this means to the body's rechargeable energy smart card. There is only about 120 g of ATP in the body (ATP + ADP and phosphate). With a rest rate of 100 watts, the equivalent of 40 kg of ATP must be cycled each day — over half your body mass. This means that each ATP molecule is cycled over 340 times per day — a robust, and smart, card indeed. (If 8 hours of hard physical labour is included, the turnover of ATP rises to 800 per day).

into fatty acids and glycerol. The glycerol is turned into glucose by the liver, while the free fatty acids fuel you in two ways: 120 g/day goes to muscles (but cannot be used by the brain or nervous tissue) while the other 40 g/day is converted to ketone bodies (such as acetoacetate) which can be used by the brain.

The fatty acids (120 g) and ketone bodies (60 g) are the prime fuels for heart, skeletal muscle, and kidney. (The ketone bodies are particularly important under starvation and other stressful conditions (but are not used during severe exercise).)

About 75 g of muscle protein is converted to glucose in the liver every day. The liver stores 90 g of glycogen and releases 144 g/day of glucose to the nervous system, and 36 g/day for blood cells, while maintaining the blood level of glucose at the required 5.5 mM.

So that is fine, but how do you get to actually use this energy; after all, you don't have an internal combustion engine. No, but you have a 'smart card' for energy transfer.

'Smart card' energy transactions

Food cannot be used directly. Instead, the food components are used to switch a chemical compound between an 'energy-discharged' state adenosine diphosphate (ADP) and an 'energy-charged' state adenosine triphosphate (ATP). ATP acts like a cash rechargeable smart card in commerce, which stores a small amount of money to be used in everyday transactions, and when depleted, is refilled electronically from a bank account.

For carbohydrates, the switch between ADP and ATP is done by three main methods: two processes in the absence of oxygen, and one in the presence of oxygen.

In the absence of oxygen, one process uses glycogen to convert ADP to ATP, and the glycogen (glucose) is converted into lactic acid. Three moles of ATP is formed for each glucose unit of glycogen molecule broken down (only two if glucose is used).

Along with a process for short start-up boosts (ATP–PCr (phospho-creatine) system), this main anaerobic glycolysis allows the muscles to operate, even when oxygen is limited and predominates in the early minutes of high intensity exercise. The glycogen or glucose is converted to lactic acid. This builds up in the muscles (1 mmol/kg to more than 25 mmol/kg), and slows down the further use of glycogen, which finally impedes muscle contraction.

The system of energy production using oxygen involves a cyclic system of chemical reactions called the Krebs (citric acid) cycle, which in turn is coupled to other reactions called the electron transport chain. (See Chapter 7 box, Murderous organofluorides.) The overall process produces 39 molecules of ATP from one molecule of glycogen (38 if glucose is used), and oxidises glucose to carbon dioxide.

With fat as the source of energy, the fatty acids must first be liberated from their bonds to glycerol by an enzyme called lipase, and the free fatty acids (FFA) can then enter the bloodstream and diffuse into the muscle fibres. One molecule of a typical fatty acid, palmitic acid (C_{16}) produces 129 molecules of ATP. (See later in this chapter.)

Fuel comparisons

On a molecule basis, a typical fat molecule produces about 3.3 times the energy of a carbohydrate molecule. Molecular comparisons are of interest when studying the mechanism of chemical reactions.

On a mass basis (rather than molecule basis) fats are just over twice as energy producing as carbohydrate. This is of interest in studying nutrition, or the contribution of food components to body energy needs.

On the basis of energy density, and efficiency of storage, fat is very efficient. Because it is hydrophobic, fat is stored 'dry'. In contrast, glycogen is hydrophilic and stored 'wet'. Physiological glycogen is 65% water. Thus in practice, body fat has six times the energy density of carbohydrate. As a 70 kg, young, lean male, the body is one-eighth fat, while a 65 kg lean young female companion has a quarter of her body mass as fat. She is more efficient.

On the basis of oxygen used (and getting oxygen into the system can be limiting), carbohydrates yield 6.3 ATPs per O_2, compared to 5.6 for fats, so carbohydrate is preferred under high intensity exercise. The body has no capacity to store oxygen, so oxygen use is an accurate measure of energy production.

Note in both cases that only about 40% of the energy available in the food fuel is captured to form ATP; 60% is converted directly to heat.

So how do we know all this? It must be quite hard to measure the rate at which each food component is used! It is hard but can be done. However it is much easier to measure how much oxygen is used and carbon dioxide produced.

The long walk
A feat of the early 1880s is unlikely to be repeated. Captain Barclay Allardice walked 1000 miles in 1000 hours, one mile in each successive hour. He walked for 40 days without more than 45 minutes rest at any one time. His nutrition would also be unlikely to find favour among professionals. It included roast fowl and spiced cider, no vegetables, but a recommended three pints of stale beer for breakfast and dinner, and definitely no supper.

Measuring which fuel is being used

The most direct way to measure energy use is to measure heat output in a calorimeter (an insulated 'room' surrounding the runner that absorbs and measures the heat output). However, this is expensive, inconvenient, and cannot measure rapid changes in energy use. So instead, the energy is estimated indirectly by measuring the amount of carbon dioxide produced, and the amount of oxygen consumed. This ratio, VCO_2/VO_2, is called the respiratory exchange ratio (RER).

For glucose ($C_6H_{12}O_6$) , six molecules of O_2 are used and produce six molecules of CO_2 (along with 38 molecules of ATP) and this corresponds to an RER of 1.0.

For fatty acids, the ratio varies a little with the acid, but for, say, palmitic acid $C_{16}H_{32}O_2$), 23 molecules of O_2 are used and produce 16 molecules of CO_2, so the RER is 16/23 = 0.70. Therefore the RER can give the ratio of carbohydrate to fat you are using. Cunning, eh?

Some care is needed. While oxygen is not stored in the body, carbon dioxide is. For example, at near exhaustion, lactate accumulates in the blood (from anaerobic energy use) and this lowers blood pH. The lower

Osler's experiment
In 1978, a world ranked marathon runner, Tom Osler, attempted to walk and run continuously for 72 hours in a sports laboratory. Chemical measurements showed that for the first hours of exercise the energy for his muscles came mainly from carbohydrate, but as the hours passed, the energy supply changed increasingly to fat stores. This occurred in spite of continued intake of milk saturated with sugar, along with a large birthday cake. In spite of an intake of almost 40 MJ during the first 24 hours, Osler was exhausted and out of energy after 70 hours, having completed 200 miles.

pH in turn converts carbonic acid to carbon dioxide, which is released in the lungs. At exercise start-up, the body goes into oxygen debt, and after completion, the oxygen demand does not stop immediately but is used to reset the borrowings from haemoglobin stores, and in removing lactic acid that was produced anaerobically.

Endurance events

The ability to exercise at high intensity without accumulating lactate (which contributes to fatigue) is a measure of ability in endurance events. Endurance events! So you reckon you're not just a couch potato. Let us look at athletic power.

If you get up and start to walk at a pace that doesn't cause shortness of breath, your muscles use mainly fatty acids. As your pace increases, so does your energy expenditure, and the amount of carbohydrate used. At a fast pace, fat can't be utilised quickly enough, and nearly all your energy demand comes from carbohydrate. But as we saw, the body storage of glycogen is modest and you can only improve this by training hard enough to keep you in the carbohydrate high-demand mode. During hard exercise, you can use up carbohydrate at a rate of 3 to 4 grams per minute, exhausting your stores in under 2 hours.

A marathon is 42.2 km long, lasts about two and one-half hours, and the body consumes approx 12 MJ of energy. Running at 20 km/h uses oxygen at 60 mL/min/kg body mass, which, for a 70 kg person, corresponds to 4.2 L/min. Burning of glucose/glycogen provides 21.12 kJ/L of O_2 used. Now multiply this by 4.2 L/min, giving 89 kJ/min, or 1.48 kJ/sec, which is 1483 watts of chemical power. The amount of useful power output is about 20–40%, or about 300 to 600 watts. (For comparison, one horsepower is 746 watts.)

A typical experiment on your 70 kg body machine would show that it varied its use of fuel with time as follows:

TABLE 4.7
Use of fuel by the body

Exercise period (minutes)	liver glycogen	liver fatty acid	muscle glycogen
40	27	37	36
90	41	37	22
180	36	50	14
240	30	62	8

You can preboost your glycogen stores by a three-day carbohydrate binge (this can give you 100 to 180 μmol/g muscle) but as an elite marathon runner, you can still only store enough glycogen in the liver for about 90 minutes of exercise to exhaustion, and sustain a power output for only 20 minutes. Fats are a pretty important energy source, as Osler discovered.

Athletes need extra energy, most of which must be in the form of carbohydrate; dehydration impairs exercise capacity, and should be offset by extra fluid intake; athletes do not generally need extra protein, vitamins or minerals.

Oils and fats, the chemistry

Fats, oils and some waxes are the naturally occurring *esters* of long, straight-chain carboxylic acids. These esters are the materials from which soaps are made (see Soaps in Chapter 3). An ester is produced when an organic acid combines with an alcohol. Whenever the alcohol is *glycerol* — glycerine, a by-product in the manufacture of soap from fat — the esters are fats or oils (see Fig. 4.5).

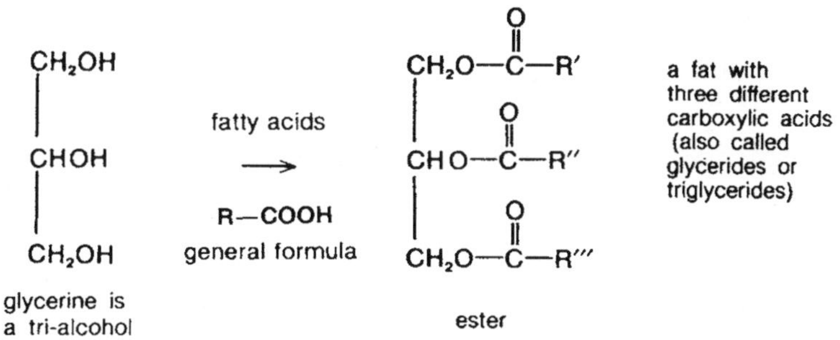

CH_2OH | $CHOH$ | CH_2OH — glycerine is a tri-alcohol

fatty acids $\longrightarrow$ $R{-}COOH$ general formula

$CH_2O{-}C({=}O){-}R'$ | $CHO{-}C({=}O){-}R''$ | $CH_2O{-}C({=}O){-}R'''$ — ester

a fat with three different carboxylic acids (also called glycerides or triglycerides)

Fig. 4.5
Esterification of glycerine to form fats

The difference between fats and oils is merely one of melting point: fats are solid or semi-solid at room temperature, whereas oils are liquids.

Since glycerol is common to all fats, whether animal or vegetable, it is the fatty acid part of the fat that is of interest. The differences between fats depend on the nature of the acid groups — the length of the chain (which controls the molecular mass) and the number and position of double bonds (unsaturation). Before proceeding further, we will briefly describe the acids. There are three groups important to this discussion: the saturated fatty acids, the straight-chain unsaturated fatty acids, and the polyunsaturated fatty acids.

Normal saturated fatty acids

Normal saturated fatty acids have the general formula $CH_3(CH_2)_n COOH$, where n is usually even, and varies from 2 to 24. The building block for producing fatty acids is the acetate ion, CH_3COO^-, hence the predominance of even carbon chains. The most common examples of normal saturated fatty acids are palmitic acid ($n = 14$), and stearic acid ($n = 16$), which is illustrated in Figure 4.9. Others are lauric acid ($n = 10$), which is dominant in coconut and palm kernel, and myristic acid ($n = 12$), dominant in nutmeg.

Even shorter-chain fatty acids (n less than 10) form a large proportion of the fats in milk fats, especially those of ruminants. Odd-numbered acids do occur, but only in traces, and then over a wide range up to $n = 23$, generally in ruminants. The unusual composition occurs because it is bacteria in the rumen that carry out the reactions for the animal. Geographic variations in products such as milk can cause temporary diarrhoea until readjustments are made by the victim's body.

Normal straight-chain unsaturated fatty acids

The most important unsaturated acids have 18 carbon atoms, usually with one double bond at the middle of the chain. If other double bonds are present, they lie closer to the carboxyl group, COOH. The double bond cannot be rotated, and so there are two distinct possible geometries, which are called *cis* and *trans* (Fig. 4.6).

cis: H(R)C=C(H)R — trans: R(H)C=C(H)R

Fig. 4.6
Cis and trans isomers

Fig. 4.7
Monounsaturated oleic acid, a component of olive oil

Oleic acid, a component of olive oil and the most abundant of all fatty acids, is

$$CH_3(CH_2)_7CH \overset{cis}{=} CH(CH_2)_7COOH,$$

(see also Fig. 4.9). The vast majority of olefinic (unsaturated) linkages in fats and oils are of the *cis* type.

The first olive oil produced in Australia received an 'honourable mention' at the Great Exhibition in London in 1851. South Australia has been our major producer, the first cuttings having been imported from Brazil in the 1830s, and in 1840, five top varieties were brought in from Marseilles.

Polyunsaturated fatty acids

The polyunsaturated acids are those that have more than one *cis*-methylene interrupted double bond, and they have the general formula:

Fig. 4.8
General formula for a polyunsaturated fatty acid

$$CH_3(CH_2)_x(CH{=}CHCH_2)_y{-}(CH_2)_zCOOH$$

where x and z range between 3 and 20, and y is usually from 1 to 4. Several polyunsaturated acids are illustrated in Figure 4.9.

The double bond, especially when of the *cis* type (see Fig. 4.6), means that the molecules do not pack together easily, which is seen in the low melting point of double bond-containing material (i.e. oils). Substances made up of shorter chains also melt at lower temperatures.

Why do animals produce fats, mainly saturated, while plants produce oils, mainly unsaturated? Plants suffer extremes of temperature and require their fats or oils to be semi-fluid even at low overnight or winter temperatures, whereas some animals can maintain a high temperature

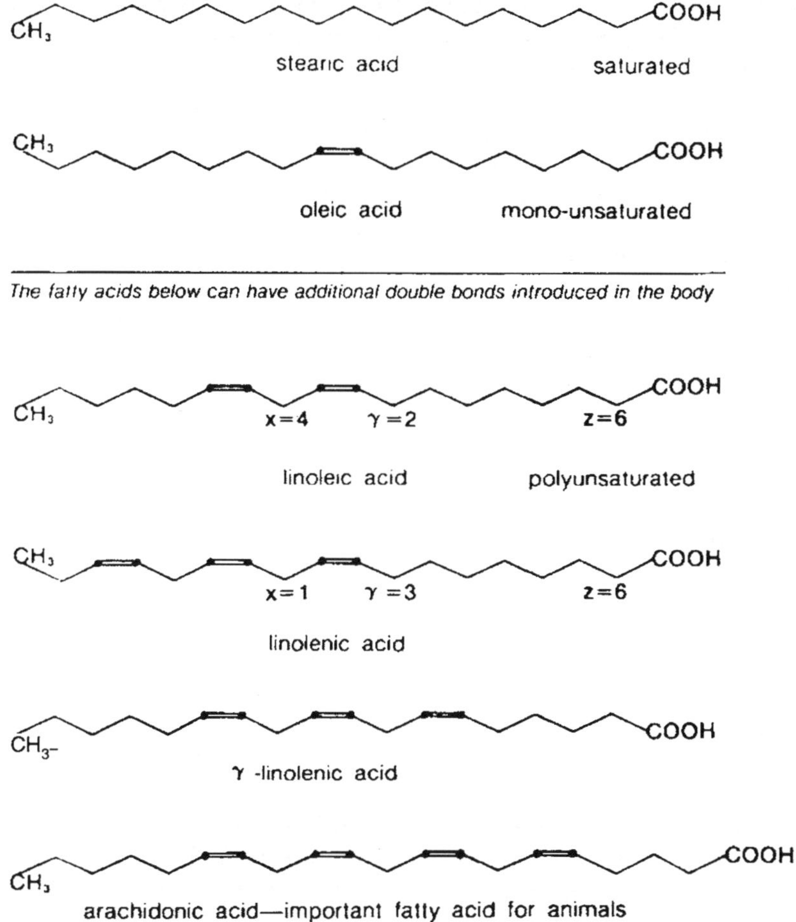

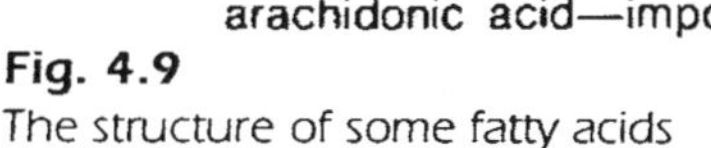
Fig. 4.9
The structure of some fatty acids

Prostaglandins
Fatty acids used to be uninteresting. Arachidonic acid is now recognised as the source in the body of a range of compounds that are critical for the health and illness of humans and animals. They are present in minute quantities, about one part per million, and often have lifetimes of only minutes in the body, yet their functions and effects are dramatic. Polyunsaturated fats are the building blocks for the important prostaglandins, and they are also alleged to be effective in lowering the cholesterol content of the blood.

through internal heating and insulation, and in the case of mammals can even regulate it. In mammals a higher melting compound is preferable because the fats also have a structural part to play and must not be too fluid. Kidney fat is solid, so as to provide mechanical support for the organ, although fat circulating in the cells is fluid, even at lower temperatures — otherwise you might get a solid casing in a cold shower.

FATS AND OILS IN ANIMALS AND PLANTS

The distribution of fats and oils varies in different plants and animals. Fats and oils appear to be a biological solution to the problem of storing, transporting, and utilising those fatty acids that an organism requires for its metabolic processes. The ester bond is quite stable, but is easily split, when required, by a specific enzyme. Lipids (a term used in biochemistry) includes fats and fat-like materials of biological origin. They are the major energy storage medium in animals (38 kJ/g, compared with 17 for carbohydrates and 23 for protein). When fats are 'burnt' in the body to produce energy, they also produce water, and more than from burning sugar. The hump of the camel is, in fact, a fat-storage unit that provides it with both energy and water. Hence the snide camel-selling trick of pumping the hump with air!

Plants, including moulds, yeasts and bacteria, synthesise both fats and their component fatty acids. Animals can synthesise much, though not all, of their fatty acid requirements, but they prefer to ingest plant foods and modify them to their own needs. Only plants are known to synthesise linoleic and linolenic acids, but animals can increase the chain length and further increase unsaturation, giving, for example, acids characteristic of fish oils, which are particularly rich in unsaturated acids, and have up to six double bonds.

Oils and fats, the products

Margarine

The origin of margarine dates from 1869, when Napoleon III proposed a competition with the aim of discovering 'for the use of the working class and, incidentally the navy, a clean fat, cheap and with good keeping qualities, suitable to replace butter'. The prize was won by a chemist, Hippolyte Mege-Mouries.

In the USA, prior to 1950, margarine manufacturers could not colour their products unless they paid an additional 22 cents per kilogram federal tax, the result of an effective lobby by the dairy industry. By changing from coconut oil to soybean oil as a raw material, the makers of oleomargarine were able to enlist the aid of soybean growers to combat the dairy lobby, and the law was repealed in 1950. The Australian states had equally obnoxious laws for equally devious political motives.

Antioxidants, flavouring (3-hydroxy-2-butanone and diacetyl, which give butter its characteristic flavour), vitamin D, and vegetable colouring (usually carotene, a source of vitamin A, which gives the colour to butter) may be added.

Hydrogenation

In general, vegetable oils are a good source of polyunsaturates, and animal fats are a poor source, although there are exceptions to this rule (e.g. coconut oil). However unsaturated fats can be saturated by adding hydrogen to the double bonds. Hydrogenation has been used since early this century. Adding hydrogen to double bonds raises the melting point of the oils, and turns them into fats. The oil is mixed with nickel powder and heated to 200°C, and hydrogen gas is passed through for up to six hours. Thus hydrogenation converts a substance with the properties of a vegetable oil into one with the properties of an animal fat; it changes a liquid into a solid. Thus linoleic and oleic acids would turn into stearic acid (see Fig. 4.9). When margarines are said to be made *from* pure vegetable oils, the emphasis could be very much on the 'from'.

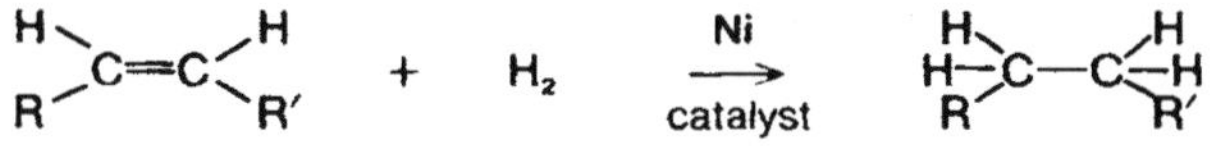

$$\mathrm{RHC{=}CHR'} + \mathrm{H_2} \xrightarrow[\text{catalyst}]{\text{Ni}} \mathrm{RH_2C{-}CH_2R'}$$

Fig. 4.10
Adding hydrogen to an unsaturated bond, which turns an unsaturated fatty acid into a saturated one

There has been increasing concern that during hydrogenation, other reactions occur in which the natural *cis*-methylene interrupted-double-bond configuration of the polyunsaturated acid has been altered, whereby a double bond shifts one place closer to the adjacent double bond. Such isomers are said to have a conjugated double-bond system. At the same time, the shifted double bond usually alters from *cis* to the more stable *trans* arrangement. The final product is about 15% *trans* fatty acids (TFA), with the total hydrogenated fat higher still.

The presence of such fatty acid artefacts in processed vegetable oils has been the subject of some controversy. Although it is true that fat in products derived from ruminant animals (beef, lamb, and dairy products) also contains conjugated and *trans* unsaturated fatty acids (arising from biohydrogenation of unsaturated fatty acids in the rumen) these are present only in small amounts. *Trans* fatty acids in processed oils such as shortenings may reach levels of over 50%.

A large prospective study of over 85 000 healthy nurses, starting in 1980, showed a highly significant direct link to increased coronary heart disease with intake of TFA. Intake of butter was not associated with the risk.

THE FATS IN OATS

Wheat does not like cold and wetness while oats does. So when oats first invaded wheat crops, it was regarded as a weed, choking the wheat. Later it was recognised for its own value. By the Bronze and Iron ages, farmers in northern Europe were growing oats as a specific crop, and from there it spread south. Much of the sustenance of oats comes from its fat content, oleic, linoleic and palmitic acids, spread throughout the kernel, except the outer layer.

This outer layer contains an enzyme, lipase, which on release (during milling) reacts with the fat, splitting out the fatty acids. These free fatty acids can impart a bitter taste, and also react with raising agents such as sodium bicarbonate, to produce soaps in baked foods like oat cakes. The oat berries are steamed before kiln drying, or the miller denatures the enzyme with heat, but this can lead to the fats going rancid if it is not done with care.

Unlike wheat, oat husk is tightly bound to the grain, and must be removed before milling through a 'dry-shelling' process carried out between revolving stones to release the kernel known as a 'groat'. Oat husks are used as low-grade animal feeds, in the brewing industry, as poultry bedding and as garden mulch. The cleaned groats are cut to make pinhead meal (or pinmeal) which goes to making porridge, or ground further to form oat flour. Cooking pinmeal in a steamer and flattening the moist plastic brew produces rolled oats. Rolled oats placed in a blender or food processor can produce small quantities of oat flour (it needs sifting and regrinding).

Source: Michael Boddy, in *The Canberra Times*.

Fake Fats

Natural fat consists of a molecule of glycerol attached to three fatty acids. Olestra™ (Proctor & Gamble) is a calorie-free fat replacement. The glycerol in natural fat is exchanged for sucrose, which attaches six, seven or eight fatty acids. This molecule is the first fat substitute that does not break down during cooking, and approval has so far been sought for its use in snack foods. However, because the Olestra™ molecule is also not broken down in the gut, this also means that it can cause stomach cramps, bloating and diarrhoea, much like some undigested sweeteners or overly high-fibre diet. However the body accommodates high fibre, but not fake fats. Like other undigested oils, for example castor and paraffin, Olestra™ can remove oil-soluble vitamins and carotenoids from the body. Proctor and Gamble say they will saturate the oil with these vitamins. Like castor oil, it can also cause anal leakage (politely known as passive oil loss).

The panel recommending approval of Olestra™ to the US Food and Drug Administration (FDA) was not allowed to judge if this will be a societal benefit. The FDA standard for approval is 'a reasonable certainty of no harm', and that is quite different from 'no evidence of harm'. To say whether this compound is needed is another question.

The iodine number

Fig. 4.11 Adding iodine to an unsaturated bond to determine the iodine value

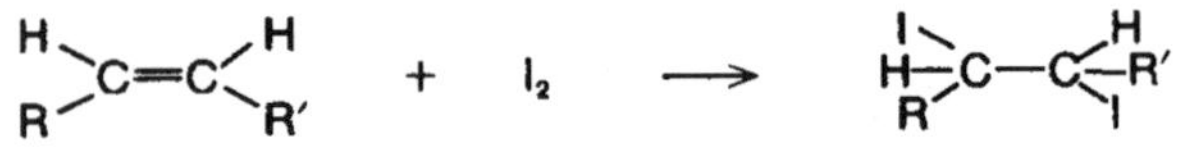

$$\mathrm{R(H)C{=}C(H)R'} + \mathrm{I_2} \longrightarrow \mathrm{R(H)(I)C{-}C(H)(I)R'}$$

The number of grams of iodine that react with 100 g of fat or oil is known as the iodine value. Some iodine values are given in Table 4.8.

Rancidity

Releasing the fatty acids from butter fat produces a complex mixture of short-chain, volatile, foul-smelling products, the cause of rancidity. With margarine, the longer chains' release of fatty acids is not enough, they

TABLE 4.8
Iodine values

Coconut oil	7–12	
Butter	26–45	Low–predominantly saturated
Lard	46–66	
Olive oil	79–88	Mono-unsaturated
Peanut oil	83–98	Hgh, because of polyunsaturated component
Cottonseed	103–113	
Safflower oil	130	
Linseed oil	170–204	

must then be 'broken' beforehand. The effect of the oxygen in the air on unsaturated fats is to add an oxygen atom adjacent to the double bond, followed by further reactions that split the chain at the double bond. However, margarine rarely becomes rancid, as commercial fats and oils have *antioxidant* added to them.

The milk of sheep and goats has an even greater concentration of short-chain fatty acids in their fats than butter (where it is about one-third). The cheeses from these milks are correspondingly stronger.

Olive oil
Olive oil was something Anglo-Celtic Australians used to buy at the pharmacist, rather than the grocer. With multiculturalism, imports have reached 18 000 tonnes, trebling since 1983. Diminished European subsidies and drought have increased prices, and encouraged development of plantations in Australia.
Varieties that are good for eating will not be best for oil, although there is a dual purpose Israeli variety called Barnea.

Heating of oils and fats

There are three important points in the heating spectrum of an oil or a fat. The first is the *smoke point,* the temperature at which a fat breaks down into visible gaseous products and thin wisps of bluish smoke begin to rise from the surface. The *flash point* occurs when brief but sustained bursts of flame start to shoot up. Higher still is the *ignition temperature,* at which the entire surface of the frying medium becomes covered with a continuous sheet of flame.

TABLE 4.9
Smoke points of some oils

Oil type	Av. smoke point range in °C	P/S ratio[a]
Safflower	246–258	6.0–7.4
Sunflower	229–252	4.7–5.2
Maize	229–268	3.1–4.2
Peanut	246–251	1.9–3.5
Soybean	256	3.7–3.9
Olive	204	0.5–0.7

[a] The P/S ratio is the ratio of polyunsaturated fatty acids in the fat or oil to the saturated fatty acids present. Mono-unsaturated acids are ignored. The range of P/S values can be affected by such factors as where the seeds are grown.

The temperature of the smoke point does not stay constant; it tends to fall with the continued use of the oil or fat (because the oil or fat decomposes, and the free fatty acids lower the smoke point). It drops about 20°C each time the fat is reused for 30 minutes at 270°C. Thus the higher the initial smoke point, the longer the fat is usable before it starts to smoke.

Heating does not significantly change the P/S ratio of polyunsaturated oils, but it causes the formation of oxidised compounds, which tend to destroy the vitamin E content, and can also cause the oils to have a tendency to polymerise, which makes them unpalatable (see Oil-drying paint in Chapter 12). Big changes to the peroxide value of oils after heating reveal how heating oxidises oils. Olive oil (which is mainly mono-unsaturated oleic acid) appears to be the most stable cooking oil. This is probably why it has been used for so long in Mediterranean countries. It needs no refining, preservatives or refrigeration.

Antioxidants

The two most common antioxidants added to consumer products during processing are butylated hydroxyanisole (BHA) and butylated hydroxytoluene (BHT). Their structural formulae are given in Figure 4.12.

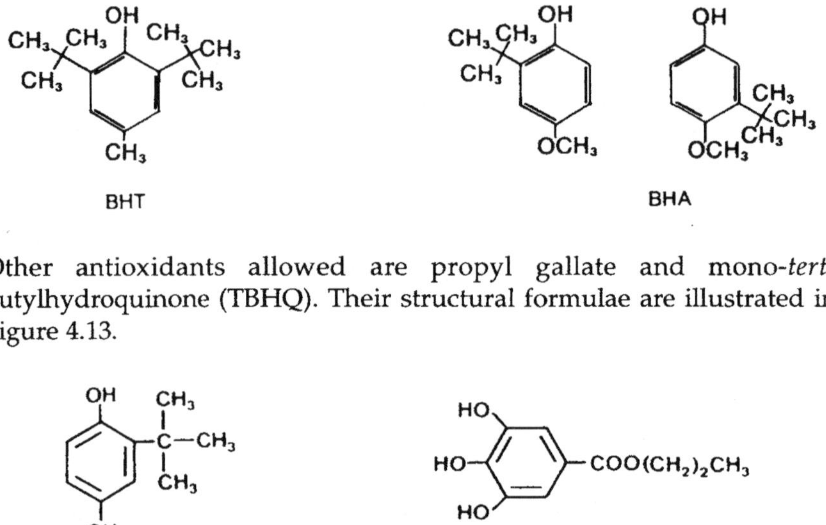

BHT BHA

Fig. 4.12 The antioxidants BHT and BHA

Other antioxidants allowed are propyl gallate and mono-*tert*-butylhydroquinone (TBHQ). Their structural formulae are illustrated in Figure 4.13.

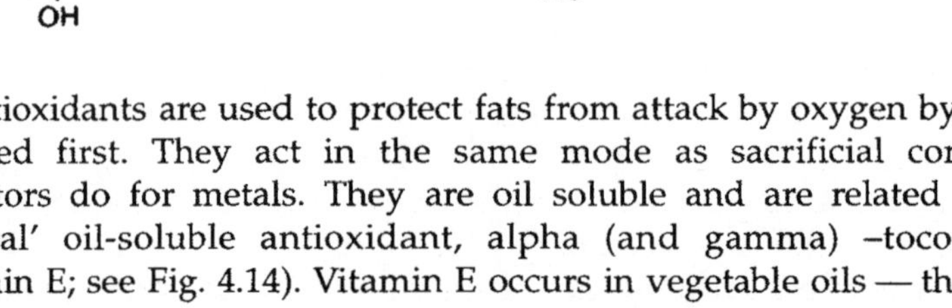

Fig. 4.13 The antioxidants TBHQ and propyl gallate

Antioxidants are used to protect fats from attack by oxygen by being attacked first. They act in the same mode as sacrificial corrosion inhibitors do for metals. They are oil soluble and are related to the 'natural' oil-soluble antioxidant, alpha (and gamma) –tocopherol (vitamin E; see Fig. 4.14). Vitamin E occurs in vegetable oils — the most important source is wheatgerm oil — and it protects them against oxidation.

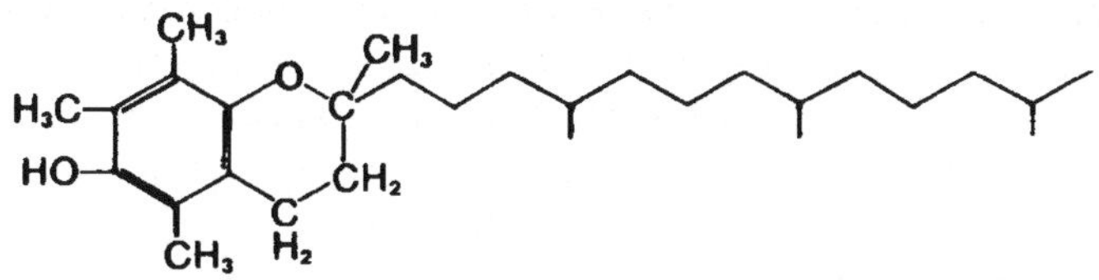

Fig. 4.14
Vitamin E

Humans have a need for vitamin E in an amount proportional to the amount of polyunsaturated fat in the diet; it is apparently required when the fats are laid down in the body, and presumably it is also an antioxidant there.

Cholesterol

Our brains are about one-third cholesterol. It acts as an electrical insulator for sheaves around the neurones. Cholesterol is also a precursor of vitamin D and the steroids, including sex hormones.

Cholesterol is not technically a fat, but a steroid related chemically to the bile acids, cortisone, the sex hormones, and vitamin D — a motley collection. Cholesterol is a necessary substance found in all the cells of the body. It is produced in the liver and it may also be taken in directly from foods of animal origin, so the level in the blood comes from two sources, both of which can vary.

About 93% of the body's cholesterol exists in cells, especially the cell membranes that encase them, where it is vital for structural support and certain biochemical reactions. The remaining 7% circulates in the blood, where problems can occur. Because cholesterol and fats are insoluble in water, they are coated with a phospholipid–protein envelope called lipoprotein. This comes in a high-density (HDL) and low-density (LDL) form. LDL carries cholesterol. Special receptors on the cell recognise LDL and let it in. If our genes code for too few LDL receptors, we are likely to have high levels of cholesterol in our blood, with the consequent health risks.

Cholesterol level
In the USA the target level of cholesterol in the blood is 200 milligrams per 100 millilitres of blood. In Europe (and Australia) the same magic number is 5.2 millimoles per litre of blood. A mole of cholesterol is 386 grams. The rest, as they say, is arithmetic.

Heart disease mortality
In countries such as Australia and the USA, where people eat large amounts of meat and dairy products — that is, their diets are high in cholesterol and saturated fats — the mortality rate from heart disease is high. In countries such as Japan, where diets are traditionally low in cholesterol and rich in the polyunsaturated fats found in vegetable oils and fish, the death rate from coronary disease is lower. But Japanese who migrate to the USA soon follow the US pattern.

Ice-cream

Ice-cream is a foam that is preserved by freezing. Four different phases can be seen under a microscope: solid globules of milk fat; air cells, which should be no bigger than 0.1 mm; tiny ice crystals, formed from freezing out pure water; leaving behind a fourth phase of water, with concentrated sugars, salts and suspended milk proteins. The ice freezes out to a point where the lowering of the freezing point caused by the concentration of the remaining solutes in the water corresponds to the freezer temperature (see also in Appendix III Eutectics). Legally the mix can be expanded with air to double its volume (called in the trade — overrun). Ice-cream with more air feels fluffier and warmer to the taste. Less milk fat means bigger ice crystals, and a coarser texture and colder taste.

Emulsifiers and stabilisers can mask the lower fat properties, and can impart a gummy, sticky quality to the product.

If ice-cream is not stored properly, the partial thawing causes the *smaller* crystals to melt, and refreezing causes the *larger* crystals to grow (this is related to the demonstration of small bubbles blowing up large ones in Chapter 6 (see Bubble politics)). This coarsening of texture from partial thawing can also occur by the crystallising out of the lactose (milk sugar) which tends to persist after the ice has melted, either in a dish or on the tongue. (Lactose occurs only in milk, and is one-tenth as soluble as sucrose).

Chocolate

Chocolate and related products begin with the beans of the cacao tree (*Theobroma cacao,* 'food of the Gods'). These grow in elongated melon-shaped seed pods, each of which contains about 25 to 40 white or pale purple beans, each slightly larger than a coffee bean. The beans are collected, heaped, covered with leaves, and allowed to ferment through the action of microbes and enzymes naturally present. This process kills the germ of the bean, removes adhering pulp, and modifies the flavour and colour (now brown). After drying, the beans are ready for export. (See Plate III, Chocolate at the Powerhouse Museum.)

For chocolate manufacture, the beans are roasted and passed through a complex set of milling processes. The heat of grinding melts the fat and produces chocolate liquor, which is composed of about 55% fat, 17% carbohydrate, 11% protein, tannins, and ash. Theobromine, the stimulant alkaloid related to caffeine, is found in amounts ranging from 0.8% to 1.7%, depending on its source. Somewhat less caffeine is also found. The solidified liquor forms the bitter cooking or baking chocolate. The fat removed from the chocolate liquor is cocoa butter. This consists mainly of triglycerides in which the middle fatty acid is oleic, and the two outside fatty acids are saturated, generally stearic or palmitic. In beef tallow, the opposite is true: the dominant triglyceride has saturated fatty acid on the inside and unsaturated on the outer positions.

Although these two fats have rather similar overall fatty acid compositions (though cocoa butter is more saturated than any animal fat!), their triglyceride composition is strikingly different, and so are their physical characteristics. The simple composition of cocoa butter leads to a relatively sharp melting point, 30°C to 35°C, which makes it attractive in chocolate. However the solid is polymorphic: it can crystallise in at least three, possibly six, different crystal forms, with melting points varying from 17.3°C to 36.4°C. Only the fifth of these (a so-called β-3 type) with a melting point of 33.8°C, is suitable). If the fat crystallises in an unstable form, it will cause problems.

A high-quality sweet chocolate might consist of 32% chocolate liquor, 16% additional cocoa butter, 50% sugar, plus flavouring. The mixture is ground to about 25 micrometres (μm) or less and then 'conched' (kneaded) for 96 to 120 hours for a high-quality product. The critical operation of controlled crystallisation follows. The liquid chocolate is cooled to initiate crystallisation, and reheated to just below the melting point of the desired crystal structure so as to melt undesirable lower-melting types of crystals. The chocolate is then stirred at this temperature for a while (tempered) and then crystallised quickly to produce fine crystals. Different sources describe the procedure differently — chocolate making is still a dark art!

For milk chocolate, the Australian Model Food Act specifies a minimum of 45 g/kg milk fat, 105 g/kg non-fat milk solids (milk sugars mainly) and 30 g/kg water-free, fat-free cocoa paste. While cocoa paste is defined as the product prepared by grinding solidified chocolate liquor containing not less than 480 g/kg of cocoa butter, the fat-free specification means there is no *minimum requirement* for chocolate to contain cocoa butter (since October 1983). There is an obvious incentive to replace the expensive (and often variable in quality) cocoa butter with a cheaper fat. White chocolate, on the other hand, does have a minimum content of 200 g/kg of cocoa butter specified, and also must contain not more than 550 g/kg of sugar, that is it can be over half sugar.

Fat bloom is the development of a new phase in a chocolate fat, causing surface disruption with large clusters (5 μm), to give the grey mould-like coating inevitably blamed on poor consumer storage. Fat bloom in chocolate is distinguished from loss of gloss, which occurs when small crystals (0.5 μm) on the surface grow into large ones, and scatter light. The growth of large at the expense of small is thermodynamically favoured. (See Chapter 6, Fig. 6.3). The use of emulsifiers and stabilisers can greatly affect the rate at which crystal changes occur in the solid state. Various additives, particularly sorbitan fatty acid esters, are used to control crystallisation and phase change in substitute chocolate. Partial glycerides are also good stabilisers. They occur naturally in high levels in palm oil, which is used in margarines to stabilise the β' crystal, which, if it transforms to the β crystal, causes the margarine to become grainy.

Waxes

In the 1970s, when the first edition of this book was being prepared, the creature that Moby Dick made famous, the sperm whale, was being hunted and was nearing extinction. Whalers since the 16th century have sought whales for their oil, which was used for lighting, as a lubricant, and to make soap. Whaling intensity increased in the 1950s and in 1960 the explosive harpoon dispatched about 60 000 of these magnificent leviathans to be chemically processed for their unique oils and waxes. A single whale can deliver 20 tonnes of oil.

There are two classes of whales. Baleen whales include humpbacks, gray, right, fin and the gigantic blue whales. Baleens have sieves instead of teeth to trap krill, shrimps, and other small creatures. Toothed whales include dolphins, porpoises, killer whales, and sperm whales. These feed on squid and large fish, catching them on their peg-like teeth.

Now the baleen whales have oils which are just like plant and animal fats, but the toothed whales contain not oil but technically a liquid wax.

Fats and oils are fatty acid esters of the tri-alcohol glycerol. Waxes are esters of long-chain ($\geq C_{16}$) alcohols with one hydroxyl group and a long-chain ($\geq C_{16}$) fatty acid. These esters are unsaturated (with two double bonds) which makes them liquid. Unsaturated liquid waxes are rare in nature and unusually stable.

Commercial drilling for petroleum oil began in 1859 and refining produced kerosene, which replaced whale oil as a lamp fuel. Sperm whaling fell sharply until the introduction of hydrogenation in 1908, which allowed the conversion of sperm oil to hard waxes, used widely for floor polish, the highest quality candles and cosmetics. The other impetus to a resumption of full-scale whaling came from a discovery in the 1930s that treating sperm oil with sulfur, (equivalent to vulcanisation of polyisoprene in rubber, see Elastomers in Chapter 9), produced a lubricant with unique properties. Unlike conventional lubricants, this one still worked at the high temperature and pressure that occurred in the newly developed automatic transmission of cars, where it was added (along with other additives) to oil at a level of 2% (half unvulcanised). This might not seem much, but in the 1980s this amounted in the US to over 6 million US gallons of sperm oil each year.

Sperm whales also produce *spermaceti* , a waxy liquid mixture of similar esters found only in a network of capillaries in its head. The main component is the ester cetyl palmitate, $C_{15}H_{31}COO–C_{16}H_{33}$, which has a melting point of 42°C to 47°C. But because spermaceti is a mixture, it does not have a sharp melting point (see also Eutectics in Appendix III), and both liquid and solid are present in the range 22°C–30°C (all solid below, all liquid above). The whale can control the partial to complete melting and freezing of this material by controlling the circulation of warm blood through it. The expansion and contraction that accompanies this melting and freezing gives the whale fine control over its buoyancy to aid in diving, surfacing, and remaining suspended mid-water.

Other natural products have now been found with similar properties to sperm whale oil, such as jojoba 'oil', which has a longer carbon chain (40 or 44 carbons). More than 90% of jojoba oil goes into cosmetics.

Beeswax. The cells of the honeycomb contain esters of C_{26} and C_{28} acids, with the C_{30} and C_{32} alcohols plus 14% hydrocarbons (mainly C_{31}). Beeswax is used in leather and furniture polishes.

Carnauba wax, the most valuable of the natural waxes, is obtained from the coating on the leaves of a Brazilian palm. Hard and impervious, it is used in car and floor polishes. It has a melting point of 80°C to 87°C, and consists of the esters of the C_{24} and C_{28} acids with the C_{32} and C_{34} alcohols. There is also a considerable amount of ω-hydroxy fatty acids, $HO–(CH_2)_xCOOH$, where x = 17–29 (ω, (omega) is the last letter of the Greek alphabet and in this application it means that the –OH is at the end of the chain). These ω-hydroxy fatty acids can form long polymer esters, which give this wax its unique properties of being hard and impervious.

Wool wax (wool grease, degras) is recovered from the scouring of wool and is unusual because it forms a stable, semi-solid emulsion containing up to 80% water — a purified product known as lanolin is used as a base for salves and ointments in which it is desired to incorporate both water-soluble and fat-soluble substances. This 'wax' consists mainly of fatty acid esters of cholesterol, lanosterol and fatty alcohols.

Candles

Yes,' I answered you
last night;
'No,' this morning, sir,
I say,
Colours seen by
candle-light
Will not look the same
by day.

(Source: 'The Lady's yes' by Elizabeth Barrett Browning)

Candles

The colour of objects depends on the light that illuminates them. Hence the importance of choosing colours for drapes, carpets, paint, etc. under the appropriate light. Note the ghastly features we have under street lamps of the sodium and mercury type, which provide only ultra-thin slices of the colours of sunlight. Light globes do not have as much blue as sunlight, and hence render colours a little softer. Candlelight is softer still. Photographers talk of the colour temperature of their flash and flood lamps, to compare these with sunlight.

The original candles were made from solid animal fats. On burning, the glycerine in the fat produced tear-gas type fumes (acroleins) which could not have made the stuffy Middle Age hovels too pleasant. Hence religious candles were required to be made of the more expensive and more pleasant beeswax. Church legislation of the 16th century decreed that candles used at Mass and the Paschal Candle should be chiefly of beeswax.

In 1811, a French chemist, Chevreul managed to split fats into glycerine and fatty acids. One of these, stearic acid, or stearin, became available for clean-burning candles and was used until the discovery of paraffin, prepared from Scottish shale in 1850 and later from crude oil.

For a candle to burn, there must be a supply of liquid wax at the base of the wick, which gets sucked up and burns as a vapour. Too little liquid, and the flame is starved; too much and the candle drips. The size of the wick in relation to the size of the candle (and the composition of the wax) are critical. Early wicks required continuous snuffing (trimming) and frequently feathered to a short stump and drowned in the wax. A major advance was the self-snuffing wick introduced by Cambacères in 1820 by plaiting the wicking, and later by chemically treating it in various salts. This makes it curl over and burn off at the edge of the flame, making snuffing unnecessary.

Candles in the Bible

Candles derive their name from the Latin candere, ' to shine' and are mentioned six times in the Bible, first in Proverbs. The hymn that is sung during the Catholic Church's Easter vigil Service of Light, the Exsultet is in praise of the Easter/Paschal Candle. In the pre-1970 version there was a stanza praising the industry of bees: 'This light is nourished by the flowing wax which the mother bee brought forth to make into the substance of the precious light.' It may have been dropped when it was realised that queen bees don't make wax!

References

Nutritional value of processed food, G. J. Walker, CSIRO Information Service, 1979 (CSIRO Division of Food Processing, PO Box 52, North Ryde 2113, Australia), and references therein.

Chemistry of cooking, D. J. Smith, *School Science Review* **58** (202), p. 25, 1976.

Impacts of the domestic microwave oven, C. Aitken and D. Ironmonger, *Prometheus,* 14(2), 168–178, Dec. 1996.

Superheating water in a microwave, R.E. Apfel and R. L. Day, *Nature,* 321, p. 65, 12 June 1986.
Killing for oil, J. Alper, Chem Matters, Oct. 1988, pp. 4–8, citing *The head of the sperm whale,* M.R. Clarke, *Scientific American,* 240, 128–41, 1979.
Intake of trans fatty acids and the risk of coronary heart disease among women, W.C. Willett, M.J. Stampfer, J Manson, et al., *Lancet* 341: 581–585, 1993.
The effects of dietary trans fatty acids, D. Kritchevsky, *Chem & Ind,* 5 Aug 1996.
Cut the kilojoules with fake fats, C. Reed, *Guardian,* Nov. 1995.
Success on a plate, R. Maughan, *New Scientist,* pp. 36–39, 25 July 1992.

Bibliography

Chemistry of organic compounds, W.B. Noller, 3rd edn, Saunders, Philadelphia, Pa., 1966. An excellent organic chemistry textbook of the more traditional type. It gives background information on industrial and agricultural aspects of chemistry.
On food and cooking: The science and lore in the kitchen, H. McGee, Unwin Hyman, London, 1987. A delightful book.
Nutrition and health, L. Pauling, supplement to the RACI Proceedings, May 1973. But see even earlier, *Vitamin C, its production and uses in the fortification of foods,* C.E. Napier, Proc. Roy. Aust. Chem. Inst., **13**, 19, 1946.
The healing factor, vitamin C against disease, I. Stone, Grosset & Dunlap, NY 1972. For the mammalian mutant argument on vitamin C.
Food: The chemistry of its components., T.P. Coultate, Royal Society of Chemistry Paperbacks, London, 1984. Good technical stuff.
The Choice guide to vitamins and minerals, R. Soothill, Australian Consumers' Association, 2nd edn. 1996. Popular paperback written by a literate science graduate.
Physiology of sport and exercise, J.H. Wilmore and D.L. Costill, Human Kinetics, Ill, 1994. An excellent reference.

CHAPTER

Chemistry in the boudoir

It was at this point that Lucy realised how dependent she was on chemicals for her natural look

In the boudoir, we spend most of our time on cosmetics, including sunscreens (sunglasses), perfumes, and deodorants.

Introduction

In 1770 a Bill was allegedly introduced into the British Parliament that read:

> . . . that all women, of whatever age, rank, profession or degree, whether virgins, maids or widows, that shall impose upon, seduce and betray into matrimony, any of His Majesty's subjects by the scents, sprays, cosmetic washes, artificial teeth, false hair, Spanish wool, iron stays, hoops, high-heeled shoes and bolstered hips, shall incur the penalty of the law in force against witchcraft and like misdemeanours and that the marriage upon conviction shall stand null and void.

After a reader's inquiry, the author went to a law library and manually searched Great Britain's *Statutes at Large* right back to Magna Carta, and found no such Act recorded. But it would be a shame to let facts interfere with such a great story. Although cosmetics have been used — by men as well as women — since prehistory, the popular general use of cosmetics is a modern phenomenon. Many new and improved cosmetics have been made available by scientific research, and extensive advertising has been an important factor in bringing them to the attention of consumers and in increasing the number of people who use cosmetics. Advertising standards are tightening up around the world. The words are becoming more subtle: 'wrinkles and fine lines will appear reduced', 'helps drainage of toxins formed around the eye'. Cosmetics companies are aware of the power of (pseudo-) science in selling, and use words such as 'active natural ingredients' and ' tonicity'.

One major company, L'Oréal, owners of Lancôme, Laboratoires Garnier, and Helena Rubenstein, has four research laboratories in Paris alone, employing 1500 researchers in cosmetics and dermatology, as well as many robots used for repetitive testing such as hair combing. They also have an enormous testing salon with about 400 seats and mirrors and a list of 9000 volunteers.

In her book *The Beauty Myth* (1990), Naomi Wolf likens the 'anti-age industry' to the diet industry in that 'universal female self-esteem would destroy (it).' Wolf estimates that urban professional American women spend up to one-third of their income on 'beauty maintenance'.

This chapter deals with the chemistry of cosmetics and related products and Chapter 6 takes a more detailed look at the basic science behind the technology, surface chemistry.

The chemistry of cosmetics

Once upon a time a clever market operator repackaged a waste chemical from an oil refinery and sold it as a personal care product. Petroleum jelly is still a big seller, but the chemistry of cosmetics has become much more sophisticated.

The market

The range of goods covered by the term *cosmetics* is very large. It includes skin-care products, eye and face make-up, fragrances, hair-care products, hand creams and lotions, nail-care preparations, bath products, deodorants, depilatories, toothpastes and mouthwashes, shaving lotions and soaps, and sunscreens. The total market for chemicals used in products was $US 4 billion in 1995 and is increasing (see Table 5.2, on page 112), mainly because of an aging population seeking youthful looks.

Australians spend about $260 million per annum, about the same amount per person as in the USA. World wide, sales have reached $20 billion.

Cosmetics are applied to the skin, so it is helpful to study the skin first.

The skin

The skin is an important organ with many tasks to perform. It encloses the body, preventing some internal materials from escaping, while

TABLE 5.1
Cosmetics time scale

BC	
5000	Eye shadow of green copper ore.
3500	Eyelids darkened and lustre of eyes increased by Egyptian women using kohl (a fine-powdered form of antimony).
3000	Balsam and perfume was a Cypriot industry of the Bronze Age.
1600	Red hair-dye in use.
200	Dye extracted from a root called rizion used for rouging the cheeks (mentioned by Theophrastus).
68–30	Asses-milk bath used by Cleopatra to improve and whiten the skin. Henna used to dye fingernails, palms of hands and soles of feet. Egyptian beauty shops sold cosmetics and perfumes.
AD	
800–1100	Perfume distilled by Arabs.
1100	Gifts of cosmetics brought back from the Crusades.
1558–1603	Reign of Queen Elizabeth 1. Toilet preparations in use. For example, perfume boxes ('sweet coffers') were a necessary item of bedroom furniture. Face powder was made from powdered marble, borax and starch. Women of the court washed their faces in wine to achieve a ruddy complexion.
1570	Mary, Queen of Scots, bathed in wine.
1610–43	Use of cosmetics encouraged by Louis XIII (France).
1649–58	Use of cosmetics discouraged by Cromwell.
1660–85	The Restoration. Cosmetics played a vital role in improving physical appearance. Charles II encouraged the use of cosmetics.
1700	Poisonous white lead (lead carbonate) to coat the face was used by many women, many of whom died as a result.
1770	Widespread use of cosmetics allegedly stimulated the introduction of a Bill into the British Parliament to protect men from being falsely attracted into marriage (see Introduction).
1780	Lavoisier delivered a paper on cosmetics to the French Academy of Science. He distinguished between those rouges made from plant and animal extracts (plant rouge is decolorised by spirits of wine; animal rouge is decolorised by alkali) and those made from minerals (not decolorised by either chemical).
1800	A great revival of interest in cosmetics early in the century during Napoleon's time.
1850	Synthetic perfumes and synthetic dyestuffs developed.
1939	World War Two caused a 75% reduction in British cosmetic manufacture. This resulted in a fall in morale and a decline in the number of factory employees. When cosmetic rooms were set up in factories, women flocked to them and production efficiency improved — proving a direct relationship between the sense of well-being imparted by cosmetics, and excellence of performance (industry advertisement).
1920s	Natural lipsticks, Tussy and Tangy.
1930s	The Cold Wave, paving the way for the home perm.
1950s	Oil of Olay, the first mass-market moisturiser.

TABLE 5.2
1995 US $4 billion cosmetic market

Category	Percentage
Skin care	40
Hair care	23
Perfume	13
Face cosmetics	5
Oral hygiene	5
Other(shaving, deodorants, bubble bath)	14

Source: Producers raise high-tech stakes in personal care ingredients, markets, E.M. Kirschner,C&EN, pp. 13–19, July 1, 1996.

allowing others to pass through, and at the same time it keeps most external materials out. It regulates body temperature by controlling the escape of water (as sweat); it regulates the penetration of sunlight by allowing sufficient light for the production of vitamin D, but not so much that the underlying tissues are damaged; and it senses and transmits information on temperature and pressure.

As Figure 5.1 shows, the skin has many components. For convenience it can be considered as four layers — subcutaneous tissue (underlying fatty material), on top of which is the dermis or true skin, which in turn is covered by an outer layer called the epidermis, on top of which is a horny layer, the *stratum corneum.*

The *stratum corneum,* the horny 10 μm outer layer of the skin, consists of about 20 layers of dead cells containing mainly a protein called keratin. These cells are overlapped like roof shingles. Water-soluble substances cannot penetrate this layer but oil-soluble substances (organic solvents but not ethanol) can dissolve the sebum (oil) that surrounds each hair follicle, and then move along the follicle. These outer cells are being continuously sloughed off. This is obvious on the scalp, where the cells stay caught in the hair and we see the excess as dandruff.

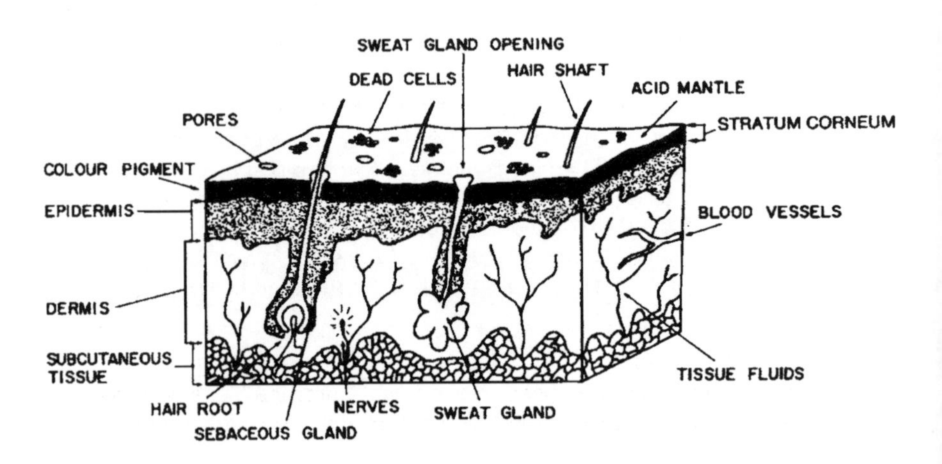

Fig. 5.1 Cross-section through the skin

Underneath the *stratum corneum* is the living epidermis, whose outer cells replenish the *stratum corneum* as it wears away. There are some chemicals that can destroy the *stratum corneum,* such as sodium hydroxide, formic acid and hydrogen fluoride. Keratinised growths (warts, etc.) are removed with agents such as silver nitrate, acetic acid, lactic acid, and resorcinol. Stronger agents are also used, such as trichloroacetic acid and phenol. Soap, detergents, and organic solvents remove the sebum, after which the skin tends to crack and then dehydrate. This can lead to chemical dermatitis and infection.

The *epidermis* contains the pigment-producing cells that determine the degree of darkness of the skin and has a deep layer of growing cells. The *dermis* contains the working elements of the skin — sensory nerves, hair follicles, blood vessels and sweat glands. Associated with each hair follicle are several sebaceous glands, which excrete, through the pores, an oily material called *sebum.* This is how the oiliness of the skin is controlled.

Sebum is approximately 50% fats, 20% waxes and about 5% free fatty acids which give the fluid a slightly acidic pH, helping a little to combat bacteria. (See Deodorants and antiperspirants, below.) This acidity would normally be neutralised by excess soap. Blackheads are formed when sebum dries in the skin ducts and blackens on reaction with the air. Pimples are inflammations caused by the irritation of sebum escaping into surrounding tissue. They are not initially caused by infection, although secondary infection can occur and make them worse.

Under normal conditions, the water content of the skin is higher than that of the surrounding air, so evaporation occurs. If this water is not replaced, the skin becomes dry. However, direct application of water to the skin does not replace the water in the skin. An oily vehicle is required to hold the water in contact with the skin and to hinder evaporation of the water.

Skin patches

The skin can carry out metabolic reactions similar to, but much weaker than those of the liver. Drugs can be administered to the skin to avoid their inactivation through other modes of administration, or to achieve a prolonged effect. Nitroglycerine ointment (2%) is applied to a patient's chest to treat an attack of angina pectoris, and this can be effective for about three hours. Seasickness can be kept at bay by applying hyoscine in a plaster affixed behind the ear. Nicotine patches are popular.

Skin penetration by drugs

In treating the skin with drugs, the great problem is to arrange for the substance to enter *into* the skin without going *through* the skin. Salicylic acid ointments (10%), to suppress chronic skin conditions (e.g. dandruff, psoriasis) are often applied daily for years. Symptoms of poisoning by salicylic acid are chronic gastritis, ringing in the ears, irregular heartbeat, and even impaired hearing. Hexachlorophene (pHisoHex), while safe for adults, has killed new-born babies. Treatment of burns in children with a 3% solution has also caused death.

Lindane (see Organochlorine compounds in Chapter 7), which is used extensively to rid children of lice and scabies, can also penetrate the skin and has caused convulsions in patients who have been too liberally treated. Boric acid (boracic acid) has also been a health hazard as an additive in talcum powder used on babies' nappies. The symptoms produced ranged from vomiting to circulatory collapse and death.

Skin types

Cosmeticians divide skins into the following types (dermatologists disagree):

- *Normal* — although 90% of children have this ideal skin, very few adults can boast of a skin that is smooth, soft, moist and of healthy appearance.
- *Oily* — shiny, with enlarged pores, a tendency to blemishes and coarse texture, and sometimes a flakiness caused by the accumulation of dried out oils.
- *Dry* — fine-textured, flaky and taut; expression lines around eyes, mouth and throat; with aging, dry skin loses elasticity and develops pronounced wrinkles.
- *Combination* — areas of both dryness and oiliness.
- *Sensitive* — florid, with tiny broken capillaries; usually fine-textured and with many of the characteristics of dry skin (a rather different definition from that used by dermatologists).
- *Blemished* — excessively oily, with blemishes such as pimples and blackheads.

The skin analysis by the beauty consultant is simple. She cleans the skin on the highest point on the cheekbone, then mixes a quick-setting latex paste and smooths it on. After three or four minutes, she peels it off and looks at it under an optical microscope. Large, irregular pores suggest oiliness; stretched out skin means it is too dry — a measure of scientific objectivity has entered the Land of Hope and Glory.

Aging

With age, the skin loses its elasticity and becomes thinner and drier. Folds appear and wrinkles fail to vanish. Sometimes changes in colour occur. The process is universal and irreversible, yet there are some people who look younger at 50 than others do at 40. Although aging is partly determined by factors such as heredity, dermatological research has shown that exposure to sunlight (i.e. to ultraviolet light) is an important contributor. Evidence of the effect of sunlight can be seen in the contrast between the leathery skin that is normally exposed, and the soft smooth skin that is usually covered. Persistent and repeated exposure to sunlight can also cause skin cancer. The incidence of skin cancer is much higher in Australia than in Britain for people of comparable heredity.

The most obvious signs of aging are wrinkles. Their cause is not known but it is probable that they are the result of changes in the dermis, or in the subcutaneous tissue. The dermis becomes thinner and less elastic, and the sebaceous glands are less active, leading to dryness. Much research is invested in understanding and trying to combat these effects.

The alpha hydroxy acids are found in nearly every skin product, even in hair care, as a wrinkle reducer. Kalaya oil, a derivative of emu oil, has shown promise in studies on mice, not only to rebuild the skin, but as a carrier for pharmaceutical and beauty care active ingredients. No wonder the emus are running scared these days!

Hair

The reason we are 'naked apes' is not that we have fewer hair follicles than our chimpanzee cousins, but that we produce fine, short hair, rather than fur. (Some simple genetic engineering is called for here!) Our heads, however, have about 150 000 individual hairs, each about 70 μm in diameter. Hairs develop in long pits in the epidermis, extending deeply into the lower active dermis. At the base of the hair follicle is a region supplied with nutrients by blood vessels called the papilla. The rapidly multiplying cells are surrounded by an inner root sheath and as they move up, they gradually fill with keratin, dry, harden and eventually die. As the hair emerges through the skin it is compressed in a roller action that seals the cuticle, with cells overlapping, shingle style, as in skin, or more accurately like a set of plant tubs or barrels stacked inside one another.

Hair yarn
The structure of the hair shaft is a complex yarn (literally). There is a tangle of keratin protein chains, about 1 nm (10^{-9} m) long, imbedded in a protein matrix. Five of these threads then twist together to form a yarn. The yarn in turn is bundled into cables. A single hair will support about 80 g. So if hairs did not pull out, a mere thousand (less than 1% of the scalp) could support a person weighing 80 kg!

The growth of scalp hair is complex and fascinating. The growing phase lasts about six years in women and three to four years in men. The length attained, if uncut, is 70 to 80 cm, and 40 to 50 cm, respectively. The growth then ceases and roots develop, attaching the hair to the follicle. The follicle shrivels and rests for three to six months, then starts to produce a new hair, which pushes the old one out. So every day, about 100 mature hairs are pushed out (150 000 follicles in a hair production time of 365 × 4 days for a man) as normal turnover. At any one time, about 90% of the hairs are actively growing. The cuticle (of shingled cells) protects the hair shaft by forming a sheath around it. If a hair is cut, the cement that holds the overlapping cells together oozes out and seals the end.

The pigment melanin, which gives skin its colour, also colours hair. Different amounts of the same pigment, and in different places, account for light blond to blue-black. Only redheads are different, in that their hair contains an additional unique iron-based pigment, which indeed makes their hair 'rusty'. The melanin-containing granules are present mainly in the cortex of even the darkest European and Japanese hair, but in Bantu hair they are present also in the surrounding cuticle. Dark hair has both more granules and more melanin per granule.

The clue to shape of a mop of hair came from a study of the adsorption of dyes on wool. Basic dyes and metal-based stains adsorb at two different regions of the fibre cortex, now called orthocortex and paracortex. The orthocortex is mainly in the form of fibrils and takes up basic dyes, while the paracortex has a matrix of protein with many exposed disulfide bridges between protein molecules that can take up the metal dye. In wool, the ortho and para forms tend to line up on each side of the fibre, which leads to the characteristic crimp of wool, because the para form crosslinks to form the inside of a bend. The most crimped human hair of some Africans has a similar structure, while the straight hair of the Japanese has the ortho and para forms arranged in concentric circles to give symmetric disulfide bridges. Other hair lies somewhere in between in structure.

A hairy tale
The strong bonding of metals by the sulfur group on the protein in hair creates an interesting forensic possibility. The metal content of a particular section of hair reflects intake when that section of hair was growing. A classic case is the death of Napoleon in 1821 in (his second) exile on St Helena. Later examination of his hair showed significant levels of arsenic, and the position on the hair gave an approximate date of exposure. Arsenic was common in dyes used for the popular green wallpapers of the time and microbes digesting the paper could disperse the arsenic into the surroundings. There is no need for a deliberate poisoning hypothesis (see Arsenic in Chapter 10).

Greying with metal
A popular preparation for camouflaging stray grey hairs works by depositing a layer of metal salts on the hair. The metal tarnishes in air to give a dull green-black result of dubious merit.

The condition, or gloss of hair depends on the outer cuticle, whose stacked transparent plates reflect the light. Chemicals, too much heat, and excessive brushing dislodge the protective tiles, reduce reflection, and give that dull appearance seen in the 'before' part of advertisements for hair products. The raised scales can catch one another and cause tangles, while the escape of moisture from the cortex causes the hair to appear dry and brittle.

The basic aim of all hair conditioners is to smooth the surface of the hair and thereby avoid all the other negative effects. While cationic surfactants (see How do surfactants work? in Chapter 3) were the original conditioners, these have been replaced by silicone-based conditioners, which are extremely good lubricants for hair. The displacement of moisture on the surface (similar compounds are used for waterproofing camping and wet-weather wear) allows hair to dry more quickly. Linear long-chain silicone co-polymers are skinny and pack onto the surface, and when you feel the hair, you are really feeling silicone. The reduction in friction also reduces static, and thus 'flyaway hair', particularly if functional groups like amines are added to the molecule.

Making your hair darker is relatively easy, particularly if you don't mind doing it fairly often. Semipermanent rinses and washes all share the property of attaching the colour to the outside of the hair. The bright fashion colours use large granules, which vanish with one washing. The finer particles of more traditional rinses bond to the scales of the cuticle and survive many shampooings. With permanent dyeing, the idea is for the dye to move through the cuticle and enter the cortex cells. Hydrogen peroxide is used to soften the cortex and thus allow very small granules to pass through. It also bleaches the melanin and so allows lighter shades. Its oxidation action clumps the granules in the cortex and thus hinders their migration out again. Hydrogen peroxide probably weakens the hair in the longer term.

The chemistry of adjusting the curl falls into groups, depending on the degree of permanence required. The simplest procedure is to wet the hair, which protonates and breaks some of the disulfide bonds holding the proteins in shape (see Silk in Chapter 11). Drying the hair while it is held in either a straighter or curlier shape than before sets it temporarily. Indeed, even heat alone on dry hair causes some changes. The slightest increase in humidity causes the hair to revert. This sensitivity of hair curl to heat and humidity is the basis of the picturesque 'weather houses' where two figures balanced on a horizontal axle are suspended on a (human) hair. The changing weather conditions twist the hair, which causes the figures to move in and out of a 'doorway' (Fig. 5.2).

A longer-lasting wave is obtained by spraying the hair with a solution of polymer. On evaporation of the solvent, the hair remains coated with a resin. (The most common polymers are poly(vinyl pyrrolidine)polymers, or co-polymers with vinyl acetate; co-polymers of maleic anhydride and vinyl acetate, etc.) However, even with heavy, thick coatings, the effect does not last long.

Fig. 5.2
Weather house

The physics of combing hair

Too much friction when combing hair pulls the skin, but too little makes the hair feel greasy. It is the friction between the hair fibres, not between the hair and comb, which is important. The average force between two contacting hairs is 30 micronewtons (μN), the coefficient of friction is 0.3, and so the frictional force between the hairs is 10 μN. Each cm of hair has about 2.2 contacts (with other fibres), their average length is 15 cm, and there are about 100 000 per head. That adds up to a lot of adhesion. By increasing the friction further, all sorts of wonderful shapes can be kept in place and this has implications for makers of setting lotions and styling mousses.

When the hair is wet, it can carry about one-third of its mass in water. This increases the frictional force, but the major effect is due to the surface tension of water. In contracting a surface to minimum area, this force acts to drag hair fibres together and triples the attractive force to 35 μN. (See also The Laplace force in Chapter 6.) With Afro and other very curly hair, there is less opportunity for drawing together, and indeed the water relaxes the curls and reduces the fibre-to-fibre contact. Such hair is about three times easier to comb wet than when dry, exactly the opposite to straight hair. Chemically straightening or perming of hair can swap between these opposite behaviours.

Chemical attack on the internal proteins is the only way to obtain a 'permanent' wave. In about 1930, scientists at the Rockefeller Institute demonstrated that the disulfide bonds that give proteins their spatial 3D macroscopic structure can be split at ambient temperature and slightly alkaline pH by the action of sulfides or mercaptans. Thioglycollic acid is used in a suitable pH buffer, together with cuticle softener. The smell of rotten eggs accompanies this process to a greater or lesser extent. Having rendered the internal structure of the hair floppy, the hair is set, the effect of the thioglycollic acid neutralised with peroxide, and the disulfide bonds reformed to hold the hair in its new shape.

Unwanted hair
Incidentally, depilatory creams and lotions work on the same principle and with the same ingredients. They soften and loosen the unwanted hair, which is then easily removed.

Baldness

Over the 25 years of this book, the author's interest has shifted from topics such as nappy rinses to baldness. The orang-utan and a particular species of macaque monkey go bald in middle age like the 40% of men who show 'male-pattern baldness'. Both the tendency to baldness and its pattern tend to be inherited. More reassuring is that in male gorillas, social dominance is strongly associated with age, and the leader of a troupe often has grey hairs on his back and flanks.

The reason for this phenomenon lies at the other end of life. In the developing foetus, both forehead and scalp contain hair follicles, and in fact form one tissue structure. At about the fifth month, the follicles on the scalp continue to grow, while those on the forehead do not. After birth, the hair follicles regress over most of the baby's body, including the forehead, and the hairs there become finer and almost invisible. In older age, under the influence of the male sex hormone testosterone, some follicles on the scalp become infantile again and the fine downy hair they produce is called *lanugo*.

And a cure for baldness? An unexpected side-effect of a drug used to treat high blood pressure, minoxidil, sold under the brand name Loniten (Upjohn Co.) caused reasonable hair regrowth in about a third of patients, downy growth in another third, and nothing in the remaining third. An ointment with the active ingredient is about to go on sale, after, presumably, the usual tests on the flanks of orang-utans.

Skin-care cosmetics

Moisturisers — These are preparations that replace the water lost from the skin, and both oil/water and water/oil emulsions are used. The dryness and reduced flexibility of the skin cannot be corrected by adding oily materials, but the skin will become more flexible when water is replaced. The skin can be protected, and skin dryness prevented or relieved, by emollient creams and lotions which slow the evaporation of water from the outer layer of the skin. Detergents cause dry and chapped skin because they dissolve some of the water-attracting components of the skin.

Hyaluronic acid
Hyaluronic acid was once described (by Ted Cleary, Pathology, Adelaide University) as 'like a piece of piano wire 400 m long folded up into a bucket . . .' It is resistant to compression, it's got space-filling properties, and it can trap over twice its mass in water'.

It is hard to believe claims that chemicals like hyaluronic acid can penetrate the skin. This material is a natural high-molecular-mass (50 000 to eight million) cellulose-like material found in umbilical cords and other body fluids.

Liposomes, on the other hand, can get into cell layers, delivering, for example, antibiotics. However they do not get through into the dermis, which is where you need the material if you want cosmetic changes to occur.

Cleansing creams and lotions — Even though adequate washing with soap and soft water will achieve the same result, there are possible advantages in using a cleansing cream to remove facial make-up, surface grime, and oil from the face and throat. The specific chemical design of a cleansing cream allows it to dissolve or lift away more easily the greasy binding materials that hold pigments and grime on the skin. Most emulsified cleansing creams can be considered as cold-creams that have been modified to enhance their ability to remove make-up and grime.

If eye cosmetics are used, they also need to be removed and mineral oil, alone, is a safe, effective agent.

Cleansers for oily skin — For the usual type of oily skin where there is no associated acne, the use of ethyl alcohol or isopropyl alcohol can provide temporary relief from excessive oil flow and the resulting shiny skin. The concentration of alcohol should not exceed 60% and preferably should be below 50%, or it could be too drying or irritating. Other modifying ingredients are included in the formula to balance the harshness of the alcohol.

Deodorants and antiperspirants

All you could ever want to know about armpits!

Most antiperspirants are based on aluminium aqueous (chloro) complexes, and these have been analysed in agonising detail with all possible techniques under every conceivable condition. Aluminium and zirconium compounds are all that is left in the market to stop perspiration, after the whole of Mendeleef's Masterpiece (the periodic table of the chemical elements) has been tried and rejected. It is not that other salts don't work, they do, such as the common ones: copper, iron, tin; the obvious ones: lanthanum and cerium (being relatives of aluminium); the esoteric ones: samarium and praseodymium. Unfortunately, while people perspired less, they expired more, which for some reason tended to provoke customer resistance!

See Plate IV, chemistry of perspiration from the Powerhouse Museum.

Aluminium salts appear to work by producing an insoluble hydroxide gel in the sweat pores and thus blocking them. Numerous tests show that the aluminium does not cross the skin barrier and enter the body, so this source cannot be accused of possibly contributing to the body load (see Aluminium in Chapter 10).

If 70% of the world's scientists work for the military, then most of the others must work for cosmetic companies. This is not surprising when we are told that in 1986 the sales of deodorant soaps, antiperspirants, and deodorants made up approximately 14 per cent of the $8600 million spent in the USA on health and beauty aids.

Perspiration was first effectively controlled by Mum™ (*c.* 1888) which contained zinc oxide (now used as a sunscreen). This neutralised the smelly acids and helped kill bacteria. In 1895 came Lifebuoy™ containing cresylic acid. It smelt like phenol (Dettol™) and replaced one unpleasant odour with another. In 1948 came Dial™ with hexachlorophene and this antibacterial survived until a French firm in 1972 inadvertently put a level of 6 per cent into a baby powder that then caused over 30 deaths.

We learn that sweat glands can be divided into two types, eccrine (E) and aprocine (A). The E's are stimulated by heat and help cool us down. The A's are triggered by emotion, and cause copious flow under embarrassing conditions (for instance, when on a lie detector). A's are

Samuel Johnson and the lady
The eighteenth century essayist, Samuel Johnson once corrected the English grammar of a lady complaining about his bodily odours with the retort: 'Lady, **you** smell, I stink'.

found on palms and soles. E's are found everywhere else. The armpits (called axillae in polite chemistry texts) have both.

If all the approximately 3 million sweat glands worked at full bore, so to speak, we would produce about 10 litres of sweat per day. Each armpit (sorry, axillary vault) has about 25 000 sweat glands and it takes only ten minutes to produce 1.5 mL after an emotional stimulus. You can see why there is an annual $1.5 billion market for antiperspirants in the USA alone.

Microbiologists love studying the axillae because their 'semi-occluded anatomy is less prone to environmental contamination'. Apparently it is hard to miss the microbes because they crowd the pit at a million to the square centimetre. An initial study shows that there is no statistically significant difference between the left and right armpit, or between left- and right-handers, or between males and females. There is, however, a big difference between those who wash and those who don't. The variety and names of the microbes suggest to a non-microbiologist that they must be a pretty ferocious lot. They work on the secretions to produce the odours. It has been said:

> Axillary odour is a mixture of many 'notes' (a technical term used in the chemical perfumery industry (see later in this chapter)) with the dominating notes identified as isovaleric acid and 5-andost-16-en-3-one and 5-andost-16-en-3-ol.

Isovaleric acid has a sweaty odour. The latter two unpronounceables are formed from body steroids, and the last is variously described as smelling like stale urine or worse. They probably do wonders for the microbe's sporting prowess. A closely related armpit steroid with a natural musk odour gained notoriety by being added to a perfume (Andron™). However, there were no reproducible(!) effects.

CATULLUS' POEM

Once upon a time, (in 50 BC), Catullus wrote a poem that foreshadowed the modern theory that microbes are responsible for the odours. The poem has been freely translated as follows:

An ugly rumour harms Your reputation.
Underneath your arms they say you keep a fierce goat which alarms all comers — and no wonder,
for the least Beauty would never bed with rank beast.
So either kill the pest that makes the stink
Or else stop wondering why the women shrink.

The hircine (L. goat-like) smell is due to another compound present in the armpits, 4-ethyloctanoic acid. Mature female goats in oestrus respond specifically to this compound, so when in range, keep your arms down unless you want an unusual experience.

Naturally the cosmetics industry has poured millions of dollars into seeing whether this response can be transferred to more affluent females. Humans can smell this acid at 1.8 parts per billion, one of the lowest thresholds for any similar compound. Unfortunately, human females find it disagreeable.

Nevertheless, experiments have been done on 'cognitive evaluation of sexual stimuli' (viewing porn photos) under the influence of various of these chemicals to see whether they change male or female responses. Some very interesting results have been obtained . . .

Widely published, on the other hand, is the experiment in which extracts of axillary odour (ethanol extracts of armpit sweat on pads) placed on the upper lips of female subjects were found to cause their menstrual cycles to approach that of the donor. This suggests that the well-established synchronisation of cycles of women living in close proximity is mediated chemically by sweat compounds.

Extracts from males appear to have a mild regularisation effect on aberrant length menstrual cycles (< 26 or > 33 days). This could be biologically significant in that fertility is correlated with normal length cycles (29.5 give or take 3 days). We must always be thankful for the initiatives of industrial research, regardless of their original motives.

Armenian proverb
When the times are tough and the road is long, it helps to remember what the Armenians say: 'The watermelon will not ripen in your armpit'. Just what chemistry have they discovered?

Sense of smell

In order to have a scent, a molecule must evaporate reasonably efficiently. This means that the molecule must not be too large, and a molecular mass under 300 is usual. The molecule must also be able to bind to the scent receptors, which, in humans, are found deep within the nose and cover an area about the size of a postage stamp. Although humans have a poor sense of smell compared to other mammals, most of us can identify, say, a strawberry, which is a complex mixture of about 300 components, at a concentration of around 10 parts per million.

Chemists use a large slow complex machine called a gas chromatograph to do the same job. In 1986, in a large smell survey, the National Geographic magazine tested 1.5 million people world wide with six scents using 'scratch and sniff' sheets. The six scents were androstenone (steroid from the fat of boars, released (in modified form) when the boar drools after sexual arousal), eugenol (oil of cloves), galaxolide (synthetic musk), isoamyl acetate (main component in banana scent), mercaptan (a smelly sulfur compound added to odourless natural gas as a warning agent), and rose. Women smelt better than men. However, 1.2% of the sample could not smell any of the scents! Only half could smell all six scents. Androstenone, found also in very low levels in human male sweat, was mooted as a possible sex attractant, but was smelt by the fewest number, and they described the smell very differently. However, it is used commercially as an aerosol spray in assessing sows for artificial insemination programs. See Plate V, The Fragrance Wheel.

Aromatherapy
Aromatherapy is based on the premise that scents affect moods and emotion at a deep level. There is even one patent published that is claimed to induce people to pay bills when the paper is impregnated with a particular substance!

Synthetic noses

The nose has around 10 million olfactory receptors, made up of 30 different types, which work in parallel to absorb and analyse a range of 'smell' molecules. A receptor molecule flips its shape as soon as an appropriate molecule is adsorbed and this causes a signal to be sent down a nerve to the brain smell centre. An analysis of all the receptors that sent signals with that breath produces a smell signal 'profile' to be checked against memory.

A prototype olfactroscope using conducting polymers of pyrrole and aniline to try to do the same job was developed by detecting the change in electric resistance (rather than shape) when an appropriate molecule is adsorbed. Aromascan™ (earlier Odourmapper) is now listed on the UK stock market, and is used to test the effectiveness of water treatment, and the freshness of food. Others detect 'unsavoury' body odours of individuals, even if masked by other aromas. The process is based on adsorption of the odours onto a resin, followed by heating to release the chemical to a gas chromatograph. A dog acts as a sniffing sensor and allows correlation of the instrument peaks with smells. The technology has promise as a security detector to allow access through a door to those who pass the sniff test.

Acetone in breath has long been used as an indication of diabetes, and for centuries Chinese doctors have used an array of smells from a patient's breath as a diagnostic tool. One instrument is being used to detect the presence of the bacterium *Helicobacter pylori,* found to cause stomach ulcers by researchers in Western Australia in the 1980s. (The usual procedure is by endoscopy.) Soda water is given to the patients, who are then encouraged to belch to provide a sample. The unit uses adsorption of gas onto vibrating quartz crystals as its means of detection.

Perfumes — (in the USA they say fragrance)

Perfume comes from the Latin *per* (through) *fume* (smoke). The first perfumers were priests who burnt resins, leaves and wood as incense. They believed that the sweet-smelling smoke carried their prayers to the Gods. The Egyptians were probably the first to use perfumes in their private lives: Cleopatra ordered the sails of her barge drenched with cyprinum in order to attract Mark Antony. When the crusaders returned to Europe, they brought new fragrances such as musk, citrus, jasmine and sandalwood. The peak in the passion for perfumes was reached by Louis XV in his 18th century royal courts.

Perfumes consist of essential oils (from *essence*) and are generally concentrated in the petals of the flowers, but can occur elsewhere. Peeling of an orange releases tiny droplets of a sweet-smelling oil located just beneath the skin. The Egyptians steeped petals in warm liquid fat to dissolve out the oils, a process called maceration. The petals were changed regularly, and when it was saturated, the fat was cooled and the oils were extracted. The French developed a gentler process called enfleurage (the hot oil decomposes some of the more sensitive components). Glass plates are coated with purified fat and the plates are placed in an enclosed chamber. Again the petals are replaced every few days. The fat is scraped off and extracted with ethanol, and this is concentrated, providing floral absolute. Steam distillation was used to make rose water, the first real perfume, in the 11th century, and is still used for most oils, including rose, spice and mint oils, sometimes under vacuum for more sensitive oils. (See Appendix III.)

The best citrus oils are produced by rupturing the oil glands in the peel and not allowing this oil to come in contact with the juice. Other methods include extraction with liquid carbon dioxide (see Caffeine in Chapter 15) and molecular distillation.

Synthetics now dominate the market, although the really top quality perfumes still rely on many natural extracts.

Picturing a smell!
'I "know" that smell but can't picture it!!' Without a visual image it is often hard to recognise a perfume. A rose laced with lavender smells like a rose. Most flowers stop making perfume after being picked. The tuberose is an exception.
See plate VI.

EUCALYPTUS AND TEA-TREE OILS

Professor Ian Rae describes a marvellous eucalyptus distillery a couple of hundred kilometres out of Melbourne at Wedderburn.

> Leaves and twigs are collected into a large tank mounted on a truck. At the distillery, on goes the lid, steam is blown in at the bottom, and away she goes. Most of the boiler fuel consists of dried gum leaves and the whole thing has a Heath Robinson look about it, with the oil finally collecting in an old Hoover Twin Tub.

See Appendix III.

TABLE 5.3
Australian eucalyptus oil production

Year	Local production (in tonnes)	Imports (in tonnes)
1947–48	900	—
1948–49	560	—
1949–50	520	—
1950–51	780	—
1951–52	775	—
1952–53	540	—
1977	200	c. 75
1987	140–160	c. 270

Back in 1788, only 10 months after the establishment of the colony, the Surgeon-General John White sent a quarter of a gallon of eucalyptus oil to England for further testing. Unfortunately the thriving industry that built up over the years has declined, and today Australia is a net importer of eucalyptus oil (see Table 5.3).

Natural perfumes
John Lambeth, perfumer at Dragoco (Sydney), led me through his library of rows of shelving full of valuable little bottles. You match the perfume to the product. You ask yourself, 'What are the trends, what is the market? What are the demographics?' The 'new-car smell' used to sell old 'lemons' is designed to match the buyer. What works for a Ford may be quite unsuitable for a Porsche. Natural is in, and the levels used are now lower.

Perfume notes
A perfume is designed in terms of three 'notes' The top notes evaporate quickly. They are the citrus and fresh components. The middle notes are slower, and include jasmine, violet, and rose — the florals — components. The base notes are slow and are woody, mossy, musky and amber.
Perfumes are spotted on smelling strips (blotting paper) and smelt at various intervals. The top note comes off in the first minutes. Smelling in 15 minutes reveals the middle notes. Smell again in 30 minutes and further components will be revealed. The base may still be evident a few days later.

Fig. 5.3
Australian tea-tree oil has many different uses.

In contrast, exports of tea-tree oil are increasing. Its main use is in medicated cosmetics. Distillation of tea-tree oil (Melaleuca alternifolia) yields mainly (+)terpinen-4-ol and related compounds. This oil has bactericidal and fungicidal uses, from dealing with Salmonella typhi, to treating tinea, acne and diabetic gangrene. It is used as a perfume toner and nutmeg substitute. It kills dry rot fungi, and is used in a variety of products ranging from water-gel fire blankets to oral contraceptives. It is also used to combat legionnaire's disease.

The largest variety of products in the essential oil industry (but not quantity) is found in Tasmania, where oils of fennel (exported to make Pernod), peppermint, lavender, caraway, parsley, boronia, blackcurrant and hops are produced.

With a fruit flavour, the question is, at what stage in the 'ripeness' is the smell of fruit 'correct'? In the fruit, it changes continuously and the traditional 'expected' smell often corresponds to 'overripe'. Chemicals used include *p*-hydroxy phenyl butanone (raspberry ketone), (see also Fruit fly attractant in Chapter 7); eugenol methyl ether (from Huon pine); Bulgarian lavender oil. Rosemary extracted by liquid carbon dioxide (see Supercritical solvents in Chapter 15) is very close to real. From 2.5 tonnes of fresh garden peas Keith Murray (CSIRO) extracted one drop of essence (pyrazines) to which the nose is instantly fatigued (nice while it lasts, though) — see also Flavours in Chapter 15.

The ever popular Eau de Cologne 4711 is just a top 'note' (lemon, orange, bergamot, rosemary). The name comes from the house number where it was made. When Napoleon occupied Köln in 1792, his troops bought up supplies for the hot road ahead. Some modern perfumes use a monolithic block, for example in Trésor: four ingredients make up 80% of the perfume, and in Stephanie: one ingredient makes up 50% of the perfume.

Methyl octin carbonate (methyl 2-nonynoate) evokes the smell of violets and motorcycles: Dior's Fahrenheit™ uses a lot of it. Coumarin, the primary ingredient of Cacherel's Lulu™, is the characteristic smell of late summer, from whose flowers and grasses it is actually derived.

Orris butter, a complex derivative of the roots of the orris, is vaguely floral in small amounts, but obscenely fleshy (like the smell beneath a breast) in quantity. Whales vomit ambergris. It floats on the surface, to be transformed by sunlight. The musk deer is farmed in China. Tonnes of musk is used each year and it is mostly synthetic (remember those pink musk sweets we had as kids? They were probably quite toxic).

The civet cat is kept in cages in Ethiopia and its glands are scraped. The concentrate smells like skatole (faeces, hope they have scraped in the right place!). It is amazingly sexy in subliminal doses. It features in Guerlain's Jicky™, introduced in 1889, and is probably the first modern perfume (and one whose market has changed over 100 years — it now has a following in gay men). Chanel No 5 was launched 1923 with a synthetic 'aldehyde' base and is still very popular. The Americans challenged French dominance in 1953 when Estee Lauder introduced Youth Dew, dropping delicate floral for a powerful oriental scent. Revlon's Charlie in 1972 was for the liberated lady who bought her own perfume, and Giorgi in 1984 was the first perfume to be advertised in scent strips in magazines.

You do not have to dabble for very long to realise that the world of smell has no reliable maps, no single language, no comprehensible metaphorical structure within which we might comprehend it and navigate around it.

> The best we can do with smells is to make comparisons. Karanal is like 'striking a flint', aldehyde C_{14} is 'like peach-skins', beta ionine is 'like latex'. Perhaps our sense of belonging to a world held together by networks of ephemeral confidences (such as philosophies and stock markets) rather than permanent certainties, predisposes us to embrace the pleasures of our most primitive, unlanguaged sense. Being mystified does not frighten us as much as it used to. And the point for me is not to expect perfumery to take its place in some nice reliable, rational world order, but to expect everything else to become like it; the future will be like a perfume.
>
> Source: Brian Eno, visiting Professor, Royal College of Art

Special aromas

Specialist aroma producer Air Products of Blackpool, UK (Fred Dale), has been supplying custom-made aromas such as gun smoke, cut grass, smell of dinosaurs and Zulu warriors (with a little poetic licence). He is looking to using smells to help the visually impared, for example in a shopping centre, using a pine aroma outside toilets, and a floral aroma outside lifts. At Christmas 1995, unsuspecting passers by were lured into the Tate Gallery with the smell of brandy, and into Woolworths with mulled wine.

John Lambeth of Dragoco (Sydney) has a brilliant 'rainforest smell' for the army of a country to our north, to mask the odour of their jungle troops.

Carmine bugs
The cochineal insect (Coccus cacti) breeds on several species of cacti (Opuntia) in Mexico. The female is the source of carmine. The cactus was imported into Israel and cultivated in the 19th century in an attempt to produce the insect for dyeing. The enterprise failed, but the cactus flourished. In Australia, it became a major pest requiring destruction, first by arsenicals and finally by biological means.

Perfume colour
Why is Christian Dior Poison™ in a dark bottle? Perhaps to give it a poisonous ambience? Maybe. More likely, it is to hide the fact that the colour changes in time, a no-no for most consumer products, particularly cosmetics. With time, a so-called Schiff's base is formed from the reaction of an amine with a carbonyl, but this has no effect on the fragrance or the colour on the skin. Again there is no skin colour change with Elizabeth Taylor's Passion™, which is actually dark blue in the bottle.

Lipsticks

The requirements of a good lipstick are: uniform, intense colour with good coverage; shiny but not too greasy; retention of form and consistency in reasonable temperatures; usable in cold temperatures without crumbling or breaking; stable to light, moisture and air; non-toxic and non-irritant; and neutral in taste.

Your tube of oil-wax base with antioxidant, preservative, perfume and colour has to perform all the above functions! Some of the properties are easier to test than others. For example, it is easy to establish the 'droop point', the temperature at which lipstick lying flat in its case will droop against the case and ooze oil or flatten out. This should be over 45°C and preferably over 50°C. The colours and dyes in lipsticks are regulated in the USA (but not in Australia) and the materials used include some that can cause skin allergy, especially on exposure to light (see also Sunscreens in this chapter). Colours used include brilliant blue, erythrosine, amaranth, rhodamine, tartrazine and eosin (tetrabromofluorescein, D&C Red No 21).

Lipsticks in their modern form were introduced after the First World War. They were coloured with carmine, a dye and acid–base indicator, made from cochineal, a small red insect, by powdering the dried insect, and extracting it with ammonia.

Indelible lipsticks were introduced in the 1920s. The dyes in these lipsticks had little colour in the tube, but became coloured on reacting with the lips and stayed on for many hours. Tussy and Tangee natural lipsticks were introduced in 1925 and remained popular until the 1950s. In the 1960s the pale, lipless look was popular and there was no need for the long-lasting qualities. However, today we are back to the twenties, but with the additional twist of colour-changing lipsticks.

The body of a lipstick is a mixture of castor oil and wax, generally beeswax or carnauba wax (popular as a car wax because of its high melting point, 85°C). The aim is to have a mixture that is thixotropic; that is, it should remain stiff in the tube but flow easily when under the pressure of applying it to the lips. Esters such as 2- propyl myristate (14 carbon carboxylic acid) are added to reduce 'stickiness'.

The colours used must be insoluble in water, otherwise you would lick them off in no time. Thus the dyes used are generally oil soluble. Water-soluble dyes such as the food dye FD&C Blue No. 1 can be used if they are precipitated together with aluminium hydroxide ($Al(OH)_3$) to form an insoluble 'lake'. This precipitate is ground and can be suspended in castor oil, but does not actually dissolve in it. Such lake dyes are often brighter and more vigorous than the original water-soluble dyes from which they were made, and have less tendency to run.

The colour-change lipsticks go back to the indelible lipstick technology of the 1920s and use a dye such as eosin. This dye is lightly coloured, but becomes red when it reacts with the free amine ($-NH_2$) groups on the protein present in the skin. In the tube, this dye is masked by a lake dye (green or blue are popular) which is the colour you see when you apply the lipstick. This colour is overtaken by the eosin as it turns red and indeed is removed when the lipstick cream is wiped off, while the reacted dye remains. As the colour-change dyes do not work well at a basic pH, the lipsticks often contain citric or lactic acid.

Exercise
Use your mood-change lipstick to test the pH of apples and potato slices or chips; it also reacts with some paper plates. Why?

Toothpaste

Toothpaste has come a long way since Hippocrates first suggested cleaning teeth with powdered marble. As long ago as 1683 Anthony van Leeuwenhoek first demonstrated the presence of bacteria on teeth. The film of plaque that bacteria form is a highly organised structure containing 10^{12} aggregated bacteria per gram of film. The film contains polysaccharides, fats and protein as well, and allows the bacteria and the acids they produce continuous access to the tooth surface.

The human mouth contains a large number of bacteria. Within minutes of the tooth being cleaned, it is coated by a thin film (called pellicle) that is derived mainly from saliva. The pellicle is colonised by bacteria which produce a sticky, gel-like substance called plaque in which to live. The bacteria in the plaque ferment sugars to produce acids, which attack the tooth enamel and eventually form a hole. Plaque builds up on the gum and causes gum inflammation (gingivitis) and later attacks the deeper tissue and bone (periodontal disease).

In the 1950s epidemiology studies suggested that fluoride protects against dental decay. As a result, tin(II) fluoride was added to toothpastes in 1955 because tin ions are known to be effective against oral bacteria. Unfortunately they react with other components, forming insoluble products. The product was replaced by sodium monofluorophosphate (MFP). Today sodium fluoride and amine fluorides are also used, and fluoride is added to many town water supplies. In all cases it is the fluoride ion that is the active constituent.

The mineral of which teeth are formed is hydroxyapatite $Ca_{10}(PO_4)_6(OH)_2$. The effect of the fluoride ion is to replace the OH^- with F^-, to give $Ca_{10}(PO_4)_6F_2$.

STPP (sodium tripolyphosphate, see The detergents, Chapter 3) was used in toothpastes for other reasons as well. Most toothpastes are sold in either plastic dispensers or laminated plastic tubes. Before this, toothpastes were sold in metal (film) tubes. Lead was used for some time but was replaced with aluminium that was often unlaminated (to save cost). Alkaline components, such as calcium carbonate, could cause the production of hydrogen and blowing of the tube. STPP at a level of 0.2 to 1.5% would prevent this reaction.

Garlic odour
Toothpaste is meant to remove mouth odours, but 'bad breath' can originate in the mouth or the lungs. When you eat garlic, your social unease comes not from bits of garlic left in your mouth but from garlic entering the bloodstream and releasing volatile sulfur vapours through your lungs. Why does an intact garlic bulb have virtually no odour? Enzymes are released when the plant is damaged, and they attack stored precursor molecules and release the gases.

FLUORIDE IN WATER SUPPLIES

The author remembers how annoyed he was on arrival in Canberra at the end of 1964 to discover that the Federal government had decided by fiat to fluoridate Canberra's water supply without consultation with the populace. The columns of the Canberra Times were filled with letters complaining about all sorts of ills. Cats had lost their fur, bathtubs were turning green (true, but this was due to excessive chlorine levels and hot weather combining to leach copper from the plumbing), teeth were discolouring overnight, and so on.

A week or so after the torrent had subsided, there was a small announcement in the paper that because of technical problems in the distribution mechanism fluoridation had been delayed. It was generally accepted that bureaucrats lacked the sense of humour to have deliberately planned this ploy.

Back in 1931, the Aluminium Company of America (ALCOA) analysed water supply of the town in which their factory was situated. They were concerned that the fluoride that they used in their smelting operations in the town of Bauxite was entering the drinking water and causing mottled teeth. They were right. The levels were between 2 and 13 ppm and the correlation matched that found in areas of high natural fluoride in water, such as in Colorado. Only when a complete survey was done on fluoride levels and mottled teeth throughout the US was the further correlation with a reduction in tooth decay discovered.

The trick then, was to find an optimum compromise in levels. Low levels of continuous application of fluoride reduce dental caries. There are a number of chemical mechanisms that suggest why this should be so. There is also considerable selected epidemiological evidence used to support the efficacy of fluoridation. However a study in Canberra itself showed that the reduction in tooth decay continued well after the ten years beyond which the effect of fluoride would have been expected to level out. Such data as there were seemed to show that the more fluoride used, the more the decay was reduced until 1 ppm is reached. The amount of improvement above this level flattened out and was not worth the increased risk of other problems. Water fluoridation at this level was based on the assumption that this was the only significant source of the element.

Water dosed at 1 ppm means taking in 1 mg of fluoride for every litre of water drunk. Such a low level of fluoride in the water supply is harmless beyond reasonable doubt for all but a minority. Among this minority are people who drink really excessive amounts of water, diabetics, patients undergoing dialysis, and people with defective kidneys. Arguments about causing cancer have been discounted, although there are problems with bone diseases at much higher levels of fluoride.

Fluoridation is generally achieved by adding solid sodium silicofluoride so as to provide 1 mg of fluoride to one litre of water, with an error margin of 10%. Is the decision taken in 1964 still appropriate? What has changed in the last 33 years? Because our methods of analysis have improved enormously, we are better aware of the background fluoride level in our water supply (0.2 to 0.3 ppm) and other dietary levels of fluoride. Processing of food and beverages with fluoridated water makes a contribution to our daily intake commensurate with that coming from fluoridated water. Other sources of fluoride are spinach, gelatin, bone meal and fish protein. Tea provides approximately 0.12 mg per cup.

We now tend to be more affluent, middle class and better exposed to public health messages. The market has changed as well. For a start, it takes effort to buy a toothpaste which does **not** contain fluoride in one form or another. The levels used are of the order of 1000 ppm. Children under five are the highest swallowers of toothpaste, and absorb about 0.3 mg per brushing.

Non-fluoridated cities, such as Brisbane, have arguably a similar level of caries to corresponding fluoridated cities. It is very likely that fluoride from these other sources, (including soft drinks imported into Brisbane from fluoridated areas) is the reason, although this is still disputed. There is thus a very good case for keeping the water used for (most) soft drinks fluoridated at the current level.

What about taking the fluoride out? Remember we are talking about a level of a material in water measuring one part per million of water. We are also not talking about a controlled laboratory experiment with small amounts of sample, but kilolitres of water passing through unattended and unserviced home filtering units. For all practical purposes, once the fluoride is in, there is no economical way of removing it. Otherwise those towns that have excessive natural fluoride in their water supplies would have reduced it to 1 ppm.

Back to the bathroom and toothpaste. It has a solid phase (a polishing agent or blend of agents) suspended in aqueous glycerol, sorbitol or propylene glycol (all polyalcohols) by means of a suspending agent (e.g. sodium carboxymethylcellulose: see also The detergents — The organic builder in Chapter 3; and Plastic or latex paints in Chapter 12. A similar chemistry is used in the froth flotation method; see Froth flotation in Chapter 6.) The mixing is done in a vacuum to avoid air bubbles and to give a consistency to the paste. Air bubbles can also lead to deterioration in flavour. If the suspension is destroyed, the product becomes watery. A small percentage of anionic detergent (sodium lauryl sulfate) is added as a foaming agent.

An Australian standard for toothpaste stipulates a number of important requirements. It prohibits the use of sugar or other readily fermentable carbohydrate. The pH range allowed is quite wide (4.2 to 10.5). When heated to 45°C and kept there for 28 days, toothpaste must not form gas, separate, or ferment. The container must withstand this temperature also, an important requirement in some parts of the country. Toothpaste should not run out of the tube when the cap is left off and the tube is left lying on its side, nor should it sink into the bristles of a toothbrush as soon as it is applied. However, it should not be so firm as to roll off the brush under normal use.

The success of fluoridated toothpastes in preventing tooth decay has resulted in a shift of focus to other problems, for example, formation of tartar (mainly calcium phosphate, $Ca_3(PO_4)_2.2H_2O$) on the surface of the teeth. Surveys in the US show that 34% of children and 20–30% of adults

Insights from the developers of Colgate Total™

Bacteria attach to tooth surfaces and form a film called dental plaque. If teeth are cleaned twice a day, and if 99% of the bacteria are killed, plaque formation will still only be inhibited for six hours. Among the most widely used agents are cationic antibacterials (see How do surfactants work? in Chapter 3) such as chlorhexidine, which have good retention on oral surfaces. However, they disturb taste perception, and cause discoloration and tartar formation. Triclosan (2,4,4'-trichloro-2'-hydroxydiphenyl ether) was found to work without side effects.

Source: Toothbrush chemistry, A Gaffar, J Afflitto and N Nabi, ChemTech, p. 38, July 1993.

TABLE 5.4
Mohs' Hardness Scale (modified)

Hardness number	Material
1	Talc
2	Gypsum
3	Calcite
4	Fluorite
5	Apatite
6	Orthoclase
7	Vitreous silica
8	Quartz
9	Topaz
10	Garnet
11	Fused zirconia
12	Fused alumina
13	Silicon carbide
14	Boron carbide
15	Diamond

have tartar above the gum line, while 60–65% have deposits below the gum line. Mechanical removal is the only available process. Tartar is only formed in the mouth, although saturated levels of the calcium and phosphate salts are also present in other body secretions — the reason may be the presence of suitable inhibitors there.

Because of consumer concern about the abrasive action of toothpaste, an optical illusion is used to disguise the abrasive in translucent toothpastes. The abrasive and the surrounding medium have approximately the same refractive index and so the abrasive particles cannot be seen. (The refractive index is a measure of a substance's ability to bend light.) If the abrasive and the surrounding medium had different refractive indices, you would be able to see the solid particles. (See also 'Hiding power, and box, Disappearing tricks in Chapter 12.)

However the abrasive function of a toothpaste is most important. In ascending order of difficulty, the paste has to remove food residue, plaque, pellicle and calculus (or tartar). Calculus is plaque that has calcified or hardened. It can come to look and feel like an extra bit of tooth. Calcification can start within two to 14 days of plaque formation. The correct hardness of a toothpaste is critical. Hardness is measured according to a geological scale called Mohs' Scale (Table 5.4). Tooth enamel is quite hard (5.5–7), but the roots are soft (3.5–5) and can be worn away if gum disease has exposed them. You can check toothpastes to see if they will scratch against glass (which rates 5.5). Toothpaste should be below 5.5, and thus not scratch glass.

Baby-care products

The skin of a baby is a much less effective barrier than that of older children (from age 3) and of adults. Because it is thinner and softer, and contains more water, material can pass through it both ways more easily. Because of the concentration of moisture and soil in the nappy (diaper) area, bacteria breed quickly, and irritants remain in contact with the skin. This is the major cause of nappy rash. In 1921 it was proved conclusively that ammonia liberated from urinary urea by the action of an enzyme present in a bacterium *(Bacillus ammoniagenes)* inhabiting the colon was a cause of nappy rash:

Other bacterial organisms act similarly on urea. These bacteria are present in the faeces under normal conditions. When the intestinal contents are of a low acidity, the number of such organisms is greatly increased, and consequently the nappy region becomes infected with the bacteria. They grow most rapidly under neutral to alkaline conditions (pH 7–9) whereas there is practically no growth at pH 6.

Nappy rash can be a frightening condition to a young mother. It happens only rarely during the baby's first months but later it can be very severe. Fortunately, in most cases, nappy rash responds readily to treatment, but it can persist or occur again.

$$CO(NH_2)_2 + 2H_2O \rightarrow CO_3(NH_4)_2$$
urea

$$CO_3(NH_4)_2 \rightarrow 2NH_3 + H_2O + CO_2$$
ammonia

Fig. 5.4
Nappy rash reactions

Manufacturers make nappy conditioners, nappy sterilisers and nappy washes to assist in combating nappy rash. Some claim to sterilise, some to sanitise, and some to soften. The compositions of two typical nappy sanitisers are given in Table 5.5. These are approximate ingredients only, true at the time of analysis.

For more on disposable nappies (diapers) see Non-woven fabrics in Chapter 11.

Baby oils are generally based on a mineral oil. They may also contain vegetable oil, lanolin, antioxidants and germicides. Products such as pHisoDerm are emulsions and lather like soap. Their pH is slightly acidic so that it matches that of the skin.

The US Food and Drug Administration (FDA) required that all labels affixed to cosmetic and toiletry products after 15 April 1977 list the ingredients in descending order of predominance. An ingredient list allows shoppers to make comparisons, and helps consumers avoid ingredients to which they are allergic or sensitive.

TABLE 5.5
Compositions of some nappy sanitisers

Product 1	Product 2	Product 3
Sodium tripolyphosphate (see Fig. 3.9)	Sodium perborate	Typical active ingredient
Sodium chloride (common salt)	Sodium tripolyphosphate	25% sodium percarbonate
Potassium persulfate	Sodium bicarbonate	
Surfactant	Sodium chloride	
Fluorescers	Surfactant	
	Optical brightener, perfume and paraffin	
This product operates on the principle that potassium persulfate and salt are stable when dry, but when dissolved, react to form chlorine bleach.	This is similar to a built laundry detergent (surfactant plus additive) but with extra perborate bleach and less surfactant.	

Sunlight on skin

The sun irradiates the surface of the earth in the wavelength range from 290 nanometres (1 nm = 10^{-9} m) in the ultraviolet through the visible into the infra-red or heat end of the spectrum. The amount of ultraviolet radiation reaching us is limited by its absorption by ozone in the upper atmosphere.

Ultraviolet light damages the hereditary material in the cell (DNA) by causing two pyrimidine bases to link as a dimer. In ultraviolet-sensitive

Suntan and social status

A historical view of sunbathing among Caucasians is revealing. For most of recorded history, white skin implied a lofty position in society. While workers, serfs and slaves spent most of their time in the sun, aristocrats sought shade by carrying parasols, wearing hats and sun bonnets, and staying indoors. However, the industrial revolution did away with the pursuit of pallor. Workers, herded in factories, spent long hours indoors. Shade became cheap, while sunlight was expensive. A suntan showed that its wearers had the wealth and leisure time to travel to places where they could get a lot of sun.

bacteria, this linkage inhibits DNA replication and thereby stops further growth. Normal human skin cells have an enzyme system which can repair ultraviolet damage by excising the dimers and closing the gap. A rare genetic disease where this enzyme is lacking makes its sufferers very liable to skin cancer.

The natural skin colour of humans evolved to match the intensity of sunlight according to the region of the earth in which they evolved. However, with mass movement of peoples, this neat balance has been upset: Caucasians in hot climates suffer from sunburn and increased skin cancer, while black-skinned people in cold climates have problems caused by insufficient vitamin D synthesis in the skin. Bedouin women, who wear veils and the traditional black robes, can also suffer this deficiency. Tanning is nature's control of the level of sunlight activity on the skin.

The ultraviolet range of the spectrum is classified into three regions, the approximate range of wavelengths (nm) being as follows: UV 'A', 320–360; UV 'B', 280–320; far UV 'C', 200–280; actinic UV, 200–320 (see Fig. 5.5).

The intensity of sunlight at various wavelengths is shown in Figure 5.6. (The wavelength sensitivity of the human eye is shown for comparison).

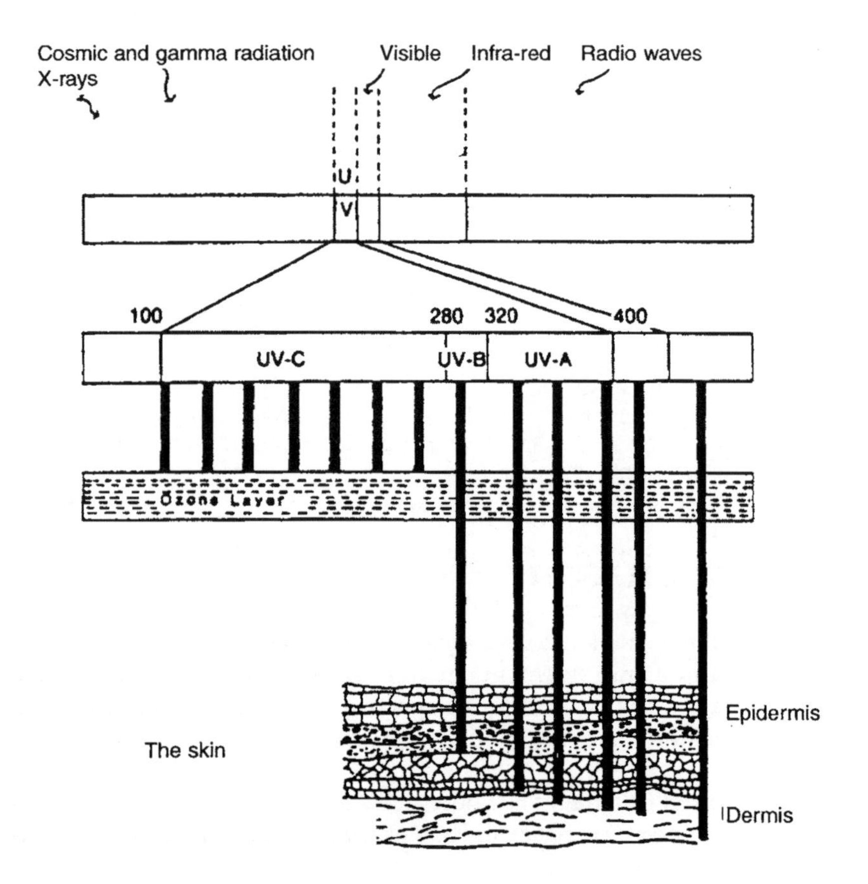

Fig. 5.5
The solar spectrum

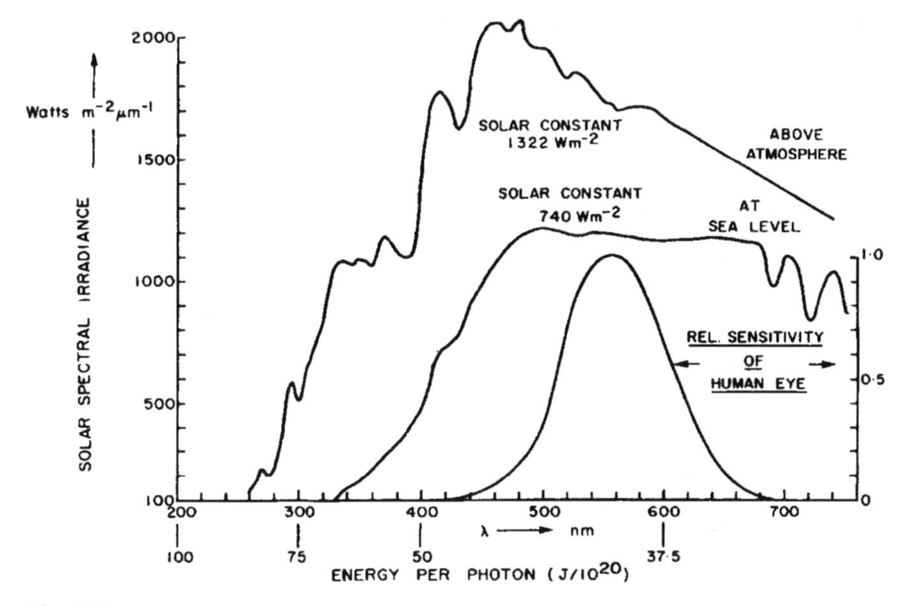

Fig. 5.6
Sunlight above the atmosphere and at sea level. Also shown is the colour sensitivity of the human eye. Note the solar irradiance P expressed in units of wavelength (λ) μm (micrometres) is watts $m^{-2}\mu m^{-1}$ and is related to that expressed in units of frequency (ν) watts $m^{-2}Hz^{-1}$ through the relationship $P_{\nu}(\nu) = \frac{d\lambda}{d\nu} P_{\lambda}(\lambda) = \frac{c}{\nu^2} P_{\lambda}(\lambda)$ where c is the speed of light (2.998×10^{14} $\mu m s^{-1}$). The conversion of either of these to the spectral energy density is given by $p(\nu)$ [joule $m^{-3}Hz^{-1}$] $= \frac{1}{c} P_{\nu}(\nu)$ and $p(\lambda)$ [joule m^{-3} μm^{-1}] $= \frac{1}{c} P_{\lambda}(\lambda)$ where c is now $2.998 \times 10^{8} ms^{-1}$.

A great deal of the UV 'B' is removed by the ozone in the earth's atmosphere (see Ultraviolet radiation and the ozone layer in Chapter 14).

The peak sensitivity of the skin occurs at about 297 nm. For wavelengths greater than 320 nm the sensitivity drops to less than one thousandth of the value at the peak at 297 nm. When the dependence of the solar radiation on wavelength is combined with the dependence of skin sensitivity on wavelength, the product curve shows a maximum at about 305 nm (Fig. 5.7, overleaf). For comments on the energy of sunlight see the Chapter 14 box on Biological effects of radiation.

Although much is argued about 'broad spectrum' effects, the increase in the intensity of solar radiation in the UV 'A' compared to the UV 'B' is much less than the factor of one thousand drop in skin sensitivity.

The tanning of our skin involves a polymer called melanin. We are all born with different amounts of it. Fair-skinned people have a little, olive-complexioned people have more, and black-skinned people have a lot.

Melanin reacts to the sun in two stages. In the first stage, pale (unoxidised) melanin granules near our skin's surface are changed by ultravio-

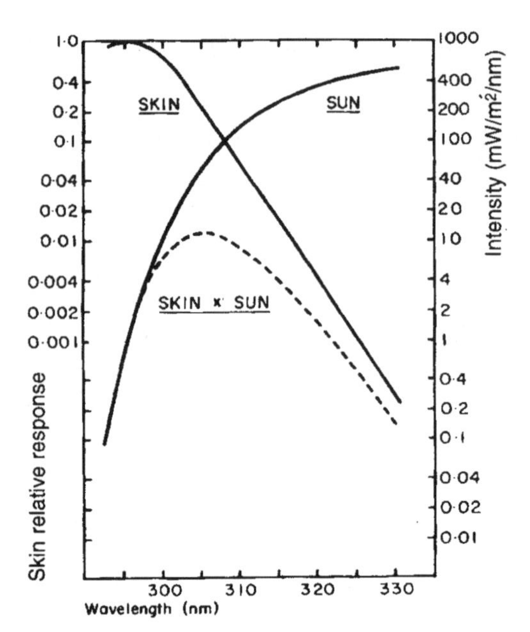

Fig. 5.7
The skin sensitivity and sun spectrum combination showing at each wavelength the equivalent intensity at 295 nm for the same sunburn result (After D.F Robertson, University of Queensland)

let light to their dark-brown (oxidised) form (Fig. 5.8, below). This gives an immediate tan, usually within an hour. It fades within a day. A more lasting tan results from the second stage. In this process, new quantities of melanin are produced from tyrosine, an abundant amino acid in our skin's protein. This second-stage tan endures for several days without further exposure. Additional sunbathing not only produces more melanin but also lengthens the polymer chains and deepens the colour.

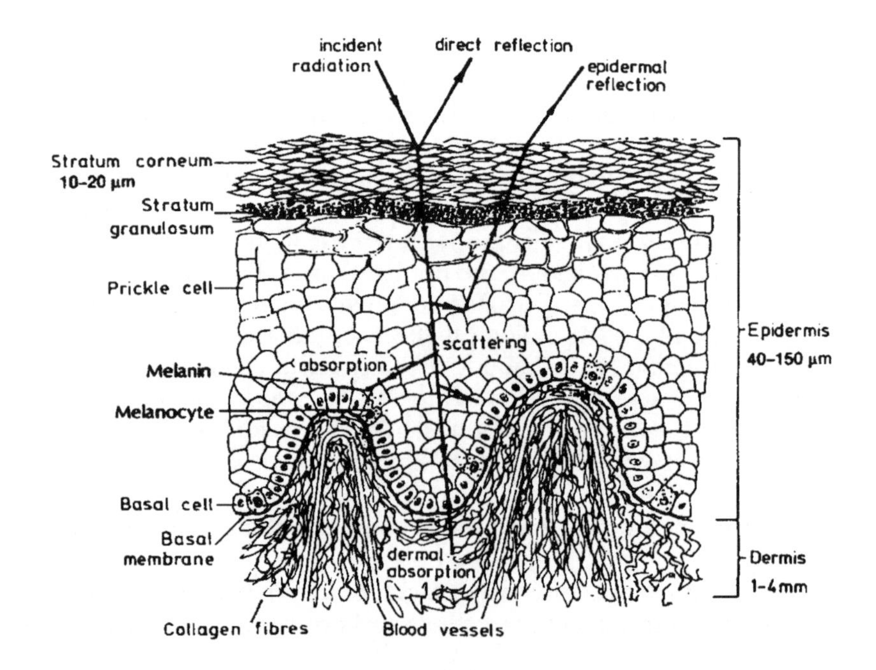

Fig. 5.8
Possible pathways for incident radiation through the human skin: release of melanin from melanocytes

(Courtesy Dr Keith Lokan, Director, Australian Radiation Laboratory, Melbourne)

Some formulations induce a suntan-like darker skin without sunlight. The tan is artificial because no melanin is involved in the process. This browning offers no protection against sunburn. A brown complexion is formed with the skin protein by the active ingredient (usually dihydroxyacetone). This material is produced commercially from glycerine using the bacterium that converts alcohol into vinegar.

Leather skin
The final effect of ultraviolet radiation is the damage to the proteins that make up the skin's connective and elastic tissue. This leads to an irreversible wrinkled, leathery and sagging skin. The word 'tanning' for the effect of sun on skin is well chosen (see Leather in Chapter 11).

Of the three types of skin cancer, the least common but most dangerous is melanoma. Death from melanoma has increased since the 1920s, and its victims are often professional or managerial workers, not workers who spend their days in the sun. In Queensland, one new case of melanoma occurs each year for every 6000 inhabitants. In cloudier Britain, the incidence is one new case per year per 37 000. The Queensland rate is 50% higher than in the US where nearly 20 000 will be diagnosed with the disease, and more than half of these will die from it (mainly men).

While melanoma is increased by exposure to sunlight, it does not necessarily occur on exposed parts of the body. Women in Australia develop melanomas mainly on the legs, while for men they occur anywhere. The rare instances of melanoma in dark-skinned people tend to be on the palms of the hand or soles of the feet, another clear example of the occurrence of melanomas in parts of the body not exposed to the source of radiation.

The 290–320 wavelength range is the active one that causes sunburn, triggers skin cancer, and produces vitamin D in the skin (which takes place through a photochemical reaction).

SKIN CANCER PROTECTION

Recently a businessman was sighted in Stockholm after a three month absence of sun, in shoes, socks and bathers, walking to work with a briefcase. Most Scandinavians can relate to that. Down-under, we have gone around fearing the Chernobyl in the sky. The Anti–Cancer Council of Victoria quietly concedes that the hole in the ozone layer will have no practical impact on skin cancer rates. The American National Academy of Sciences has found that the chance of getting skin cancer increases by 1% every 10 km a person moves towards the equator. It also estimates that a 1% decrease in ozone (in the ozone layer) increases the chance of skin cancer by 2%. So even if, as some doomsday scientists have predicted, the ozone hole increases UV rays by 10%, that would be equivalent to swimming at Bondi Beach rather than at Wollongong, 100 km to the south. While sunscreens undoubtedly save our skin, some active ingredients are absorbed straight into the bloodstream and could increase the risk of liver cancer.

Photos from the war years shows crowds of pedestrians on their way to work, all wearing hats. In the 1950s the fashion changed and hats disappeared. Now they are slowly coming back again, and that is good, because the increase in skin cancer twenty to forty years later is probably due to that change in fashion, rather than any ozone hole. A constant tan and the avoidance of binge burning is probably a good thing.

Source: Prof Bruce Armstrong — a leading Australian skin cancer researcher.

Sunlight transmission

It is interesting that glass does not transmit much light with wavelength below 350 nm. If you sunbake behind a window, the main effect is reddening of the skin by heating. The same is not true of perspex, which does transmit light of shorter wavelengths. For sunburn, it is also important to take reflected light into account. Typical reflection values for 300 nm radiation from specific surfaces are: snow, 0.85; dry sand, 0.17; water, 0.05; grass, 0.025 (the higher the index, the greater the reflection).

Sunscreens

From the above argument, it follows that, except for the immediate short-lived tanning response, it is not possible to screen selectively for true tanning while protecting against sunburn and skin cancer. Pigmentation requires preceding sunburn (which can be kept at a low level) to trigger the process. Sunscreens are used to lower the dose of light received by the skin to the point where tanning may have a chance to catch up. The amount by which the dose must be lowered will depend on the intensity of the light and susceptibility of the skin.

Sunscreens that are formulated to give a quick tan, but to suppress sunburn (and true tanning), will absorb light more strongly in the 290–320 nm range than in the longer wavelength range. This can be checked by measuring an absorption spectrum. For the suppression of sunburn it is only the absorption in the 290–320 range that is of interest. PABA(p-aminobenzoates) show an optimum ratio of quick tanning transmission to sunburn suppression, but problems with PABA allergy means that it is being replaced mainly by chemical actives such as octyl methoxycinnamate (OM). Other chemicals such as octyl salicylate, oxybenzone (2-hydroxy-4-methoxybenzophenone), homosalate (salicylic acid 3,3,5-trimethylcyclohexyl ester) are also used.

Inorganic materials such as ultra-fine titanium dioxide and zinc oxide powders provide non-allergenic alternatives. Sun damages hair as well as skin; the tryptophan degrades, disulfide bonds are broken, and the hair surface gets rougher.

An Australian standard for sunscreens AS 2604 was issued in 1983 (and modified in 1986, 1997) after an extensive six-year survey of the literature and critical discussion. At the beginning of this exercise the complexities of devising a reliable, yet not too expensive test for a sunscreen were hard to imagine. What is most important is that an adequate amount of sunscreen be used. The quality of the spreading agent is vital. If the emulsion or solution does not spread to form a continuous, coherent, stable film on the skin, but breaks up into globules, you could come out of the sun looking like a micro-dotted Dalmatian!

Because of the importance of adequate application of sunscreens, their sale in small packs and at high prices, encouraging sparing use, is counterproductive.

People's skins vary a little in the range of ultraviolet to which they are sensitive and so a quoted 'protection factor' can vary for different people for some sunscreens but not for others. The most effective sunscreens are zinc and titanium oxide creams. Although they may not look very attractive, at least you can tell they are still there, which is not the case with most screens. Some sunscreens are supposed to be swim-resistant, but they will wash off eventually, and it should be noted that fresh water dissolves sunscreens more effectively than salt water.

Many parameters are important in a sunscreen. The ability to absorb or reflect radiation is perhaps the easiest parameter to measure and it used to be quoted for commercial products. A sunscreen must be chemically and photochemically stable, otherwise its absorption ability changes with time. It must be soluble in the cosmetic base but insoluble in water or perspiration.

Alexander's rag-time band

The story is told that Alexander the Great made use of the fact that some colours are photochemically unstable and bleach rapidly in sunlight. Because his commanders didn't have watches to synchronise their attacks, he gave them bleachable coloured rags to put around their arms so that they could measure time during the day. Thus came into being Alexander's rag-time band!

Evaluation of screening

After about 20 minutes' exposure to the midday sun, an average untanned white skin will be affected by sunburn, although the actual reddening will not appear until after about six hours. This reddening will still be visible 24 hours later. The exposure needed to give this effect is called the minimum erythemal dose (MED) and it depends on the intensity of the radiation and the time of exposure. By comparing the time necessary to produce this MED on unprotected skin to that needed to produce it on skin protected with a standard amount of sunscreen, it is possible to give a protection factor (PF) for the sunscreen (independent of the absolute intensity of the radiation):

$$PF = \frac{\text{exposure duration for minimum erythema in } \textit{protected} \text{ skin}}{\text{exposure duration for minimum erythema in } \textit{unprotected} \text{ skin)}}$$

A protection factor of 10 means that if a sunscreen is used, a person can stay out in the sun about 10 times longer than without a sunscreen, and achieve the same effect. The protection factor should be proportional to the quantity of UV light transmitted through the layer of sunscreen on the skin. That is, if the sunscreen has a transmittance (T) of 50%, it should provide a PF of 2. Conversely, a PF of 10 should correspond to a transmittance of 10%. When such a screen is tested *in vitro* (on an absorption instrument) the actual transmittance is closer to 10^{-8}%! The most probable reason for this enormous difference is that the roughness of the skin can reduce the film thickness from the value calculated on the basis of the amount of material used divided by the surface area (normally about 20 mm) to about one-tenth of that (average of 2 mm). With an assumption of this much thinner film, the PF and transmittance are in better agreement. It is problems such as this that have led to the rejection of laboratory measurements on transmittance of sunscreens to actual testing being carried out on human backs *(in vivo)* to determine the protection factor (Fig. 5.9, overleaf). For a variety of technical reasons, the preferred sources of radiation are artificial (xenon and mercury UV sources) rather than sunlight. The largest source of error is the accuracy and uniformity of the application of the sunscreen.

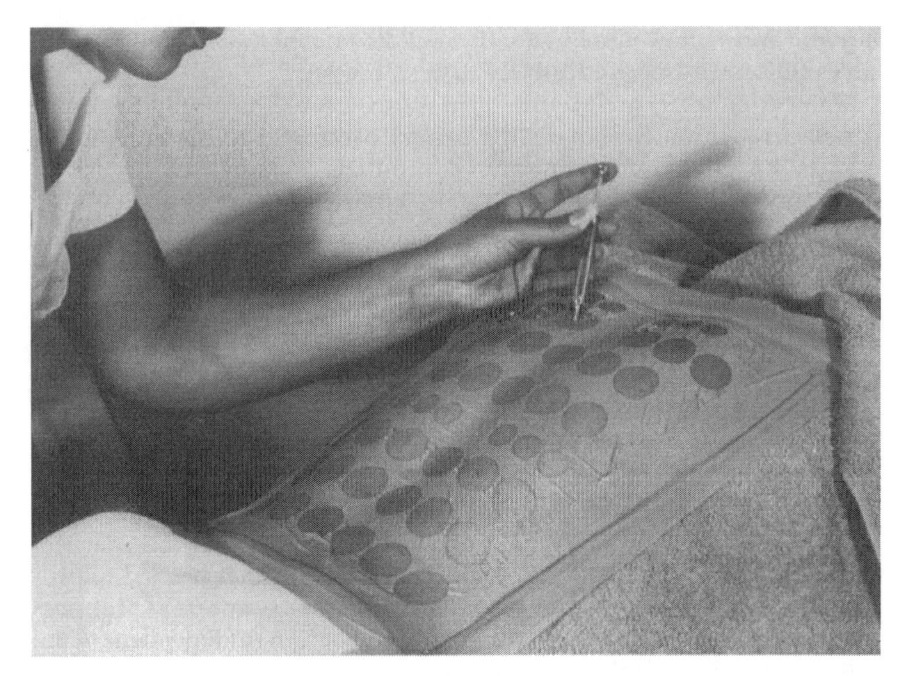

Fig. 5.9
In vivo testing of sunscreens

(Courtesy of Greiter, A.G.)

LIGHT TRANSMISSION

The amount of light a material will let through (at a given wavelength) depends on the thickness of the material, d(cm), the concentration of the material, c(g/L) and a property of the material called the absorption coefficient, a. The relationship between the light transmitted I, and the light falling on the surface I_o, is given by Beer's Law:

$$\log I/I_o = -a.c.d$$

I/I_o is called the transmittance. It can be expressed as a percentage, $100\ I/I_o$ called the per cent transmittance:

$$\%T = 100\ I/I_o$$

Thus a 1% transmission requires I/I_o to be 0.01 and then $\log I/I_o = -2$ and a.c.d. = +2. For a thickness of 0.01 mm, a.c. = 2000. Thus knowing the characteristic value of a for a material allows c, the necessary concentration, to be determined. Because the relationship is logarithmic, a 0.1% transmission requires 1.5 times (not 10 times) the concentration required by a 1% transmission. Conversely, halving the concentration increases the transmission from 1.0% to 10%. What we have said about concentration is equally applicable to path length (thickness) d. These two parameters occupy equivalent positions in the equation. The percentage of light transmitted increases and decreases exponentially with both variables.

Other examples where this relationship applies are the thickness (normally given as weight) of window awnings used to provide shading from sunlight and the thickness of thermal insulation for 'shading' of heat radiation.

UV fluorescence
The predominant emission of the ultraviolet (mercury fluorescent) lamps used in discos occurs at 365 nm. This wavelength causes the cornea of the eye to fluoresce, and that is why these lamps appear fuzzy when you look at them. No damage is allegedly caused to the eyes at this wavelength. The same wavelength stimulates blue fluorescence from the whiteners used to launder shirts (see Fluorescers in Chapter 3).

There has been considerable argument as to whether a PF greater than 15 is any more than just a marketing gimmick. Few cosmetic preparations are used as extensively as sunscreens, or cover such a large area of skin. If one estimates the total skin area of an adult (which is about the same area as the area to be covered) to be about a square metre, and assumes that a sunscreen lotion with an active concentration of 1.5% is used, then about 0.3 g of lotion will be deposited on the skin. This increases to 2.2 g for a lotion with 11% active concentration.

When sunscreens containing benzoate or salicylates (≈5%) are applied to the skin, these chemicals can be detected within 30 minutes in the urine of the user. Those people who need to avoid salicylates in food should also avoid their absorption from sunscreens.

Sunglasses

In the *Canberra Consumer* (No. 63, September 1978) there was a report giving the results of optical transmission measurements on a wide range of sunglasses supplied by a local pharmacist. The most surprising result was that a number of sunglasses had widely differing optical properties in the right- and left-hand lenses in the ultraviolet region. Some sunglasses showed evidence that the colouring had faded with time or exposure while on display, and one pair showed a greater reduction in the visible light transmission than in the ultraviolet region.

The problems with sunglasses arose when the 'sun glass' was replaced by 'sun plastics'. Glass absorbs ultraviolet light (as we saw in the discussion of sunscreens) and some infra-red light. Plastics on the other hand have variable optical properties in these regions (but see Infra-red radiation and the greenhouse effect in Chapter 14).

The pupil of the eye opens up in response to a reduction of visible light intensity caused by absorption of light by a plastic sunglass lens. However, if the plastic cuts back disproportionately less ultraviolet or infra-red radiation than it does visible light, the overall exposure of the eye lens (and retina) to these radiations will be increased. Excessive ultraviolet and infra-red radiation can cause eye damage. However, the eye protects itself quite adequately in normal sunlight if no sunglasses are worn, and for comfort in the sun, hats are far better anyway, and give protection for the face as well.

Sunglasses of any type are dangerous for night driving.

The transmission of ultraviolet light at different wavelengths in the eye is shown in Figure 5.10 (overleaf).

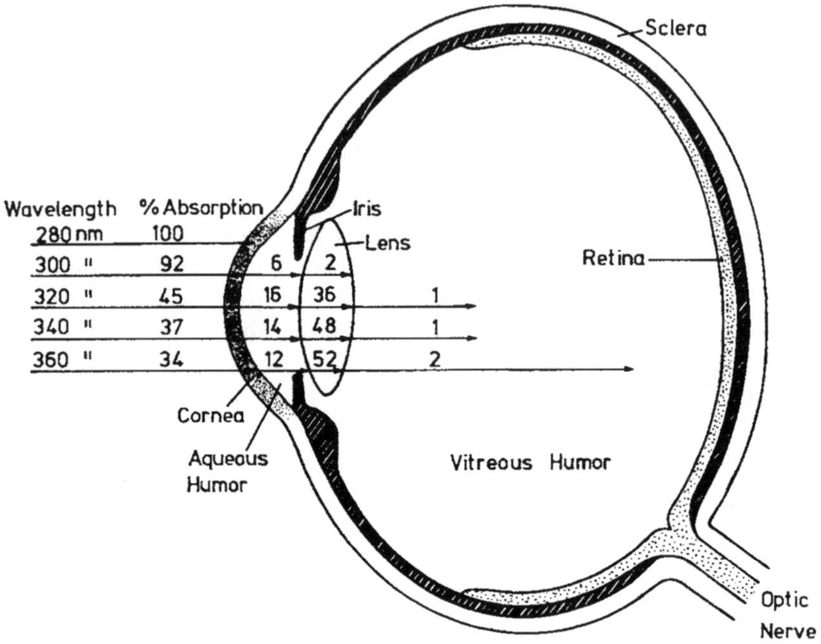

Fig. 5.10
Ultraviolet radiation absorption in the eye. The values given are the percentages of the incident radiation absorbed in each layer of the eye.

(Courtesy of Dr Keith Lokan, Director, Australian Radiation Laboratory, Melbourne)

References

Only skin deep, E. Geake, *New Scientist*, 25 Dec 1993, pp. 52–54,

A hair piece for Christmas, J. Cherfas, *New Scientist* **19** (26) December 1985.

Antiperspirants and deodorants, Volume 7 in a series Cosmetic Science and Technology, edited by K. Laden and C.B. Felger, Marcel Dekker, New York, 1988.

Web of scent, Eric Albone, *Chemistry Review*, p. 19, Jan 1996.

Nose by any other name, V. Hook, *Chem Brit*, p. 513, July 1995.

Perfume, M. Linner-Luebe, *ChemMatters*, pp. 8–11, Feb 1992, and references therein.

Brushing with confidence, A. Gaffar, S. Nathoo, *Chem Brit*, p. 51, May 1996.

Producers raise high-tech stakes in personal care products markets, E.M. Kirschner, C&EN, pp. 13–19, July 1 1996.

Bibliography

Physical foundations of perfumery, L. Appell, *American Perfumer and Cosmetics*, reprints of a series of ten articles, Dec. 1970. Very classical physical chemistry put to work. Not for the uninitiated.

Perfumery, practices and principles, R.R. Calkin and J. Stephan Jellinek, J. Wiley Interscience, NY, 1994. Industrial reference.

CHAPTER 6

Chemistry of surfaces

Surface chemistry impinges on so many aspects of consumer science that it acts as a link to many other chapters in this book. This chapter goes beneath the surface and explores some of the physical chemistry needed for much else in the book.

Introduction

Surface chemistry is the basis of detergency as discussed in Chapter 3. Our own surfaces mean a lot to us, as the discussion in Chapter 5 suggests. If you prefer eating cream or ice-cream to eating solid butter or margarine, and prefer both to solid lard, you owe it all to surface chemistry, see Chapter 4.

Soil behaviour is determined by the surface properties of particles, as are the methods for clarifying your swimming pool from suspended soils, Chapter 8. Many toy materials, like Slime, depend on particle surface properties for their fun characteristics. Adhesives are all about surface energy, and paint formulation, spreading and setting into a film is subtle surface interaction (Chapter 12). Extracting oil from depleted oil-fields uses microemulsions, and fine flour can be used as diesel fuel, Chapter 14. Much of food technology of taste and feel revolves around

The surface
'Superficial' means pertaining to the surface. It has a derogative connotation because it implies the ignoring of important aspects below the surface. However surfaces are very important in themselves and often determine many aspects of chemical and physical processes.

getting it right, surface-wise (Chapter 15) not to speak of winning metals from their ores by froth flotation (this chapter).

This, then is another linking chapter.

Consider a cube with 10 cm sides. It has a surface area of 0.6 m^2 (600 cm^2). If you cut this cube up into cubes with 1 cm sides (1000 cubes) the surface area increases to 6 m^2. Further subdivision to cubes with 1 mm sides increases the surface area to 6000 m^2. Such a large surface area can completely change the chemical reactivity of a material. The surface area of the porous particles of activated charcoal used in cleaning up chemicals and upset stomachs is quite enormous, maximally 800 m^2/g.

A lump of coke is quite stable in air, but if you suck up the fine carbon powder used in some photocopiers into a vacuum cleaner, you can cause quite an explosion. Flour in flour mills can ignite spontaneously.

POWDERS TO BURN

Dragon's Breath, available from magicians' supply houses, is a powder which when sprayed into a fire burns with a large billow of flame. It is composed of lycopodium, a fine yellowish powder of club-moss spores (also available from laboratory supply houses). Lycopodium in bulk does not burn.

Pyrophoric iron can be made by heating $FeCO_3$ with a small amount of glucose. This drives off carbon dioxide, leaving FeO which is very unstable. If the powder is poured through the air it ignites spontaneously. Use very small amounts only.

A nice (but dangerous) experiment demonstrates the rate of reaction as a function of particle size which involves the reaction of potassium permanganate (Condy's crystals) with oxidisable organic material such as glycerol or glycol (brake fluids, etc.). The delay before the reaction starts is a function of particle size. This reaction is the basis of the aerially dropped capsules used in controlled burning in forests and it is also the so-called 'insurance' reaction, where people burn down unprofitable businesses using the delayed reaction in an attempt to establish an alibi. (See also Chapter 8.)

Small is better

If you look at a glass of soda or beer, you see bubbles of gas rising in the liquid because the liquid can flow. At the top of the beer there is foam in which gas is dispersed as very small bubbles in a liquid. The gas does not move up and, in fact, you can't really distinguish the liquid and gas states. You can cut the foam with a knife and examine a slice in the colloidal state. Incidentally, the stabilising agent in *real* beer is the protein derived from grain. However, when large amounts of adjunct are used (such as sugar, starch, potatoes) a protein extract of seaweed is needed to maintain a reasonable head. The protein 'protects' the air particles by denaturing at the interface. Next time you are at the beach you can lament all that seaweed going to waste that could have formed foam on your beer.

Talking of seaweed, there is a theory that the infamous Bermuda triangle, where ships sank without warning, might have been subject to a high level of bubbles released from rotting vegetation below. These would lower the density of the seawater, and perhaps reduce the buoyancy of a boat sufficiently to allow it to sink (so it's a nice theory!).

A foam is used in greenhouses to prevent heat loss during the night. Each evening, a water foam is injected into the thin space between the double walls of some greenhouses. The foam cuts down the convection currents that normally move the air around and cause it to cool (but greenhouses don't use the greenhouse effect, see Chapter 14). At sunrise, the foam collapses. The next night the drained liquid is pumped up into a foam again.

If you are selecting a dry battery (Leclanché cell) in the supermarket, you may well be bewildered by some of the choices. The difference may be only in the nature and area of the surface of one of the components, manganese dioxide, and zinc. The larger the surface area (i.e. the finer the powder) the more extensive the reaction, and hence the greater the current that can be drawn.

Let's look at some other examples. Why is cream more palatable than butter, which in turn is more palatable than lard? Lard is solid fat, whereas the butter is an emulsion of *water* in *fat,* which provides the fat with a large surface area with which to react with our digestive juices. Cream is an emulsion of *fat* in *water,* which provides an even greater surface area for the fat. Making butter from cream involves inverting a fat-in-water emulsion to a water-in-fat emulsion by denaturing the protein that 'protects' the fat globule and prevents it from coalescing. Our digestive detergents — the bile acids — form emulsions with fat and so increase the rate of reaction.

Emulsions

The basis of many products, ranging from cosmetics to chicken soup, is an *emulsion.* An emulsion suspends two materials one within the other that do not normally dissolve in each other. Because they do not mix, the components are called phases. For oil and water to form an emulsion an *emulsifier* must be added (Fig. 6.1, overleaf).

The purpose of the emulsifier is to reduce the difference in surface tension (i.e. the mutual repulsion) between the two phases. For example, in making mayonnaise from oil and vinegar, egg yolk acts as an emulsifier. Emulsifiers reduce the interfacial tension between the two immiscible phases; they allow the oil and water to 'wet' each other.

$$\begin{array}{l} CH_2OH \\ | \\ CHOH \\ | \\ CH_2OCO(CH_2)_{16}CH_3 \end{array}$$

glycerol monostearate

a 'fat' with two of the fatty acids missing; a *mono*glyceride.

$$R-C_6H_4-(OCH_2CH_2)_nOH$$

polyoxyethylene alkyl phenol

a non-ionic detergent
e.g. triton X-100, teric X10

Fig 6.1
Typical emulsifiers used in cosmetics

HLB, the hydrophilic-lipophilic balance

Selection of the correct emulsifier is critical. Emulsifiers are characterised on a scale called HLB, the hydrophilic-lipophilic balance (hydrophilic = water loving; lipophilic = oil loving). In this system the relative affinity of an emulsifier for the oil phase is expressed as a number ranging from 1 to 20. Propylene glycol monostearate has a low HLB number — it is more at home in the oil phase. Polyoxyethylene monostearate $CH_3(CH_2)_{16}COO(CH_2CH_2O)_n H$, which has a long polyoxyethylene chain with lots of polar oxygen atoms, has a high HLB value and is quite at home in water. In general, emulsifiers with HLB values of 3 to 6 will produce emulsions of water dispersed in oil, whereas those with HLB values of 7 to 17 give emulsions of oil in water. Cosmetic emulsions can be either dispersions of oil in water — o/w — or water in oil — w/o (see Fig. 6.2). (See also Chapter 4 box, The Quintessential Mashed Potato)

But it is not quite as simple as that. A scientific investigation spanning time, space and discipline raged for many years, engaging the vital question of what contributed to the stability of the oil in water, *sauce Béarnaise*. A major contribution was made by Jearl Walker in his 'Amateur Scientist' column of *Scientific American*. It is described also by Harold McGee in *On Food and Cooking*. Kirk Othmer's *Encyclopedia of Chemical Technology* is an authoritative source. 'The most stable emulsion systems usually consist of two or more emulsifiers, one portion having lipophilic tendencies (fat attraction), the other hydrophilic (water attraction) . . . only in relatively rare instances is a single emulsifier suitable.' Now egg yolk contains cholesterol and lecithin. Cholesterol stabilises w/o emulsions; lecithin

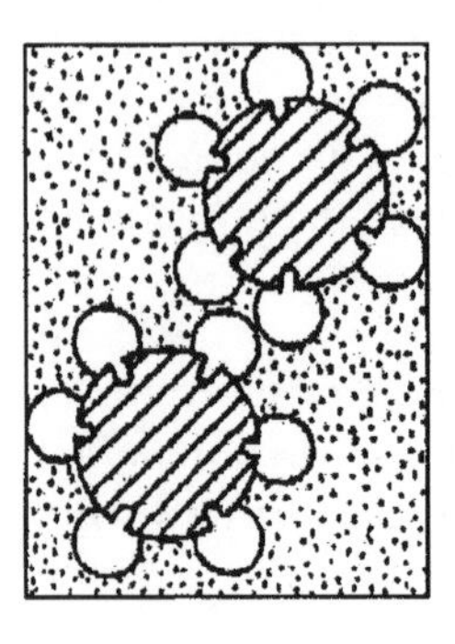

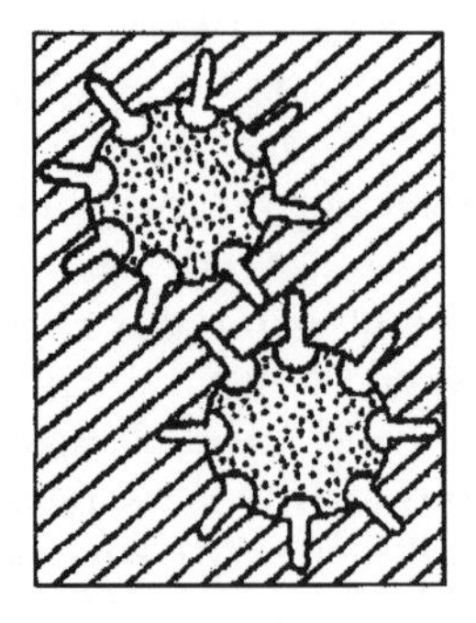

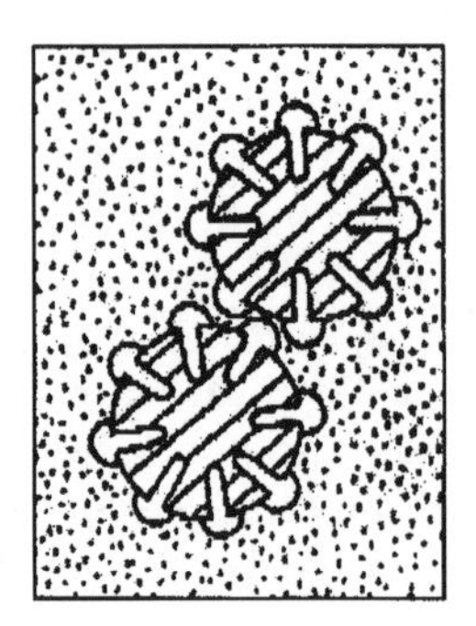

Fig. 6.2
(a) Water in oil requires a low HLB emulsifier
(b) Oil in water requires a high HLB emulsifier
(c) A wrong emulsifier can mean collapse of the emulsion

Source: On food and cooking, H. McGee, p. 357, Unwin Hyman Ltd., 1987

stabilises o/w emulsions; both are needed, but the balance must be perfect. Fresh eggs have such a balance, but as the egg ages, the lecithin slowly deteriorates, but the cholesterol remains unchanged — the sauce collapses. Beating in extra lecithin (but be careful of the source) can do a resurrection job. Back to the kitchen!

The effect of the two different types of emulsions is quite different. As mentioned above, cream is an emulsion of fat in water, while butter is an emulsion of water in fat. Beating of cream denatures the protein protecting the oil droplets, and the emulsion 'inverts'. In cosmetic creams, water evaporates from an o/w emulsion, causes cooling and leaves a film of the oily ingredients (e.g. oils, waxes, emulsifiers, humectants). On the other hand, a w/o emulsion permits direct immediate contact of the oil phase with the skin. No cooling effects occur because evaporation of the emulsified water is much more gradual. These are 'warm' emulsions, in the sense of their apparent effect on skin temperature. Geometry requires that the minimum concentration of the continuous phase be at least 26% of the total.

In an emulsion, the finer the particle size is, the more stable the emulsion, and the higher the *viscosity* (resistance to flow). (The lower the viscosity of the emulsion, the more runny it is.) Large particle size increases the tendency for the particles to coalesce and hence, finally, for the emulsion to separate into the two phases.

The easiest way to test which type of emulsion is present is to measure electrical resistance. Most oils have a much higher electrical resistance than aqueous solutions. The w/o type is relatively non-conducting compared to the o/w type. A simple resistance meter is often all that is required for this test, or a light-emitting diode and two electrodes can be used with a battery. Another test is to use an oil-soluble dye (e.g. fuchsin) which will spread and colour the surface only if the oil is in the continuous phase (i.e., w/o). Conversely, water-soluble dyes (e.g. food colours or methyl orange) will colour only o/w emulsions.

Rubber latex

A most important emulsion in the Third World is that produced by the rubber tree — an emulsion of rubber latex in water. It is collected from the trees in the early morning when the flow is greatest. The flow is stimulated by a synthetic plant hormone, chlorethylpropionic acid. The latex is coagulated (curdled) with acetic acid (vinegar) and the excess water removed in several stages to produce mats, which are sun dried. These are bought by traders, who take them to a factory where the rubber is macerated and pushed into blocks. The whole process is highly labour intensive and uses a potentially renewable resource.

Foams

Australia's most famous foam is probably not from a 'tinny' but the pavlova, named after the ballerina Anna Pavlova. It is vital to add a *pinch* of salt and a *squirt* of vinegar. However, too much of either spells disaster in the form of a collapsing crust and a panic journey to a cake shop. Too much vinegar lowers the acidity (pH) which removes the charge of the protecting proteins (pK_a of COO^- is 4.8); too much salt collapses the atmosphere of counter-ions (Debye–Hückel layer) around the charged proteins. Both of these excesses remove the mechanisms that keep the bubbles of air apart, and hence keep the foam stable.

An old cook's tale recommended copper bowls for beating eggs into a better, creamier foam. In fact, experiments show that under identical conditions, whipping of egg whites in a glass bowl produces a grainy and dry-looking foam after 10 minutes, whereas beating in a copper bowl

Experiment

- Place a generous squirt of detergent in the bottom of a 250 mL cylinder. Add 50 mL of hydrogen peroxide (H_2O_2) and mix well.
- Add a spatula-load of potassium dichromate. Leave cylinder in a tray to catch overflow.
- A hot purple foam develops quickly which turns back to orange. Ventilate well.

produces a stiff and smooth foam after 20 minutes. Apparently a certain amount of copper is released and this reacts with the egg protein to stabilise the partially denatured protein film that constitutes the foam. So if your pavlova has collapsed, here is the chemistry to explain it. Your guests will surely be impressed.

Froth flotation

Australia's most *important* foam is neither beer head nor pavlova but the process used for separating crushed mineral ores using *froth flotation*. The first successful large-scale flotation process in the world was developed by Charles Potter (from Balaclava, Melbourne) working at Broken Hill. It was discovered from the observation that the bubbles of hydrogen, produced when ore from accumulated dumps containing 20% zinc was dissolved in sulfuric acid, carried particles of zinc blende to the surface, where they could be skimmed off (Victorian patent 18775 (1901)). Full-scale operations were in place by 1905. In the years that followed, a number of important further innovations were developed, including the use of xanthates as flotation collectors, by Keller and Lewis in the USA in 1925.

Most ores consist of a number of mineral species, some of which are valuable and some worthless (gangue). Froth flotation is a technique for separating ground particles based on the different properties of their surfaces. If a material that is not wetted by water (such as finely ground candle wax) is placed in water and air is bubbled through, the bubbles will attach themselves to the wax particles and lift them to the surface. Air 'wets' the surface of wax better than water does.

Some minerals have 'waxy' surfaces and are naturally floatable. These are graphite, talc, sulfur, molybdenite (MoS_2), orpiment (As_2S_3). Most minerals are fairly wettable by water and won't float. If we can *selectively* render particular minerals 'waxy', then we can float those minerals. Originally pine oil was used as both a waxing agent and a frothing agent. Today more selective materials are used.

Fire-fighting foams

Firefighters use foams for fire control, particularly in enclosed spaces, because people can still breathe, even when completely immersed in a foam. The foam also traps the products of the fire, such as noxious chemicals, coal dusts, or radioactive products.

There are two categories of fire-fighting foams, chemical and mechanical. The earliest chemical foam was developed in the 1880s to combat coal and oil fires. It involved the reaction between sodium bicarbonate and aluminium sulfate powders when water is added. The reaction forms carbon dioxide gas which is trapped in bubbles. This chemical foam is expensive, difficult to produce and use, and does not reseal well when disrupted. Mechanical foams are created by bringing together three components; water, foam liquid, and air, and foaming with a turbulent gas or pressure source.

Typically, low-expansion foams have 3% concentrate with 97% water and are expanded with air to a foam-to-liquid ratio of 4 to 12 (max. 20). The first of these in the 1930s used a protein source such as animal horn, fish by-products, blood, and even vegetables, which was digested in acid or alkali to produce hydrolysed protein. This mixture gave a dense thick, viscous (slow drain-out) and very stable foam that resisted burn-back (being burnt up by the fire). However, there are disadvantages. Fuel can saturate and then coat the foam, and burn on top of it. The solids in the concentrate settle out, so it has a short shelf-life. The protein foam is not compatible with some dry chemical agents. Interestingly, protein foams do not have an aqueous film, and so can expose flammable liquid and its vapours to fire.

In the 1960s, fluorocarbon surfactants were added to the protein. The film now became non-wetting to organic fuel and so the fuel could no longer form a film on top of the foam and burn. This property also allowed the foam to be injected into the bottom of fuel tanks, where it floated up through the fuel and extinguished a fire at the top.

A synthetic foaming agent, plus fluorocarbon surfactants and stabilisers, formed the basis of aqueous film-forming foam (AFFF). This combination rapidly covered the fire (fast knock-down) with the added advantage of forming an aqueous film over the fuel after the foam receded, thus suppressing vapour formation. The AFFF blanket has to be constantly monitored, and fresh foam applied where necessary. A fire produces heat and that heat produces an updraught and smoke column. This hot mix can move with the equivalent of an 80 km/h wind. Into this inferno is pumped a foam which will be carried away, but AFFF foam is less aerated, heavier than the earlier ones, and copes better. AFFF has a low surface energy and acts as a wetting agent, allowing weather better access to a fire source. AFFF can be pre-mixed into a water tank and as a last resort the tank foam can be used in 'dump-and-go' mode. AFFF is compatible with dry chemical agents and airports often use the two as a 'married' nozzle system at airports.

In the 1980s film-forming fluoroproteins (FFFP) were developed which have further improved properties.

'Polar' solvents (see Solubility in Chapter 1) such as methyl ethyl ketone or acetone can also burn and cause fires. These fuels will destroy the standard foams by mixing with water and 'pulling' it out of the foam (which is 97% water). By adding a special polymer to the standard foam, which is water soluble but not miscible with the polar solvent, a tough polymeric self-healing layer is formed around the foam, and acts as a physical barrier. These are the multipurpose synthetic foams used at 3% level for conventional non-polar fuels (at which level no polymer film forms) and 6% for polar fuel fires.

Detonation foams

A Bunsen burner (or better an oxy-acetylene torch) is adjusted to give a good flame and then the mains-gas control is used to turn the gas off and on, so that the burner adjustments are not changed. This gas mixture is then used to produce a foam by bubbling the gas into a soap solution contained in a large shallow plastic bowl. The foam filled with the right combustion mixture can be set alight with a **long** taper, and the bubbles will explode. This method was used by the British army to detonate land mines on the Falkland Islands after the war with Argentina. The foam can absorb up to 90% of the pressure of a detonation.

Environmental effects

The fire at the Sandoz warehouse on the Rhine in 1987 brought a fundamental change in German fire legislation which became the European

Bubble politics
Soap bubbles are blown on a T-piece with three taps so that one bubble is bigger than the other (Fig. 6.3). The input tap A is closed and the other two taps are opened so that air can pass from one bubble to the other. Does the big bubble blow the little bubble up until they are equal (socialist) or does the big bubble get bigger and the little get littler (capitalist)? (You're wrong — they are capitalists!).

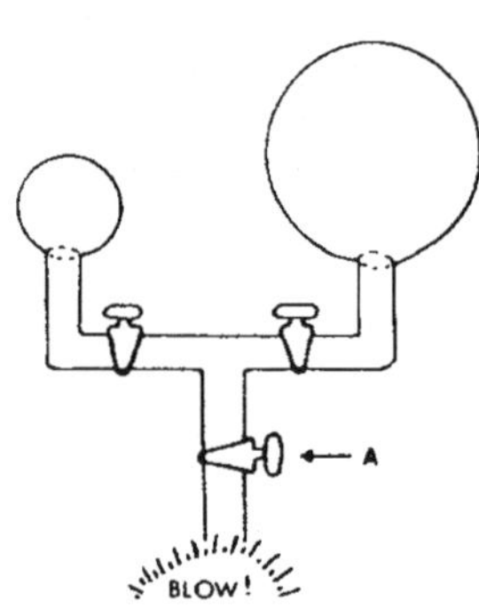

Fig. 6.3
Blowing soap bubbles with a tee-piece

norm. If the fire-fighting foam runs off into a river, the different foam types have different effects on aquatic life. Protein-based foams are generally more benign than those based on synthetic surfactants by factors ranging from 9 for fish to 40 for crabs. FFFP gave the best environmental result for protein-based foams. The natural-based protein foams are also biodegraded more efficiently.

The equation behind being small

What is a surface anyway? It is what separates two *phases* — generally gas (g), liquid (l) and solid (s). This gives five types of surfaces: s/s, s/g, s/l, l/l, l/g, noting that all gases are miscible (i.e. mix in all proportions). One of the characteristics of surfaces is that their production requires energy. You have to do work to create a surface (e.g. to blow a soap bubble). It is not so obvious, but equally true, that you need work to create a solid surface. That is why surfaces spontaneously contract — that is, reduce the amount of surface. Soap bubbles contract, and lots of small crystals in a crystallising dish dissolve and allow big crystals to become bigger. A small soap bubble contracts to allow a large bubble to grow (see Figure 6.3).

The *smaller* the bubble, the *higher* is the internal pressure. Actually the French mathematician Laplace is responsible for the precise relationship. The excess pressure inside a bubble, ΔP, is equal to twice the surface tension γ, divided by the radius r of the bubble:

$$\Delta P = 2\gamma/r$$

That raises an interesting question. When you form a bubble, say in a beaker of boiling water, the initial pressure required when $r = 0$ is infinite! In fact it is very difficult to boil very pure water in a very clean, blemish-free beaker. If you observe boiling water in a beaker closely you will see that tiny air bubbles are released from flaws in the beaker. Into these air bubbles, water vapour rapidly moves to form an expanding bubble of steam. The same flaw will continue to supply nuclei of air for some time. In a perfect beaker, with no nuclei, the water heats up and leaves the beaker in one go in a process called 'bumping'. A boiling chip (porous pot) is added deliberately to provide nuclei and prevent bumping. (Anti-bumping granules are available from chemical supply houses but they don't work on airline flights!)

The lability of small compared to large is true of 'solid' bubbles (i.e. drops). Small drops coalesce to form larger drops. Small drops have a higher vapour pressure than large drops. Special efforts are needed to keep small stable with respect to large, for example, water drops in clouds, in emulsions, and in precipitated crystals.

SHAKING THE BOTTLE

Carbonated soda water (Coke, etc.) is bottled with carbon dioxide gas under about 3 atm pressure (45 psi, 300 kPa), greater than in a car tyre. On unscrewing the top of the bottle, the pressure of gas in the headspace is released (gradually through a vertical track in the screw thread). If the top is immediately replaced, the pressure gradually builds up again to (almost) the original pressure, as equilibrium is set up again between the gas in the liquid and gas in the headspace. (In spite of reacting to form carbonic acid, carbon dioxide in water obeys Henry's law to 5 atm, and to a good approximation up to 10 atm.) If the closed bottle is shaken, there is no change in headspace pressure, but plenty of nuclei for gas bubbles are created. If the bottle is then opened, escaping gas pushes the contents out (e.g. launching with champagne). We are dealing with a kinetic effect; calm open bottle, slow, gradual escape of gas; shake and open, fast release and setting up a new equilibrium (i.e. flat drink). The Laplace equation tells us that bubbles do not form easily unless given a nucleus. However, the equation also tells us that the excess pressure inside a bubble once formed, decreases as the bubble expands (from more carbon dioxide moving into the bubble from the liquid).

Source: Will that pop bottle really go pop? An equilibrium question, D. Deamer and B. Selinger, J. Chem. Ed. **65**(6): 518, 1988 . . . don't always believe what you read in New Scientist!

Fission drop
The liquid drop was the model used in explaining nuclear fission. As a nucleus becomes larger, the charged repulsion of the positive protons increases and 'lowers the surface tension'. The nuclear drop then reverses the coalescing process and breaks up into two drops (fission).

Bubbles in water, and the 'bends'

If you place a plastic straw (spoon) in a glass of soda water (pop, Coke, etc.), bubbles form immediately on the straw. Now the gas in the soda has been added under pressure so, you are releasing gas from non-equilibrium. However, even if you fill a plastic glass with ordinary water (ungassed) fresh from the tap, and let it stand for half an hour or so, very fine air bubbles will form spontaneously on the surface. (Glass also works, unless it is exceptionally clean (hydrophilic).) In contact with a normal atmospheric pressure of air at 15°C, water dissolves 7 mL of oxygen and 13.5 mL of nitrogen per litre at equilibrium. However, in a manner not yet properly understood, hydrophobic surfaces help release some of this dissolved air to form bubbles. The tissues in the body provide many hydrophobic surfaces, easing bubble release.

At 10 metres depth, which is equivalent to about 2 (extra) atm pressure, divers don't need to worry about decompression sickness for the usual length of a dive. At greater depths (pressures) the danger from the release of nitrogen gas from the blood as bubbles, on surfacing, increases and dive charts classify the events and the remedial action needed.

Laboratory experiments on the effect of salts on bubbles

In the laboratory, *in the absence of hydrophobic surfaces,* compression up to about 180 atm, followed by decompression, will not cause bubbles to be formed. If, in the experiment, hydrophobic surfaces are added, there is a (delayed) bubble formation at much lower pressures.

Vince Craig, an ANU researcher, passed nitrogen gas from a cylinder through a frit at the bottom of a 1 metre cylinder half-full of water. The fine bubbles formed at the bottom quickly coalesced to form bigger ones. Then he added saturated salt solution through a tube to the bottom of the cylinder to give a solution that is about 1M NaCl when mixed. Now the bubbles stay fine, the foam is more stable and longer, and is opaque — quite a dramatic difference — from just adding salt.

So salts in the plasma would protect against the bends by inhibiting the coalescence of fine bubbles into bigger ones which could cause blockages. Without the salt in the tissue, driving up a modest hill could give you the bends. The composition of body salts happens to occur just at the point where (from the experiments with the bubbles in the water column) maximum coalescence prevention is achieved. This surprising result indicates a possible link between salt levels and susceptibility to decompression sickness, as small bubbles present in tissue do less damage than big bubbles. Life as we know it could probably not have evolved out of the sea until the oceans had become sufficiently salty for this effect to be manifest. Some salts are effective, and others totally ineffective; there is a clear pattern, but again no explanation.

Bubbles in champagne

Consumers allegedly prefer smaller, slow-moving bubbles in their effervescent wines, but such a 'quality of the bead' criterion is a bit of an illusion, in spite of its inclusion in the sensory analysis sheets proposed for sparkling wines:'. . . like all enduring totems, wine supports a diverse mythology, which is not confounded by its own contradictions' (R. Barthes, *Saponides et Détergents, Le Vin et Le Lait,* in 'Mythologies', 38–40 & 83–86; Editions du Seuill, Paris, 1959).

Careful observation of bubbles in effervescent wine or beer shows that bubbles start only from a limited number of specific sites in the glass, and each site emits bubbles of a uniform size and frequency. Different sites can vary. The bubbles expand as they move up through the liquid, rapidly reaching a 'terminal' velocity, as (anti-)gravity acceleration, that is buoyancy, is opposed by the inertia and viscous drag of the liquid. With time, the expansion on rising also drops, because there is less gas left in the surrounding liquid.

Sparkling wines will often produce high-frequency streams (10–20 per second) with little expansion during ascent because of the depletion of dissolved gas in their path. As the gas content of the liquid is reduced, the number of sites decreases, and their bubble frequency (but not size) decreases.

The distinctive appearance of effervescence in sparkling wines is mainly due to the presence of 10–12% ethanol, which lowers the surface tension of carbon dioxide and raises the viscosity of the liquid.

Foam in wine is found only on the surface and is short lived for a number of reasons. The foam is destroyed by the evaporation of the alcohol. Thus the foam is more stable in a closed shaken bottle than in the poured glass. (Compare this with covered saucepans of boiling liquids which froth over more than when uncovered, and beer froths more in a closed stein.)

Carbon dioxide, being quite soluble in water, diffuses quickly between bubbles and destabilises a foam. Air, or nitrogen, is much less soluble in water and therefore forms more stable foams (hence the use of nitrogen in Guinness packs).

Finally, surface active agents help stabilise foam, and they are found in higher concentrations the 'closer' the product is to the natural biological origin. Compare the stability of foam by shaking closed containers of red and white, still and sparkling wines, beer, spirits, fruit juices, black tea, soft drinks, milk, skim milk, and others.

An open unshaken bottle of effervescent wine has carbon dioxide under 3–6 atmospheres, but it does not rush out (with or without spoons etc. across the mouth) because prolonged storage under pressure has thoroughly wetted the internal glass surface and left no sites for bubble formation. A soon as the contents are poured into a clean dry glass, streams of bubbles will form tiny discontinuities (scratches, etc.) where the liquid does not wet the glass.

In effervescent drinks without alcohol the bubbles tend to be larger before detachment (the surface tension is higher, 0.07 J/m^2, compared to 0.05 J/m^2) the speed of rising is faster (lower viscosity) and the larger bubbles tend to flatten and wobble during their ascent.

In plastic bottles (PET and others) there is a strong attraction between the bottle and the bubble, which therefore has to be larger before detaching. In a (plastic) glass of aerated water the bubbles can become quite large and yet remain attached to the side of the glass as they rise.

TEARS ON WINE, LEGS ON PORT

Proverbs 23

31: 'Look not thou upon the wine when it is red, when it giveth this colour in the cup, when it moveth itself aright'.
32: At the last it biteth like a serpent, and stingeth like an adder.
33: Thine eyes shall behold strange women, and thine heart shall utter perverse things.

While verses 32 and 33 no doubt provide the usual good advice on the effects of excess alcohol, it is verse 31 that has the physical chemistry. The

Experiment
Make up a set of alcohol–water mixtures across the full range, and pour them into identical clean wine glasses. Swirl the mixture, connoisseur style, and after a short period measure the height of the 'ring' above the bulk level for each composition and plot the height (in mm, 0–8 scale) against the composition.

basic explanation for the phenomenon of alcoholic drinks of sufficient strength moving, as a thin film, up the sides of a clean wineglass and then forming 'kissing tears', is as follows.

Water will wet clean glass, and surface energy will cause it to rise up the side a little. Alcohol plus water will do the same, but with the additional effect that the alcohol will evaporate from the wine more quickly than the water. This will happen much faster in the thin film up the side of the glass than in the bulk liquid. Loss of alcohol raises the surface tension in the film and causes it to climb further up the side of the glass, and drag fluid of lower surface tension from the bulk with it. During its climb, the leading edge of the film becomes heavy, drops back and forms finger-like projections with large drops at the end. These finally tumble down as 'tears' or 'legs'. These tears have difficulty in re-entering the bulk (the tears 'kiss' the bulk, and retract) because the surface energy of the tear is higher than that in the bulk (because of greater evaporation of alcohol in the tear). Rather than merging with the bulk on touching, the opposite happens, and the tear tends to draw liquid **from** the bulk. The tear has great difficulty in returning to the bulk liquid and just drops, kisses it, and retracts.

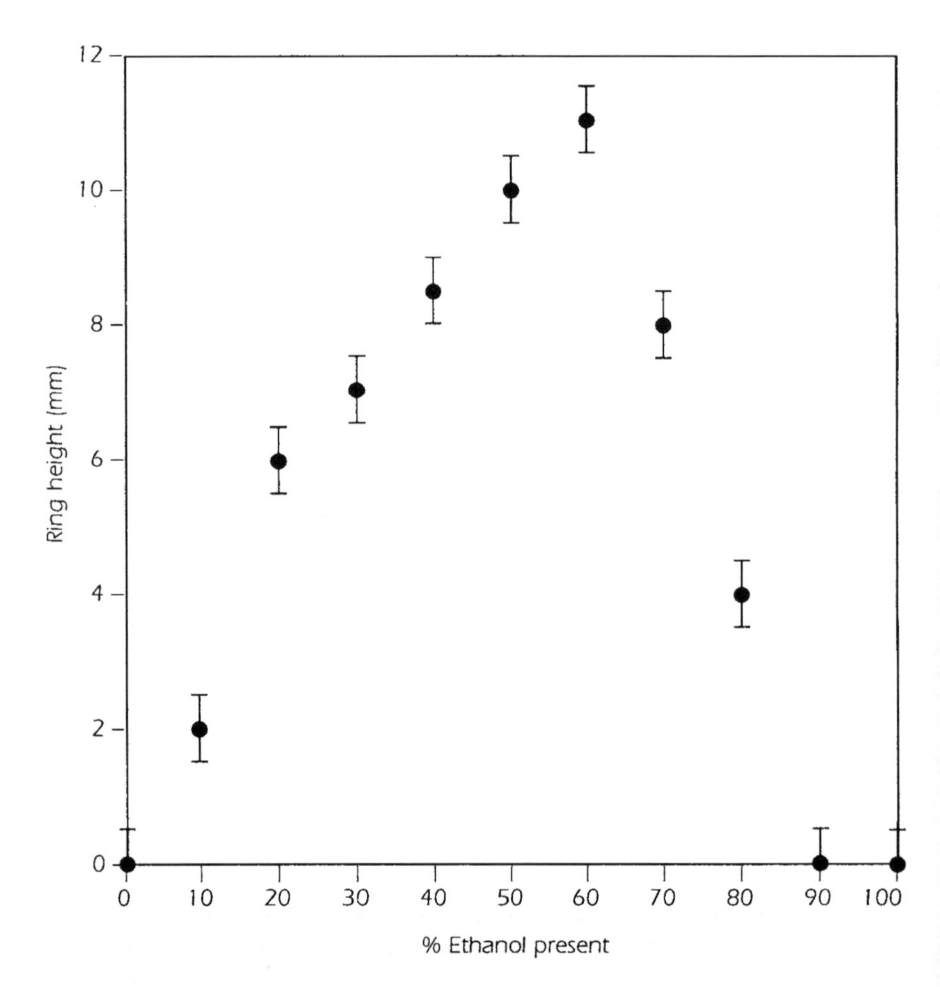

Fig. 6.4
Plot of ring height for ethanol/water mixtures

Contrary to the quotation, it works just as well on white wine and pure alcohol/water mixtures!

Assigning the effect to 'slow-draining, viscosity-enhancing glycerol in wine' (Peter Atkins, Molecules, Scientific American Library, 1987, p. 52) is thus unsustainable.

Bubbles and viscosity

A simple device for measuring the relative viscosity of liquids is to include an air bubble in a glass tube filled with a range of different liquids (compare water, sunflower oil, washing-up liquid, motor oil, and bubble bath). When the tubes are simultaneously inverted, the bubbles are now on the bottom, and there is a bubble race to the top of each tube. The speed of the bubble is inversely proportional to the viscosity of the liquid. Injection of air bubbles into the base of a column of high viscosity silicone oil gives a slow motion display revealing interesting hydrodynamic details of the relative speeds of different size bubbles.

Opening a tap at the base of a column of water causes the water to spin. Spinning causes a pressure drop at the centre of the surface and the surface curves downward and eventually forms a vortex or air tube. The faster water exits, the deeper is the vortex. A cylindrical column of water from a tap breaks up into droplets when the slightest disturbance causes a curvation in the surface along the cylinder length. As you lower the water pressure, the smooth column of water becomes shorter.

Columns of water formed as dew along the threads of a spider's web break up into consecutive microscopic small and large drops. (*Source*: Techniquest Science Centre, Cardiff, UK).

Bubbles and noise

When making a drink by dissolving a solid such as instant coffee, bubbles are formed. When the side of the cup is clunked vigorously with a spoon, the pitch drops several octaves with time. This appears to be correlated with the dispersion of the bubbles. One can use cold concentrated detergents in a small stout coffee jar to demonstrate the effect. It is probably due to the change in the speed of sound in a liquid with and without bubbles.

Steam bubbles arise in the region of a kettle element where the water is being superheated. These bubbles rise into the cooler water, then collapse and send out a shock wave. The process repeats itself cyclically and causes 'singing'. As the temperature of the water rises, the bubbles travel further before collapsing, and the pitch drops. Just before the kettle boils, the singing abruptly stops.

When you pour liquid nitrogen into a Dewar flask, the sizzling sound caused by the cooling of the container maintains a steady volume until a

Stop slop
You can devise an interesting experiment using a swelling clay from cracked, dry clay pans (montmorillonite or fine-grained illite). Pack the clay in tubes with a fitted filter base. Add sodium chloride to one, and calcium chloride to the other; then pass water through. Note the different rates at which the water passes through. This explains the effectiveness of the practice of throwing gypsum ($CaSO_4.2H_2O$) into pig pens to reduce the sloppiness.

crescendo heralds the end of the cooling. Why? Gases are very much less-efficient conductors of heat than liquids. When the container is relatively very hot the liquid nitrogen in contact with it is in the vapour phase and cooling is slow. As the container cools, it reaches a point where liquid nitrogen comes in contact, and the rate of cooling increases rapidly, as does the sizzling sound. (See also Catalysis in Chapter 14.)

Micro-emulsions

Emulsions such as mayonnaise are *thermodynamically* unstable. This means there is a natural tendency for the suspended droplets to coalesce and reduce the overall surface area, and hence surface energy. The two components then separate into two layers.

We know that adding surfactants to water reduces the surface tension. There are also surfactants that, when added to oil, lower its surface energy. By choosing very effective agents we can reduce the surface energy between oil and water to zero — or even negative! Then the whole situation will reverse and the surface area will tend to *increase* and we find spontaneous emulsification to a *thermodynamically stable*, so-called *micro-emulsion* (in which the droplet size tries to become as small as possible, limited by the amount of surfactant added, which limits the total surface area possible). Micro-emulsions of oil in water are used experimentally in tertiary oil recovery where about 70% of the oil remains in the well after it has been pumped 'dry'.

Suspending small particles

The 'behaviour' of soils, particularly clay soils, depends on the *surfaces* of the soil particles. The physical properties and behaviour of a soil are strongly influenced by the relative proportions of calcium, magnesium, sodium and potassium adsorbed onto the surfaces of its clay minerals, as well as the amounts of organic matter, oxides and carbonate present. (See Soils ain't just soils in Chapter 7.)

The best agricultural soils have a low proportion of 'exchangeable' sodium. As the proportion of exchangeable sodium increases (particularly if there is more (exchangeable) magnesium than calcium), soils become difficult to work. In surface soils with a high exchangeable sodium content, rain breaks down aggregates to a dispersed layer of soil in which water-conducting cracks are blocked, and this leads to waterlogging. On drying, the clay sets hard, shrinks and cracks.

In washing clothes we have exactly the opposite problem. We want to *disperse* clay particles and hold them in suspension. So the opposite tactic is used. Washing soda ($Na_2CO_3.10H_2O$) is added so that *sodium* ion can displace the *calcium* ion. Modern synthetic detergents use polyphosphates or zeolites, which 'tie up' calcium ion and remove its influence from the system (see The detergents — Inorganic builders' in Chapter 3).

When a swimming pool is full of suspended clay washed in by rain, we can coagulate the fine particles by going one better than using calcium. We use aluminium (which has a charge of three compared to calcium, with two, and sodium with one). The addition of alum (a salt of aluminium) 'flocculates' the clay into large particles and allows it to be caught in the filter, and so the pool can be clarified again. (See also Experiment 6.1 below). Some of the cationic detergent algacides sold for swimming pool maintenance (which, incidentally, were originally developed for clearing yeast from beer lines) can cause problems by the positively charged surfactant chain attaching itself to the negative clay particles and keeping them in suspension.

EXPERIMENT 6.1 Clarification of river water

Equipment

Jam jars, potash alum, clay soil or river water.

Procedure

Muddy river water contains soil held as a colloid in the water. Colloidal particles are generally charged with either positive or negative charges, and it is the repulsive forces between charges of the same sign that keep the colloid particles suspended in the solution.

If compounds that form ions are added to the colloidal solution, the positive ions will gather around any particles that are negatively charged (and negative ions around particles that are positively charged) thereby effectively screening the repulsive forces between the colloidal particles. The reduction in mutual repulsion allows the particles to clump together and thus settle out of the water (flocculation).

Many colloidal suspensions are found to carry negative charges; thus ionic compounds that yield multicharged positive ions are particularly effective flocculating agents. The most commonly used is potash alum, $KAl(SO_4)_2.12H_2O$, which can be purchased at many pharmacies, photo shops, or swimming pool supply shops. It is sometimes used to clarify city water supplies (see Aluminium and Alzheimer's disease in Chapter 10).

Fill a jam jar with water and add a small quantity of clay soil to the jar. Shake the jar and then allow the heavy solid material to settle out by leaving it to stand for a few minutes. (You could use creek or river water instead of the clay suspension.) Pour some of the muddy water you have prepared into two similar jam jars. To one of the jars add a small quantity (about a teaspoonful) of potash alum and shake the jar.

Allow the two jars to stand. Check the clarity of the water in the jars from time to time. You should notice the colloid suspension in the jar to which you have added the alum begin to clarify as the colloidal particles flocculate and settle to the bottom of the jar.

Test other ionic compounds for their effectiveness as flocculating agents. Consider the relative rates at which given quantities of the compounds clarify the water.

The Laplace force

Why are bread crumbs easier to brush off a plate than toast crumbs? Why does warming the plate first, help? The reason is the Laplace force (see p. 148). Water is released from the toast onto the plate and its surface tension holds the crumbs down (Proof: using a pre-warmed plate allows the water to evaporate.) This effect is also the reason for having sand for sand castles somewhat wet, but not too wet. When shipping coal in bulk carriers, just the right amount of water is needed to damp down the dust; enough so that it mostly evaporates by the time the ship reaches its destination. The same force makes it hard to comb straight hair when it is wet; see The physics of combing hair in Chapter 5.

Warning

Do not try to drink the clarified water. Extra stages are undertaken at drinking-water treatment works to ensure the purity of the reticulated water supply. The clarified water you have prepared is not suitable for drinking.

Coating with a water repelling layer

The froth flotation process mentioned earlier under the heading Foams is an example of coating a material to make it hydrophobic, in that case, so that it will hold onto air bubbles.

Hair conditioners and laundry wash final rinses for wool both work by coating fibres with a waxy outer layer one molecule thick. As we saw in Chapter 3, the active ingredient is a cationic detergent with a positive charge (see also Chapter 8) that is attracted to the negative charge on the fibre. Glass also has a negative charge and is made waxy by these detergents. It follows that ground silica and silicate minerals can be floated using these detergents. On the other hand, minerals with positive-charged surfaces will be made waxy with negatively-charged soaps and detergents, just as plastic plates are affected in the same way.

A surface can be modified more effectively by changing the surface with a chemical reaction rather than just relying on attraction of opposite charges. For floating sulfide ores of copper, lead, nickel and zinc, sulfur-containing compounds such as thiocarbonate (xanthate) or thiosulfate (photographic 'hypo') are used to change the surface chemically.

The children's toy material called Magic Sand (Wham-O Manufacturing Co., California: US Patent 3,562,153) is just ordinary sand that has been treated with a dye and coated with a hydrophobic silicon coating. When Magic Sand is placed in water it can form underwater columns and other structures, but it is perfectly dry on removal. The 'cohesion' of Magic Sand under water is a form of phase separation that reduces the surface area of the water–sand interface — exactly the opposite effect that we try to achieve by using detergents so that soil is wetted by water and de-aggregates. The material was originally developed for large-scale clean-up of oil-on-water pollution. Why would it be effective for this purpose? What advantage would it have over the use of detergents?

Isn't it annoying when a baked cake adheres to the cake tin. This tends to happen when the water in the dough wets the metal surface and then dries during baking, depositing dissolved materials that then bond the product to the metal. Oiling of the surface is intended to prevent wetting and hence adhesion. The degree of wetting by a liquid on a solid can be measured by studying the contact angle that the liquid and solid form (see Fig. 6.5). For further details on adhesion, see Adhesives in Chapter 12.

Metals are characterised by high surface energy (this replaces the older term surface tension) ranging from 0.5 to 5 J/m^2, while water has a value of 0.07 J/m^2 and oil about 0.02 J/m^2. A material with a low surface ener-

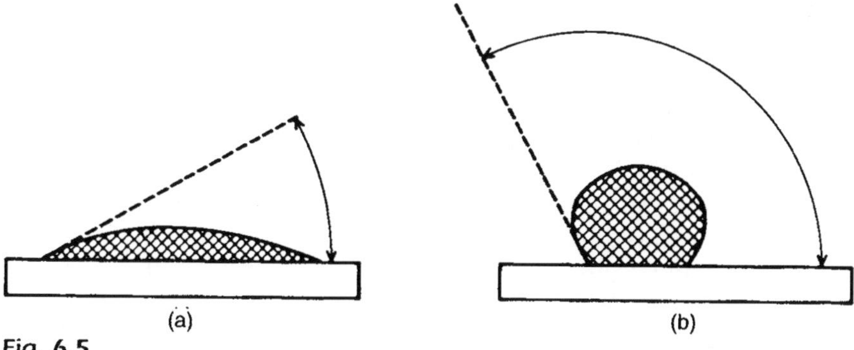

Fig. 6.5
Contact angles of a liquid on a solid surface: (a) small contact angle — a high degree of wetting; (b) large contact angle — a low degree of wetting.

(From Food Research Quarterly **39**, pp. 30–33, 1979.)

gy will wet a material with higher surface energy. Both water and oil will wet metal. Oil will spread on water to a greater or lesser degree, but water won't spread on oil. Commercially a lecithin/white oil mixture is used to coat aluminium baking trays to prevent 'stickers'. Trays contaminated with polymerised processing oil when they are made may stick and need to be scrubbed with a solvent. The build-up of polymerised oil on bread-baking pans is a major problem as it increases the incidence of sticking loaves and stained bread.

Clumping in cups
Add water to two polystyrene (or other plastic) cups, one to within a few cm of the top, the second to the point of overflowing. Place small (≈ 1 cm) pieces of polystyrene foam on the surface. In cup 1, the pieces move to the edge; in 2, they move away. If a pencil point is used to shove them against their inclination, the rejection is vigorous. (Twigs and leaves in ponds clump the same way). Contrary to New Scientist, water does not wet or rise up a plastic cup wall; quite the opposite.
The reason for the move to touch the side is that it reduces plastic foam surface touching the water. In the overflowing bulging cup, the pieces can't move to the edge because that would mean floating downhill. The pieces clump in both cases because that reduces the total amount of hydrophobic (plastic) surface exposed to hydrophilic water.

Source: New Scientist, p. 89, Dec 21/28, 1996.

The physics of muesli

An hourglass filled with sand shows that sand flows like a fluid through the orifice but packs like a solid on the bottom to form a heaped pile. It does this reproducibly so that for hundreds of years people have used the device as a clock. However, when farmers store grain in huge silos, these occasionally collapse unpredictably; it never happens with liquids.

When you think about it, much of our food comes in granular form and while this may not seem a big deal for the consumer, a manufacturer shifting large quantities through machines and packaging lines must face up to the physics. An average spoonful of sugar may contain between a thousand and a billion grains, depending on whether it is plain granulated, castor, or icing sugar. When you pour, shake, or compress a powder, the particles all jostle each other and readjust their positions all at the same time (but do not exchange energy the way molecules in a liquid do). So funny things can happen to packets of muesli on the way to the supermarket shelf.

If you pour granulated coffee or muesli into another container, it occupies a volume that decreases with tapping or light shaking, down to a constant *bulk density*. But if you look at your package of muesli, where are the large flakes, raisins and nuts, and where are the fine broken bits?

Shaking or vibration quickly lets small particles fall through to the bottom and at the same time helps segregate and group particles of similar size, perhaps by blocking a larger particle returning to its original spot. This segregation will make a product non-uniform, and may cause aesthetic and performance problems.

If you place lycopodium powder onto a vibrating plate, it spontaneously runs into small heaps. Michael Faraday described this *convective grain flow* in 1831 as follows: 'the particles of the heap rise up the centre, overflow, fall down upon all sides, and disappear at the bottom, apparently proceeding inwards'. Salt grains vibrated at 50 Hz form a slope that may be inclined at 20°C. Sand dunes show this on a grand scale. Adding more sand to a dry sand pile overextends the maximum slope and an avalanche occurs to re-establish it.

References

You can't keep a good bubble down, J. Harris, *New Scientist*, 24/31 Dec. 1987.

Effervescence in sparkling wines: The sequel, John Casey, *Aust Grape & Winegrower*, Annual Tech Issue, pp. 37–46, 1995.

Effect of electrolytes on bubble coalescence, V. Craig, R. Pashley, B. Ninham, *Nature:* 364, pp. 317–319, 22 July 1993.

The physics of muesli, G. Barker and M. Grimson, *New Scientist*, pp. 23–26, 26 May 1990.

Firefighting effects on the environment, R. Whitely (Ansul UK), *Fire International*, pp. 36–37, Aug/Sept 1994.

Fire prevention 259, J. Brittain (Angus Fire), pp. 24–27, May 1993.

Bibliography

The science of soap films and soap bubbles, C. Isenberg, 1972, Tieto, reprinted Dover, 1992. A classic on the physics of soap bubbles.

CHAPTER 7

Chemistry in the garden

In the cut-throat world of tomato competitions

Step into my garden and examine the soils, fertilisers and pesticides. We also look at sex attractants (for pests) and a sexy murder.

Introduction

'Until about 10 000 years ago, with the invention of agriculture and the domestication of animals, human food supply was limited to fruits and vegetables in the natural environment, and game animals. The sparsity of naturally grown foodstuffs was such that the earth could sustain no more than about 10 million people; by the end of the 20th century there will be 6 billion.

That means that 99.9% of us owe our existence to agricultural technology and the science that now underlies it; plant and animal genetics and behaviour; chemical fertilisers, pesticides and preservatives; ploughs, combines and other agricultural implements; irrigation; and refrigeration in trucks, railway cars, stores and homes. Many of the most striking advances in agricultural technology — including the 'green revolution'— are products of the 20th century'.

Source: Carl Sagan, prize-winning author and astrophysicist, shortly before he died, 20 December, 1996.

URBANISATION

'They paved Paradise and put up a parking lot', song, Joni Mitchell.

Agriculture enabled the first human aggregations; people congregated to cooperatively grow crops where the land was most fertile. Until methods of preserving food were developed, food growing and residence had to be close. When the settlements cover the productive land, horticulture is pushed out to more marginal land, and chemical inputs increase. In 1900 a mere 14% of the world's population lived in cities: by 2000 half will. About 63% of Australians live in the capital cities, and another 20% in other towns, supporting our claim to be the most urbanised country in the world. Sydney and Melbourne were richly provided by close market gardens until as recently as 20 years ago, but free market forces have turned these into sterile suburbs. Subdivision, often seen as the farmer's superannuation, has more recently been dubbed 'the final crop'. Urbanisation has alienated the food supply, so that people no longer know what food seasons are, or where certain produce actually comes from.

Source: Cherry Ripe, columnist, Australian newspaper.

Soil acidity

On the east coast of northern Queensland soils were laid down from an anaerobic mangrove swamp-type terrain. The reducing conditions caused the formation of iron sulfide, FeS. When these soils are now ploughed or mined, the aeration causes oxidation of the sulfide, ultimately through to sulfuric acid, and thus they become acid. When later limed to increase the pH, excess lime can hydrolyse organophosphate pesticides (see below) added to treat soil pests such as nematodes which attack sugar cane. Thus the organophosphates are made ineffective.

Soils ain't just soils

If you stick your hand in a bit of dirt in the backyard you will come up with a handful of something that will differ widely in physical and chemical composition, depending on where you live, and how keen a gardener you are. Let us think of the soil as we might consider a wine.

Colour

Organic matter darkens any soil, so topsoils are usually darker than subsoils. Light or grey soils are often leached of their hydrated iron oxides, which give the yellow through orange, brown and red colour to soil. The greenish-grey colour of some waterlogged soils is due to the reduced ferrous form of iron.

Texture

Soil particles range in size and these are defined in Table 7.1.

The typical feel or texture of a soil depends on the distribution of particle size (see Bubbles and the physics of muesli in Chapter 6), and especially the clay content. Field texture is determined by working the moistened soil in the hand and comparing its feel with that of a series of standard soils. A triangular graph (Figure 7.1) can show the definition of soils based on the proportions of three components.

Water percolates readily through sands with little silt or clay, so saltier water can be tolerated in irrigation; but fertilisers leach out quickly too (slow release formulations should be used).

TABLE 7.1
Particle dimensions

Particle	Equivalent diameter mm	Surface area cm^2/g
Gravel	>2	
Coarse sand	2–0.2	23
Fine sand	0.2–0.02	90 to 230
Silt	0.02–0.002	450
Clay	<0.002	8 000 000

'Aah . . . now **that's** a top soil'

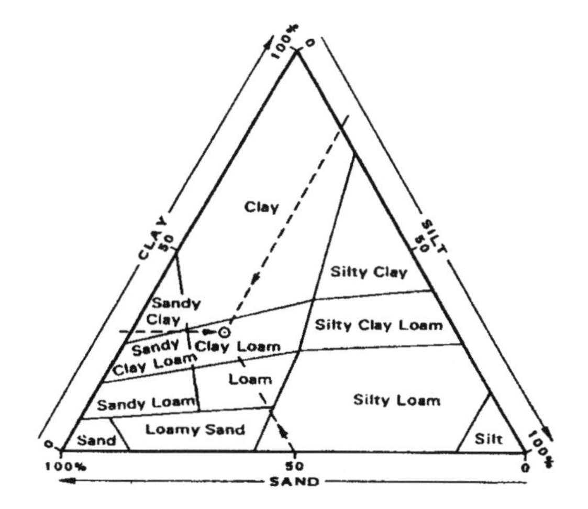

Fig. 7.1
Some grades of soil textures: The clay loam soil represented by ⊙ has 50% sand, 19% silt, and 31% clay.

(From Soils, CSIRO, Melbourne, 1977, p. 12)

Generally clay soils are thus difficult to cultivate because they become sticky when wet and set hard on drying. The reason for this is that they lack sand-size particles, and contain too little humus and too much sodium. They are great for sealing earth-fill dams, but not for draining septic tank effluent. The type of mineral present is also very important. Adding sand to clay can open up the structure, but, as Figure 7.1 indicates, amounts of the order of 60% are needed to make loams from clays. This is impractical, except for small areas. The problem of excess sodium can be solved by adding calcium ion in the form of gypsum or lime. The calcium ion pushes the sodium ion off the clay particles and allows them to aggregate into larger clumps which do not pack as tightly, allowing easier circulation of air and water. A typical application rate is 0.5 kg/m^2, but it must be spread over the whole surface.

To test your soil for excess sodium, drop an approximately 6 mm diameter fragment of soil into a glass of distilled water (rainwater or defrost refrigerator water). Leave it to stand for 24 hours. If the fragment is surrounded by a milky-brown halo, the soil will certainly respond to gypsum.

SOIL CARBON

Soil contains a vast amount of carbon — as much again as the total in all terrestrial vegetation. The carbon content of soil can vary from under 1% in a sandy soil to over 40% in peat soil. Carbon dioxide is released naturally when soil organic matter is oxidised by microbes. The amount depends on temperature, aeration, and moisture, and ranges from 80 to 550 g of carbon per m^2 of ground per year. Draining and cultivation of flooded soils allows oxidation of an estimated extra 150–180 Mt/a of carbon worldwide. Reduction or elimination of tillage avoids the aeration of the upper layer (and also its mineralisation). About 30 Mt/a might be saved, as well as 10 Mt/a reduction from less use of fossil fuel.

A more effective approach is to increase carbon storage in the soil. Reforestation of old agricultural land stores carbon in the ratio of 76% in the trees, 13% in the litter, and 10% in the soil.

Mineralogy

Sand and silt contain a variety of minerals that are similar in physical properties, although they differ chemically. This is not true of clay minerals, whose two major types, kaolinite and montmorillonite, have very different physical properties. Because clay minerals have crystal shapes with a very large surface-to-volume ratio, the shape and chemical bonding between particles determines the physical behaviour.

Montmorillonite clays are called reactive clays, and swell by as much as threefold as water molecules enter between the sheets of atoms in their crystal lattice. They cause engineering problems for buildings because they shrink at different rates on drying out. Trees can cause cracking of buildings by drying out the soil under the foundations. The physical condition of the soil in relation to plant growth is called tilth.

Chemistry

Soils provide an anchor for plant roots, and also provide the chemical elements necessary for growth. The elements nitrogen, phosphorus, potassium, calcium, magnesium, and sulfur are called the major elements, while manganese, boron, iron, copper, zinc, molybdenum, sodium, chlorine, and cobalt are called trace elements. Other elements are not essential for the plants, but are taken up and are essential for animals feeding on the plants. The availability of elements depends not only on the amount present but also on the form in which it is present, the rate it is released from the minerals, and the pH of the soil. Positive ions are held on negatively charged clay and humus particles. Excess potassium can prevent the uptake of magnesium, and vice versa. Even with adequate zinc in the soil, excess superphosphate can induce zinc deficiency. Excess manganese can inhibit cobalt uptake, essential for the bacteria that fix nitrogen in the root nodules of legumes.

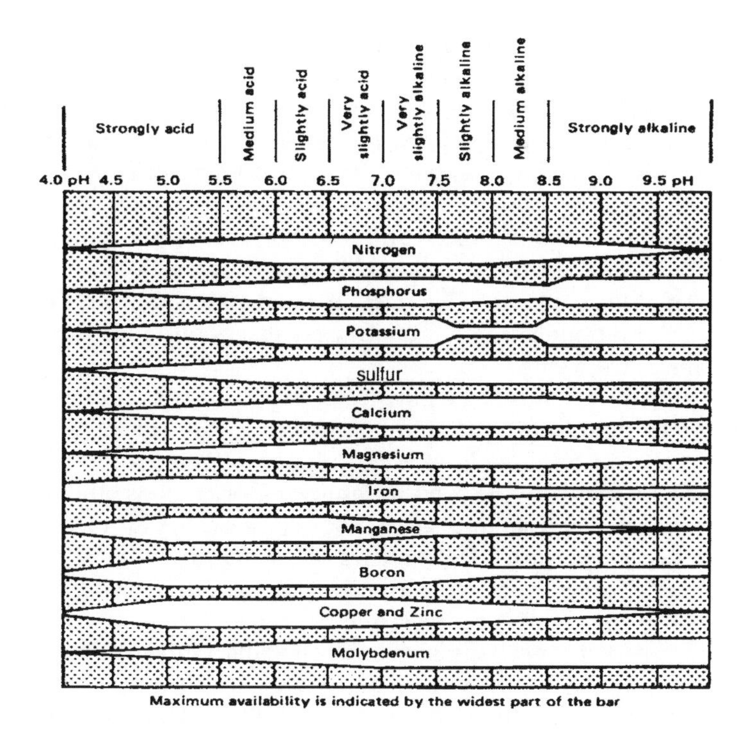

Fig. 7.2
The availability to plants of nutrient elements varies with soil pH.

(From Soils: an outline of their properties and management, Discovering Soils 1, CSIRO, Melbourne, p. 20, 1977.)

The pH of the soil has a critical effect on element availability, as seen in Figure 7.2 (above). Thus heavy liming will reduce the availability of iron and manganese, while ammonium sulfate causes the pH to drop and can cause molybdenum deficiency, and possibly manganese toxicity.

Food for plants

What really *is* a cabbage? We see the top as a symmetrically packed set of leaves, while a worm would see a root system exploring the soil. Both leaves and roots 'catch' food for the whole plant. If you dry a 2 kg cabbage in an oven, driving off all the water, you will be left with about 160 g dry weight. Burning will reduce this to a grey-white ash of about 12 g, and mainly carbon dioxide gas will be given off. The ash is only 0.6% of the original mass of the cabbage, and contains a wide variety of elements.

Carbon dioxide in the air is present at a level of only 0.033%, and yet this is the sole source of organic carbon in plants and animals. One form of fertilising in greenhouses is artificially increase of the carbon dioxide level to around 0.1 to 0.15%. While this tiny amount of carbon dioxide is directly usable, the much greater amount of nitrogen (78% of the air) cannot be used by plants until it is 'fixed' as ammonium cation or nitrate anion. Micro-organisms such as *Rhizobium* that live in the nodules on the roots of legumes fix the nitrogen equivalent of at least 10 million tonnes of ammonium sulfate every year.

Asparagus odour
In 1891, M. Nencki distilled the urine of four individuals who had eaten 7 kg of asparagus and deduced that the compound contained sulfur, suggesting methanethiol (methylmercaptan). A bit less than half the population produces this odour from asparagus and it is controlled by a single gene. With more sensitive analytical equipment, the individuals used in modern experiments had to eat only 100 g of asparagus. The compounds are S-methyl thioacrylate and S-methyl 3-(methylthio) thioproprionate.

Phosphorus is vital to virtually all processes of life, particularly in the processes of energy usage. Australian soils are nearly all deficient in this element. Agricultural products remove and export it from the soil on cropping, for example, 3 to 6 kg/ha for cereal grains, and 5 to 15 kg/ha for hay and lawn clippings. As we transfer nutrients from soil to sea via sewage, we need to provide a continuous input to the soil. Superphosphate is made from imported and local rock phosphate, and sulfuric acid. When added to the soil, some is 'fixed' by soil minerals, and some is available to plants. The fraction fixed decreases with each ad dition. Blood and bone and other organic fertilisers provide slow-release formulations of nutrients.

Potassium is important in the balance of sap minerals and is involved in extension of stems and thickening of the outer walls in the outer cells of the plant. Its absence leads to accumulation of sugars and nitrate in plant tissues. Sugars provide food for attacking organisms ,and so make the plant more susceptible to disease. Potassium occurs in micas and on the surface of clay particles. Deficiencies occur mainly in the wet coastal regions. Good sources are wood ash (hence the name 'potash'), flue dust from cement works, seaweed, and urine.

Calcium is needed for cell division and cell walls. Only sandy, acid soils in high rainfall areas are likely to be deficient.

Magnesium is needed to produce chlorophyll, and hence is vital for photosynthesis. Deficiencies are the same as for calcium, but high doses of potassium can cause magnesium deficiencies.

Sulfur is essential for the amino acids cysteine and methionine, and so is essential for protein production. Sulfur is also a constituent of many flavours and odour components in plants, for example brussels sprouts, cabbages, and onions. The smell of urine after eating asparagus is due to sulfur compounds.

Sulfur loss is a problem in sandy soils. It is readily available from sulfate, and often the response to superphosphate is as much due to the sulfur as to the phosphorus. Near the sea, sulfate from seaspray can be a major source. Burning of fossil fuels adds sulfuric acid to rain, and in highly industrialised countries, amounts of 30 kg/ha can reach the soil each year from this source and acidify the rainwater to such an extent that trees and other plants are damaged. In Europe acid rain has been reduced to such an extent that in some areas there is now a sulfur deficiency!

Trace elements

Iron is essential to the function of the chloroplasts in photosynthesis. Iron is present in abundance in all soils as oxides and silicates. However, other soil components can interfere with its availability, particularly lime, either added or naturally occurring as limestone. *Lime-induced chlorosis* is the name given to the disease this causes in such plants as citrus, soft fruits, beet, spinach, peas, and in many native plants transferred from their natural acid-soil environment to an alkaline soil.

Manganese is important for reasons that are not well understood. It is present in soils as insoluble oxides formed from soluble compounds by bacteria (particularly in alkaline soils).

Copper is essential to the production of enzymes, critical to plant cell function. Deficiency occurs in acid sands, loams, and gravelly soils. Copper toxicity can occur through the use of copper-containing sprays, such as Bordeaux mixture.

Zinc is involved in producing the plant hormone auxin responsible for stem elongation and leaf expansion.

Boron is important for the uptake of calcium. Boron levels are high in sea water, and so also in soils with a marine sediment origin, sometimes to a toxic level. Liming can reduce boron toxicity.

Molybdenum is needed by the bacteria that fix atmospheric nitrogen and so is particularly important for legumes. In addition, it is needed to form proteins from soluble nitrogen compounds. Molybdenum is associated with iron minerals. It dissolves more rapidly in alkaline soils.

Cobalt is also needed by the nitrogen-fixing bacteria (as well as directly by some plants). Thus cobalt deficiency, like molybdenum, will be seen in legumes as a nitrogen deficiency. Cobalt in soils may be deficient, or just mopped up by manganese oxides.

NUTRIENT DEFICIENCIES IN PLANTS — A QUICK GUIDE

Symptoms appear first in the OLDEST leaves

Nitrogen	General yellowing; stunting; premature maturity
Magnesium	Patchy yellowing; brilliant colours especially around edge
Potassium	Scorched margins; spots surrounded by pale zones
Phosphorus	Yellowing; erect habit; lack-lustre look; blue-green, purple colours
Molybdenum	Mottling over whole leaf, but little pigmentation; cupping of leaves and distortion of stems
Cobalt	Legumes only, as for nitrogen
(Excess salt)	(Marginal scorching, generally no spotting)

Symptoms appear first in either the OLDEST or YOUNGEST leaves

Manganese	Interveinal yellowing; veins pale green, diffuse; water-soaked spots; worst in dull weather

Symptoms appear first in the YOUNGEST leaves

Calcium	Tiphooking; blackening and death
Sulfur	Yellowing; smallness; rolled down; some pigmentation
Iron	Interveinal yellowing; veins sharply green; youngest leaves almost white if severe
Copper	Dark blue-green; curling; twisting; death of tips
Zinc	Smallness; bunching; yellow-white mottling
Boron	Yellowing margins; crumpling; blackening; distortion

Synthetic fertilisers
Nutrients removed from the land through cropping must be replaced. Australia uses 3–4 million tonnes of fertiliser per year. The industry has a turnover of $1000 million and employs about 2500 people. Fertilisers are applied annually to about 30 million hectares of farmland but account for only 5% of the farm input costs; the amount used per hectare is much less than is normal in North America or Europe because of the large broad-acre component in Australian agriculture.

Source: What's wrong with my soil? Discovering soils, No. 4, CSIRO, Melbourne, p. 4, 1978.

Composting

Composting is really just a method of speeding up the natural process of rotting under controlled conditions. The essential ingredients are organic materials, micro-organisms, moisture, and oxygen (and a little soil). The organic matter is the food for the bacteria and it must meet their nutritional requirements by having a suitable carbon–nitrogen ratio.

Moisture The moisture content of the heap is very important. Below 40%, decomposition does not occur. Above 60%, airflow is reduced and the heap becomes anaerobic. The microbes in a reasonable-size heap go through many cubic metres of air a day. The right moisture content of 55% feels damp but not soggy, like a squeezed sponge.

Temperature If you place a thermometer in the compost heap you will notice some interesting changes. The working microbes generate heat and in two or three days the temperature will rise to 55–60°C (bigger heaps get hotter and are better in winter). At about 40°C there is a change in shift as the starting microbes, that like the same temperatures as we do, flake out, and the work is taken over by other microbes that like higher temperatures. Compost heaps are turned to aerate them and this causes a temporary drop in temperature of 5–10°C. Temperatures higher than 60°C start to kill off bacteria and the process slows down until the heap cools again.

pH The pH of the heap starts off slightly acidic from the cell sap. As fermentation starts, the acidity increases and the pH drops further. In the hot-heap stage, the production of ammonia causes an increase in pH, and the heap becomes alkaline. Finally the pH drops to almost neutral as the ammonia is converted back to protein and the natural buffering capacity of humus takes control. Lime causes loss of ammonia and is best added to soil, and not to compost heaps.

Microbiology In the early stages of decomposition, acid-producing bacteria and fungi dominate, consuming the sugars, starches, and amino acids. The high-temperature bacteria decompose protein, fats, and hemicelluloses (similar to cellulose but composed of mannose and galactose as well as glucose — see Fig. 15.12). The high temperature fungus *Actinomyces* decomposes the cellulose. An important function of the high-temperature stage is to kill parasites and dangerous organisms, as well as weed seeds. Much of the carbon of the original heap is 'burnt' by the microbes in their life processes and escapes as carbon dioxide. The dry weight of the heap is reduced by 30–60%, and the volume drops by about two-thirds.

If you look inside a compost heap that has only partially completed its task, you will often notice that the materials have turned white or grey-white. This is because *Actinomyces* has been hard on the job. If the compost heap dries out, these white spores can disperse, and this can be irritating, so keep the heap moist.

If green grass is piled in a pit, and air is sealed out with a layer of plastic sheeting or soil, then *anaerobic* bacteria dominate, and the main acids produced are lactic and acetic acids. No other processes occur. The pH is around 4 to 5. This product is called *silage.* The grass has been pickled or preserved by fermentation in the same way as peat and peat moss, and is used as fodder.

From compost to gunpowder

1788 was a great year for European discoveries (apart from Australia). A prize was awarded by the French Academy of Science for the best essay on '. . . the most advantageous way of making saltpetre, (the critical component in gunpowder).' The essay was to draw up a detailed program which set out '. . . precisely what was known about the formation of saltpetre (also known as nitre), what books were to be consulted, and what experiments were to be tried' — a bit like a modern grant application.

In 1788 that most famous of French chemists, Antoine Lavoisier (discoverer of oxygen, and destroyer of the Phlogiston Theory of combustion) had just completed the task of reorganising France's gunpowder production as head of the Gunpowder Commission. He turned his headquarters, the Arsenal in Paris, into the centre of scientific discovery for the next twenty years. All the great scientists came, including Watt and Franklin.

One of the major reasons for the end of the Seven Years War with Prussia in 1763 was that the French needed to buy gunpowder at very high prices from the Dutch because the Prussians were well ahead in nitre technology. The main component, nitre (KNO_3), was found in soil that had been trodden by farm animals and mixed with excreta. The other components, charcoal and sulfur, were no problem, although grinding and mixing them was hazardous.

The French saltpetre makers were a rough lot. They could claim free lodgings and transport, they had the right to dig anywhere, but could be bought off by those who could afford to protect their stables, sheepfolds, pigsties, etc. They had rights to cheap wood. They became rich, while their country ran out of explosives.

The dung converted the urea in the excreta to impure ammonium nitrate. This was dug out, dried, and mixed with lime and wood ashes, the latter containing potash (hence the name potassium). The mix was leached with boiling water and the resulting solution boiled to concentrate it to the point at which ordinary salt (sodium chloride) crystallised out first.

EXERCISE

Consider the equilibrium: $KCl + NaNO_3 \rightleftharpoons KNO_3 + NaCl$.

From the solubilities below, see if the production of nitre is feasible.

Solublity of various salts (g/100 mL)
NaCl = 35.7 (cold 0°C), 39.1 (hot, 100°C);
$NaNO_3$ = 92 (cold, 25°C), 180 (hot, 100°C);
KNO_3 = 13.3 (cold 0°C), 247 (hot, 100°C);
NH_4NO_3 = 118 (cold, 0°C), 871 (hot, 100°C)

Why do you think that the solubility of salt is so independent of temperature?

(See also Nitre and nitrate in Chapter 1 and Solubility at the sea in Chapter 2.)

In Petroleum oil fractionation (Chapter 14) we see that additives are used in petrol to slow the combustion chain reaction and prevent knocking. For the opposite reason salt crystals had to be *removed* from the nitre because (as we now know) sodium causes a fast rather than slow explosion in the gunpowder, which would destroy the gun rather than the enemy. Salt is deliberately added to today's nitrate-based blasting powders when used in coal mines, so as to shorten the explosive chain reaction and thus prevent any methane gas present from having time to ignite before the explosion is over.

Proofing spirits
Wet powder was put to good use by customs officers who once added alcoholic spirits to gunpowder and touched the mixture with a lighted taper. If there was sufficient alcohol (about 50%) the gunpowder/alcohol/water mixture would still burn, the spirits were declared as 'proof', and duty was payable. That term has survived to present times.

After the salt crystals were removed, the solution was allowed to cool, whereby crystals of fairly pure potassium nitrate separated out. It was important to obtain potassium rather than sodium nitrate because the latter is hygroscopic, that is, it absorbs water from the air and in its damp state will not burn reliably. Hence the advice from Oliver Cromwell to his soldiers during the English Civil War (1653–1658), 'Put your trust in God, my boys, but keep your powder dry'.

Lavoisier organised artificial nitre beds along Prussian lines, which reproduced the conditions in the stables. Ditches were dug and roofed. Animal excreta, decaying vegetable matter, and old mortar or chalk were mixed. Some sources suggest the army was required to urinate in shifts to fill the urea quota. For his troubles, Lavoisier later lost his head in the French Revolution and his widow then lost hers to another chemist, Count Rumford, later Lord Benjamin Thompson, whom she married. He set up the Royal Institution of Great Britain, later made famous by Michael Faraday and Humphrey Davy. Rumford went on to bore cannons for the Minister of War in Bavaria, and provide yet another fascinating chemical story, this time on the history of understanding of the concept of heat.

(Source: *Of nightsoil, nitre and soups*, B. Selinger, *Canberra Times*, p. 12, 2 April, 1988.)

Pesticides and alternatives

Insects may be small in size but that's all. Insects represent about 76% of the total animal mass in the world today, and probably have always been the major form of animal life. Egyptian hieroglyphics mention the frightening effect of locust swarms, which can have a mass of 15 000 tonnes. Locust swarms are highly visible, but most insects are much less conspicuous. If you examine a few square metres of typical sheep-grazing soil, and count all the grass grubs, the chances are that the weight of the grubs eating the grass from below will be greater than the weight of the sheep eating it from above. The development of agriculture and the concentration of growing plants was a marvellous step forward — for insects. It concentrated their feeding and breeding. The storing of produce, and the keeping of reservoirs of water for domestic use, also helped insects.

If you are intent on following the advice of your gardening books, and the helpful hints in the gardening section of your newspaper, and wish to continue the chemical warfare in the garden, perhaps this section will help you understand what you are using and help to ensure that it is the insect rather than you that ends up knocked out.

A pesticide is a material that is selective to a degree in killing a pest. Pesticides are classified by their intended use; for example, insecticides, fungicides, herbicides (weedicides), rodenticides, and acaricides (mites, ticks, spiders, etc.). This list is by no means exhaustive and some classifications overlap. These major divisions are subdivided according to chemical type (generic group), mode of action or specificity, and these subdivisions are again not mutually exclusive.

Historical perspective

Control of insects by chemicals goes back to antiquity. The fumigant value of burning sulfur is mentioned by Homer, and Pliny the Elder was aware of the use of soda and olive oil for seed treatment of legumes and the use of arsenic to kill insects. Marco Polo reported the use of mineral oil to treat mangy camels in 1300 AD. On Sir Francis Drake's third voyage to the New World in 1572, lemon juice was used against mosquitoes.

The Chinese were known to be using arsenicals and tobacco extracts (nicotine) in the 16th century, while at that time, the people in South America were using rotenone to stun and harvest fish. Tobacco extracts of nicotine were used in the 18th century, and in the early 1800s, mixtures of mercuric chloride and alcohol became fashionable for bedbug control, as well as sulfur for mildew in England.

Prussia declared phosphorus paste the official rodenticide in 1845 and extended it to cockroaches in 1859. The year 1867 saw the introduction of Paris green, and the year 1878, London purple (both arsenical insecticides). They were used in the USA to check the spread of the Colorado beetle and this led to the introduction of probably the world's first pesticide legislation (the Food, Drug and Cosmetic Act (Pure Food Law)), in

1906, and the Federal Insecticide Act in 1910. The US set maximum residue limits (tolerances) for arsenic in apples in 1927.

Lime sulfur was introduced in Europe in 1885 to control downy mildew on vines. In 1896 a French grape grower noticed that the leaves of yellow charlock growing nearby turned black and realised that it was due to the Bordeaux mixture he was using. Thus was born the first chemical herbicide. Iron sulfate was soon shown to kill dicotyledonous weeds, but not cereal crops when both grew together.

The organomercury compounds that had been used to combat syphilis were found in 1913 to be effective in protecting seeds from insect attack. Tar oil used to control aphids by killing their eggs on dormant trees was introduced between the wars. The control of weeds in cereals using dinitro-orthocresol was patented in France in 1932 and thiram was the first dithiocarbamate fungicide (1943). In 1938 the first organophosphate insecticide TEPP was discovered, and also *Bacillus thuringiensis* was first used as a microbial insecticide.

Entokil to Rentokil
Rentokil started out as Entokil and was named by the Professor of **Ento**mology at Imperial College, London, as an insect killer in the 1920s. The 'R' was added later for trade mark reasons.

During World War Two, organophosphorus compounds were discovered in Germany, while the phenoxyacetic acid(2,4-D) types of weedicide were developed in the UK. After the war came the soil-acting carbamate herbicides in the UK and the organochlorine insecticide chlordane in the USA and Germany. Carbamates were introduced in Switzerland as insecticides soon afterwards. The period 1950 to 1955 saw the introduction of urea as a herbicide, captan and glyodin as fungicides, the first synthetic pyrethroid, allethrin, and the insecticide malathion. Then from 1955 to 1960 came the triazine herbicides from Switzerland.

The important systemic fungicide benomyl and soil-acting herbicide glyphosate are US products of the late 1960s, but relatively few actual new *groups* of pesticides have appeared in the past few decades. In 1962, Rachel Carson published her landmark book, *Silent Spring*, that gave the first warning of the environmental effects of persistent pesticides like DDT. Yet ten years later, in 1972, the Australian Academy of Science in a report 'The use of DDT in Australia', still recommended the continued use of DDT, albeit with the 'usual suspect' statements; 'search for alternatives and confirmation of reported non-target effects'. It was this report that triggered the author's activism and led to the lecture course and finally this book. See Fig. 7.3 on p. 177.

The first photostable pyrethroid, permethrin was introduced in 1973.

In general, the demand for chemical pesticides increases with increasing scarcity of land and with better access to markets. Pesticide use is high on deciduous fruits, vegetables, cotton and cereals, and more moderate on citrus fruits, tropical fruits, cocoa, coffee, and tea.

In the mid-1980s, developing countries accounted for about 20% of the global consumption of pesticides (40% of this in East Asia, including China).

In humid tropical countries, pest generations may follow each other without being reduced by low temperatures or aridity. Pest pressure is high, and fungal infection is strong. Low labour costs do not make herbicide use competitive with manual weed control. Pesticide use in the developing world was about 530 000 tonnes (active ingredient) in 1985, down from 620 000 in 1980, but then increased by 1% per year throughout the world. Since 1990 there has been a slow worldwide decline in amount of chemical active sold.

In Australia, the 1993/94 figures for pesticide usage (based on sales) are as follows ($AUS mill.): Paraquat $51.1; Diquat $38.9; Chlorpyrifos $37.5; Diazinon $23.0; Fenthion $6.8; Chorfenvinphos $4.1; Fenitrothion $2.2; Moncrotophos $1.1; Cyanazine $0.79.

Insecticides

Insecticides are sometimes divided *functionally* into:

- stomach poisons (require ingestion);
- contact poisons (absorbed through the cuticle);
- fumigants (gases).

Chemically the more important groups of insecticides are:

- the inorganic group;
- organochlorine compounds (chlorinated hydrocarbons), such as DDT, aldrin;
- organophosphorus compounds (e.g. parathion) and carbamates, which have similar action (e.g. Aldicarb, Methiocarb, Pirimicarb);
- plant extracts (botanical), for example rotenoids, pyrethrins and their synthetic analogues, the pyrethroids.

The inorganic group are almost exclusively stomach poisons, which are active only after ingestion, and are thus restricted mainly to chewing insects. They are not very effective against sucking insects such as aphids and mosquitoes. Inorganic pesticides are the most persistent.

Other classes of chemicals used either alone or in conjunction with insecticides are attractants and repellents, and synergists.

Inorganic insecticides

Typical inorganic insecticides are usually heavy metal compounds, particularly of lead, mercury, arsenic, and antimony, although some others, such as fluoride salts (e.g. NaF), sulfur and polysulfides, and borax, have limited use.

Lead arsenate ($PbHAsO_4$) is a typical heavy metal compound. It is toxic and extremely persistent. It is water insoluble and thus is effective only by ingestion. The lead ties up essential sites on enzymes and is non-specific. It is not readily absorbed by plants on contact. Sodium arsenite was once used as a cattle dip to kill ticks.

Sodium fluoride (NaF) and *cryolite* (Na_3AlF_6). These compounds liberate fluoride ion, which precipitates Mg^{2+} as fluorophosphate and upsets magnesium-dependent enzymes. It is non-specific and toxic to animals.

Borax ($Na_2B_4O_7$). This is used as a cockroach and ant poison but is not very toxic to mammals.

Elemental sulfur (S) and *lime sulfur* (CaS_n, $n \approx 5$). Lime sulfur is a solubilised form of sulfur (a polysulfide). Its mode of action is through aerial oxidation to SO_2 which is an effective fungicide and acaricide, but is of limited use as an insecticide. It is one of the safer fungicides.

Copper compounds. Two copper compounds are widely used by home gardeners: Bordeaux mixture, a combination of copper sulfate and lime; and Paris green, copper aceto-arsenite, which is very poisonous to humans. Copper sulfate causes vomiting and promotes its own elimination — feedback, so to speak!

Timbers impregnated by the pressure pot method with various brands of copper–chromium arsenite (CCA) appear greenish in colour. Pine logs treated this way are used as fence posts and in children's playground equipment. The arsenic in CCA-treated timbers is strongly locked into the wood. Toxic doses would require the ingestion of about 10–20 cm^3 of treated wood and it is readily excreted if absorbed. The only place where a problem might arise is when such treatment is taking place, or the timber is being sawn and livestock have access to the sawdust. Exhaust ventilation should be provided when sanding or finishing treated timber, and the urine of regular workers should be analysed for arsenic content. Dermatitis can result from chronic skin contamination.

The whitish crystals or powder-like substances that appear on newly purchased timber are mostly sodium sulfate, which is harmless. Timber that has been aged properly after treatment, and washed, should not show this effect.

WARNING: Burning of treated timber or sawdust is dangerous because arsenic is released as a vapour.

Natural insecticides

There are several well-known insecticides of plant origin. Nicotine is a potent insecticide. The rotenoids (sold as Derris Dust) are not persistent and must be sprayed every three days. They are also used to poison fish. Garlic oil has been shown to be effective against the larvae of mosquitoes, house flies and other pests.

Some of the oldest and best known of the natural insecticides are those that make up the class of compounds known as pyrethrins, which are extracted from the pyrethrum flower, a daisy-like member of the *Chrysanthemum* genus, which originated in Persia (Iran) but was being cultivated in Yugoslavia by the mid-nineteenth century. Pyrethrins have a remarkable 'knock-down' effect on flying insects, but a second poison is needed to kill the insect. They have extremely low toxicity to warm-blooded animals. Currently, the major production is in East Africa. They are expensive and photosensitive and thus commercially unreliable. Synthetic pyrethroids are now used widely on fields because they are not photosensitive.

Limonene (citrus peel oil) is the latest addition to the botanical insecticides (although Drake was using it in 1572). It is effective against fleas, lice, mites, and ticks. It has a similar action to pyrethrum. Oil extracts of the neem tree contain an active ingredient *azadirachtin* (a nortriterpenoid of the limonoid class) which is a light-green powder with a garlic-like odour. It has a broad range of insecticidal, fungicidal, and bactericidal activity, as well as insect growth regulator qualities.

MURDEROUS ORGANOFLUORIDES IN PASTURES AND POISONS

Gibb Bogle was a charming man. He was also a brilliant physicist at CSIRO Radiophysics Laboratory next to the Chemistry Department at Sydney University. In 1962, he and the wife of a colleague died in a park after leaving a party under what the local constabulary called 'mysterious circumstances'.

The cause of death was never discovered, but there was suspicion that a metabolic poison such as fluoroacetate was involved. The Bogle–Chandler case has gone down as one of Australia's most bizarre unsolved mysteries.

Fluoroacetate blocks the Krebs (citric acid) cycle in the metabolism of animals, by replacing normal acetate at a critical juncture. (See Sports chemistry and biochemistry in Chapter 4.) Just as a washing machine with a blocked pump will go on and overflow with water, so the Krebs cycle goes on and builds up citric acid. The muscles (usually the heart muscle first) are deprived of energy, and fail. At the time of the Bogle–Chandler case, a metabolic poison was not considered by Rolle Thorpe, the investigating Professor of Pharmacology (and incidentally, the founder of the consumer movement in Australia). This particular poison is known commercially as Compound 1080, and is used to bait rabbits, dingoes, and foxes.

During World War Two there was a rumour that the Germans were going to use a volatile derivative as a war gas. Some research was carried out at Cambridge aimed at countering this, and a successful gas mask was developed. It later turned out that the rumour was wrong. Further studies at Cambridge detected low non-toxic levels of this and related compounds in tea, so humans were apparently not the only species to have learnt how to make and use 1080.

Security?
Gibb was the co-supervisor for the author's Master's thesis, and on the basis of comparing his charm with the security value of his research into masers, I believe it is less likely that the deaths were instigated by agents of some foreign secret service, and more likely by a femme-très-fatale.

From the hallowed halls of Cambridge the scene shifts to the dead heart of Australia, between Alice Springs and the Queensland border. Here graziers have lost untold numbers of stock over the last century from toxic pastures, causing so-called Gidyea poisoning. Many owners have walked off their land; others have developed great skills in assessing the danger from the look of the landscape. In Western Australia, many of the pea flowering plants contain the toxin as well.

It took a century to understand the problem. The 'poison country' was first identified in the 1890s but the toxin was not found until the 1960s. The answer came as a very cunning piece of plant chemistry. Certain plants, notably Acacia Georginae, could take up low levels of ordinary fluoride from the soil (particularly over limestone and dolomite), convert it to fluoroacetate, and concentrate it in the pods at a sufficient level to give cattle and sheep heart failure. In South Africa there was even one plant that could build up a level of one per cent, one leaf of which was sufficient to kill a bullock.

Now Keith Gregg at the University of New England has taken a gene from a soil bacterium which can detoxify the compound and inserted it into another bug that can live in the rumen of stock (from a Canadian white-tail deer). There is some concern that this approach may lead to immunity in pest animals that are currently susceptible to 1080, but this is unlikely as the bug lives only in the forestomach of ruminants and would not persist in rabbits, dingoes and like pests.

Note: There is no relation between the toxicity of fluoroacetate and fluoride. (See Toothpaste in Chapter 4).

Organochlorine compounds

Bubonic plague is estimated to have killed 100 million people in the 6th century, topped up with another 25 million in the 14th century. Typhus killed 2.5 million Russians during the First World War and was a critical factor in the collapse of the Balkans and in the defeat of the Russian armies on the Eastern front. Malaria has long been *the* most fatal and debilitating human disease. As late as 1985 three million people died in an Ethiopian epidemic, and in 1968, Sri Lanka (then Ceylon) suffered more than one million cases.

What do these diseases have in common? They are all spread by insects. The flea (plague), the louse (typhus), and the *anopheles* mosquito (malaria) have all been controlled since 1939 with DDT (1,1-bis (4-chlorophenyl)-2,2,2-trichloroethane). In the developed world, DDT has been replaced by other chemical, medical, and physical means.

The discovery of the insecticidal properties of DDT in the Swiss laboratories of Geigy Pharmaceuticals in 1939 was an event of major importance. Before this discovery the main insecticides available were natural products. DDT is a highly effective insecticide, both by contact and by ingestion, and is of very low toxicity to mammals. Oral LD_{50} is 300–500 mg/kg; dermal LD_{50} is 2500 mg/kg. It is practically odourless and tasteless, and it is chemically stable, which is now seen as a major disadvantage. It can be made in a one-step reaction from low-cost raw materials and therefore was cheap.

Increasingly, many insects developed resistance to DDT by the natural selection of surviving members whose enzymes can detoxify the DDT to a substance known as DDE, which is flatter than DDT, and the change in shape alters the toxicity. The chemically stable DDT and its metabolic breakdown product DDE tend to accumulate in the fat of birds and fish because these creatures are at the end of the biological chains.

The mosquito eradication campaign on Long Island, New York, showed levels of DDT in the sea water of 3×10^{-6} mg/kg (non-toxic); in the fat of plankton, 0.04 mg/kg; in the fat of minnows, 0.5 mg/kg; in the fat of needlefish, 2 mg/kg; and in the fat of cormorants and osprey,

25 mg/kg. This progression is called biological magnification. The level in the birds is biologically active and is believed to upset the metabolism of the female hormone oestrogen. It has been alleged that this has resulted in eggs being produced with very thin shells that break easily.

Humans are also at the end of a food chain and organochlorines will concentrate in fatty tissues, including the milk of nursing mothers. Decades after the removal of most organochlorines from use in Australia, the residues, although decreasing, are still found, see Tables 7.2 (below) and 7.3·(overleaf).

TABLE 7.2

Amount of organochlorines (μg/kg fat) in breast milk of women in Victoria, Australia

Organochlorine	Number*	Mean	SD	Median	Range
p,p'-DDE	60	960	892	632	150–390
p,p'-DDT	58	225	177	176	6–960
Aldrin	3	20	20	nr	nr
Dieldrin	26	159	306	48	13–190
Heptachlor epoxide	18	61	39	51	5–150
HCB	59	411	1350	69	16–7600
α-HCH	56	71	67	46	6–270
β-HCH	54	345	680	184	1–4400
Lindane	46	108	101	73	3–480
δ-HCH	43	121	107	115	4–510
Oxychlordane	48	130	140	74	5–540
trans-Nonachlor	20	77	98	33	9–380
Tri-CB	36	126	258	42	7–220
Tetra-CB	9	34	27	24	2–88
Penta-CB	25	256	282	120	8–1000
Hexa-CB	10	77	63	60	22–230
Hepta-CB	12	7	11	2	0.2–33

nr = not relevant; CB = chlorinated biphenyls
* Total sample number = 60
Numbers, means and medians are for all samples greater than the detection limits.

Source: Persistence of organochlorines in breast milk of women in Victoria, Australia, P.M. Quinsey, D.C. Donohue, and J.T. Ahokas, Key Centre for Applied Nutritional Toxicology, RMIT University, GPO 2476B, Melbourne, publ. Food Chem. Toxic, **33** (1): 49–56, Elsevier Science Ltd, 1995.

The organochlorine pesticides are so called because they have one or more chlorine atoms attached to the carbon atoms, replacing hydrogen. DDT was the first, the most dramatic in terms of human life saved, and the safest in terms of immediate short-term effects. As already mentioned, the organochlorines are cheap to make. Often only a couple of chemical steps are needed, including the famous Diels–Alder reaction (hence dieldrin and aldrin). As well as meeting an important human need, the organochlorines solved an industrial problem. At the very base of an emerging chemical industry is the process of passing electricity through a solution of salt to produce caustic soda and chlorine, just as

Supertermite
Of the 200 or so termite species in Australia, one stands out in size and diet — Mastotermes darwiniensis Froggat. It can grow to about 15 mm — three times the size of other termites. It is found north of the Tropic of Capricorn and, although it prefers a diet of timber, it has been known to attack plastic cables, cow-dung pads, paper, wool, corn, bagged salt, ivory, bitumen, pebbles, ebonite, lead, and even billiard balls. You can actually hear it in the walls of your house if it has decided to have a snack there. What a super recycler! The actual digestion is done by a complex of single-cell organisms that live symbiotically in the termite's gut.

TABLE 7.3

Median concentrations of organochlorine pesticides in adipose tissue*, Western Australia, 1970–1991

Survey	DDT	Dieldrin	HCB	Heptachlor	Chlordane
1970	2.75	0.19	—	—	—
1988	2.40	0.20	0.50	nd	nd
1991	0.72	0.04	0.14	0.03	0.02

* Adipose tissue concentrations as mg/kg extractable fat.
nd = not detected.

Adapted from The Medical Journal of Australia, Vol. **158**, 15 February 1993.

occurs in salt-water swimming pool electrolytic chlorinators, except that the concentration of salt is very high (see Chapter 8). The demand for caustic was very high for making soap and paper, and more recently bauxite processing in the manufacture of aluminium. The demand for chlorine was low. Organochlorines, and later chlorinated plastics such as PVC, restored a profitable balance.

Lindane/gammexane/BHC (benzene hexachloride) This substance, known by many names, was first discovered in 1825 by Faraday. It is made simply by the addition of chlorine to benzene in the presence of light. A mixture of nine isomers is formed, but the only active one is the γ-isomer, which forms only 13–18% of the mixture. The insecticidal properties of the γ-isomer were discovered in 1943. Detoxification is accomplished by elimination of HCl. Lindane is still accepted as a treatment for head lice in children. The following related compounds, however, are now considered too dangerous to be used for any purpose.

Aldrin, chlordane, dieldrin, heptachlor, endrin These compounds form a closely related group of insecticides, and are all made by a common final step, a Diels–Alder reaction with hexachlorocyclopentadiene. Chlordane and heptachlor are less toxic than dieldrin or aldrin.

The use of organochlorine pesticides peaked worldwide in the 1970s (Rachel Carson's *Silent Spring* was published in 1962) and has declined ever since. See Fig. 7.3. Chlordane and heptachlor are not banned in Australia but legal uses have virtually disappeared, except against *that* termite in the Northern Territory. However, the products are no longer available from its only Western manufacturer.

Chlordane and heptachlor are included on a list of persistent organic pollutants (POPs) (see Chapter 2) that have been identified by the United Nations Environment Program (UNEP) for possible international phase-outs on their production and use. POPs are subject to long-range global transport and are highly persistent, semi-volatile, highly toxic, and liable to bioaccumulate. Both pesticides are currently included under the Prior Informed Consent (PIC) procedure, an international program designed to notify governments about (and give them an opportunity to prevent) certain hazardous pesticide imports. PIC is currently voluntary, but

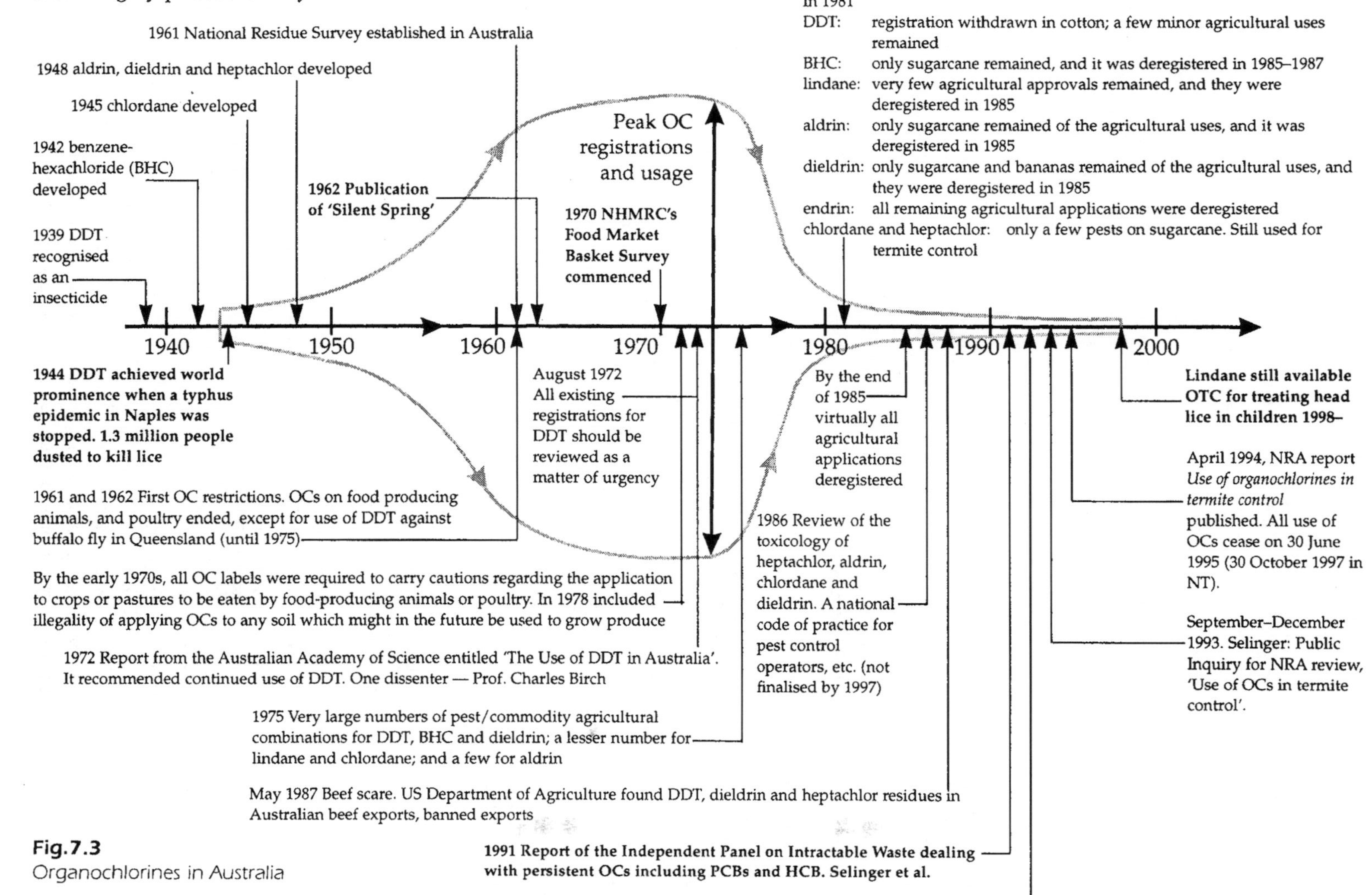

Fig.7.3
Organochlorines in Australia

Data from National Registration Authority and Department of Primary Industries and Energy governments are negotiating to make the procedure a legally binding instrument.

Organophosphorus insecticides

The organophosphorus insecticides are a wide variety of compounds with a tremendous range of activity, persistence, specificity, and function. All have the general formula (RR'X)P = O (Fig. 7.4), where R and R' are short-chain groups, and X is a leaving group, especially selected so that it is easily removed from the molecule, either directly or after a reaction in the body. This group is built in so that the persistence of the substance is reduced.

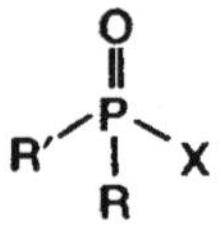

Fig. 7.4 Organophosphates. P has valencies 3 or 5. It is the 5-valent state that is of importance for insecticides.

The five compounds depicted in Figure 7.5 are among the best known organophosphorus insecticides. Some are very easily and cheaply made.

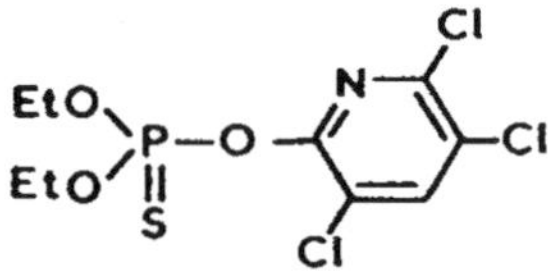

Common name: *chlorpyrifos* (BSI, ISO-E, ANSI, ESA, BAN)
Chemical names:
O,O-diethyl *O*-3,5,6-trichloro-2-pyridyl phosphorothioate (IUPAC)
O,O-diethyl *O*-(3,5,6-trichloro-2-pyridinyl) phosphorothioate (CA)

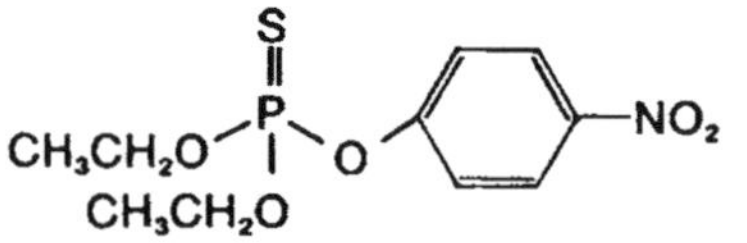

parathion (code number E 605)
oral LD_{50} 3–6 mg/kg
dermal LD_{50} 4–35 mg/kg

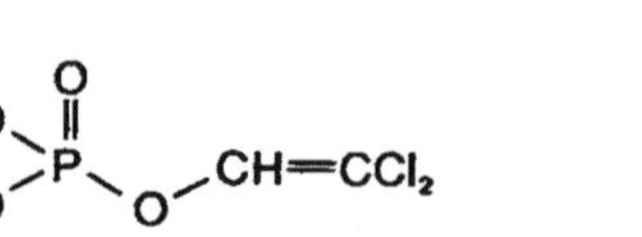

dimethoate (Rogor)
oral LD_{50} 200–300 mg/kg
dermal LD_{50} 700–1150 mg/kg

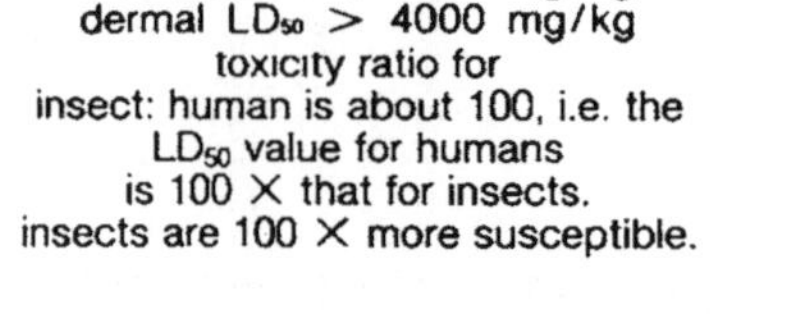

maldison
(Malathion)
oral LD_{50} 1400–1900 mg/kg
dermal LD_{50} > 4000 mg/kg
toxicity ratio for insect: human is about 100, i.e. the LD_{50} value for humans is 100 × that for insects. insects are 100 × more susceptible.

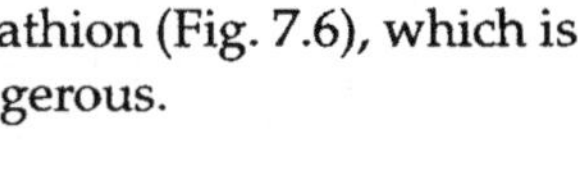

dichlorvos (Shelltox strips)
oral LD_{50} 25–30 mg/kg
dermal LD_{50} 75–900 mg/kg

Fig. 7.5 Some organophosphorus insecticides

An example is the production of parathion (Fig. 7.6), which is still used because it is cheap — but it is also dangerous.

$$P_2S_5 + EtOH \longrightarrow (EtO)_2P(=S)SH \xrightarrow{Cl_2} (EtO)_2P(=S)Cl$$

$$(EtO)_2P(=S)Cl + \text{4-}NO_2C_6H_4O^-Na^+ \longrightarrow \text{parathion}$$

Fig. 7.6
Synthesis of parathion

The other important characteristic, apart from ease and cheapness of production, is persistence — how long the material persists before chemical degradation. For crop-spraying insecticides, this is specified by a withholding period — the number of days between application and harvesting. (See MRL in box Mevinphos on p. 184.)

$t_{½}$ = time for half the material to disappear
$t_{99}\%$ = time for 99% of the material to disappear

Thus, for TEPP the LD_{50} oral/rat value is 1 mg/kg, which is highly toxic. The half-lives are:

$t_{½}$ (25°) = 6.8 hours
$t_{99}\%$ (25°) = 45 hours — quickly detoxified

For *parathion*, the LD_{50} oral/rat is 3–6 mg/kg — highly toxic.

$t_{½}$ (28°) in water = 120 days — very persistent

Parathion is used as both the methyl and ethyl ester. It has probably been responsible for more deaths than any other pesticide. The methyl ester has not caused many deaths but this is probably because of lower usage.

Malathion like parathion, is persistent but has low *mammalian* toxicity. The LD_{50} oral/rat is 1300 mg/kg, because mammals and insects detoxify it differently, and with very different efficiency (see Fig. 7.7). (For trade mark reasons, malathion cannot be used as a generic name in Australia — maldison is used instead.)

Dichlorvos (dimethyl dichlorovinyl phosphate — DDVP) is relatively volatile (low molecular mass), broad spectrum, but rather toxic to mammals: LD_{50} oral/rat is 30 mg/kg. It is not very persistent: $t_{½}$ (25°C in water, pH 7) is eight hours. This compound is used in enclosed slow-release formulations in cockroach baits.

All these organophosphorus insecticides are contact poisons that are rapidly absorbed through the insect cuticle. However, they can also be absorbed through human skin during spraying operations. One way to solve this problem is to use *systemics*. Systemic insecticides are absorbed into the plant (e.g. through the leaves) and are ingested by sap-sucking insects, along with the plant juices (but they do not harm the plants

Run, it's a cereal killer!
Alternatively, for a cockroach poison, take four or five parts wheat flour, one part sugar, plus two heaped tablespoons of boric acid. Mix into a light dough with milk; form into little balls.
In Madras, cockroaches would gnaw the dead skin of children until they hit raw flesh and there was a scream of pain!

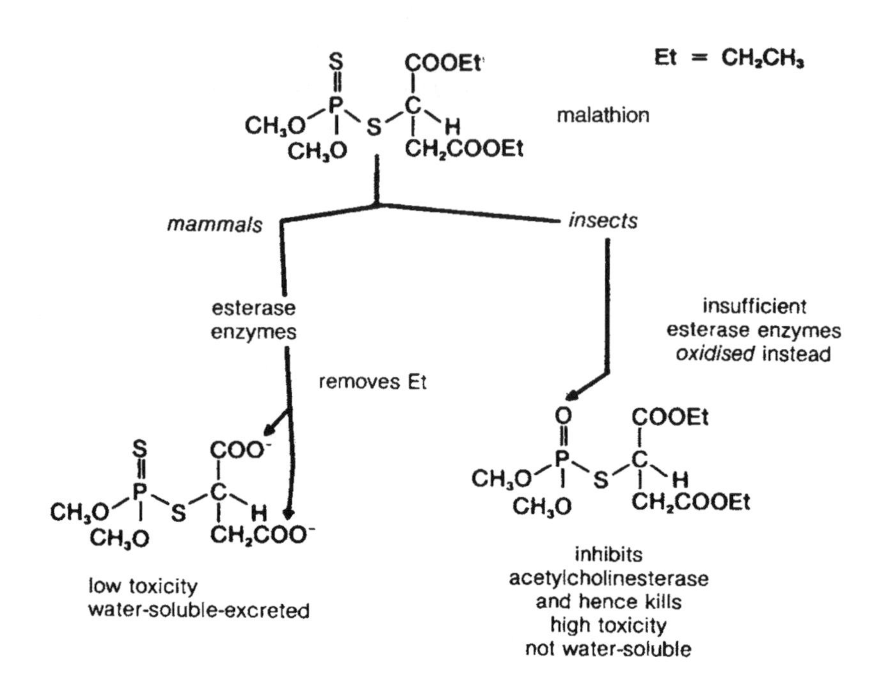

Fig. 7.7 Detoxification of malathion

Fly swats
Flies tend to flit indoors and then, attracted to light, whiz up and down the window. This is a good moment to swat them. The insect has a marvellous reflex action and can respond in milliseconds. However the nervous system is 'hard-wired'. It has no brain, but produces automatic responses to events. It is not programmed to respond to two threats at a time. Presto! Use two fly-swats, moving the hands to and fro slightly to confuse the response. Then pounce with both hands simultaneously! Exit fly, scientifically.

because they do not have a nervous system). They are equally effective against both chewing and sucking insects and are administered as a spray, or to the plant roots as granules, or injected into the plant trunk. Typical organophosphorus systemics are demeton and dimethoate.

Demeton (Systox) is a mixture of two components:

$(EtO)_2P(=S){-}OCH_2CH_2SEt$ (Demeton O), and
$(EtO)_2P(=O){-}SCH_2CH_2SEt$ (Demeton S)

It is a broad-spectrum, persistent systemic, very toxic to mammals: LD_{50} oral/rat is 9 mg/kg. It was commonly used by home gardeners, and cases of poisoning have resulted from absorption through the skin.

Dimethoate (Rogor) is also broad spectrum, but is much less persistent than demeton because it is detoxified by the plant fairly rapidly. It is formed by the reaction:

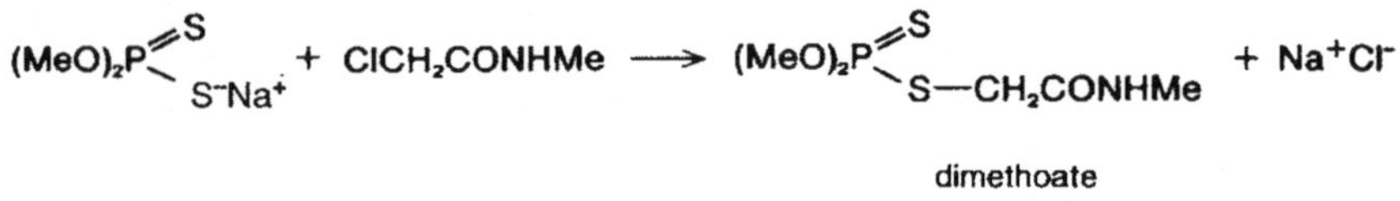

$$(MeO)_2P(=S)S^-Na^+ + ClCH_2CONHMe \longrightarrow (MeO)_2P(=S){-}S{-}CH_2CONHMe + Na^+Cl^-$$

dimethoate

Fig. 7.8
The dimethoate reaction

Comparative toxicities of organophosphates and organochlorides are given in Tables 7.4 and 7.5. The longer the bar, the less toxic is the chemical. Source: The pesticide book (see Bibliography).

TABLE 7.4

Acute oral LD_{50}s for rats and dermal LD_{50}s for rabbits for some organophosphate insecticides

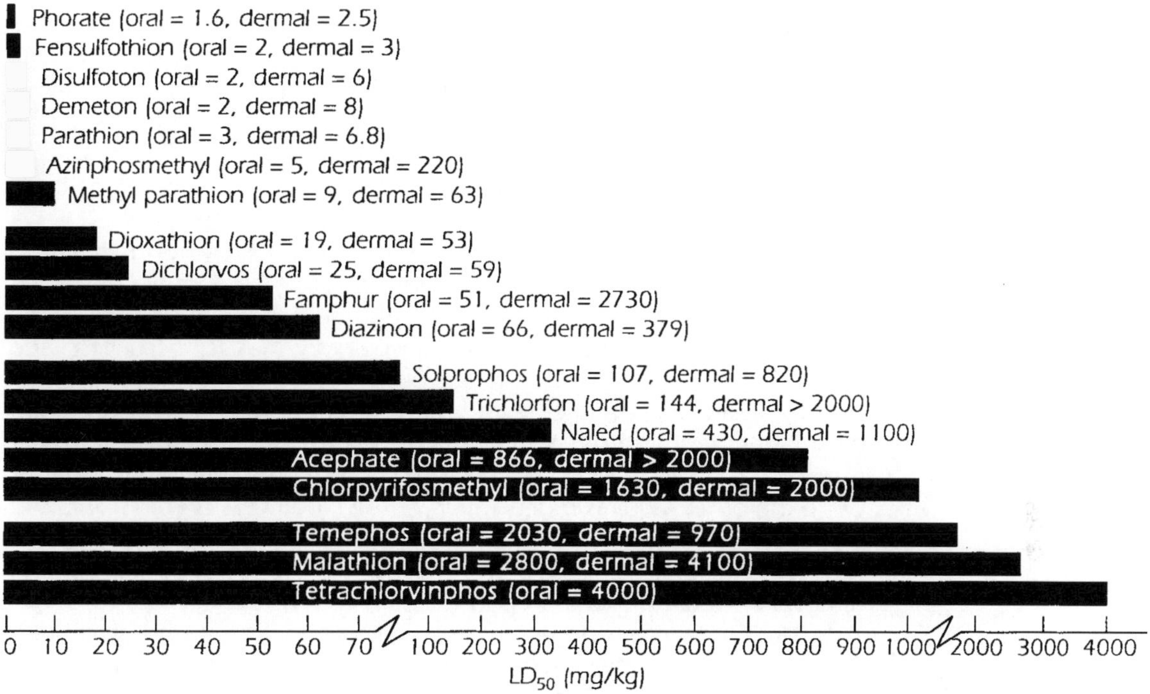

Phorate (oral = 1.6, dermal = 2.5)
Fensulfothion (oral = 2, dermal = 3)
Disulfoton (oral = 2, dermal = 6)
Demeton (oral = 2, dermal = 8)
Parathion (oral = 3, dermal = 6.8)
Azinphosmethyl (oral = 5, dermal = 220)
Methyl parathion (oral = 9, dermal = 63)

Dioxathion (oral = 19, dermal = 53)
Dichlorvos (oral = 25, dermal = 59)
Famphur (oral = 51, dermal = 2730)
Diazinon (oral = 66, dermal = 379)

Solprophos (oral = 107, dermal = 820)
Trichlorfon (oral = 144, dermal > 2000)
Naled (oral = 430, dermal = 1100)
Acephate (oral = 866, dermal > 2000)
Chlorpyrifosmethyl (oral = 1630, dermal = 2000)

Temephos (oral = 2030, dermal = 970)
Malathion (oral = 2800, dermal = 4100)
Tetrachlorvinphos (oral = 4000)

0 10 20 30 40 50 60 70 100 200 300 400 500 600 700 800 900 1000 2000 3000 4000

LD_{50} (mg/kg)

TABLE 7.5

Acute oral LD_{50}s for rats and dermal LD_{50}s for rabbits for some organochlorine, carbamate, botanical, pyrethroid, and formamidine insecticides

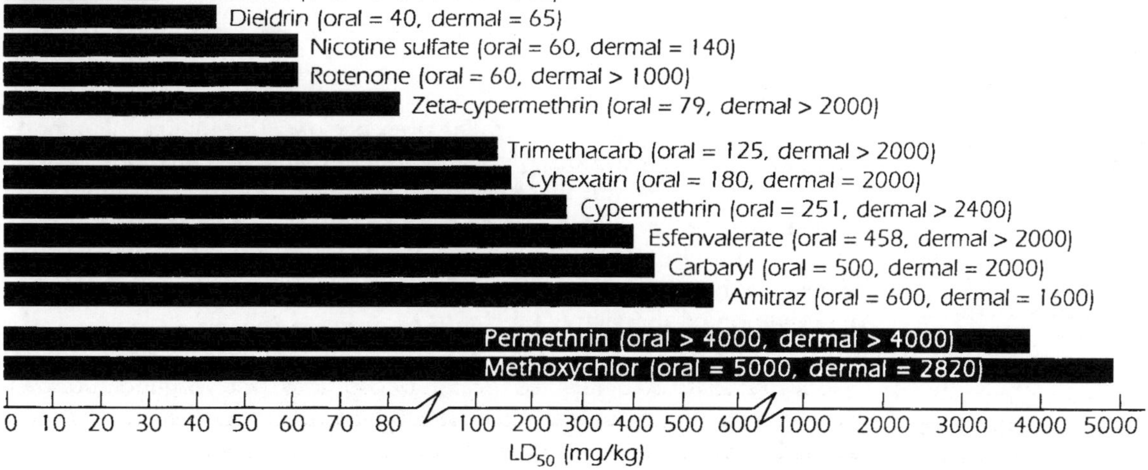

Aldicarb (oral = 0.9, dermal = 5)
Endrin (oral = 3, dermal = 12)
Oxamyl (oral = 5, dermal = 710)
Carbofuran (oral = 8, dermal = 2550)

Methomyl (oral = 17, dermal = 1000)
Endosulfan (oral = 18, dermal = 74)
Bendiocarb (oral = 34, dermal = 566)
Dieldrin (oral = 40, dermal = 65)
Nicotine sulfate (oral = 60, dermal = 140)
Rotenone (oral = 60, dermal > 1000)
Zeta-cypermethrin (oral = 79, dermal > 2000)

Trimethacarb (oral = 125, dermal > 2000)
Cyhexatin (oral = 180, dermal = 2000)
Cypermethrin (oral = 251, dermal > 2400)
Esfenvalerate (oral = 458, dermal > 2000)
Carbaryl (oral = 500, dermal = 2000)
Amitraz (oral = 600, dermal = 1600)

Permethrin (oral > 4000, dermal > 4000)
Methoxychlor (oral = 5000, dermal = 2820)

0 10 20 30 40 50 60 70 80 100 200 300 400 500 600 1000 2000 3000 4000 5000

LD_{50} (mg/kg)

DANGERS OF ORGANOPHOSPHATES

Poisoning by organophosphates may occur if they are inhaled, swallowed, or come in contact with the skin. The first symptoms of poisoning arise within minutes, or at the most, hours after exposure. They are usually headache, fatigue, giddiness, saliva formation, sweating, blurred vision, pin-point pupils, chest tightness, nausea, abdominal cramps, and diarrhoea. If someone who has been using one of these insecticides develops these symptoms, **they should seek medical help immediately.**

When absorbed, these pesticides inactivate an enzyme in the body called acetylcholinesterase. This enzyme is normally responsible for returning to normal a compound (acetylcholine) used to activate muscles. When the enzyme is inactivated, spasms in involuntary muscles occur first, and overactivity of certain glands can take place. A second enzyme, called non-specific esterase, occurs in the bloodstream. Monitoring of blood cholinesterase of exposed persons has been used as a safety measure. After exposure, several days can be needed to restore normality in one enzyme, while the other requires regeneration of red blood cells and this occurs at 1% of the normal value per day.

Treatment involves inducing vomiting (for swallowed poison) and removal of clothing and thorough washing of skin.

In 1937, while carrying out research on organophosphorus insecticides, Schrader, in Germany, discovered the nerve gases. The first was called tabun (Fig. 7.9). These compounds will kill humans as well as insects. A 0.2 mg splash on the skin is fatal; in fact, every accident the workers had was fatal. Although outlawed by definition by the Geneva Convention of 1925, tabun was available for use in the Gulf War in 1984.

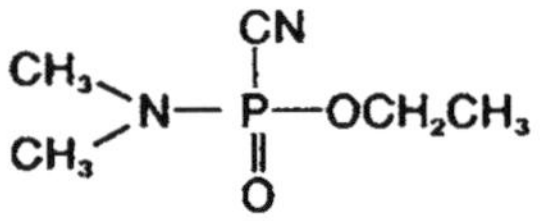

$(CH_3)_2N{-}P(=O)(CN){-}OCH_2CH_3$

Fig. 7.9
Tabun

Plate VII illustrates the handling of toxic nerve gas weapons following the Gulf War.

The persistent organophosphorus insecticides are mostly thiophosphates (Fig. 7.10) (thio = sulfur). The presence of sulfur makes them more resistant to decomposition by water, and lengthens their shelf life as well as their persistence. But these thiophosphates are not reactive enough to be toxic. Except for insects, living organisms lack enzymes that can convert the thiophosphates to their toxic reactive phosphate analogues, which are the active insecticides. This is an example of *biological priming* or activation of an inactive precursor.

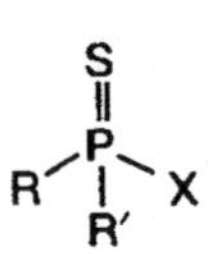

$R{-}P(=S)(R'){-}X$

Fig. 7.10
Organothiophosphates

Water solubility for most of the contact organophosphorus insecticides is very low (typically 0.000 01%) but the water solubility is improved by selecting a suitable group, X, which changes with time and allows the compound to become water soluble.

Systemics are necessarily fairly persistent to allow time for the slow plant transportation process.

Carbamates

Carbamates act in a similar way to the organophosphates. They are based on carbamic acid (Fig. 7.11).

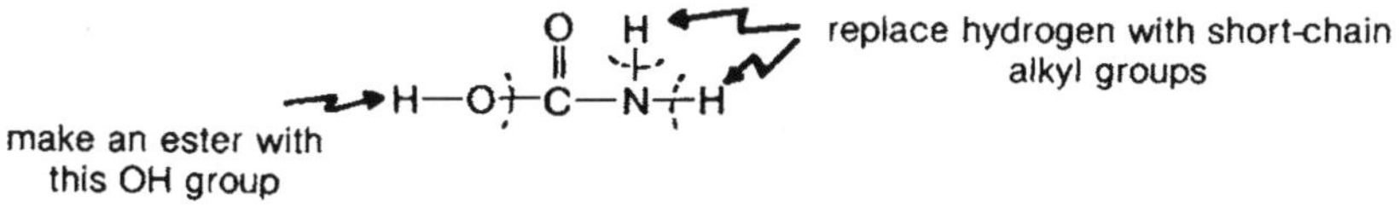

make an ester with this OH group

H—O—C(=O)—N(H)—H

replace hydrogen with short-chain alkyl groups

Fig. 7.11 Carbamic acid and derivatives

Dealing with new concerns about older chemicals

Chemicals are under continuous review, and in Australia this is done by the National Registration Authority for Agricultural and Veterinary Chemicals (NRA). The first cycle of review involved five active chemicals found in 64 products. These were parathion, parathion ethyl, mevinphos (see below), endosulfan, and atrazine. In the case of parathion, the data submitted weighed over half a tonne. In a second cycle, NRA is looking at seven active chemicals in 213 products. Then there are special reviews triggered by particular concerns, which currently involve 161 products.

SECOND CYCLE NRA REVIEW

The seven chemicals currently under review in a second cycle are:

chlorpyrifos, (widely used organophosphate): concerns with health and worker exposure, water pollution, and effects on wildlife.

diazinon, (used in dog washes and on lawn grub killers): concerns with groundwater contamination, bird kills, worker exposure

fenitrothion, (broad range insecticide, outdoor foggers, grain protectorant): concerns with chronic effects on workers, acute on wildlife

monocrotophos, (kills mites): worker exposure and bird hazard

chlofenvinphos, (cattle and sheep dip): worker exposure — overseas concerns

dichlorvos, (pest strips, cat collars, parasite treatments for animals) acute toxicity, potential carcinogen

demeton-S-methyl, (systemic insecticide) reports from US of acute poisoning — only one product in Australia.

The review is by experts in the area of human health, occupational health and safety, environment, efficacy, and trade issues.

Sometimes, as in the case of the organochlorines and mevinphos, active chemicals are removed completely from the market, but mostly their use is further restricted, or the personal and environmental protection requirements are changed.

For example, examination of a commercial insecticide used by market farmers (but never available to home users) in the first cycle review resulted in the following decision (NRA March 1997):

> 'All non-essential uses of the insecticide mevinphos are withdrawn, and limited use on *Brassica* crops is allowed until the end of 1998'. . . .'Only suitably trained and qualified people will be allowed to purchase, handle, and use mevinphos.'

MEVINPHOS

Mevinphos (BSI, ISO, ESA), $C_7H_{13}O_6P$, is a broad spectrum organophosphorus insecticide. Trade names are Phosdrin, Duraphos, Mevidrin, Mevinox, and OS-2046. Phosdrin™ (Cyanamid) is registered in Australia for the control of a variety of insects on broccoli, cauliflower, cabbage, and brussels sprouts. It is particularly effective in controlling Diamond Black Moth.

The active chemical has two IUPAC names:

> 2-methoxycarbonyl-1-methylvinyl dimethyl phosphate, and
> methyl 3-(dimethoxyphosphinoyloxy)but-2-enoate,

and one CA name:

> 2-butenoic acid, 3-[9dimethylphosphinoyl)oxy]-,methylester; CAS reg. no. 7786-34-7 ((Z)-+ (E)-isomers; 26718-65-0 ((E)-isomer; 338-45-4 ((Z)-isomer).

MeO, MeO, P(=O)–O–C(Me)=C(H)(CO–OMe) **cis**

MeO, MeO, P(=O)–O–C(Me)=C(CO–OMe)(H) **trans**

Fig. 7.12
Cis and trans forms of mevinphos

Like all organophosphates, mevinphos is stable in acid to neutral conditions, but rapidly hydrolysed in alkaline conditions. 50% hydrolysis occurs at pH 6 in 120 days, at pH 7 in 35 days, at pH 9 in 3 days, and at pH 11 in 1.4 hours.

Environment

Mevinphos degrades rapidly in natural systems with a half-life of < 1 day in aerobic soil; because of strong adsorption onto soil, no leaching is expected. It is extremely toxic to aquatic invertebrates (yabbies, etc.), birds and mammals.

Public health

Because of its high acute toxicity it is a hazard to public health and so it is not registered for domestic use. The greatest risk is therefore through food intake. The acceptable daily intake was set at 0.002 mg/kg/day (see below). The recommended ADI (see Chapter 2) is 0.0008 mg/kg/day based on a NOEL (see Chapter 2) of 0.015 mg/kg/day for cholinesterase inhibition in a human study, and using a 20-fold safety factor.

Occupational health and safety

The occupational health and safety risk is assessed by a model called the predicted operator exposure model (POEM). To reach a level equivalent to the dermal $LD_{50,}$ an average 60 kg worker would need to be contaminated with 250 mg of mevinphos, (or 0.23 mL of Phosdrin concentrate or 325 mL of working strength spray (0.07%) active).

In order **not** to show **any** effect from exposure, that is be below the no observed effect level (NOEL) for skin, the exposure must be less than 0.005 mL active chemical per day.

In order **not** to show a **toxic** effect from exposure, the exposure levels must not exceed 5 mg of mevinphos, (or 0.005 mL (5 μL) of Phosdrin or 6 mL of working strength spray). None of these calculations includes a safety factor. Improvement of worker practices cannot sufficiently reduce exposure to protect them adequately.

Residues

The current Codex MRL (maximum residue limit) is 1 mg/kg (broccoli, brussels sprouts and cauliflower). The MRL in Australia is 0.25 mg/kg (for vegetables) but there are no data on actual residue levels.

Trade

It has become an essential chemical in integrated pest management (IPM) strategies (particularly as a backup for failure of other chemicals) and as part of the rotation of chemical groups to reduce resistance. It is important for controlling Diamond Black Moth, as this has become increasingly resistant to other organophosphates, carbamates, and endosulphan. New active chemicals are coming on stream that will be able to replace mevinphos.

Recommendation

Mevinphos has the advantage of disappearing very quickly after use and so protects the final consumer better than most products. However, it creates an unacceptable risk to the farm worker.

Source: National Registration Authority on Agricultural and Veterinary Chemicals, 1997.

Fungicides

Interestingly, it is the fungi (rather than bacteria, viruses, or insects) that have had the most devastating effect on our crops. Potato blight (Irish famine), wheat rust and smut, (ergotism in Europe), and wine mildew, are a few examples with severe historical consequences.

Contact fungicides are designed to attack the fungus directly before it penetrates the plant, and inorganic chemicals were the earliest ones used. The use of sulfur goes back 2500 years. Bordeaux mixture (lime and copper sulfate) was introduced in 1885. In 1934 the iron salt (ferbam) and zinc salt (ziram) of dimethyldithiocarbamic acid were introduced. A related compound, captan, is widely used on fruit trees, and works by liberating thiophosgene ($CSCl_2$) inside the fungus. Nabam is a related compound. The chemical 8-hydroxyquinoline (Quinolinal, Oxine), which needs iron to activate it, is used for mould-proofing outdoor equipment.

Treatment of plants with systemic fungicides is more difficult than treatment of animals with oral drugs because plants lack true circulation or excretion mechanisms. The compound benomyl (Benlate) was introduced in 1968 with great success. It has been particularly useful for protecting stored seeds (used for planting) for which purpose it has replaced the very hazardous mercurials and hexachlorophene, which have killed people who have eaten the seeds by mistake.

Garlic extract
A sulfur-containing antibiotic in garlic is known to be active against a variety of bacteria and fungi and garlic juice has been shown to inhibit the growth of fungus spores on nutrient plates in the laboratory. Garlic extract dramatically reduces disease in germinating bean seeds inoculated with fungus spores, and also increases growth of the root system. So, eat salami and breathe on your roses!

Herbicides

Herbicides are chemicals used by humans to kill unwanted vegetation. The compounds can be completely non-selective, killing every species of plant, or very selective, killing only certain plants.

The use of chemicals that selectively destroy plants had a very slow beginning. During 1895–97, copper sulfate solution was found to kill a weed (charlock) without causing injury to crops, and in 1911, dilute sulfuric acid was used in a similar situation. Dinitro-*o*-cresol was soon found to be useful as well. The 1930s saw the introduction of chemicals that mimic the natural auxin (plant hormone) indoleacetic acid. These were the phenoxyacetic acids, which, unlike the natural auxins, were not destroyed by the plant. They were also very effective in setting unfertilised fruits and promoting root growth. Their use in excess showed that they were very effective herbicides. The most used of these is 4-chloro-2-methylphenoxyacetic acid (MCPA); it is used to kill weeds in cereal crops (Fig. 7.13).

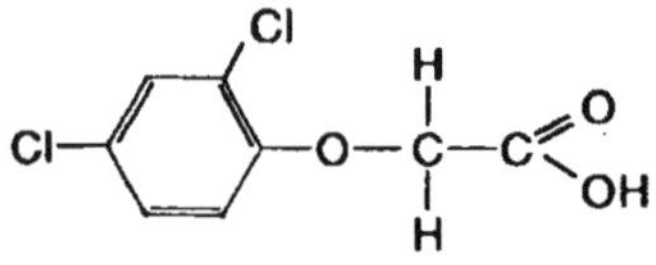

2,4-D
oral LD_{50} 400–500 mg/kg
dermal LD_{50} 1500 mg/kg

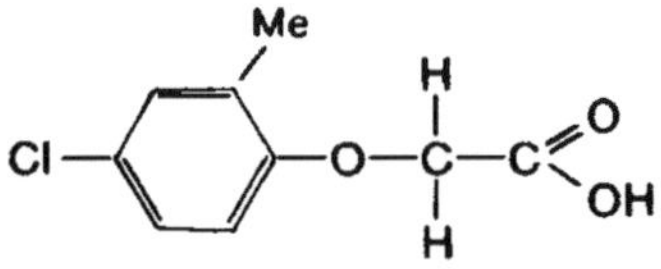

MCPA
oral LD_{50} 800 mg/kg
dermal LD_{50} > 1000 mg/kg

Fig. 7.13 2,4-dichlorophenoxyacetic acid (2,4-D) and 4-chloro-2-methylphenoxyacetic acid (MCPA), Methox, Methoxone, and Tuloxone

The selectivity for killing dicotyledons but not monocotyledons (when equal amounts are *absorbed* by weeds and cereals, the cereals remain unharmed) is explained by the differences in their growing shoot structure.

It was later found that orthocarboxylic acids often behave in a similar manner. Picloram (Fig. 7.14) was introduced in 1963 and found to be more potent than 2,4-D.

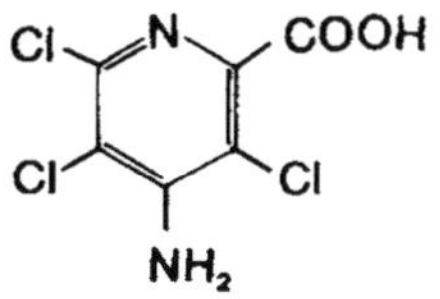

Fig. 7.14
Picloram

The compound 2,4,5-trichlorophenoxyacetic acid was found to be able to deal with such noxious weeds as privet and blackberry and thus became an urban and domestic herbicide (Fig. 7.15).

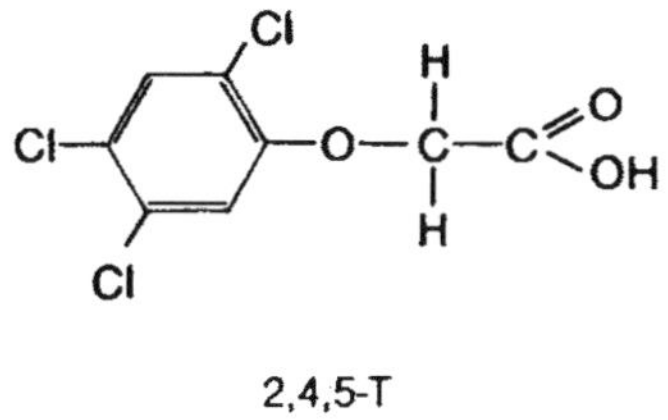

2,4,5-T

Fig. 7.15
2,4,5-trichlorophenoxyacetic acid (2,4,5-T)

The chemical 2,4,5-T was implicated in causing birth defects in animals. It turned out that the cause of the problem was a small quantity of an impurity — a dioxin formed during manufacture (Fig. 7.16) — for which LD_{50} guinea pig is 0.0006 mg/kg. It is fat soluble and may be concentrated like DDT. It is no longer used.

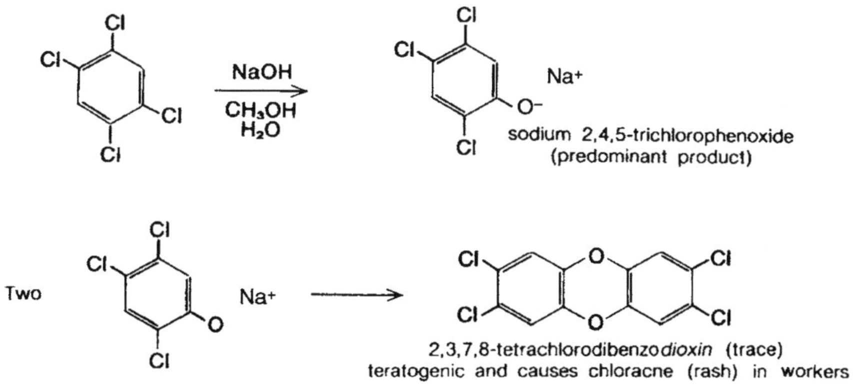

NaOH, CH_3OH, H_2O

sodium 2,4,5-trichlorophenoxide (predominant product)

Two ... Na^+ →

2,3,7,8-tetrachlorodibenzo*dioxin* (trace) teratogenic and causes chloracne (rash) in workers

Fig. 7.16
2,3,7,8-tetrachlorodibenzodioxin (trace)

The redox indicator, methyl viologen, has been used by chemists since 1933, but its herbicidal properties were not discovered until 1955 (by ICI), when it was renamed paraquat. Both uses depend on its ability to form a stable, free radical. (A molecule with a single unpaired electron is called

Genetic labelling rules

The rules agreed to in the European Union (EU) for labelling genetically modified products are as follows: A 'genetically modified living organism' such as tomato, containing seeds, must be labelled. Labels are also required if a novel substance can be detected in the food. Thus labelling is required for genetically modified soya beans, or the solids left when they are crushed, because they contain novel DNA. On the other hand, oil pressed from the beans, or the meat of the animals that ate the solids, are not required to carry labels.

Novel foods must be evaluated for safety by the national authority of one member country before they can be sold anywhere in the EU, but permission may be refused in a country that has concerns. The rules do not apply to additives, flavourings, or to purified products such as sugar from modified sugar beet.

a radical. Such a species is very reactive and chemically ferocious.) Paraquat, diquat and cyperquat are quaternary ammonium herbicides (related to cationic surfactants) and are unique in that they kill by contact with the plant foliage and cannot enter the plant via its roots because they are very strongly adsorbed onto clay and soil particles (not true for hydroponics). The 'quats' are used as chemical ploughs to kill weeds between crops, or just before a crop emerges.

The quats are moderately toxic. Small amounts can irritate the skin or, if inhaled, can cause nose bleeding, but such amounts are rapidly and completely eliminated from the body. However, larger amounts cause a terrible death. The symptoms are minimal for several days, but later a non-cancerous proliferation of lung cells occurs which continues long after all traces of herbicide have disappeared, leading to respiratory failure. There is no known antidote.

Most herbicides generally act on only one target point in the biochemistry of the plant, which causes downstream effects that eventually kill the plant. Thus herbicides can be grouped, depending on their mode of action, and this grouping is important when trying to cycle products to avoid resistance, and when crops are biogenetically engineered so as to be resistant to a particular herbicide to be used on weeds around them.

Use of biotechnology — An example; herbicide-resistant crops

Bacteria resident in the soil of heavily sprayed fields have evolved to degrade the herbicides used there and scientists have taken the genes (that have been selected in the microbes to cope with this assault) and inserted them into the crop that researchers desire to protect from the herbicide. A similar approach is used for biological degradation of hazardous chemical sites by using microbes that have evolved there naturally (sometimes with further genetic modification).

Resistance to herbicides is only one of the first of numerous traits under development. Plants that produce their own pesticides is another. Bt cotton (e.g. Ingard™) has an introduced gene from a bacterium that makes a natural pesticide. This natural pesticide is rapidly degraded by light so spraying is effective for only a short time. Having the pesticide expressed for most of the growing season in the plant is thus much more efficient, but on the other hand it makes it locally 'persistent', with the associated potential for resistance that this threatens. Bt cotton will feature widely in the clothing sold at the Sydney 2000 'green' Olympics.

Other developments involve genetic changes to provide resistance to frost, drought, and disease; and foods that contain more starch, or a different mix of proteins, and so on into the technological future. However, we are ignorant of the potential problems this new approach might surprise us with. New technologies bring tremendous benefits, but invariably there is a price to pay somewhere, sometime. We must remain vigilant and be prepared to continually question scientific salespersons promoting genetically engineered 'snake oil'.

TABLE 7.6
Molecular site of action of commercial herbicides (there is some overlap in action).

Site of action	Some actives	Some trade examples
Inhibits branched-chain amino acid biosynthesis	Sulfonylureas, imidazolines	Glean, Classic, Permit, Escort
Inhibits aromatic amino acid biosynthesis	Glyphosate (more details below)	Zero, Roundup
Inhibits glutamine biosynthesis and halts carbon fixation. (This molecular site is also found in mammals, but they don't fix carbon.)	Glufosinate or phosphinothricin (when obtained from a natural fermentation).	Ignite
Disrupts cell mitosis. Causes swelling and stunting of roots and stops water transportation	Dinitroanilines; N-phenyl carbamate; DCPA, Chlorothal dithiopyr, pronamide	Trifluranil propham, Dacthal, Dimension Stampede
Blocks natural plant hormone (auxin) sites	2,4-D; trichlopyr, dicamba, picrolam; clopyralid	—
Inhibits photosynthetic electron transport (system 2)	s-triazines (atrazine); phenyl ureas	Aatrex Cotoran
Produces photosynthetic toxic superoxide radical (system 1)	Paraquat, diquat, (more details above)	Avenge
Interrupts photosynthesis by producing a toxic porphyrin (protox inhibitors)	Acifluoren, lactofen, oxyfluorfen, formesafen	Blazer, Cobra, Goal
Stops production of cellulose	Dichlobenil, bromoxynil	Casaron, Brominal
Stops production of lipids (fats)	Diclofop, sethoxydim (cyclohexadione)	Hoelon, Poast,

Source: diverse

Glyphosate and synthetic pyrethroids
Glyphosate (Roundup™, Zero™) is strongly adsorbed onto soil particles and rendered inert. Calcium and magnesium also deactivate it, so water hardness should be less than 100 mg/L (as calcium carbonate), or 80 mg/L (as calcium). Synthetic pyrethroids behave in a similar manner. They will kill fish in clean but not muddy water. Sensitivity of some tadpoles to glyphosate formulations has turned out to be due to the spreading agent, not the chemical active.

The molecular site of action of some herbicides is indicated in Table 7.6.

The world's most popular herbicide is glyphosate (*N*-[phosphonomethyl]glycine), the active ingredient in Roundup™, and Zero™, and is used to kill weeds after they have emerged. Glyphosate has an extremely low toxicity to mammals, birds, and fish.

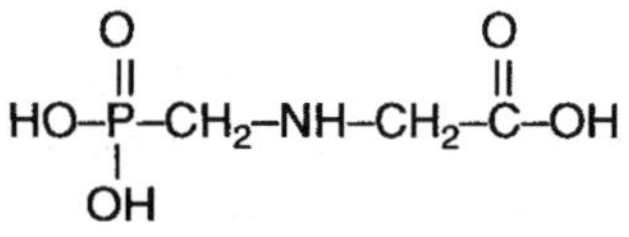

$$HO-\overset{\overset{O}{\|}}{\underset{\underset{OH}{|}}{P}}-CH_2-NH-CH_2-\overset{\overset{O}{\|}}{C}-OH$$

Fig. 7.17
Glyphosate

It is not surprising that this herbicide is one of the first for which biotechnology has been used to protect important crops — Roundup Ready™ genes for soya beans. While this will result in an increased use of this herbicide, the use of others (often less benign) will decrease, and so will soil erosion, because of the minimum tillage regime it allows. Soybeans are capable of producing the greatest amount of protein per unit of land of any major plant or animal source used as food, and they have an unusually good amino acid profile for a plant (i.e. contain more of the essential amino acids). The oil is the major edible oil used in the USA.

There are still many arguments about this use of biotechnology, and in particular about the possibility of creating resistance (which is virtually unknown for glyphosate). There are serious issues of social equity (particularly in poorer countries) about the way agriculture is being 'tied-up' by the combination of herbicide and food crop seeds.

Some politics of herbicides[1]

Australia is a major trading country, and exports represent 14% of the gross domestic product, compared to the USA (7%) and Japan (9%). Of this trade, agricultural products represent about 30% of export earnings, and for wheat, cotton, and sugar cane, about 80% of the production is exported. As a measure of the problem of weeds to crop production in Australia, herbicides represent about 60% of agricultural chemical sales (excluding animal health care). Herbicides have been replacing mechanical control of weeds, reducing labour (and farm accidents), and reducing soil erosion. However, a legacy from the Vietnam war, where herbicides (in particular Agent Orange, a mixture of impure 2,4-D and 2,4,5-T) was used indiscriminately, has clouded attitudes (see earlier editions of this book for a wider discussion).

Sugar beet production in Europe is heavily subsidised for political, historical, and emotional reasons. The crop needs extensive weed control, conventionally one pre-emergent and four post-emergent treatments. The average yield is 8.5 tonnes/ha (raw sugar). The (1996) value of the crop was $2500 per ha, and a substantial fraction of this is (subsidised) herbicide cost. The 2 million hectare sugar production in the European Union alone requires 8000 tonnes of herbicide per annum, with consequential environmental problems. A transgenic sugar beet, resistant to glyphosate, could substantially reduce overall use of herbicide, but also increase the spread of resistance to non-productive beet species. (The answer of course is to buy sugar from efficient (tropical, wide-acre, mechanised) sugar cane producers like Australia and Cuba!).

In terms of value, the Netherlands is one of the top three world exporters of agriculture. It has achieved this position by developing highly intensive and sophisticated systems of primary production, supported

---

[1] *Herbicide-resistant crops and pastures in Australian farming systems*, ed. G.D. McLean and G. Evans, Bureau of Resource Sciences, workshop 15–16 March 1995, AGPS, Canberra, 1995.

by large inputs of pesticides. The pesticide reduction program adopted by the Netherlands imposes new and more demanding environmental criteria relating to persistence, mobility, and toxicity to aquatic organisms, and to soil biota. The target of a 50% reduction of pesticide use in the Netherlands, Sweden, and Denmark seems feasible.

The 'green revolution' dwarf wheat varieties have transformed China and India into the second and third largest wheat producing countries in the world. These varieties had a high grain-to-straw ratio and this tripled yields. This in turn made it economic to use fertilisers and irrigation. However, the dwarf varieties cannot compete with grass weeds and so herbicides were then needed. Unfortunately, both India, and then China used only one herbicide (isoproturon) and in the case of India — 2000 tonnes per annum. This soon caused weed resistance that was predictable from earlier results in England, Germany, Israel, and Spain. While there might be some short-term solutions with other herbicides, genetic modification of the wheat is probably the only longer-term hope.

Fair comment
Because chemicals allowed the dramatic worldwide increase in food production per hectare, it would appear that they are absolutely essential to this expansion. Indeed, numerous studies show that banning some or all herbicides and pesticides would have a draconian and unacceptable effect on production. But is this really a fair comment?

If the agricultural research community had put as much resource into non-chemical improvements in agriculture as it has into chemicals over the last 50 years, then an environmentally and socially more benign production regime might well be in place. Weed control by rotation, tillage, and other management practices is a public good. Unlike a product such as chemicals, it cannot easily be embedded into products that become property rights, and can thus be sold. World War Two gave the chemical industry an opportunity to produce chemicals en masse, and after the war it saw a future in switching large-scale production to fertiliser and pest control (as well as of course to other areas such as plastics). Once an industry develops along a particular path, and invests huge amounts of money, alternatives tend to be pushed aside. (Another example is the use of high-temperature incinerators for waste treatment.)

However, command economies such as in communist countries (where the direction is dictated from above rather than from the sum of individual choices) have had a worse track record for making wise choices. We are better off lobbying and improving our current system. Recent environmental pressures have boosted the importance of integrated pest management (IPM), and this is returning us part of the way to the alternative (non-chemical) approaches.

The production of mass chemicals has had its share of disaster, as the following case illustrates only too clearly.

Bhopal

Until a cloud of poisonous gas enveloped Bhopal on 2–3 December 1984, few people would have heard of this Indian city, or of methyl isocyanate. Union Carbide was manufacturing carbaryl (Sevin), a carbamate pesticide (see Carbamates, p. 183). The process proceeds in several stages (Fig. 7.18). Carbon monoxide is produced by the reaction of petroleum coke with oxygen. Carbon monoxide is reacted with chlorine to produce phosgene. Phosgene is reacted with (mono) methylamine to produce hydrogen chloride and methyl carbamoyl chloride. Methyl carbamoyl chloride is then pyrolysed to produce methyl isocyanate. The final step is to react methyl isocyanate with 1-naphthol to produce carbaryl, as shown in the figure.

1. $2C + O_2 \longrightarrow 2CO$
2. $CO + Cl_2 \longrightarrow COCl_2$
3. $COCl_2 + CH_3NH_2 \longrightarrow HCl + CH_3NHCOCl$
4. $CH_3NHCOCl \longrightarrow CH_3NCO + HCl$
5. $CH_3NCO + OH \longrightarrow$ OCONHCH$_3$

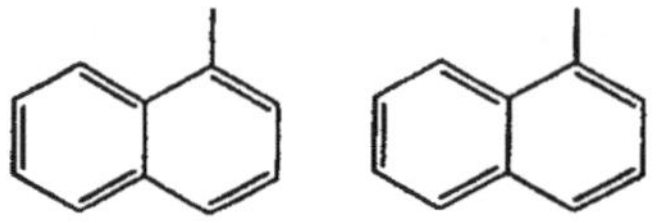

Fig. 7.18 Stages in the production of carbaryl

At the opening of the plant in 1977 only this last step was carried out, using materials imported from the US plant. In 1980, all the reactions were done at Bhopal.

All the gases involved are very poisonous. While little was known about the toxicity of methyl isocyanate itself, under the action of heat it forms hydrogen cyanide. Phosgene ($COCl_2$) is a more potent war gas than chlorine. Carbon monoxide is toxic. Altogether (literally) this was not a very pleasant soup to spread over 40 km^2, and 200 000 people. Thus there was considerable confusion as to what the actual poison was, and therefore how to treat its victims.

At first, doctors in Bhopal thought the gas was from a leak, and therefore phosgene, and based their treatment accordingly. They argued that on a cold December night, phosgene (b.p. 8°C) was more likely to escape than methyl isocyanate (b.p. 39°C). Phosgene was present in the methyl isocyanate as a stabiliser (200 to 300 mg/kg) to slow the reverse (breakdown) reaction during storage.

Damage to crops for kilometres around the factory was attributed to phosgene because it was not expected from methyl isocyanate.

But phosgene, like chlorine, slowly fills the lungs with fluid and would not kill people immediately, as was patently obvious in Bhopal. Autopsies revealed that the blood of victims was a dark, cherry red, and that lungs and other organs were also red. If fluid in the lungs had caused death by suffocation, then the blood would be bluish. Hydrogen cyanide was suspected, for which the antidote is sodium thiosulfate (photographic hypo). It reacts with cyanide to produce thiocyanate, which is excreted in the urine. This is a safe compound to administer, even if the diagnosis is wrong, but its use was forbidden because it was not proved that cyanide was involved. However, by giving this treatment and monitoring for thiocyanate it can be proved that cyanide (above the tiny amount that is present in the body naturally) has been absorbed. Cyanide does not accumulate in the body and so treatment must be fairly rapid.

When sodium thiosulfate was finally administered after the Catch-22 argument was resolved months later, marked improvements were found, probably because of damage still present in the body from the methyl isocyanate gas itself.

When Union Carbide published its report in March 1985, no mention was made of the nature of the gases that escaped from the tank.

Scaling up a chemical process. Many reactions that test as safe on a small scale can become dangerous when scaled up. This may have been part of the problem at Bhopal. Heat dissipation is of critical importance to scaling up chemical plants.
Place hot water in a small round-bottom flask and in a big one. Add a thermometer to both. Note the difference in cooling rates. Heat dissipation is a function of the area of the surface, while heat production or content is a function of the volume of the unit. The area increases as the square of the radius but the volume increases as the cube of the radius.

Biological control

Instead of using chemicals against plant or insect pests, an alternative is to use their natural (or imported) enemies. Probably the best known and most successful application of this approach was against a plant — the prickly pear — in southern Queensland and northern NSW. During the late 1920s, caterpillars of moths brought in from Argentina literally ate their way through some 25 million hectares of prickly-pear infested country. The Cactoblastis Memorial Hall near Chinchilla in central Queensland, on the Warrego Highway, commemorates the little grub's feat.

From the time they were introduced by the First Fleet, cattle have been upsetting the ecological balance in Australia with their dung, which gives bush flies, and the blood-sucking buffalo flies, copious breeding places. Australia has native beetles that bury dung, but they evolved to cope with the pellet-like droppings of the native marsupials, not the massive and sloppy offerings of cattle. Dung beetles from Africa have lived on the droppings of large plant-eating animals for millennia and appear to be finding the local drop quite digestible. They were introduced into Australia in the early 1970s.

Fig. 7.19
Spraying prickly pear wth arsenic

Insects have been used to control *Lantana,* a fungus has been found effective against skeleton-weed, and a bacterium is being used commercially to control caterpillars.

Sex attractants

One of the most fascinating approaches to insect control involves *pheromones.* These are chemicals excreted externally, and may serve to mark a trail, to send an alarm, or to attract a mate. The last, the sex attractants, are usually excreted by the female to attract males. These compounds are detectable in extremely low concentration by males, and can be used to lure males into traps, or to disorient them. Field tests have shown that the sex attractant of the gypsy moth is effective at amounts of 0.000 000 000 0001 g (10^{-13}) in the field. It is interesting that the first claim, in 1961, for having discovered the structure of the attractant was wrong. Research in this area is difficult. In 1967 researchers used the abdominal tips — which contain the glands that produce the sex attractant — of hundreds of thousands of female gypsy moths to isolate a minute amount of attractant, and it was synthesised three years later (see Fig. 7.20).

Fig. 7.20
Sex attractant of the gypsy moth

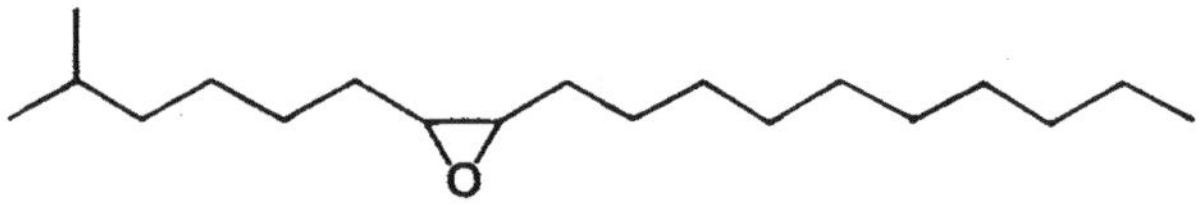

The original suggestion that the message was chemical was published as *The social Life in the insect world,* by J. H. Fabre, T. Fisher, Unwin, London, 1912.

The sex attractant for the silkworm moth (a C_{16} alcohol) inspired the Canberra poet, A.D. Hope, as follows:

II THE PERFUME

O Chloe, have you heard it,
This news I sing to you?
It's true, my lovely bird, it
Is absolutely true.
A biochemist probing
Has caught without a doubt
The Queen of Love disrobing
And found her secret out.

Poetic licence
The silkworm moth can only walk. For other species, such as the gypsy moth, the tale is true.

What drives the **Bombyx mori**
To fly, intrepid male,
Lured by the old, old story
Six miles against the gale?

The formula, my Honey,
Is now in print to prove
What is, and no baloney,
The very stuff of love.

At Munich on the Isar
Those molecules were found
Which everyone agrees are
What makes the world go round;
What draws the male creation
To love, my darling doll,
Turns out, on trituration
To be an alcohol.

A Nobel Laureatus
Called Adolph Butenandt*
Contrived to isolate us
This strong intoxicant.
The boys are celebrating
And singing at the club.
Here's Bottoms up: to mating,
Since Venus keeps a pub!

My angel, O, my angel,
What is it **you** suffuse,
What redolent evangel,
What nosegay of good news?
What draws me like a dragnet
And holds and keeps me tight?
What odds! my fragrant magnet,
I shall be drunk tonight!

Source: 'Six songs for Chloë' by the Canberra poet Alec Derwent Hope, in Selected Poems Angus & Robertson 1966, p. 243.

*See page 357

So whether it's moths or pigs (after truffles) — sex hormone molecules rule, OK.

Some sex attractants are simple, others are complex. In some complex mixtures, the *ratio* of chemicals is important. The sex attractant for the common house fly is now known to be (at least) a two-compound system. One component is $CH_3(CH_2)_7CH{=}CH(CH_2)_{12}CH_3$, which is fairly easy to make. The use of sex attractants can control insect pests in the mobile stage of their life cycles.

FRUIT FLY ATTRACTANT

In the 1950s, Alan Willison (Concord, Sydney) worked for a small chemical company interested in fragrances and had synthesised p-hydroxybenzyl-acetone (4-(4-hydroxyphenyl)-butane–2–one), as part of that work. He spilt some of that material on his shoes and placed them on a bench in front of a window to let the material evaporate. Numerous flies arrived and he recognised them as fruit flies, and did some work at home comparing the parent compound with some derivatives, with his wife as fly observer and unpaid assistant. He also contacted Harry Friend in the NSW Department of Agriculture.

There was a cautious paragraph in the 1959 Annual Report of the NSW Department of Agriculture:

> 'A locally-produced chemical, para hydroxy benzyl acetone, prepared by Mr A. Willison of Concord, together with some analogous compounds, was tested in a limited way. It showed some value as a male attractant for Queensland fruit flies.'

The 1960 Annual Report enthused for a whole ten paragraphs, and described the Caneite traps designed by Union Carbide laced with a range of modified attractants and malathion, plus a little Alacor (a mixture of PCBs) to reduce evaporation of the organophosphate insecticide. This established the feasibility of male fly eradication for the first time.

In 1961, a US patent (2974086) was taken out on Cue-lure which is the acetate of Willison's lure. This derivative has the advantage that it is not as easily oxidised as the straight phenol, but hydrolyses to it in the field (and also presumably got around the 'prior discovery' by Willison, assuming that the US Patent Office bothered to check Australian publications at the time).

This chemical is also known as raspberry ketone, or phenol-2-butane-2-one, and is found naturally at a level of about 0.2% in raspberries, and 2% in synthetic raspberry flavour.

Source: Tom Bellas, from a discussion in August 1996 with Mrs Willison

---

Mammals as well as insects produce chemicals that can be detected by their scent by other mammals of the same or different species. Some of these substances can be found in sweat, breath, urine and faeces. These can provide information to other animals, much like a urine analysis does for a doctor, and can act as sex attractants, alarm signals, recognition signatures, or used for marking trails and territory. Some mammals also have specialised scent glands (see Scents and perfumes, Chapter 5).

Juvenile hormones

Juvenile hormones control the rate of development of the larval stage of insects and are switched off to allow development of the adult. Application of a mosquito juvenile hormone will keep mosquitoes in the harmless larval stage. Synthetic hormones that are much more potent than the natural compounds have been produced. This technique is effective for insects which are pests in the adult stage. It is not effective against insects that are pests in the larval stage (e.g. the caterpillars that are a pest in farmers' fields).

Sterile males

In some cases, male insects are sterilised by irradiation from a radioactive source and then let loose in the insect populations. The females of many insect species mate only once, so it is statistically possible to wipe out a species in an area quickly. This technique has been used successfully against the fruit fly in an *isolated* area. The biggest control program was one in Texas against the screw worm (a fly), and the 'barrier' to prevent the screw worm moving north has now been pushed down south of Mexico.

Resistance

The main reason for the development of new insecticides and the search for biological control measures is the development of resistance by insects to present methods. Some insects are resistant to organochlorine, organophosphorus, and carbamate compounds — at least at levels that do not give problems of residues in crops. One method of developing resistance depends on the Darwinian concept of the survival of the fittest. If an insecticide wipes out 99% of a population of insects, then the 1% that survive contain those best suited to deal with the poison, and it is these that breed the next generation. So insecticides carefully breed resistant insects!

Insect repellents

On a less drastic level, there are several products that will keep insects away, at least from you. In the past, insect repellents contained strong-smelling oils such as citronella. These products kept away friends as well as the insects. It was later found that *contact* with the product was needed, not smell. In a study of more than 7000 chemicals, the US Department of Agriculture found that only a few were really effective repellents. *N*,*N*-diethyl-*m*-toluamide, (deet), (Fig. 7.21), was found to last twice as long as any other. Dimethyl phthalate (DMP), (Fig. 7.22) is also very effective, but it is a rather good solvent for plastic watch 'glass' and spectacles as well (see also Vinyl polymers in Chapter 9). Other compounds found by the study to be effective were ethyl hexanediol (E-Hex) and Indalone (registered trade name for butyl 3,4-dihydro-2,2-dimethyl-4-oxo-2*H*-pyran-6-carboxylate).

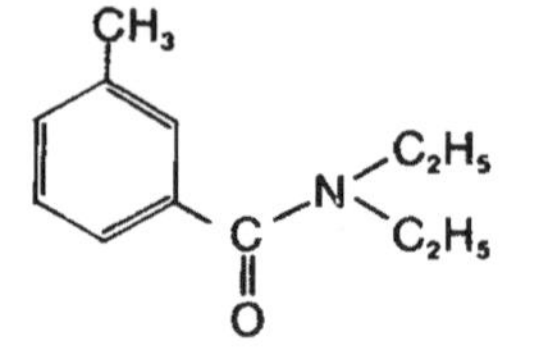

Fig. 7.21
Deet

a phthalic symbol!

Fig. 7.22
Dimethyl phthalate (DMP)

Australian salute
One point that is not adequately emphasised is that you should know which insect you are intent on repelling. The Australian bush fly, Musca vetustissima, is a notorious nuisance to people throughout Australia. (The so-called Australian salute consists of one hand brushing flies away from the face.)

On fabric, repellents are not rubbed off as quickly as they are from skin and will last for days instead of hours, but they can stain permanently and damage some synthetics and plastics.

It has been known since 1947 that effective mosquito repellents such as DMP are of no use, and that deet, which is one of the best mosquito repellents, is of little value. An aerosol with an effective repellent was introduced to the Australian market in 1961 (Scram™ — David Gray Co.). It contains 5% di-*n*-propyl isocinchomeronate, whose persistence is extended by adding the pyrethrum synergist *N*-octyl bicycloheptenedicarboximide. The use of such aerosols a few times a day prevents the flies from *settling* but not from momentary contact to test the site. Other Australian manufacturers market aerosols containing this repellent, along with others to repel mosquitoes, sand flies, etc. If it's bush flies you're after (or not after!) then 'isocinch' plus synergist is what to look for on the can.

References

A series of publications from the CSIRO (1977–1979) was used in the section on soils, now out of print.

The use of DDT in Australia, Report No. 14, Australian Academy of Science, Canberra, Feb. 1972.

The chemistry of pesticides: their metabolism, mode of action and uses in crop production, K.A. Hassal, *Verlag Chemie*, Weinheim, 1982, 1990.

Herbicide-resistant crops and pastures in Australian farming systems, ed. G.D. McLean and G. Evans, Bureau of Resource Sciences, workshop, 15–16 March 1995, AGPS, Canberra, 1995. Good basic science, far beyond the title.

HCB waste: Background and issues paper, Report prepared for ANZECC, B. Selinger, ANUTECH, Canberra, July 1995, 257 pp., comprehensive data on organochlorine levels in breast milk in Australia, as well as much else.

Bibliography

Gardening down-under, K. Handreck, CSIRO publications, 1993.
Pesticide index; index of chemical, common and trade names . . ., H. Kidd and D. Hartley, RSC, London, 1988.
The agrochemicals handbook, ed. H. Kidd and D.R. James, 3rd edition, Royal Society of Chemistry, London, 1991. Short summary, and chemical structure, on all major and most minor pesticides. Looseleaf and updatable.
The pesticide book, G.W. Ware, Thomson Publications, Fresno CA, USA, 4th edn. 1994. An excellent textbook and basic reference for a variety of users.
Pesticide application methods, G.A. Mathews, Longman, Essex, UK, 1979, minor update in 1982.
Pesticide formulation and adjuvant technology, ed. C.L. Foy and D.W. Pritchard, CRC Press, 1996. Publication of the first ever international symposium on the topic. Can recommend the book to professionals.
Recommended common names of pesticides, Australian Standard AS 1719 — 1994, 132 pp.
Pesticide Action Network North America, Updates Service, WWW site at http://www.panna.org/panna/
Environmental Research Foundation, P.O. Box 5036, Annapolis, MD, 21403, Internet: erf@rachel.clark.net

Light reading

The crumbling of cricket pitches', J.R. Harris, *The Australian Scientist*, April 1961, pp. 173–178. From the days when CSIRO did non-commercial public benefit science.

Chemistry of swimming pools

When was the last time you checked the salt level?

The view of any Australian city from the air presents myriad blue spots, some green. These are backyard swimming pools. As well, there are numerous municipal pools. They require maintenance. This needs chemistry.

Swimming pool chemistry

A remarkable variety of chemical concepts can be taught using the swimming pool as the medium. These concepts include pH, acids and bases, buffers, equilibrium, oxidation–reduction, properties of chlorine, electrolysis, photochemistry, mole concept, and Faraday's law. On a more practical level a number of preliminary questions can be asked.

Question 1

Are you aware of any problem associated with pool chemicals? Can you remember if there are warnings on the containers, and what they are? Any newspaper reports of fires?

Question 2

Why do we need to chlorinate our swimming pools?

Answer: Even water that is not meant to be drunk must be kept free of micro-organisms. When the same water is used over and over again, it quickly becomes contaminated with bodily excretions and garden debris. We use a strong oxidising agent such as chlorine to sterilise the water.

Question 3

What happens when chlorine is bubbled through water?

Answer: $$Cl_2(aq) + 2H_2O \rightleftharpoons HOCl^- + Cl^- + H_3O^+$$

When chlorine is added to chemically pure water, a mixture of hypochlorous (HOCl) and hydrochloric (HCl) acids is formed. At ordinary water temperatures, this reaction is essentially complete within a few seconds. In dilute solution and at pH levels above about 4, the equilibrium (in the equation) is displaced to the right, and very little dissolved Cl_2 exists in solution. In normal practice (except for the concentrate from solution-feed chlorinators) the amount of chlorine supplied to water does not produce a concentrated solution of such strength as to yield such a low pH. The oxidising property of the chlorine is, however, retained in the HOCl it has formed, and this provides the principal disinfecting action of chlorine solutions.

Hypochlorous acid ionises (i.e. dissociates), in a practically instantaneous reaction, into hydrogen and hypochlorite ions (note that the reaction is reversible), the degree of dissociation depending on pH and temperature:

$$H_2O + HOCl \rightleftharpoons H_3O^+ + OCl^-$$

Question 4

Why don't people chlorinate their pools with chlorine gas?

Answer: Chlorine is a yellowish green, dense gas. Chlorine gas bubbled through the pool water is an obvious means of treatment. However, it is inconvenient to use as a gas because it causes rapid corrosion of metals, and destruction of plastics. It is also a dangerous gas which attacks the mucous membrane linings of the eyes, nose, throat, and lungs. It causes the lungs to fill with fluid, and the victim drowns. During the First World War, chlorine gas was used as a chemical weapon.

Question 5

Is there a more convenient way to chlorinate a pool?

Answer: All forms of chlorine, such as pool chlorine (70% available chlorine), calcium hypochlorite, $Ca(OCl)_2$, and sodium hypochlorite, NaOCl, chlorine gas, etc. ionise in water and yield hypochlorite ions:

$$Ca(OCl)_2 + H_2O \rightleftharpoons Ca^{2+} + 2\,OCl^- + H_2O$$
$$NaOCl + H_2O \rightleftharpoons Na^+ + OCl^- + H_2O$$

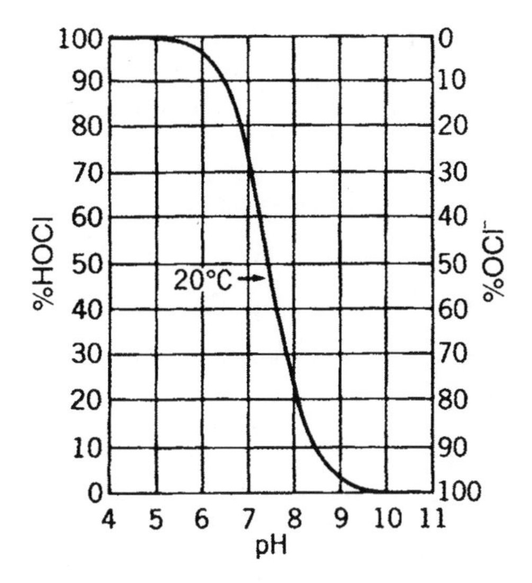

Fig. 8.1
Distribution of HOCl and OCl^- in water at indicated pH levels

The hypochlorite ions also establish equilibrium with hydrogen ions, depending on the pH. Thus, the *same* equilibria are established in water, regardless of whether elemental chlorine or hypochlorites are employed. The important distinction is the resultant pH, and, hence, the relative amounts of HOCl and OCl^- existing at equilibrium. Chlorine tends to decrease the initial pH, and hypochlorites tend to increase it.

Question 6

What is this effect of pH?

Answer: Hypochlorous acid is weak, and dissociates poorly at pH levels below about 6; thus, chlorine exists predominantly as HOCl at low pH levels. Between pH 6.0 and 8.5 there occurs a very sharp change from undissolved HOCl to almost complete dissociation. At 20°C above a pH of about 7.5, hypochlorite ions (OCl^-) predominate, and they exist almost exclusively at levels of pH around 9.5 and above (Fig. 8.1, above).

The HOCl is much more effective in killing bacteria than OCl^- because the negative charge on the ion hinders it entering into the bacteria.

Question 7

How do you measure the amount of chlorine in water?

Answer:

$$HOCl + 2I^- + H_3O^+ \longrightarrow Cl^- + I_2 + 2H_2O$$
$$I_2 + 2S_2O_3^{2-} \longrightarrow S_4O_6^{2-} + 2I^-$$

In the laboratory, the amount of oxidant is measured by its ability to liberate iodine from acidified iodide solution. A sample of water can be titrated with standard iodide solution. The iodine released is detected by a blue colour formed with a fresh starch indicator (Experiments 4.2, 4.3).

Exercise
Note that the reagent is only oxidised to tetra-thionate by the iodine, whereas with chlorine it is oxidised to sulfate. Calculate the oxidation number of sulfur in these various compounds.

The amount of iodine released is determined by back titration with sodium thiosulfate (hypo). Potassium iodide releases iodine with chloramines as well, so these are included in the titre if present. See Question 15.

This titration is still the standard method for determining free chlorine, although other more convenient tests are used in practice at the pool.

Question 8

What does 'available chlorine' mean, and why is chlorine gas 100% available when it isn't?

Answer: When chlorine gas dissolves in water it forms one molecule of hydrochloric acid and one molecule of hypochlorous acid. Only the latter is 'active' so only half the chlorine is usable. Nevertheless, chlorine gas is set at 100%. So compounds for which all the chlorine in solution is active will have percentages twice the value based on composition.

Table 8.1 (below) gives some typical values of available chlorine. Note the entry for $Ca(OCl)_2$, which is 99.2% for the pure material. This is generally quoted as 100% and raises some confusion. $Ca(OCl)_2$ produces two moles of active chlorine compared to only one from Cl_2. However, it has just over twice the molecular mass (a ratio of 143: 71). So, on a mass basis, both materials are equally effective. Materials releasing other oxidising agents when dissolved in water are measured on the basis of the same redox reaction. Chlorine dioxide thus gives a value in excess of 200%.

TABLE 5.6

Per cent available chlorine of various products

Material	Available chlorine %
Cl_2 (chlorine)	100[a]
Bleaching powder (chloride of lime, etc.)	35–37
$Ca(OCl)_2$ (calcium hypochlorite)	99.2
Commercial preparations	70–74
NaOCl (sodium hypochlorite) (solution 100%)	95.2
Commercial bleach (industrial)	12–15
Commercial bleach (household)	3–5
ClO_2 (chlorine dioxide)	263.0
NH_2Cl (monochloramine)	137.9
$NHCl_2$ (dichloramine)	165.0
NCl_3 (nitrogen trichloride)	176.7
$HOOCC_6H_4SO_2NCl_2$ (Halazone)	52.4
CONClCONClCONCl (trichloroisocyanuric acid)	91.5
CONClCONClCONH (dichloroisocyanuric acid)	71.7
$CONClCONClCON^- Na^+$ (sodium dichloroisocyanurate)	64.5

[a] by definition.

Source: Chlorinated bleaches and sanitizing agents, W.H. Sheltmire, Ch. 17 of ACS Monograph, Chlorine, Reinhold Publishing Co., New York, 1962.

Question 9

Are those the only ways of obtaining chlorine in a pool? What about electrolysis?

Prior electrolysis
Sometimes the electrolysis of salt is carried out beforehand, and a solution of sodium hypochlorite, NaOCl (liquid bleach), is sold for addition to the pool. Sodium hypochlorite is not stable as a solid.

Answer:

cathode	$2Cl^- \longrightarrow Cl_2(g) + 2e^-$
anode	$2H^+ + 2e^- \longrightarrow H_2(g)$
followed by:	$Cl_2(g) + H_2O \longrightarrow HOCl + H^+ + Cl^-$

Note: e^- in a half-equation represents an electron that is transferred. When the two half-equations are added, the electron cancels out. The symbol (g) means that the substance is a gas.

If salt is added, it is possible to generate the hypochlorous acid continuously in the pool by using an electrolysis cell. With this method of chlorination, the pool water will also gradually become acidic, and may need some alkali added to adjust the pH. Commonly used alkalis are sodium carbonate and sodium bicarbonate.

Question 10

Why not drop the pH of the pool well below 7 if the oxidising strength of HOCl increases by doing so?

Answer: For a start, more acidic solutions will corrode the pool components. In marblesheen and tiled pools, the corrosion is even greater, and the recommended pH range for these is 7.4–8.0. In addition, calcium chloride is added (100–200 mg/L) to counter the removal of calcium salts from the grouting.

More interesting, though, are the reactions of chlorine with ammonia and ammonia-like compounds that are formed from organic waste to form chloramines.

$$NH_3 + HOCl \rightleftharpoons NH_2Cl + H_2O \qquad (\approx 1 \text{ minute})$$
$$NH_2Cl + HOCl \rightleftharpoons NHCl_2 + H_2O \qquad (\text{slower})$$

The chloramines and dichloramines also react with each other:

$$NH_2Cl + NHCl_2 \rightleftharpoons N_2(g) + 3HCl$$

The oxidation of chloramines in total is therefore given by:

$$2NH_3 + 3HOCl \rightleftharpoons N_2(g) + 3HCl + 3H_2O$$

Further additions of chlorine beyond the break point can result in the formation of nitrogen trichloride:

$$NHCl_2 + HOCl \rightleftharpoons NCl_3 + H_2O$$

This gives rise to the so-called 'smell of chlorine' because it escapes easily from the water when the water is stirred up. It causes severe eye irritation.

A point of major confusion is whether a test for chlorine in water measures the free residual chlorine or the free chlorine, *plus* the chloramines (combined chlorine). The lower the pH, the more readily chloramines form. Above a pH of 7, their formation is minimal. This is another reason for keeping the pH above 7.

Question 11

What is superchlorination?

Answer: When chlorine in its various forms is added to water, it is used up in oxidising any material for which it is a sufficiently strong oxidising agent (iron II, sulfide, nitrite, etc.). After this demand is satisfied, chloramines (combined chlorine) are formed from reaction of chlorine with organic nitrogen compounds. Additional chlorine (at pH greater than 7) will be used up in oxidising these chloramines to nitrogen gas and nitrate ion. The break point occurs when all this material has been oxidised (Fig. 8.2, below). Further additions of chlorine are not used up, and remain as residual chlorine, ready to react with any material now added to the pool. This process of superchlorination carried out at regular intervals involves precisely this piece of chemistry. Strong oxidation can also be carried out using potassium monopersulfate. This will also oxidise any chloride back to chlorine (see also Nappy sanitisers, which use the same reaction, in Baby-care products in Chapter 5).

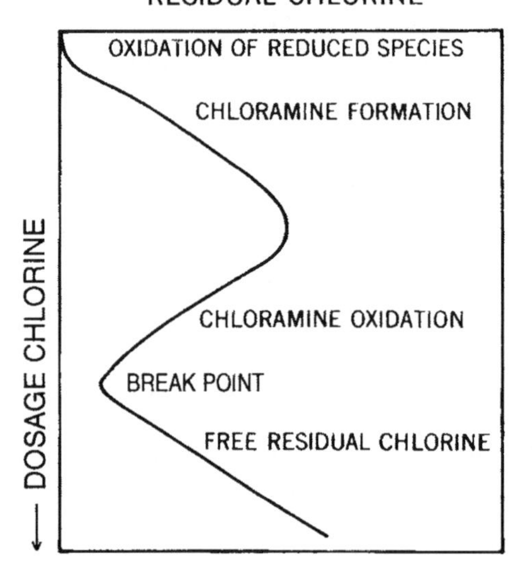

Fig. 8.2 Relationship between chlorine dosage and residual chlorine for break point chlorination.

Question 12

How do we adjust the pH of a pool?

Answer: We now see the need to control the pH of the water in a swimming pool. We can measure the pH with a suitable indicator that changes colour at the pH in which we are interested. The indicator is itself a weak acid that shifts from one colour to the other, just as hypochlorous acid shifts from HOCl to OCl^-. That is, we choose an indicator with the same pK_a as hypochlorous acid. The pH of natural water is generally about 6 because dissolved carbon dioxide from the air forms carbonic acid, and lowers the pH below the neutral value of 7. The continuous addition of hypochlorite powder will gradually raise the pH so addition of acid may be necessary after some time.

Solids are easier to store and use than liquids, so the 'acid' generally used is sodium hydrogen sulfate, $NaHSO_4$, although hydrochloric acid is supplied for salt-water pools with electrolysers. Why?

Alkalinity is a term used as a measure of the buffer capacity of water, or the degree of resistance to change in the pH of water on the addition of strong acids or bases. If the alkalinity is too low, pH control is difficult, because the pH is sensitive to small amounts of acid and base. From the volume of water in the pool, calculate how much 0.1 M hydrochloric acid needs to be added to change the pH by one unit if there is no buffering capacity.

Sodium bicarbonate in amounts of 80 to 120 mg/kg is added as a buffering agent. The pH of sodium bicarbonate in water (8.4) lies between the pK_a values of the first and second dissociation constants of carbonic acid (6.35 and 10.33).

The term alkalinity comes from the common use of sodium bicarbonate as a buffer and its alkaline pH of 8.4. However, acid buffers can also be used and buffer capacity is a better term than alkalinity. Very high buffer capacity makes it difficult to change the pH when you want to because very large amounts of acid or base are then required. Sodium bicarbonate is also used as a base to raise the pH, but sodium carbonate is more effective, and more common. Addition of sodium bicarbonate to water gives a pH of 8.4, whereas addition of sodium carbonate to water gives a pH of 11.6. Addition of acid changes the pH according to the titration curve (Fig. 8.3).

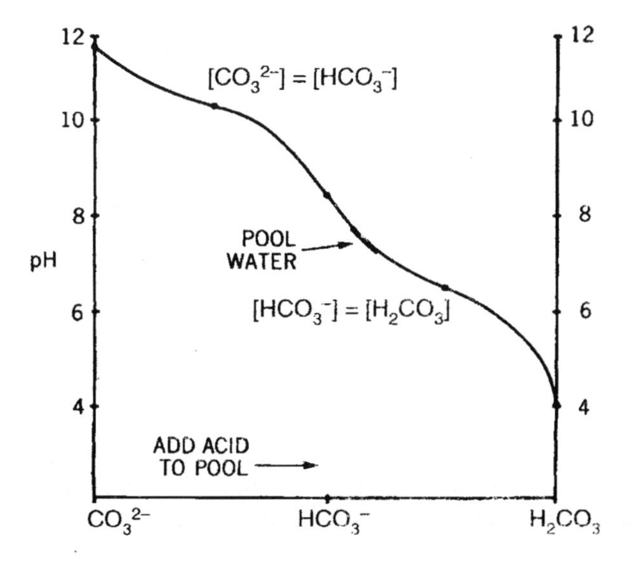

Fig. 8.3 Titration of sodium carbonate with hydrochloric acid.

Question 13

What do we mean by more powerful oxidising agent?

Answer: The ability of a material to oxidise is measured by the standard half-cell reduction potential (given in volts). This is an equilibrium value given for very specific conditions, and so gives only a general indication for a practical situation. The larger the potential is, the stronger the oxidising agent. The standard electrode potentials, $E°$, are just two points on a potential versus pH curve ($E° = 1.49$ at pH = 0, $E° = 0.94$ at pH = 14). The equations and Figure 8.4 (below) show that HOCl is a stronger oxidising agent than OCl^-.

	$E°/V$
$HOCl + H_3O^+ + 2e^- \rightleftharpoons Cl^- + 2H_2O$	1.49
$ClO^- + 2H_2O + 2e^- \rightleftharpoons Cl^- + 2OH^-$	0.94

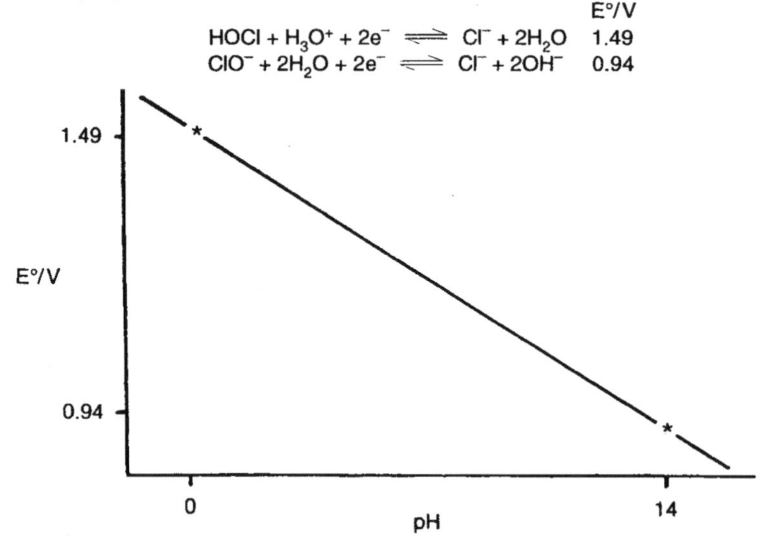

Fig. 8.4 Reduction potential of chlorine in water as a function of pH

Question 14

Is it dangerous to store pool chlorine near materials that can burn, because it is a strong oxidising agent?

Answer: Yes! Hypochlorite is chosen for sterilising water, including swimming pools, because HOCl is a strong *oxidising* agent. If the material comes in contact with something that is able to be oxidised (i.e. burnt, but with chlorine, not oxygen, as the agent) then a fire can occur.

Hypochlorite powder + brake fluid ⟶ flame.

WARNING: The experiment described here is ***very dangerous.***

Note that chlorine gas is released when solid hypochlorite powder comes in contact with moisture (at any pH). The constant escape of chlorine as a gas means the reaction continues to maintain equilibrium, and thus produces more chlorine gas.

Question 15

How do we measure chlorine levels?

Answer: It is important to ensure that excess 'pool chlorine' is present in the pool after all the organic material has been oxidised. Some method of measuring free chlorine is needed. The level of chlorine in water is a difficult analysis to carry out satisfactorily. A number of commercial test-kits are available.

Question 16

Why is chlorine lost from pools in sunlight?

Answer:

$$OCl^- \xrightarrow[\text{light}]{\text{UV}} Cl^- + \tfrac{1}{2}O_2(g)$$

Fig. 8.5
Photolysis of chlorine

Chlorine is rapidly lost from pool water in the presence of sunlight. It is estimated that about 90% of the chlorine consumed in pools is the result of this reaction (Fig. 8.5), which is an example of *photolysis* (chemical reaction brought about by the energy of light). Shallow pools require frequent addition of chlorine.

The ultraviolet range in sunlight is from 290 nm to 350 nm and all the chlorine species absorb in this region, with the OCl^- (pH 7–8) showing the strongest absorption (see Fig. 8.6).

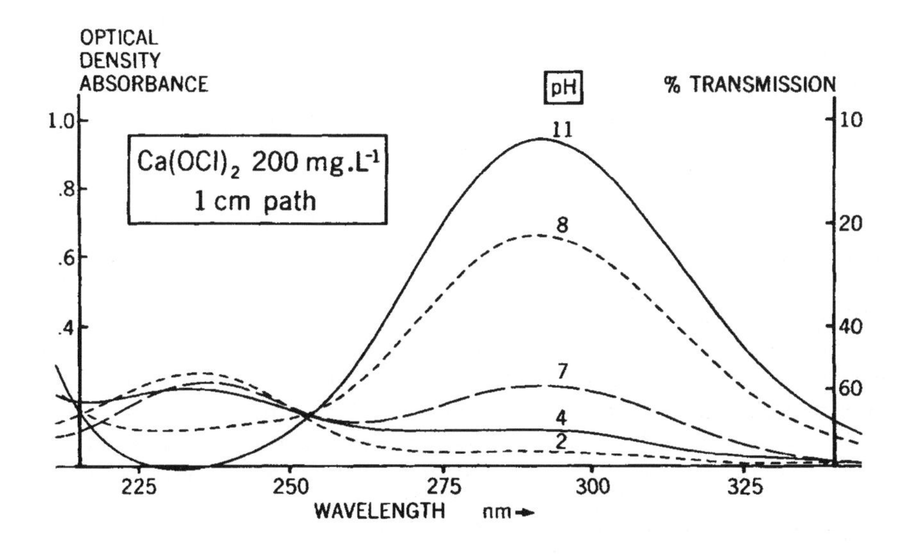

Fig. 8.6
Ultraviolet absorption spectrum of pool chlorine for the equivalent of 1 m depth at recommended concentrations

Question 17

What do we do to stop loss of chlorine in sunlight?

Answer: There are compounds such as cyanuric acid (Fig. 8.7, overleaf) which can be used to stabilise chlorine in swimming pools.

Cyanuric acid is made by heating urea or some of its derivatives. (It is a selective herbicide, very toxic to barley and radishes.) It exists as two tautomeric forms in equilibrium — an alcohol and a ketone.

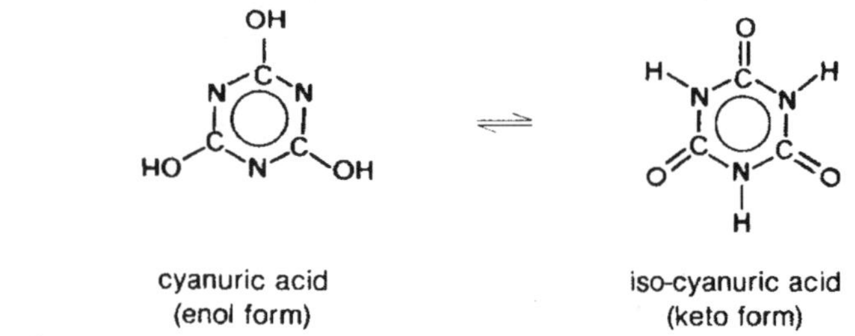

cyanuric acid (enol form) ⇌ iso-cyanuric acid (keto form)

Fig. 8.7 Cyanuric acid exists as two tautomeric forms in equilibrium

Question 18

How does cyanuric acid work?

Answer: Cyanuric acid reacts with chlorine to give dichloro(iso)cyanuric acid in a chemical equilibrium (see Fig. 8.8).

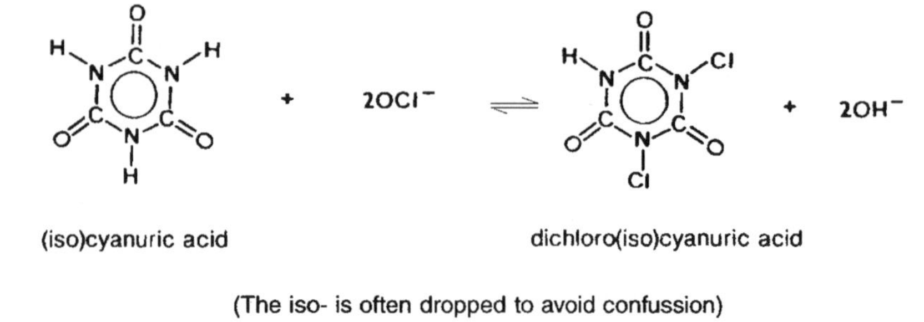

(iso)cyanuric acid + $2OCl^-$ ⇌ dichloro(iso)cyanuric acid + $2OH^-$

(The iso- is often dropped to avoid confussion)

Fig. 8.8 Equilibrium of cyanuric acid with chlorine and dichlorocyanuric acid

As chlorine (as OCl^- or HOCl) is used up in the pool, more OCl^- is released from the dichloro(iso)cyanuric acid to re-establish equilibrium. Thus a constant amount of chlorine is maintained in the pool until all the dichloro(iso)cyanuric acid has lost its chlorine.

The general commercial chemicals are trichlorocyanuric acid and sodium dichlorocyanuric acid. Which would you expect to be more water soluble, and why? These compounds are not photosensitive. The same effect can be obtained by adding just cyanuric acid, and the chlorine from another (cheaper) source. This produces the chlorocyanuric acid as shown in the reaction. The amount of cyanuric acid is kept constant in the pool between 30 and 80 mg/L by the initial addition of any of the three compounds. From then on, only hypochlorite is added (to keep the level of free chlorine at 2 mg/L). The *equilibrium* in the reaction shown *stabilises* the concentration of OCl^-. What happens if excess cyanuric acid is used? (Consider the equilibrium, and the effect on OCl^-.) The chlorocyanuric acids do not absorb the sun's ultraviolet light, and thus their use retards the loss of chlorine through sunlight.

Question 19

Can we measure the level of cyanuric acid in a pool?
Answer: To determine the level of cyanuric acid in the pool, you can carry out a reaction with melamine (Fig. 9.2) to form a salt that precipitates and scatters light (see Fig. 8.9). The amount of turbidity is proportional to the amount of cyanuric acid. The turbidity is measured by the depth of solution required in a standard (Nessler) tube to just obliterate an object at the bottom of the tube.

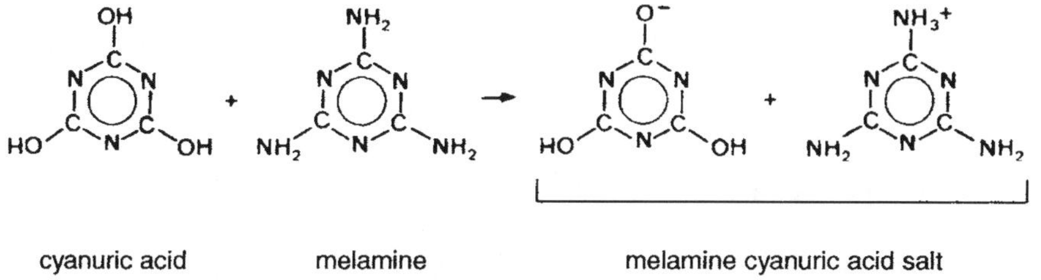

cyanuric acid + melamine → melamine cyanuric acid salt

Fig. 8.9 Salt formation by cyanuric acid and melamine

Interestingly, anionic surfactants are measured by a similar salt formation, using methylene blue. The salts formed are sufficiently lyophilic (non-polar portion overwhelms ionic contribution) to transfer into organic solvents, and to be estimated by the depth of colour in the organic solvent. (See The physics of plastics in Chapter 9.)

Question 20

What about algae?

Answer: In addition to the need to sanitise the water against body waste and bacteria, algae can also be a problem. Small concentrations of copper or silver ions produced by electrolysis can be effective. Silver's bacteria fighting properties have been known for centuries. The ancient Phoenicians learnt that silver containers kept their water pure on long voyages.

Silver ionisers are based on technology developed by NASA for the Apollo missions, and 220 parts per billion (ppb) was found to prevent bacterial growth during storage. However the Australian Standard does not accept that these are effective in pools.

Materials similar to cationic surfactants are also used as algacides (see How do surfactants work? in Chapter 3). Clarification of muddy pool water is carried out with a flocculant, commonly a highly charged ion such as Al^{3+} from alum. (As well as coagulating clay suspensions, aluminium ion is used in sticks to coagulate blood from shaving cuts, and in antiperspirants to coagulate sweat from pores under the arm. See also Deodorants and antiperspirants in Chapter 5 and Chemistry of surfaces, Chapter 6.)

Question 21

What about phosphates and algae?

Lanthanum toxicity
Lanthanum's toxicity is very similar to that of barium. LD_{50} for $LaCl_3$ is 5g/kg, while for sulfate it is 500g/kg.

Answer: If algae grow in a typical swimming pool that is being chlorinated, then it will contain at least 0.5 ppm of phosphate. With light (and nitrogen) this could produce a kilogram of algae. (The chemical composition of Black spot algae is typically 98% organic, 2% phosphate, 1% calcium.)

Aluminium salts can precipitate phosphate but more is removed if lanthanum is used (lower solubility product). About 10 ppb of lanthanum sulfate can be effective in reducing the phosphate to below the level needed to sustain algae. (Inventor: Starver pool treatment™, Innvez Pty Ltd, Canberra.) This idea cannot be used for fish tanks. Fish are very sensitive because their lungs hang out in the water! (Use an ion exchange resin instead.)

Bibliography

The explosive reaction between swimming pool chlorine and brake fluid, K.P. Kirkbride and H. Kobus, *J. Forensic Sciences*, May 1991, pp. 902–907. (The authors are forensic scientists in South Australia.)

CHAPTER

Chemistry of hardware

I'd like you to meet your new replacement . . . man of Kevlar

In this chapter we move into the hardware store and discuss two of the most important materials from which consumer items are made, plastics and glass.

Plastics

Early plastics

In the early days of the industry there was concern that the word plastic undervalued all the wonderful properties of the new materials, and a competition was held to find a better one. One suggestion was PUCCA standing for phenol, urea, cresol, casein and acetate, then the basic raw materials. However *plastic* stuck, getting an official extra 's' in 1951 to become *plastics*.

Plastics are materials that can be moulded. The earliest true plastic (other than clay) was probably hoof and horn, a thermoplastic material based on the protein keratin. The Worshipful Company of Horners is first recorded in the 13th century making snuff boxes and drinking beakers. The beakers were made from oval cross-sections, heated and straightened over wooden formers. Horn was also flattened into thin sheets to produce transparent sheets in windows and in lanterns (*lanthorn*, see Oxford English Dictionary). It was the ability to produce moulded products more quickly and with less skill, and consequently more cheaply than artisan carvings, that spurred the expansion of the plastics industry. Tortoiseshell is also based on keratin and was another attractive raw material for moulding.

In the Indian subcontinent, the lac insect (*Tacchardia lacca*) was harvested in its larval stage to produce shellac for the varnish and paint industries. Shellac is dissolved in methanol and dries to a hard thermoplastic film that cross-links in time to become less soluble. With suitable fillers, shellac was also used as a mouldable material. The ability of shellac to take the fine detail of a mould meant that it was used until after World War Two in 78 rpm gramophone records.

Molecules strung together like paper clips

Since the first edition of this book, synthetic plastics have become more complex and sophisticated and the straightforward distinctions between different types and uses have become somewhat fuzzy. The discussion that follows tries to maintain some of that simplicity, but the exceptions are as common as the rule!

When molecules are strung together like a string of paper clips, in one or three dimensions, the resulting compound is called a *polymer* or *macromolecule.* Some natural members of this group are cellulose, rubber, keratin and chitin. Some synthetic members include the plastics, fibres and rubbers. The small, basic building-blocks are called *monomers.* Differences in the chemical constitution of the monomers, in the structure of the polymer chains, and in the interrelation of the chains, determine the different properties of the various polymeric materials.

There are many ways of classifying plastics. One method is by division into two groups: those polymers that extend in only one dimension (that is, those that consist of linear chains); and those polymers that have cross-links between the chains, so that the material is really one three-dimensional giant molecule (an example is given in Fig. 9.1, p. 216). The first group of plastics are called *linear polymers,* and are *thermoplastic;* that is, they gradually soften with increasing temperature and finally melt because the molecular chains can move independently. An example is polythene, which softens at about 85°C. The second group are the *crosslinked polymers,* which are *thermosetting.* They do not melt on heating, but finally blister, the result of gases being released, and char. An example is bakelite.

One readily available linear polymer is made from milk. The milk protein casein is separated out (using rennet from calves' stomachs, or acid) and moulded, and then cross-linked (hardened) with formalin. Casein plastic was used to make buttons and knitting needles.

HISTORY OF THE DEVELOPMENT OF SYNTHETIC POLYMERS

1839 Styrene polymerisation observed.
1843 Caoutchouc (native rubber) vulcanised.
1844 Linoleum produced.
1864 Christian Schonbein prepares cellulose nitrate by treating paper with nitric acid.
1865 Alexander Parkes prepares the first plastic — parkesine — composed of cellulose nitrate, vegetable oils and camphor.
1869 John Hyatt patents celluloid, an improvement on parkesine.
1880 Acrylates first prepared in the lab. Related to Perspex.
1897 Casein–formaldehyde polymer.
1897 Urea–formaldehyde first prepared in the lab.
1898 Einhorn prepares polycarbonate in the lab.
1905 Schnitzenberger prepares cellulose acetate.
1907 Leo Baekeland patents bakelite. Produced from phenol and formaldehyde, this was the first wholly synthetic plastic.
1911 Cellophane developed by Brandenburger.
1912 Ostromislensky produces PVC.
1915 Dimethylbutadiene rubber produced.
1918 Urea–formaldehyde patented.
1921 First plastic injection moulder developed.
1922 Staudinger shows that rubber is a chain of isoprene units.
1924 Poly(vinyl alcohol) first prepared.
1927 Perspex first prepared in the lab.
1928 Du Pont develops 'superpolyamide', now known as nylon.
1930 Poly(vinyl acetate) developed. Epoxy polymers produced from phenolpolyalcohols.
1931 Neoprene (polychloroprene), poly(vinyl acetal) and polyisobutylene first produced.
1932 ICI develops polyethylene.
1933 Roy Plunket discovers Teflon, polytetrafluoroethylene.
1935 First extruder for thermoplastic polymers developed.
1936 ICI produces Perspex commercially.
1937 Polyurethane developed.
1939 Polyethylene produced commercially.
1939 Melamine formaldehyde begins to replace urea–formaldehyde because of superior properties.
1940 Polyethers (acetyls) first prepared.
1942 Saran fibres (poly(vinylidene chloride)) are produced.
1943 Silicones become commercially available.
1943 Epoxy resins are developed.
1950 Orlon (poly(acrylonitrile)) developed.
1956 Polycarbonates commercially produced.
1960 Polyacetal first prepared.
1965 Poly(phenylene oxide) produced.
1965 Polysulfone developed.

1969 Polybutylene terephthalate developed (BT) by Celanese (later acquired by Hoechst)
1977 PET introduced for soft-drink bottles

Source: CSIRO Science Education Centre

In 1907, when Leo Baekeland, a Belgian working in the United States, was looking for a shellac substitute, he discovered the first artificial plastic — bakelite — by mixing phenol and formaldehyde. Bakelite is a good electrical insulator and is still used for power plugs, points and switches, and for electric jug lids, etc. It is still the most important cross-linked plastic because it is cheap and rigid (but it is available only in dark colours).

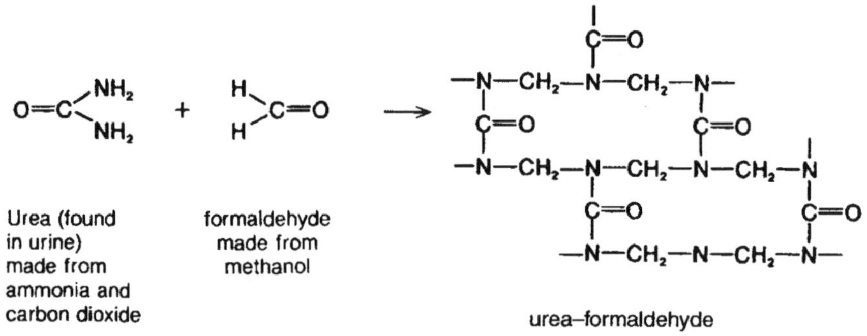

Urea (found in urine) made from ammonia and carbon dioxide

formaldehyde made from methanol

urea–formaldehyde

Fig. 9.1 Three-dimensional giant molecule of urea–formaldehyde

Urea–formaldehyde (Fig 9.1) is usually formulated as a moulding powder with purified wood pulp filler. Because the refractive indices of the resin and cellulose are similar, there is little light scattering at the phase boundaries, and so it is easy to produce pastel-coloured products (see Toothpaste in Chapter 5; Solid paints in Chapter 12, and Plates VIII and IX). They are less water and heat resistant than the phenolics.

In melamine ware (used for dinner ware) the compound melamine (Fig. 9.2) is used instead of urea.

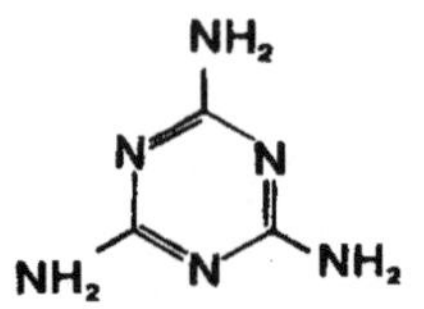

Fig. 9.2 Melamine

This gives a product that combines the good properties of the other two resins. Melamine formaldehyde is water resistant, can be produced in pastel colours and can be kept continuously at above 100°C (dishwasher proof).

Some interesting economics emerge. Urea–formaldehyde resins dominate the adhesives market because they are cost effective. However, when a filler is needed, the phenol formaldehyde can tolerate cheaper fillers, so the moulding powder market is dominated by the phenolics. When a water-resistant chipboard is required, or a moulding in a pastel colour, you have to pay extra for the urea resin. If you want a plastic that is both water resistant and pastel coloured, you have to buy melamine, which is more expensive.

Laminated plastics and veneers (e.g. Formica and Laminex) are made by impregnating several sheets of materials (usually paper or cloth) with plastic, then pressing the sheets together and hardening them in the oven.

Formica
Note: Use of formica as an adjective is free, but the noun Formica is a trade mark.
Origin: as a resin substitute 'for mica', used as an insulation material in electrical boards.

Addition polymerisation

The *linear* polymers can be made by joining a sequence of monomers, but here the possibility for cross-linking in two or three dimensions does not exist. The simplest and most widely used is polyethylene, shortened to polythene (. . . $-CH_2-CH_2-CH_2-$ . . .). When the chain length exceeds about a thousand units, the material becomes relatively rigid and can be used in a variety of ways. The linear molecular chains are partly tangled with each other to form amorphous (without structure) regions that impart strength and a higher melting point to the material. Polyethylene is made by joining together monomers of ethylene in a process called *addition polymerisation*, according to the reaction:

$$R{-}O{-}O^{\bullet} + CH_2{=}CH_2 \rightarrow R{-}OOCH_2{-}CH_2^{\bullet}$$

$$+ \text{ next } CH_2{=}CH_2 \rightarrow R{-}OOCH_2{-}CH_2{-}CH_2{-}CH_2^{\bullet} \quad \text{etc.}$$

Fig. 9.3 Addition polymerisation

where $R{-}O{-}O^{\bullet}$ is a free *radical* and acts as an *initiator.* (When a molecule has an odd electron it is called a radical, and this is indicated by a dot at the top right of the atom where the electron is possibly localised. The final product of the reaction is polyethylene, $(CH_2)_n{-}CH{=}CH_2$, termination of the chain often occurring by a reversal of the initiation step. About every tenth unit has a CH_3 branch. Polythene is fairly inert, but biodegradability can be built in by increasing the number of double bonds, which gives microbes somewhere to chew (see Biodegradability in Chapter 3). Polyethylene is the most widely used plastic. It was discovered in the early 1930s and gained prominence during World War Two as an insulator for high-frequency cables used in radar installations.

Under normal conditions, using the usual catalysts, the spatial arrangements of the branches in polymer products are random. Such polymers are called *atactic* (an example is given in Fig. 9.4).

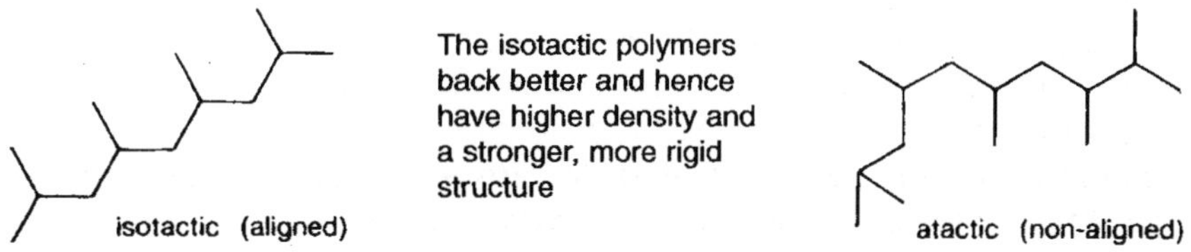

isotactic (aligned)

The isotactic polymers back better and hence have higher density and a stronger, more rigid structure

atactic (non-aligned)

Fig. 9.4 Polypropylene

Ionic polymerisation

When a suitable acid (A) is used to initiate the reaction, the process becomes a particular type of addition polymerisation called *ionic polymerisation*. The reaction proceeds as follows:

$$A + CH_2{=}CH_2 \longrightarrow \bar{A}{-}CH_2{-}\overset{+}{C}H_2$$

$$\bar{A}{-}CH_2{-}\overset{+}{C}H_2 + CH_2{=}CH_2 \longrightarrow \bar{A}CH_2{-}CH_2{-}CH_2\overset{+}{C}H_2$$

Fig. 9.5
Ionic polymerisation

This process allows the use of suitable catalysts to give products in which the branches are arranged in an orderly manner. These plastics are called *isotactic* polymers. An example, polypropylene is illustrated in Figure 9.4 (above). For polyethylene, this form of polymerisation has the advantage of reducing the number of branches that are formed. The process uses lower temperatures and pressures than radical polymerisation and consequently little cross-linking or branching occurs. The truly linear chains pack together to give a *high-density* polymer that has a high melting point. By using anionic catalysts mounted on a crystalline solid, the geometry (stereochemistry) and shape of the product can be determined.

The molecules in linear isotactic polyethylene can line up with one another very easily to yield a tough, high-density compound that is useful for making toys, bottles, etc. The polyethylene with irregular branches is less dense, more flexible, and not nearly as tough as the linear polymer, because the molecules are further apart.

The factor that affects the rigidity of the polymer is the packing of the molecules, which is seen as variations in *density* and *molecular mass* — the number of monomer units combined to give a polymer. With increasing molecular mass there is a corresponding increase in strength, toughness and chemical resistance.

A third basic parameter is molecular mass *distribution*. Because of the statistical nature of polymerisation, polymer molecules show a variation in mass (see *non-ionic* detergents in How do surfactants work, Chapter 3). Although the chemical properties of a polymer remain much the same regardless of mass, its mechanical properties are affected by variations in the mass of its molecules. If most of the molecules of a resin fall within a very narrow mass range, products made from it will have better mechanical properties than those made from resins whose molecules vary over a large range of masses. It is the balance in these three characteristics that provides the variation in properties of the different plastics formed from the same basic monomer.

Certain properties can also be controlled by adding small amounts of other monomers to the ethylene monomer. Vinyl acetate or ethyl acrylate in low-density resins, and hexene or butene in high-density resins, will increase branching — thus increasing flexibility and elasticity. Finally, polyethylene can be changed from a linear polymer to a cross-linked polymer by incorporating particular reagents. Additives used in polyethylene include slip agents (to decrease frictional properties), anti-block agents (to prevent sheets of molecules from sticking together), and antioxidants.

Most polymers undergo oxidation and photo-initiated degradation, and this can be retarded by antioxidants. Low-density polythene requires only a very small amount of antioxidant. High-density polyethylene, polystyrene, and in particular, polypropylene, are much more sensitive both during processing and on exposure to the environment. Oxidation causes the links in the polymer chain to break, which lowers the molecular mass of the polymer and makes it less tough. It also causes cross-linking of the molecules, which, although resulting in a higher molecular mass, in this case decreases toughness and makes the plastic more brittle. To combat oxidation, various materials are used: primary antioxidants, BHT and BHA (see Fig. 4.12), and other materials. Polypropylene used in, for example, washing-machine agitators, requires antioxidants capable of withstanding hot detergent. High molecular mass (> 800) derivatives of BHT and BHA are used.

Vinyl polymers

If one of the hydrogen atoms in the ethylene molecule is replaced by chlorine, *vinyl chloride is* formed. If two hydrogens on the same carbon atom are replaced, *vinylidene chloride* is formed. If we replace a hydrogen in ethylene with acetate (from acetic acid) we form *vinyl acetate*. These monomers can all be readily polymerised. Their molecular structures are given in Table 9.1 (overleaf).

Vinyl polymers and co-polymers make up one of the most important and diversified groups of linear polymers. This is because PVC (poly(vinyl chloride)) can be *compounded* to produce plastics with a wide range of physical properties. PVC is used to make products ranging from guttering and water pipes to the very thin, flexible surgeon's gloves.

Polymerisation of vinyl chloride to PVC can be carried out in three different ways. In the *suspension* process, droplets of monomer are dispersed in water and polymerised. In the *mass* process, special agitation is used to polymerise liquid vinyl chloride monomer without water present. The commonest method of making PVC is to disperse vinyl chloride in water as an *emulsion* (using surfactants) and using catalysts and heat. The monomer is polymerised to solid particles of polymer that emerge as a suspension in water. This is centrifuged and dried. A large molecular mass means a stronger and more rigid polymer, but this type of polymer is more difficult to work. Vinyl acetate is also polymerised by emulsion polymerisation because most PVA (poly(vinyl acetate)) is used in making emulsion (latex- or water-based) paints. Plasticiser and pigment are added. When the paint is applied to a surface, the water evaporates and leaves a polymer film containing pigment and plasticiser.

About 80% of all plasticiser production is used in PVC, and the amount can form up to 50% of the finished product. The esters of

TABLE 9.1 Vinyl polymers

Monomer	Monomer structure	Polymer	Main uses
A. Common polymers			
Ethylene	$H_2C{=}CH_2$	Polythene low density high density	Bottles, tubing, sheets and other moulded objects
Vinyl chloride	$H_2C{=}CHCl$	Poly (vinyl chloride) (PVC)	Raincoats, shower curtains, gramophone records, garden hose, rigid clear bottles, swimming-pool liners
Propylene	$H_2C{=}CH{-}CH_3$	Polypropylene	As for polythene, and carpets (isotactic)
B. Specialised polymers			
Vinyl acetate	$H_2C{=}CH{-}O{-}C({=}O){-}CH_3$	Poly (vinyl acetate) (PVA)	Adhesives, latex paints
Vinylidene chloride	$H_2C{=}CCl_2$	Poly (vinylidene chloride) co-polymer with PVC in Saran	Some clinging wraps, some freezer bags
Tetrafluoroethylene	$F_2C{=}CF_2$	Poly (tetrafluoroethylene) (PTFE), Teflon	Bearings, gaskets, non-stick pan lining, chemical resistant films
Styrene	$H_2C{=}CH{-}C_6H_5$	Polystyrene	'Rigid foams', moulded objects, electrical insulation
Methyl methacrylate	$H_2C{=}C(CH_3){-}C({=}O){-}O{-}CH_3$	Poly (methyl methacrylate), Perspex, lucite, plexiglass	'Safety glass' but PV Butyral is the adhesive in Triplex and Pilkington's windscreen glass
Acrylonitrile	$H_2C{=}CH{-}C{\equiv}N$	Poly (acrylonitrile)	Orlon, acrilan textile fibres
C. Mixed polymers			
Styrene and acrylonitrile		SAN	Latex paints, plastic plates, etc.
Acrylonitrile, butadiene and styrene		ABS	Rigid: telephone sets, shoe soles, automobile parts

phthalic acid are most commonly used, particularly di(2-ethylhexyl)-phthalate (DEHP), also called dioctylphthalate (DOP), about which there are environmental concerns. There are several reports of PVC plasticisers in sealants used in greenhouses inhibiting growth (of brassicas — the cabbage and related vegetables).

PLASTICS IN THE BUILDING INDUSTRY

Of the over 30 million tonnes of plastics produced in the US in 1992, 20% or 6 million tonnes was used in the building industry, second only to packaging. Of this, PVC is still the dominant material. Roughly 75% of PVC use is in the building industry, and of this, 45% is used for pipes (replacing copper). In Australia, 0.2 million tonnes is used, 80% in the building sector, of which about 10% is used for cable insulation.

PVC can be pulled through cracked cast iron pipes and heat sealed to repair the pipes internally. Pipes are now produced with a hard inner and outer shell but with a foamed PVC inner core. Further chlorination of PVC gives a pipe material that can be used at high temperature.

About 5% of the PVC market is for electrical insulation on wires and cables. Phenolics come second and are used for electrical devices and in laminating plywood, followed by urea and melamine resins.

Polyethylene is used in natural gas pipes and wiring conduits (halogen-free retardants can be added) and also increasingly for water pipes. Because of its low-temperature flexibility, polyethylene could replace PVC in sewer pipes.

Other materials likely to make inroads into the use of PVC in building are modified polycarbonates and iodated polymers.

Better Windows
PVC has been replacing wood, and then aluminium, in window sills and frames. In turn, acrylonitrile–butadiene–styrene (ABS) is moving in on PVC in this market because of its better high-temperature stability.

Heavy metal compounds are used as heat stabilisers in the manufacture of PVC articles. Lead is commonly used, but could be replaced by calcium/zinc at a modest extra cost and a great extra benefit to the environment on disposal of the plastic. The lead either ends up in the ash or volatilised in the flame. Fires, either accidental or deliberate incineration, also generate dioxins in the flue gas and flue ash from chlorinated material and PVC is one contributor. High-temperature incinerators can be designed to reduce the gas emission to very low levels, but are expensive, unpopular, and not always available, particularly for hospitals dealing with medical waste. Only cable insulation is recycled, or more correctly 'respiralled' into less demanding products such as hose cores and mud flaps. Recycling of building products is critical to the long-term acceptance of PVC.

Polystyrene is an amorphous, transparent polymer (the bulky phenyl groups inhibit crystallisation) that softens at 94°C and so cannot be sterilised. With a sharp hit, it gives a metallic ring. Products made from it like drinking glasses, sparkle and are quite attractive because it has a high

Biodegradable credit cards
Greenpeace is using a Visa credit card manufactured by Monsanto (from a process developed by Zeneca) with a plastic called Biopol. It is produced by fermenting grains, starch or sugar with microbes. Under the right conditions the plastic is biodegradable. Less than 1% of PVC of the 18–20 Mt produced each year is used for the credit cards. In the UK there are about 90 million PVC credit cards in circulation and at least 14 million are thrown away each year.

refractive index (1.6). (See also Hiding power in Chapter 12.) The brittle nature of the polymer can be overcome by adding 5 to 10% of butadiene monomer to give 'impact polystyrene', but then light scattering occurs at the phase boundaries and the product is opaque. Polystyrene is flammable, softened by many solvents, and light sensitive. Polystyrenes that contain UV absorbers are used as fluorescent light diffusers.

Because of the wide gap between the glass transition temperature ($T_g = 94°C$) and the melting point ($T_m = 227°C$), polystyrene is a pleasure to process (see Glass transition temperature later in this chapter). Injection moulding is used for making bottles and jars. Extrusion moulding is used to produce sheeting, which can then be thermally shaped to make refrigerator linings and three-dimensional contour maps. Rigid polystyrene foam accounts for about 15% of this plastic's use and a considerable amount of this is used in packaging. Co-polymerisation with acrylonitrile to give styrene acrylonitrile (SAN) provides a slightly better product.

Polystyrene burns with a hot smoky flame and melts while burning. If the molten, burning polymer falls on the skin, it tends to stick and cause severe burns. Because polystyrene dissolved at high concentrations in petrol forms a gel, this mixture is used in the modern formulation of that nasty weapon called napalm. The name napalm comes from the original gelling agent for petrol — the aluminium salt of naphthenic and palmitic acids (the aluminium salt of household soap also works well).

Condensation polymerisation

Another form of polymerisation involves the joining together of monomers by removing a small molecule in the joining process. This is called *condensation polymerisation*. An example of such a process is the formation of a *polyamide*. The amide bond is illustrated in Figure 9.6.

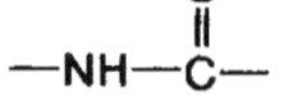

Fig. 9.6
The amide bond

$$-NH-\overset{\displaystyle O}{\overset{\|}{C}}-$$

The original polyamide was 6,6–nylon, which was developed as a replacement for silk in parachutes. The reaction is given in Figure 9.7.

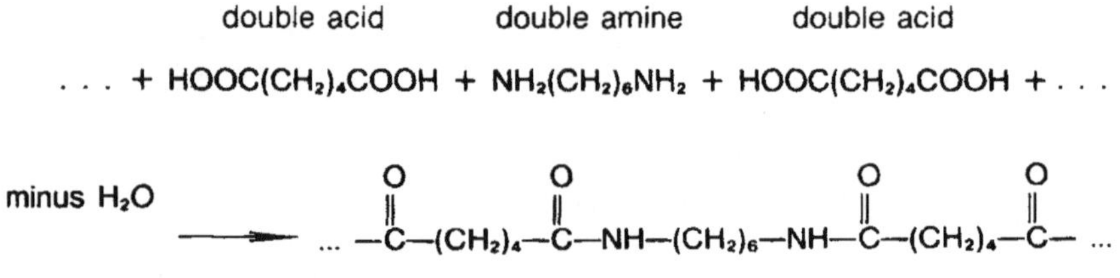

Fig. 9.7
The 6,6–nylon reaction

$$\underset{\text{}}{\ldots +} \underset{\text{double acid}}{HOOC(CH_2)_4COOH} + \underset{\text{double amine}}{NH_2(CH_2)_6NH_2} + \underset{\text{double acid}}{HOOC(CH_2)_4COOH} + \ldots$$

$$\xrightarrow{\text{minus } H_2O} \ldots -\overset{\displaystyle O}{\overset{\|}{C}}-(CH_2)_4-\overset{\displaystyle O}{\overset{\|}{C}}-NH-(CH_2)_6-NH-\overset{\displaystyle O}{\overset{\|}{C}}-(CH_2)_4-\overset{\displaystyle O}{\overset{\|}{C}}- \ldots$$

Stretching aligns the chains, and additional weak (hydrogen) bonds between the chains strengthen the fibre. The two different monomers each have six carbon atoms — hence the polymer is called 6, 6-nylon.

Why not have *one* monomer with *two different* functional (or end) groups? Using NH_2–$(CH_2)_5COOH$ (illustrated in Fig. 9.8), the polymerisation product is 6–nylon.

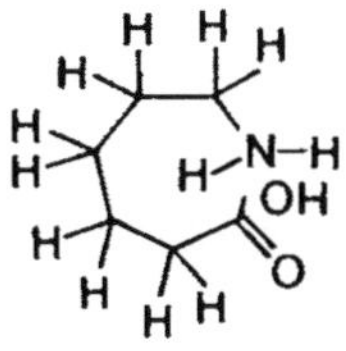

Fig. 9.8
6-nylon monomer

The amide bond is similar to that formed when proteins (which are polymers) are formed from individual (but not identical) amino acids. (However, in protein, instead of there being only one type of monomer, there is a choice of about 20 amino acids to string together; that is, the paper clips can be different.) The shorter the –CH_2– chain the more hygroscopic the nylon is. For industrial applications (nylon bearings, etc.) long –CH_2– chains are used to reduce water absorption.

The sharing of a hydrogen between an oxygen on one chain and a nitrogen on another (a hydrogen bond) is what gives nylon its unique fibre-like properties. (See Fig. 9.9, overleaf.)

The locations of these atoms 'are just right' to link up every one of them in nylon 6,6. You can draw the structure of nylon 6,10 and show that it will also 'line up'. But try something arbitrary like nylon 7,7 and it won't line up.

Nylons are among the toughest plastics: they can withstand repeated blows, and they also have the advantage of low friction. Nylons also resist many solvents, but they are soluble in formic acid and phenols. (See also From mess to millions, the story of nylon in Chapter 11).

By condensing terephthaloyl chloride with p-phenylene diamine, a polymer with an amide bond connecting aromatic groups is produced, called an aramid. In this case the compound is known as Kevlar (see Fig. 9.10).

Kevlar is an early example of a liquid crystal polymer in which the rigid chain of the molecule maintains its rigidity in the melt, or in solution. Kevlar has replaced asbestos in many applications. It is used as cord for large tyres, reinforcement for bullet-proof vests and in helmets. In combination with epoxy, it is used for canoes and as a cloth in boat sails. Nomex is a related polymer used for fire resistant cloth for firefighters and racing car drivers. (See Spider silk in Chapter 11.)

By forming an ester bond instead of an amide bond, we obtain *polyesters*. (Compare the ester bond in Fig. 9.11 (overleaf) with the amide bond in Fig. 9.6.) The reaction is shown in Figure 9.12 (see p.225). The polyester product of this reaction has numerous trade names. Examples are Terylene and Crimplene in the United Kingdom, Dacron in the United States, and Trevira in Germany. Polyester film material has unusual strength and electrical resistance (mylar film, magnetic recording tape, frozen-food packaging).

Kevlar man
On a weight for weight basis, fibres made from Kevlar are stronger than steel.

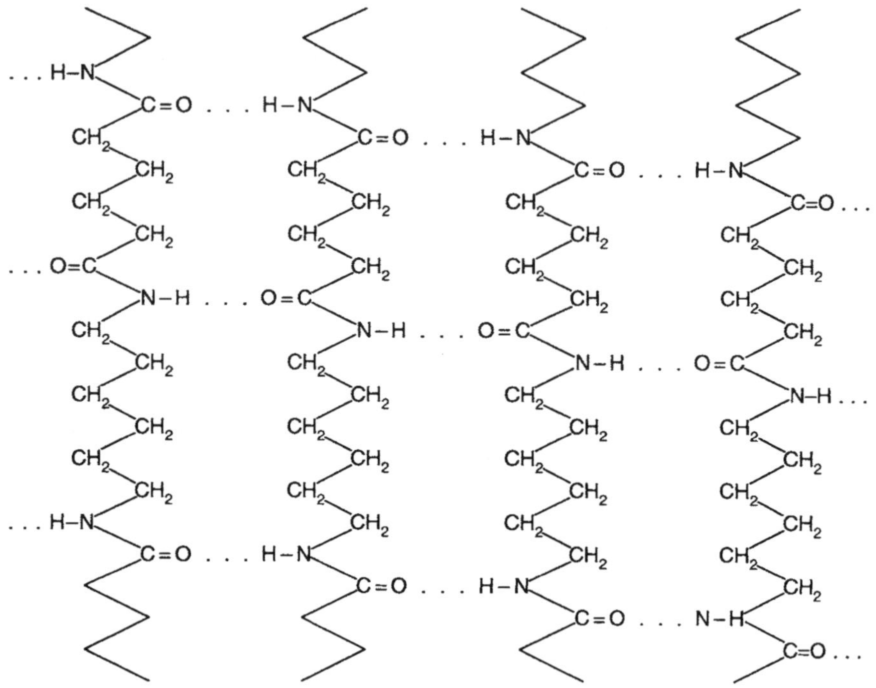

Fig. 9.9
Bonded chains of nylon 6,6. Each CO and each NH group is located conveniently to bond to its counterpart in the adjacent molecule, effectively locking the individual chains together in a filament.

(J. Cross, Australian chemical resource book 1990 (9) p. 83).

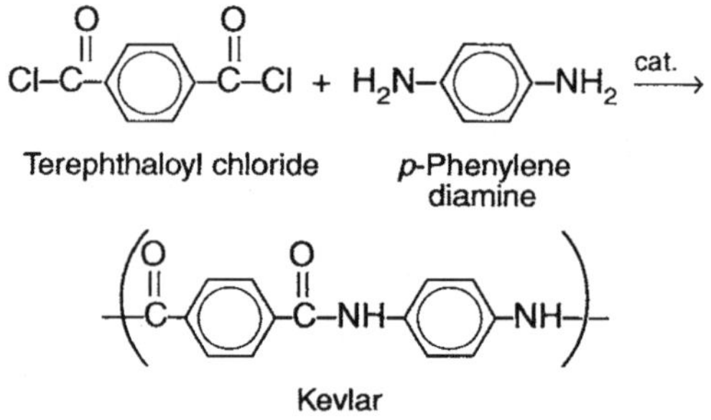

Terephthaloyl chloride   *p*-Phenylene diamine

Kevlar

Fig. 9.10
Production of Kevlar

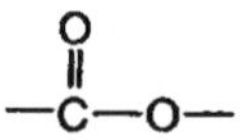

Fig. 9.11
The ester bond

An unsaturated polyester resin consists of a linear polyester whose chain contains some double bonds, together with a monomer, such as styrene, that co-polymerises with the polyester to provide a cross-linked product. It is, in effect, a *cold-curing, thermosetting* plastic.

The formulation has inhibitors (to prevent reaction during storage), UV absorbers, extenders, thixotropic agents, etc. Curing is brought about with a free radical initiator. Small heat-sensitive objects (e.g. electronics components) can be 'potted', and large objects such as boats and swimming pools can be made without having to use huge ovens. As well as the fibreglass layering technique, which is a slow process, injection

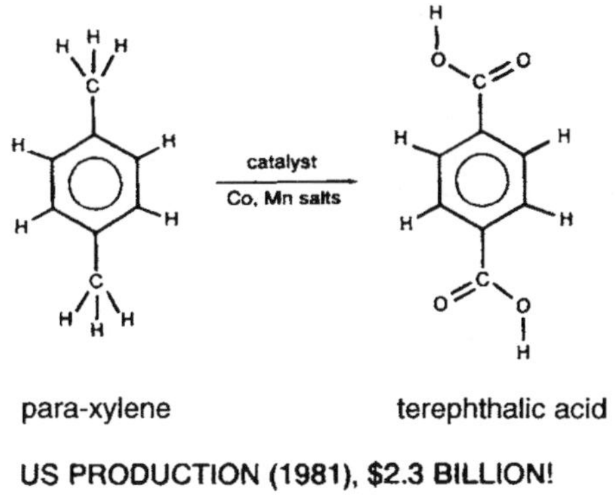

catalyst
Co, Mn salts

para-xylene → terephthalic acid

US PRODUCTION (1981), $2.3 BILLION!

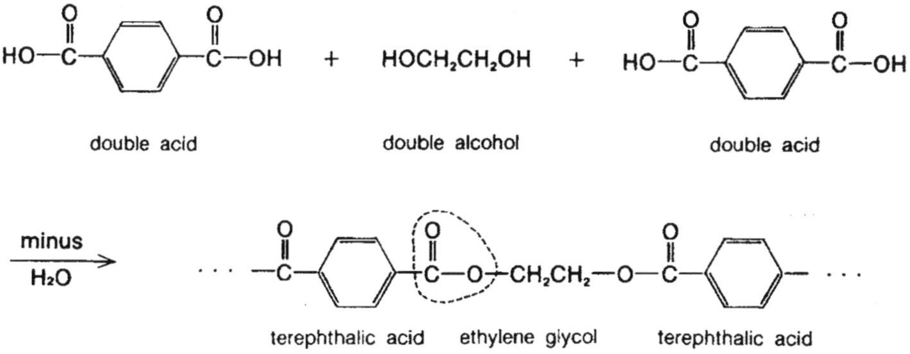

HO—C(=O)—C_6H_4—C(=O)—OH + $HOCH_2CH_2OH$ + HO—C(=O)—C_6H_4—C(=O)—OH

double acid | double alcohol | double acid

minus H_2O → ··· —C(=O)—C_6H_4—C(=O)—O—CH_2CH_2—O—C(=O)—C_6H_4— ···

terephthalic acid | ethylene glycol | terephthalic acid

Fig. 9.12
Polymerisation to a polyester

moulding, which is much faster, can be used by mixing the polyester and short glass fibres into a 'dough'. This overcomes the problem of delamination in poorly fabricated fibreglass products.

For short-run production of custom car bodies, fibreglass is more economical than metal stamping. A rapidly increasing use of fibreglass is in the manufacture of simulated wood for doors, and imitation marble. Products as diverse as pistol grips, explosion barriers for TV tubes, and bowling balls are made from polyester. A cunning way of providing a smooth skin is to include an insoluble thermoplastic additive (such as Perspex) that migrates to the surface of the mould and layers out as a smooth 'skin' that may be readily decorated.

Epoxy resins are dealt with in Chapter 12 (under Adhesives). They adhere better to metals than polyesters do. See also PET below (under Recycling).

If each monomer has only two functional groups, the resulting polymer is a linear chain. If there are more than two functional groups on the monomers, cross-linkage can occur and this will result in a more rigid lattice. An example is the reaction shown in Figure 9.13.

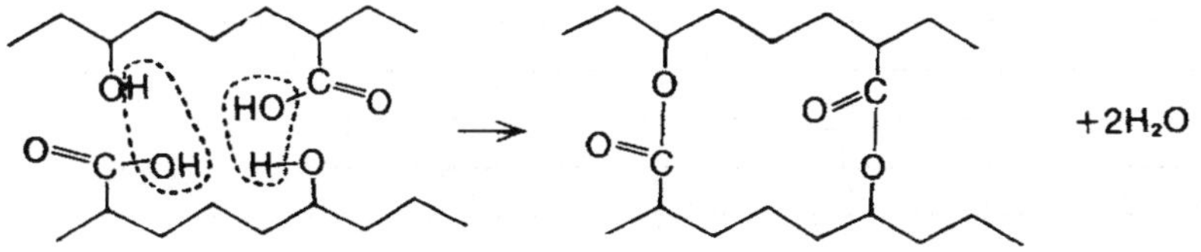

$+2H_2O$

Fig. 9.13
Cross-linking in a polyester polymerisation

Cross-linked polymers of this type are known as *alkyd* resins. They are used in paint enamels and in the manufacture of false teeth.

Molecules containing the isocyanate group, –NCO, will react with other molecules containing, say, an –OH group to give a *urethane* linkage (Fig. 9.14), which is similar to the amide bond in nylons. Polyurethanes can be formed by using bifunctional molecules (Fig. 9.15).

Fig. 9.14
The urethane linkage (compare Fig. 9.6).

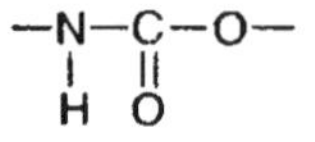

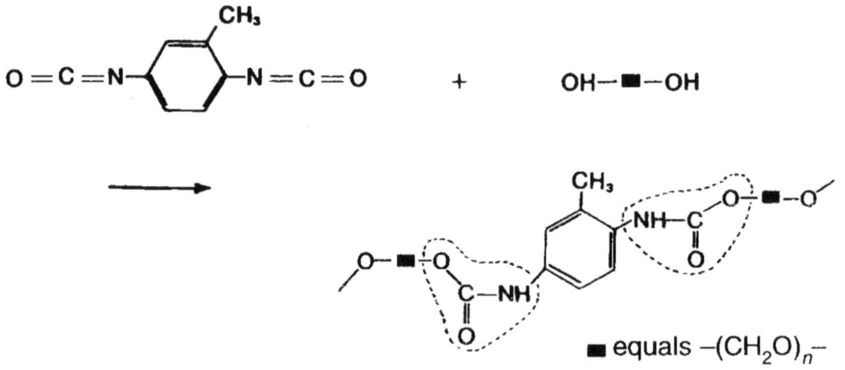

–N–C–O–
H O

Fig. 9.15
Polymerisation to polyurethane

■ equals $-(CH_2O)_n-$

Because the isocyanate monomer decomposes with water to form gaseous CO_2, judicious amounts of water can be added during polymerisation to form polyurethane foam rubber. Heating of polyurethanes produces unpleasant vapours containing nitric acid (HNO_3), nitrogen dioxide (NO_2), and even hydrogen cyanide (HCN). A burning pillow will quickly fill a room with thick, dense, toxic fumes.

Elastomers

Natural rubber (Fig. 9.15) contains a linear polymer of isoprene, $CH_2=C(CH_3)\ CH=CH_2$, in which all the –CH=CH– groups are *cis* (see Fig. 4.6).

Fig. 9.16
Natural rubber

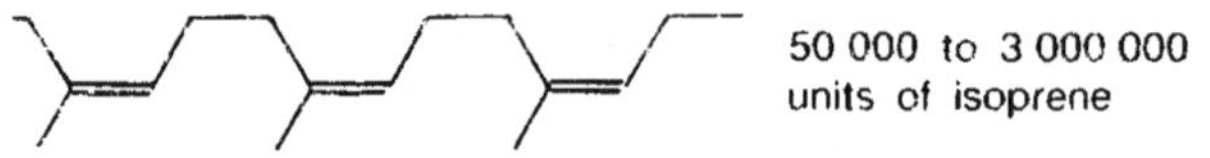

50 000 to 3 000 000 units of isoprene

Strictly speaking, the property of being elastic means the degree to which a material returns to its original shape after stretching. However, in general usage, it also means the degree to which a material can be stretched. The polymer chains in elastomers are elastic in the second sense because the chains can be unravelled without coming apart.

Elasticity, in the first sense, was improved by cross-linking with sulfur (as discovered by Goodyear), and this produces *vulcanised* rubber. Synthetic rubbers are produced from related monomers. Polymerisation of butadiene (CH_2=CH–CH=CH_2) using a sodium (Na) catalyst produces Buna rubber. The monomer chloroprene, CH_2=C(Cl)–CH=CH_2, on polymerisation gives neoprene, an oil-resistant rubber. Co-polymers such as SBR (styrene–butadiene rubber) are produced in even greater quantities for vehicle tyres (90% of Australian-made tyres). To produce all *cis* polymers (which have much higher elasticity than those which are not entirely *cis)* a special stereo-regulating catalyst must be used. (The trans-isoprene polymer also occurs in nature in various tropical trees, and is called gutta-percha.) Silicone rubbers are thermally stable, resist oxidation, are chemically inert, are flexible over a wide range of temperatures (–90 to 250°C) and are virtually non-stick.

STRETCHING RUBBER

Rubber is a typical polymer and is not very 'elastic' in the Hooke sense of stress being proportional to strain. Stretching aligns the random chains, and that temporarily crystallises (and toughens) the material (rubber tyres do not suffer from crack propagation). Take a thick rubber band, condom, or finger of a pure rubber glove (thin, surgical, latex). Stretch it quickly and place it against your lips to sense the temperature. It has become hot. Keeping it stretched, allow it to cool back to room temperature over half a minute or so. Then let it suddenly contract to its natural length. Sense the temperature again with your lips. It has cooled. Well, anyway, it is great way to break a lull in the conversation.

Working hard to align the random chains (on stretching) reduces the entropy of the system and increases the entropy of the surroundings in the form of heat given off (see Theory of heat engines in Chapter 14, and Appendix VIII). Spontaneous contraction means the aligned chains randomise. This increases their entropy and causes absorption of heat from the surroundings, (i.e. cooling). Polymers thus behave in the directly opposite way to metals. Rubber contracts on heating and expands on cooling.

Warm band
Try warming a stretched rubber band (e.g. with a weight on the end) with a hair dryer. Well, Robert never got his 'Hooke' into that piece of science.

Polymer foams

Polyurethane and polystyrene foams, and rubber foam latex, dominate the market. The glass transition temperature of rubber is so low that only flexible foams are possible; it is so high for polystyrene that only rigid foams are possible. However, polyurethanes can be formulated to provide either type. These foams can be cured *in situ*, which gives them great versatility in the packaging industry and in producing items such as shoe soles.

Plants eat formaldehyde
NASA scientists have found that spider plants are best for removing formaldehyde, the possible cancer agent found in many homes, from the air. They say that an average home could be kept completely free of formaldehyde gas by installing 70 spider plants. The researchers, working on an enclosed ecological life-support system project, looked at a number of house plants but the spider plant was five times better at absorbing the gas than any rival. Nobody knows quite how it does it. They recommend one plant for every 2.5 square metres in homes or offices.

Source: New Scientist, 8 November 1984.

The foaming agent is variable. In foam rubber, the latex is mixed at very high speeds and the foam contains air. PVC foam is produced when a 'blowing agent' that decomposes to a gas on heating is included. Polystyrene beads are impregnated with pentane (C_5H_{12}), which boils at 36°C.

Urea–formaldehyde foam

Under certain conditions some polymers can break apart into their component monomers. Examples are poly (methyl methacrylate) which reverts to the monomer on distillation, and polyurea–formaldehyde, which reverts to its monomers, to a small extent, in the presence of acid. Other polymers do not decompose into their monomers but into other products. An example of this is poly(vinyl chloride). The only monomer released by this polymer is the one that has been left behind from the initial reaction forming the polymer. On decomposing, the polymer forms hydrochloric acid.

Polyester embedding resin

Apart from being used to make fibres such as Dacron and Terylene, some polyesters are used as a clear, crystalline, embedding resin. This resin is supplied as a viscous liquid of the already linear-polymerised unsaturated polyester dissolved in styrene monomer. The catalyst generally used is methyl ethyl ketone peroxide, which cross-links the styrene to the double bonds in the polyester to give a three-dimensional structure (see Fig. 9.17). The amount of cross-linking can be controlled by using different proportions of saturated and unsaturated polyester.

EXPERIMENT 9.1 Tests on plastics

Equipment
Samples of plastics
Copper wire in cork holder
Large beaker with water
Spatulas

Method
When burning plastics:

- Use cm square pieces of your samples, and hold them with tongs or a wooden peg.
- Experiment in a well ventilated place because of the fumes produced.
- Hold the burner or candle at an angle so that any drops of molten plastic fall onto a non-flammable mat.

1. Copper wire test
Heat a copper wire of length about 10 cm (stuck into a cork as a holder) in a gas flame until any yellow or green colour disappears. Press the heated

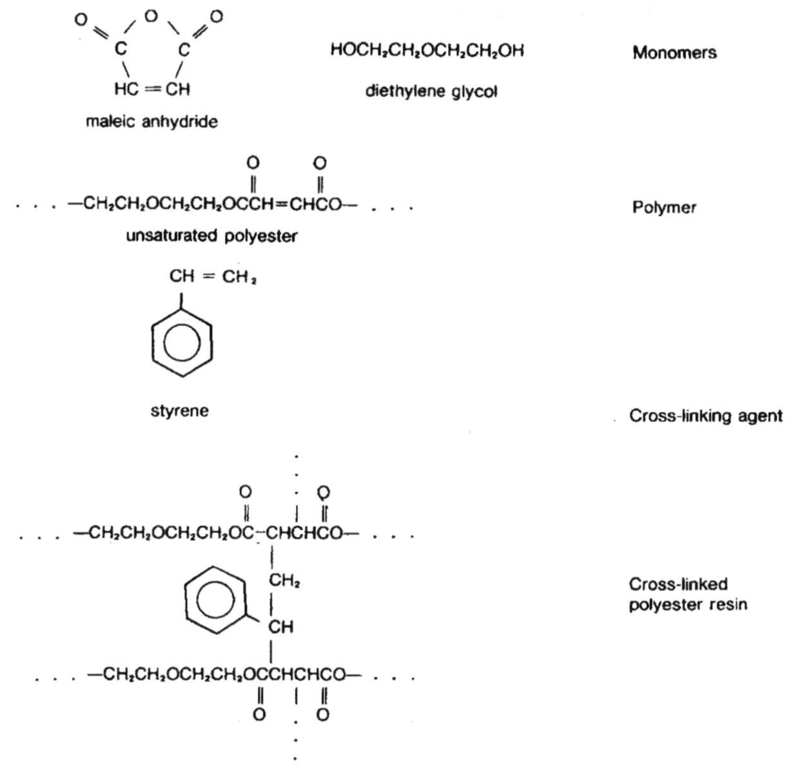

maleic anhydride

$HOCH_2CH_2OCH_2CH_2OH$

diethylene glycol

Monomers

$\ldots -CH_2CH_2OCH_2CH_2OCCH{=}CHCO- \ldots$

unsaturated polyester

Polymer

$CH = CH_2$

styrene

Cross-linking agent

$\ldots -CH_2CH_2OCH_2CH_2OC-CHCHCO- \ldots$

CH_2

CH

$\ldots -CH_2CH_2OCH_2CH_2OCCHCHCO- \ldots$

Cross-linked polyester resin

Fig. 9.17
Cross-linked polyester

Experiment 9.1 (continued)

wire into the plastic sample and then put the wire with a little molten plastic on it back into the flame. A green colour indicates that the plastic contains a halogen — probably chlorine in poly(vinyl chloride) (PVC) or poly(vinylidene chloride) (PVDC). Before repeating the test with another plastic, again heat the copper wire until the green colour disappears.

It should be noted that an additive may contain a halogen (chlorine, bromine, or iodine) that gives rise to a positive result. Also cyanide (from, say, Orlon) may give a positive result.

2. Density

Some polymers are less dense than water and hence will float. These are polyethylene, polypropylene, styrene–butadiene and nitrile (some types). It is essential for the sample (not a foam type) to be properly wetted and pushed below the surface, and then released. The presence of large amounts of additives can change the density.

3. Feel

Poly(ethylene) and poly(tetrafluoroethylene) have a waxy feel not possessed by other polymers. Clean the surface to remove grease or plasticisers.

4. Heating tests (in a laboratory)

A small piece (0.1 g) of the material is placed on a clean spatula (nickel spoon-like object) previously heated to remove traces of combustible material. It is then gently warmed, without ignition, over a small gas flame (with minimum colour) until it begins to fume. The sample is removed from the flame and the fumes are tested with moist litmus paper (red and blue) to determine whether they are acidic, alkaline, or neutral. Check the odour as well. The sample is now moved to the hottest zone of the small gas flame and the following points noted:

(a) Whether the material burns, and if so, how easily.
(b) The nature and colour of any flame (a very sooty flame generally indicates an aromatic polymer, but may result from carbon black filler).
(c) Whether the material continues to burn after removal from the flame.
(d) The nature of any residue.

Table 9.2 considers some of the possibilities. It was compiled from uncompounded polymers. Thus highly plasticised PVC may not be self-extinguishing.

The physics of plastics

Plastics can be subdivided roughly into groups on the basis of their mechanical properties. One way is by measuring their rigidity as a function of temperature (Fig. 9.18, p. 232). In another method the *stress* versus *strain* curves for plastics place them approximately into different categories. Some of the plastics not mentioned in the previous section fall into one of these categories — *hard and tough.*

Plastics that are hard and tough have a high tensile (elongation) strength, and stretch considerably before finally breaking. The top performers in this grade are called *engineering plastics.* They are relatively expensive and have specialised applications. *Polyacetals* are very abrasion resistant, and resist organic solvents and water. Hence they are used in plumbing to replace brass or zinc in shower heads, valves, etc. Furniture castors, cigarette lighters, shavers, and pens are also often made from polyacetals as they give a non-stain, satin finish.

Polycarbonates are often used instead of glass because they are transparent and retain their dimensions and resistance to impact, even when subjected to a wide range of temperatures. Babies' bottles, bus-shelter windows, plastic sheeting for roofing and telephone dials are just a few examples. Because of their fire resistance, they are used in firemen's masks, interior mouldings in aircraft, and in electronic equipment. In sporting equipment they are found in helmets used by cricketers and motor cyclists, as well as in baseball helmets and snowmobiles. The most common plastic optical lens material is a poly-diallylcarbonate (Columbia resin CR39) and related materials (CR64, EX80), which are often used for embedding (of flowers, etc.) and casting.

TABLE 9.2
Burning test on plastic

Plastic	Flame colour	Odour	Other features
1. The material burns but extinguishes itself on removal from the flame			
Casein	Yellow	Resembles burnt milk	
Melamine–formaldehyde	Pale yellow with light blue-green edge	Formaldehyde and fish-like	Very difficult to ignite, alkaline fumes
Nylon	Blue with yellow tip	Resembles burning vegetation	Melts sharply to clear liquid which can be drawn into fibre
Phenol–formaldehyde	Yellow	Phenol and formaldehyde	Very difficult to ignite
Poly(tetrafluorethylene)	Yellow	None	Burns with extreme difficulty, chars very slowly, acidic fumes
Poly(vinyl chloride) Poly(vinylidene chloride)	Yellow with green base	Acrid	Acidic fumes
Urea–formaldehyde	Pale yellow with light blue-green edge	Formaldehyde and fish-like	Very difficult to ignite, alkaline fumes
2. The material burns and continues to burn on removal from the flame			
Acrylonitrile–butadiene–styrene	Yellow with blue base, smoky	Burning styrene	
Alkyd	Yellow, smoky	Pungent, unpleasant	
Cellulose acetate	Yellow	Acetic acid (vinegar)	Acidic fumes
Cellulose nitrate	Yellow	Possibly camphor on gentle warming	Burns at a very fast rate, may explode
Epoxide	Orange-yellow, smoky	Acrid	
Ethyl cellulose	Pale yellow with blue-green base	Resembles burning wood	Drips on ignition
Poly(acrylonitrile)	Yellow	Cyanide initially, then resembles burning wood	
Poly(carbonate)	Yellow, smoky	Phenolic	Difficult to ignite initially
Poly(ethylene) Poly(propylene)	Yellow with blue base	Resembles burning candle wax	Becomes clear when molten
Poly(methyl methacrylate)	Yellow with blue base	Methyl methacrylate	
Polystyrene	Yellow with blue base, very smoky	Burning styrene	
Polyurethane	Yellow with blue base	Acrid	
Poly(vinyl acetate)	Yellow, smoky	Vinyl acetate	Black residue
Poly(vinyl alcohol)	Yellow, smoky	Unpleasant, sweet	Black residue

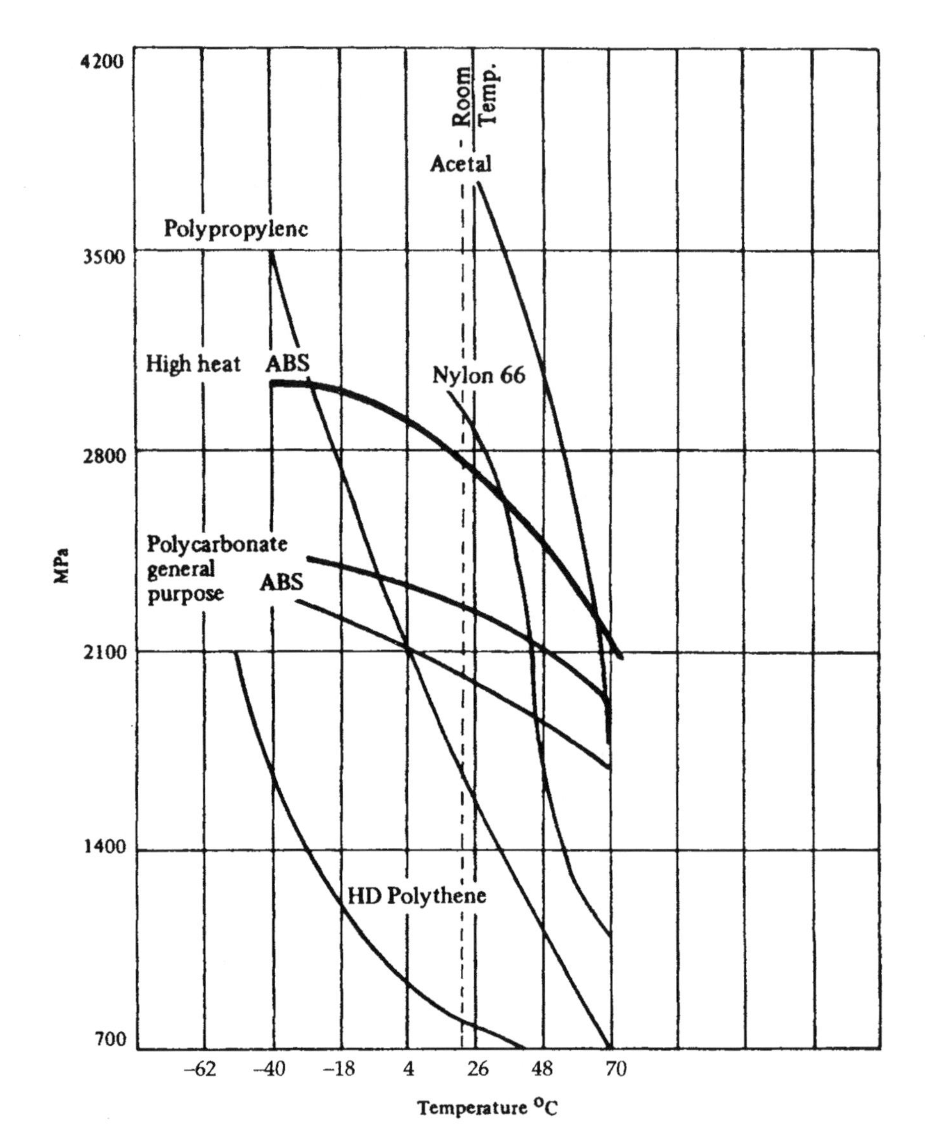

Fig. 9.18 Comparison of flexural modulus (rigidity) versus temperature for various plastics.

(From Know your plastics, Plastics Institute of Australia, 1980)

Polycarbonates are produced by an interesting process called phase transfer catalysis. The two components of the polymer are each in separate non-mixing phases — one in water and the other in an organic solvent — so that they do not react directly. A quaternary ammonium positive ion (a short, cationic surfactant — see How do surfactants work? Chapter 3) forms a neutral salt with the negative ion of the water-soluble component, and takes it across into the organic layer where it can react with the component there, releasing the quaternary ammonium ion, which can then go back into the water. (Such a salt was mentioned in Chapter 8 (Question 19) in regard to measuring cyanuric acid levels in swimming pools.) An example of this procedure is given in Figure 9.19.

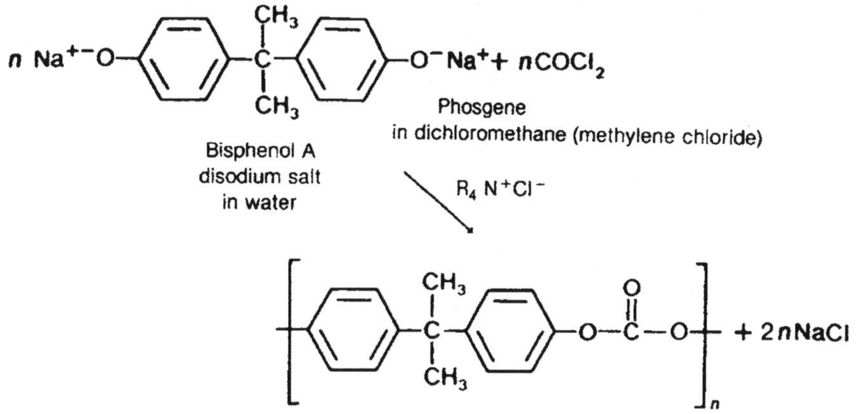

$$n\ Na^{+-}O-C_6H_4-C(CH_3)_2-C_6H_4-O^{-}Na^{+} + n\,COCl_2$$

Bisphenol A disodium salt in water

Phosgene in dichloromethane (methylene chloride)

$$\xrightarrow{R_4N^+Cl^-}$$

$$\left[-C_6H_4-C(CH_3)_2-C_6H_4-O-\overset{O}{\overset{\|}{C}}-O-\right]_n + 2n\,NaCl$$

Fig. 9.19
Formation of polycarbonate

One plastic that is not considered an engineering plastic but is still hard and tough is *high-density polyethylene* (HDPE). However, during injection moulding, stresses are set up inside the material which cause bonds to rupture. These cracks propagate (see below) until visible cracks appear, and the surface of the polymer roughens. This is called *environmental stress cracking* (not to be confused with personal problems at work!). If small amounts of propylene as co-monomer are added, this problem is reduced. Polypropylene itself has superior properties, but is expensive and light sensitive.

Moulding of plastics allows new forms of design, such as the one declared in US patent 2 487 400 (1949), lodged by one Earl S. Tupper (Fig. 9.20).

> . . . the invention herein provides a sealing enclosure for containers in the form of a hollow finger-engageable stopper having elasticity and flexibility with a slow rate of recovery to provide a non-snapping and noiseless type of cover which is applicable to the lip of a container by hand conformation and removable therefrom by a peeling-off type of procedure.

The patent does *not* specify that you can buy Tupperware only at special parties!

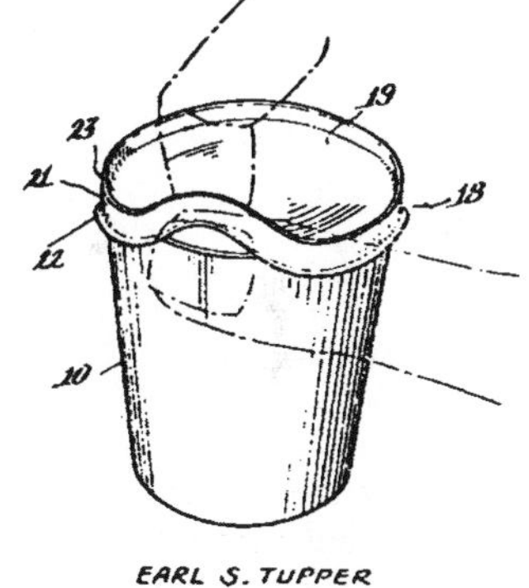

Fig. 9.20
Tupperware

An interesting plastic system is *acrylonitrile–butadiene–styrene* (ABS). This is in fact a two-phase system in which a styrene–butadiene rubber is dispersed in a glassy styrene–acrylonitrile (SAN) matrix. (See also Part I Appendix III.) Without fire-retardants, ABS burns. It has a high gloss and keeps its shape, and finds uses in children's toy block-building kits, housings for radios, calculators, telephones, computers and the better quality refrigerators. It is also used in lawnmower housings, high-quality pipes and fittings, luggage, and the top layers of skis. ABS tends to be used as a replacement for metals (e.g. telephone dials).

Nylon has excellent mechanical properties, resists solvents, and is an ideal material for gears and bearings that cannot be lubricated. About 50% of all moulded nylon fittings go into cars in the form of small gears (wipers), timing sprockets, and all sorts of clips and brackets. A special nylon was developed for the important use of sealing the side of steel cans to replace solder. The nylon seal is mechanically stronger and does not present problems of lead release. The nylon used has a very long chain (C_{36} dibasic acid) which increases its adhesive power. It bonds directly to steel and does not require a tin surface, unlike solder, and the seam itself is a simple overlap of the two ends of the steel rather than a soldered join of the two edges.

Fluoroplastic (e.g. PTFE or Teflon) is very hard and tough, is also non-flammable, and has unique resistance to chemical attack. However, it lacks dimensional stability and is subject to creep. The non-stick frypan is the classic example of this plastic's chemical resistance. For a long time, no way was found to make fibres from PTFE, until one day a researcher got so impatient, he just yanked apart a rod that had been heated in its centre. It formed multiple fibres and quickly gave us Goretex™ clothing.

Plastics for packaging

Cellophane — regenerated cellulose in plasticised film form — is, of course, the original manufactured packaging film, and it still enjoys wide usage because of its strength, clarity, non-toxicity and moderate cost. Cellulose ester plastics, such as the acetate and butyrate, generally plasticised by phthalic esters, are outstanding for their toughness, clarity and gloss, but cost considerably more. Their use in packaging is restricted to applications where a glistening appearance is important for marketing. They find considerable application in blister packaging and as see-through windows for boxes. (Table 9.3 shows how permeable to water vapour they are.)

Package tax

France puts a tax of a quarter of a centime on every package. There are of the order of 10^{11} items of packaging disposed of each year, from 50 million people using about 2000 items per person which is equivalent to about 6 items per day. This collects 2.5 billion francs per annum, about half a billion dollars.

Of the wide variety of commercial organic polymers, several major types occupy a commanding position in packaging (see Table 9.4, p. 238). Low-density polyethylene is fairly flexible and is outstanding for its extensibility, toughness, chemical resistance, and low cost. Its limitations are low strength, inability to recover after stretching, low melting point, translucency, and sensitivity to oxidation — it requires the addition of hindered phenolic antioxidants (see Antioxidants in Chapter 4) for stability during processing. It is particularly useful as film packaging, for tubing, and for squeeze bottles.

In contrast, high-density polyethylene is more rigid, stronger, less extensible and less tough, higher melting, and more opaque. It finds its major use in blown bottles. It is ued in laboratory ware because it can be autoclaved. Polypropylene is stronger still, and even more rigid, less tough, higher melting, and is much more sensitive to oxidation, requiring multicomponent antioxidant systems to stabilise it during processing and use. Shrink films are formed by stretching the film, which orientates the molecules. Upon heating, the film shrinks back to its original size. In this way, close-fitting wraps can be achieved.

Clear, unmodified 'crystal' polystyrene has found wide use in drinking glasses and as vials for pharmaceutical tablets. Oriented polystyrene foam is used in packaging of fresh produce. Rubber-modified polystyrene, generally called 'impact styrene' or even just 'polystyrene' is the most widely used plastic for moulded and heat-pressed food packaging because it is reasonably rigid, strong, tough, chemically stable, and relatively cheap.

In contrast to other packaging plastics, the use of polyethylene terephthalate (PET) has expanded ever since it appeared in the market in 1977. It is excellent for carbonated beverages — light-weight, shatter-resistant, and heat stable. Polyester drawn in one dimension gives a strong orientated fibre (Terylene™, Dacron™) and about 65% goes into this use. Another 17% goes into bottles using blow moulding which orientates the plastic in two dimensions, and strengthens it.

This PET sheet is both strong enough to resist the pressure of carbon dioxide (see Shaking the bottle in Chapter 6) and prevent it diffusing through. (The thread in the top of the bottle has a vertical groove so that gas pressure is released slowly on unscrewing.) However, the sharp bend needed to make a flat base weakens the plastic too much. The first PET bottles had round bottoms and a second plastic (black polyethylene) as a cup on the bottom to provide the base. Current bottles mould a bottom with five convolutions, all of which have rounded edges, eliminating the second plastic and making recycling easier.

Two-dimensional orientation of this plastic does something else; it coincidentally makes it birefringent.

EXERCISE

Cut out a square from the side of a PET bottle. Place it between two crossed pieces of Polaroid material (i.e. transmitted light extinguished). Note that where the material sits, the combination of all three sheets lets through some light again!

More correctly, light waves are a set of travelling helixes; unpolarised light is an equal mixture of left- and right-handed helixes. A birefringent material has a different refractive index in the two dimensions, that is, a different speed for left- and right-handed helixes, slowing one down with respect to the other. When polarised light passes through, this differential results in a rotation of the plane of polarisation. So crossed Polaroids are no longer crossed when material in between has caused a rotation. The amount of rotation depends on the degree of orientation of the polymer molecules, thickness of the film, etc. Polaroid sheet is just a piece of birefringent material that, in addition, absorbs light in one of the directions strongly, while transmitting much of the other. Polaroid is made from plastic by adding a dye (of the type with a long chain molecule) before stretching, so that the dye gets orientated as well.

Other birefringent clear plastics that can be used are the clear polystyrene rectangle from a window envelope or the crispy transparent cellophane wrapping found on some products such as confectionery, potato chip, cigarette, and disk packets.

(See also McFourier, B. Selinger, Oxford University Press, 1991.)

Picket fence?

The simple picket fence explanation for the action of a Polaroid sheet as letting through one direction of polarisation and stopping the one at right angles, is fine, as long as you haven't done this one extra experiment!

PET bottles have replaced PVC for drink bottles because of the health fears from PVC monomer and PVC additives (see earlier editions of this book). The caps and seals are still HDPE with an ethylene–vinyl acetate gasket or liner.

Recycling

In the United States, about 7% by weight of the municipal waste stream is plastics; it can be from 12–20% by volume, most of which is from packaging. Burning for fuel value recovery is an option for the non-chlorinated plastics. Recyclers grind the bottles, use blown air to separate the fines and paper, and then use a sink/float in water to separate the PET (density 1.33–1.52) from the HDPE (0.94–0.97). PET from diverse sources is routinely recycled into moulded products, geotextiles, and carpet, as well as pillow filling and outdoor wear. Mixtures of crystalline polymers such as polyolefins, however, have different melting points; LDPE 110°C, HDPE 120°C, and PP (140°C). When such a mixture is melted without intensive mixing, crystallites of minority components will separate out, which are unsightly and may cause weaknesses in the new product. In practice HDPE is the only olefin seriously recycled. Milk bottles can be reground, cleaned and recompounded into pellets for new products which are not displayed too obviously, such as car washer fluid containers, bumpers, and boot liners.

Another possibility is to depolymerise polymers back to monomers, and either remake them into polymers or find other uses for the molecular fragments to be used elsewhere. Polyolefins can be cracked to yield the original monomer(s), plus higher olefins. PET is easily depolymerised by heating in glycol, or even water at high temperatures and appropriate pressures. They can then be reacted to form thermoset resins in another life. Halogen-containing polymers such as PVC release HCl (not monomer) on heating and become brittle and discoloured.

It should not be forgotten that each new polymeric material, along with its positive qualities, may introduce unexpected problems. If the problems are foreseen, they can be tested. If not, we tend to learn from experience. Those developing new packaging should never be content with meeting the present specifications drawn to cover known problems.

Potential problems of plastic containers

Desorption: The author was heavily involved in the 1970s with the problem of migration into food of PVC monomer and plasticisers. The foods involved were those with oil-solubilising components such as butter and margarine, vegetable oils, fruit drinks (because of citrus oils) and high-alcohol drinks (see earlier editions of the book). Industry eventually agreed to Standards Australia setting up a committee, FT/8, which produced a new standard, Plastics for Food Contact Use, AS2070. It has since been under continuous development. PVC has been largely replaced for food contact by other plastics.

Photodegradation: Most plastics exhib^it varying degrees of degradation on prolonged exposure to sunlight. Poly(methyl methacrylate) (Perspex) and poly (carbonate) are exceptions. The ultraviolet and blue parts of sunlight are sufficiently energetic to cause polymer bonds to break. One way of retarding this effect is to add a compound that will absorb the radiation and convert it efficiently to heat. Derivatives of benzophenone are used (particularly in polypropylene) because of their high absorbance of light in the 290 to 400 nm range (see Sunscreens in Chapter 5).

Permeability of plastic films: One of the important properties of plastics when used as flexible films is their permeability to various gases. (See Table 9.3.) This is particularly true when they are used for wrapping food, because if they are permeable to oxygen from the air, the food will spoil. Oxygen diffuses through most plastics about four times as fast as nitrogen; carbon dioxide diffuses through most plastics about 25 times as fast as nitrogen.

Polythene shrinkage
When heated high-density polythene and low-density polythene melt to form liquids with the same density. So, when these liquids are poured into a mould and allowed to cool and freeze solid, the high-density polymer shrinks much more than the low-density one.

TABLE 9.3
Gas permeability of plastics[a]

Name	Permeability to			
	Oxygen	Carbon dioxide	Water[b]	Nitrogen
Cellulose (PVDC-coated)	0.06	0.4	20	0.09
Cellulose acetate	7.8	68	75 000	2.8
Rubber hydrochloride	0.3	1.7	240	0.8
Polyethylene (low density)	55	352	800	19
Polyethylene (high density)	10.6	35	130	2.7
Polypropylene	23.0	92	680	—
Poly (vinyl alcohol)	0.3	low	soluble	—
Poly (vinyl chloride) (PVC)	1.2	10	1560	0.40
Poly (vinylidene chloride) (PVDC)	0.053	0.29	14	0.0094
Polystyrene	11.0	88	12 000	2.9
Polyamide (nylon)	0.38	1.6	7000	0.10
Polyester	0.3	1.53	1300	0.5
Chlorotrifluoroethylene	0.10	0.72	2.9	0.03

[a] The permeability P is the volume of gas in cc which passes through a square centimetre of plastic surface 1 mm thick in 1 sec for a pressure difference of 1 cm height of mercury — i.e. the lower the number, the slower the gas moves through the plastic film. Figures given are $P \times 10^{10}$ cc cm^{-2} mm^{-1} sec^{-1} (cm Hg)$^{-1}$.
[b] Water at 90% relative humidity and 25°C.

Source: After Sacharow, Drug Cosm. Ind. **97** (3), 359, 1965.

Glass transition temperature (Tg)

We will see in Chapter 10 that atoms pack together as spheres to form ordered crystals with sharp melting points. For molecules of diverse shapes and attractions to surrounding species it is sometimes difficult to pack symmetrically into a neat crystalline solid. Sucrose is a somewhat elongated molecule, but that would not be a problem if it were not for its

TABLE 9.4
The big six (with the code identifying the polymer.)

Polymer	Monomer	Properties of polymer
Polyethylene (LDPE) 4 LDPE	Ethylene $\mathrm{H_2C{=}CH_2}$	Opaque, white, soft, flexible, impermeable to water vapour, unreactive towards acids and bases, absorbs oils and softens, melts at 100–125°C, does not become brittle until −100°C, oxidises on exposure to sunlight, subject to cracking if stressed in presence of many polar compounds.
Polyethylene (HDPE) 2 HDPE	Ethylene $\mathrm{H_2C{=}CH_2}$	Similar to LDPE, more opaque, denser, mechanically tougher, more crystalline and rigid.
Polyvinyl chloride 3 V	Vinyl chloride $\mathrm{H_2C{=}CHCl}$	Rigid, thermoplastic, impervious to oils and most organic materials, transparent, high impact strength.
Polystyrene 6 PS	Styrene $\mathrm{H_2C{=}CH{-}C_6H_5}$	Glassy sparkling clarity, rigid, brittle, easily fabricated, upper temperature use 90°C, soluble in many organic materials.
Polypropylene 5 PP	Propylene $\mathrm{H_2C{=}CH{-}CH_3}$	Opaque, high melting point (160–170°C), high tensile strength and rigidity, lowest density commercial plastic, impermeable to liquids and gases, smooth surface with high lustre.
Polyethylene terephthalate 1 PET	Ethylene glycol $\mathrm{HOCH_2CH_2OH}$ Terephthalic acid $\mathrm{HOOC{-}C_6H_4{-}COOH}$	Transparent, high impact strength, impervious to acid and atmospheric gases, not subject to stretching, most costly of six.

hydroxyl groups which have an unnaturally high affinity for water, forming hydrogen bonds. So a concentrated solution of sugar forms viscous molasses, and growing of crystals of sugar is quite difficult. For long-chain polymer molecules, the formation of crystals can be even more difficult. Sometimes an encouraging 'yank' while cooling can at least partially align some of the long molecules. For one molecule to move it must have the cooperation of its neighbours.

To understand this, try to move a single cooked spaghetti strand in a pile on a plate without disturbing its neighbours. It's hard enough when its uncooked! When a plastic is cold, this cooperative movement is unlikely, and the polymer appears solid and brittle, like ordinary glass. With heat, the molecules move more, and eventually sufficient co-operation is available for the glass to 'melt' at the glass transition temperature T_g. It is not a 'real' melt, just a marked drop in viscosity. For partially crystalline polymers, there is both a real melting point, T_m (of the crystals) and a T_g (for the amorphous regions).

IMPORTANCE OF GLASS TRANSITION TEMPERATURE TO EVERYDAY ACTIVITIES

Natural chewing gum from chicle is based on gutta-percha which is naturally plasticised by triterpenes. Commercial chewing gums are based on PVA. To remove chewing gum from carpet, it is first frozen with ice to bring it below its T_g, and it then comes off like a solid soil rather than sticking like a soft adhesive (see Table 3.4, A guide to stain removal).

Simple plastic garbage tins made from recycled materials invariably have a raised T_g because the mix will hinder molecular movement. In cold climates, the temperature drops below the T_g and this leads to brittle bins which crack easily on bouncing, while the council-supplied auto-collection bins, tailor made for the task, do not.

Cotton is a cellulose polymer with a T_g of 225°C, so the fabric made from its fibres keeps its shape. It absorbs water because the water molecules can slip in between the polymer molecules and this plasticises the cotton and lowers the T_g. Wrinkles can be removed with a hot iron, but for best results, use water as well, because it increases the plasticising effect. When flat, the water is removed by heating and its absence raises the T_g and 'freezes' in the new shape. That is how to iron a cotton shirt correctly.

Ironing?
Well actually we should be 'aluminiuming', but language is very slow to catch up with technology.

Unlike cotton, nylon fibres will melt onto the iron. Nylon's T_g is 50°C, so a cooler iron is fine. Polyester has a T_g of 69°C and much the same applies.

Steam is also used in ironing wool, but here the reason is to break the disulfide bonds that keep wool fibres in shape and then allow them to re-form. The same is true for silk (T_g is 162°C) but because of its smooth filament fibres, water has difficulty penetrating — it is best to put it moist in the fridge so that the water has time to soak in. For analogous chemistry, see kneading of dough (Fig. 15.19), and Experiment 9.2, below.

EXPERIMENT 9.2 Permanent press for wool

Equipment

Permanent crease solution (3% sodium bisulfite or sodium metabisulfite solution containing a little detergent); two wool samples; small sponge; watch glass; detergent; steam iron.

Glass is rigid
There is a widespread opinion that glasses are supercooled liquids and therefore have a finite viscosity at ordinary ambient temperatures. Stories are told of glasses flowing under their own weight; of ancient window panes thicker at the bottom; of glass that has sagged in storage. Other explanations must be found because glasses of commercially useful compositions are in fact rigid solids at ordinary temperatures.

Source: F.M. Ernsberger in Glass: science and technology, D.R. Uhlmann and N.J. Kreidle, Eds; Vol. V, Chapter 1; Acad: New York, 1980.

Method

Pour some of the permanent crease solution into a watch glass and sponge a line of solution down the centre of one of the wool samples. Then crease the sample along the sponged line and press the crease in with the steam iron for about 30 seconds. Using the steam iron again, press a similar crease in the untreated sample. Immerse both samples in warm water (about 70°C) containing a small amount of detergent. Then check to see if the treatment has produced a 'permanent crease'.

Human hair is a protein polymer very similar to wool. A 'permanent wave' in human hair is much the same as a permanent crease in a skirt or a pair of trousers.

Glass

A mixture of silicon dioxide and metal oxides forms glass, which is a disordered solid but not a supercooled liquid. Glass may crystallise over a period of many years and then become more brittle, but pieces of ancient Egyptian glass have remained uncrystallised for 4000 years. Egypt is one of the few places on earth where sodium sesquicarbonate, $Na_2CO_3 \cdot NaHCO_3 \cdot H_2O$ (trona) occurs as a natural mineral.

Talking of Egypt, while serving as a scientific adviser to Napoleon with the expedition to Egypt in 1799, Claude Berthollet contemplated these Egyptian mineral deposits. He decided that they must have been formed by the action of concentrated salt solutions on limestone.

$$\underset{\text{saturated brine}}{2NaCl(aq)} + \underset{\text{limestone}}{CaCO_3(s)} \rightarrow Na_2CO_3 \cdot 10H_2O(s) + CaCl_2(aq)$$

Fig. 9.21
Action of brine on limestone

This is the direct reverse of the well-known laboratory precipitation reaction in which sodium carbonate solution will precipitate calcium carbonate from an aqueous solution of calcium chloride

Berthollet postulated correctly that the reaction had been very slowly reversed in these evaporating deposits because of the very high concentration of sodium chloride.

The conversion of salt to soda is of tremendous commercial importance and one of the oldest pieces of industrial chemistry still in use.

SOLVAY PROCESS

The salt fields of Adelaide's Dry Creek in the northern suburbs produce 775 000 tonnes of salt per annum. Fifty million tonnes of sea water is pumped into 4000 hectares of evaporation and crystallising ponds.

Harvesting takes place in the last week of March for about 90 days until the winter rains start. The Dry Creek fields were originally developed to provide unemployment relief during the Great Depression and employed 650 men who used shovels, wheelbarrows and planks to transport salt over the swampy ground. Today there are 19 permanent, and five casual employees.

The company, Penrice, exploits this deposit and is one of the top five producers of soda ash in the world, and the only one in Australia (it supplies product to the value of almost $100 million). The soda is extracted from the salt using the Solvay process developed in 1863. Soda ash is used to produce glass, detergents for metal refining, and for water purification.

The overall reaction is: $CaCO_3 + 2NaCl \longrightarrow Na_2CO_3 + CaCl_2$.

The natural direction of this reaction is backwards. To make the process go forward, a roundabout series of steps involving at least five individual reactions, and six chemical intermediates, is needed.

One crucial step is when carbon dioxide is forced under pressure into a concentrated cold brine solution which has been saturated with ammonia. This adds ammonium ions and bicarbonate ions to the sodium and chloride ions already present.

$$NH_3(g) + CO_2(g) + NaCl(aq) + H_2O(l) \longrightarrow NaHCO_3(\text{precipitates}) + NH_4Cl(aq)$$

The least soluble combination of ions is sodium bicarbonate, and this precipitates and is separated in rotary drum dryers. This anhydrous product is called light soda (density 0.5 tonnes per cubic metre) and is hard to handle. The liquor is fed to the ammonia recovery plant where it is liberated with lime to leave calcium chloride. Lime kilns produce both lime and carbon dioxide for the process.

Sodium bicarbonate is decomposed to sodium carbonate in heated rotary driers at about 300°C and the carbon dioxide released is recycled. The ammonia is regenerated and recycled by decomposing the ammonium chloride formed. The process is very heat-energy-intensive.

Sodium carbonate solid is hydrated to monohydrate crystals for easier handling. It then has twice the density. Washing soda is produced by recrystallising the monohydrate from water to form the decahydrate.

Washing soda will dehydrate spontaneously (efflorescence) back to the monohydrate under dry conditions and powder is often seen to form on the surface of washing soda, even inside the packet (this depends on the permeability to water vapour of the packaging, see p. 237, Permeability of plastic films).

Some of the waste concentrated calcium chloride liquor finds use as a medium for passive storage of solar heat, some as a drilling mud for the oil industry (and as an ice and snow melting salt in cold climates) but mostly it is waste.

This process is important worldwide. The town called Solvay, along the west shore of Onondaga Lake near Syracuse in the USA, was once very grateful to this industry. The plant has since closed and the town has been left with a calcium chloride pollution problem.

Berthollides

Berthollet went too far and insisted that you could also change the actual composition of the compounds by varying the proportions of the reaction mixtures. In the ensuing fight with Louis Proust, the Law of Definite Proportions was established. Some fifty years later, exceptions to this rule were found in some inorganic compounds such as metal oxides and sulfides, and led to his rehabilitation with coining of the term berthollides.

Exercise:

Adding nine extra molecules of water of crystallisation before selling the product on to the retailer provides what additional percentage profit?

With temperatures up to 1000°C in charcoal furnaces, the ancients discovered that mixtures of sand, limestone and sodium carbonate fuse to form glass, in which all the crystalline order of the added minerals has been lost. With only sodium carbonate and sand, a glass is also formed, but this is soluble in water:

Fig. 9.22 Waterglass

$$Na_2CO_3(s) + SiO_2(s) \xrightarrow{heat} \underset{glass}{Na_2SiO_3} + CO_2(g)$$

This material is sodium silicate (waterglass) which was used as an inorganic builder in detergents (see Chapter 3), for preserving eggs, in fireproofing materials, including paper, and making 'chemical gardens'.

CHEMICAL GARDENS

Put a centimetre or so of sand in a big jar (say 500 mL). Make a 50:50 mixture of sodium silicate (waterglass) and pour it onto the sand to almost fill the jar. Put the jar in a place where it won't be disturbed. Drop in whatever crystals of metal salts you may have (must have insoluble hydroxides); iron sulfate, copper sulfate, alum, Epsom salts, and so on. The crystals develop 'shoots' some of which are directed upward by small bubbles. In a few hours you will have a silicate jungle fully grown. The metal hydroxide skin formed around the crystal is permeable only to water and not the salt. The water diffuses in to balance the concentration inside and out, bursts the skin, which then re-forms further out. (See also Plate XXVIII.) For sources of material see Appendix X.

(The oral tradition of recipes for wonderful little experiments like this is being lost. Reference must be made to the 1960s when 'real kids had chemistry sets'; perhaps instead their experiments are now on illicit drugs! See for example, The secrets of chemistry: How to set up a home laboratory — over 200 simple experiments, R. Brent, Paul Hamlyn, London, 1965).

When the glass is made with calcium as well, it becomes insoluble in water and much harder. A typical glass contains (by weight) 70% silicon dioxide (SiO_2), 15% sodium oxide (Na_2O), and 10% calcium oxide (CaO), with 5% other oxides. As we saw, the sodium and calcium are added as carbonates and lose carbon dioxide to form the oxides in the glass. The glass formed in this process softens at 650°C.

There are two distinct constituents of glass. The non-metal as an oxide is usually silicon, but it can be boron, aluminium, or phosphorus. This is the network former. The network modifiers are usually sodium, potassium, calcium, and magnesium. Without the metal oxides, silicon oxide or silica is a crystalline material called quartz (although fused silica can be made at 1700°C). In silica glass each silicon atom is surrounded tetrahedrally and rigidly by four oxygen atoms, and each of these is linked with some variation in angle to other tetrahedra. These linkages are rigid below the glass transition temperature (T_g) and can break and reform above T_g. When ionic oxides of, say, sodium or calcium are added, they open the Si–O–Si bridge and slip in between, thereby weakening it; this lowers the melting point by about 1000°C. In borosilicate-type glasses, BO_3 groups substitute for some of the SiO_4 groups. The glass transition temperatures for different glasses as measured and as calculated are as follows.

TABLE 9.5
T_g for glasses

Glass	T_g°C (measured)	T_g°C (calculated)
Silica glass	≈1200	—
Pyrex	550	350
Window glass	550	270

Glass kept at ambient temperature has the properties of a disordered solid and will not flow. Instructions for storing glass relate to protecting it from breakage, not from bending. Antique window glass at the time of production was non-uniform in its thickness and was probably and sensibly installed thicker side down. Unless we can find some instruction manuals from that period, we won't know for sure.

Particular metal oxides are used for particular effects. If we take a set of wine glasses of different composition and try their 'ring' on striking, we can see the effect of different solute or modifier ions on the properties. Cheap (easy to work) soda glass has no ring. The sodium ions can hop around the glass more easily than the lead ions, and this hopping efficiently converts the vibration of the struck glass into heat and damps out the ringing. Lead crystal has a beautiful ring, as well as a high refractive index and more brilliance.

Cobalt gives a blue glass, manganese gives purple, chromium gives green, and copper gives either red or blue-green. Borax beads can be made to illustrate glass colouring at a lower temperature than for conventional glass (see Table 9.6, p. 245). The natural colour of glass (best seen edge on) is green to yellow because it contains iron impurities. By adding manganese dioxide, MnO_2, the coloured iron (II) is oxidised to the paler iron (III). Solarisation of old manganese-doped glass transformed it into blue 'desert amethyst glass' (see also Quartz below). Modern glasses turn a dirty brown.

In 1912, the Corning Glass Company in the USA found that adding 10% to 15% of boron oxide, B_2O_3, gave a glass that was more shock-resistant. In Jena, Germany, such borax glasses were also on sale. This borosilicate glass (Pyrex) was found to be resistant to chemical attack from virtually everything except hydrofluoric acid (HF). A glass composed of 96% SiO_2, 3% B_2O_3 and 1% of another oxide produced Vycor, which could withstand very sudden changes in temperature (e.g. it did

Sodium silicate solution

Salt crystal begins to dissolve

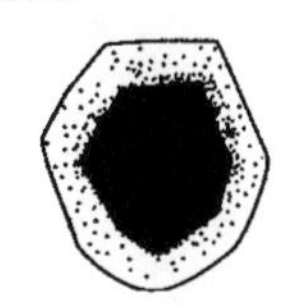

Shell of insoluble hydroxide forms around crystal — this membrane is permeable to water not salt solution

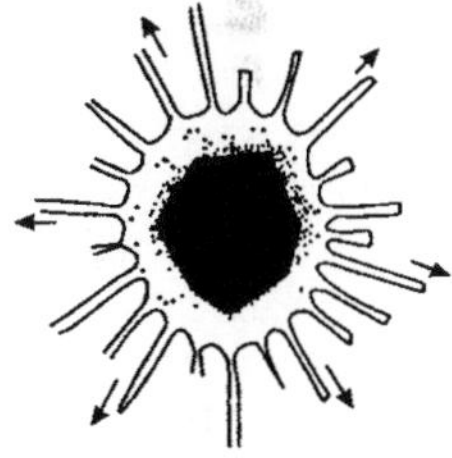

The osmotic pressure finally ruptures the membrane causing growth of fine hollow tubes

Fig. 9.23
Diagram illustrating the mode of growth of a 'silicate garden'

(Courtesy of D.D. Double.)

Verre anglais
To claim that something as French as champagne is dependent on foreign technology is heresy but true. Bottles made of glass strong enough to be of service to wine — let alone withstand the pressure of bubbly — were an English invention. Firing of glass using coal-fired furnaces provided higher temperatures than the charcoal method. These dark-green or dark-brown bottles had less flaws, and were able to withstand wine fermenting in a bottle. They were known as verre anglais. They made their way to France around 1670. Wine was not generally bottled at that time in France, but kept in barrels. It may have been taken to the table in bottles, but these were covered in wicker (much like Italian Chianti bottles today) to prevent breakage.

not break when plunged red hot into cold water). Vycor found use in the glass panels of oil heaters and for scientific work, as it was a cheaper alternative to fused quartz.

Aluminium oxide gives aluminosilicate glass, which is used in glass reinforcing fibres.

Glass can be 'chemically' toughened by rapid cooling and dipping into potassium chloride (KCl) solution. This replaces some of the sodium atoms near the surface with the larger potassium ions, and puts the surface under compression stress (see Chemistry and design below). Titanium compounds are also used. They allow the production of lighter glass bottles.

An interesting development was the incorporation into glass of silver chloride (AgCl), which darkens on exposure to the ultraviolet component in sunlight (just as it does in a photographic film). The darkening is the result of the silver ions (Ag^+) converting to metallic silver (Ag) by picking up an electron. This colour is lost again in the dark, and so we can produce photochromic sunglasses.

Accidental overheating of this photosensitive glass at Corning led to the glass becoming permanently opaque and apparently useless. However, it was quickly found that the glass was now virtually unbreakable. While glass can crystallise on its own (devitrify) over centuries or at 600°C in a matter of months, with silver (titanium, phosphorus, etc.) the process is speeded up, occurs in hours, and yields a very strong crystal (with the dislocations locked in — see Chemistry and design, below).

Small changes in composition can cause large changes in properties. A formulation of 74% silicon dioxide, 16% aluminium oxide, 6% titanium dioxide, and 4% lithium oxide doesn't expand on heating. Some *glass ceramics* do not expand on heating; some have high resistance to fractures. Glass ceramics out-perform glass in almost every respect except one, transparency.

New formulations for glass bring surprises. A glass that has phosphorus oxide instead of sodium and calcium dissolves very slowly. It is being tested as a medium for the controlled release of other substances such as copper into ponds for controlling the disease schistosomiasis (bilharzia). Then there is the continuing debate about the type of glass that should be used to store nuclear waste material.

Quartz

Naturally occurring crystals of quartz were probably among the first pieces of jewellery. Quartz is common and durable. Many of the biblical precious stones were forms of quartz, including the popular amethyst. Ancient intaglios and cameos from Rome and Greece were carved from agate, a form of quartz whose crystalline structure is only visible when magnified (cryptocrystalline). Rock crystal, a clear form of quartz, and rose quartz, containing titanium, are often carved.

TABLE 9.6
Colour test using borax beads

Metal salt[a] (solid)	Oxidising flame Hot	Oxidising flame Cold	Reducing flame Hot	Reducing flame Cold
Copper	green	blue	—	opaque red-brown
Iron	yellow-brown	yellow	green	green
Chromium	yellow	green	green	green
Manganese	violet (amethyst)	amethyst	—	—
Cobalt	blue	blue	blue	blue
Nickel	violet	red-brown	grey	grey
Gold	rose violet	rose violet	red	violet
Tungsten	pale yellow	—	green	blue (but blood red if iron is present)

[a] Minerals work quite well.

Borax beads
The free end of a platinum wire is coiled in a small match-head size loop. The loop is heated in a Bunsen flame until red hot and then quickly dipped into powdered borax, $Na_2B_4O_7 . 10H_2O$. The adhering solid is held in the hottest part of the flame; the salt swells as it loses water of crystallisation, and shrinks on the loop, forming a colourless, transparent, glass bead. The bead is moistened and dipped into the finely powdered sample, allowing only a very small amount of material to adhere (otherwise the colour to be formed will be too dark. See Table 9.6.). The bead is then reheated in the oxidising (outer purple) or reducing (inner blue) part of the flame.
After each test, the bead is removed by heating it again to fusion, and then flicking it off the wire into water. The platinum can be cleaned by melting a pure borax bead and running it up and down the wire. A white ceramic tile can be used instead of a platinum wire.

Imitation jewellery was generally made from glass and called paste, (derived from the Old French and late Latin *pasta* meaning small square medicinal lozenge).

The usual 'crown' glass has a refractive index between 1.47 and 1.51, compared to quartz of 1.54–1.55. Measurement of refractive index is not simple, but noting the visibility of edges of a specimen in water, or a liquid of higher refractive index, can be used (See Plate VIII). Flint glass containing lead has a higher value 1.57–1.77. Quartz has a Mohs' hardness of 8 (see Table 5.4), while various glasses range between 5 and 6.

Quartz forms crystals with six-sided prisms and six-sided triangular faces on ends. Nearly always one end is broken where it was attached to a cluster or host rock. The faces are flat (except for the parallel striations or ridges at right angles to the prism faces) and the edges between the faces are sharp.

Moulded glass imitations tend to have slightly concave faces with rounded edges and show scratches from dust (which contains quartz). Glass often has bubble inclusions (round or elongated), or swirl marks which look 'syrupy', while quartz can include other crystals such as thin needles of rutile or show feathery shapes and cracks.

Any colour variation in quartz (common in amethyst) arises from changes in composition during crystallisation from liquid (see Alloys in Chapter 10) and so appears as an internal smaller crystal within the larger one, that is the edges of colour change are parallel to the crystal faces.

When a quartz crystal is viewed along its length, it appears to have three-fold rather than six-fold symmetry because one set of faces has grown much faster. The long prism faces are seldom equal in width, and so the cross-section appears distorted rather than a regular hexagon. Carved and other 'unnatural' crystals generally fail to capture these aspects.

Colour in quartz
Dark brown or black 'smoky' quartz occurs when one silicon atom in 10 000 has been replaced by aluminium and the crystal is then irradiated with X- or gamma rays. The colour disappears on heating to around 400°C and reappears again on re-irradiation. Natural quartz nearly always contains enough aluminium to colour with irradiation. With iron (as Fe III) instead of aluminium, irradiation produces the purple colour of amethyst. Heating causes bleaching to a pale yellow (citrine) and re-irradiation restores the purple. Heating under reducing conditions gives Fe II and 'greened' amethyst (both in natural Brazilian and in synthetic material) which can be re-irradiated to purple.

Chemistry and design

A light breeze, an expanse of water and a few hours to spare — why not go sailing? The physics of sailing has remained unchanged since the first human being ventured onto water. However, the materials from which boats and sails are made have changed dramatically. The chemistry of new materials has encouraged new designs that have drastically reduced costs and enormously increased efficiency.

The great division in technology has always been between metals and non-metals. Their properties are clearly very different (we have seen some explanations) and the mental processes involved in using them are also very different. The history of Western technology has very much been the history of steel.

The price of steel *dropped* by a factor of more than 10 during the reign of Queen Victoria. This was the most important event of its kind in history until perhaps today, when something similar is occurring with micro-electronics.

For generations, engineers had no idea at all why steel and concrete behaved in the different ways they do. However, they described their behaviour in minute detail and filled many unreadable books with their results. These 'properties' of materials were essential to design, but useless in logically devising new materials.

It is a mistake to exaggerate the virtues of traditional design, even though the workmanship may have been excellent. The wheels really did keep coming off coaches, and wooden ships always leaked, quite unnecessarily, because shipwrights did not understand shearing stress.

The understanding began with the work of Robert Hooke and with what is now known as Hooke's law. Hooke did not suffer unduly from modesty (see also Chapter 11) and staked his claim to priority in a number of fields by publishing in 1676 'A tenth of the hundredth of the inventions I intend to publish', among which was an anagram which revealed *Ut tensio sic vis* — 'as the extension, so the force'.

It took over a century to move from Hooke's law to Young's modulus *E*, which described the elastic or stiffness properties of a *material* rather than the properties of an *object* (of undefined dimensions). Thomas Young published the idea of his modulus in a rather incomprehensible paper in 1807 after he had been dismissed from his lectureship at the Royal Institution, London, for not being sufficiently practical. Thus was born the most famous and most useful of all concepts in engineering: the modulus *E* determines the deflection of structures when loaded. The stiffness of a material is not a measure of its strength. A biscuit is stiff but weak; steel is stiff and strong; nylon is flexible and strong; raspberry jelly is flexible and weak. The two properties together describe a solid about as well as you can reasonably expect two parameters to do.

Strength is usually thought of as the stress needed to pull materials completely apart, although materials are most often used in compression. When a squat column of material is compressed, it can squash out sideways like plasticine. Copper behaves in this way. Brittle materials like stone and glass will explode sideways into dust and splinters. A long, slender column may buckle like a walking stick or tin can.

STRESS IN BUILDINGS

In the biblical tale of the people of Shinar wishing to build a tower (of Babel) that would reach the heavens, it is said that they were thwarted by God replacing their single language with a host of disparate tongues, thus destroying the coordination of their efforts. However, divine intervention was unnecessary as the finite strength of interatomic bonds would have stopped their efforts to reach the infinite, fairly early on.

The strongest material available at the time was granite, which, like all polycrystalline substances, fails by shear at 45° to the direction of compression. Granite has a density of 2.7×10^3 kg/m^3 and a failure strength of 20 MN/m^2. A stress of 1 kg/m^2 is about 10 N/m^2. The 45° introduces an extra factor of two and the danger point for shear failure for the tower would be reached at a height of 1445 m, only a very short way to heaven!

The 528-floor skyscraper designed for Chicago in 1956 was to be one mile high — 1609 m. Even though it would have been reinforced with steel, the structure would have failed. The height limitation through the stress induced by the weight of material explains why the larger a planet is, the lower are the mountains — increased gravity means failure at lower heights.

When crystalline materials are stressed with moderate loads for long periods, the materials can 'creep'. Creep is a gradual change in shape, and depends on the stress and temperature. It starts slowly and builds up to a faster steady rate. This was demonstrated by E.N. da C. Andrade, who found that for many materials it became important at temperatures of about half the (absolute) melting point. Lead melts at 600.6 K (327.4°C) and so the critical temperature is 300.3 K (27.1°C). Thus at ambient temperatures, lead in say a plumbing joint that is under stress will creep, whereas copper (m.p. 1356 K, 1083°C) will not.

Source: Cambridge guide to a material world, ed. R. Cotterill, Tech. Uni. Denmark, Cambridge University Press, 1986.

During the 1920s an engineer named Griffith was working at the Royal Aircraft Establishment at Farnborough. He asked, 'Why are there large variations between the strengths of different solids? Why don't all solids have the same strength? Why aren't they much stronger? How strong *ought* they to be anyway?' These were regarded as quite silly questions.

Liquids such as water have surface tension — or surface energy as it is correctly called (see Chapter 6). This is a skin-like property of the surface that allows drops to form, and insects to walk on water. Solids also have surface tension. Just as it is possible to calculate the weight of the largest

Cutting glass
Glass is cut by using a shallow but sharp scratch on the surface and then breaking it along the stress line. The sharper the angle of the cut, the greater is the stress.

insect than can walk on water, so it turned out to be quite simple to calculate the stress needed to separate atom layers inside a material. The problem was that most materials came nowhere near these theoretical results. Griffith did some experiments on glass fibres because glass was easier to handle than steel. However, when his superiors became aware of his work (after a fire wrecked his laboratory) he was immediately transferred to other duties.

Glass was an ideal material for theoretical study. The idea that strain must overcome surface energy to cause a break arose very naturally from the belief that glass was a solidified liquid.

Another fascinating result was that the strength of glass fibres increased enormously with *thinness* — by a factor of about 100! In fact, thin glass fibres are stronger than the strongest steel, and close to the theoretical limit of strength. Bulk materials are actually much weaker than expected because they have local high-stress points. Chemical corrosion also occurs most readily at points of high stress.

METAL FATIGUE

The Australian novelist Neville Shute published No Highway in 1948. It is a story about a researcher who was obsessed with metal fatigue in aircraft components and the drama centres around the storyteller's efforts to have this man's experiments taken seriously, and forestalling a projected set of sudden failures after a fixed number of flying hours for the then current commercial Reindeer fleet of Transatlantic airliners. Around page 4 (of an almost 200 page novel) he said:

> 'Fatigue may be described as a disease of a metal. When metals are subject to an alternating load, after a great many reversals the whole character of the metal may alter and this change can happen very suddenly. An aluminium alloy that has stood up quite well to many thousands of hours flight can suddenly crystallise and break under quite small forces, with the most unpleasant consequences to the aeroplane.'

(His explanation of fatigue being due to crystallisation was incorrect because the failure is now known to be due to extension of small fractures.)

Neville Shute was an aircraft engineer turned novelist and his novel was technically prescient in relation to the series of British Comet aircraft which fell out of the air in the mid-1950s. At the time, these disasters did little for confidence in air travel in general, and British engineering in particular.

Source: Porcelain to steel: A fractured tale, L.M. Brown, Royal Institution Proceedings, 62, pp. 279–305, 1990.

The chemical molecular picture of solids now provides the theoretical framework for explaining the real strength of materials and the methods by which weakness may be overcome or compensated for. By making glass fibres thin enough, the room for surface cracks to form was reduced and the material became stronger. From this grew the research that led to the development of ultra-strong whiskers of metals and non-metals. Purity of material was of great importance because impurities would sit at so-called grain boundaries (the surfaces between pure microcrystals packed together) in the crystal structure and cause weakness. Thus when sea water freezes, the ice formed is substantially fresh because the crystallisation forces out the salt. The wrong impurity in an alloy can completely ruin its strength by being forced into grain boundaries.

An interesting application of this principle of migration of impurities to grain boundaries is in the use of glycol antifreeze in car radiators. While it is true that glycol lowers the freezing point and postpones freezing, the more important aspect is that less than the calculated amount of coolant is required, because when the mixture does eventually freeze, it forms mushy ice without much mechanical strength and is unlikely to burst the radiator or head (see Antifreeze in Chapter 14).

So the worst feature in an engineering material turns out to be not lack of *strength* or lack of *stiffness*, desirable as these properties are, but lack of *toughness*, that is, lack of resistance to the propagation of cracks. The lack of strength or stiffness can be overcome by adequate design considerations.

Strain induced in a material by a sudden blow can propagate through the material and cause very high local stresses far removed from the impact point. The material has to resist the effect of these stresses. A shell fired at armour plate, but not penetrating, can throw off a 'scab' on the inside of the plate that can bounce around causing great damage. The reason for the internal head-band in crash helmets is not to ventilate it but to cushion shock waves, and prevent damage at the back of the skull, opposite the usual point of impact. The trick of 'opening' a bottle of wine by giving it a sharp hit on the bottom often breaks the top off, for the same reason.

Wooden sailing ships were used to explore the world. Large ships were built from naturally curving wood, chosen to have the right shape. The watertight skin and deck were laid lengthwise, at right angles to a closely spaced framework of ribs. The grooves between the timbers were filled with *oakum* (made by picking old rope to pieces in the prisons and work-houses) driven in by mallet and chisel. Decks were 'payed' with hot pitch and the surplus chipped off when cold. (The Devil, incidentally, was a particular seam, hence the expression 'the Devil to pay and only half a bucket of pitch!') With this design the shear stresses were not taken up and the strains were left to the caulking, which always leaked and occasionally fell out. Generally the crew became exhausted from having to continually pump water. Up to 1914, Norwegian shipowners were still making a living by buying up British sailing ships that had become too leaky with age and converting them to run with windmill pumps. The quality of ropes and spars was awful and this created problems, particularly when ships needed to go into the wind. Bligh's crew on the

Dinky toys and Hornby trains are being affected by 'lead rot'.

From the mid-1930s, Meccano began making toys from a zinc alloy called Mazak, which was zinc containing 3 or 4% aluminium and 1 to 2% copper. It gave better detail than the lead it replaced. However, scrap containing lead was occasionally thrown into the pot and as little as 0.008% lead or cadmium will cause problems. The lead migrates to the metallic crystal grain boundaries, causing a change in crystal structure which leads to an uneven expansion of up to 10% in these areas. Cracks appear in the models and ultimately they fall apart. Wheels on prewar Hornby trains would turn to dust; plane wings would curl up. During the war, Meccano had to make precision die-casts for the Ministry of Defence and after the war their Mazak (and other alloys) were of high quality.

Source: Tragedy in Toytown, Andy Coghlan, New Scientist, 21/28 Dec. 1996, pp. 38–41.

The 'Torrens'
The last major wooden passenger sailing ship seems to have been the **Torrens**, which carried passengers to Adelaide in 1903.

Bounty, for example, mutinied after one appalling attempt to beat around Cape Horn in which the ship almost fell apart, forcing Bligh to turn around and go right around the world the other way in order to reach the Pacific. With the discovery of gold in California, the trip around the Horn was in great demand. It is clear that the West was largely won with better rope. Hemp is still very popular, but for a different part of the plant.

Slow ships, and lack of facilities ashore, meant that the fouling of ships' bottoms by weed and worm was a very serious problem. Copper sheathing was introduced in 1770 and the slow release of small amounts of copper ions proved highly toxic to marine organisms. So successful was this piece of chemistry that it delayed the replacement of wooden hulls by iron because the copper sheathing on iron set up an electrochemical cell in sea water (a battery) which rapidly corroded the iron. In order to save copper, Sir Humphrey Davy applied blocks of zinc to the copper-sheathed hull of the frigate *Samarang* in the 1820s. The idea was to give cathodic protection to the copper in the same way that zinc does for iron in galvanised steel. The experiment was a great success in preventing the copper from dissolving, but then the copper failed completely to kill the growth of fouling. This brings us to a discussion of metals in more detail in Chapter 10.

References

The new science of strong materials or why you don't fall through the floor, J. E. Gordon, 2nd edn, Penguin, Harmondsworth, Middlesex, 1982. Brilliant and easy reading.

MegaMolecules, H.G. Elias, Springer Verlag, Berlin 1985, 202 pp. Dilettantish and delightful.

Glass: past elegant, future thin, J. Emsley, New Scientist, 8 December 1983. Elegant.

Plastics in building and construction, M.S. Reisch, *Chemical and Engineering News, 30* May 1994, pp. 20–43. Practical.

Crystals in the marketplace, A.L. Rennie, *Aust. Science Teachers, J.*, **35**(4), p. 57–59, Nov 1989 . Take it with you to the markets.

Bibliography

The development of plastics, S.T.I. Mossman and P.J.T. Morris, Royal Society of Chemistry, Publication 141, London, 1994. Very English.

Textbook of polymer science, F.W. Billmeyer, John Wiley, NY, 3rd edition, 1984. For the serious.

Industrial organic chemicals, H.A. Wittcoff and B. G. Reuben, John Wiley, NY, 1996. The wider picture.

A history of rubber, P. Mason, ABC, Sydney, 110 pp., 1979. Timeless.

Know your plastics, Plastics Industry Association, 2nd edition, Melbourne, Australia, 1992. Useful compilations.

Physics and chemistry of color: The fifteen causes of color, K. Nassau, Wiley NY, 1983. A lateral look at colour, including crystals.

Chemistry of metals

In spite of the rise of plastics, metals are still the mainspring of consumer products. In this chapter we start with the properties of pure metals, then look at a very few of the huge variety of their alloys, and finally at the biological importance of metals as both essential elements and pollutants.

Introduction

Metals are crucial to our technology. A glance at the financial pages of a newspaper reveals the price of the major commercial metals. Table 10.1 (overleaf) gives the values from the fourth edition of this book in 1988 and those quoted for this edition. The relativities have not changed much, and given that the *prices are in the dollar value prevailing at the time* (i.e. excluding the effect of a decade of inflation) the prices have dropped substantially. This is due to the level of economic activity, but also partially due to the replacement of some uses of metal by other materials such as plastics, see Table 10.2. Countries that rely heavily on mineral exports (like Australia) have suffered. The static demand for wool (and wheat) has not helped either. Indeed the Australian dollar is basically a measure of the value of commodities such as metals.

Around 1990, Australia could no longer afford its one and two cent coins; the currency had devalued so much that the metal content was too

Banana republic
The first sign of a banana republic is when the currency is (close to being) worth more when melted down!

TABLE 10.1
Metal prices

Metal	US$/lb (1988)	US$/lb (1997)	US$/kg (1997)
Platinum	8190	5840	12 850
Gold	6370	5760	12 672
Palladium	1920	2016	4435
Silver	100	77	170
Nickel	—	3.3	7.32
Tin	3.36	2.7	6.0
Copper	1.18	1.03	2.28
Aluminium	1.06	0.73	1.61
Zinc	0.70	0.51	1.13
Lead	0.40	0.32	0.71

Source: Financial Review, 20 Jan. 1997

TABLE 10.2
Material volume of world production of various commodities (millions of tonnes)

Material	1913	1938	1950	1960	1970	1980	1989
Plastics	0.04	0.3	1.5	5.7	27.0	40.0	80
Aluminium	0.7	0.5	1.3	3.6	8.1	11.2	23.1
Zinc	0.8	1.4	1.8	2.4	4.0	4.8	—
Copper	1.0	1.8	2.3	3.7	6.1	8.4	9.4
Pig iron	53	88	153	241	448	480	570
Rubber (synthetic)	—	0.01	0.5	1.9	4.5	7.7	10.1
Rubber (natural)	0.12	0.92	1.9	2.0	2.9	3.7	5.2
Fibres (synthetic)	—	—	0.12	0.65	4.5	8.4	11.6
Cotton	—	5.2	6.0	7.1	7.7	9.1	18
Wool	—	1.6	1.7	2.1	2.2	2.2	2.3

Source: The Development of plastics, S.T.I. Mossman and P.J.T. Morris, Roy. Soc. Chem., London, p. 1, 1994. Synthetic fibres exclude semi-synthetic polymers. Most metals indicated are primary refined products.

Exercise
Play with some pieces of different metals such as copper, soft iron, zinc, aluminium and magnesium. They look different: magnesium, aluminium and zinc are silvery; iron is dark; while copper is salmon-brown. The really striking difference is in the way they behave physically, when we hammer, stretch or twist them. Why?

expensive! Check the 1988 prices of copper to see why. A 1 cent coin (97% copper) weighed 2.60 g, and a 2 cent coin 5.20 g.

There is a great deal we can learn about metals from an inquisitive look around our shops. In a supermarket we can buy aluminium foil. In the hardware store we can buy lead and sometimes copper foil, but we cannot buy zinc foil, even though zinc is cheaper than copper! Nickel, chromium and iron can come as wire, but never their close neighbour, cobalt. One type of brass can be machined, while the other must be cast like bronze.

The answer can be obtained using a heap of ping-pong balls, polystyrene balls, or marbles. Rack some together tightly in one layer, as at the start of a snooker game. Then place another layer on top of the first, in the indentations created by the first layer. The two layers are of course not directly on top of one another. Then place a third layer on top of the second. Wait! There are two ways we can do this. They can be placed so that the third layer is directly over the first layer, or they can be placed so that all three layers are each in a different position. In the first case, we can continue to pack layers alternately, while in the second case we repeat the pattern every three layers. We can call these different packing arrangements *ab* and *abc*. The packing efficiency for the *ab* and *abc* arrangements is the same, and they are the most efficient possible. In both cases, each sphere is surrounded by 12 nearest neighbours. For equal-size spheres, geometry shows that in both cases the spheres occupy 74% of the total volume.

The best way to obtain a feel for the structures is to prepare triangular planes of balls (just as in snooker) and stick them together. (Polystyrene balls can be stuck with dichloromethane solvent.) Make the triangles of decreasing size. Then place one triangular layer on top of another. If you pile up the layers in an *abc* arrangement, you will form a pyramid on the triangular base. In fact it is a tetrahedron where three different faces have the atoms in layers just like the base layer (See Plate X(a).) There are thus four directions in which the balls are layered. On the other hand, if you pile the layers up *ab* as in Plate X(b), the layers will jut in and out alternately as you go up. If you look down from the top to the base you will see channels going through the whole structure in position *c*, which has been kept clear. There will be no new planes of atoms formed, only the horizontal ones you have been placing.

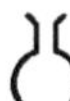

Experiment
If your geometry is not up to it, you can 'use your marbles'; pack them into an empty ice-cream tub, and fill it with water to a level near the top of the marbles. Then pour the water out, and measure the volume. Tip out the marbles and fill only with water. Compare the volumes. (Error due to 'edge effects' at the bottom, sides and top will be small if the marbles are small compared to container size.)

Metals (or alloys) that pack *abc* are much easier to shape than those that pack *ab*, because it is the planes of atoms that move and make metals ductile and malleable. Metals found in foils and wire are almost exclusively those that pack as *abc*. An exception is tungsten.

Tungsten lamp filaments
Survival of tungsten filaments in incandescent lamps depends on a deliberate impurity (potassium) which forms micro-bubbles in the filament and prevents the growth of microcrystals across the whole diameter of the filament. Heating at grain boundaries would then allow the whole filament to fail. This increases the life of the lamp from a few minutes to 1000–3000 hours.

The *abc* structure is called face-centred cubic close packing (*fcc*), while the *ab* structure is called hexagonal close packing (*hcp*). The reasons for these names can be seen by studying the models more closely. The way different metals pack is shown in Table 10.3 (overleaf).

Another, but less efficient, way of packing spheres involves surrounding a sphere with only eight nearest neighbours. Geometry shows that, for equal-size spheres, only 68% of the volume is occupied. This packing arrangement is called body-centred cubic (*bcc*). It does not have quite as many slip planes as *fcc* but many more than *hcp*. Among metals, *fcc* and *hcp* are about equally as common, while *bcc* is about half as common. Other structures constitute about 10% of the total.

Tin

Tin is the Oscar Wilde of metals. It is a little uncertain as to its status as a 'real' metal and shows some outrageous behaviour. Because tin is available as a foil, you would predict that it packs *abc*, and it does, almost. Tin is soft and weak; it melts at 232°C. Lead added to tin lowers the melting point even further, and gives us pewter. On the other hand, a small amount of tin added to copper strengthens it enormously, and this

TABLE 10.3
Interatomic distances of common metals (298 K)

Metal	Structure	Interatomic distance (nm)
Ag	fcc	0.40857
Al	fcc	0.40496
Au	fcc	0.40782
Cd	hcp	0.29793
Co	hcp	0.25071
Co	fcc (> 417°C)	0.35447
Cr	bcc	0.38848
Cu	fcc	0.36146
Fe(α)	bcc (< 912°C)	0.28665
Fe(γ)	912°C < fcc < 1400°C	0.36467
Mg	hcp	0.32094
Mo	bcc	0.31470
Ni	fcc	0.31470
Pb	fcc	0.49502
Pt	fcc	0.39236
Sn[a]		
Ti(α)	hcp (< 880°C)	0.29506
Ti(β)	bcc (> 880°C)	0.33065
W	bcc	0.31652
Zn	hcp	0.26650

[a] Tin (Sn) has three crystalline structures:

α-Sn ↔ (13.2°C) β-Sn ↔ (161°C) γ-Sn ↔ (232°C) Sn (liq)

α-Sn: 'grey' (diamond structure), density is 5.75
β-Sn: 'white' (metallic-tetrahedral), density is 7.31
γ-Sn: also metallic

Source: CRC Handbook of Chemistry and Physics, 77th edition, 1996/7, 12-18

Winter, best time to lose trousers
Pure tin is not much used today, but historically tin 'pest' has had its consequences. Napoleon's army lost their trousers in the freezing winter of 1812 when their tin buttons crumbled away. Another cold Russian winter was said to be responsible for the rumour that the Russian hoard of silver had turned to powder in 1867–68. It was actually tin.

piece of technology that science now makes understandable gave us the Bronze Age, which lasted a millennium and allowed the development of civilisation as we know it.

When a piece of pure tin is bent backwards and forwards near your ear, it can be heard to produce a plaintive 'cry'. This is due to the presence of 'twinned' crystals in the metal, which you are 'cruelly' separating. Tin also suffers from a thermodynamic 'disease'. Below 13°C the atoms of tin very slowly change their packing arrangements. In practice you need a much lower temperature to see this happen in a reasonable time. The new structure is *not a metal structure* at all (tin is as closely related to the non-metal semiconductors such as germanium as it is to the metals) and tin in this form is grey and crumbly (see note at left).

Contrary to popular belief, tin (plated) cans survive long periods at low temperature very well (such as in a fridge), because of reinforcing by the underlying steel (see Fig. 10.1). However, lead solders (see p. 225) with a high tin content once used in sealing fuel tanks, do not survive and

Fig. 10.1
Can of roast veal, 1824 (Courtesy of the International Tin Research Institute, UK)

early expeditions to cold climates, such as Antarctica, failed through loss of fuel rather than food.

We see tin mainly in tin-plated cans. Unless they are lacquered internally, a little tin dissolves in the natural vegetable acids and gives, say, tomato soup its special colour and flavour. One cause of a can 'blowing' is the production of hydrogen from the acid reaction with the tin. Now, tin is just above hydrogen in the activity series, and therefore does not readily react with dilute acid. Do the food acids have special properties? Well yes, they do. They are strong complexing agents for metals and so keep the concentration of free metal ion in solution low. By Le Chatelier's principle, more tin will dissolve and be complexed and so on. A maximum level of 250 mg per kilogram has been set for tin in food, based on taste, not toxicology. Tin gives the soup a sharp metallic taste; in fact when they make generic soups, the companies often put the same product into lacquered cans. The supermarket has demanded the same product for its house brand and gets it; the consumer buys the branded one because the no-frills product doesn't taste as good. There is no accounting for taste.

Aluminium

ALUMINIUM OR ALUMINUM?

The spelling can be traced back to ancient Rome, being derived from the Latin alumen meaning 'light' because of the brightening effect that alum (called alumen by Pliny) had on dyeing processes. Humphrey Davy who had produced many active metals such as potassium, sodium, barium, strontium, calcium, and magnesium by the electrolysis of the fused salts, failed with Al. Sir Humphrey intended to call the metal alumium, but later linked it to its ore, alumina, and changed it to aluminium — keeping the traditional metal ium ending. Alcoa in the US switched to the um ending in July 1888 for reasons obscured in company minutes.

Source: G.B. Kaufmann & M.L. Mathews, Ed Chem, pp. 36–39, March 1990.

Helium discovery
Helium was first discovered in the sun through analysing sunlight. It was thought to be a highly charged ion of iron, and as a metal was thus given the ium ending. Chemists' mistakes remain in the language to haunt them forever.

Case study — magnesium

An early experiment that most chemistry students carry out is the burning of magnesium in air. The aim of the experiment is to measure the mass increase that occurs when oxygen is added to magnesium. Seldom is the question asked: how does the *volume* change, and why?

Magnesium, with a density of 1.74 g/cm^3, oxidises to MgO with a density of 3.58 g/cm^3.

$$Mg + \frac{1}{2}O_2 \longrightarrow MgO$$

The percentage volume change occurring in the reaction is negative, that is, –19%. *Less* volume is required for MgO than Mg, even though oxygen has been *added!*

The ionic Mg^{2+}–O^{2-} bond is so strong that it pulls the ions close together (0.21 nm, centre to centre). In magnesium, the metallic Mg–Mg bond is weaker (with a centre to centre distance of $\approx$ 0.32 nm). The stronger bonds in the oxide lead to a higher melting point, 2800°C, compared to that of metallic magnesium, 650°C.

This highlights the major difference between the metal magnesium and the ceramic magnesium oxide, and the connection between the metals and ceramics (discussed earlier). See also Experiment 10.1, Tarnishing and corrosion.

Alloys

Metallic mixtures are generally very complicated. Because of the limited solubility of one metal in another, we are often dealing with several solid phases in the one piece of metal. Alloys are studied by means of equilibrium phase diagrams and an example of one for tin and lead is given in Appendix III.

Metals are often quenched (sudden cooling) so as to deliberately prevent equilibrium from being established.

The following discussion of alloys makes use of Table 10.3 which gives atom size and crystal structure. With regard to crystal structure, like dissolves like more readily. The closer in size the atoms of the two metals are, the more soluble in each other they will be.

Copper–zinc alloys are known as brass

Copper is *fcc*, whereas zinc is *hcp*. Up to about 35%, Zinc can be added to copper without changing the *fcc* structure. Copper–zinc alloys are known as brass.

About 5% Zn alloy is known as gilding metal. It has a golden colour and is used for cheap jewellery. Australian one- and two-cent coins were 97% copper, 2.5% zinc, and 0.5% tin, as are the English one and two pence coins. US one cent pieces since 1982 consist of zinc (+0.8% Cu) covered by a thin copper shell, and weigh just over 2.5 g. The overall composition has 2.5% copper. Before 1982, the cent was made from pure copper. In 1943 the cent was a zinc-coated steel coin.

- The 10% Zn alloy is known as *commercial bronze* (cheaper than real tin bronzes)
- The 15% Zn alloy has a red tint and is known as *red brass*.
- The 30% Zn alloy (70:30) has good mechanical properties and is the military's favourite brass because it can be shaped by cold-pressing for cartridge and shell cases.

Admiralty *gun metal* is 88% copper, 10% tin, and 2% zinc, really a bronze, and no longer used for guns. A further 0.04% arsenic reduces the preferential loss of zinc from the alloy, dezincification, which is often seen in brass tap fittings exposed to chlorinated water supplies. These fittings sometimes also have lead, a source of water contamination.

Copper–tin alloys are known as bronze

Tin is neither *fcc* nor *hcp*. A maximum of about 14% of tin will dissolve in copper at high temperature, but as little as 2% at equilibrium at room temperature. Above about 2% , two phases are generally present. Most cast binary bronzes contain about 10% tin. Phosphorus is used to remove oxygen (as P_2O_5 in the slag) but the residual further increases the strength to give phosphor bronzes.

Copper–aluminium alloys are also called bronze

Both copper and aluminium are *fcc* but the aluminium atom is larger than the copper atom. Up to about 10%, aluminium will dissolve in copper to give an aluminium *bronze*, with a pleasing 18 carat gold look. The aluminium oxide inhibits oxidation of the copper. Australian one- and two-dollar coins are 92% copper, 6% aluminium, and 2% nickel, while the English £1 and £2 coins are 70% copper, 24.5% zinc, and 5.5% nickel.

Copper–nickel alloys

Copper and nickel are completely soluble in each other in both the solid and liquid states — they are both *fcc* and the nickel atom is slightly smaller than the copper atom. However, when cooling from the melt, the composition of the single solid phase formed varies with the amount solidified, and the alloy still consists of a number of cored crystals of varying composition. Suitable annealing can remove this.

Release of copper ions from the metal is used to protect marine surfaces against barnacle growth, but the sheathing corrodes quickly. When the change in weight of an alloy is plotted against its nickel concentration, it is found that the copper-rich alloy loses weight, but the nickel-rich one gains weight from barnacle growth. A good compromise is 70:30 and is used in marine condensers. (See Chemistry and design in Chapter 9.)

German silver is stiffer
German silver is stiffer than real silver (larger Young's modulus, see Chemistry and design p. 246) and consequently creates too long a resonance time for quality flute notes. Real silver (less elastic, hence damping the sound more) is just right, but unfortunately more expensive

A solution of copper, nickel, and zinc was once known as German silver, now known as nickel silver. It is widely used for plumbing fixtures and as a base of silver plate. It is a solid solution, provided that the combined nickel and zinc content is below about 36%. Monel metal is two parts nickel to one of copper and is a strong alloy.

Australian 5, 10 and 20 cent (silver) coins are 75% copper and 25% nickel, as are the English 5p, 10p and 50p coins. The English 20p coin has more copper and a slightly different colour; it is 84% copper and 16% nickel.

Tin–lead alloys

A 60:40 tin–lead mixture is called electrician's solder. It is a mixture of two solid phases and melts and freezes sharply, like a pure metal, at a lower temperature than either tin or lead. Such a sharp melt for a mixture is called a eutectic. This is useful for electricians, who do not want their heat-sensitive components to be exposed longer than necessary. A 30:70 tin–lead mixture is called plumber's solder. It melts and freezes slowly over a range of about 70°C, staying mushy for some time. This is useful for plumbers (L: *plumbum*, lead) who need time to move a joint properly into place. *Pewter* is tin with up to 35% added lead, which lowers its melting point by up to 40°C.

Leadless pewter
While this lowering was important in older times, today the increased awareness of the toxicity of lead has meant that pewter (at least for food purposes) tends to be tin with copper, to give a tarnished appearance but with no lead content.

Shape-memory alloys

Nitinol is a nickel–titanium alloy that has a 'memory effect'. Above a temperature called the transition temperature T_c, the alloy is stiff, and below that temperature it is soft. For an alloy to have shape memory it must have crystal structure that can shift in and out of a configuration called martensite (see A very special alloy, below). Below T_c, the shape can be changed completely. Taken above T_c, the atoms move back in a co-operative manner to the original shape, like soldiers in a formation making a coordinated turn. Each atom, through interatomic attraction, remembers where its neighbours were. The change back is correlated, 'military', no-bonds-broken reversal, and happens at the speed of sound (in that material). This is in marked contrast to the general movement of atoms when metals are bent, which introduces irreversible defects, such as in the work-hardening of copper on repeated bending. While such transitions had been noted in many alloys, including brass, it was not until nickel–titanium was studied that any interest was aroused. (The name Nitinol comes from nickel–titanium National Ordnance Laboratory.)

By varying the nickel–titanium ratio and by adding small amounts of other elements, T_c can be set at any temperature within the range –200 to +140°C.

The heat engine (see Fig. 10.2) consists of thin Nitinol wire, looped around one plastic pulley and one brass pulley. The bottom edge of the brass pulley (in contact with the wire) is heated in hot water to 60–65°C, while the rest of the machine is at ambient temperature (25°C).

Rubbery metals
Consumer products that regularly get sat on, now often use a shape-memory alloy at well above its transition temperature. When bent and released, the metal springs back immediately with a rubber-like response. Examples are the aerials on mobile phones and some metal-framed spectacles.

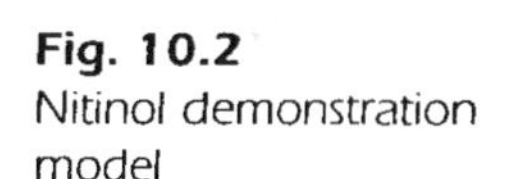

Fig. 10.2
Nitinol demonstration model

Once started (with a bit of encouragement) the device can be taken out of the water and keeps running in thin air for a couple of minutes (until the brass cools down). An important version of nitinol alloy with about 50% each of nickel and titanium is Biometal™ (Toki Corporation) and is used in products called Space Wings and Kinetic Butterfly that flap their wings with temperature change. The first patent for converting low temperature heat straight into mechanical energy with such alloys was issued in 1966. Other uses are in self-opening stents for keeping arteries from collapsing, self-opening fasteners for rooting an artificial tooth in a jaw, and heat-sensitive switches for trash cans used in national parks. If the contents catch fire, the can closes and later reopens on cooling.

See also Theory of heat engines in Chapter 14. Source: *Shape-memory alloys*, L.Mc. Schetky, *Scientific American*, pp. 68–76, Nov. 1979.

Noble metal alloys

The noble metals (Au, Ag, Pt, and Pd) are all *fcc* with very similar size atoms. They all form solid solutions. The 24 carat gold used in the Australian proof $100 coin is 99.99% gold. The 22 carat gold in the (1995) $100 uncirculated coin is 91.67% Au, 4.17% Ag, and 4.17% Cu, while the 22 carat gold in the $200 coin is 91.67% Au, 8.33% Ag, and a trace of Cu. The gold content of both 22 carat alloys remains the same.

Pre-1996 $10 coins are sterling silver: 92.5% Ag, 7.5% Cu, while the newer ones are fine silver, 99.5% Ag.

A very special alloy — iron and carbon

Metals were brought into use throughout history in a sequence determined by the increasing difficulty of winning them from their ores, that is, in a sequence which parallels their thermochemical properties. Whereas bronze melts between 900 and 1000°C (within the reach of an ordinary wood fire) pure iron melts at 1535°C, which, for a large part of our history, was out of reach of human technology.

When an iron ore is heated in a charcoal fire, two reactions occur — reactions that together ushered in the iron age.

$$3Fe_2O_3 + 11C \longrightarrow 2Fe_3C + 9CO(g)$$

The oxygen in the ore is removed as carbon monoxide gas and the iron reacts to form iron carbide (cementite), a compound with 6.7% carbon. Some pure iron is formed at the same time.

$$Fe_2O_3 + 3C \longrightarrow 2Fe + 3CO(g)$$

Now the crux of the matter is that iron and iron carbide (cementite) are mutually soluble and form a *eutectic*. As we saw, the mixture of the two compounds has a *lower* melting point than each of the components. Whereas pure iron melts at 1535°C, the mixture with, say, 4% overall carbon, melts at only 1150°C.

Liquid melt
The reason that iron can be smelted at all is that the two reactions occur together; that they produce two components; and these happen to form a liquid melt, at a reasonable reachable temperature.

Cast iron contains about 4% carbon, which is as much carbon as iron will hold. As well as carbon, other impurities are present because the original ore is not pure. These are removed by adding a 'flux' in the form of limestone, which combines with the other metals to form a low-melting glassy slag. This material is sometimes made into fibres (rock wool) for home insulation.

Crude cast iron is thus a mixture of metal and non-metal crystals, which makes it brittle when cold but malleable above 250°C, because at that temperature the crystal dislocations can move.

Cast iron is 'wrought' at 800–900°C by hammering. The hammering mechanically squeezes out most of the slag and solid impurities and also burns out most of the carbon. The heated iron forms an oxide layer (FeO) on the surface. When the iron is beaten and folded over like pastry, the surface layer can react with the bulk material.

$$Fe_3C + FeO \longrightarrow 4Fe + CO\ (g)$$

This long and laborious process was repeated perhaps one thousand times to produce almost pure iron with a few streaks of inclusions such as slag, which provide the traditional wavy patterns of wrought iron. To produce swords, the wrought iron had to be 'case hardened' by inducing retention of a certain amount of carbon, but only on the surface. For best results the 'steel' was quenched by cooling it suddenly in a liquid. The mechanism of this process is now fully understood and is related to the packing of the spheres.

Below 910°C iron metal has the *bcc* structure. This form of iron is called *ferritic* and it is magnetic. Above 1400°C, iron again has a similar ferritic structure, but between 910 and 1400°C, iron has the *fcc* non-magnetic *austenite* structure (see Fig. 10.3)

Sphere packing
In 1831, Gauss calculated the percentage occupancy as being $\pi/(3\sqrt{2}) = 74.05\%$. Simple geometry shows that if the spheres have a radius of, say, 1.0, then for each sphere in the crystal there will be eight holes of radius 0.155, two holes of radius 0.225 and one hole of radius 0.414.

Fig. 10.3
Left: body-centred cubic structure with some interstices filled, e.g. low-temperature (ferritic) form of iron with space for carbon. Right: face-centred structure with interstices filled, e.g. high-temperature (austenite) form of iron with space for carbon

Many structures of metals can be rationalised on the basis of little atoms fitting in the spaces provided by packed big atoms.

Carbon can dissolve in the holes between the iron atoms, as well as combine with iron chemically to form the compound cementite (Fe_3C). More carbon can dissolve in the high-temperature form of iron (austenite) than in the low-temperature form (ferritic).

If austenite steel is cooled slowly, a banded structure consisting of layers of pure iron (ferrite) and iron carbide (cementite) forms. This steel is tough and strong, but not particularly hard. However, if the austenite steel is cooled very quickly, the carbon atoms cannot move quickly enough from the holes between the iron atoms to react with the iron atoms and form cementite. They therefore remain trapped in a ferrite structure, but now without sufficient room. This means that we have pro-

Cooling steel
This can be demonstrated by dropping water on an electric hot-plate or hot iron and watching the water droplets bounce around. With urine, however, crystals form on the steel and break up the steam barrier. (Don't try it, the smell is awful!)

duced a structure in which dislocations cannot move. This quenching process is called *martensite hardening* and provides a very hard ferrite steel.

Biological fluids used to be preferred to water as the quenching agents. For example, prisoners are alleged to have been sacrificed in the production of Japanese samurai swords, but only people's urine was used in Europe. Technological innovation of this type often waits a long time for chemical explanation.

It is important that the steel be cooled as quickly as possible to prevent the carbon atoms from moving. When water is poured on hot metal, steam is formed, which prevents the liquid from touching the metal, and the transfer of heat from the metal to the water is poor. The nitrogen in urine (as urea) can also react with iron to form hard needle crystals of iron nitride (Fe_2N) and also nitrogen molecules can enter the vacant holes in the same way as carbon atoms and help to pin the dislocations. The modern method uses several days of urea or ammonia treatment on the hot metal.

Tempering of steel at between 220 and 450°C oxidises the carbon in the steel, which softens the steel to make it more ductile. Quenching and tempering are alternated and the art is to provide just the right balance of hardness and ductility for the purpose.

When a process is scientifically understood, art is replaced by science and skilled artisans are replaced by automated processes.

KNIFE EDGE

On a regular Dial-a-Scientist talk-back radio show I was once asked the following by a bushman on a mobile phone, out on a tractor: 'Why had he been told that to judge a knife that will take a good edge, breathe on it first, and compare how long the breath stays on before evaporating?'.

Observation: The longer, the sharper.

Explanation: Soft steels (almost pure iron) will conduct heat better than high carbon steels which have lots of grains of iron carbide. So, if the breath stays on, the heat from the metal is not getting to it too quickly. Low thermal conductivity — hard metal. (However, stainless steels have a much lower heat conductivity (25–30%) but do not sharpen easily (see questions at the end of this section).)

Reason for the explanation: Grains, precipitates, and small crystallites all impede movement of dislocations, and hence the steel is hard. They also scatter waves of electrons and waves of atoms vibrating, and so reduce both the electrical conductivity and the conduction of heat. If you look up a catalogue of steels you will find that hardness and thermal conductivity run in parallel.

Experiment:
Expose different metals to a brief contact with steam (from a teapot), so that the initial condensed layer does not depend much on the sample. The slower evaporation of the harder metals does show up. Compare the blade and the handle of a stainless-steel knife. Better controlled conditions would be preferable!

The military history of Europe has in many ways been a history of improvements in iron and steel production. Cast iron was a poor material for any purpose involving tension, such as cannons. The main armaments of HMS *Victory* were 32-pounders weighing 3 to 4 tonnes. (These cannons have been replaced in the ship's museum by wooden replicas because otherwise the ship would collapse under their weight.) The *Victory*'s opponent at Trafalgar, the *Redoubtable*, surrendered after two of her guns burst. Even by the time of the Crimean War the guns had still not improved much and the Light Brigade was probably more at risk from its own guns.

Henry Bessemer had the brilliant idea of blowing air through liquid pig-iron to remove carbon and other impurities as their oxides, and so 'convert' the iron to steel. The addition of manganese, the subject of another man's patent (Mushet), ensured the success of the Bessemer process. The original process would have failed because sulfur impurities react with iron to form iron sulfide (FeS), which can dissolve in *molten* iron but will crystallise out of the *solid* iron (like salt from freezing sea water) to settle at the boundaries between the iron crystals, and considerably weaken the steel. Addition of manganese changes FeS to MnS, which is insoluble in liquid iron, and so passes into the slag.

Stainless steel

The rusting of iron and steel occurs because an incoherent oxide layer is formed, that is, the iron oxide crumbles off. This is a major drawback.

Chromium atoms are about the same size as iron atoms and form the same crystal structure (*bcc*). When alloyed, the chromium atoms just replace some of the iron atoms. Chromium forms coherent oxide layers that prevent further attack. At least 12% chromium is needed to give corrosion resistance to iron, and 18% is usual. Chromium has another effect on iron; it stabilises the low-temperature *bcc* (ferritic) structure even at high temperatures. Add much above 12% chromium and no *fcc* (austenite) structure is formed at any temperature, and for this reason iron chromium (Fe–Cr) steels resemble most non-ferrous alloys, that is they show no transformation hardening by quenching, no grain refinement by heat treatment, and are non-magnetic. Binary Fe–Cr alloys free of carbon are really stainless irons, not stainless steels.

To make a real steel again, nickel is added. When nickel, which has an *fcc* structure, is also alloyed to iron, it stabilises the *fcc* (austenite) structure of iron. The popular cutlery stainless steel is generally type 302, or 18:8 (18% Cr, 8% Ni).

Like iron, chromium also forms a carbide (actually a mixed carbide with iron) but nickel does not. Chromium diffuses to grain boundaries (where the carbon is) to form the carbide phase at these sites. This depletes chromium near the grain boundaries. The level of chromium can then drop to below the 12% needed for corrosion resistance. In a corrosive environment, the low chromium area is anodic to the rest of the grain and pitting occurs. To reduce attack and corrosion of stainless steels at the grain boundaries, the carbon content can be lowered, or those metals added which preferentially react with carbon, such as niobium or

Stainless steel content
Stainless steels contain chromium for corrosion resistance, and nickel to give the austenitic crystal structure of a good steel, with some carbon and other elements added. Cold working of stainless steels to harden them can cause them to revert and become magnetic (ferritic).

Copper release
Adding 1.5% copper and a trace of niobium provides a slow release of copper, which reduces bacterial contamination (golden staph and E. coli) very significantly. Washing machine drums appear to be the first commercial application, but medical and food equipment cannot be far behind.

titanium. The chloride ion attacks the oxide layer and this can be reduced with added molybdenum.

Chromium is the basic element for forming the passive oxide layer and preventing corrosion. Contrary to intuition, strongly reducing conditions aid corrosion, while oxidising conditions prevent it by strengthening the oxide layer. Anything that blocks oxygen access to the surface will aid corrosion: grease, oil, dirt, hand prints, dust, and salt crystals. Architectural stainless steel structures must be washed as routinely as the windows, and for more critical reasons.

If you remove a paper clip from a sheaf of papers that has been left around for some years, it shows rust marks on the paper *underneath* the clip, where the clip was in contact with the paper. This effect has little to do with the chemistry of the paper but is due to oxygen being excluded from the metal at the point of contact. Why does steel rust where the oxygen level is *reduced*? (The study of electrochemistry and corrosion explains all this but we must leave at this stage.) But what does that tell you about the way you should store your stainless steel cutlery? Loosely, and with no greasy spots, so that there is uniform easy access to air all around. A small cathode, that is a stainless steel fastener into an aluminium or galvanised steel plate, is protected by the aluminium or zinc having the greater tendency to corrode. Conversely, an aluminium or zinc plated fastener in a stainless steel sheet can lead to significant corrosion of the fastener.

Questions

- Test your stainless steel 18:8 cutlery with a magnet and compare the knife handle and blade. Sometimes they are made from two materials welded together. Even when they are one piece, the blade tends to be magnetic, while the handle is less or not at all. Why?
- Stainless steel pots and pans often have copper bases or at least copper with a thin cover of steel. Why?

EXPERIMENT 10.1 Tarnishing and corrosion

Equipment

Alfoil sheet, tarnished silver, sodium bicarbonate (baking soda), aluminium soft drink cans.

Procedure — Part A

Sulfur compounds abound as a result of natural and human activities. With silver, they form black silver sulfide, which adheres strongly to the silver surface. (This is particularly noticeable on silver spoons used for eating eggs because eggs contain protein with sulfur.) The silver sulfide can be polished off, or dissolved off by solutions containing ammonia, with resulting loss of silver. The best method uses an oxidation–reduction reaction to reverse the corrosion process.

Put some hot water in a pan (it can be aluminium) and add a few teaspoons of sodium bicarbonate ($NaHCO_3$, baking soda) to the water. Wrap the silver to be cleaned in aluminium foil. Leave a small opening in the foil and put the foil and silver in the solution. Ensure that the foil is completely covered with solution and that no air bubbles are trapped inside the foil (otherwise those bubble contact points will not be cleaned). All of the silver utensils must touch the foil at some point. The silver will be cleaned in about 1–5 hours, depending on the degree of tarnishing. Some commercial aluminium alloys are sold for this purpose, with a claim of higher efficiency.

The redox processes that occur are:

$$Al(s) \longrightarrow Al^{3+}(aq) + 3e^-$$
$$2e^- + Ag_2S(s) \longrightarrow 2Ag(s) + S^{2-}(aq)$$

The overall reaction is:

$$3Ag_2S(s) + 2Al(s) \longrightarrow 2Al^{3+}(aq) + 3S^{2-}(aq) + 6Ag(s)$$

Procedure — Part B

A common method used for packaging food and beverages is to use tin-coated steel, or aluminium cans. These forms of packaging are chosen because both tin and aluminium readily oxidise to form a protective oxide coating.

Take two aluminium soft-drink cans and cut the tops off them. Fill both cans with water and to one can add a few teaspoons of sodium bicarbonate, and to the other add a few teaspoons of vinegar (acetic acid). Allow the cans to stand for a few days.

You will notice that the can containing the vinegar shows no evidence of corrosion, but the can containing sodium bicarbonate has corroded. The aluminium oxide formed on the surface of the aluminium is stable in the acid solution, but in the basic solution formed by the sodium bicarbonate, the oxide dissolves. The aluminium surface is therefore unprotected in basic solution. This explains why the sodium bicarbonate was added to the water in Part A — to keep the aluminium surface clean and free from the aluminium oxide. It also explains why soft drinks can be stored in aluminium cans — soft drink solutions are acidic.

Note

Try the experiment in Part B using tin-coated steel. Be sure you choose a can which has only a tin coating. Some cans are coated with a protective lacquer coating as well. Also try a set of experiments in which you have scratched the tin surface to reveal the underlying steel.

EXPERIMENT 10.2
Magnesium pencil sharpener

Introduction

You probably thought that pencil sharpeners are for sharpening pencils. However, there is a metal version that has many other uses, at least for

Magnesium from the sea
The concentration of magnesium ion in sea water is
= 1.3 mg/L
= 5.4×10^{-8}M.
The solubility product for $Mg(OH)_2$ is
$K_{sp} = 1 \times 10^{-11}$
It is much less soluble than $Ca(OH)_2$
($K_{sp} = 4 \times 10^{-6}$).
Lime is therefore added to (concentrated) sea water to 'mine' magnesium. Can you do the calculations? (See also Exercise: Solubility at the sea in Chapter 3.)

creative chemists. It's a Staedtler™ and is made from a magnesium alloy with 92.7% Mg, 2.2% Al, 0.5% tin, and 0.1% manganese. If you toss it into some salty water, the steel blade with magnesium base acts as a corrosion cell. Bubbles of hydrogen are seen to come off the steel blade and the metal turns fleetingly black and then a white gel precipitate forms. Overnight the sharpener dissolves, leaving only the blade (like the smile of Alice's Cheshire cat). The magnesium has acted as a sacrificial cathode. The precipitate is magnesium hydroxide, which is quite insoluble, and is used as milk of magnesia for upset stomachs (after eating US army rations — see below). As 3.8 g of magnesium is 0.16 of a mole, complete dissolution should yield 3.5 L of hydrogen, and in a 1 L PET bottle, a few atmospheres pressure — **Take care!**

Magnesium strip is used as a time fuse to hold down the buoys that indicate the positions of lobster pots beneath the surface of the ocean. The corroding fuse releases the buoy for the pot at a roughly predetermined interval, depending on the thickness of the strip. This keeps the location of the pots secret from the prying eyes of poachers (or fisheries inspectors).

The attack by chloride on magnesium becomes marked if the impurity content exceeds a certain tolerance limit (0.017% for iron and only 0.0005% for nickel). Magnesium sacrificial anodes contain 6% aluminium, 3% zinc, and 0.2% manganese.

INSTANT HEAT

The US army developed a Meal, Ready to Eat (MRE) for the Desert Storm war in Iraq. A heat-producing plastic sleeve Flameless Ration Heater is placed around the aluminium food pack and 30 mL of water is put inside. After 12–15 minutes, you have food at about 60°C. The protective cardboard is peeled off before use.

Good catalyst
Steel is a particularly good catalyst for the reduction of water to hydroxide, and steel balls are used in the industrial conversion of mercury–sodium amalgams produced in brine electrolysis cells to give caustic soda. (The steel blade on the sharpener (Experiment 10.1) is thus fortuitously ideal!)

The heat pad is a composite of a powered alloy of magnesiun and iron sintered with HDPE to form a porous matrix. In theory, 24 grams or one mole of Mg ($\Delta H_f = 0$) releases 354 kJ on reacting with two moles of water ($\Delta H_f = -285 \times 2$) to form one mole of $Mg(OH)_2(s)$ ($\Delta H_f = -924$kJ) plus one mole of hydrogen gas ($\Delta H_f = 0$).

Without losses, this would be enough to bring one litre of water to boiling, from 20 to 100°C. The reaction also releases 22.4 L of hydrogen per mole of magnesium, which can be burnt to give water again, plus even more heat ($\Delta H_f = -285$ kJ) (To heat water from 20°C to 100°C takes 4.18 J/deg./mL × 80° = 335 kJ/L.)

However magnesium metal normally resists corrosion, so salt is added. The chloride from the salt breaks down the oxide layer, and the iron forms an electrochemical cell. The magnesium hydroxide product is not very soluble, and flocculates and then precipitates out: $Mg(OH)_2$: sol. 4 mg/L.

(1 oz = 28 g; ΔH_f MgO is –601.6 kJ/mol (–15 kJ/g), ΔH_f $Mg(OH)_2$ is –924 kJ/mol (– 16kJ/g).)

Source: A flexible electrochemical heater, W.E. Kuhn, K.H. Hu, S.A. Black, US patent 4,522,190; 11 June 1985.

EXPERIMENT 10.3

You will need a battery wall clock that (normally) runs off a single AA battery, and a juicy orange.

Unscrew the blade of the pencil sharpener and discard it. Clip a copper wire lead onto the sharpener body, which will act as one electrode. The clip must not come in contact with juice. Clip another copper wire lead to a piece of copper strip (about 1 to 2 cm wide) for the other electrode. Wrap some filter or toilet paper around the copper strip and insert both electrodes into the orange (made squishy in the centre) as close as possible, with just the paper separating them. The lead from the magnesium is attached to the negative terminal of the battery holder, and the lead from the copper strip to the positive terminal.

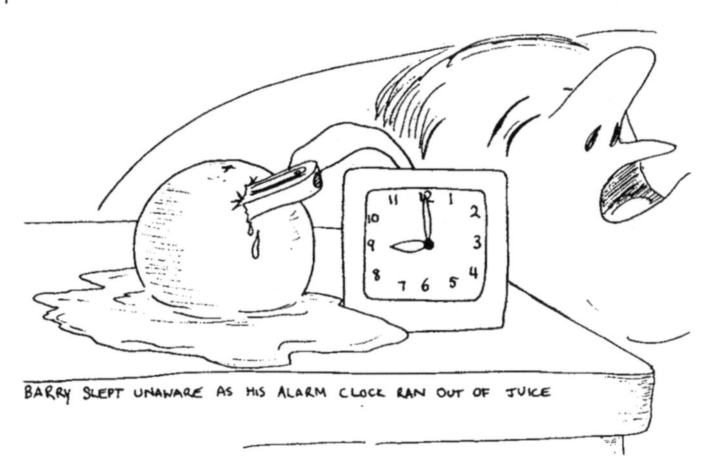

Under standard conditions and under zero load, the theoretical potential should be 2.37 V in acid (pH = 1).

$$Mg \longrightarrow Mg^{2+} + 2e \qquad E_o = 2.37 \text{ (1 M } Mg^{2+})$$
$$2H^+ + 2e \longrightarrow H_2 \qquad E_o = 0.00 \text{ (1 M acid)}$$

In practice the voltage is much lower, particularly under load (it drops during the clock tick), and an average current of about 3 mA is typical.

Biological effects of metals and metalloids

As a very broad generalisation, the elements found on earth have been utilised by the body, or are kept out by various biological ploys developed during evolution. The elements we use in bulk are the lighter ones: the metals Na, K, Mg, Ca; and non-metals C, N, O, P, S, and Cl.

We use members of the first transition row of the periodic table such as Fe, Mn for oxidation/reduction processes, but most use of Co and Ni was dropped by life forms when oxygen appeared on earth. There is still one single use of Co; in cyanocobalamin (vitamin B_{12}); of Ni in urease; and of Cr in glucose control. Zn, however is most important. Further down the periodic table, the heavier elements are not used to any extent, and many are toxic and must be kept out.

Cd is toxic because it competes with Zn and Ca. The body gets rid of Cd by deposition in the kidney and liver. As the animals we eat do the same, we should avoid eating offal from older animals, which have had time to accumulate lots of it.

We require some elements only in minute (trace) quantities and some are very toxic indeed at higher concentrations, like selenium and fluorine.

The heavy metals are widely used in industry. When released into the air, or into rivers, they distort the naturally occurring distribution in which we have evolved. Quite apart from the amount, the changing pattern of use has a dramatic effect on how biologically available the metals are. Acid rain leads to acidic waters, and this releases metals such as aluminium from an insoluble and unavailable form. Zinc production, and use of superphosphate, has released cadmium. This cadmium is absorbed passively by potatoes, and concentrated in oysters.

The availability of elements is very critical because our need for traces of some elements, such as selenium, is very low. Seals and dolphins from many parts of the world show a correlation between the accumulation of mercury and selenium which is not found in the fish on which they feed. The uptake and need for copper depends on the level of zinc and molybdenum to which we are exposed, and vice versa. Such subtle interactions need to be teased out further. Modern technology is heavily dependent on the use of these materials, but in many instances they can be replaced.

Essential selenium
Selenium is the most toxic element known to be essential to mammals. It is also known that selenium compounds have a protective effect against the toxic action of mercury.

Metals in medicaments

Examples include the use of

Li	in depression,
Cu, Ag	in precipitants,
Au, Hg	in germicides,
Al, Zr	in astringents,
As, Sb	as parasiticides,
Se, Te	as topical agents (anions), and
Cl, Br, I	as oxidising germicides.

In what form are they dangerous?

Depending on the physical and chemical form of the metal, the substance can be either very poisonous or completely harmless. *Liquid* mercury, as in a thermometer bulb, or in a mercury dental amalgam, is harmless, but long exposure to the small amount of *gaseous* mercury given off by the liquid (but generally not the amalgam) is readily absorbed into the lungs and can lead to poisoning. The toxicity of *inorganic* mercury compounds depends on how soluble they are in the body fluids. The use of *soluble* mercuric nitrate in the felt-making process led to the traditional notions of 'hatters' shakes' and 'mad as a hatter', and also accounts for the use of the term 'hatter' for solitary, acutely shy, goldminers who recovered their gold by distillation of the gold–mercury amalgam in crude stills.

The most toxic forms of mercury are now recognised to be the organo-mercury compounds — in particular, alkyl-mercury, which is produced from inorganic mercury by micro-organisms present in the bottom of waterways. These compounds, while not soluble in water, are very soluble in fat, and hence are stored in the body, with an average residence time of about 70 days, in contrast to inorganic mercury, whose residence time is about six days. Thus, with continuous exposure of low levels, it is possible to build up concentrations in the body of 10 times the amount of organic mercury, compared to inorganic mercury, at the same level of exposure.

While soluble barium compounds are very poisonous, barium sulfate, which is used for increasing the contrast of internal organs for X-rays, is so very insoluble as to be harmless. Metallic lead is not considered particularly dangerous in itself, although water collected from lead-lined roofs once was a considerable problem. Some forms of lead used in paint are very soluble in body fluids and hence are toxic. Copper is an essential metal, but water passing through copper plumbing in Australian hospitals was found to produce up to 1 mg/kg of copper compound in dialysis fluid, which could add about 10 times the dietary intake of copper to the copper stored in the liver. Domestic copper water pipes normally present no problems.

Cadmium compounds are used as pigments (lemon, yellow, orange, maroon) in ceramic glazes, paints, and plastics. (See Appendix X.) Cadmium is absorbed from food and water slowly, so something like 98% of the cadmium in food is excreted within 48 hours. However, cadmium that is absorbed remains, so cadmium levels build up from zero at birth. Absorption from the air is much more efficient (10–15%). Cigarette smoke contains cadmium — 20 cigarettes contain an average of 30 μg, of which about 70% is extracted into the smoke. There is a good correlation between stored cadmium levels and high blood pressure, and injection of cadmium chloride into animals can apparently cause irreversible damage to the testes. More details on cadmium are given later in this chapter.

Fig. 10.4 Misleading poison.

(From New Scientist, 9 January, 1975, reprinted with permission).

Trace elements — essential, non-essential or toxic?

Research work has gradually added to the list of essential trace elements and it is now believed that 14 trace elements are essential for animal life: iron, iodine, copper, zinc, manganese, cobalt, molybdenum, selenium, chromium, nickel, tin, silicon, fluorine, and vanadium. Probably other elements will be added to this list as experimental techniques are further refined. Five of the 14 elements — nickel, tin, silicon, fluorine, and vanadium — have more recently emerged as essential nutrients in the diets of laboratory animals following the introduction of ultra-clean environments, and the use of pure crystalline amino acids and vitamins. Much remains to be learnt about the metabolic functions of the 'newer' trace elements and their practical significance in the health and nutrition of humans.

No surprise
It should not be surprising that minute quantities of some elements are essential for the health and well-being of people, for the human race evolved in an environment containing most of the known elements.

Plants extract elements from the soil in which they grow, and some of these elements are required for the plant's nutrition. We receive our trace elements from our food, and from the air. Processing and packaging may increase the undesirable trace elements, while depleting essential trace elements. The environmental relationship is illustrated by the fact that a person's body burden of sodium and potassium follows fairly closely the levels of these elements in the earth's crust.

In the following more detailed discussion of some elements, there is no significance in the selection of elements.

Aluminium

Aluminium toxicity and Alzheimer's disease

There is evidence of short-term toxicity of excess aluminium. In England, a population of about 20 000 individuals was exposed for at least five days to increased levels of aluminium sulfate, accidentally placed in a drinking water facility. Case reports of nausea, vomiting, diarrhoea, mouth ulcers, skin ulcers, skin rashes, and arthritic pain were noted. No lasting effects on health could be attributed to the known exposures from aluminium in the drinking water. It was concluded that the symptoms of this massive overexposure were mild and short lived.

There is no doubt that aluminium is neurotoxic. Renal dialysis patients exposed to aluminium in the dialysis fluid, for example, will suffer dementia (but not Alzheimer's disease).

Biological transport of aluminium

While aluminium is toxic to nerves, aluminium salts have difficulty passing through the gut lining. In the acid stomach, the aluminium is there as Al^{3+}, and charged ionic species cannot pass through. The small intestine is alkaline, and aluminium is present as the insoluble hydroxide, which physically cannot pass through. However, there are some complexes of aluminium, notably citrate (citric acid in fruit juices, etc.) that are neutral, and can pass through the gut.

Sources of aluminium

We ingest small amounts of aluminium in various forms. Adults consume between 2.5 and 13 mg of aluminium daily from food and beverages. Drinking water is generally clarified using alum as a flocculating agent. This precipitates out suspended clay (a source of insoluble aluminium) and leaves behind extra soluble aluminium, which contributes about 0.2 mg/day (up to 0.4 mg/day) in the drinking water, while absorption in the lungs for non-occupationally exposed people is up to 0.04 mg/day. On the other hand, two average size antacid tablets may contain in excess of 500 mg Al.

Citrate is a normal dietary breakdown product in the gut, as well as a common food (beverage) component. The aluminium content of citric beverages in aluminium cans has been reported to rise at up to 0.9 mg/L per year of storage. The presence of citrate can raise blood plasma levels of aluminium independently of the level of intake.

Some aluminium cooking pans corrode more easily than others, and release more aluminium into cooked foods and boiled water, particularly if scoured with steel wool or gritty powder. Acidic foods attack the protective oxide layer. Wine and egg yolk sauces are known to be discoloured in aluminium pans.

Aluminium in water

Most of Melbourne's water is not treated with alum, while most of Brisbane's water *is* treated. Melbourne's water should be below 0.05 mg/L soluble Al (< 0.1 mg/L total), while in Brisbane regions it should be below 0.10 mg/L soluble (0.2 mg/L total). The US Environmental Protection Agency (for aesthetic reasons) sets a secondary maximum limit for aluminium in drinking water ranging from 0.05 to 0.2 mg/L (depending on water source) designed to prevent precipitation of aluminium salts in the distribution system. The aluminium speciation at the reservoir may not correlate well with that at the household tap

Water filters
Water filters remove particulates and so reduce aluminium in raw water, but not treated water.

because of precipitation or complexation followed by release along the reticulation path.

Aluminium in water
Water is a minor source of aluminium in the diet and unlikely to contribute significantly to body absorption.

The chemistry of aluminium compounds in water is straightforward. Aluminium occurs as two main species in water. In mildly acidic solution pH < 5.5, it occurs as the aluminium ion attached to six water molecules, $Al^{3+}(H_2O)_6$. At pH > 7, Al is present as aluminates ($Al(OH)_3$) of varying degrees of insolubility depending on what else is present that can complex the aluminium (K_{sp} for Gibbsite (the pure hydroxide) is about 10^{-9}). The highest concentration for aluminium occurs at near neutral pH, and is about 3×10^{-11} mol/L or 0.8 μg/L. Fluoride forms strong complexes with aluminium, and at concentrations of 10^{-4} to 10^{-5} mol/L, the AlF_2^+ and AlF_3 are the most common species.

Aluminium in food

Some food components, for example aluminium compounds are used in processed foods — aluminium phosphate (541 — see Appendix IX) and aluminium silicate (554). They are both acid soluble and are used for pH stabilisation, emulsifying, thickening, rising, and anti-caking. (In bread the range can be from 350 μg/L up to 13 000 μg/kg in pumpernickel.)

Antiperspirants

Most antiperspirants contain aluminium chlorhydrate (see Deodorants and antiperspirants in Chapter 5), which is known to be absorbed into the bloodstream, but correlation with Alzheimer's disease is not established.

Bio-availability of aluminium

There is little correlation between total aluminium ingested and the amount taken up into the bloodstream (i.e. bio-availability) because of the way it is complexed. For example, tea has high aluminium levels (1000 to 3000 μg/L) but very little is absorbed because it occurs as complexes with polyphenols which are poorly digested. The efficiency with which aluminium is taken up from water decreases in the following series: lemon juice > orange juice > wine, coffee > tomato juice > beer > tea, milk.

Aluminium readily substitutes for calcium in the bones, and bone acts as a passive reservoir, so the Al level is a good measure of long-term exposure, ranging from a normal 3.3 μg/g dry mass to in excess of 200 μg/g for persons on a high aluminium diet.

Arsenic

Arsenic is very common. It ranks twentieth in abundance among the elements in the earth's crust. It is found in igneous and sedimentary rocks, particularly with sulfide ores. Arsenic is released into the environment through industrial emissions from smelting, and from burning fossil fuels. The most common form of arsenic is as inorganic compounds. However, methylation by microbes produces methyl arsines and organo-arsenic acids.

Main uses of arsenic compounds

Before the advent of DDT, arsenic was one of the principal pesticides. Arsenic is also toxic to plants and so was used as a weedkiller. (See Inorganic insecticides in Chapter 7.) In fact it will generally kill the plants long before it accumulates enough to be a health hazard to humans. (Copper and nickel are similar.) Therefore arsenic levels in food used to be much higher than they are today, as very little arsenic is now used in agriculture.

There appears to be a level of intake above which arsenic accumulates in the system, and below which the body appears to be able to excrete all the arsenic ingested. Normal intake is about 0.007–0.6 mg/kg body mass per day. It appears unlikely that any problem should arise with this element: arsenic poisoning is only likely in individuals occupationally at risk, or who happen to be the victims of attempted murder. (See History of drug control in Chapter 13.)

The biggest hazard from arsenic in soil is eating the soil (licking dirty fingers, etc.). Because these materials are absorbed much more efficiently through the lungs, inhalation of dust could be a potential hazard.

The 1992 NHMRC/ANZECC guidelines for taking action on contaminated sites from a direct (i.e. eating) health perspective set 'trigger levels' for taking action as follows:

As	100 ppm	20 ppm (environmental)
Pb	300 ppm	
Cd	20 ppm.	

Note that the environmental level for arsenic is much lower because any more than 20 ppm kills plants. On the other hand, cadmium is not toxic to plants, and so cadmium in soils can be a health hazard in crops such as potatoes and lettuce (see below).

Organic arsenicals are widely used as additives in stock food because these compounds appear to stimulate growth and improve food utilisation. Animal diets typically contain 40 μg/kg arsenic, at which level there is some accumulation in edible tissues, usually less than 0.5 mg/kg in muscle and 2.0 mg/kg in liver.

Arsenic is used in wood preservatives (see copper chromium arsenite in Inorganic insecticides, Chapter 7) and small quantities are used in glass making and electronics as GaAs transistors and photovoltaic cells.

Salvarsan, Ehrlich's compound 606 (arsphenamine) was the first drug capable of successfully and reasonably safely treating the sexually transmitted disease syphilis. That was in 1910, and it remained in use until the 1940s when replaced by penicillin. Salvarsan continued to be used to treat animal parasites, for example canine heartworm.

'Gosio gas' is trimethyl arsine (Me_3As). It is the volatile poisonous arsenic compound produced by moulds growing on wallpaper containing the arsenic pigment Scheele's green (copper arsenite).

Napoleon and arsenic

One of the theories about Napoleon's death is that he was poisoned by trimethyl arsine. Arsenic accumulates in keratin found in hair and fingernails.

The high level of arsenic in marine animals such as rock lobster is organically combined, and arsenobetaine ($Me_3As^+CH_2COO^-$) is its most common marine form. It is also found as arsenocholine ($Me_3As^+CH_2CH_2OH$). We rapidly excrete arsenic in this form, apparently unchanged. The no observable effect level (NOEL) is very high (10 g/kg).

The most toxic form of arsenic is As^{3+} because this reacts with SH^- groups found in vital enzymes. As^{3+} poisoning can be treated with BAL British anti-Lewisite, developed to combat the arsenical war gas Lewisite; see Fig. 10.5).

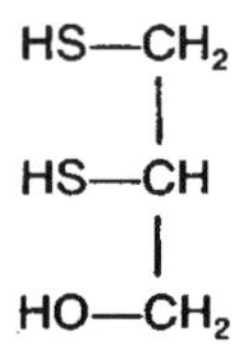

Fig. 10.5 BAL (British anti-Lewisite)

Cadmium

Cadmium looks like zinc but, on bending, the coarse-grained cadmium gives a crackling sound similar to that given by tin. It is used for plating, bearing alloys, soldering aluminium (Cd/Zn), orange-red glaze in ceramics, and in nuclear reactors to absorb neutrons, especially in control rods. It is used in nickel oxide cadmium (Nicad) accumulators. Organo-cadmium compounds are used as PVC stabilisers and as mould-release agents for plastic articles. Oysters grown in unpolluted water may contain as little as 0.05 mg/kg of cadmium, whereas oysters grown in polluted water may contain cadmium in excess of 5 mg/kg.

Large concentrations of cadmium have bizarre effects. In 1955 the Japanese reported an affliction called 'Itai-Itai Byo'. The name mimics the cries of sufferers, and has been translated as 'Ouchi-ouchi disease'. It takes a long course of increasing painfulness, beginning with simple symptoms such as 'joint pains', and ending with total and agonising immobility as the result of skeletal collapse. Cadmium leads to bone porosity and to the total inhibition of bone repair, so the load-bearing bones of the skeleton suffer deformation, fracture, and collapse. In Japan the disease has been associated with rice and soya (in the range 0.37–3.36 mg/kg dry weight) in the local diet, and was also known to occur in workers engaged in the preparation of cadmium-based paints.

Cadmium is chemically related to zinc (they are in the same column of the periodic table, see Fig.1.1) and is found together with zinc in nature and in zinc products. The zinc/cadmium group is also related to the magnesium/calcium group, so that is why cadmium (and strontium) can interact so effectively with calcium in the bones. The mutual replaceability of metals depends both on their chemical similarity and the size of the *atomic ion* (see Table 10.4).

TABLE 10.4
Ionic radii in picometres (10^{-12} metre)

Group IIA			Group IIB		Group IV	
Ca^{2+}	Sr^{2+}	Ba^{2+}	Zn^{2+}	Cd^{2+}	Hg^{2+}	Pb^{2+}
99	113	135	74	97	100	121

Zinc and cadmium each inhibit the other's absorption and retention in the body, perhaps because they compete for similar protein-binding sites. High zinc intake may reduce the toxicity of cadmium, and conversely, high cadmium intake with marginal zinc deficiency may aggravate the deficiency. Unfortunately the residence time of cadmium in the body is very long.

Cadmium in fertilisers

See Plate XI, a and b, Cadmium in fertilisers.

Chromium

Chromium deficiency in humans appears to reduce tolerance to glucose. (Chromium is a co-factor of insulin, essential to proper glucose metabolism.) Chromium deficiency may result from a grossly deficient diet. The chromium requirement of humans is very difficult to estimate because little is known of its form in food or its biological availability. Meat appears to be the best source of chromium in the diet, as it may contain several μg/kg. Yeast-leavened bread is also an important source of chromium.

Cobalt

Cobalt is thought to be unique because it is the only trace element shown to be physiologically active in one particular form only, namely cyano-cobalamin, or vitamin B_{12}. Cobalt nutrition in humans is thus primarily a question of the source and supply of vitamin B_{12}. Ruminants, on the other hand, can utilise dietary cobalt directly because the microflora of the rumen convert cobalt into vitamin B_{12} (cyano-cobalamin). Hence sheep have been given metallic cobalt pellets, which remain in their rumen. All ordinary diets contain considerably more cobalt than can be accounted for as vitamin B_{12}. Cobalt intake is about 0.15 to 0.6 mg/day. Levels of 25 to 30 mg/day can be toxic to humans.

Cobalt was implicated as the cause of heart failure in beer drinkers consuming about 12 L a day. The cobalt had been added to the beer in concentrations of about 1.2 to 1.5 mg/L to improve foaming. At this level

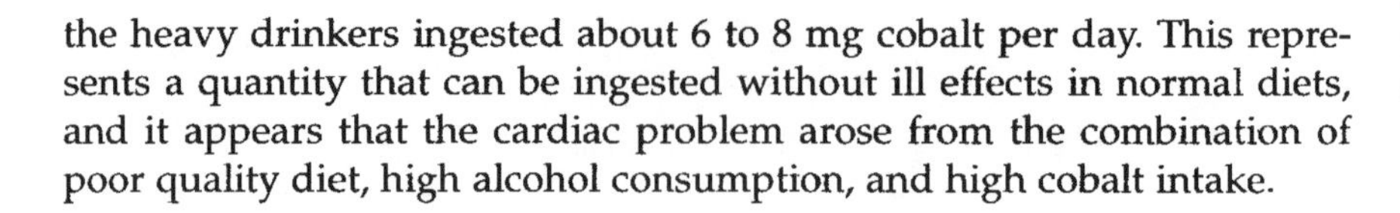

the heavy drinkers ingested about 6 to 8 mg cobalt per day. This represents a quantity that can be ingested without ill effects in normal diets, and it appears that the cardiac problem arose from the combination of poor quality diet, high alcohol consumption, and high cobalt intake.

Kinky hair in infants
The condition commonly known as 'kinky hair' in infants is caused by a genetic defect causing inefficient copper absorption.

Copper

Copper deficiency has not been reported in adults, even in areas where copper deficiency is severe in grazing animals. But copper deficiency is implicated in anaemia in infants from impoverished communities where the diet is based on cow's milk. Infants' diets containing, less than 50 μg of copper per kilogram body mass per day have resulted in copper depletion and produced clinical lesions. (Copper is a component of several amine oxidases and it is possible that, in some animal species, defects in the synthesis of vascular elastin and of collagen in the bones and connective tissue are the result of copper depletion causing a decline in amine oxidase activity.) Sulfur in the diet in the form of sulfides can markedly decrease copper absorption, and cadmium concentrations in the order of 3 μg/kg can adversely affect copper utilisation. Infants appear to require 50 to 100 μg/kg per day, and adults about 30 μg/kg per day.

Liver, oysters, many species of fish, and green vegetables are good sources of copper, but milk and cereal products are poor ones. Indeed, copper is an undesirable constituent of milk, fats, and fatty foods as it acts as a catalyst, and promotes rancidity of fats even at a very low concentration.

Copper in ice blocks
Following illness in children immediately after eating ice blocks, 285 samples were analysed for copper. A small number had levels of 43 to 80 mg/kg — levels that are sufficient to cause vomiting. This is a recurring problem and arises because some manufacturers use tinned copper moulds. The citric acid in ice blocks will dissolve copper from de-tinned areas of the moulds if the mixture is left in the moulds for excessive lengths of time.

Copper is frequently implicated in food poisoning, and copper in water at a concentration of about 20 mg/L is often the problem. Under certain conditions, drinks from machines dispensing carbonated beverages may contain high concentrations of copper. Water from a copper water service may also contain high copper concentrations (up to about 70 mg/kg) if allowed to stand for some days without flushing.

Copper bracelets

Is there anything in the so-called myth that copper bracelets have therapeutic value? This question has been investigated in Australia by W.R. Walker in the Department of Chemistry at the University of Newcastle. Through letters in newspapers about 300 sufferers from arthritis were contacted (half of whom previously wore copper bracelets) and were randomly allocated for a psychological study. This involved wearing copper bracelets and placebo bracelets (anodised aluminium resembling copper) alternately. The results showed that for a statistically significant number of subjects, the wearing of a copper bracelet appeared to bring some therapeutic benefits. Copper from the bracelet is found to dissolve in sweat and the amount lost was about 13 mg/month. If this were absorbed into the system it would amount (over 12 months) to more than the usual copper level in the body. The level in sweat was about

'Could I 'ave a copper pair for me arthritis?'

Fig. 10.6
'For me arthritis' (From Newcastle Morning Herald, 18 September 1976)

500 mg/kg from the bracelet, but, if sweat is left in contact with copper turnings for 24 hours, the concentration rises by a factor of 100 (turns blue). The skin is permeable to some materials, and copper does move through.

Lead

Lead was used in ancient times by the early Egyptians to glaze pottery. It was also used by the Romans to make water pipes and to store wines (this being a possible source of lead poisoning). Lead tetra-ethyl additive for petrol accounted for about 10% of the world lead consumption in 1970, with the USA consuming about 0.26 megatonnes per year for this purpose. Most of the bromine produced from sea water is used to produce ethylene dibromide, which, along with lead tetra-ethyl, is added in order to exhaust the lead from engine cylinders as volatile lead dibromide. (For each molecule of lead tetra-ethyl there are 0.5 molecules of ethylene dibromide, and 1.0 molecule of ethylene dichloride, in the usual antiknock additive; see also Petroleum oil fractionation in Chapter 14.)

The toxicology of lead is complicated. Inorganic (Pb^{2+}) lead is a general metabolic poison and is accumulated in humans. It inhibits enzyme systems necessary for the formation of haemoglobin (levels of urinary δ-aminolaevulinic acid are monitored to indicate this interference). In particular, children and young people appear liable to suffer permanent damage. Lead can replace calcium in bone, and so it tends to accumulate there.

Lead alkyls (organo-lead) such as lead tetra-ethyl are even more poisonous than Pb^{2+}, and are handled quite differently in the body. Lead tetra-ethyl causes symptoms mimicking those of conventional psychosis. There is little or no elevation of blood lead, so correct diagnosis is difficult in the absence of suspected exposure. Yet leaded petrol has been almost a 0.1% hydrocarbon solution of this compound. The danger of

Lead risk
An unpleasant feature is that lead may be remobilised long after the initial absorption; for example, under conditions where calcium is suddenly needed by the body, such as during feverish illness, cortisone therapy, and also in old age. It can also cross the placental barrier, and thereby enter the fetus.

using leaded petrol for degreasing or cleaning is generally not appreciated. Humans have evolved in the presence of a certain amount of lead — it averages about 10 mg/kg in the earth's crust. We know that over 2 megatonnes of lead is mined each year, in comparison with the 180 kilotonnes estimated to be naturally mobilised and discharged into the ocean and rivers.

Typically an adult living in a city who breathes air containing, say, 3 mg lead/m^3, and who respires about 15 m^3 of air per day, will absorb 20–25 μg of lead per day (about 50% efficiency in absorption). A normal daily diet containing about 300 μg of lead will result in about the same amount of absorbed lead (5–10% efficiency in absorption). Fallout of airborne lead also adds to the lead present in food and water. Even the present average levels of lead in the blood of adults in industrial countries are not far below those that can lead to obvious clinical symptoms. The greated danger of exposure to lead comes from flakes of old lead paints and improperly fired lead glazes (see Plate XIV, Household lead test kit, and Appendix IV).

Mercury

The first recorded mention of mercury was by Aristotle in the fourth century BC, when it was used for religious purposes. Earlier still, vermilion (cinnabar, HgS) is known to have been used as a decorative warpaint (cosmetic). Paracelsus (1493–1541) introduced the treatment of syphilis with mercury. In 1799, Howard prepared mercury fulminate as a detonator for explosives.

With the exception of iron, almost all other metals can be *amalgamated* (alloyed) with mercury. Sodium amalgams are formed in electrolysis cells used for producing chlorine and caustic soda. Many mercury compounds are used as industrial *catalysts*. Dental amalgam is also prepared with mercury (see Chapter 12). For this application, the mercury and alloy must amalgamate upon mixing, to form a smooth paste within 90 seconds. Within three to five minutes it should set to a carvable mass and remain so for 15 minutes. Within two hours it must develop sufficient strength, hardness, and toughness to resist biting and chewing stresses. It must expand to maintain a good marginal seal, but not so much as to overstress the tooth. It must not produce toxic or soluble salts, or tarnish, or produce significant amounts of mercury vapour.

Irreplaceable mercury
Mercury often cannot be replaced by any other metal (nor can it replace other metals); therefore its unique properties have proliferated its uses.

The *equilibrium* vapour pressure of mercury (20°C, 13 mg/m^3 of air) is about 200 times the recommended atmospheric concentration. In an amalgam, the equilibrium vapour pressure is greatly reduced and ventilation prevents build-up to equilibrium.

One of the biggest mercury poisoning disasters, at Minamata (Japan) in 1953, was the result of effluent of a poly(vinyl chloride) factory using inorganic mercury. The inorganic mercury waste was *converted* in the sludge at the bottom of a bay by anaerobic microbes (oxygen-depleted conditions) to organic mercury, and entered the food chain in its most deadly form. The preferential methylation of mercury by methane-producing anaerobes results in most aquatic animals containing monomethylmercury, the concentration being a rough guide to the species', position in the food chain. Thus shrimps, which are low in the chain, generally contain less than 0.05 mg/kg mercury, and sharks, which are at the top of the chain, contain levels often in excess of 2 mg/kg. Marlin and swordfish can contain around 16 mg/kg.

Selenium

Selenium may well prove to be one of the most important elements since it has been shown to be essential in animal diets, although many toxic effects have been described. No pathological conditions in humans appear to have been identified as resulting from selenium deficiency, but selenium *reduces* the toxicity of methyl mercury, and selenium deficiency may reveal an underlying heavy-metal toxicity.

Selenium and sulfur can replace each other in certain chemical structures and reactions, but sulfur cannot replace selenium in its essential nutritive role. The selenium intake varies widely throughout the world, and blood levels accordingly vary from about 0.8 μg/mL in Venezuela (selenium-rich area) to 0.07 μg/mL in Egypt (selenium-poor area).

Tin

Tin in its inorganic form is generally regarded as non-toxic but the attachment of one or more organic groups to the tin atom produces biological activity against most species, and this is greatest when the number of attached groups is three (R_3SnX). If the chain length of the *n*-alkyl group is steadily increased, the highest toxicity to mammals is attained when R = ethyl. Tributyl-tin compounds, on the other hand, show a high activity against fungi and are used as fungicides in wallpaper pastes. They are less dangerous to mammals. A combination of quaternary ammonium salts ($R_4N^+X^-$) with tributyl-tin oxide gives a water soluble formulation. Organo-tin fungicides have also been incorporated into marine paints; they protect the surface from marine growth but cause sex changes in mussels. Triphenyl-tin compounds are also toxic to fungi. Increasing the chain further reduces biological activity, and the tri-*n*-octyl-tin compounds are of low toxicity.

The largest single application for organo-tin compounds is as stabilisers for poly(vinyl chloride) plastic, and sulfur-containing organo-tins are unsurpassed in their ability to confer heat resistance to the plastic. Several dioctyl-tin compounds are allowed in PVC used for food packaging.

Zinc

Zinc deficiency was once regarded as important only in the practical nutrition of pigs and poultry, and therefore was only of indirect importance to humans. Zinc deficiency has now been demonstrated to be a public health problem in several countries. Inadequate zinc in the diet results in growth failure, in sexual infantilism in teenagers, and in impaired wound healing. Zinc-responsive growth failure has been observed in Egypt and Iran, where a major constituent of the diet is unleavened bread prepared from high extraction wheat flour. In the USA the same phenomenon has been observed in young middle-class children if they consume little meat (less than 30 g/day). The average adult requires 15 mg or more of zinc per day and lactating women require about twice this quantity, but the biological availability of zinc is related to the type of food consumed. For example, zinc availability in cereals and vegetables appears to be lowered by the complexing action of some cereal components, and this appears to be largely responsible for the zinc deficiency in Egypt and Iran. Studies indicate that only about 20% to 40% of the zinc in a mixed Western diet is actually available for absorption.

A good source of zinc is yeast, which may contain about 100 mg/kg, but meat is the most important source in a normal diet. High levels of zinc in food are undesirable. For example, a large number of Sydney schoolchildren were ill when they consumed a cordial that had been stored overnight in a galvanised container and had developed zinc levels of about 500 mg/kg.

Zinc in oysters
The food that contains most zinc is oysters, some of which contain more than 1000 mg/kg.

Oysters and zinc

In 1972 the CSIRO reported on the level of zinc in oysters collected from the river Derwent on which the city of Hobart, capital of Tasmania, is situated. Downstream from the Electrolytic Zinc Co. refinery, these oysters concentrated so much zinc from the water that consumption of as few as six oysters could cause vomiting from zinc poisoning. High cadmium and copper were also found, as well as fish with high levels of mercury, the latter probably from a paper mill further upstream. Matters were not helped when on 5 January 1975 the ore-carrying ship, *Lake Illawarra,* on its way to the zinc refinery, crashed into the bridge across the Derwent and cut the bridge in two. Apart from the economic and social upheaval that this severing of the city caused, thousands of tonnes of heavy metal-containing ore then lay on the bottom of a deep part of the harbour from where it was not salvaged.

In 1988, the metal levels of most oysters outside the estuary were within the legal Tasmanian limits, but these are set to ensure that most oysters are inside the limit. This leads us to ask what really are natural and safe levels?

Analysis of oysters at Port Davey in the remote southwest of the state, away from direct industrial and urban pollution, still showed mean levels of 1000 mg/kg of zinc, but the levels in mussels were mostly below

Amazing! Eighteen oysters before chundering!

Fig. 10.7
Zinc Point Casino (The first legal casino in Australia was Wrest Point in Hobart.)

the old standard of 40 mg/kg. This is not entirely explained on the basis of known bio-accumulation rates (Table 10.5) but may have to do with the high tissue level of calcium in the oyster, which could suppress the availability of zinc for its enzymes. The mussel has most of its calcium in the shell. The interaction of metals in different species raises the whole question of how best to use these organisms as biomonitors of pollution.

A final point raises the interesting question as to how the oyster itself survives such a massive load of toxic metals (it obviously has survival

TABLE 10.5
Average abundance of certain trace elements in the earth's crust and sea water, and enrichment factors in selected sea organisms (all values in mg/kg)

Element	Earth's crust	Ocean water	Oceanic residence time (years)	Enrichment factors		
				Scallop	Oyster	Mussel
Be	2.8	0.000 001	150	—	1 000 000	—
Ag	0.1	0.000 1	2×10^6	2 300	18 700	330
Cd	0.2	0.000 05[a]	5×10^5	2 300 000	320 000	100 000
Cr	100	0.000 6	350	200 000	60 000	320 000
Cu	55	0.003	5×10^4	3 000	14 000	3 000
Mn	950	0.002	1400	55 500	4 000	13 500
Mo	1.5	0.01	5×10^5	90	30	60
Ni	75.0	0.002	1.8×10^4	12 000	4 000	14 000
Pb	12.5	0.000 03	2×10^3	5 300	3 300	4 000
V	135.0	0.002	1×10^4	4 500	1 500	2 500
Zn	70.0	0.005	1.8×10^5	28 000	110 000	9 000
Hg	0.1	0.000 05	4.2×10^4	—	100 000	100 000

[a] 0.000 02 to 0.0008 ppm in some NSW waters.

Source: S.R Taylor, Geochem, Cosmochem Acta **28**, 1273 (1964).

value, in that the more inedible they are, the better their chance of reproducing!). Oysters apparently have amoeba-like cells reminiscent of our own white cells, in which little membrane-covered bags (vesicles) store copper and zinc. These beasties scavenge among all the tissues of the oyster in much the same way as macrophages (also large amoeba-like cells) scavenge around our lungs for dust particles, and ingest them to prevent damage to other cells. (This does not always work; for example, asbestos fibres of critical size kill these cells and they are again released — hence the danger of this commonly used mineral.) The oyster appears to remove some of the metal-loaded scavengers slowly through its gills.

Source: *Oysters and zinc—the Derwent revisited*, R. Beckmann, *Ecos* 50, Summer 1986–87.

References

The new science of strong materials, or why you don't fall through the floor, 2nd edn., J.E. Gordon, Penguin, Harmondsworth, Middlesex, 1982. Brilliant and easy reading.

Orange juice clock, P.B. Kelter, J.D. Carr & T. Johnson, JCE, **73**(12), p. 1123, 1996.

Shape-memory alloys, L.McC. Schetky, *Scientific American*, pp. 68–76, Nov. 1979.

Bibliography

Out of the fiery furnace: The impact of metals on the history of mankind, R. Raymond, MacMillan Australia, 1984. From a television production of the ABC. Brilliant.

The toxic metals, A. Tucker, Earth Island Ltd, 1972. An inspiration.

Trace elements in human and animal nutrition, E. J. Underwood, 4th edn., Academic Press, New York, 1977. Classic.

Metal ions in biological systems, ed. H. Sigel, Vol. 20 *Concepts on metal toxicity,* Marcel Dekker, NY, 1986.

Metals in the environment, Ed. H.A. Waldron, Academic Press, New York, 1980.

Introduction to material science, B. R. Schlenker, Jacaranda Press, Queensland, 1986.

Industry publications from Int. Council on Metals and the Environment (ICME), Ottawa, Canada. E-mail info@icme.com. Balanced presentations:

Metals in the global environment: Facts and misconceptions, I. Thorton, ICME, Ottawa, Canada, 1995.

Persistence, bioaccumulation of metals and metal compounds, Parametrix, Inc, ICME, Ottawa, Canada, 1995.

Chemistry of fibres, fabrics and other yarns

The first wash-n-wear fabric

From hardware to 'software', this chapter deals with many of the same materials, polymers, but in a different form. We start with their natural precursors.

Natural fibres

Wool

Wool comes from the coat or fleece of domesticated sheep. Wool fibres have a natural crimp, which stops them lying close together when made into yarn or cloth. The pockets trap air, which acts as an insulator. Wool fibres can be bent up to 20 000 times before breaking and this flexibility makes wool a great material for carpets.

One of the key properties of wool is its ability to absorb large amounts of moisture — from the air or from sweat. This absorption leads to (hygral) expansion when wet, and reversion to the original dimension when dried. Wool can absorb up to 35% of its own weight without actually feeling wet. Wool and absorbed water have an ancient connection.

DID GIDEON GET FLEECED?

'And Gideon said unto God, "If thou wilt save Israel by mine hand, as thou hast said,

'"Behold, I will put a fleece of wool in the floor; and if the dew be on the fleece only, and it be dry upon all the earth beside, then shall I know that thou wilt save Israel by mine hand, as thou hast said."

'And it was so: for he rose up early in the morrow, and thrust the fleece together, and wringed the dew out of the fleece, a bowl full of water.

'And Gideon said unto God, Let not thine anger be against me, and I will speak but this once: let me prove, I pray thee, but this once with the fleece; let it now be dry only upon the fleece, and upon all the ground let there be dew.

'And God did so that night: for it was dry upon the fleece only, and there was dew on the ground.'

[Judges 6:36–40].

A woollen fleece is about 90% air and so has an extremely low thermal conductivity and heat capacity, which is why wool is so comfortable. This means that as the temperature falls during the night, a fleece left outside overnight will cool more quickly than the ground, and indeed fall below the dew point (at which water condenses from the air) while the surrounding ground is still above the dewpoint. A five kilo fleece would collect about three litres (3 kg) of water.

Calculations on assumed values for a primitive fleece suggest about two million fibres per kilo of fleece, providing a surface area of about 60 square metres per kilo (fibres assumed to be about 16 cm long, 0.05 mm diameter and density 1.32).

Deposits of dew up to 3 mm have been recorded in Israel. Assuming only the average figure of half that (1.5 mm), and a penetration of dew into the outer 1.5 cm of fibres (not matted together), a five kilo fleece would collect about three litres (3 kg) of water. The fibre surfaces are water-repellent and the water droplets would coalesce and run into the centre of the fleece and be protected against the evaporation that would remove the dew from the surrounding ground when the sun rose. (Sand grains are used in deserts to collect water in the same way as this fleece).

The second test is also explicable. A true dew will deposit from the air only when a set of precise conditions are fulfilled simultaneously. There must be a clear sky, high humidity in the surface layers of air, moderate wind (4–10 kph), and a good radiating surface that is thermally insulated. A slight shift from these conditions can prevent true dew formation, but water droplets can still appear on grass and leaves. This 'false dew' can occur either by condensation of water rising from the soil, or by 'guttation', a process whereby drops of water resembling dew are exuded at the tips of grass, or around the edges of leaves. Such processes would provide a dry fleece and 'dew' on the surroundings. A light real dew of about one-tenth of the previous night would be totally absorbed by the internal fibres of the fleece (assuming it had dried out after the previous night).

If Gideon had been a skeptical physical chemist he would have tested these theories by examining the fleece on the first morning after all the dew had disappeared from the surroundings, and on the second morning before the dew had disappeared. Instead, he was a true believer and went on to smite the Midianites . . . Hmm!

Source: C.M. Giles, J. Chem. Ed. **39**:584 (1962).

Wool causes itchiness in many people by mechanical stimulation of the pain receptors just below the skin's outer layer. It is not an allergic reaction. The prickle sensation increases with the diameter and length of the fibre projecting from the fabric surface. These can be lessened by the finishing process, such as by brushing, or raising to increase 'hairiness', or flat setting to press fibres parallel to the fabric surface.

Silk

The most important commercial silk comes from the silkworm, which spins a cocoon that is 78% silk protein and 22% silk glue. The glue is softened in hot water and the silk unravelled. Of the 3000–4000 metres of fibre in a cocoon, only about 900 are usable, the tangled fibres on the outside of the cocoon and inner core being processed separately along with damaged cocoons to be made into *spun* silk. *Wild* or *tussah* silk comes from uncultivated silkworms. *Weighted* silk contains metallic salts, which make the silk cheaper and more drapable, but less serviceable. Silk is weakened by sunlight and perspiration, and is yellowed by strong soap, age and sunlight.

The Chinese emperor Huang-Ti is believed to have been the originator of commercial production in 2600 BC and it was kept secret until some Christian monks smuggled out the silkworm eggs in AD 555. In AD 827, Sicily was captured by the Arabs and became a centre for silk weaving. In the 16th century the industry became established in Lyons, France. Three centuries later, economic disaster was about to strike in the form of a disease of the silkworm. This led to an urgent search for a synthetic, and Chardonnet silk made from cellulose nitrate was the first artificial silk. However, it had the unpleasant habit of bursting into flames near ovens — understandable, as it was after all, almost chemically identical to gun cotton!

Rayon fared better because it is a synthetic cellulose fibre made from wood pulp treated with caustic soda which is then treated with carbon disulfide to give a viscous orange solution, hence giving the name *viscose* to this product. Viscose is spun through spinnerets into sulfuric acid to precipitate the fibres. Because of its silky appearance it was called artificial silk. Like the analogous first stage chemistry in the production of paper pulp, the production of rayon can be an environmentally messy process.

Unlike synthetic filaments, which are produced at the required thickness, silk is not used as a single fibre. During the unwinding of the

Hooke on artificial silk

Already back in 1665, when Robert Hooke was extending springs, he was also thinking chemistry. He was attracted to the wonders of silk thread that was stronger (on a weight basis) than his metal springs. 'And I have often thought that there might be a way found out, to make an artificial glutinous composition, much resembling, if not full as good, nay better, than that Excrement, or whatever other substance it be out of which, the Silk-worm wire-draws his clew.' Possibly to his chagrin, fibres do not obey his law of stress being proportional to strain, except at low extensions, from which the modulus of elasticity is calculated.

TABLE 11.1
Strengths of fibres

Fibres	Modulus of Elasticity GPa	Tensile strength MPa = N/mm^2	% dry elongation	Resistance to abrasion	Resistance to alkali	Resistance to acid
Silk-like						
Natural silk	7–10	350–600	20–25	****	o	o
Nylon 6, 6	6	800	19	******	*****	o
Polyester	15	900	20	*****	***	****
Kevlar	150	2800	4	*****	****	**
Wool-like						
Sheep wool	1–3	150–200	30–40	**	o	o
Acrylic	5	280	35	***	**	****
Cotton-like						
Cotton	6–11	250–800	6–8	**	******	o
Rayon	9	340	25	**	******	o
Hemp	29	850	2	**	******	o

The more * the better
o = not satisfactory

Source: Adapted from Megamolecules, H.-G. Elias, Springer-Verlag, Berlin, 1985.

cocoon, several filaments are joined together to form a weavable fibre. The wet glue resets, to bind them together as the thread dries. This makes silk strong, and one of the best threads. Its tear resistance may reach 600 MPa, compared to wool of 200 MPa. Cotton is as strong as silk but cannot be stretched as much (8% compared to 25%).

Silk is much stiffer than wool in spite of both being proteins made from amino acid chains. Parts of the silk chain are very simple and can pack densely (crystalline) and this gives silk its strength, but it also contains complex chain sections that pack irregularly and these account for its easy extensibility.

The same trick of varying crystallinity has been adopted in making synthetics. In nylon 6, a low crystallinity (> 15%) is used for shopping bags, higher for tyres (75–90%), and highest (> 90%) for fishing lines and rock-climbing ropes, while 20–30% is chosen for women's underwear, and 60–65% for hosiery.

Spider silk

Spider silk also has an interesting history. In 1710, the French physicist Reaumur (of temperature scale fame) calculated that it would take 27 648 female spiders to produce one pound of silk. An 1897 delegation from China presented Queen Victoria with a gown made from spider silk. Silk from both the US black widow spider and British garden spiders was used for the cross-hairs on telescopic gunsights and binoculars.

Spider silk is stronger than cable steel of the same diameter, 50 times more reliable than the human tendon, and harder to snap than rubber. It is composed of mainly glycine and alanine, but the secret is in the microstructure, as revealed partially by X-ray diffraction. The material is excreted as a liquid from the spider and it is not the temperature change or contact with air but the act of pulling itself that creates the strong fibre as the spider drops from the ceiling to the floor. 'It is as if Batman were to abseil down the outside of the Empire State building on a rope spontaneously spun from treacle squirted out of a bagpipe under his arm. (See References.)

Bio-distortion
Most tarantulas are too heavy for draglines. In the film 'Arachnophobia' the producers apparently taped fishing line to the spider! Shame on them for such bio-distortion.

The golden orb weaver spider *Nephila claipes*, found in Panama, can do a 300 metre dragline of silk in one go. It produces seven types of silk from seven glands specialised for the web, each with unique mechanical, biochemical and functional properties, wrapping insects, and attaching the web strands to trees, etc. In a web, there is as much as 30 metres of silk, different segments having different properties. Orb web spiders often rebuild daily because their delicate webs weather rapidly and the sticky patches need replacement.

Whereas Kevlar can stretch 4% before breaking, spider silk can stretch 15%. It is designed to elongate under sudden load; a spider's web catches insects by converting the prey's kinetic energy into heat. Because the strands are not elastic, they don't rebound to throw the prey out again. This is very useful to the military — the silk has a low 'glass transition temperature' (–50 to –60°C) (see Chapter 9) which allows it to operate at low temperatures while retaining good dynamic properties which are needed for sudden dynamic loads such as in a parachute. The military have been trying to clone the spider silk-producing genes into bacteria, which can then be grown in vats.

Cotton

Cotton comes from the protective covering of the seeds of the cotton plant *Gossypium*. Either on its own, or in mixtures, cotton provides most of the world's clothing. Because cotton is stronger when wet, it will withstand regular washing. Cotton quality depends on fibre length, fineness, colour, and lustre. It can be used for hot or cold weather garments from the finest gauze to denim jeans. Most Australian cotton is exported, and cloth, clothing, and textiles are imported.

Virgin harvest
The rumour that cotton is best harvested by sacred virgins at the new moon is only partly wrong. Cotton is the seed hair of a bush and if the capsules are opened too early in full sunshine, light and oxygen can photochemically oxidise the cellulose and break the chains resulting in poorer quality. As for the virgins, well, they are part of many recipes.

Linen

Linen is produced from the fibrous materials in the stem of the flax plant. It is the oldest textile fibre and was used to wrap Egyptian mummies. During the 1700s and 1800s, linen was used by labourers and peasants as smocks for work clothing, as it was strong, washable and durable. Linen does not lint, but does not drape well; it has poor resistance to flex abrasion and may crack or show wear along seams and edges where fibres are bent. While linen absorbs moisture strongly, a natural wax keeps moisture on the surface and it dries quickly, making it a popular choice for tea-towels and, well, linen.

The industrial revolution made cotton much easier to process than linen, and so today, linen is expensive and considered a prestige textile.

The fibre makes the fabric

Silk fibres are continuous (filaments) and can be a few thousand metres in length, while those of wool are shorter (staple) fibres of maximum length about 13 cm. The two halves of the wool fibre are slightly different in their chemistry and adsorb water differently. This accounts for the bending and crimping of wool, a trick that is copied in synthetic fibres. Modern textile technology can provide polyacrylonitrile fibres for wool-like properties, polyamides with high strengths for silk-like fibres, polyester for good shape retention for suits, water-absorbing fibres for sports people, elastic, water-repelling, soft, flame resistant fibres, and so on. See Coloured fabrics in Plate XII.

Fabric nomenclature

When the material in a fabric is given a name, it can refer to a number of different things. It may be the name of the *type* of fibre — nylon or acrylic, for instance. It may be the *brand* name of that particular fibre; for example, Crylon is Courtauld's nylon, and Orlon is Du Pont's acrylic. It may be the brand name of the *yarn* into which the fibres are usually formed before the fabric itself is made; for example, Agilon is the brand name used for a particular stretch-nylon yarn. It may be the brand name of a special *treatment,* for example, Koratron is one for a durable press finish. It may be the brand name of the *fabric,* for example, the Moygashel range can be woven from natural or synthetic fibres. Or — a more recent development — it may be the brand name used for the fabric in a particular type of *garment.* Thus Tricopress is the brand name used only on approved shirts and pyjamas made from Bri-Nylon, ICI's brand name for its nylon.

In France and the USA, fibre brand names can be used only alongside the name of the type of fibre. Different brand versions of the same fibre type vary somewhat, but their similarities are very much greater than their differences. Some characteristics — notably shrinkage — depend very much on the way the fabric is made up.

The type of yarn is also important. Yarn can be bulked to make it feel softer and warmer. Crimplene is the yarn made by bulking Terylene (polyester). It is crimped (coiled or crumpled) to give bulk or stretch, or is given stretch characteristics in other ways, including twisting. Helanca covers a range of crimped or twisted yarns, mainly of polyester or nylon.

Synthetic fibres

Acrylic fabrics
The major contributions of acrylic fabrics are their wool-like qualities and their easy-care properties.

Acetate is made from cellulose, but is considered a synthetic fibre because of the significant alteration in properties. It is closely related to rayon. The fabric dissolves in nail polish, paint remover, and some perfumes (e.g. Acete from Du Pont).

Triacetate is similar to acetate but has lower strength when wet, and has low resistance to abrasion. It is much more heat resistant, and hence shows durable crease and pleat retention, dimensional stability, and resistance to glazing during ironing. It can be machine washed, tumble dried, and ironed at temperatures up to 230°C (e.g. Arnel and Tricel from Celanese).

Acrylic is polyacrylonitrile. Brand names include Acrilan, Creslan, Orlon, and Zefran (all of which have 10% to 14% of other monomers added to improve dyeing), Verel (which is 40% to 50% vinyl chloride), Dynel (more than 50% vinyl chloride), and Belson, Cashmilon, Courtelle, Crylor, Darvan, Dralon, Teklan and Vonnel. Compared to wool, acrylic fabrics are stronger, easier to care for, softer, do not felt, and provide more warmth for less weight. However, they pill worse than wool (i.e. gentle rubbing forms small but unsightly nodules or pills as the surface fibres are raised) and the pills do not break off. Acrylic does not bounce back in the same way as wool, and garments do not retain the snug fit of wool. They are also more flammable. Acrylic is not attacked by common solvents, bleaches, dilute acids or alkalis, and is resistant to weathering.

Spandex (polyurethane) usually occurs as a 5% to 10% contribution by weight in blends with other fabrics such as wool, cotton, linen, nylon or silk, without changing the look and feel of the basic fibre, but it contributes properties of stretch and recovery. For example, in upholstery, the fabric can stretch to conform to the contours of furniture. The fibres are used in lightweight garments and swimsuits. Note that they are vulnerable to bleaching agents such as chlorinated swimming-pool water.

Modacrylic is a fibre composed of less than 85% and more than 35% by weight of acrylonitrile units. Dense, fur-like fabrics are produced by combining fibres with different heat shrinkage capacities, to form a surface pile that resembles the hair and undercoat fibres of natural fur. The use of 15% to 65% vinyl chloride or vinylidene chloride co-monomer reduces the flammability of the fabric.

Polyamide occurs in various types: Nylon 66, Nylon 6, Nylon 11 (Rilsan), Antron, Bri-Nylon, Caprolan, Enkalon, and Nylon 4.

Nylon is one of the strongest of all synthetic fibres in common use. It neither shrinks nor stretches on washing, and is abrasion resistant. It is, however, degraded by ultraviolet light and is useless for curtains. White nylon garments yellow with age. Nylon pills easily. Its use in garments has disappeared and it is now used where its strength is important (e.g. in tyres).

Polyolefins occur as polyethylene with brand names Courlene, Nymplex, and Pylen; and polypropylene with brand names Drylene, Herculon, Marvess, Merkalon, Moplen, Polycreast, Pylene, Reevon, and

Ulstron. Polyethylene and polypropylene are the lightest in weight of all fibres and are difficult to dye. They are generally used in blankets, carpets, upholstery, and also apparel. When mixed with wool they provide better thermal insulation than wool alone, but when used as a filler in quilted pads (and not treated with wash-resistant antioxidant) they can catch fire in a tumble drier. Polypropylene has virtually replaced sisal in cheap ropes (nylon and polyester for high quality). It is also the fibre used in synthetic grass, webbing, and carpet backing.

Polyester is made under the brand names Dacron, Diolen, Fortel, Kodel, Tergal, Terlenka, Terylene, Tetoron, Trevira, and Vycron. It does not shrink or stretch appreciably in normal use; heat-set pleats and creases last well, and water-borne stains may be quickly and simply removed. Like nylon and acrylics, polyester tends to pick up static charge, which attracts dust particles, and it has an affinity for oils, fats and greases. It has a high density (1.38 compared with 1.14 for nylon and 1.18 for acrylics) and so the cost per unit area of cloth is high. The close-packed rigid structure without highly polar groups makes it difficult to dye. However, the stress–strain curve of the polyester fibre can be varied to match that of other fibres with which it is to form a blend, and it was the development of this technology that made possible the polyester–cotton blends now so widely used.

Rayon occurs as two types, both of which are made from cellulose; cuprammonium and viscose. In recent years several 'new' rayons, called polynosic rayon, have been developed with greater wet strength. Rayon is one of the cheapest 'synthetics' and is easily blended.

Chloro-fibres are poly(vinyl chloride) and co-polymers. Brand names are Vinyon, Geon, Krekalon, Movel, Pe-Ce, Rhovyl, Saran, Teviron, Tygan, and Velon.

Vinal, poly(vinyl alcohol), poly(vinyl acetate) and co-polymers, are made under the brand names Kuralon, Mewlon, and Vinylon.

Nylon discoverer
Just three weeks before the basic patent on nylon had been filed, the discoverer of nylon, Wallace Hume Carothers, suffering from one of his increasingly frequent attacks of depression caused by his conviction that he was a scientific failure, drank lemon juice containing potassium cyanide. (Why the lemon juice? A careful chemist to the end.)

From mess to millions

In January–February 1939, a consumer product hit the US market. It is without equal in its impact before or since. Nylon stockings were exhibited at the Golden Gate International Exposition in San Francisco, and were sold first to employees of the inventor company, E.I. Du Pont de Nemours. Some months earlier, nylon had made its first debut in a less-published manner as 'Exton' bristles for Dr West's toothbrushes. On 15 May, 1940, nylon stockings went on sale throughout the USA, and in New York City alone, four million pairs were sold in a matter of hours.

At this time, chemists believed the the tarry messes, often obtained in the bottom of their flasks and called polymers, consisted of the simpler monomers held together by vague physical attractive forces. The German chemist Hermann Staudinger believed true chemical bonding was responsible, but the whole chemical establishment was opposed to this idea, as the largest molecule then known contained no more than about 200 atoms. In his farewell lecture in 1926 in Zurich, Switzerland, there was so much acrimony that Staudinger is reputed to have nailed his lecture notes to the door of the theatre and, emulating Luther, yelled 'Here I stand'. By the time he received his Nobel prize for his studies 27 years later, the chemical establishment had relented. The chemical establishment is an unforgiving lot.

'TWAS THE NIGHT TO MAKE CRYSTALS

'Twas the night to make crystals, and all through the hood,
Compounds were reacting, I'd hoped that they would.
The hood door I'd closed with the greatest of care,
To keep noxious vapors from fouling the air.
The reflux condenser was hooked to the tap,
And the high vacuum pump had a freshly filled trap.
I patiently waited to finish my task,
While boiling chips merrily danced in the flask.
Then from the pump there arose such a clatter
That I sprang from my chair to see what's the matter.
Away to the fume hood! Up with the door!
And half of my product foamed out on the floor.
Then what to my watering eyes should appear,
But a viscous black oil which had once been so clear.
I turned the pump off in a terrible rush,
And the oil that sucked back filled the line up with mush.
The ether boiled out of the flask with a splash,
And hitting the mantle went off with a flash.
My nose turned quite ruddy, my eyebrows went bare,
The blast had singed off nearly half of my hair.
I shut the hood door with a violent wrench,
As acid ate holes in the floor and the bench.
I flushed it with water, and to my dismay,
Found sodium hydride had spilled in the fray.
And then e'er the fire got way out of hand,
I managed to quench it with buckets of sand.
With aqueous base I diluted the crud,
Then shovelled up seven big buckets of mud.
I extracted the slurry again and again,
With ether and then with dichloromethane.
Chromatographic techniques were applied
Several times 'til the product had been purified.
I finally viewed with a satisfied smile,
One half a gram in a shiny new vial.
I mailed the yield report to my boss,
Ninety percent (allowing for loss).

'Good work', said the boss in the answering mail,
'Use same conditions on a preparative* scale.'

Source: Professor John F. Hansen, Department of Chemistry, Illinois State University, USA, American Chemical Society, December 1978.

It was while Carothers was investigating this theory of polymers that he discovered (in 1930) that the polymerisation of chloroprene yielded an interesting polymer called neoprene, which was later to save the war for the US.

Carothers then worked on polyamides. These are polymers made by stringing together pairs of two different monomers. If you think of a polymer as a string of paper clips, then for a polyamide, the paper clips are of two different types where the length of each paper clip can be changed independently. He found the best polyamide for a textile fibre in 1935, which was elastic, stronger than silk, and inert to moisture and solvents. The two monomers had 5 and 10 carbon atoms respectively, so it was called Fibre 510. Du Pont looked for cheaper starting materials and so Fibre 66 came into being.

What's in a name
The commercial name for this product as used for stockings was suggested as norun because it was more resistant than silk to laddering. There were problems, and so norun was spelt backwards to read nuron. However, this was too close to neuron and could be construed as a nerve tonic. An Asian intonation to nülon ran into trade mark problems and a change was made to nilon. English speakers differed in their pronunciation of this, so, to remove ambiguity the name finally became nylon.

Come Pearl Harbor, Japan was in a position to strangle the American war effort by the loss of two vital natural products, rubber and silk (for parachutes). Japan was the major supplier of silk, and rubber came from South-East Asia. Carothers had already invented neoprene or synthetic rubber (which was superior to the natural product in many applications, particularly if exposed to oil). Nylon was to replace silk for parachute cloth, 'flak vests', tow ropes, and tyre cords. All nylon production was commandeered, and nylon stockings were donated by thousands of patriotic women to the war effort.

Today about half the chemists in the US work on the preparation, characterisation, or application of polymers. And as for nylon, the objects in regular use that use this versatile material include hosiery, textiles, paint brushes, fishing lines, runners, nets, tennis racquets, upholstery, sewing thread, tyre cords, ropes, films for boil-in-the-bag, sails and rigging, gears, oil seals, casings, hoses, combs, zip fasteners, hinges, syringes, spectacle frames, and ski bindings.

The versatility of nylon (and the later polyesters) is due to the fact that it is made from two paired monomers. By varying the length of the two monomers, the properties can be changed from a reasonably water-friendly fabric, suitable for body contact, to a tough engineering material, replacing metal in gears.

Mixed fibres

In a blended fabric, two or more fibres are blended before spinning them into yarns. In a combination (or union) fabric, individual yarns composed

---

* Preparative scale — larger than the experimental amounts just used.

of one fibre are combined during weaving with yarns composed entirely of another fibre.

Cotton and rayon, for example, are combined with other fibres to increase absorbency and comfort, decrease static build-up, improve dyeability, and reduce production costs. Acrylics improve softness and warmth without adding weight. Nylon adds strength. Polyester contributes several properties to blends, including wash-and-wear qualities of abrasion resistance, wrinkle resistance, and dimensional stability. Acetate improves drapability and texture.

The quantities of blend needed vary considerably: 15% nylon improves the utility of wool; 10% elastic fibre gives strength properties to clothes; addition of 30% modacrylic reduces the flammability of acrylic carpets.

Non-woven fabrics

Many cultures have traditionally made fabric direct from fibres without first producing thread to weave into cloth, as in papermaking. Hats made from wool felt are a classic example.

Today, non-wovens fabrics can be found in the interlining of suits, parkas, padding in bras and shoulder pads, disposable nappy liners, as well as doona (duvet) filling, upholstered furniture and car carpet. They are found in pot scourers and carpet underfelt. Filters often contain non-woven fabric for gases, such as in vacuum cleaner bags, air-conditioners and on a bigger scale, the smoke filters on coal-burning power stations. They are used for liquids in tea bags, wine pressing, and on a larger scale for civil engineering projects to stabilise earthworks, or as covers in agriculture. Non-woven felt, coated with silicone, covers the upper surface of the space-shuttle orbiter. Non-woven fabrics are the fastest growing area of the textile industry.

Non-woven pads of titanium are used in hip implants to allow the bone to grow between the fibres and through the mesh. Calcium alginate (from seaweed) is used in non-woven wound dressings, which react with the fluid from the wound to form a gel.

Most non-woven fabrics are made by forming fibres into a more or less tangled web, and then sticking them together. This can be done by mixing the fibres with water and depositing them on a moving screen as in traditional papermaking, or by combing the fibres in a carding machine, or else blowing them onto a moving screen. The fabric is then bonded (sometimes several layers) with a liquid adhesive in a bath, print, or spray process. A typical rayon (cellulose) cleaning cloth (e.g. Chux™) will have been made by carding or combing the short fibres and blasting holes with jets of water (which also helps tangle the fibres). This then allows the cloth to pick up dust and be absorbent. It will have been glued in a pattern to leave most of the surface suitably exposed.

Thermoplastic extruded polymer filaments are also used in the same way but can be bonded by melting (rayon would burn first). The dots of the melt points can be seen as a pattern on the surface. A disposable surgical gown is typically made from polypropylene; spun outer layers give strength, sandwiching an inner layer of molten and blown non-woven fibres to act as a barrier to microbes. Non-woven polypropylene is also used as the liner in sanitary napkins and nappies (diapers).

Needle punching is used to entangle a thick fabric used for geotextiles.

Exercise
Make a little light frame out of something suitable like grass straw and 'trawl' repeatedly for spider webs in the bush. Having captured sufficient web in the frame, 'fix' the resultant non-woven fabric by dipping it in diluted milk. When dried, the web fabric should be strong enough to take paint from an artist's brush. **Do not include the spider! repeat: Do not include . . .**

Fabrics and flammability

The large sheets of butcher's paper that are often used for children's paintings can be a fire hazard. Paper can be rendered non-flammable by soaking it in alum, $K_2SO_4Al_2(SO_4)_3 \cdot 2H_2O$. The protection afforded by the alum can be simply demonstrated. Write on paper with alum solution, allow it to dry, and then heat the paper carefully (over a warm stove); the dry invisible alum letters become visible as dark carbonised areas. The alum dehydrates the (polyalcohol) cellulose by acting as a proton donor to form H_2O from the –OH groups of the cellulose. Alum is representative of substances that are inert at normal temperatures, but become active when heated. Sodium silicate (waterglass, see Chemical gardens in Chapter 9) also works.

The serious nature of clothing fires was aptly summed up in a National Bureau of Standards (1973) report: 'If your house catches fire, you will probably escape with your life and your skin. If your pyjamas catch fire, you will probably lose your skin and possibly your life.'

Carpets

The criterion of a hard-wearing carpet is simple—it should stand up to wear without looking shabby. Where you put a carpet and how much use it gets is very important in assessing its life. Underlay also helps a carpet survive. The two most important questions to ask when assessing the quality of a carpet are 'how much pile?' and 'what is it made of?'

The pile *density, height* and *weight* will tell you how much pile there is. Density is the thing to look for first. Fibres wear more if you tread on the sides rather than the ends. The more tightly the fibres are packed, the more likely they are to stay upright and the better the carpet will wear. Provided the pile is equally dense, the higher the pile, the better. The type of fibre used is also important. To test a carpet:

- Bend the carpet sample back on itself. You can compare samples by seeing how much they gape. The more easily you can see the backing through the pile, the less dense the pile is.
- Tug at a few tufts to see if they are firmly anchored.
- Look at the backing. A closer weave lasts longer and the threads should be straight and at right angles. A rigid backing will help to keep the carpet in shape.

Underlay is used for the following reasons:

- It forms a cushion between the carpet and the floor, so the carpet wears evenly. This is particularly important if the floor is uneven.
- It protects the backing from rubbing and rotting.
- It stops any dirt that comes through the floor boards from soiling or abrading the pile.
- It provides extra insulation.
- It makes the carpet feel softer and thicker.

The greater the likely traffic, the heavier the underlay needs to be. Do not use foam with underfloor heating as it can disintegrate. Carpets with heavy foam secondary backing (like many tufted carpets) do not need a separate underlay. This seems to save money but of course you cannot use the underlay again if you want to change the carpet. There is also the danger that if the underlay disintegrates before the carpet, or if it separates from the carpet backing, the carpet will wear unevenly and more quickly.

Excellent standard
Standards Australia has prepared an excellent standard for the classification and terminology of textile floor coverings (AS2454, 1981) in which the various terms are also defined pictorially.

Leather

The Australian leather industry

In 1794, an attempt was made to tan kangaroo skins but it failed. By 1803 the first two successful tanneries for cattle hides were established in Sydney. Demand increased steadily, so that in 1945–46 Australia had 152 tanneries or related establishments employing 5022 people. In that year the industry treated 3.8 million cattle and calf hides, 2.5 million sheep skins, and 0.8 million goat hides. However, the industry soon declined. Synthetics moved into traditional leather areas such as shoe soles, suitcases, and motor vehicle upholstery. Imports increased, and by 1978 employment in the industry had been halved.

In 1978–79 (the last year for which figures are available) 90 to 95% of cattle and sheep hides were exported. France was the major buyer of woolly sheep skins, while cattle hides went to Japan and Poland. Hardly any goat-skin leather is now produced in Australia and imports have increased, particularly from India. Interestingly, India provides a substantial air freight subsidy for its export of finished leather products and does not allow the export of raw hides and skins (except fur skin and skins of 'stray' dogs) to keep the price to its own manufacturers below world parity. However, many more of our exports are now only semi-processed. The total value of raw hide and skin exports was $373 million in 1978–79.

Tan from oak bark
The word 'tan' derives from the crushed bark of an oak tree which was one of the first materials found to be capable of converting skin to leather.

Leather making is an ancient art that originated in the Mediterranean region about 3500 years ago. It is a process used to preserve hides, skins, and furs. It was not until the late 19th century that it was realised that the tanning was brought about by chemical agents called 'tannins' present within the oak bark (see below).

Tannic acid
Tea leaves contain a tannin, tannic acid, as the brown colouring material. In adding milk to tea, the tannins and milk proteins react and this results in neutralisation of the astringent taste of the tea. Tannins appear in beer from the barley, and in wine from grape skins. The tannins react with protein to form a (non-biological) haze, as occurs in home brewed beer and wine. Wines and beers are clarified with a 'fining' agent such as protein (e.g. gelatine) or a colloid such as bentonite. See experiment on tannic acid, Chapter 8.

Some production realities

Only 50–60% of the original hide is transformed into leather. The rest is disposed of. Both the manufacture of leather and its main product (shoes) are labour intensive, so production has moved into the developing world. Because half the selling price of leather is bound up in cost of hides, and these are traded on the commodity market, there is considerable uncertainty in the trade.

Leather is still ahead of its synthetic competitors in its ability to absorb and release more than 20% of its mass in water vapour (e.g perspiration). It is permeable to air and a good heat insulator.

Leather is mainly made from the hides of animals bred for meat (cattle, sheep and goats). Pig skin goes to sausage and gelatine production. Smaller volumes come from kangaroos, crocodiles, snakes, fish and birds.

All the chemicals used in tanning have to penetrate the surface of the hide in order to react with it. Consequently each stage in the production of leather takes some time. All tannins have several properties necessary for their action as tanning agents. They are all water soluble, and they precipitate proteins from solution.

Vegetable tanning Vegetable tannins are mixtures of large polyphenols (molecular masses 500–3000). They are extracted from the bark, roots, wood, leaves and fruit of wattle, mangrove, oak, eucalyptus, pine and willow trees, and the woods of oak, chestnut, poplar and quebracho trees. If a skin is placed in dilute acid, the non-crystalline, polar regions of the collagen will be penetrated by the acid. The peptide linkages within this region will be protonated, resulting in breakage of the hydrogen bonds within this region. The collagen then swells, allowing the tannins to penetrate. The phenol groups within the tannins are protonated at a much lower pH than the peptide linkages, and so the pH is controlled to a point where the peptide linkages are protonated and the phenol groups are not. The tannins thus form strong hydrogen bonds to the collagen, cross-linking the fibrils in many places. The process is complete when about 50% by mass of tannin has been absorbed. The leather is then resistant to bacterial attack, is much stronger, and will repel water. No resistance to heat is achieved.

Chrome tanning Chrome tanning is performed in a basic solution ($pH > 7$). The processes involved are still uncertain but the key step is believed to be the production *in situ* of some insoluble chromium (III) hydrated hydroxides. These either form hydrogen bonds to the collagen fibrils, or react with them in some way. The final result is again cross-linking of the fibrils. The process provides resistance to bacterial attack and shrink resistance to boiling water, but provides no resistance to water penetration. Water resistance is achieved by adding vast quantities of waxes and other substances after tanning is complete.

Synthetic tanning (syntans) The most important syntans are condensation products of formaldehyde with one of the following classes:

naphthalene sulfonic acids, phenols, sulfonated phenols, diaryl sulfones, urea, melamine, and dicyanodiamine. They were introduced around the 1920s and were developed during World War Two. They may act as partial or complete replacements for natural tannins.

The all-leather shoe — the use of leather

The shoe is a very good example of the use of leather. There are four different pieces of leather within a shoe — the sole, the shoe upper, the inner sole and the laces.

The shoe sole must be strong and stiff, with high resistance to abrasion and water. A heavy leather is used, made from vegetable-tanned cowhide. The shoe upper, on the other hand, must be light and flexible, but still strong and water repellent. It is generally made from calfhide that has been chrome tanned. Often a plastic 'patent' paint is applied to the outside surface to improve the finish, change the colour, and add water resistance. An inner sole leather must be light and spongy, being able to absorb perspiration and not be destroyed by it. A combination tanning is often used, where the leather is half-tanned by vegetable tanning, and the process is completed by chrome tanning. Lastly, a shoelace must be strong and very flexible and have resistance to water. Chrome tanning is invariably used to tan either calf or kangaroo skin for this purpose.

Because dry collagen will contract rapidly and irreversibly when heated to above about 70°C, the dry-shrinkage temperature of all leathers is about 70°C. When wet vegetable-tanned leather is heated, the same occurs. However, when wet chrome-tanned leather is heated, shrinkage does not occur until the temperature reaches 95 to 100°C, or the leather dries, in which case shrinkage occurs at 70°C again. This goes some way to explaining what happens when wet shoes are placed near a fire or oven to dry. At 90°C the wet shoes may dry properly but, as soon as they do, they start to shrink because they are above 70°C (greater than the dry-shrinking temperature). Because the thin uppers dry and shrink fastest, the shoe curls upwards. Even at 70°C the sole leather will shrink long before the shoe dries, and the shoe will curl down. In either case the result is not a pretty sight.

EXPERIMENT 11.1
An experiment with dyeing

Equipment

Multifibre fabric or individual fabrics
Set of suitable dyes
250 mL beaker
Bunsen burner on a tripod with a wire gauze.

Introduction

Dyeing involves immersing fabrics in a bath of dye at a suitable temperature for a period of time. The dye molecules are attracted to the fabric, partly as a result of their large size and partly because of the presence of particular molecular groups or charges which interact with the fabric. Hydrophilic (water-loving) fibres such as wool and cotton swell appreciably in water and allow entry of water-soluble dyes. The relatively few polar groups in synthetics make them hydrophobic (water-fearing) and allow little swelling. Polyester is thus difficult to dye with water-soluble dyes.

Method

A multifibre fabric woven from wool, Orlon, Dacron, nylon 66, cotton and acetate is dyed with a mixture of chlorazol scarlet (a direct dye — large molecule), dispersal yellow, (a disperse dye — non-ionic) and edicol blue (an acid dye — negatively charged).

The dye bath contains 0.05 g chlorazol scarlet, 0.03 g dispersal yellow, and 1.25 g edicol blue (40%) in 100 mL of water. Other dyes of a similar nature will also work.

The dye bath is brought to boiling temperature and a small piece of fabric is added. After about 10 to 25 minutes the hot liquid is poured off and the murky-coloured fabric is rinsed several times in a large volume of cold water, squeezing between washes, and then allowed to dry. Plate XIII shows schematically how the dyes interact with different fibres.

Principle

Wet wool has NH_3^+ and COO^- groups. In acid solution the COO^- is neutralised to COOH and the fibre is positively charged, thus attracting mainly the negatively charged sulfonic acid dye. That is why wool can be used to extract (legal) food colours in jelly beans, because these should all be acid dyes.

Cotton is attracted to the direct dyes, which are similar to acid dyes but consist of much larger planar molecules with groups that form strong hydrogen bonds with cellulose. Cotton fibres are cellulose, and the OH groups give a negative charge in water and so salt is often added to reduce repulsion between like charges.

Acetate is a cellulose-based polymer, sufficiently modified to warrant its designation as synthetic. This fabric serves directly as a solvent for the dye. It thus attracts the non-ionic yellow (disperse) dye.

Like wool, nylon is positively charged in water, so it also takes up the blue acid dye. It takes longer for the larger red direct dye to be adsorbed (this is speeded up at a lower pH of 3, given by adding acetic acid). Nylon is less polar than wool and so takes up the yellow non-ionic dye as well, becoming green.

Dacron is a trade name for polyester, while Orlon is a trade name for polyacrylonitrile (sometimes with 10% other monomers present to aid in dyeing.)

References

Megamolecules, H.-G. Elias, Springer-Verlag, Berlin, 202 pp., 1985.
Spider webs and silks, F. Vollrath, *Scientific American*, pp. 52–58, March 1992.
Warding off bullets with a spider's thread, Stephen Fossey, *New Scientist*, 14 Nov. 1992.

Bibliography

Kirk–Othmer encyclopedia of chemical technology, M. Grayson, Ed., 3rd edn., Wiley, New York, 1983. One of the best resources.
The Textile Institute, 10 Blackfriars St, Manchester UK M3 5DR. A major source of publications and information.
Cauchu: the weeping wood: A history of rubber, P. Mason, ABC, Sydney, 1979.
A material world, publication from the Powerhouse Museum, for the exhibition, 9 Nov. 1990–30 June 1991, 72 pp. The exhibit and publication are really useful.
Fibre to fabric, Australian edition, R.E. Griffith, M. D. Potter, B. P. Corbman, 400 pp., McGraw Hill, Sydney, 1970. A classic.
Textile care and the consumer, S. Pyott, 150 pp., Longman, Melbourne, 1985. Consumer friendly.
Textiles in perspective, B. F. Smith and I. Block, 430 pp., Prentice Hall, NJ, 1982. One of very many texts.

Plates III, IV, VI, XII, XXVII are all courtesy of the Sydney Powerhouse Museum.

Plate I
Stick and space-filling models for aspirin. Each model has its uses for visualising the shape of this molecule and the way the atoms in the molecule can move (see p. 15).

Plate II
Model of the structure of zeolite (see p. 50)

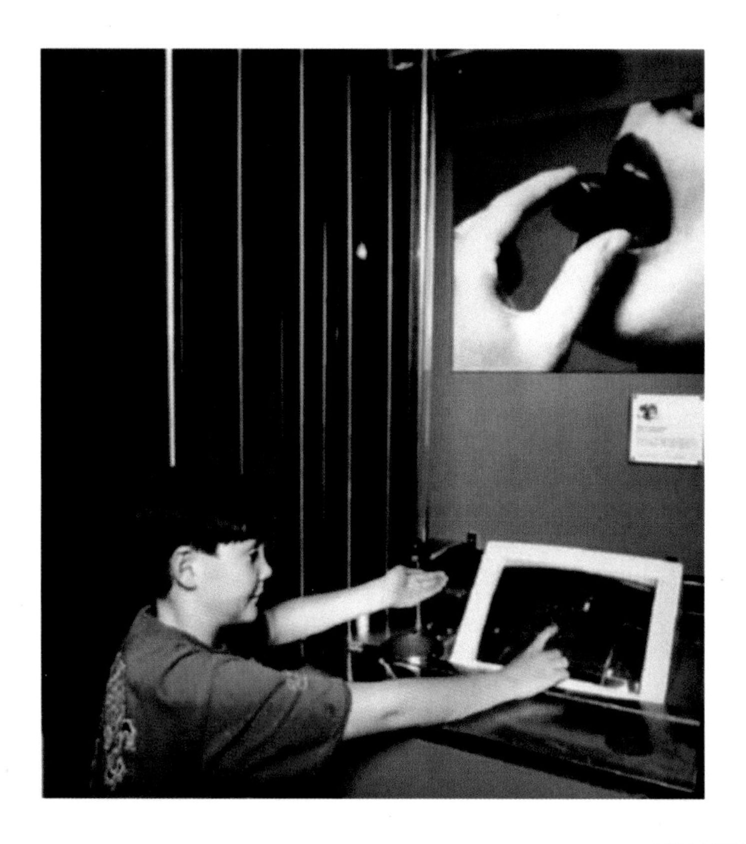

Plate III
Powerhouse Museum display of the processing of cacao beans to produce chocolate (see p. 104)

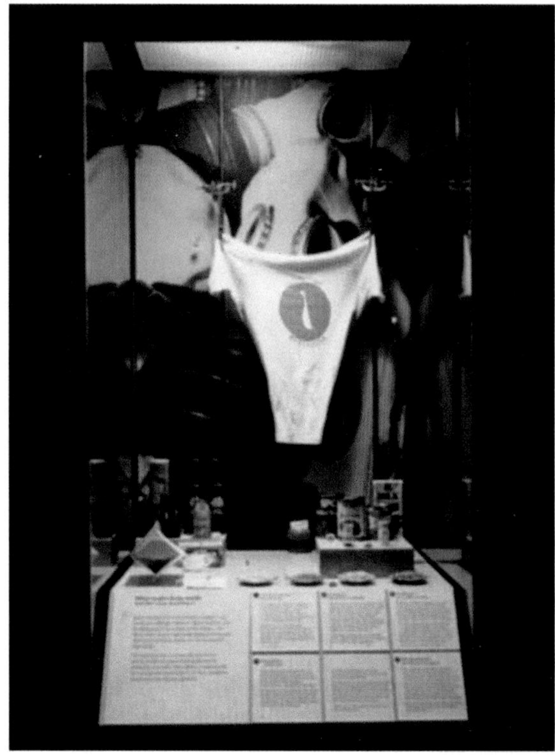

Plate IV
Powerhouse Museum display of the chemistry of perspiration (see p. 119)

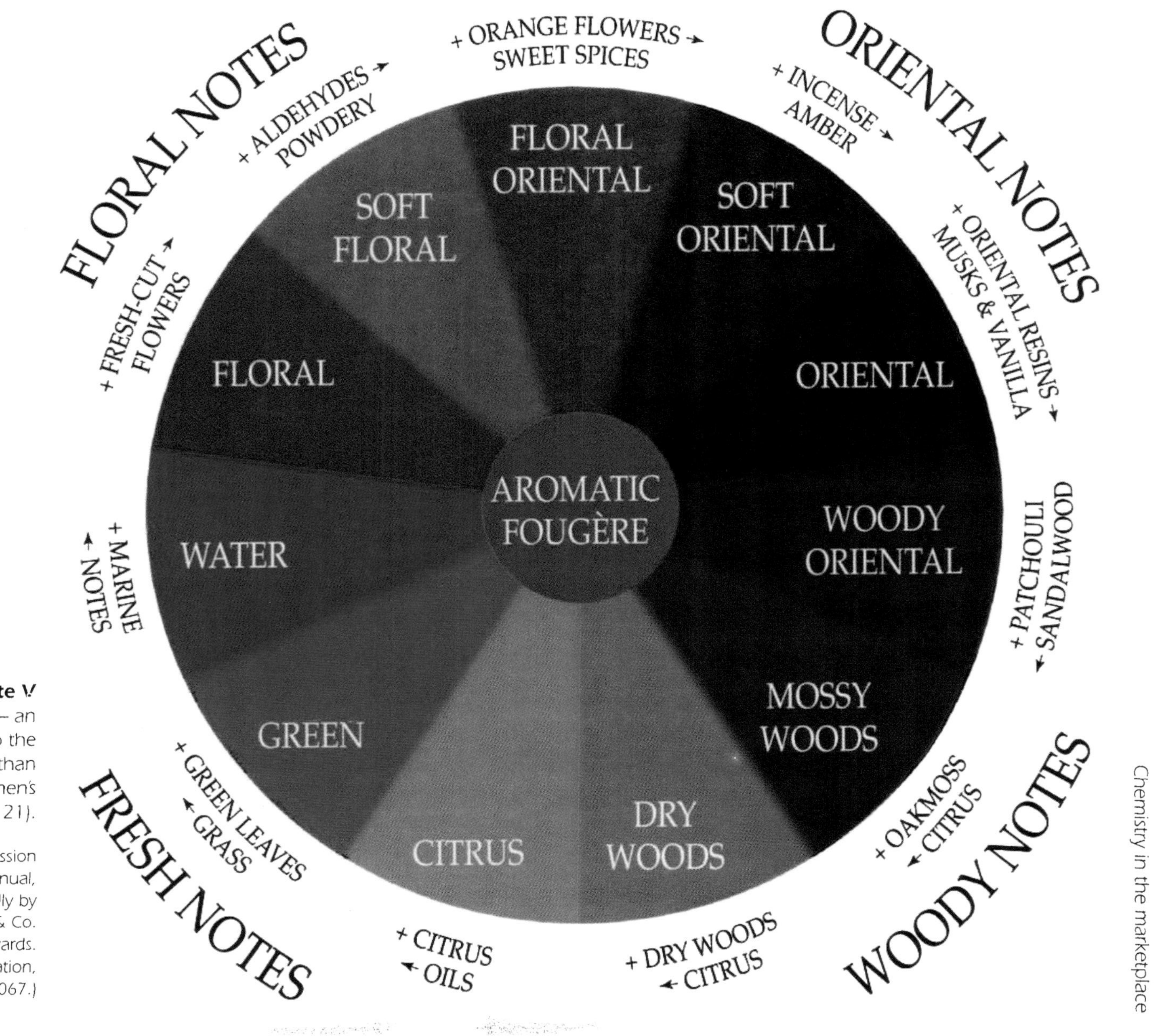

Plate V

The fragrance wheel — an international guide to the classification of more than 1300 women's and men's fragrances (see p. 121).

(Reproduced with permission from The Fragrance Manual, published annually by Michael Edwards & Co. For further information, fax +61 2 9546 8067.)

Plate VI
(left)
As sweet as a rose — appreciating a perfumed bloom (see p. 123)

Plate VII
(below)
The handling of toxic nerve gas weapons following the Gulf War (see p. 182)

Plate VIII
(left)
Disappearing glass tubing. The liquid matches the refractive index of one type of glass but not the other. No light is scattered when there is no refractive index difference (see pp. 216 and 245).

Plate IX
(below)
Unbalanced U-tube. The liquid poured in the two arms of the U-tube settles at different heights. The reason for this is that there is a different liquid on each side! However they are matched for refractive index so the meniscus between them is not visible. The liquids have different densities. The two liquids are revealed by adding iodine crystals, which dissolve in one liquid but not in the other. (See pp. 216 and 310.)

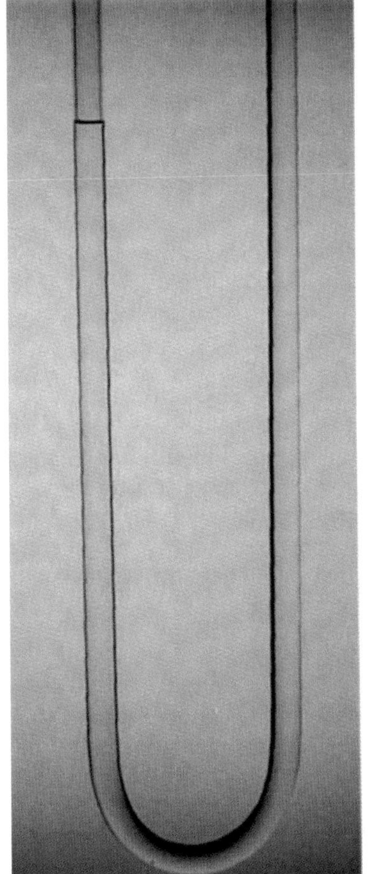

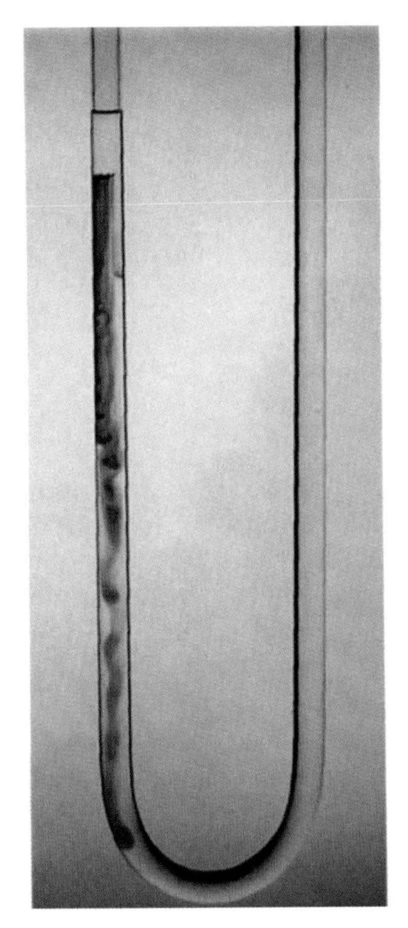

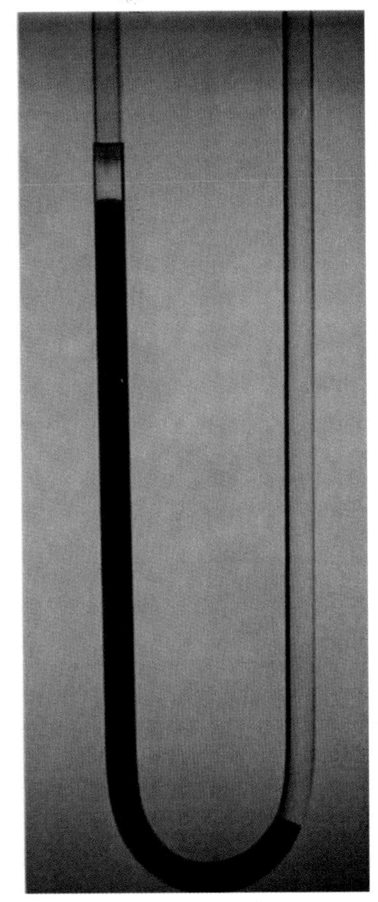

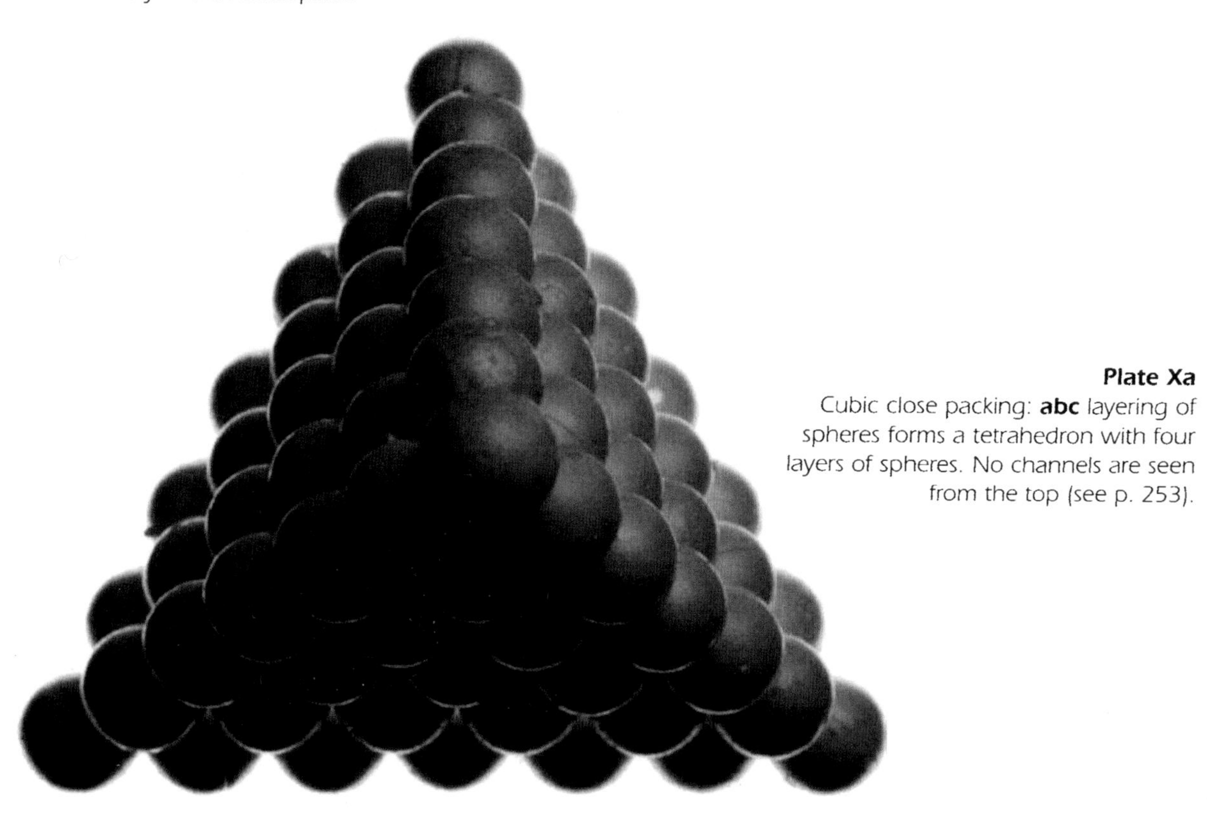

Plate Xa
Cubic close packing: **abc** layering of spheres forms a tetrahedron with four layers of spheres. No channels are seen from the top (see p. 253).

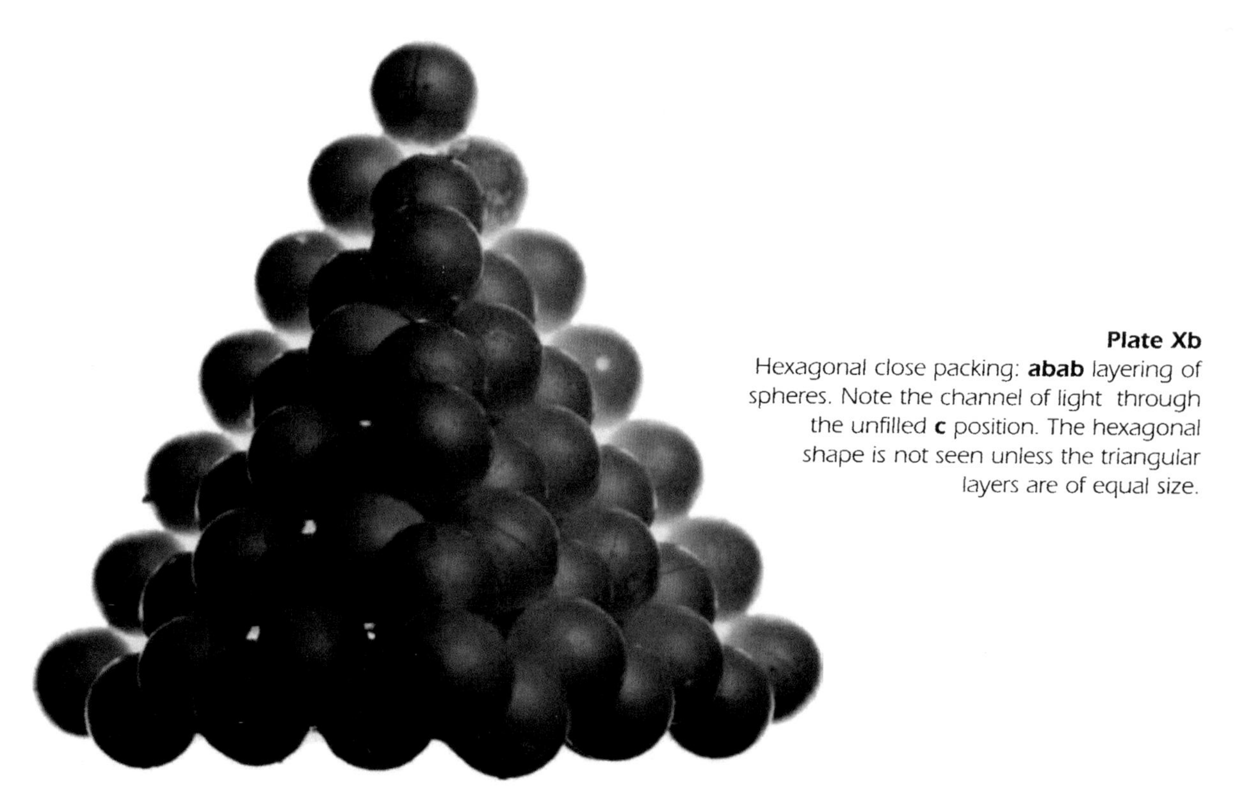

Plate Xb
Hexagonal close packing: **abab** layering of spheres. Note the channel of light through the unfilled **c** position. The hexagonal shape is not seen unless the triangular layers are of equal size.

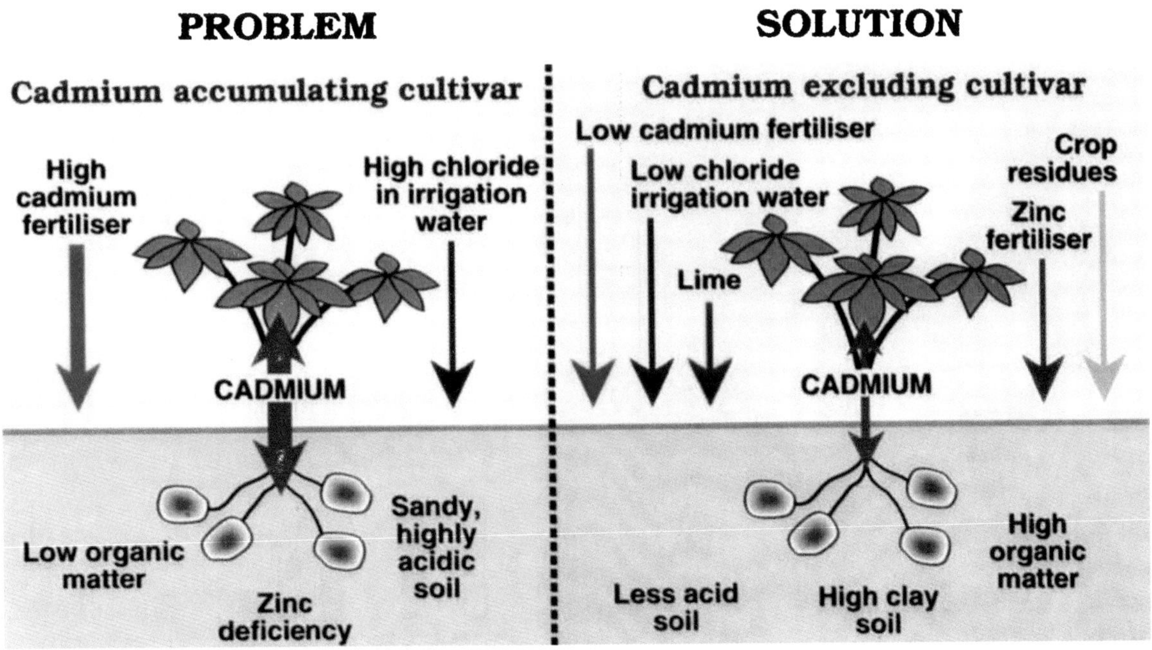

PROBLEM

Cadmium accumulating cultivar

High cadmium fertiliser

High chloride in irrigation water

CADMIUM

Low organic matter

Zinc deficiency

Sandy, highly acidic soil

SOLUTION

Cadmium excluding cultivar

Low cadmium fertiliser

Low chloride irrigation water

Lime

Crop residues

Zinc fertiliser

CADMIUM

Less acid soil

High clay soil

High organic matter

Plate XIa

Managing cadmium effectively means implementing a range of practices as a total system. In paddocks where tuber cadmium concentrations are already high, the impact may be small in the short term, but sound management will be essential to assist control of long-term cadmium levels (see p. 275).

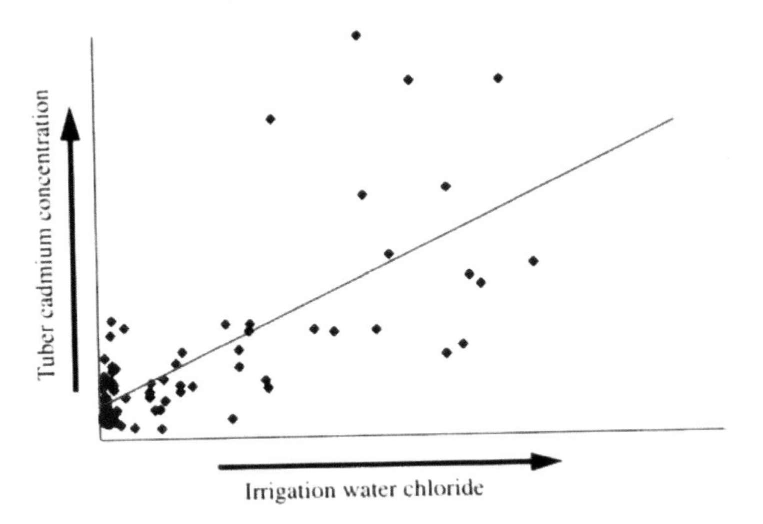

Plate XIb

Tuber cadmium versus chloride concentration in irrigation water at 150 trial sites in southern Australia

Plate XII
(left)
Chemistry gives colour to fabrics (see p. 288).

Plate XIII
(below)
Dyes and how they interact with different fibres (see p. 298)

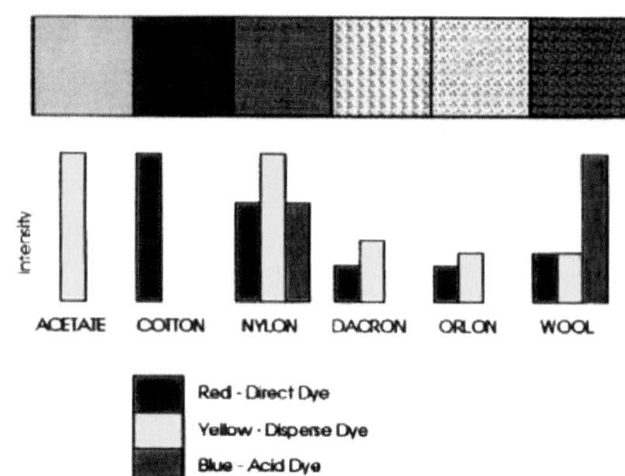

TABLE 1

Chlorazol Scarlet	Dispersol Yellow	Edicol Blue	3 Dye Mix
Red	llow	e	
Pink	llow	e	
Pink		e	

Plate XIV
(right)
Renovation of buildings poses a hazard from lead paints. A household lead test kit is shown (see p. 312).

Plate XV
(below)
Engineering peak performance (see p. 376).

Plate XVI
The Freiburg house — an energy-efficient solar house (see p. 382).

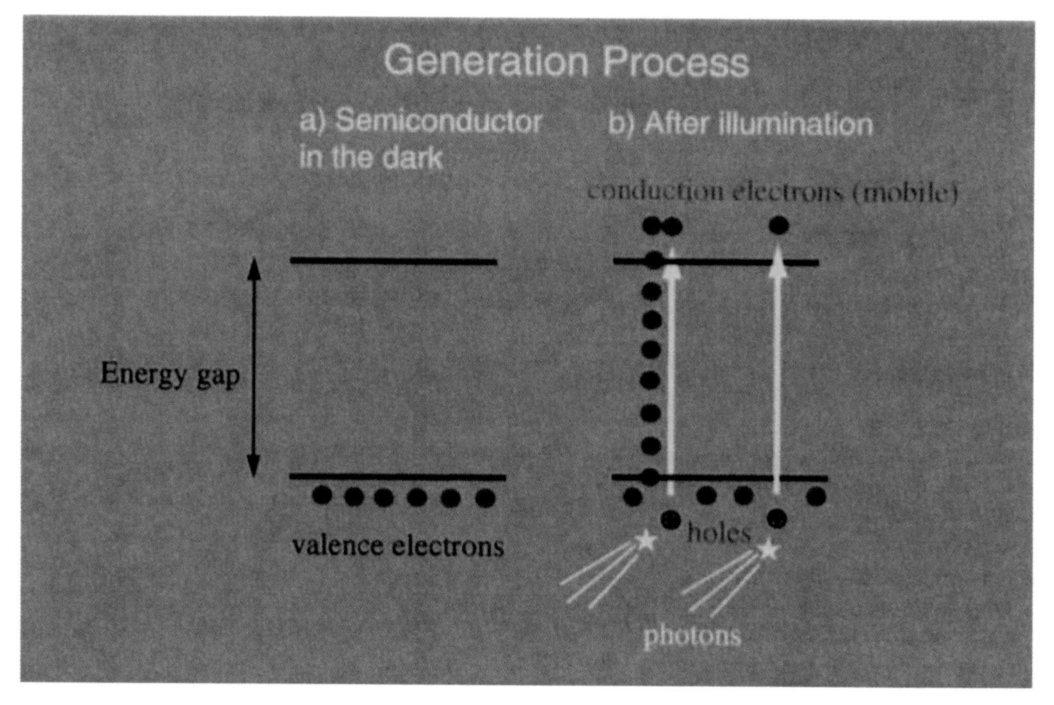

Plate XVII
Photons of visible sunlight 'promote' electrons to a higher level (see p. 388).

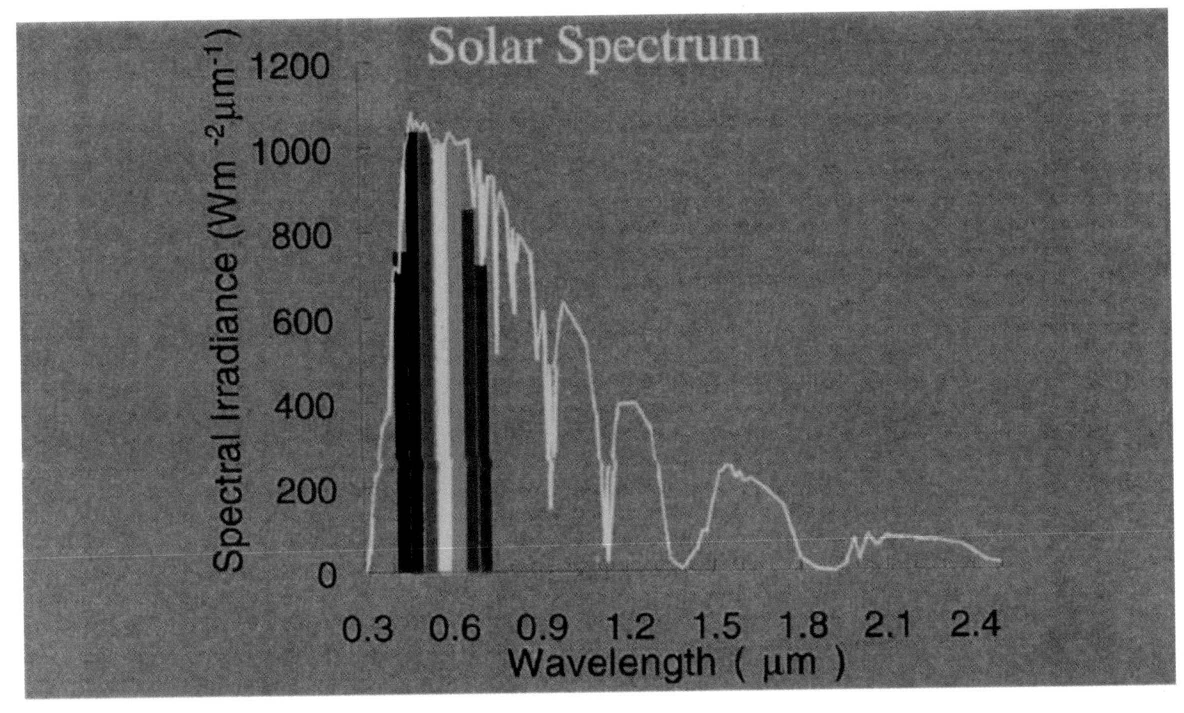

Plate XVIII
The solar spectrum (see p. 388).

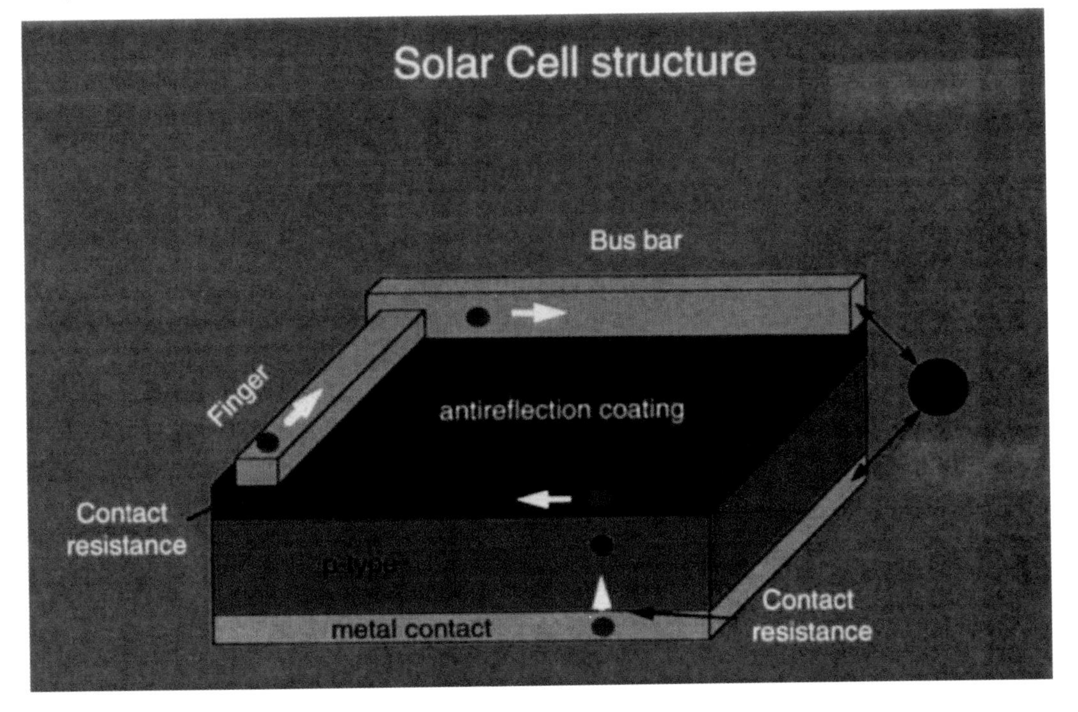

Plate XIX
Solar cell structure. The current depends on the surface area exposed to light, and its intensity (see p. 388).

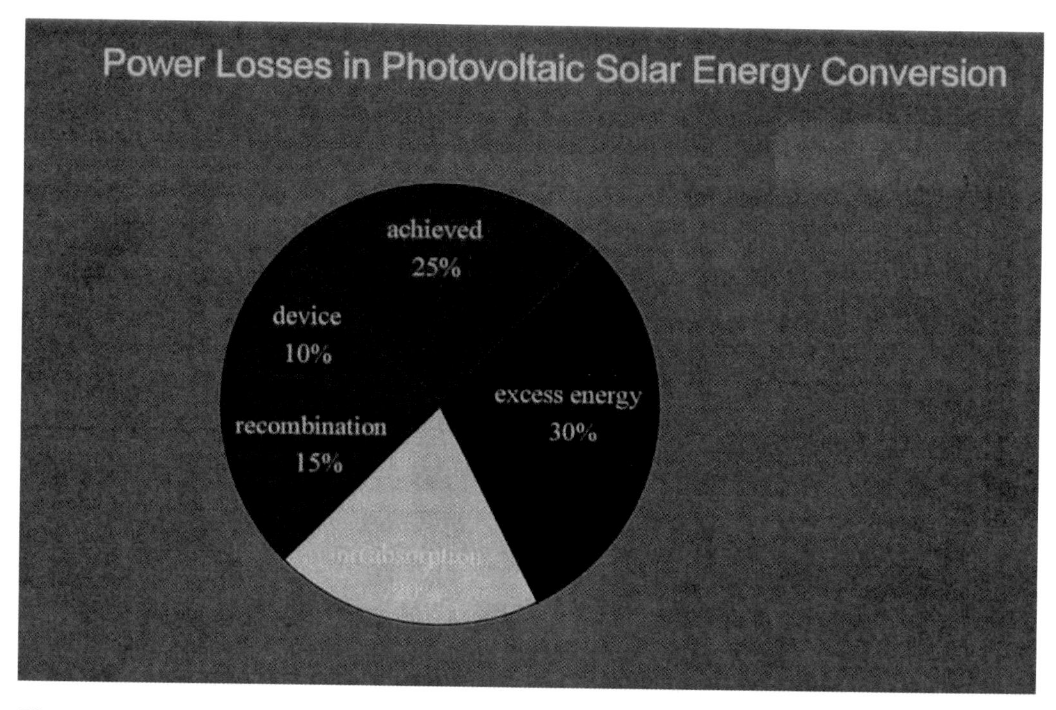

Plate XX Power losses in photovoltaic solar energy conversion (see p. 388).

Plate XXI The car and team from the Ginninderra College (ACT) competing in the Darwin to Adelaide solar car race (see p. 389).

Plate XXII
(right)
The big bang — the combustion chamber designed for a reproducible explosion of hydrogen and oxygen to illustrate flammability limits (see p. 405).

Plate XXIII
(below)
The copper metal suspended over a trace of liquid acetone is kept red hot by the heat released at the surface of the copper from catalysing the oxidation of acetone vapour (see p. 411).

Plate XXIV
Distribution of sources of radiation in the UK, as identified by the National Radiation Protection Board (see p. 474).

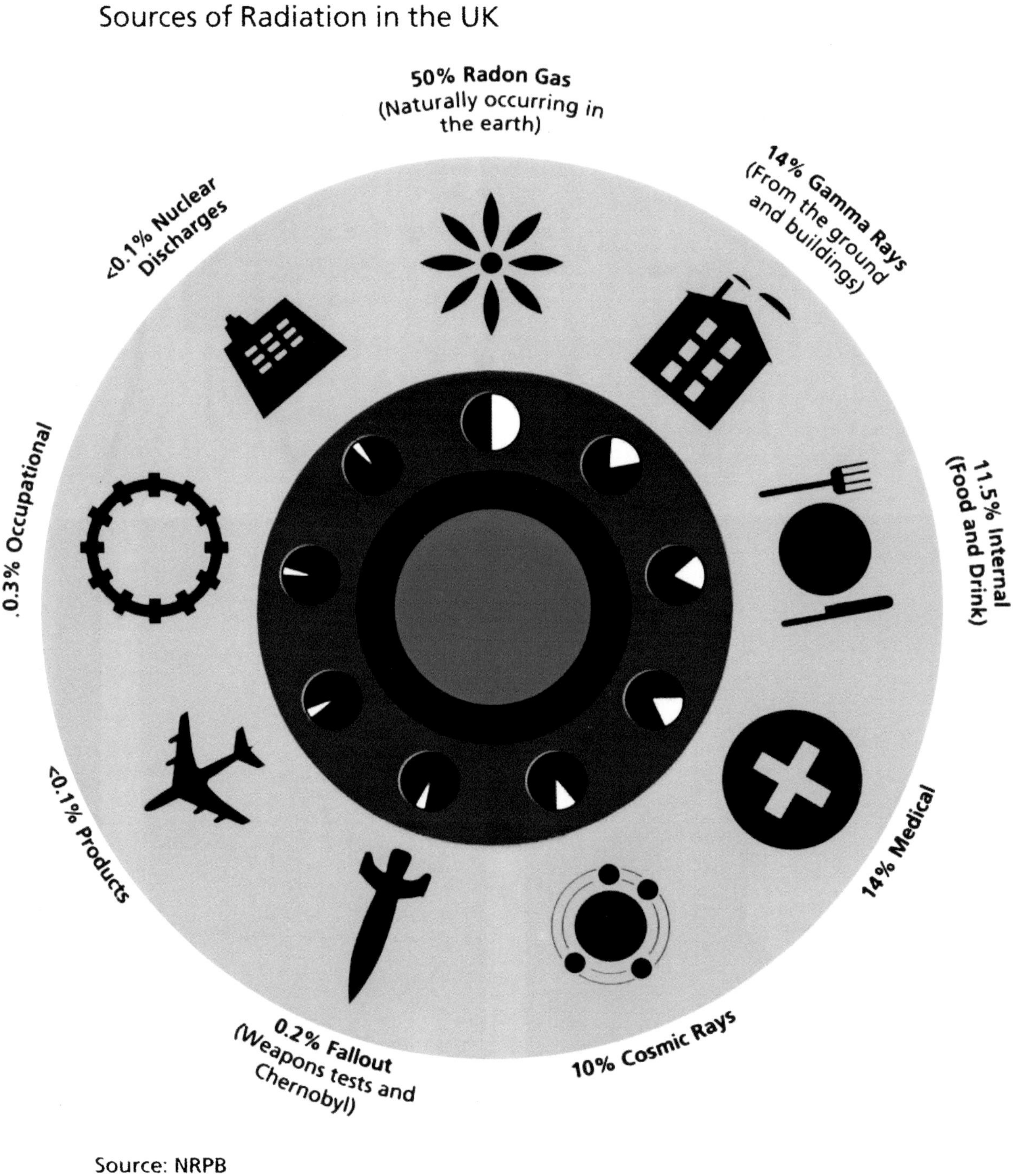

Sources of Radiation in the UK

50% Radon Gas
(Naturally occurring in the earth)

14% Gamma Rays
(From the ground and buildings)

11.5% Internal
(Food and Drink)

14% Medical

10% Cosmic Rays

0.2% Fallout
(Weapons tests and Chernobyl)

<0.1% Products

0.3% Occupational

<0.1% Nuclear Discharges

Source: NRPB

Plate XXV

Structures of (left to right) water, ammonia and methane. The geometry of methane is a tetrahedron. In ammonia one hydrogen is replaced by a lone pair of electrons, in water two hydrogens are replaced. The angles between the remaining atoms change a little as a consequence (see p. 486).

Plate XXVI

The Selinger family; Michael, Adam, Veronique and author Ben, fronting an H_2O structure

Plate XXVII
Safety is paramount in chemical experiments (see p. xi)

Plate XXVIII
Growing a chemical garden — in vitro (see p. 242).

Chemistry of paints, adhesives, enamels and concrete

John realised he should have read the label

From paints to concrete this chapter deals with various protective treatments of a variety of surfaces.

Paints

Before the First World War, practically all pigments, linseed oil, turpentine and varnishes were imported for bulk sale to tradesmen, who mixed their own paint as required. The mid-1920s saw the development of ready-to-use hard gloss paints and nitrocellulose lacquers, and the whole marketing pattern changed — the decorator carried 'colour cards' to arrange colour schemes with his customers, and ordered ready-to-use paints from the manufacturers for immediate delivery. The mass-media advertising of the 1950s and 1960s, together with the development of paints that were easier to apply, and improved painting techniques, have led to the situation today where 75% of homes are painted by their owners.

Until the early 1950s, the *vehicles* used in paints were principally natural polyunsaturated oils such as tung, fish, and linseed oils. Linseed oil was thinned with turpentine and pigmented with white lead (i.e. basic

Paint quality important to artists
Leonardo da Vinci may not have been the greatest painter of his day but his superiority in the technology of paint making ensured that his works survived, while those of many of his contemporaries did not. His technical competence was of course shown in many other spheres. The protection of a film of paint is remarkable, considering it is only about 25 to 50 µm thick. Compare this to the thickness of newsprint, which is about 75 µm.

lead carbonate, $Pb(OH)_2 PbCO_3$) and tinted with one of a small range of colouring agents.

Alkyd resins were introduced into the industry during the 1940s and have since become the basis of nearly all 'oil'-based paints (e.g. house paints, alkyd enamels, undercoats, and primers). Water-based paints, known as latex paints, plastic paints, etc., based on poly(vinyl acetate) (PVA) or acrylic resins, were introduced during the 1950s and became increasingly popular. At first these water-based paints were recommended only for interior use, but later formulations suitable for external use became available as well. The introduction of easily dispersible pigments (universal stainers) during the 1960s meant that the retailer needed to keep only white paint bases, and add small quantities of universal stainers to obtain a wide range of colours.

Function

The basic function of a paint, the protection of a surface from the action of light, water and air, is achieved by the application of a thin, resistant, impervious, flexible film to the surface. The film usually contains pigments to hide and decorate the surface. Thus paints have two basic components:

- The *vehicle* — the liquid part of the paint, which polymerises in some way to provide the bonding and protective film.
- The *pigment* — a solid, suspended in the vehicle, which is opaque, scatters light, and colours the film.

The protective film

There are two ways of establishing the flexible film on the surface to be protected:

- Apply the monomer and allow it to polymerise *in situ*. The old oil-drying paints operated on this principle, but curing times tended to be long.
- Apply a suspension of a high molecular mass polymer in a solvent, and allow the solvent to evaporate. This method is utilised in modern plastic paints. Both oils and alkyds 'cure' (convert from liquid to solid form) by the same mechanism.

Oil-drying paint

The monomer in oil-drying paints is linseed oil or some other unsaturated oil (tung, soya, castor, menhaden, etc.). These oils are mixtures of long-chain, unsaturated fatty acids such as linoleic acid (see also Fig. 4.9).

$$CH_3\text{–}(CH_2)_4\text{–}CH{=}CH\text{–}\mathbf{CH_2}\text{–}CH{=}CH\text{–}(CH_2)_7\text{–}COOH$$

The point of attack by oxygen on exposure of the oil to air is the methylene $-CH_2-$, (shown in bold type) between the double bonds. This leads to the formation of peroxides, and eventually to polymerisation by cross-linking at the active oxygen sites. The cross-linking can continue until there are two links for every double bond in the original oil, which is far beyond the point required for good film properties, and causes paint breakdown.

The old oil-based paints were slow drying (several days between coats), tended to weather, and were not suitable for metals.

The Romans had a word for it

On some tubs of polyunsaturated margarine, one of the ingredients listed will be linoleic acid. This is one of the polyunsaturated fatty acids we eat in the hope of reducing the cholesterol deposited in our arteries (see Chapter 4). Linseed oil and linoleic — can you see the connection? The only edible oil the Romans knew about was olive oil. The Latin for oil, oleum (our use: oleum for oily fuming sulfuric acid), gives us oleic acid, the main component of olive oil. So linoleic is the oil of linseeds. What about the word linoleum? Good old 'lino' was made by pouring linseed oil over hessian (with cork and a few resins etc. thrown in) and allowing it to harden in air (like putty).

Plastic or latex paints

Plastic or latex paints contain highly polymerised resins. Solutions of such polymers are too viscous to be readily applied, so the paints are formulated as emulsions of the resin solution in water. These emulsions are known in the industry as *latexes*, after the natural, milky, emulsified juices of certain plants (e.g. rubber and dandelions). Latex paints dry quickly and lack a persistent odour; they are water soluble (hence spills can be cleaned up easily, and brushes and rollers can be washed in water); the vehicle is non-flammable and non-toxic; and they can be used to paint on damp surfaces. The paints may contain poly(vinyl acetate)(PVA), based on the monomer vinyl acetate, or acrylic polymers based on the monomer methyl methacrylate (Fig. 12.1).

$H_2C{=}CH{-}O{-}C({=}O){-}CH_3$ $\qquad$ $CH_2{=}C(CH_3){-}C({=}O){-}O{-}CH_3$

Fig. 12.1
Vinyl acetate and methyl methacrylate

Interior paints may also contain PVA or styrene–butadiene resins (Fig. 12.2) and they have high proportions of pigment.

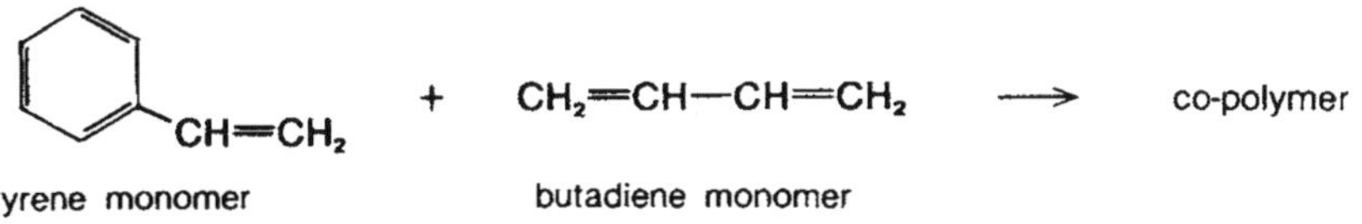

$C_6H_5{-}CH{=}CH_2 + CH_2{=}CH{-}CH{=}CH_2 \longrightarrow$ co-polymer

styrene monomer $\qquad$ butadiene monomer

Fig. 12.2
Styrene and butadiene monomers

To improve washability and adhesion, up to 15% of drying oil or alkyd resin is added. One of the alkyd resins used for this is Glyptal which is formed by the reaction shown in Figure 12.3.

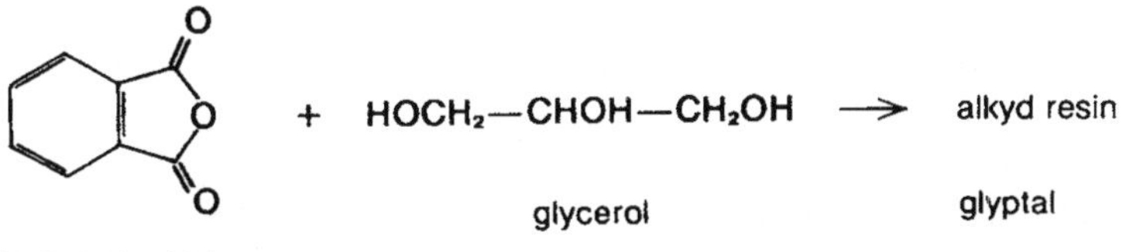

phthalic anhydride $+ HOCH_2{-}CHOH{-}CH_2OH \longrightarrow$ alkyd resin

glycerol $\qquad$ glyptal

Fig. 12.3
Formation of Glyptal

Plastic paints contain a number of additives to stabilise and thicken the emulsion (sodium methacrylates, carboxymethyl cellulose, clays, and gum arabic) to assist in dispersion of the pigment (tetrasodium pyrophosphate, and lecithin), to reduce foaming, and to preserve the paint. In more humid areas, anti-mould agents become essential. Polymerisation of alkyd resins with polyamides gives a thixotropic (non-drip) paint which has a gel-like consistency when standing, but which flows under stress when applied with a brush. The degree of thixotropy must be modified so that the brush marks can level out before the paint re-gels. The opposite effect, an increase in resistance to flow with stress, is called *dilatant*.

Fun with fluid flow

The change in flow properties with applied pressure can be exploited in making fun materials.

Slime: Slime™ is the name given to a material made commercially (Mattel Co.) from guar gum and crossed-linked with borax. Like Silly Putty (made from silicone), it belongs to a class of materials that do not obey the usual laws of viscosity and are called non-Newtonian fluids. This group is called dilatant and becomes stiffer when stressed quickly. This contrasts with materials like quicksand and paint which become less viscous under stress and are called thixotropic (more details below).

Precautions
Keep Slime off furniture and clothes; Slime becomes dirty and may become mouldy after a few days. Discard Slime in a rubbish bin (not down the drain, which it can block). Store it in secure plastic (zip-lock) bags. **Do not allow it to escape!**

A low stress, such as slow pulling, allows Slime to flow and stretch and you may even be able to form a thin film; a high stress such as pulling sharply will cause a break. If Slime is poured from its container and the container is then tipped upward slightly, the gel will self-siphon. Try cutting the pouring stream with scissors. Hitting a small piece of Slime with a light hammer will not cause splashing or spattering and the material will even bounce to a small extent. If pushed through a tube, Slime will emerge with a swell (known as die swell in the plastics extrusion trade).

This delightful material can also be made from PVA.

There are a number of recipes for do-it yourself versions.

Recipe 1

Squeeze a very fine stream of carpenter's glue (wood glue) into a container of saturated borax solution. Stir the mixture, then remove the glue and knead it into a ball. try bouncing it, in contrast to pulling it out slowly.

Addition of flourescein dye gives a ghoulish green which glows in the dark under ultraviolet light.

Source: Chem 13, Nov. 1995.

Recipe 2

PVA bags are used for hospital laundries so the whole bag is thrown in the wash without anyone touching the contents. These bags dissolve in hot water and can be used to make a 'slime'.

Hydrolysed polyvinylalcohol is used for packaging solid pesticides so that they can be thrown into spray tanks to dissolve without operator contact. These bags dissolve in cold water **but must never be used if they once had contents.**

The PVA dissolves in water without the polymer breaking down.

Do-it-yourself Slime

Use a PVA laundry bag; cut into squares of 20 cm x 20 cm weighing about 1.4 g. Make up a 4% solution of borax in water (max. solubility is 63 g/L). Add an optional food colour. Measure out 25 mL of water; (heat to 60°C). Add one strip and stir to dissolve. Add 5 mL of the 4% borax solution. Remove the material from the cup and knead until it becomes firm and loses its stickiness.

Source: David Katz, notes 'Chemistry of toys', 6th edn.

How does it work?

Borax is sodium tetraborate, $Na_2B_4O_7 . 10H_2O$, which has a complex structure in the solid and in solution. Borax hydrolyses in water to form boric acid, $B(OH)_3$, a weak monobasic acid, which in turn forms an equilibrium with the borate ion, $B(OH)_4^-$.

$$B(OH)_3 + 2H_2O \rightleftharpoons B(OH)_4^- + H_3O^+$$

The borate ion is tetrahedral which allows it to form a three-dimensional cross-linked network with the polymer chains. Water molecules are trapped in the network that is 95% water. The structure is hydrogen bonded, breaks down on warming, and reforms on cooling.

The boric acid–borate buffer equilibrium has a pK_a of 9.2 and the aqueous solution has a pH of 9.

Dilute acid shifts the equilibrium to the left and destroys the gel. This is very useful for cleaning up glassware crusted up with dried Slime! Addition of dilute alkali restores the gel, but excess makes a permanent jelly that will no longer flow.

Source: School Science Review, 73, (202), Sept. 1991; *Journal of Chemical Education*, **63** (1), 57–60, 1986.

Recipe 3

You can demonstrate the same dilatant effect in the kitchen with a mixture of 2:1 (by volume) of cornflour and water. (Corn (maize) flour is almost pure starch, while wheat flour contains about 10% protein, so cornflour congeals better and makes better sauces.) Starch and cellulose are both polymers of glucose but with different geometry by which the units link together. This slurry has a low viscosity to slow stirring with a spoon, but will instantly thicken to an almost solid paste if stirred quickly.

OTHER FUN MATERIALS

There are a number of other toys based on this type of behaviour. For example, A Bad Case of Worms (Mattel Toy Co.) is made from a Shell isoprene polymer that is a thermoplastic elastomer, plasticised and tackified. When the worms are thrown against a smooth, clear surface such as a wall they stick, but after a while the worms start to 'crawl' down the wall. The rate of progress depends on the cleanness of worm and wall. The worm is soap-washable.

'A magic octopus' works in the same way and is probably a styrene butadiene co-polymer. Excess plasticiser can leave an 'oily' residue that is difficult to remove.

To make cornflour dilatant
Mix in a bowl, cup or glass three measures of cornflour — finely ground maize (not wheat) — with two measures of water. Then add more water, up to about one more measure, until the mixture becomes rich and creamy. (Try pulling your finger quickly out of a wine glass containing this mixture — it's weird!) Place it in a bowl. Put your hand in it slowly and withdraw slowly — no worries. Withdraw fast and the bowl comes as well.

Shrinky Dinky
These are plastic films on which you draw pictures. They are then placed in an oven for four minutes at about 165°C. The film shrinks to about one-third all round, thereby becoming nine times thicker. The films are made from a polystyrene film that has been extruded under stress to become oriented in both planes (directions).

These are some examples from David A. Katz of Cabrini College, PA, USA.

Notes on viscosity

To understand a little more about viscosity, consider the following experiment. Take a tall glass with a small amount of honey in the bottom. If the glass is tipped on its side with the open end facing slightly downward, the honey will begin to flow. The molecules closest to the wall will be attracted to the surface and will be held back from flowing. These molecules will in turn slow down the molecules in the next layer trying to flow past. The process of each layer slowing down the next layer extends throughout the fluid. The fluid moves in layers like a pack of cards being pushed along a table by a flat hand on top of the pack. The card in contact with the table hardly moves at all (see Fig. 12.4, below).

If the forces between the molecules are weak, this drag does not extend very far from the surface, but if the forces are strong, the drag is transmitted further into the liquid. Increasing temperature reduces the attraction between molecules and hence reduces the viscosity of liquids.

For non-Newtonian fluids the viscosity also depends on the rate of shear, as we saw, sometimes increasing with shear (dilatant), sometimes decreasing (thixotropic), sometimes behaving in a complex combination of both, depending on the way the force is applied and for how long. The mechanisms responsible for the different behaviours can be quite diverse.

Have you ever experienced the 'shear' pleasure of running along the wet sand near the water's edge at a beach? The sand feels as solid as a footpath and your feet hardly mark it at all. If you stand for even a short time you will sink in a little and leave an obvious, temporarily dry, footprint. Running on dry sand is more difficult because the sand collapses rapidly under your feet. When water is added to dry sand it acts as a lubricant and the sand grains pack together much more efficiently. However, the sudden shear force of the feet of a runner on wet sand

Fig. 12.4
Pushing a pack of cards

forces the grains past each other, forming cavities which do not have time to be filled with lubricating water. The resultant friction hinders movement of the grains. If, on the other hand, you stand still, the water can move in, lubricate, and hence allow the sand beneath your foot to reshape. (See also Chapter 6.) Although this behaviour is similar to that of Slime, the mechanism here relates to the behaviour of the macroscopic solid particles, rather than polymer molecules in solution.

The natural shape of polymer molecules is like a random coil. This shape results from the effect of entropy. *Random* has a special meaning here. It means that there are many more possibilities for throwing down a chain that looks tangled in roughly the same way than putting it down in a stretched out manner. So being coiled in a variety of ways is the most probable shape a polymer molecule will take (see Appendix XIII). For example, the molecules in rubber are randomly coiled. When rubber is stretched, the molecules are unwound. The elasticity of rubber comes from the tendency of the molecules to return to the random coil shape — the condition of maximum entropy, see Fig. 12.5 below. (See also Stretching rubber in Chapter 11.)

In Slime, a rubber-like polymer molecule is dissolved in a solvent. Shear forces can unwind the coils and the molecules then have a greater chance of entwining, and this increases the viscosity of the solution temporarily. With the shear force removed, the molecules move back to random coil, and this moving back is manifest macroscopically as the elastic behaviour seen in the experiments on Slime. A stream of Slime pouring out of a container will be stretched under its own weight, and when you cut it, the upper cut portion *re-coils* (literally) elastically. You cannot cut a stream of honey.

Thixotropic fluids are easier to understand. They all have some loose secondary structure which binds them sufficiently to give the initial high viscosity. You don't have to get exotic paint formulations to see the effect; just take out your tomato sauce bottle from the pantry. If you tap or shake the bottle, the shear forces break the secondary structure and lower the viscosity to the point where the contents may flow so well that they spread themselves around the kitchen. It is the weak hydrogen bonds in the starch gel that provide the secondary binding in tomato sauce.

Tomato Sauce
You bang and bump
and shake the bottle:
first nothing comes,
and then a lot'll.

Source: Anon — deathless verse.

Shear thinning is used in many food and cosmetic products. When making a sandwich by spreading a solid onto bread, the viscosity at the

Fig. 12.5
Randomly coiled and stretched rubber molecules

Stirring soup
Next time you are stirring a bowl of tomato soup, watch what happens when you suddenly stop stirring. It shows Slime-like elasticity, and may recoil with a slow reverse spin.

shear rate of about 100 mm per second (appropriate for a hurried early morning lunch maker) must not be much higher than 1–3 Pa s if the bread is not to tear. Some idea of the magnitude of viscosities (in Pa s) is given by the following: marshmallow creme 1000, honey 10, motor oil 0.1, water 0.001, and air 0.000 02. Peanut paste (butter) is much harder to spread than margarine because its viscosity is much higher at the shear rate of spreading. Other products with similar shear-thinning properties to margarine are mayonnaise and mustard.

Ink held in a ballpoint pen is much more viscous than honey, but on shearing with the ball in the nib it flows onto the page, setting when the pen has passed by. Higher temperatures lower the viscosity, and ballpoints can leak under these conditions. More seriously, higher temperatures make engine oils less viscous and the friction protection they offer falls dramatically. Polymers are added to increase the viscosity and so compensate for this change with temperature (see — Lubricating oils in Chapter 14). The children's fun material 'Silly Putty' is a silicone polymer called a polyborosiloxane and has more complex time-dependent viscosity properties.

A silicone consists of chains of Si–O–Si–O–Si–O– with methyl groups attached to the silicon. However, in this polymer, about 5% of the silicon atoms are replaced by boron and these boron atoms can cross-link weakly with oxygen atoms in other chains — like Slime.

Toothpaste viscosity
Imagine toothpaste with the properties of honey! It would be too viscous to squeeze from the tube, but not viscous enough to stay on the brush.

Solid paints

Paints consisting of 100% solids (such as epoxies and polyesters) are sprayed onto metal surfaces and held electrostatically. They are then heated to cause the material to flow and coalesce into a continuous film, at the same time causing chemical cross-linking in three dimensions. These products find use in coating of aluminium frames and heavy-duty machinery.

Chalking

Under the influence of sunlight, oxygen and water, paints tend to degrade by breakdown of the long chain of the polymeric binder into shorter chains. This process continues until pigments at the surface of the paint film become liberated as 'chalky' residue (Fig. 12.6). This erosion

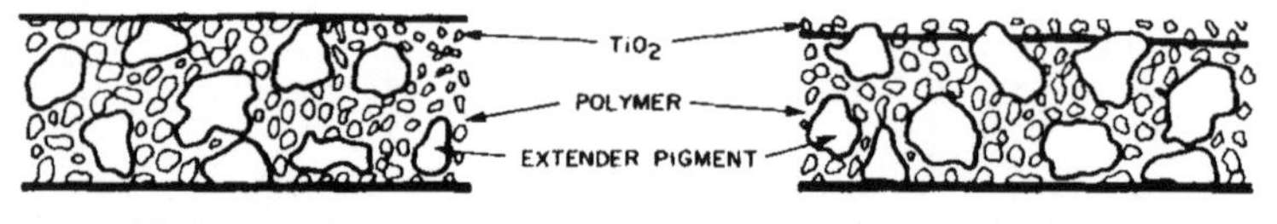

Fig. 12.6
Schematic representation of a dried paint film. The paint film degrades by exposure to sunlight, water and air. The pigment particles are liberated at the surface, giving the appearance of a chalky residue.

From F. L. Floyd, 'Emulsion polymers in coatings', Chem. Tech., 484, August 1973, with permission of the American Chemical Society)

continues until the entire paint film has weathered away, a process requiring anything from one to 20 years. Although this erosion is advantageous in white paints as it provides a self-cleaning attribute, it is most undesirable in coloured paints, because it gives the impression of fading. (Coloured pigments are lost from the chalky layer more rapidly than white ones, resulting in a gradual fading.) Resistance to chalking can be improved by changing the monomer used to produce the polymer.

Flexibility

Polymeric substances are glass-like in their mechanical behaviour as a function of temperature. Like glass (silica), an inorganic polymer with SiO_2 repeat units, these polymers do not 'melt' in the sense that crystalline materials do. Rather, they exhibit a characteristic 'glass transition' region (see Chapter 9) in which they pass from a hard, brittle substance to a soft, rubbery, elastic substance. Hardness is at a maximum below the glass region and flexibility at its greatest above. These two properties must be balanced.

Hiding power

The hiding power of a paint is associated with the difference in refractive index of pigment and polymer. The reflectivity F is an approximation for the hiding power:

$$F = \frac{(n_1 - n_2)^2}{(n_1 + n_2)^2}$$

where n_1 is the index of refraction of pigment and n_2 is the index of refraction of polymer. Hiding power is therefore maximised by minimising the refraction index of the polymer. With the design of translucent toothpastes (Chapter 5) and translucent urea–formaldehyde filled resins (Chapter 9, p. 218), the refractive index difference is minimised so as to reduce light scattering.

DISAPPEARING TRICKS

The index of refraction of Corning 7740 (Pyrex™) glass is 1.474, while that of pure glycerol is 1.4746 (at 20°C). When two pieces of glass tubing, one soda and the other Pyrex, are placed in pure glycerol, the soda tube is seen shifted by parallax, while the Pyrex 'disappears' completely from view (see Plate VIII). Glass blowers use this effect to distinguish glass types.

Another experiment is to pour two immiscible liquids of different densities but the same refractive index, on either side of a U-tube. The system stabilises, with the liquids on each side at a different height. By matching the refractive indices, the interface cannot be seen. Use glycerol (to which ≈ 6–7% water has been added) in one arm, and paraffin oil (with about

Grease spots
A grease spot (butter, for example) on paper becomes transparent because the refractive index of the grease matches that of the paper fibres. What could be simpler?

8% decane added to match the refractive index) in the other arm. (You may need to measure the refractive index of the glycerol mixture and adjust the amount of water added to obtain an exact match.) The interface is completely invisible. A close inspection of the interface shows that it is curved, and seems to slope away from you, no matter from what angle you view it.

Plate IX shows such an experiment, but with CCl_4 and some iodine crystals instead of hydrocarbon for visibility. When the iodine was completely dissolved, the author was surprised to see a short column of virtually uncoloured liquid on top of the carbon tetrachloride. What had happened was that during the demonstration the tube had been shaken backwards and forwards in response to a suggestion from the audience to prove that there was no careful layering. Because of the invisible interface, the author had not noticed some glycerol moving across to the other arm. Actually it is a good idea that it is there, because it prevents the evaporation of toxic carbon tetrachloride vapours. The demonstration is done more safely with hydrocarbons.

Pigments

The traditional white pigments added to paints were white lead (basic lead carbonate, $Pb(OH)_2PbCO_3$) and lithopone (30% ZnS + 70% $BaSO_4$). Today, the major white pigment is titanium dioxide (TiO_2) which exists in two forms: *rutile* and *anatase.* Rutile has the higher refractive index (2.73) and hence the higher obliterating power but reflects a light that is slightly yellow. Besides its other advantages, TiO_2 also promotes the self-cleaning of the paint by chalking. Paints sometimes contain mixtures of the two forms in varying proportions to achieve a compromise of their best features. Examples of coloured pigments are given in Table 12.1.

Particle size

Levelling (disappearance of brush marks after paint application) improves with increasing particle size. If the particles are large, they are encapsulated by the polymer, rather than being stuck together. This improves levelling because the flow properties are improved. Gloss, on the other hand, declines with increasing particle size.

Solvents

When replacing one solvent with another which has less toxic or environmental problems, their important physical parameters have to be matched. The transition has been from organic solvent to water emulsion for paints, in the printing industry, and in the electronics industry.

For the last decade, about 80–90% of all decorative paints have been water based. The enamel paints for the trim and wet areas such as

TABLE 12.1
Coloured pigments

Colour	Inorganic	Organic
Black	carbon black, copper chromate/MnO_2	
Yellow	Pb, Zn, Ba chromates, cadmium sulfide, iron oxides (ferrite yellow)	azo (arylamide) yellows, nickel azo yellow
Blue/violet	ultramarine, Prussian blue, cobalt blue	phthalocyanin blue, indanthrone blue, carbazole violet
Green	chromium oxide, mixtures of yellow and blue	phthalocyanin green
Red	red iron oxides, cadmium selenide, red lead, chrome red	azo pigment dyestuffs and toners, (e.g. toluidine red — fades badly in tints), quinacridones, perylenes

Source: Based on Paints and plastic coatings, W.M. Morgans, Education in Chemistry **10**(1): 12, 1973.

Paint procedures
Many consumer products are made from prepainted steel strip, called 'coil coating' because the strip is transported as a giant, painted coil. When refrigerators, freezers and washing machines are stamped and pressed out of this steel, there is a tremendous stress applied to the paint. Because different parts of each item come from different parts of the coil, there has to be a high level of consistency in paint colour. In the older process of painting after manufacture, it was easy to keep each item looking uniform and it did not matter that no two looked exactly the same!

bathrooms need to be hard and mould resistant and are often still organic-solvent based, but this is no longer necessary. Water-based paint generally still has small amounts of other components so that the tinter will dissolve; the smell is often from ammonia added as a preservative. However, there are now some virtually solvent-free products on the market for many applications.

Industrial paints for cars and refrigerators have traditionally been organic-solvent based, but this is also changing to water base (e.g. Toyota — Altona, Victoria). This is not only good for the workers, but also means that water-based paint can be used for a touch-up job if you later have an encounter of the too-close kind.

Even better is to apply a paint without solvent at all. A coloured powder is used and an electrostatic generator gives it, and the metal you are going to coat, an opposite charge (like rubbing a comb and picking up paper). The resin sticks to the metal and is put through an oven to melt it evenly, and it then solidifies. Powder coating technology is now more widely used.

Some industrial paints have remained solvent based. These include the polyurethanes and epoxy resins, and situations where fast drying is required. The industrial colour-bond roofing metal is one example. The application of epoxy resin followed by polyurethane coatings to floors is another. Because the epoxy resins tend to be sticky and stiff, the applicators often use more solvent xylene, sometimes 20% instead of the recommended 5%. The xylene penetrates and is sealed in by the thinned resin. To make it worse, the required 24 hours between polyurethane coats to allow solvent vapour escape is often shortened to as little as four

Floor hazard
Misuse of solvent-based coatings for floors can lead to serious health problems for occupants.

hours. As well as solvent xylene, isobutyraldehyde is formed from the reaction of the cross-linking agent with moisture. The isocyanates themselves are potent sensitising agents.

Paint removers must act to remove the paint film, or to swell and soften it so that it can be easily removed by scraping, or flushing with water. The liver toxin, dichloromethane (methylene chloride) has been in large measure replaced by polyethylene glycol esters and ethers. These have excellent solvent properties for organic substances, and many of them are miscible with water. They have low volatility and relatively high viscosity. While the low molecular weight members are toxic, for example, 2-methoxy ethanol (methyl cellusolve), the high molecular mass ones are not.

Toxicity

The toxicity of a paint depends on two factors:

- *The chemical composition.* For example, compounds of metals such as lead, cadmium, antimony and barium are generally poisonous.
- *Solubility.* A substance that is *insoluble* in body fluids, even though it is a compound of a metal whose soluble salts are toxic, will pass through without injury. For example, barium sulfate is used to provide contrast in X-ray images, and is taken as a barium meal. However, white lead is *soluble* in body fluids, rendering it highly toxic. Legislation was passed after World War Two to prohibit the use of 'soluble' lead in paints, but they are still present in old houses. A lead test kit is shown in Plate XIV.

Thin primers
Primers are always thin because the thinner the layer, the more the adhesive forces overcome the cohesive forces.

Red lead and zinc chromate are used in primer coats; they protect metal surfaces against corrosion. The molecular contact must be close, hence the need for a very clean surface and, preferably, a large surface area. The latter is achieved by sand-blasting to provide ridges and valleys. However, multiple coating increases the labour.

Special-purpose paints and coatings

Heat-resistant paints Ordinary paints blister, then char and disintegrate at relatively low temperatures. Coatings for ovens, heaters, stills, engines, turbines, etc. must withstand much greater heat. Paints with silicone vehicles, metallic powders (Al, Zn, Sn to reflect and conduct heat away) and heat-resistant pigments (Cr_2O_3, Fe_2O_3, C, TiO_2, CdSe) are widely used for specialist applications. Polymers of alkyds with saturated fatty acids give non-drying paints that can be baked on, and are called oven or baked enamels. These must be contrasted with vitreous or porcelain enamel (see Vitreous enamels later in this chapter).

Fire-retardant paints Addition of a variety of compounds to paint renders it less flammable. These compounds (phosphates, tungstates, borates and carbonates) decompose on heating to give gases that do not support combustion, and hence tend to extinguish flames. Alternatively, a substance that fuses on heating to give a glass-like layer on the surface can be added to the paint. A third approach is to use non-flammable constituents — silicones, chlorinated resins, or mineral powders. Water-based paints are usually non-flammable before they are applied, but become flammable when dried out.

Anti-fouling and insecticidal paints Anti-fouling paints have much application in marine construction. They usually contain inorganic poisons (mainly copper and mercury salts) or organic molecules (such as pentachlorophenol). More recently, organo-tin groups have been directly incorporated into the polymer. These have been found to be very effective against marine organisms but have caused mutations in them (see Tin in Chapter 10).

Luminous paints These fall into two groups:

- *Fluorescent* paints absorb ultraviolet radiation and re-emit it as visible light only while being irradiated. They contain zinc and cadmium sulfides, together with organic dyes.
- *Phosphorescent* paints are irradiated with ultraviolet light and continue to glow in the dark for some hours after the irradiation has ceased. Phosphors include ZnS (green, yellow, orange) or CuS and SrS (bluish). Additional salts can be used to change the colour.

EXPERIMENT 12.1
Preparation of ferric tannate

Equipment

Tea bags, saucepan, two glass jars, a pad of steel wool, a small quantity (100 mL) of vinegar, about 1 mL of a 3% hydrogen peroxide solution, filter funnel, cotton wool.

Procedure

Prepare a solution containing tannic acid by adding boiling water to fresh tea leaves or a tea bag. Make the solution equivalent in concentration to strong black tea. One cup of solution will be sufficient.

To prepare a solution containing iron (III), boil a small quantity ($\approx$ 100 mL) of vinegar to which a pad of steel wool has been added (the steel wool should not contain any soap). Allow the solution to simmer for 5 to 10 minutes and then strain the solution through a filter funnel containing a loosely fitting plug of cotton wool.

When the solution is cool, add about 1 mL of hydrogen peroxide solution. The colour of the solution should now be a dark brownish-red, indicating the presence of iron (III).

To produce ferric tannate, add a quantity of the tannic acid solution (say 10 mL) to a roughly equal amount of the solution of ferric ion. The solution should turn black, due to ferric tannate being produced. Ferric tannate is the major ingredient of many black inks.

Reactions

$$2H^+ + Fe \longrightarrow Fe^{2+} + H_2$$

$$2H^+ + 2Fe^{2+} + H_2O_2 \longrightarrow 2Fe^{3+} + 2H_2O$$

$$Fe^{3+} + \text{tannic acid} \longrightarrow \text{ferric tannate}$$

Notes

1. Hydrogen peroxide is available from pharmacies and is used for cleaning wounds. It is a strong oxidising agent and should be handled with care. It is also poisonous if swallowed, so it must be used carefully. Before adding the peroxide to the solution, make sure the solution has cooled to near room temperature. Don't spill the peroxide, because it is a powerful bleach.
2. Ferric tannate may stain, so be careful to avoid contact with clothing.

Superabsorbants

Superabsorbants have revolutionised the personal care industry over the past decade and nearly half a million tonnes of these materials are produced every year. Traditional absorbency resulted from water moving through and filling the open space between fibres; more absorbency required more open space between fibres and therefore more bulk. Superabsorbent polymers absorb the liquids and form a gel that prevents the water being squeezed out. In baby nappies (diapers), and in incontinence pads, they absorb up to 30 times their weight in urine and retain it under pressure (gravity, squeeze). This has led to thinner, more comfortable, nappies and drier babies. In tampons, their efficiency is compromised by the increased risk of bacterial growth and the possibility of toxic shock syndrome.

The material is a polymer of partially neutralised acrylic acid; sometimes an acrylic–starch co-polymer, formed by a free radical reaction in water. The rubbery product is cut into small pieces and dried in a hot oven and then milled to produce granules. The two important properties of the final product are its swelling capacity, which decreases with increasing cross-linkage, and its elasticity, which increases. A proper balance is needed. Modification of the shape of the granule can change the ease with which liquid is absorbed. Once water enters the material, to give the acid groups a chance to ionise, the chains in the gel are driven apart by electrostatic repulsion. Absorption is reduced strongly in the presence of cations such as calcium, which weakens their use in horticulture (to hold water in soil).

Adhesives

Nebuchadnezzar used bitumen in the construction of buildings in Babylon about 3500 years ago, and it may have been used on the Tower of Babel (see Chapter 9, p 247). The ancient Egyptians used gum Arabic (hence the name), egg white and animal glues for furniture. Papyrus sheets were joined into folios with flour and water paste. By the 16th century, animal glue was used for furniture again (skills having been lost in the Dark Ages). Apart from some improvements, there was little further development until the 20th century.

Efficient binding
In pure strength (apart from flexibility) the bindings, lashings and sewings of primitive people and older seamen were more efficient than metal fastenings. In fact the wood screw forms the least efficient of all joints.

In the First World War, plywood bonded with casein and blood albumen was important for boat and aircraft construction. In 1920, flying-boat hulls were sewn together with copper wire. World War Two saw the rapid development of synthetic adhesives. By 1981 the total western European consumption of adhesives was estimated to have reached two million tonnes dry weight, at a value of £Stg 1300 per tonne. An example shows why. Boots were once held together by a combination of sewing and nailing. Today, 90% of the construction uses adhesives.

The term 'adhesive bonding' can be used to refer to a variety of operations — sealing envelopes, applying bandages, and repairing torn paper with cellophane tape are but a few common examples. Adhesion involves the fastening together of solid materials by a thin, generally continuous, intermediate layer.

Today's hardware shops provide a bewildering assortment of packages and types of adhesives that involve different types of adhesive bonding (see Table 12.2). More elaborate surface preparation is required than for mechanical fasteners, which function quite well with dirty surfaces. The mechanical strength of most adhesives takes time to develop and pressure is generally required to ensure the necessary close contact. Adhesive bonding can be 'irreversible' in the sense that the structure can be disfigured or destroyed if pulled apart.

Glued joints
The strength of a glued joint depends mainly on its width, not its area, because of the way stress concentrates.

Any two solids can be glued together if we can find a liquid that will wet them both and then harden. Wood glued with water which is then frozen passes most of the tests in the specifications for wood adhesives! Carpenter's glue can be viewed as a variant of ice, in that the melting point is raised to a more practical temperature (70–80°C).

Casein glues and moisture

Casein glues (from milk protein) used to be very popular but, as casein is more or less a mixture of lime and cheese, on becoming moist it deteriorates and its last hours are like those of a dying Camembert.

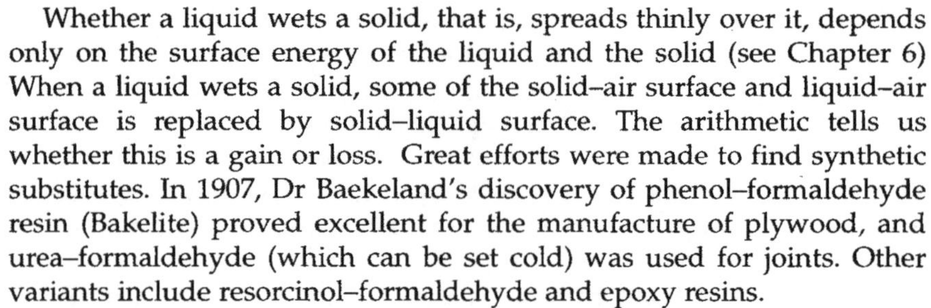

Whether a liquid wets a solid, that is, spreads thinly over it, depends only on the surface energy of the liquid and the solid (see Chapter 6) When a liquid wets a solid, some of the solid–air surface and liquid–air surface is replaced by solid–liquid surface. The arithmetic tells us whether this is a gain or loss. Great efforts were made to find synthetic substitutes. In 1907, Dr Baekeland's discovery of phenol–formaldehyde resin (Bakelite) proved excellent for the manufacture of plywood, and urea–formaldehyde (which can be set cold) was used for joints. Other variants include resorcinol–formaldehyde and epoxy resins.

Adhesives make simple, strong and cost-effective joints. They are attractive for both, children and engineers. When many simple parts are fixed together simultaneously rather than consecutively, phenolic and epoxy adhesives are chosen because of the cost savings. An example is the European Airbus. A complete adhesive-bonded section of a McDonnell Douglas YC15 aircraft 12 metres long was constructed and tested for 120 000 hours without the catastrophic failure that would have occurred with a riveted structure. However, unlike traditional ways of making joints, such as welding, riveting, bolting or brazing, there is no established method of predicting the strength of a bonded joint, although the aerospace industry has obviously developed some confidence in them for complex structures involving many materials.

Let us attempt to define what we want in an adhesive. There are many sticky materials that we would not consider to be good adhesives — chewing gum, for instance. To be predictable and consistent, an adhesive must stick more strongly to the substrate(s) than to itself, that is, it must be stronger in adhesion than cohesion. Thus a bonded joint should fail only in the breaking apart within the adhesive, or within the components, but not at the interface between the component and adhesive.

Epoxies

Epoxies are brittle and 30 times stronger than two-part acrylics, which are weaker, more flexible and stretch before failing.

The strength of a joint depends on a number of factors, including the thickness and strength of the components. The strength is not simply proportional to the overlap, and a point is often reached where there is no increase in strength with further overlap.

As aircraft are not subject to very rough treatment, the brittleness of epoxy adhesives is not a critical factor. However, for wider uses, incorporation of a little elastomeric material acts to absorb energy and stop crack propagation. This matrix polymer is often acrylate and is applied as two separate components, one to each side of the joint. This has the added advantage of dissolving grease from an improperly cleaned surface, which would otherwise have hindered adhesion.

For less demanding situations, solvent-based adhesives are used (footwear, laminating, flooring, roof tiles, upholstery and trim of motor vehicles). Natural rubber in an organic solvent has now been all but replaced by a range of polymers such as polychloroprene. As with paints, solvent-based adhesives are being replaced because of the cost of the lost solvent, and because of its environmental and occupational health effects. Hot-melt adhesives have been successful. The co-polymers of ethylene and either vinyl acetate or ethyl acrylate are used, for example in the DIY handguns for the handyman. However they melt again at high temperatures and so are not recommended for the kitchen table.

Adhesive failure occurs when the surfaces have not been prepared properly, or when there is a large difference in surface energy between the surface of the substrate and the adhesive. Liquids wet only surfaces that have higher surface energies (the solid equivalent of liquid surface tension). Polar adhesives with molecules that attract each other strongly, and are thus liquids of high surface energy, do not adhere well to Teflon or polyethylene. Cohesive failure often occurs along weaknesses caused by gas bubbles left behind from volatile by-products of the solidification process. A thinner glue line, obtained by using less adhesive, can be the answer in this case.

The process of bonding is of two types — physical and chemical. Physical bonding occurs when the process is one of cooling or solvent evaporation. Starch and animal glues and poly(vinyl acetate) (PVA) emulsions (white glues) all solidify by the evaporation of water. They cannot be used where exposure to water is likely to occur. Chemical bonding results from a reaction that changes the nature of the adhesive. An example is the cyanoacrylate systems, which cure rapidly with traces of moisture as catalyst.

It can be seen from Table 12.2 (overleaf) that some adhesives can be multi-purpose.

Poly(vinyl acetate) (PVA) glues are milky liquids, generally sold in squeeze bottles. They consist of a latex of polymer, fillers and plasticisers in water and are analogues of rubber latexes. They keep very well in a sealed container. Their setting time is short for porous substrates such as paper and cardboard (two to three minutes) but longer for non-porous ones such as polished woodwork and ceramics (12 hours). Surplus glue can be wiped off with a wet rag before it has set. The glue line is transparent and the set glue is resistant to hydrocarbon solvents (oil, grease, etc.). Formulations with an acid hardener have better water resistance. The joint softens if heated, and normally has poor creep resistance.

Plastic glues are glues made to join plastics, and are generally clear solutions. They set by loss of solvent and the joint is not very strong. They are suitable for light-weight applications such as for models, books, ceramic ornaments and leather.

Some plastics require special treatment. Vinyl plastics (e.g. inflatable toys, swimming pool liners, and some upholstery) can be joined with vinyl kits. Polystyrene cements are available for polystyrene toys. Polyethylene is very non-polar and so some polarity has to be introduced by heating it in air with a flame (but not so as to melt it). The surface has become suitably polar if water will spread on the plastic. It can then be bonded with most flexible adhesives, except water-based ones (e.g. not PVA). Casein and starch dextrose glues work well on polyethylene and polypropylene bottles.

Synthetic rubber glues include the elastomeric (contact, pressure sensitive) adhesives and silicone rubber cements. They are stronger than natural rubber. Whereas neoprene rubber adhesives lose solvents in the setting process, the others absorb moisture to set.

Removing wine labels

A 3% caustic soda solution will remove wine labels. (A strong adhesive is used to prevent the labels washing off in wet Eskys.)

TABLE 12.2

Guide to use of adhesives

	Wood	Metal	Rubber	Flexible plastics*	Rigid plastics*	Fabric and leather	Paper and card-board	China, glass and ceramics	Masonry
Masonry	solvent rubber#	epoxy resins	poly-urethane	poly-urethane	poly-urethane	contact cement	PVA glue	epoxy resin	epoxy resin
China, glass, and ceramics	epoxy resin	epoxy resin	poly-urethane	vinyl glue	epoxy resin	contact cement	acrylic glue	epoxy resin	
Paper and cardboard	PVA glue	acrylic glue	acrylic glue	vinyl glue	acrylic glue	contact cement	PVA glue		
Fabric and leather	PVA glue	contact cement	contact cement	vinyl glue	contact cement	contact cement			
Rigid plastics*	epoxy resin	epoxy resin	poly-urethane	vinyl glue	cyano-acrylate				
Flexible plastics*	vinyl glue	vinyl glue	poly-urethane	vinyl glue					
Rubber	poly-urethane	poly-urethane	poly-urethane						
Metal	epoxy resin	epoxy resin							
Wood	PVA glue								

* Does not apply to polyolefins, e.g. polyethylene or polypropylene, which need to be flamed (oxidised) first.

Block co-polymer solvent-borne building adhesive, e.g. Liquid nails™.

Source: Generic adaptation from Selley's Quick Adhesive Guide. **Use the manufacturer's own guides in practice.**

CHARACTERISTICS OF ADHESIVES

- **Clear household glues:** set by solvent evaporation and suitable for most substrates except polyolefins.
- **Contact cement:** A more specialised solvent, evaporation based, for gluing laminates to benchtops, or plasterboard to timber frames. Clamping is recommended.
- **Epoxies:** Almost universally two-pack adhesives, except polyolefins.
- **PVA paste:** Solvent based, filled adhesives, for use in areas where substrates are subject to water contact. The unfilled PVA latex is the standard wood glue. Cross-linked PVA gives a bond that is stronger than most woods.
- **Acrylic adhesives:** Emulsions of acrylic polymers, usually filled, and used as carpet glue and for other interior use where water contact is low.
- **Resorcinol–formaldehyde:** A specialised two-part woodworking adhesive for marine aplications, or where hot water or steam is used. Both parts must fit well together, and be smooth,clean, and dry.
- **Urea–formaldehyde:** A two-part system used for wood, and suitable for gluing laminated benchtops.
- **Cyanoacrylates:** Rapid setting, most suitable for non-porous smooth, close-fitting surfaces such as metals and most plastics.
- **Isocyanate adhesives:** Specialised flexible adhesives that cure by reaction with moisture to form a polyurethane polymer. They give molecular bonding to metals, rubber, and wood. Release of gas during curing is a disadvantage, and clamping or taping is recommended. In contrast to cyanoacrylates (above), reaction with water is slow, requiring two hours for initial setting, and 24 hours for maximum strength. Isocyanates are toxic and sensitising agents in certain individuals.
- **Silicone adhesives:** These will stick to metal, masonry, wood, glass, and most plastics. They provide a strong, flexible, water resistant bond, stable under hot conditions.

Epoxy resins

Epoxy resins were introduced commercially just over 30 years ago. The term is applied to a whole family of resins, and combinations of resins and curing agents, whose properties can vary widely. An epoxy resin is defined as a molecule that contains more than one epoxy group (Fig. 12.7) and is capable of being converted into a thermosetting plastic with curing agents or catalysts.

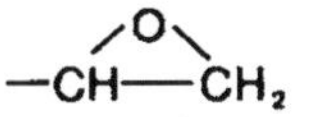

$$-CH-CH_2 \text{ (with O bridging the two carbons)}$$

Fig. 12.7
The epoxy group

Epoxy resins provide good adhesion to a wide range of materials, including metals, wood, concrete, glass, ceramics and many of the plastics. This is because polar groups are present in the cured resin. Since no water or other by-products are liberated during the curing of epoxy resins, they exhibit very low shrinkage. They can be formulated to withstand very high temperatures, and they are chemically very resistant. The most commonly used curing agents are polyfunctional amines in the form of an amine adduct (to reduce the volatility of the objectionable amines). The (new) Australian Parliament House has the granite facade fixed with epoxy resins.

Ultrafast-setting adhesive

An interesting glue sold for home use was first introduced by Kodak for industrial purposes (Eastman 910). It is a one-component system and depends on the polymerisation reaction of a monomer, *methyl 2-cyanoacrylate* (Superglue), shown in Figure 12.8.

$$H_2C{=}C(CN){-}C({=}O)OCH_3$$

methyl 2-cyanoacrylate

$$H_2C{=}C(CH_3){-}C({=}O)OCH_3$$

methyl 2-methacrylate
(forms Perspex)

Fig. 12.8
Acrylates

It is an exceedingly strong adhesive, very fast setting (10 seconds to two minutes) and bonds a wide variety of materials. It can be based on the methyl, ethyl or butyl cyanoacrylate monomer, with the addition of a suitable stabiliser. A weak base such as adsorbed water causes rapid anionic polymerisation. It gives very strong bonding, provided the gap is less than 0.5 mm. The bonding is brittle and has poor resistance to warmth and moisture, but this adhesive is suitable for electronic and light electrical units. It has also found favour with the police for 'developing' latent fingerprints on certain materials. Doubts have been expressed about its safety as a consumer product because of the toxic vapour (tolerance 2 ppm in air) and fast setting time. The glue adheres very strongly to the skin and is difficult to remove.

Formaldehyde resins

The group of formaldehyde resins are related to Bakelite — the original plastic. They have a number of specialist applications. Urea–formaldehyde is used in low-stress veneer applications such as joining Formica to wood, and in particle board, while resorcinol and phenol–formaldehyde are used in marine applications and outdoor furniture. Another use of phenol–formaldehyde is in brake linings and clutch facings where a dry abrasive material such as asbestos, or its more recent replacements, which must withstand high temperatures, is bonded to a substrate.

Vitreous enamels

Vitreous enamels (called porcelain enamels in North America) are alkali borosilicate glasses formulated to have temperature expansion coefficients slightly lower than the metal base to which they are applied. (The temperature expansion coefficient is the amount by which a material becomes larger when it is heated.) When the hot ceramic is applied to the hot metal and then cooled, the metal contracts more, and keeps the enamel compressed. This takes advantage of the high compression strengths, glasses. A ceramic glaze is similar to a vitreous enamel except that the substrate for the ceramic glaze is non-metallic (e.g. clay), generally has a low expansion coefficient, is weaker under tension, and is brittle. Vitreous enamelling and ceramic glazing involve the same process.

Contrary to what one might expect, vitreous enamel is not a thermal insulator, but is a relatively good heat conductor when applied in thin coats, and is thus useful in ovens. *Paint enamel* is a completely different material. It is a normal paint that is dried by baking at a relatively low temperature to remove traces of solvent.

Dental amalgam

Tooth enamel is composed mostly of a hydroxyapatite mineral, $Ca_{10}(PO_4)_6(OH)_2$, with many other elements in small amounts (3% in total). Protein adsorbed on the surface largely prevents attack from bacteria on the outer enamel, but they can diffuse through to attack inner surfaces. Fluoride can replace in part the hydroxide in the enamel and form a much tougher mineral.

When a tooth develops caries (a cavity) the area is drilled to remove decaying matter, and filled. The material used to fill teeth must be malleable, must adhere tightly to the tooth and thus exclude air and saliva, and must be able to withstand abrasion, corrosion, discoloration and the force of chewing. The material used must therefore satisfy quite a variety of requirements.

The search for suitable materials is fascinating and today's standard dental alloy is made up of the following metals: silver 66.7–74.5%, tin 25.3–27.0%, copper 0.0–6.0%, zinc 0.0–1.9%. The major component has the approximate formula Ag_3Sn. Zinc serves as an oxygen scavenger during manufacture. The dental alloy is mixed with mercury (1:1 weight ratio) just before packing into the cavity.

Tooth shock
A piece of aluminium foil placed on a mercury amalgam filling can give a galvanic shock (at greater than 30 mA passing through the tooth).

$$Ag_3Sn + Hg \longrightarrow Ag_2Hg_3 + Sn_8Hg + \text{unreacted } Ag_3Sn$$

The phase most readily attacked by saliva is Sn_8Hg. Saliva acts on the Sn_8Hg and releases Sn^{2+}. The process is accelerated if precipitating or complexing reagents are present, such as sulfide (from eggs) or citrate (from citrus fruit). Such corrosion accelerates if there is contact with a less reactive metal such as gold from an inlay on an adjoining tooth.

A certain amount vaporises during the first few hours of a filling and mercury can be detected in the urine. The older copper amalgam released vapour for many months after a filling and was very likely a health hazard (see 'Metals in food — Essential elements — Mercury', Chapter 10).

Portland cement and concrete

Ever since people first started to build, they have felt the need for some cementing material to bind stones and other aggregates to form strong walls and floors, and to give them a smooth appearance. The Assyrians and Babylonians are known to have used moistened clay for this purpose. The Egyptians used the mineral gypsum, which had been calcined or 'burnt', mixed with sand to make the mortar used in constructing the pyramids. Gypsum is calcium sulfate dihydrate ($CaSO_4.2H_2O$). When heated to 121–132°C, each mole of gypsum loses 1.5 moles of water, which leaves the hemihydrate $CaSO_4.\frac{1}{2}\,H_2O$. On mixing with a small quantity of water, the hemihydrate is converted slowly to the dihydrate, serving as a binder in the process. The hemihydrate is known as plaster of Paris. It received its name from work carried out by French chemists in the late 18th century on the mechanism of the setting of calcined gypsum by reaction with water.

The Greeks produced lime by 'burning' limestone and found that the action of the lime admixed with sand was enhanced if some volcanic ash was also used in the mixture. The Romans improved such mixtures by utilising pozzuolana, a volcanic deposit in Italy, which had proportions of alumina and silica nearly optimum for use with lime to make an effective cement.

The quality of building cements fell with the decline of Rome, and the art of cement making was practically lost until the middle of the 18th century. It was not until 1824 that Joseph Aspdin, an English bricklayer,

Portland cement
The product was named Portland cement because, after hardening, it resembled a natural limestone quarried at Portland, England.

found that the addition of volcanic ash to the lime could be avoided if a limestone containing a relatively high proportion of clay was used and the calcination was carried to incipient fusion.

Many limestone deposits contain appreciable proportions of clay, such 'clayey limestones' being known as cement rock. Since in such deposits the limestone and clay are intimately mixed, the early practice was to use cement rock itself if it had the approximately proper proportions of ingredients to make a good cement. The process for making Portland cement is relatively simple, but the chemistry of cement manufacture is highly complex. An example of the overall chemical composition of a high-quality Portland cement is given in Table 12.3.

TABLE 12.3
The overall chemical composition of Portland cement

Mineral	Formula	Per cent
Lime	CaO	62.0
Silica	SiO_2	22.0
Alumina	Al_2O_3	7.5
Magnesia	MgO	2.5
Iron oxide	Fe_2O_3	2.5
Sulfur dioxide	SO_3	1.5
Other	—	2.0

Although a cement can have the correct overall *chemical composition* (as listed in Table 12.3), unless the chemicals are present as the *proper compounds,* it will be worthless as a cementing material. The proper compounds in a good cement are tricalcium silicate, dicalcium silicate, tricalcium aluminate, and tetracalcium alumino-ferrate. After the calcination step, the magnesium oxide remains largely as such, there is some free lime, and the sulfur is present as calcium sulfate. The basic calcium oxide reacts mostly with the acidic silica and sulfur trioxide, and with the amphoteric alumina and iron III oxide.

The system containing cement, water, sand, aggregate (gravel) and air is called *concrete.* Concrete's most important engineering property is its *compressive* strength (i.e. strength against compression). Its *tensile* strength (i.e. strength against stretching) is only about one-tenth of its compressive strength. The compressive strength of concrete is normally 'specified' at a minimum of 20 MPa, but with better water control, a value of 30 MPa is obtainable. The more water used, the weaker the concrete, but the easier it is to work. The amount of water needed increases as the grind of concrete used becomes finer, because of the larger surface area that needs to be wetted. The finer grind creates more pores in the concrete and the concrete is weaker.

Cement hardening theories
The first theories of cement hardening were put forward by Le Chatelier (1893) and Michaelis (1893), both famous for other contributions to physical chemistry.

Both the aggregate and the cement paste have higher tensile strengths than the concrete they form — suggesting that the weakest part of a hardened concrete is at the interface between cement paste and aggregate and sand. It is because of these properties that steel mesh is used to reinforce concrete and why some structures are *prestressed* to ensure that the concrete is compressed rather than stretched.

High alumina cement called *cement fondu* is particularly interesting because it provides high performance and fast setting (used for military airfields during the war).

Cement fondu
Cement fondu lacks durability, as was discovered when it was used for cheap housing after the war.

Calcium chloride can be added to cement to obtain rapid hardening. It accelerates the hydration, and the heat of hydration causes the concrete to become very hot.

Limestone is often used as an aggregate in concrete for sewerage pipes. The acid waste attacks the aggregate preferentially and increases the life of the concrete.

Why does concrete crack?

Concrete changes its volume for a number of reasons. Expansion can be caused by the presence of undesirable compounds (e.g. MgO and CaO). Alkalis can dissolve some of the aggregate containing amorphous silica or carbonate minerals. Shrinkage can be caused by carbonation (reaction with carbon dioxide in the air).

The problem of changes in volume caused by freezing of water (ice has 9% greater volume than water) is overcome by using surfactants to entrap air in the concrete, thus making room for expansion.

A survey of buildings carried out in 1979 in Sydney indicated that 70% of buildings less than 15 years old displayed noticeable durability problems, and those less than five years old were in even greater trouble. Expansive forces, generated by rusting steel reinforcement, chip pieces of concrete from the surface, disfiguring it and ultimately creating a structural weakness. Carbonation converts calcium hydroxide into calcium carbonate and lowers the pH. This in turn increases the corrosion potential of the steel (see Experiment 12.3 on p. 324).

Concrete mixtures

The ingredients of concrete are listed in the following order: cement, sand, and gravel by volume. For example, 1:3:6 is a lean mixture; 1:2:4 is used for stronger structures, reinforced cement or when concrete is used under water; and 1:1:2 is used when the concrete is exposed to sea water (and the cement must be free of lime and alumina). The amount of water required to combine chemically with the cement is about 16% by weight, but, for efficient mixing, a greater amount than this must be used.

Sugar inhibits the setting of concrete and moreover weakens it. A tanker of sugar syrup is often kept on hand for dealing with large-scale spillage of concrete. The exact mechanism of this process is unknown, although some sugars (e.g. sucrose and glucose) are particularly effective.

EXPERIMENT 12.2 Strength of cement with changing water content

Equipment

Moulds (from milk cartons), 100 mL measuring cylinder, 400 mL beakers, cement, water, plastic wrap, steel balls, hammer.

Procedure

Prepare five moulds by cutting off the tops of some cardboard milk cartons. Label them 1,2,3,4, and 5. Half-fill a 400 mL beaker with dry cement/sand mixture. Fill a 100 mL measuring cylinder with water. Slowly add water a little at a time, with stirring, to the initially dry mixture. Continue adding water until the mixture becomes a thick paste. If too much water is added the mixture will suddenly become very thin and you will need to start again. When the mixture is thick with no pockets of dry material, pour it into the first cardboard mould and scrape as much of the material out of the beaker as possible. Smooth the surface of the cement in the mould. Wipe the inside of the beaker with a paper towel and rinse it well with water.

How much water is left in the cylinder? Record the amount of water used. Dry the beaker and repeat the experiment, but this time with 20% less water. Repeat the experiment again, but with another 20% reduction in water. Repeat again with 20% **more** water than the first mixture. Cover all four moulds with plastic wrap to prevent evaporation. Prepare one more sample identical to the first one you made, but leave this one uncovered. Leave all the moulds in a warm place for two days.

After two days, examine the samples. Do the surfaces look different? Scratch the surfaces with your fingernail, a nail, or point of a file. Note any differences. Try dropping a steel ball or marble from the same height onto the surface of each sample. **Wear safety glasses.** The harder the surface, the greater will be the bounce height. Tear the cardboard off the samples and try a 'reproducible' hit with a hammer, of increasing intensity.

Record the order of surface hardness by both methods, and the resistance to breaking. What effect does the amount of water have? What difference does keeping cement under wraps make to the surface?

Experiment
What is the effect of adding 1% sugar to an otherwise strong setting mixture?

There are other parameters you might wish to explore, for example, the ratio of sand to cement. Try mixtures varying from 50 mL sand plus 150 mL Portland cement, to 50 mL of cement plus 150 mL of sand.

Source: The chemical world, C. LeR. Darlington and N.D. Eigenfeld, Houghton Mifflin, Boston, 1977, p. 370.

EXPERIMENT 12.3 Alkalinity of concrete

Equipment

Steel reinforcing (or nails), jars, phenolphthalein indicator solution.

Procedure

Concrete is cement plus gravel, and is often reinforced with steel rods. The steel does not corrode because of the alkaline nature of the cement. Demonstrate this protection by placing pieces of steel reinforcing that has been cleaned of rust with sandpaper into two half-filled jars of water. Into one jar, also place some broken pieces of concrete. Seal the jars and leave them for a week. The rod without the concrete should corrode faster.

Reactions

In exposed buildings, carbon dioxide from the atmosphere slowly penetrates the surface of concrete and reacts with lime, $Ca(OH)_2$, converting it to limestone, $CaCO_3$, the reverse of the process by which cement is made. This reduces the alkalinity of the surroundings of the steel, which then can rust. The oxides and hydroxides of iron have a larger volume than the original iron, and this expansion cracks the concrete.

Obtain a broken piece of concrete where the surface has been exposed to the atmosphere for a while, for example from a building site. Break the piece open, and wet the whole new surface with some phenolphthalein indicator solution. A pink coloration will show the area of high alkalinity inside, with a rim of untinted concrete around the edge. The extent to which the gas penetrates depends on the size of the pores in the concrete; the more water originally used, the larger will be the pores.

Source: CSIROPRAC: Source book for science teachers, experiment 35, CSIRO and Australian Science Teachers' Association, 1985]

Volume on oxidising

Contrast the expansion of rusting iron to the reduction of volume when magnesium oxidises to magnesium oxide. (See Case study — magnesium in Chapter 10.)

Bibliography

Cambridge guide to a material world, Ed. R. Cotterill, Tech. Uni. Denmark, Cambridge University Press, 1986.

Surface coatings, Oil Colour Chemists' Association Australia, and Australian Paint Manufacturers' Federation, 2nd rev. edn., NSW University Press, Sydney, 1983. Two vols. A very clear exposition.

Solvents and human health, Total Environment Centre, Sydney, conference proceedings, Feb. 1996, 90 pp.

Chemistry in the medicine cabinet

In this chapter we have a peek into the medicine cabinet, but can afford to consider only some of the items. We look at the development of drugs, how they are controlled, and the industry that produces them. We explore some of the chemical logic in their design, and finally some undesirable and illicit examples.

BASIC RESEARCH BEATS BUREAUCRATS' BETTER MOUSETRAPS

'High technologies are the technologies of crisis medicine, the only tools we can apply, in our ignorance, when we cannot get to the real root of the disease. Through a deeper understanding of disease, gained through research, we will find a way to prevention or cure.'

'In the early 1950s each summer had its nightmare aspects. Mothers literally kept their children home from the cinema on a Saturday afternoon for fear of poliomyelitis. Once you had the disease, nothing could be done. The virus took its relentless toll of paralysis, and health officials then worried seriously about how society could afford the cost of keeping the worst affected individuals for a lifetime in an iron lung.

Funding priorities
If governments had set up committees in the 1950s to determine national research and development priorities for funding, developing a better iron lung would have been high on the agenda. Luckily for my generation such committees did not yet exist.

Then came Enders, Salk, and Sabin — and the miracle of poliomyelitis vaccine. Within five years, the scene had been transformed.

Source: Lewis Thomas, president of the Memorial Sloan Kettering Cancer Center in New York, quoted by Sir Gustav Nossal

---

Development of medication

One of the earliest records of human medication is to be found in an Egyptian papyrus dating from 1550 BC. During the 19th century the isolation and examination of the active principles of plant drugs were developed, while the 20th century has heralded the manufacture and distribution of potent synthetic drugs. Early medicaments were mainly alkaloids (complex plant chemicals containing nitrogen); the toxic properties of mandrake, opium poppy and *nux vomica* seed are due in each case to alkaloids — atropine, morphine and strychnine, respectively.

The first synthetic drugs were ether, chloroform, chloral hydrate (1869), phenacetin, acetanilide (1887), aspirin (1899) and procaine (1905). From this point there was an increase in synthetic drug production — an increase strongly reinforced by the discovery of the organic arsenicals as a treatment for syphilis, and of the antibacterial action of certain dye-stuffs. This second discovery led in turn to the development of the sulfonamides, which encouraged firms already engaged in the dye industry (the main chemical industry at the time) to move into pharmaceutical preparations. Table 13.1 shows this development.

TABLE 13.1
The rise of potent drugs

1785	Withering uses digitalis for heart disease
1803	Serturner isolated the first 'active principle' (morphine) from the poppy
1818	Strychnine isolated
1820	Quinine isolated from cinchona bark
1831	Atropine isolated from deadly nightshade
1846	Ether first used as an anaesthetic in surgery by Long and Morton
1844	Nitrous oxide used as an anaesthetic
1847	Chloroform used as an anaesthetic
1867	Amyl nitrite used for angina
1876	Salicylates found to be pain killers
1884	Cocaine found to be a local anaesthetic
1898	Barbital used as a sedative
1899	Aspirin synthesised by Dreser and marketed by Bayer
1902	Adrenaline isolated from the adrenal gland
1903	Veronal (barbital — the first barbiturate) (see below)
1905	Organic arsenic compounds used for the treatment of syphilis
1907	Ergot alkaloids found to counteract adrenaline
1910	Ehrlich discovers salvarsan as effective against syphilis
1911	Vitamin studies started

1912 Phenobarbital used as an anti-epilepsy treatment
1916 Heparin used as an anticoagulant
1921 Insulin isolated by Banting and Best and used to treat diabetics
1929 Penicillin discovered by Fleming — no medical use because of crudeness of product
1933 Vitamin C is synthesised
1935 Activity of the sulfonamides discovered and used on humans
1936 Pethidine, the first synthetic narcotic, synthesised
1937 Antihistamines discovered
1937 Curare introduced as a muscular relaxant
1941 Penicillin first used on an Oxford policeman who was suffering from multiple suppurating abscesses and osteomyelitis and was in the terminal stages of a generalised infection. Although after four days of the treatment his condition showed an astonishing improvement, supplies of the drug had run out (in spite of recycling it from his urine) and he eventually died.
1943 Diphenhydramine becomes the first practical antihistamine
1948 Chlortetracycline discovered, followed by oxytetracycline (1950) and tetracycline (1952). Chloramphenicol discovered
1950 Development of the first major tranquilliser, chlorpromazine
1952 Reserpine, the first drug for high blood pressure was isolated from a plant Synthetic anti-inflammatory steroids
1957 The tricyclic imipramine used as an antidepressant
1960 Oral drugs for mature onset diabetes

What's in a name?
Sometime in 1903, famous synthetic chemist, Professor Emil Fischer, was at a railway station in heated discussion. 'My train leaves in half an hour for Verona', he said. That settled it. The trade name for the first of the barbiturates was to be Veronal™. Later, when the French sought a licence to make the drug, the German patent holders (worried that quality might be compromised) insisted that the French use a different trade name, but ending in the generic 'al' for barbiturates. The French were to be warned not to compromise German quality — 'Gardez-vous!'. We could imagine a garbled telegraph message. It would explain the name the French gave their product, Gardenal™. C'est logique, n'est-ce pas?

Source: Jose Rappaport, unpublished memoirs as a chemistry student in Paris.

The history of the industry

The first pharmacopoeia (list of drugs) was published in London in 1618, but the first to regulate standards for drug purity was published in 1846. In 1960 the first international edition was prepared by the World Health Organization (WHO).

Drugs were traditionally taken by mouth, but in 1853, the hypodermic syringe was invented. It allowed drugs to be injected directly into the bloodstream and so take rapid effect (e.g. morphine). Later the hypodermic syringe was used to administer drugs that are destroyed in the gut, such as insulin. Anaesthesia also became possible by injection, removing the limitation of having to use gases. Preparations for injection had to be sterile, and preferably made up in single-dose containers, and fabrication moved out of the hands of pharmacists to the drug manufacturer. Another important step was the development of the compressed tablet (or 'tabloid' as it was first known).

The international pharmaceutical industry started after the First World War and grew quickly to multinational proportions. By the 1950s, painstaking research was giving way to a new era of commercialism, aimed at achieving the greatest possible consumption of pills and medicines.

By the end of the 1950s, difficulties were becoming apparent. Really new drugs were becoming harder to find. The areas of therapeutics that remained were less easy to find drugs for. Some diseases afflicted only a comparatively small number of people who were generally in hospital, and were not likely to prove a source of great income to manufacturers, even if better curative agents were found.

In 1976, the chemical and allied products industries in the USA had sales of about $100 billion of which the pharmaceutical industry contributed about $15 billion and employed about 30% of the total technical staff. In research and development the pharmaceutical industry spent over $1 billion out of a total of $2.85 billion. Companies can spend about 10% of sales revenue on research and up to 30% on advertising and promotion. The after-tax profits of the pharmaceutical industry were 11.2%, compared to 7.5% for the chemical industry overall. In spite of declining profits, the pharmaceutical industry is still one of the most profitable and research intensive areas of the chemical industry.

The production of pharmaceuticals can be classified into five groups. *Fermentation*, used for antibiotics such as penicillin, streptomycins, tetracylines and modification of steroids, is probably the most important in terms of dollar earnings, but, on a weight basis, *synthetic chemistry* is dominant because it produces the heavy-use products such as aspirin, tranquillisers, and antihistamines. *Animal extracts* provide insulin and hormones, while *biological sources* lead to vaccines and other serums. *Vegetable extracts* provide alkaloids such as opium, quinine, atropine and the precursors of steroid drugs.

An interesting technological comparison can be made between the pharmaceutical and petrochemical industry. In the petrochemical industry, sophisticated catalysts, but simple processes, are the rule. On the other hand, in the pharmaceutical industry, multistage processes and expensive reagents are more common. The petroleum industry invariably uses hydrogen as a reducing agent, whereas the pharmaceutical industry can afford to use the sodium metal–liquid ammonia, Birch reagent (which was the crucial step in producing the contraceptive pill).

Because smaller quantities are involved, and high levels of purity are required, the pharmaceutical industry generally uses batch production rather than continuous processes. The exception is aspirin!

Drugs have three sorts of names. A precise name for a chemical substance in an internationally agreed system has been devised to provide a *systematic* name. Because this is unwieldy, subject to typographical error, and highly forgettable, there is a simple, uniform 'official' or 'non-proprietary' name, often called the *generic* name or international non-proprietary name (INN), approved for a chemical likely to prove worth while as a drug (see Fig. A1.17 in Appendix I). The third name, the *trade* name, is a creation of the manufacturer.

The arguments for generic labelling are basically that it provides a simple, unambiguous, international label for drugs, and doctors are more likely to know exactly what substance they are dealing with. A trade name effectively disguises the constituents of a combined preparation, thus increasing the chances of a patient receiving something that contains ingredients which (to that person) are potentially hazardous.

Many preparations are marketed under different names in different countries (for trade mark reasons). The same name may refer to different

TABLE 13.2
Generic and trade names for some tricyclic antidepressants

Generic name	Trade names
Imipramine	Tofranil, Melipramine
Desipramine	Pertofran
Trimipramine	Surmontil
Chlomipramine	Anafranil, Placil
Dothiepin	Dothep, Prothiaden
Doxepin	Deptran, Sinequan
Amitriptyline	Tryptanol, Tryptine, Amitrip, Amitrol, Endep
Nortriptyline	Allegron, Surmontil

Many have about 20 other trade names throughout the world.

Source: Pharmaceutical benefits, and MIMS, Australia, May, 1997

drugs in different countries — which can have unfortunate results for travellers. Some examples are Anotox, Avlon, Benol, Bilagen, Bilitrast, Cedrox (which can be vitamin C or aspirin!), and we are only up to C in the alphabet. The pharmaceutical industry's reply to this is that preparations may differ in their bio-availability (the rate of release of active ingredient). It is true that manufacture of the pill in which the drug is contained has a critical role in the stability of the drug and the rate of its release. In Australia, poor quality control in this aspect led to the development of standard testing methods by the National Biological Standards Laboratory (initially for digoxin in 1966).

Generic labelling ensures that molecular manipulations are obvious because the names will be related. Compare the lists of generic and trade names for some tricyclic antidepressants (non-selective monoamine re-uptake inhibitors) in Table 13.2.

A set of similar sounding trade names may be coincidental, and does not necessarily mean that there is a relationship between the drugs. An example is the popular antidepressant, fluoxetine (Prozac, Lovan, Zactin) that belongs to a different class of compounds (serotonin re-uptake inhibitor), and in fact adversely interacts with the tricyclics, diazepam and other drugs.

Polymorphism
An interesting aspect of bio-availability is polymorphism (different solid-state structures for the same chemical). Chloramphenicol palmitate exists in two forms, only one of which is biologically active. In the other, the long hydrocarbon chain is possibly 'wrapped around' the outside of the crystal, preventing it from dissolving. There are other cases as well.

The industry in Australia

Australia provides about 2% of the world pharmaceutical market. Discovering, testing and gaining approval, initially by the US Food and Drug Administration (USFDA) for new drugs is expensive, of the order of $200 million. The company gets a patent on the drug for 20 years, but the clock started when the patent needed to be issued, not when the drug is approved, and this can cut substantially into the protected period. The company will market the drug under a brand name; for diazepam it was Valium™. After the patent has expired, the producer of a generic drug will use another name, say Antenex, and try to break into the market with a lower price to combat the established name.

In Australia, in 1995/96 more than 160 million doctor's prescriptions were presented at pharmacies (including pharmacies in hospitals). Of these, about 40 million were paid for directly by the customer. The balance of about 125 million were subsidised by the Federal government at a cost of $2.2 billion, the patient usually paying a maximum of $20 per prescription, which adds up to about $480 million per year. (See Appendixes VI and VII). Pensioners and other social security recipients pay less, and these account for 80% of the prescriptions filled. This does not include drugs that are sold over the counter without prescription. The rapid rise in the cost of the Pharmaceutical Benefits Scheme (PBS) has been attributed to several factors. These include an aging population, changes in patient expectations, more medical practitioners, and promotions by the pharmaceutical companies, which now include drug-pushing videos in doctors' surgeries.

Many drugs are heavily subsidised. In 1996, a course of insulin cost $121 but the patient paid $20. For a course of interferon, the cost is $475 and the charge is still only $20.

Drugs in the less regulated, more competitive, US market are much more expensive. Prices of Australian prescription pharmaceuticals are in total about 50% of world average.

The Factor *f* scheme was set up in 1987 to compensate drug companies for the low prices negotiated to be paid by the government for drugs under the PBS. By the time it expires in 1999 it will have allocated more than $1 billion to the industry and helped it to increase investment. As a result, research and development (R&D) spending in Australia has increased from $42 million in 1986–87 to $227 million in 1995/96. In 1987 exports were worth $165 million and R&D $7 million. In 1995, exports were $600 million and R&D $70 million, most of it in packaging and formulation from imported ingredients. Corresponding imports were $1.1 billion.

When a drug is out of patent, a pharmacist can prescribe a cheaper generic drug in place of a branded one unless the doctor specifically states 'no substitution' on the prescription. Even though companies lose patent protection after 20 years, they maintain control over their brand name, established over the same period. Valium™ is better known than diazepam.

Doctors' representatives have argued that elderly patients, in particular, might become confused because they take their pills by colour and shape. However, these elderly patients often come to the doctor from hospital, where they would invariably have been given generic drugs with several repeats to take home, only to be confused by doctors switching them over to 'brand' drugs.

Until a few years ago, patients paid a maximum price for a drug on the PBS, irrespective of cost, so they had no idea of its real cost. Now the government sets its contribution on the basis of the lowest equivalent, and requires the patient to pay the premium if the doctor insists on an expensive brand. This has had little effect so far.

Doctors oppose generic prescribing for many reasons but there is a very strong suspicion that they depend on the major companies for the sponsorship of medical educational seminars and (overseas) conferences and other fringe benefits. These big-brand companies would discontinue this financial support if doctors ceased to specify their brands. This is a marketing mechanism by which the major companies can keep control over their product after the patent has lapsed.

The industry in the UK

Some UK estimates in 1983 found that profits on top-selling drugs are so great that switching from a single-brand-name heart drug to its unbranded equivalent would save £14.5 million. There are about 4000 brand-name drugs on the market which are protected by patent but can be substituted, except for about 30 for which there is no good substitute. Patents on products developed during the 1960s are now running out and within five years about 90 of the top 100 drugs will be out of patent. It is fascinating to watch the trade advertising change for a drug that is reaching this point. Suddenly it is no longer so wonderful, as its owners try to move sales to another drug with a longer patent protection and simultaneously spoil the pitch of competitors waiting for the current protection to lapse. In the UK, drug companies are entitled to compensation for losses on branded products that are out of patent by raising the price of protected drugs.

Regulating profits
The Pharmaceutical Price Regulation Scheme actually regulates profits and not prices, which is bad for consumers, but does have a certain economic rationale.

The industry in the USA

In September 1984, the US president signed into law the Drug Price Competition and Patent Restoration Act. This Act permits simplified approval for generic versions of post-1962 products that are no longer protected by patents, and whose safety and effectiveness have already been demonstrated. The increased competition brought about price reductions. Valium was approved by the Food and Drug Administration (FDA) in November 1963 and the patent held by Hoffmann–La Roche ran out in early 1985. In September 1985, Valium acquired three generic duplicates to join the other top-selling eight prescription drugs with generic competitors.

Sales of nine of the 10 top-selling prescription drugs in the USA in 1984 totalled $US2.1 billion. Forty million Americans also spend something like $500 million annually on antihistamines for hay fever (allergic rhinitis). This thus affects one in six citizens. (See Table 13.3, overleaf.)

Drugs consumed worldwide

The top 20 selling drugs worldwide (by sales) are shown in Table 13.4 (overleaf). Most of them are still under patent protection (or just out) and so are selling at high prices.

TABLE 13.3
A list of prescribed drugs (outside of hospitals) for the USA 1992

Rank	Generic name	Typical trade name	Use
1	Amoxycillin	Amoxil	Antibiotic
2	Estrogenic hormone	Premarin oral	Hormone replacement
3	Ranitidine	Zantac	For ulcers
4	Ethynyloestradiol	Ortho-N 7/7/7-28	Oral contraceptive, steroid
5	Hydrochlorothiazide	Dyazide	Diuretic
6	Codeine	Tylenol/codeine	Analgesic (opioid)
7	Salbutamol	Ventolin	For asthma
8	Digoxin	Lanoxin	For heart failure
9	Penicillin V	V-cillin	Antibiotic
10	Frusemide	Lasix oral	Diuretic
11	L-thyroxine	Synthroid	Thyroid treatment
12	Propoxyphene	Darvocet	Analgesic (opiate)
13	Triamterene	Dyazide	Diuretic
14	Enapril	Vasotec	Lower blood pressure
15	Alprazolam	Xanax	Benzodiazepam type relaxant
16	Erythromycin	E.E.S.	Antibiotic
17	Naproxen	Naprosyn	Anti-inflammatory (non-steroid)
18	Nifedipine	Procardia	Calcium channel blocker
19	Diltiazem	Cardizam	Calcium channel blocker
20	Norethindrone	Ortho-novum	Oral contraceptive, steroid

Source: *Data from Reuben, Table 42.2, p. 906 (See References).*

TABLE 13.4
Twenty top-selling drugs worldwide

Rank	Generic name	Typical trade name	Use
1	Ranitidine	Zantac (Glaxo, UK)	For ulcers (H_2 receptor antagonist)
2	Nifedipine	Adalat/procardia (Bayer, Germany)	Calcium channel blocker
3	Enapril	Vasotec (MSD, US)	For high blood pressure
4	Captopril	Capoten (Bristol-Myers, USA)	For high blood pressure
5	Lovastatin	Mevacor (MSD, US)	For high cholesterol
6	Verapamil	Cardizem (Marion, US)	Calcium channel blocker
7	Epoetin alfa	Epogen/Procit (Amgen, US)	For anaemia
8	Omeprazole	Losec (Astra, Sweden)	For ulcers (proton pump)
9	Salbutamol	Ventolin (Glaxo, UK)	For asthma
10	Cefaclor	Ceclor (Eli Lilly, US)	Antibiotic
11	Diclofenac	Voltaren (Ciba-Geigy, Switzerland)	Anti-inflammatory (non-steroid)
12	Iohexol	Omnipaque (Hafslund/Nycomed, Norway)	X-ray contrast agent

TABLE 13.4 (continued)

Rank	Generic name	Typical trade name	Use
13	Acyclovir	Zovirax (Wellcome, UK)	Antiviral
14	Ciprofloxacin	Cipro (Bayer, Germany)	Antibiotic
15	Cimetidine	Tagamet (SmithKline-Beecham, US/UK)	For ulcers (H_2 receptor antagonist)
16	Fluoxetine	Prozac (Eli Lilly, US)	Antidepressant
17	Amoxycillin-clavulanic acid	Augmentin (SmithKlineBeecham, US/UK)	Antibiotic
18	Naproxen	Naprosen (Syntex, US)	Anti-inflammatory (non-steroid)
19	Atenolol	Tenormin (Zeneca, UK)	For angina (beta blocker)
20	Ceftriaxone	Rocephin (Roche, Switzerland)	Antibiotic

The 1992 sales, from 1 to 20, ranged from $US billion 3.44 to 0.86. Data from Reuben, Table 42.1, p. 905, citing Lehman Bros.

Five of the top six are for heart disease, and three are for ulcers (including the top ranking drug). Four of the drugs are antibiotics, but special ones, because the conventional ones are cheap. Two are for arthritis. There is only one antidepressant, a class of drugs that has gone out of favour for mass prescription. There is a single drug for asthma, a disease whose incidence is rising in the developed world.

The most widely consumed drugs, on the other hand, each annually about 50 000 tonnes, are aspirin, paracetamol (acetaminophen in the US) and vitamin C, all sold over the counter.

Drugs that are widely prescribed by doctors are more important in terms of public health. (See Table 13.5, p. overleaf.)

Heart drugs are the most prescribed in the developed world because they are taken for the rest of a sufferer's life. Antibiotics are the most important worldwide. Drugs affecting the central nervous system are next (including aspirin and paracetamol) but the psychotropic drugs such as the benzodiazepams (e.g. Valium), antidepressants, and appetite suppressants have been dropping since the 1980s because of problems of addiction and tolerance. Consumption of asthma drugs is increasing.

Many more ulcer treatment drugs have become available and prescribing has increased. The Australian discovery that many stomach ulcers are caused by a bacterium has radically changed treatment.

The non-steroid anti-inflammatories are widely prescribed for pain in arthritis, along with antihistamines for hay fever sufferers, hypoglycaemics for diabetes, and L-thyroxine hormone for thyroid sufferers.

Steroids are used, not only for contraception, but also for asthma and some anti-inflammatories.

TABLE 13.5
A list of prescribed drugs (outside of hospitals) for Germany 1990

Rank	Generic name	Typical trade name	Use
1	Mineral supplements	Calcium Sandoz	Diet supplement
2	Paracetamol	Ben-u-ron	Analgesic
3	Diclofenac	Diclophlogont	Anti-inflammatory (non-steroid)
4	Salicylicate ointment	Dolo-Arthrosenex	Anti-inflammatory (non-steroid)
5	Nifedipene	Adalat	Calcium channel blocker
6	Ambroxol	Mucosolvan	Expectorant
7	Heparin	Thrombareduct	Anticlotting agent
8	A-P-codeine	Gelonida NA	Analgesic
9	Vitamins	D-Fluoretten	Diet supplement
10	Xylometazolin	Olynth	Nasal decongestant
11	L-thyroxine	L-Tyroxin, Henning	Thyroid
12	Hydrochlorothiazide	Esidrix	Diuretic
13	Metoclopramide	Gastrosil	Anti-emetic
14	Isosorbide mononitrate	Ismo	Coronary vasodilator
15	Triamterene hydrochlorothiazide	Dytide H	Diuretic
16	Glibenclamide	Euglucon	Lowering cholesterol
17	Acetylcysteine	ACC Hexal	Expectorant
18	Aspirin	ASS Ratiopharm	Analgesic
19	Beta-acetyldigoxin	Novodigal	Cardiac glucoside
20	Doxycycline	Supracyclin	Antibiotic

Source: Data from Reuben, Table 42.3, p. 908.

TABLE 13.6
Prescriptions per person per year (1990–91).

Country	No. of Prescriptions
France	38.3
Italy	19.6
USA	15.5
Spain	15.0
Germany	12.0
Ireland	10.0
Belgium	9.5
UK	8.5
Australia	7 (1996 — PBS)
Denmark	6.2
Netherlands	4.8
Japan	cost only availalable

Source: Data from Reuben, Table 42.6, p. 911.

It is fascinating to examine the prescription differences between countries, for which there are not many obvious reasons (Table 13.6).

Drugs are cheap in France, and the number of prescriptions per person is double that of Italy and treble that of most other countries. The extent to which a patient visiting a doctor expects a prescription seems to set the scene for the level of prescribing. In Italy, patients come out with a prescription in 95% of their visits; this drops to 70% in the UK. There are major differences in prescribing requirements between countries. Diet supplements appear popular in Germany but are available over the counter in the US, and thus are not recorded. (See Table 13.5 above.)

A major difference in approach is that in the USA and the UK there is a requirement for demonstrated efficacy in prescribed drugs, whereas in Germany, France and Italy it is common to prescribe drugs agreed to have mainly a 'comfort' rather than a therapeutic value. However, the much higher rate of prescription of antibiotics in the US and UK probably reflects a comfort approach as well, as they are often given for conditions that would heal themselves anyway, or for viral infections (for which they don't work). The downside of this practice is to breed bacterial resistance to the antibiotic, and its loss as a weapon.

More and more drugs are released for sale over the counter, in order to help reduce the cost of the drug benefit schemes.

History of drug control in the USA

60 years ago, American drug manufacturers could produce drugs and sell them without trying them out first on animals or people, or clearing the products with the US federal government. Until 1938 the onus was on the government to show that a particular drug was adulterated or misbranded. The US Food and Drug Administration (FDA) sought new legislation via a 'chamber of horrors' presentation of cases, such as that of the top executive who died in agony from a non-prescription drug then widely available: radioactive radium water.

None of the horror stories prevailed over the manufacturing–advertising lobby until children started dying. In October 1937, physicians reported to the American Medical Association the deaths of six patients from a liquefied version of a new formulation of the then-new wonder drug, sulfanilamide. Because pills are hard to swallow, a small company found a good solvent for the drug and put it on the market without testing it. Elixir of sulfanilamide killed 107 people, mainly children, before supplies were recalled. Sulfanilamide was such a new drug that no-one was sure whether it was the drug or the solvent that was toxic. Dr Francis Kelsey (then a graduate) found it was the solvent, diethylene glycol, also used as an antifreeze additive. (Ethylene glycol was the adulterant used in the Austrian wine scandal in 1986.)

The tragedy of the elixir of sulfanilamide led to the 1938 Food, Drug and Cosmetic Act, requiring drugs to be cleared for safety before they go on the market. Later, as an employee of the FDA, Francis Kelsey stopped thalidomide from being sold in the USA, in contrast to what happened in most other countries. The procedure now used in the USA is outlined in Table 13.7 and Figure 13.1 (overleaf).

Australian thalidomide legislation
It took the thalidomide disaster to make Australia pass corresponding legislation 25 years after it was passed in the USA.

TABLE 13.7
How experimental drugs are tested in humans

Phase	Number of patients	Length	Purpose	Per cent of drugs successfully completing the phase[a]
1	20–100	Several months	Mainly safety	70
2	Up to several hundred	Several months to 2 years	Some short-term safety, but mainly effectiveness	33
3	Several hundred to several thousand	1–4 years	Safety, effectiveness, dosage	25–30

[a] For example, of 100 drugs for which investigational new drug applications are submitted to the FDA, about 70 will successfully complete Phase 1 trials and go on to Phase 2; about 33 will complete Phase 2 and go to Phase 3; 25 to 30 will clear phase 3 (and, on average, about 20 of the original 100 will ultimately be approved for marketing).

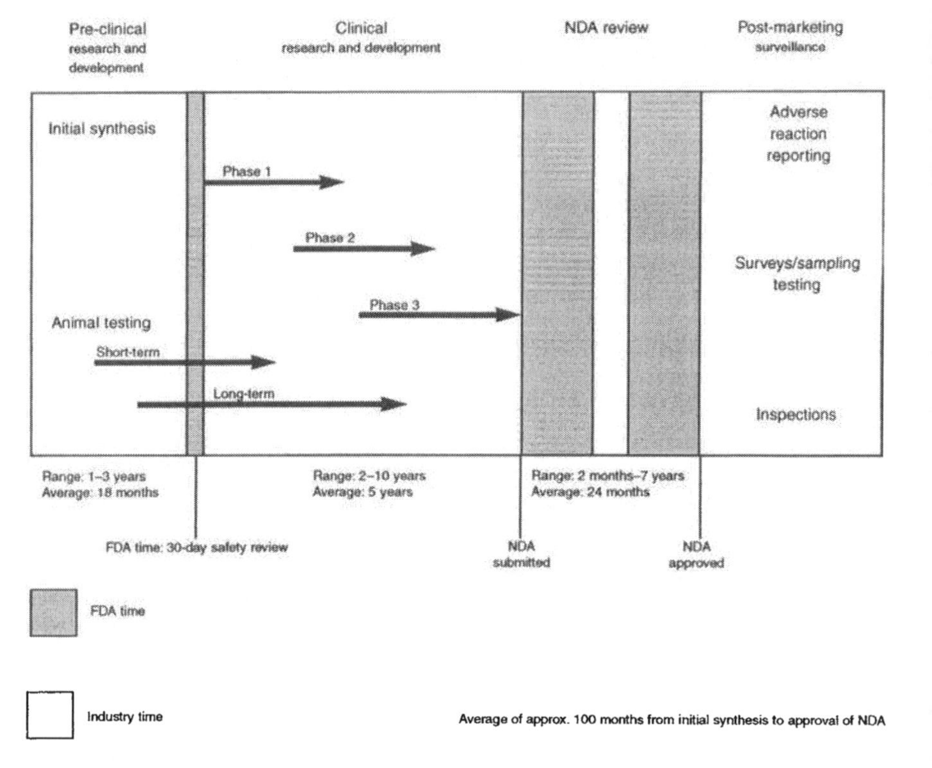

Fig. 13.1 New drug development in the USA (From From test tube to patient: New drug development in the United States, an FDA consumer special report, January 1988)

History of drug control in the UK/Australia

Legislative control of drugs and poisons has changed over the years to reflect changing social conditions, including philosophical approaches to individual freedom versus community benefit, the level of education, and a better understanding of the nature of the poisons themselves.

The economic subjugation of women once meant that killing husbands was one of the few ways of obtaining an estate, and poisoning was a common female *modus operandi*. The level of concern over poisonings in Britain in the mid-19th century mirrored that currently on gun control in the USA and elsewhere.

The early 19th century British criminal legislation was aimed primarily at dealing with murder and other crimes of physical violence, and it wasn't until the Arsenic Act of 1851 that a chemical matter came under control. It was opposed on the grounds of restrictions on personal freedom (a freedom that also allowed the random disposal of sewage, the selling of adulterated food, and so on). It became accepted that one way to reduce poisonings in the community was to restrict the availability of commonly used chemicals to legitimate users. A second way was to label the containers with enough information to allow the user to make a personal assessment of the hazard and risk of use.

In the 1800s (in Australia) dangerous chemicals were available to the public as raw materials from pharmacies, grocery stores, and produce stores in bulk, or taken in bulk and sold in small packages and boxes. These secondary containers were often flimsy paper bags, usually unlabelled, and often also used for food and beverages. This resulted in many recorded poisoning incidents.

The early Australian legislation in the 1870s required the seller to be a 'fit and proper' person, and the purchaser to sign a poisons register. The legislation covered commonly used pesticides such as arsenic and strychnine as packaged raw materials. It aimed to rectify the social problems related to abuse or criminal misuse of substances; laudanum (tincture of opium), chloral hydrate (hypnotic) used to drug victims of assault, heroin, children's 'soothing syrups' (containing opium, mercury), and so on.

In Western Australia, the Sale of Poisons Act of 1879 was expanded in 1899 to include 'all poisonous vegetable alkaloids and their salts, such as digitalis, laudanum, *nux vomica*, colocynth pulp, etc., as well as a wide range of proprietary medicines, which had earlier been exempted'.

History of scheduling
A historical view of scheduling gives a succinct social history of substance abuse.

Gazettal of further drugs in 1923 under the Act reflected public health concern over the expanding development and public use of toxic pharmaceuticals, and saw the scheduling of adrenal gland extract, acetanalide, amyl nitrite, barbiturates, benzamine, heroin, nitroglycerine, picrotoxin, pituitary and thyroid gland extracts, sulfonal, formalin and paraldehyde, and others.

The 1930s saw the addition of phenylaminediamines (12/4/35); amidopyrine (20/11/36), benzedrine (8/7/38), chloroform in toothpastes (14/10/38), benzedrine derivatives (7/7/39), sulfonamides (7/7/39), metaldehyde (22/12/39), pethidine (8/9/44), and DDT (21/9/45; 2/11/45; 6/9/46). The prescription-only drugs gazetted on 13/4/49 were barbiturates, Benadryl, Benzedrine, ergot, penicillin, and sulfonamides.

By the 1940s the advent of new drugs such as the sulfonamides which required 'good clinical practice' brought the 'prescription only' category into law around the world. It was seen to be easier to control the behaviour of professionals, such as doctors and pharmacists, than individual citizens.

The Australian Poisons Scheduling Committee was formed in 1954 and aimed to restrict and grade access on a 'need to use' basis. It was asked to separate the therapeutic from the commercial/domestic chemicals. Initially the committee just reacted to new circumstances but it has now become part of the assessment procedure prior to the release of new products. Scheduling decisions do not take economic factors into consideration but concentrate purely on public health. It is argued that the movement of drugs from prescription to OTC (over-the-counter) sale saves government subsidy, but this can have a social cost, as exemplified by the cases of thalidomide, mixed analgesics, and bromural sedatives.

Early legislation
In some of the earliest legislation (e.g. South Australia) provision was made to penalise householders who stored poison carelessly at home, but this was later recognised as unenforceable. However, the provision remains for pharmacists and storekeepers.

Introduction of the organophosphates in the 1950s with their much higher human toxicity than the earlier organochlorines, created a further need for regulatory control of pesticides. This area, and the general community access to chemicals is creating new demands. The National Registration Authority for Agricultural and Veterinary Chemicals has divided pesticides into groups, some of which are restricted to persons trained as pest control operators or accredited by a farm chemicals training program, or the like.

Source: John. P. Gregan, NHMRC, Canberra.

Consumer product information

There was a time not long ago when the information provided to the patient on a medicine could be fitted onto a generous size postage stamp. Even that minuscule disclosure was then covered up by the pharmacist's own label. All that has changed. Information is now provided in standardised modular form to the consumer, as well as more technically to the doctor.

The purpose of consumer product information (CPI) is to help consumers use medicines appropriately, and supplement the counselling activities of doctors and pharmacists. The test of their efficacy is that for literate consumers (i.e. those who can read): over 90% should be able to find information on the CPI quickly and easily, and over 90% of those who find the information should be able to understand and act on it appropriately. Therefore over 81% of literate consumers should be able to use a CPI appropriately. It takes between 30 and 60 hours to prepare a CPI, and 3 to 6 months to do all the necessary iterations of testing. The basis of the CPI is the product information approved to accompany the drug for the prescriber.

CONTENT OF CONSUMER PRODUCT INFORMATION (CPI)

The CPI must contain the following:

1. **Direction to discuss:** A direction to patients to discuss any aspect with the health professional.
2. **What is the purpose?** The expected effect of the drug, the health problem that it is intended to alleviate, for example, 'this medicine lowers high blood pressure. If this condition is not treated it can lead to more serious health problems such as . . .' The group (or the common type of activity) to which this drug belongs. Its habit-forming potential.
3. **Non-approved uses:** An explanation that the doctor may prescribe 'non-approved' uses if he/she feels it is appropriate.
4. **Before you use it:** Why not to use this medicine, such as previous reactions. Precautions for certain groups (elderly, children, pregnancy, liver or kidney problems). Expiry date or visible signs of deterioration. 'Before you start to use . . .', 'If using for the first time . . .', 'Do not use if . . .',
5. **Using this medicine:** The dosage, the method, (combining instructions with explanations can help consumers comply), the frequency, the duration of treatment, the expected effect, what to do if a dose is missed, how to stop treatment (prematurely) if this is a problem. What to do about an overdose.

6. **While you are taking the medicine:** Things you must do: 'Have your blood pressure checked regularly'. Things you must not do: 'Drive a car'; 'get up quickly'. . . Some other possible problems: 'may interact with some foods'; 'cause a sensitivity to sunlight'. Other side effects grouped as mild, serious, or very serious.
7. **After using it:** cleaning, storage, and disposal.
8. **Product description:** Dose form or strength, active ingredients, and manufacturer. What it looks like (colour, shape, and number of tablets).

Scheduling of poisons

The National Health and Medical Research Council (NHMRC) *Standard for uniform scheduling of drugs and poisons* has eight active schedules for drugs and chemicals which the state governments have largely adopted into their appropriate Acts.

Chemicals are *not* scheduled on the basis of a universal scale of toxicity, although this is obviously a most important input. Other factors are the need for the substance, the purpose of use, actual safety in use, and the potential for abuse. Chemicals are classified by inclusion in a set of eight schedules (2–9) according to the degree of control recommended to be exercised over their availability to the public. Some chemicals are deemed not in need of control (Appendix A of the Schedule). Others are deemed not to have any legitimate use in therapeutics and are listed in Appendix C of the schedule.

Old issues replaced
In the second edition of **Chemistry in the marketplace** criticism was levelled at the quite unclear distinctions between some of the schedules. Some criticism remained in the third and fourth editions, but many of the old issues have now been tackled and new ones have arisen in the scheduling of pesticides.

When listed in increasing order of the restrictions placed upon them, the order of the schedules from least to greatest is:

No restrictions < 5 < 6 < 2 < 3 < 7 < 4 < 8 < 9 < banned.

Chemicals for therapeutic use (drugs) fill schedules 2, 3, 4 and 8 in a progression of increasing strictness of control. The one substance can appear in several schedules, depending on its concentration in the product, the total amount in a pack, and a number of other factors. This is what makes scheduling difficult.

For agricultural, domestic, and industrial chemicals, schedules 5, 6 and 7 provide increasing control. The National Registration Authority (NRA) registers chemicals for agricultural and veterinary use (ranging from flea collars and vaccines, to garden sprays and genetically engineered cotton). The NRA applies the schedule requirements to labelling.

Most industrial chemicals are controlled by Worksafe Australia (National Occupational Health and Safety Commission). Worksafe also applies the schedule requirements to labelling.

Schedule 9 contains the prohibited substances, available only for special research projects.

STANDARD FOR UNIFORM SCHEDULING OF DRUGS AND POISONS

Schedule 1. has disappeared after much ridicule.
Schedule 2. Poisons for therapeutic use that should be available to the public only from pharmacies or equivalent, but no prescription is required.
Schedule 3. Poisons for therapeutic use that are dangerous, or are so liable to abuse as to warrant their availability to the public being restricted to supply by pharmacists or equivalent (medical, dental, veterinary practitioners).
Schedule 4. Poisons that should, in the public interest, be restricted to medical, dental, or veterinary prescription or supply, together with substances or preparations intended for therapeutic use, the safety or efficacy of which requires further evaluation.
Schedule 5. Poisons of a hazardous nature that must be readily available to the public, but require caution in handling, storage, and use.
Schedule 6. Poisons that must be available to the public, but are of a more hazardous or poisonous nature than those classified in Schedule 5.
Schedule 7. Poisons that require special precautions in manufacture, handling, storage, or use, or special individual regulations regarding labelling or availability (because of exceptional danger, or deficiencies in toxicological information).
Schedule 8. Poisons to which the restrictions recommended for drugs of dependence by the 1980 Australian Royal Commission of Inquiry into Drugs should apply.
Schedule 9. Poisons that are drugs of abuse, whose manufacture, possession, sale, or use should be prohibited by law, except for amounts that may be necessary for medical and scientific research conducted with the approval of Commonwealth and/or State health authorities.

Mixtures of drugs must comply with the (more stringent) individual schedule requirements. The requirements for labelling warnings, first aid, and packaging and so on, are selected from a comprehensive set of standardised focused statement modules. This provides a consistency between products, and allows a familiarity to be built up, with a reasonable degree of specificity for each situation. The cartoon created for earlier editions of *Chemistry in the marketplace* must have had some effect. It shows a frightened little girl looking up at a large elephant labelled 'If swallowed seek medical advice' (see Appendix V).

Advertising

A person shall not include any reference to a poison included in Schedules 3, 4 or 8 in any advertisement, except in genuine professional or trade journals, or other publications intended for circulation only within the medical, veterinary, dental, or pharmaceutical professions, or the wholesale drug industry.

Reading the schedules

Chemicals are now being scheduled individually using approved names. In order to have more flexibility and clarity, schedule entries are increasingly expressed in either positive or negative terms. Thus selenium is in schedule 6 *only if* one of four conditions applies. Fluorides on the other hand are also in schedule 6 *unless* exempted by one of the exempting clauses. However other legislation may apply to the exempt category.

- S2 is seen on drugs that are not safe enough to be sold by other retail outlets.
- S3 is seen on drugs that are potent, but are needed so regularly by sufferers of specific ailments that requiring a prescription each time is considered unwarranted. The pharmacist must provide adequate instructions, written or verbal, for the use of the drug.
- S4 is seen on drugs that are 'prescription only'. This is one of the largest schedules

The schedule contents change regularly with increased information. In 1973, enterovioform (clioquinol), once a common treatment for traveller's diarrhoea, was scheduled S3. Agitation by consumer organisations led to discreditation of this drug and it was rescheduled S4 in 1975. More agitation followed, and as a drug for internal human use it was scheduled as a *prohibited substance* S9. It was then promoted for good behaviour to S7 (Fourth edition). For this edition, it has had a further promotion from schedule 7 to schedule 4, but only for use on skin.

Various antihistamines in various preparations are spread through S2, S3 and S4.

- S5 for the consumer contains 'hardware' and pesticide items. It contains a variety of chemicals used around the home — from epoxy resins, turpentine, and kerosene to moth balls, antifungal agents in wallpaper paste, and some pesticides.
- S6 has a similar range to S5, but includes veterinary preparations, and has more restrictions on labelling and packaging.
- S7 is seen on more dangerous chemicals, or larger packs, or concentrations mainly for professional use, or chemicals needing special rules; for example vinyl chloride and acrylonitrile monomers (see Vinyl polymers in Chapter 9).
- S8 is for addictive drugs — morphine, pethidine, etc. that are legal.
- S9 is for illegal drugs (or concentrations of legal drugs making them illegal). The S9 schedule has some illicit designer drugs, but there is great difficulty in defining them adequately from a chemical–legal point of view.

In the second edition of this book, the odd bedfellows in S9 were noted: cannabis, heroin, and thalidomide. In spite of its criminal history, thalidomide has legitimate use and in the third edition its removal from this list was noted, as well as its rescheduling in S7 for use in treating leprosy. In the fourth edition, one of its optical isomers, (+) thalidomide was scheduled as S4 (restricted in one of the appendixes to treating leprosy).

Analgesics

The following may be sold freely (e.g. in supermarkets): plain aspirin in tablets of less than 325 mg in packets of not more than 25 (or 16 @ 500 mg), or powders of less than 650 mg in packets of not more than 12; plain paracetamol (or plain salicylamide) in tablets of less than 500 mg in packets of not more than 25, or powders of less than 1 g in packets of not more than 12. Dosages or packs greater than these are scheduled S2 and are available only in pharmacies. If you want one big pack rather than lots of small packs, the preparation is scheduled S3. Mixtures of analgesics (including mixtures with caffeine) are scheduled S4 and require a doctor's prescription. Codeine in mixtures with concentrations in excess of 30 mg per dose are scheduled S8.

Chemical synthesis normally gives equal mixtures of optical isomers, and no attempt is made to separate them unless the use demands it. The original British patent 768 821 obtained by Chemie Grünenthal describes the preparation and uses without reference to isomers.

Thalidomide was developed by Chemie Grünenthal of Aachen, Germany, in the mid–1950s as a sedative to prevent nausea during pregnancy. In 1961 it was found to cause birth abnormalities and was banned. Thalidomide has shown valuable effects as an inhibitor of replication of the AIDS virus, to clear oral canker sores in AIDS and other patients, and to ease sores in leprosy patients. So the investigations in 1979 at the University of Bonn in which the drug was resolved into its two optical isomers (see Amphetamines, below) with the report that only one isomer was teratogenic, suggested that the other isomer might be quite a safe drug. However, investigators at St Mary's Hospital in London had shown in 1967 that both isomers are teratogenic in rabbits, and in 1984 at the University of Münster it was found that even when separated, the isomers interconvert (racemise) in rabbits or in test tubes, so the question of one isomer being safe becomes academic.

Exemptions

Appendix A of the schedule exempts chemicals from scheduling (for undisclosed reasons). They may be covered under other jurisdictions, or they may require a warning without other restriction. It includes ceramics, chemistry sets (labelled in accordance with AS 1647), electronics, explosives, food, glazed pottery, matches, motor fuels, polymer pigments, inks, breathalyser tubes, timber, and wallboard.

Pharmacies

Australia has one pharmacy for every 3000 people, compared to one per 4500 in the USA, one per 5000 in Britain, and one per 15 000 in Holland (where the system is to 'allocate' people to a pharmacy). The number of wholesale outlets per pharmacy is about one in 17 (Australia), one in 53 (Britain), and one in 31 (Holland) .

At the local pharmacy you can probably buy footwear, sunglasses, health and leisure equipment, toys and electronic goods, photographic equipment and processing, cosmetics, perfumes, toiletries and toothpaste, as well as over-the-counter medicines and, of course, prescriptions. What is the history behind this situation?

In the 1930s pharmacists enjoyed a virtual monopoly over drugs and patent medicines. Suddenly department stores began marketing, more cheaply, items such as toothpaste and analgesics, which had previously been available only at the pharmacy. At the same time the pharmacist's professional skills in compounding preparations from basic materials became less and less necessary as prepackaged medicines began to dominate the field, particularly when a completely new generation of therapeutically potent drugs (the sulfonamides and penicillin) appeared in the 1940s.

Pharmacy ownership
Anyone can own a medical practice and hire a doctor to operate in it, but pharmacies cannot be owned by non-pharmacists.

At this time the British chain pharmacy, Boots, wanted to enter the Australian market. With their advantage of scale and efficiency, the threat to the local community pharmacy could not be ignored. The argument that standards would drop was rejected by a NSW government inquiry headed by Justice Browne. Nevertheless laws were passed prohibiting the *establishment* of chains (thereby allowing the continuing existence of the Soul Pattinson chain), limiting the ownership of pharmacies to one per person (who soon had to be a pharmacist), setting up licensing boards, and so on. Some controls have since been relaxed and in NSW there can now be up to three pharmacies in partnership, while in other States a pharmacist can own more than one pharmacy. The argument for restricting ownership is that the more commercial nature of the pharmacy business makes owner interference and coercion into non-ethical practices more likely.

'Chemists only' items
The Trade Practices Commission (as it was then) had outlawed formal agreements on 'chemists only' items outside the Poisons Schedules, although some companies still have such a 'policy'.

The intervention of government in the pharmaceutical distribution system both as a monopoly buyer and third party subsidiser of pharmaceutical consumption has helped turn pharmacists into shopkeepers, and encouraged a certain style of practice with high-volume dispensing, and little professional intervention. In a typical pharmacy, prescription medicines account for 45% of turnover, non-prescription 15 to 20%, and approximately 40% of pharmacy turnover comes from cosmetics and toiletries, etc. for which they enjoy about 60% of the market. These average levels fluctuate wildly with the location of the pharmacy. The turnover from pharmaceuticals can vary from as little as 20% to as much as 80%. Within the non-prescription area, pharmacists hold 25% of the photoprocessing market, but competition is increasing. Many items are now available through supermarkets, such as antiseptics, bandages, toothpaste, and analgesics, which previously were not.

Do the increasing professional qualifications required of pharmacists (over the last 20 years it has gone from an apprenticeship with some technical training to a degree course at university) make sense, when less than 3% of prescriptions are now compounded by the community pharmacist? On the other hand, the hospital pharmacy has increased in importance (representing about 15% of pharmacists) and the need for individualised medication makes greater demands on skills.

Non-prescription consultations
It is estimated that pharmacies have about 40 million non-prescription consultations every year compared to 60 million general practitioner consultations.

As the supermarket encroached on the pharmacy, the pharmacy encroached on the medical practitioner by providing information but not 'advice' as this would involve a demarcation dispute with the medicos. In the 1950s, people used the community pharmacy extensively for minor medical matters and treatments, rather than worry the busy doctor (a visit would normally turn out to be more expensive anyway). The structure of our medical and pharmaceutical benefits in the 1980s now makes a visit to the doctor with a consequent prescription a much cheaper option. Recent modifications to the scheme may tend to reverse this trend to some extent.

Against this background, an association of accredited pharmacists has been established to provide consultant pharmacy services to special-care areas and to the public. Some of this service will be paid for.

Another interesting development has come about through hi-tech developments in clinical screening kits. The legal semantics here are again interesting. A *diagnostic* kit is something used by medical practitioners, and they are jealous of the term. Use of a *testing* kit in some states requires a licence (cost approximately $1000). So the pharmacist is left with a *screening* kit. They are chemically the same! The fingerprick devices allow blood glucose, cholesterol, triglycerides, uric acid, etc. to be measured (with a reflectance meter) without the need for a syringe sample. Blood sugars, for example, provide much more useful information than urine levels. The results are of clinically useful accuracy. Pregnancy tests (screens) are also now much more user friendly.

In the USA, pharmacies have moved into the hi-tech home-care services, providing dialysis, cytotoxic drug delivery, etc., which would otherwise be carried out in a hospital. A high-level laboratory in the pharmacy is required, and in Australia the move in this direction has not been followed. However, computers have moved in, and as well as the usual management tools (such as stock control, accounting, invoicing, all aspects of the dispensing routine, and autobilling by disk or modem of benefit payments), a database warns of drug allergies, drug–disease interactions, and scaled drug–drug interaction. Because of the confusion in drug naming there are warnings when dispensing errors are likely — a so-called PDL message (PDL is Pharmaceutical Defense Ltd, which handles the indemnity insurance).

Chemistry of drugs

A discussion of the chemistry of drugs could easily fill a book on its own. For a start, it has a fascinating history. Modern research on the mode of action of drugs is exciting and potentially very useful in paving the way for a more scientific selection of effective substances. It is perhaps not realised how much we still rely on naturally occurring substances — either directly, or with some modification — for our pharmaceuticals. Even where it is possible to synthesise a drug such as morphine, it is often more economical to produce it from a plant and then perhaps add a few trimmings. A substance as complex as insulin is only synthesised once! This is done because synthesis is the ultimate proof of the correctness of the structure of the compound. However, as new and better chemistry is developed, the synthetic route to a drug can again become interesting and competitive.

This is a huge subject and this section concentrates particularly on the drugs that affect behaviour. They have become a consumer item in the sense that their use is widespread, and some of the preparations are available without a doctor's prescription. They also have the greatest potential for abuse. However, let us first deal with antibacterial and antibiotic drugs.

Antibacterial drugs

Drugs can be classified by what they do; for example, analgesics (pain deadening), sedatives and tranquillisers (reduce anxiety), stimulants, anti-depressants, hallucinogens, etc. But from the point of view of understanding, the relation between structure and activity is more useful. Let us consider the very first of the antibacterial drugs — the *sulfonamides,* which were found to be effective against the 'cocci infections' caused by the bacteria *streptococci, gonococci,* and *pneumococci.*

The basic compound is called sulfanilamide (Fig. 13.2).

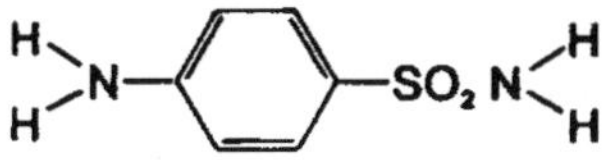

Fig. 13.2 p-aminobenzene-sulfonamide

A whole family of derivatives can be built up on this compound by modifying the molecule in a manner that either changes its potency, or reduces side effects or toxicity. Thus, if we write a general formula for a sulfonamide (Fig. 13.3) as:

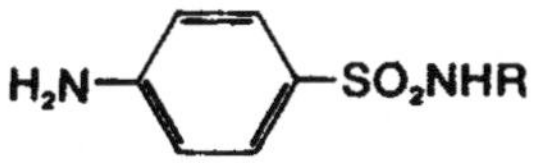

Fig. 13.3 Sulfonamide

then, in sulfathiazole (Fig. 13.4):

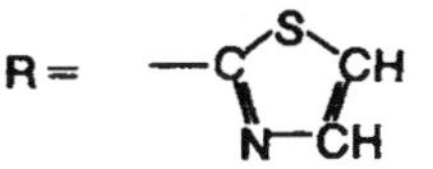

Fig. 13.4 Sulfathiazole

and in sulfadiazine (Fig. 13.5):

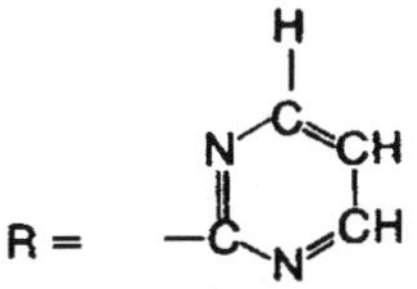

Fig. 13.5 Sulfadiazine

These are then members of the sulfonamide family, or generic group. It appears that the effectiveness of these drugs depends on maintaining the basic structure and shape of the molecule, and you might wonder why. One of the essential growth compounds for most bacteria that are susceptible to the sulfonamides is *p*-aminobenzoic acid.

The fact is that bacteria absorb a sulfonamide 'by mistake' because in shape and charge distribution it is similar to *p*-aminobenzoic acid (Fig. 13.6, overleaf), and then they cannot metabolise it. It fits into the cell machinery, but doesn't come out; that is, it blocks the active sites of the cell.

The sulfonamides, as chemicals, have been known for a long time but their medical value was discovered in 1932 by Domagk only by accident

Fig. 13.6 Sulfanilamide and p-aminobenzoic acid

sulfanilamide

p-aminobenzoic acid

while looking at a series of azo dyes he used for attenuating the virulence of experimentally used bacteria. Why are the sulfonamides active against bacteria and not against people? Well, *p*-aminobenzoic acid is used by bacteria to produce folic acid, which they need, just as we do. By blocking the enzyme that carries out the first step, the bacteria get no folic acid, and bacteria cannot absorb folic acid from their food. Humans do not synthesise folic acid, but can obtain it from their food, and so the sulfonamides cannot deprive them of it.

DISCOVERY OF SULFA DRUGS

The story of the first sulfonamide is interesting. Gerhard Domagk, working for IG Farbenindustrie, patented an azo dye, Prontosil, on Christmas Day, 1932. However his experiments with the dye on infected mice, and safety tests on rabbits and cats, were not published until they appeared in the Deutsche Medizinische Wochenschrift of 15 February 1935. The paper describes how 12 mice infected with streptococcal bacteria stayed lively for eight days, while of 14 controls, 11 survived for one day only, two for two days, and only one survived three days. This paper was followed immediately by another from the local hospital with enthusiastic descriptions of clinical tests on human patients over a period of more than two years.

Domagk had also used Prontosil to cure his daughter of a bacterial infection. Domagk noted that the chemical was ineffective in vitro and 'only works as a true chemotherapeutic agent in a living organism'. This was a problem because the antibacterial properties of the separate sulfanilamide entity were recognised back in 1919 (although not in animals) and therefore prevented the patenting of Prontosil (or, for that matter, the sulfanilamides themselves). He may have delayed publication in the hope of finding another compound that could be patented. In November 1935 a publication by researchers at the Pasteur Institute in Paris showed that the sulfanilamide entity alone was found effective in humans (as well as in vitro).

The author vividly remembers taking the huge M & B wonder tablets (May & Baker's sulfa drugs) as a child in the late 1940s and being required to drink copious quantities of water at the same time. The kidney damage caused by earlier products (see box, Sensitivity to drugs determined by genes, later in this chapter) meant that cautious doctors still insisted that patients drink many pints of water per day while on the tablets.

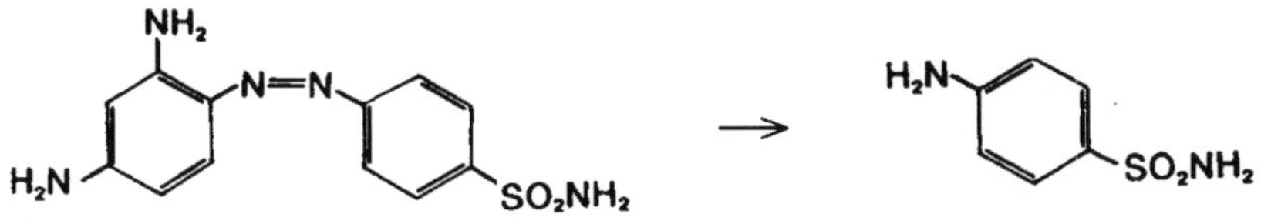

Prontosil → sulfanilamide

Fig. 13.7
The action of Prontosil

(See also the section on azo food dyes in More on additives — Colouring matter' in Chapter 15.)

Antibiotics

Antibiotics are substances produced by micro-organisms within themselves, which, when excreted, interfere with the growth or metabolism of other micro-organisms — a sort of chemical warfare on the microbe scale. In 1929, Fleming discovered a mould of the *Penicillium* genus that inhibited the growth of certain bacteria. The active compound, which he was unable to isolate, was called penicillin (Fig. 13.8). Florey and Sharp later performed the decisive clinical tests and were responsible for the success of the drug.

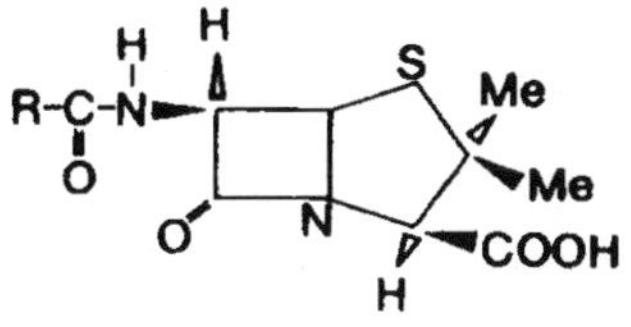

Fig. 13.8
Penicillin

The R-group can be varied by adding molecules to the nutrient solution in which the mould is growing, to produce many different penicillins (the original penicillin was a mixture). The mode of action of penicillin was determined in 1962. Penicillin interferes with the building up of the cell wall (which is continuously being digested and rebuilt) and the cells of certain bacteria are very much more sensitive to this interference. Penicillin is effective against a series of bacteria called Gram-positive (which take up and hold a certain stain or dye) but not against Gram-negative bacteria (in which the stain is washed out). (H.C.J. Gram was a Danish bacteriologist 1853–1938.) In addition, many bacteria have developed, or can develop, the enzyme or biological catalyst penicillinase, which can destroy penicillin (-ase often means an enzyme which is related to the compound or chemical reaction immediately preceding it).

Guinea pigs
In order to balance the argument on safety testing of drugs, it should be pointed out that penicillin is quite toxic to guinea pigs. During the war there was no time to carry through an adequate testing program on animals. Just as well! Mind you, virtually everything is bad for guinea pigs, poor defenceless creatures.

Analgesics and opiates

In order to see how sensitive structure is in relation to pharmacological activity, consider the alkaloids associated with opium. The alkaloid morphine was first isolated from the latex of the opium poppy (*Papaver somniferum*) by a German tinker, Sertürner, in 1803, although the ancient Babylonians probably used crude opium to relieve pain 5000 years ago. Its addictive properties were known from early times. In 1832, another alkaloid, codeine, was isolated from opium. Although codeine has only about one-tenth of the potency of morphine, its prolonged use in low doses can cause physical dependence. In 1898, morphine was acetylated to produce diacetylmorphine, or heroin, which was quickly realised to be even more addictive than morphine. Figure 13.9 illustrates the structure of morphine and some of its derivatives.

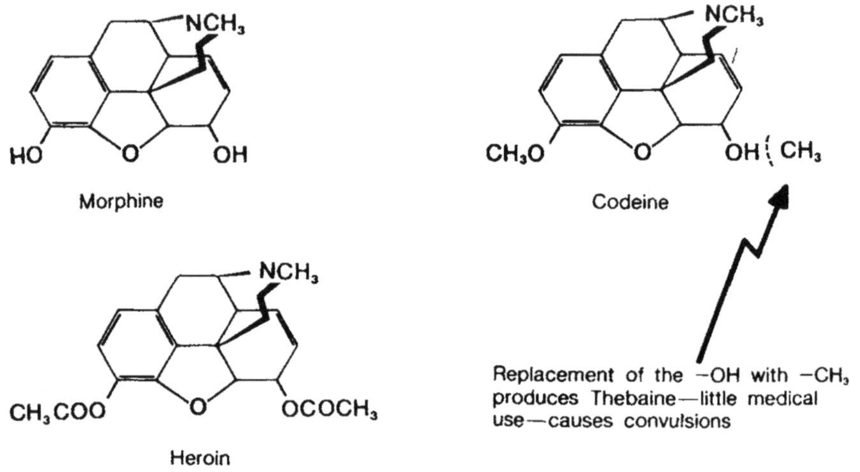

NCH_3 HO O OH
Morphine

NCH_3 CH_3O O OH ⁞ CH_3
Codeine

NCH_3 CH_3COO O $OCOCH_3$
Heroin

Replacement of the –OH with $-CH_3$ produces Thebaine—little medical use—causes convulsions

Fig. 13.9 Morphine and its derivatives

The first potent synthetic analgesic to be prepared that did not depend upon opium for its prime source was discovered quite by chance in 1939 by Eisleb and Schaumann during a search for atropine-like activity. The substance, pethidine, seems only vaguely related to morphine, but if the molecule is drawn to show a particular conformation, the relationship becomes apparent (Fig. 13.10).

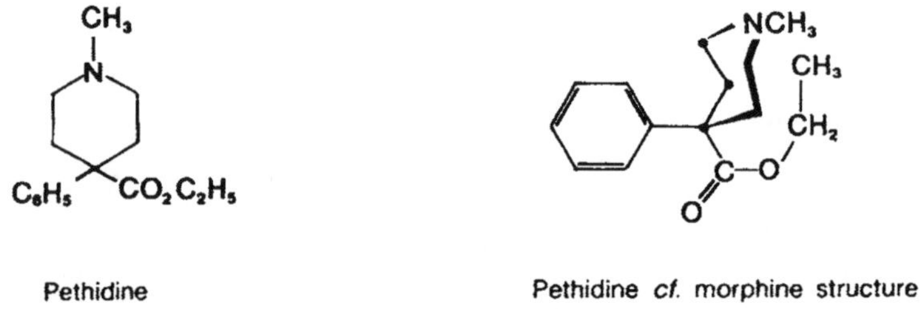

CH_3 N C_6H_5 $CO_2C_2H_5$
Pethidine

NCH_3 CH_3 CH_2 C O O
Pethidine *cf.* morphine structure

Fig. 13.10 Pethidine

In 1946, the first member of an important new group of synthetic analgesics (based on 3,3-diphenylpropylamine) was introduced under the name methadone. Again the structural relation to morphine can be detected when the flexible methadone molecule is rearranged (Fig. 13.11).

As long ago as 1915 a simple derivative of codeine was prepared in which the methyl group attached to the nitrogen ring was replaced by

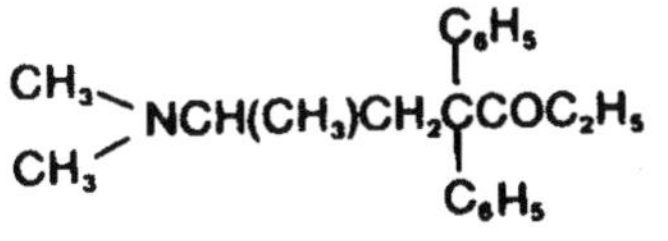

$(CH_3)_2NCH(CH_3)CH_2C(C_6H_5)_2COC_2H_5$

methadone
only the laevo form is active

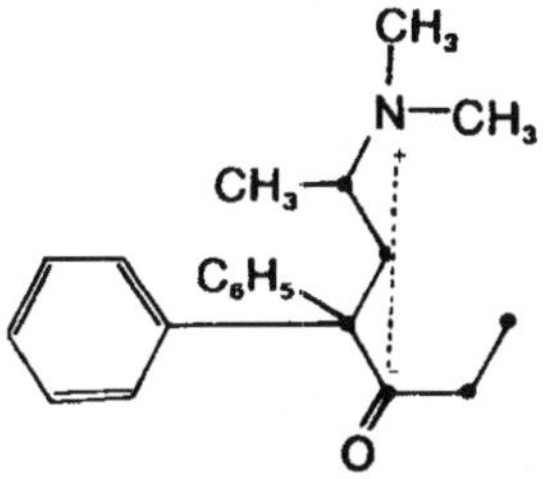

methadone *cf.* morphine structure

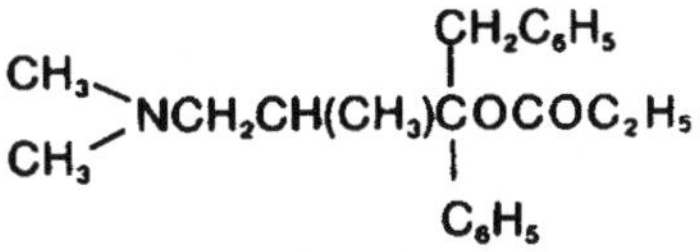

$(CH_3)_2NCH_2CH(CH_3)C(CH_2C_6H_5)(C_6H_5)OCOC_2H_5$

dextropropoxyphene

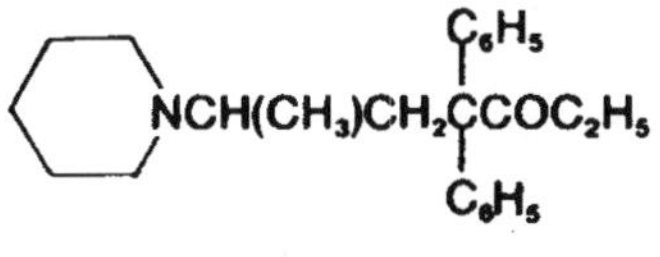

dipipanone

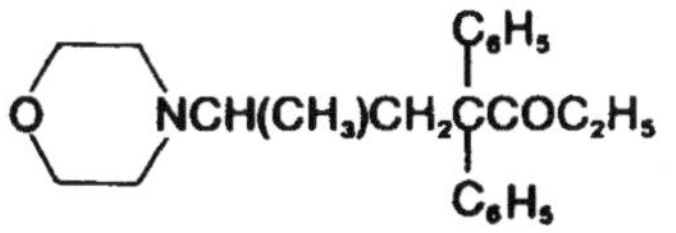

phenadoxone

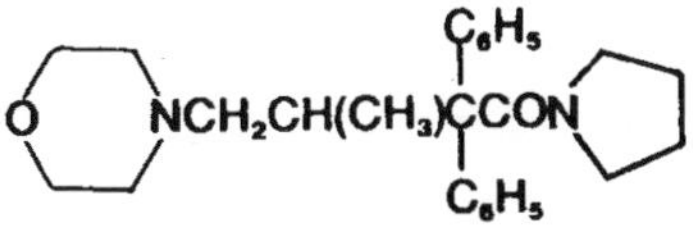

dextromoramide

Fig. 13.11 Methadone and structurally related compounds

another alkyl group. Although this compound seemed itself devoid of any analgesic properties, it was noted that it antagonised the properties of codeine. In 1941 a corresponding transformation was effected on morphine to give a substance that was named nalorphine (Fig. 13.12).

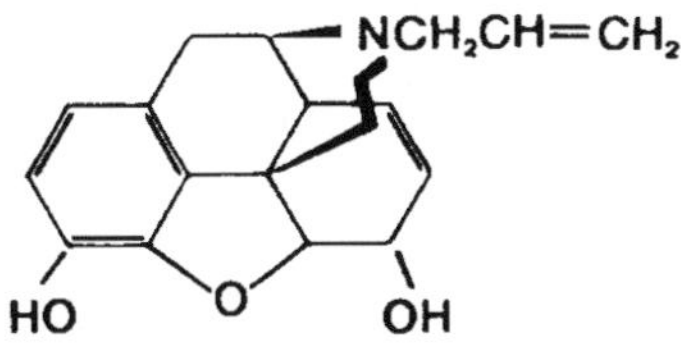

$NCH_2CH{=}CH_2$ HO OH

Fig. 13.12 Nalorphine

The first and obvious use of nalorphine was therefore to treat cases of poisoning by morphine and its derivatives. However, it has also proved very useful for diagnosing cases of addiction.

Aspirin

Experiments have shown that both salicylic acid and acetylsalicylic acid (aspirin) can breach the protective barrier in the stomach and cause stomach bleeding. For most people the bleeding produced is trivial — from half to two millilitres after two tablets; however, for some people it can be hundreds of millilitres, with such people requiring emergency hospitalisation. In acid solution, aspirin is un-ionised (Fig. 13.13a), fat soluble, and can diffuse through the stomach's protective barrier.

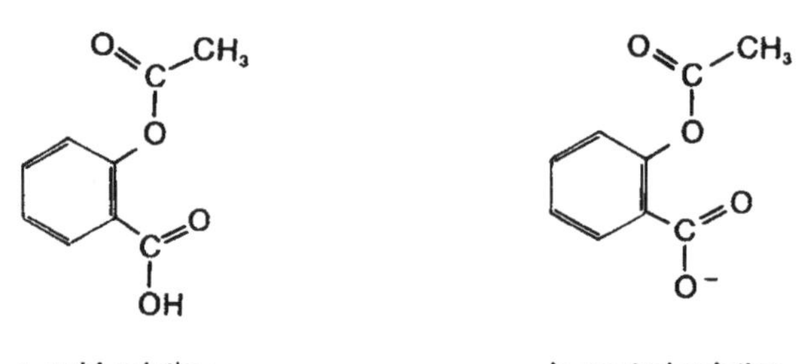

Fig. 13.13
Aspirin in acid and neutral solutions

Once through, it is in a neutral environment — it ionises and then cannot pass back again. (Compare this with the effectiveness of HOCl versus OCl^- as a bactericide in swimming pools: see Chapter 8.) The rate of diffusion is enhanced by alcohol, even when the contents of the stomach have a low acidity. The cocktail party story of aspirin and alcohol being potent is seen to be well founded. Such cooperative action is often called *synergism*.

The related methyl salicylate (Fig. 13.14), which has the common name *oil of wintergreen*, is used externally to ease the pain from rheumatism and strained muscles.

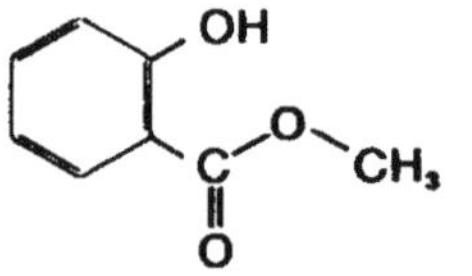

Fig. 13.14
Methyl salicylate

Aspirin that is kept too long begins to hydrolyse to salicylic acid, which is not well tolerated by the human body; the crystals can be seen on an old tablet.

Soluble aspirin is either the sodium or the calcium salt of normal aspirin. These salts immediately form aspirin in the acid stomach in the form of *fine* crystals and possibly cause less gastric distress. Some analgesics, such as Panadol/Tylenol, contain *p*-acetylaminophenol (4-hydroxyacetanilide, acetaminophen, paracetamol), which is comparable to aspirin as a pain reliever (Fig. 13.15), but is gentler to the stomach.

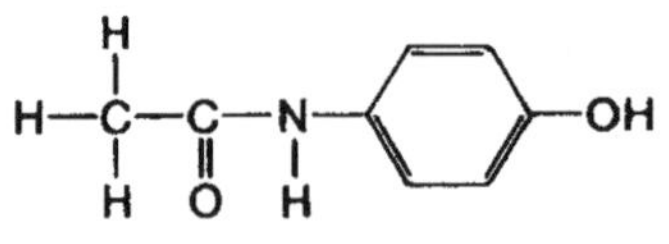

Fig. 13.15
Paracetamol/acetaminophen

Potency

Potency is compared in two ways. Chemists consider a mole-for-mole effect between drugs. Physicians compare them on a dose-for-dose basis — where in fact the number of molecules in a dose will be different for different drugs. In order to maintain a steady level of the drug in the body, the dosing must take into account the half-life of the drug in the body (see Table 13.8). Both of these approaches have validity in their

TABLE 13.8
Half-lives of various drugs in the human body

Drug	Time(hours)
Acetylsalicylic acid (aspirin)	0.25 ± 0.03
Amitriptyline	16 ± 6
Amoxicillin	1.0 ± 0.1
Bromide ion	168
Caffeine	4.9 ± 1.8
Chlorpromazine	30 ± 7
Cimetidine	1.9 ± 0.3
Clonazepam	23 ± 5
Cocaine	0.71 ± 0.26
Dapsone	28 ± 3
Desipramine	18 ± 6
Diazepam (Valium)[a]	18 ± 6
Erythromycin	1.8 ± 0.6
Flurazepam	1.5
Imipramine	18 ± 7
Isoniazid — fast acetylators	1.1 ± 0.1
— slow acetylators	3.1 ± 1.1
Methadone	35 ± 12
Morphine	3.0 ± 1.2
Nicotine	2.0 ± 0.7
Nitrazepam	26 ± 3
Nitroglycerine	2.3 ± 0.6 minutes.
Nortriptyline	31 ± 13
Phenobarbital	99 ± 18
Phenylbutazone	56 ± 8
Protriptyline	78 ± 11
Streptomycin	5.3 ± 2.2
Sulfadiazine	7.0 ± 3.9
Tetracycline	10.6 ± 1.5
Theophylline	9.0 ± 2.1
Thiopental	9.0 ± 1.6

[a] You must also consider the active metabolites, for example, desmethyldiazepam (62 ± 16) and oxazepam (7.6 ± 2.2).

Source: A. G. Gilman, L. S. Goodman, T. W. Ball and F. Murad, Goodman and Gilman's pharmacological basis of therapeutics, 7th edn, Macmillan, New York, 1985, Table A-11-1.

particular usages, but the distinction must be kept in mind (see also The dose in Chapter 2).

Amphetamines

Amphetamines were once used to treat obesity, mild depression and narcolepsy (a tendency to fall asleep at any time), and certain behavioural disorders in children. The last is the only current use. Amphetamines are pep pills. Ordinary doses of 10 to 30 mg per day provide a feeling of well-being and increased alertness.

Caffeine in athletes
Caffeine is such a common stimulant and ingredient in over-the-counter drugs that a limit has been set by the International Olympic Committee of 12 μg/mL in the urine of athletes at any time. This level is reached by imbibing about one cup of strong filter coffee.

Amphetamines were used by the military during World War Two to boost concentration during patrols and were legally available to students cramming for exams in the 1950s. Their use is now more restricted, and some derivatives are very dangerous. Athletes take them because they increase the energy-producing capacity of muscles. Ephedine is similar but weaker and was used by Diego Maradona in the 1994 soccer World Cup, leading to his expulsion.

These amines are weak bases and are ionised in the mildly acid urine, and then cannot 'escape' back through membranes into the tissue again (see aspirin earlier in this chapter, and pool chlorine in Chapter 8).

Amphetamine and epinephrine (see Fig. 13.16) are *optically active.* This means that there are two compounds with *exactly the same formula* but whose structures are mirror images of each other, and cannot be superimposed. If you look at your hands, they are pretty much the same shape, but you can't place one hand in an identical position on top of the other. However, if you hold them parallel, one acts as the image of the other in an imaginary mirror placed between them. In fact, pairs of chemicals related in this way are called left and right handed (!) or, using the Latin, *laevus* and *dexter*, *l*- and *d*- for short. It is actually a bit more complicated than this — but isn't it all? Strange as it may seem, compounds differing only in this way can be biologically very different in their activity. *Benzedrine* is a 50:50 mixture (racemic) of the *d*- and *l*-amphetamine, but, as the *l*- form is less active on the central nervous system, pure *d*- or *dexedrine* is obviously nearly twice as potent.

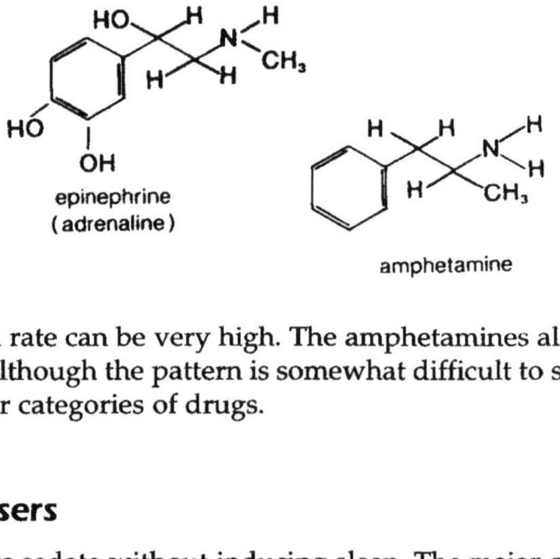

epinephrine (adrenaline)

amphetamine

Fig. 13.16
Biogenic amines

The death rate can be very high. The amphetamines also form a family of drugs although the pattern is somewhat difficult to see and tends to overlap other categories of drugs.

Tranquillisers

Tranquillisers sedate without inducing sleep. The major ones are used in the treatment of schizophrenia by blocking dopamine receptors in the brain. Many of them are based on a compound called phenothiazine (Fig. 13.17) and its derivatives (Table 13.9).

Derivatives of phenothiazine that retain the sulfur atom, but not the nitrogen atom, are the thioxanthine tranquillisers, for example flupenthixol (Fluanxol) and thiothixene (Navane).

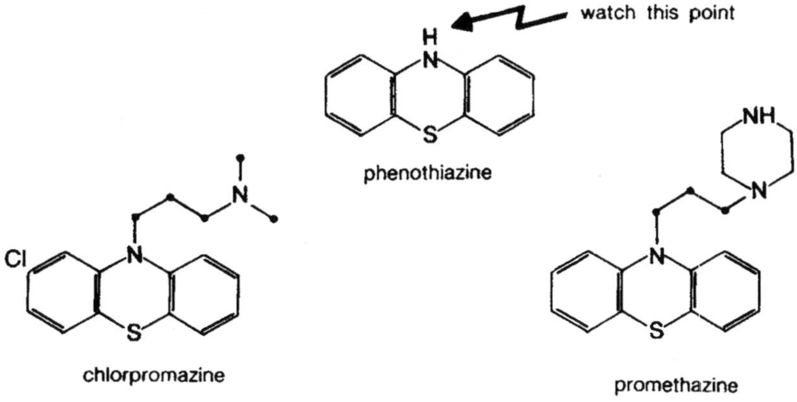

watch this point

phenothiazine

chlorpromazine

promethazine

Fig. 13.17
Phenothiazine and two of its derivatives

TABLE 13.9
Series of phenothiazine tranquillisers with different types of substituent

Generic name	Trade name
Aliphatic series	
Chlorpromazine	Largactil, Protran[a], Promacid[a]
Promethazine	Phenergan[a], Meth-Zin[a], Progan[a], Prothazine[a], Avomine[a]
Piperidine series	
Thioridazine	Melleril, Aldazine
Pericyazine	Neulactil
Piperazine series	
Prochlorperazine	Stemetil, Compazine[a], Anti-Naus[a]
Thiopropazate	Dartalan[a]
Fluphenazine	Anatensol
Fluphenazine decanoate	Modecate
Trifluoperazine	Stelazine, Calmazine[a]

[a] Many not available in Australia

If the sulfur and the nitrogen atoms of phenothiazine are replaced by –CH=CH– and –CH– respectively, one of the derivatives is protriptyline (Fig. 13.18). Compare it with the tricyclic antidepressants in Figure 13.19 (overleaf).

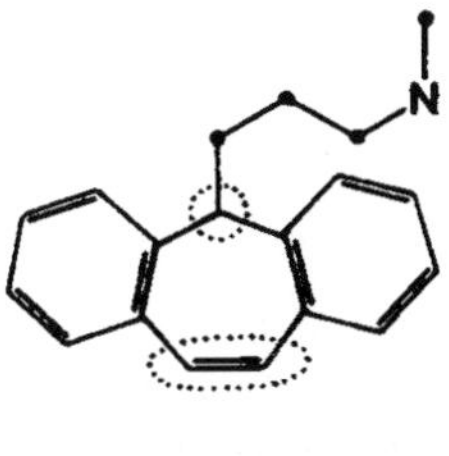

Fig. 13.18
Protriptyline

All these compounds are used to relieve the symptoms of schizophrenia and reduce the likelihood of relapse, and they affect the brain stem rather than the cortex. Their use has profoundly modified the

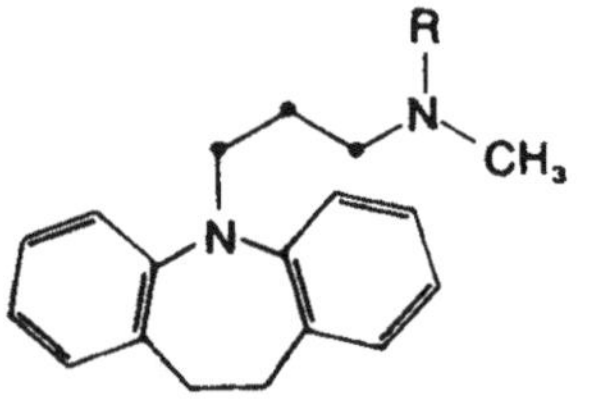

imipramine, where R = $—CH_3$
desipramine, where R = —H

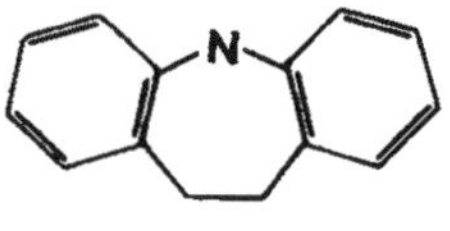

dibenzazepine

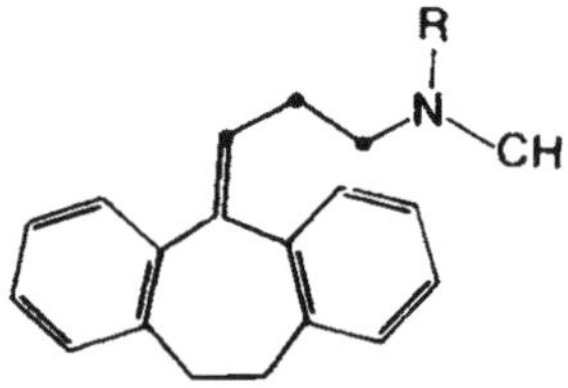

amitriptyline, where R = $—CH_3$
nortriptyline, where R = —H

Fig. 13.19 Tricyclic antidepressants

problems of the mental hospital, but they do carry a high incidence of adverse reactions.

MINOR TRANQUILLISERS

The most common of the minor tranquillisers are built up on a benzodiazepine nucleus. Four are illustrated in Figure 13.20.

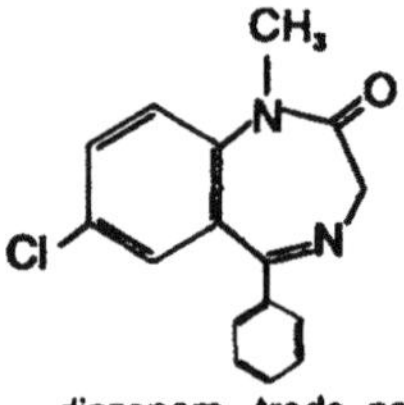

diazepam, trade name: Valium

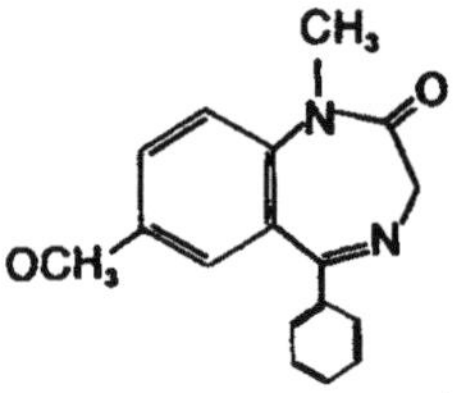

oxazepam, trade name: Serenid

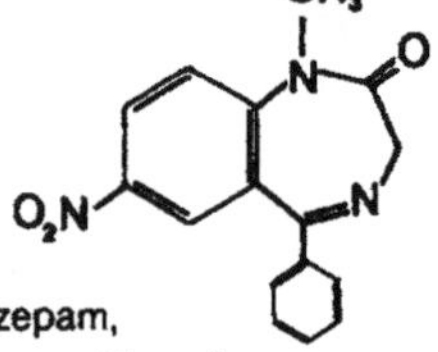

nitrazepam,
trade name: Mogadon
(sleeping pill)

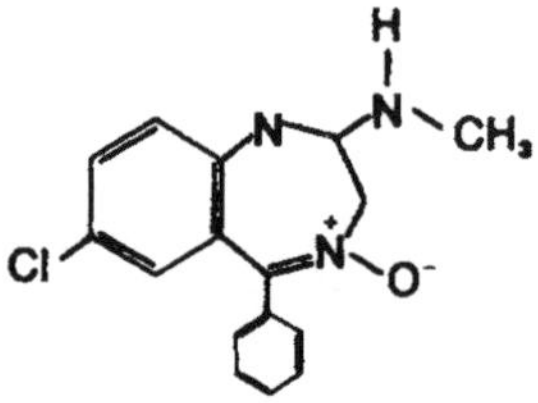

chlordiazepoxide, trade name: Librium

Fig. 13.20 Minor tranquillisers

Librium was used in the treatment of neuroses, behaviour disturbances, alcoholism, and as premedication for anaesthesia. Valium is used to reduce symptoms of anxiety. The differences relate to how fast they metabolise to the fast-acting actual drug — nordazepam. Valium loses the 1-methyl group, while Librium hydrolyses the 2-methylamino group to an oxygen.

Antihistamines

Many people are allergic to pollen, stings, dust, etc. An allergen is a substance that initiates the allergic response. It is usually a protein, but is sometimes a polysaccharide. For a person with pollen allergy, a pollen grain enters the nose and clings to the mucous membrane. The nasal secretions acting on the pollen grain release the grain's allergens and other soluble components, which penetrate the outer layer of the mucous membrane. By a series of events that are not well understood the allergen forms a complex with an antibody of a type that is present in unusually high concentrations in allergic persons. The complex is responsible for the release of the allergy mediators, one of the most potent being histamine (Fig. 13.21). Histamine is formed by the breakdown of the amino acid histidine; it accounts for many of the symptoms of hay fever and other allergies.

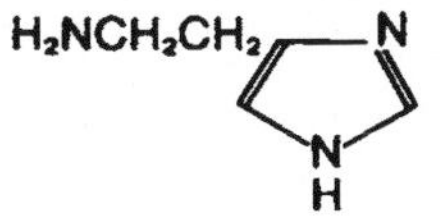

$H_2NCH_2CH_2$ N N H

Fig. 13.21
Histamine

Antihistamines are most widely used for treating allergies and there are more than 50 types available. Many contain, as does histamine, an ethylamine group, $-CH_2CH_2N=$. These drugs compete with histamine for the receptor sites normally occupied by it on cells, and thus prevent it from causing allergic reactions. An example of a well-known antihistamine is Polaramine (Fig. 13.22).

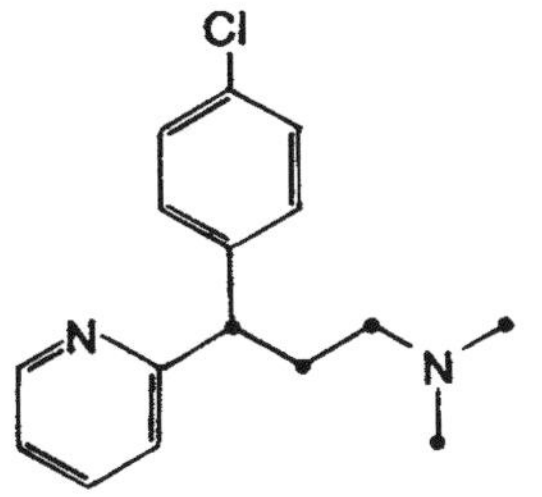

Cl N N

Fig. 13.22
Dexchlorpheniramine (Polaramine)

Note that some of the tricyclic antidepressants have antihistamine effects as well, because they also contain the ethylamine group.

Hormones used in sport

The first drug to be taken seriously for boosting athletic performance was the male hormone testosterone, produced in the testes. This hormone was made synthetically by the German chemist Butendant in 1935, and led to determining its role in male sexual development, stimulating growth through protein synthesis, and the production of red cells from bone marrow. The Nazis in particular tried to exploit its properties during World

Urine tests for UK footballers
Footballers in the UK are regularly tested without warning. Because most drug molecules tend to be non-polar, they are extracted by shaking a sample of the urine with a non-polar solvent and allowing the two layers to separate (see Solubility in Chapter 1). Diethyl ether, or (in the tropics) tertiary butyl methyl ether, is often used. Solvent is removed from the extract and the residue is analysed.

War Two by giving it to their soldiers serving in Russia in the hope of producing a 'master race' of soldiers.

However, testosterone is rapidly absorbed by the intestine and passes straight to the liver, which destroys it, so it is almost inactive by mouth. So why not inject it into the muscle? As a fat-soluble drug, injections were given in oil which is painful, and the drug also quickly 'leaked' from the muscle into fat surrounding the injection site, entered the blood, and was again destroyed. A little thought directed chemists into making a modification to the molecule, turning an OH group into an ester, and thereby making the molecule more fat soluble still, and keeping it out of the bloodstream for longer. Body enzymes hydrolysed the ester back to the hormone and allowed it to work. A whole series of so-called 'anabolic' steroids were made which are not destroyed by the usual enzymes and thus stay in the body for a long time.

At the Munich Olympics in Germany in 1972, about 70% of the competitors admitted taking these performance enhancers and they were banned for subsequent Olympics. Not only were they decreed to be unfair, but also these male hormones caused acne, premature baldness, and increased heart attacks in men, as well as masculinising women who took them. Both men and women can suffer heart conditions, liver cancer, and addiction, while teenagers can have their growth stunted.

Chemists came to the rescue by introducing the sensitive analytical technique of combined gas chromatography with mass spectrometry (GCMS) to detect one ppm in the urine of any unnatural metabolites of the anabolic steroids.

However, many steroids and their metabolites exist as complexes, and subtle chemistry is needed to cleave them before analysis.

By going back to the first synthetic, testosterone ester, only the natural metabolite would be detected. But normal testosterone always comes with a partner (isomer) called *epitestosterone* and in the blood of normal people they are in the ratio T:E of 6:1. The International Olympic Committee (IOC) declared that any athlelte with a ratio in excess of this would be banned. With funding, chemists can synthesise almost anything, including in this case, epitestosterone. Thus T and E can be given in the 'right' ratio and presto, cheating is back again.

Now production of T is controlled by a protein produced in the pituitary gland located at the base of the brain, called 'luteinising hormone', (LH). LH, and the T it produces, regulate each other in a feedback loop and so the amount of T is stabilised in adult men (>18) to approximately 0.6 μg/L. If extra T is given artificially, the amount of LH drops below normal. LH is detected by another chemical technique called radio-immunoassay. The battle continues into extra time, and one wonders whether athletes are now taking extra LH to keep the level normal. More in the next edition!

Hallucinogenic drugs

Hallucinogenic (or related psychotomimetic) drugs derived from various plants and fungi have been used from time immemorial. They were used for religious purposes, and for festivals and orgies. The use of the emetic toadstool *Amanita muscaria* extends over thousands of years. The Aztec and Mayan cultures used the peyote cactus, from which mescaline (Fig. 13.23) is derived.

T/E ratio

The median T/E ratio in the Caucasian population is about 1.5; hence the IOC ruling that a ratio in excess of 6:1 is taken as positive. Recent research shows that in the Oriental population the median ratio is 0.5, with some as low as 0.1. These athletes could take testosterone and never exceed the limit! (Note: we are talking about ratios, not concentrations.)

Source: Professor Prapin Wiliarat, who is the Deputy Director of the Thai Doping Control Centre, preparing the laboratory for the Asian Games in December 1998.

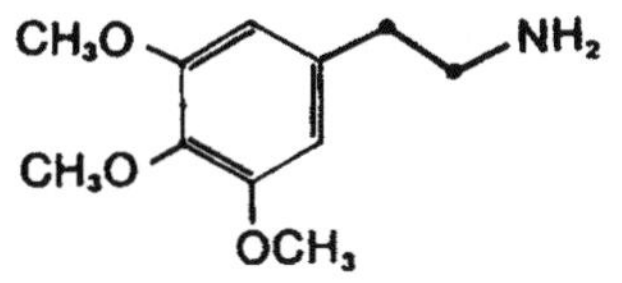

Fig. 13.23
Mescaline

They also used the psilocybe mushroom (or sacred mushroom Teonanacatl), whose active principle is psilocybin (Fig. 13.24), which is about 30 times as potent as mescaline.

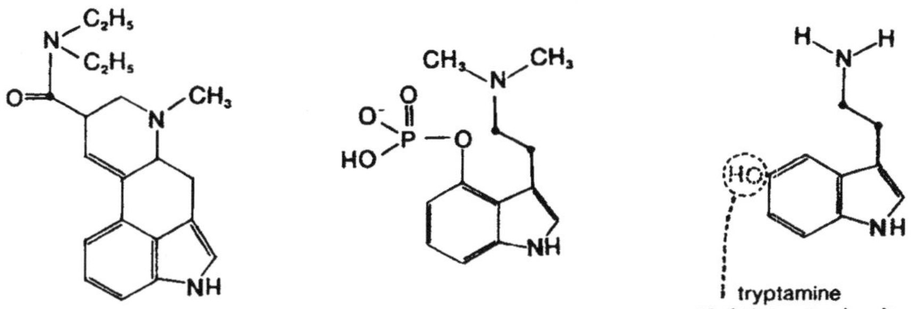

lysergide (LSD_{25})

psilocybin

tryptamine (5-OH tryptamine is also important)

Fig. 13.24
Some hallucinogens based on tryptamine

Similar mushrooms are found in Australia. From a plant called ipomoea (morning glory) the Mexican Indians obtained a substance similar to lysergic acid (Fig. 13.24), and from the plant *Datura stramonium* (thorn apple) they obtained scopolamine (hyoscine) and atropine. Other plants used in Central and South America contained cocaine (Fig. 13.25), and there is at least one hallucinogenic animal, a caterpillar found in bamboo stems.

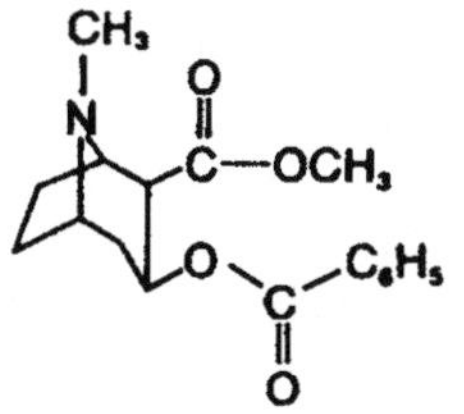

Fig. 13.25
Cocaine (blocks nerve transmissions)

Mescaline
Mescaline was made famous by the author Aldous Huxley. It is classed as a catecholamine, along with amphetamines, to which it is structurally related.

Lysergic acid diethylamide (LSD)

LSD is classed as an indoleamine. It is one of the most potent drugs known, and doses as low as 20 to 25 μg (1 μg = 10^{-6} g, 1 millionth of a gram) are capable of causing marked effects in susceptible individuals. It was at one stage believed to cause chromosome damage when taken in large doses, but this is now disputed. Lysergide (Fig. 13.24) was discovered, in a chemical sense, by the Swiss chemist Albert Hoffmann in 1938 when he routinely tasted some of the compound while investigating a modified ergotamine as an improved drug for childbirth.

Clandestine manufacture is usually from ergot alkaloids, to yield lysergic acid, to which the diethyl groups are easily added. Ergot itself is found on many plants, particularly rye. An ergot alkaloid is used to induce uterine contractions. In ergotamine, the diethylamino group is replaced by a peptide (a mini-protein).

Cannabis

The use of cannabis (marijuana, hashish, Indian hemp) has a long history. The most active ingredient in the extract is tetrahydrocannabinol, (THC) (Fig. 13.26).

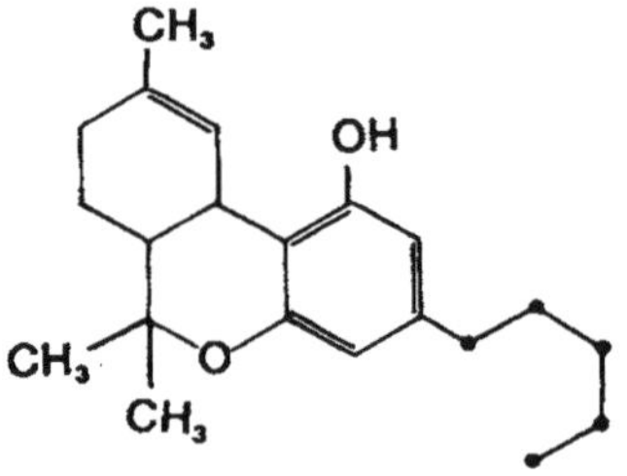

Fig. 13.26 Tetrahydrocannabinol (THC)

THC can be obtained from the fruiting or flowering tops of the cannabis plant, whose cultivation is banned in Australia. There are many ingredients in cannabis other than THC and their long-term effects are unknown. There is evidence to suggest that, chemically if not socially, it is less harmful than tobacco, except when driving a car.

Mode of action of drugs

To understand the effect of drugs on humans, we naturally have to explore the chemical and pharmacological action of the chemicals involved. However, the effects of drugs are strongly influenced by the personal and social environment. Traditional drugs used in traditional ways often cause few problems. Opium and cocaine are good examples. In a different legal and social climate their effects can be disastrous.

Physical dependence on opiates is a complex issue, which is not treated in this book. It is worth noting the chemistry. Opiates reduce pain, aggression, and sexual drive, hardly the stuff to make violent criminals. (They do, however, make zonked car drivers.) The Chinese opium smokers were blissful and peaceful. Our social structure has made these drugs illegal. This provides a certain social attraction to some, and also sets a high price for them. This in turn sets the scene for crime and corruption. One way of attempting to separate out the strictly biochemical from additional social effects is to look at animal (especially mammal) experiments.

People's inclination to take drugs is shared with that of other mammals, which show patterns of self-administration that are strikingly similar to those found in human users of the same drug. This seems to rule out any specific mental or physical weakness in the human addict. Animals will press a lever more than 4000 times to receive repeated injections of cocaine. When given free access, they will self-administer high daily doses that cause severe toxic effects and even cause the animals to mutilate themselves. With cocaine (and amphetamines) the animals alternate between periods of self-imposed abstinence, and periods of drug taking, and they generally die from starvation and poisoning. If saline solution is substituted, there is a burst of rapid lever pulling for several hours, then abruptly all responses cease and are not resumed. When morphine is provided, the behaviour is very different. The animal will raise the dose it is giving itself gradually over a period of weeks, then keep the rate at a steady level that avoids excess toxic effects on the one hand, and withdrawal symptoms on the other. When saline solution is then substituted, the animal will continue to press the supply lever at a slow but steady rate (except during the peak of withdrawal) for weeks on end, showing that the addiction remains. Although the behavioural patterns in humans are influenced by social and psychiatric factors as well, the basic similarity is worth bearing in mind.

Cocaine and amphetamines belong to a group of drugs that mimic the natural substances that stimulate the central nervous system (CNS). They cause an elevation of mood, a sense of increased strength and mental capacity, and less need for sleep or food. (Natives living high in the Andes chewed the leaves of the coca bush for generations for just this purpose.) The cocaine is converted from the hydrochloride salt to the free base with alkali, and extraction with organic solvents. Absorption from the lungs is then increased dramatically. The drug is highly addictive. Given a choice of cocaine and food, monkeys select cocaine (over a period of eight days). Continuous access causes weight loss, self-mutilation, and death within about two weeks. Whereas the effects of cocaine fade quickly, because the esterase enzymes in the bloodstream quickly hydrolyse it, the effects of amphetamines last for hours. This has been demonstrated in double-blind experiments where subjects could not tell with which drug they had been injected until some time had passed.

'I have an earache . . .'

2000 BC: here: eat this root.
AD 1000: that root is heathen; here, say this prayer.
AD 1850: that prayer is superstition; here, drink this potion.
AD 1940: that potion is snake oil, here, swallow this pill.
AD 1985: that pill is ineffective; here, take this antibiotic.
AD 2000: that antibiotic is unnatural; here, eat this root.

SENSITIVITY TO DRUGS DETERMINED BY GENES

While 90% of Asians (Japanese and Chinese) are fast acetylators, only 40 % of Americans (both black and white) acetylate drugs fast. Acetylation is often

Acetylation
Individuals fall into two genetic groups; those who acetylate drugs like sulfonamides quickly, and those who do so slowly. The amount of this enzyme produced is caused by a single difference in one chromosome.

the first step in metabolising and thus deactivating a drug and so **slow** acetylators are exposed to **higher** levels of a drug given at the **same** dose.

The acetyl derivative of sulfathiazole is not very soluble. It tended to block kidney tubules and led to many deaths. It was replaced by sulfadiazine para. The same acetylating enzyme deactivates some carcinogens, for example aromatic amines such as benzidene and o-tolidine, used in dyestuff man-ufacture and as analytical reagents in the detection of blood (and once for chlorine levels in water). Slow acetylators are at higher risk of bladder cancer from these chemicals. In a study at ICI in the UK, former dyestuff workers with bladder cancer were examined. Twenty-two were slow acetylators. Genetic probes allow the detection of such differences. Other enzymes are related to the detoxification of some pesticides. Genetic screening will have profound effects on our attitudes to occupational health and the grounds on which legal compensation claims are liable to be based.

A parallel genetic difference exists in regard to alcohol dehydrogenase, which converts alcohol in the liver to acetaldehyde.

Drug tolerence and lethality
The therapeutic index is the ratio of the toxic dose to the effective dose. The larger this factor, the greater is the safety in the use of the drug. The therapeutic index is dependent on two types of drug tolerance:
- **pharmokinetic tolerance** — a tolerance due to changes in the concentration of the drug in the body caused by changes in liver activity, and
- **pharmodynamic tolerance** — a tolerance caused by the receptor (where the drug acts) requiring more drug, while the concentration for receptor poisoning may not change.

Most Caucasians oxidise ethanol only slowly in the first stage to acetalde-hyde. Most Japanese and Chinese, on the other hand, generally have a gene that codes for an enzyme that oxidises alcohol faster. As a result, a few sips of ethanolic beverage bring a deep red colour to their cheeks and an unpleasant tingling caused by the acetaldehyde, while Caucasians tend to get drunk instead because the level of alcohol stays high for longer.

The further oxidation to acetic acid occurs at the same rate in both fast and slow acetylators.

Tolerance

There are aspects of drug taking that have a definite chemical basis. One of these is tolerance. Drug users often develop tolerance to the chemical they are using. This can occur for a number of reasons. In the case of the opiates, most of the tolerance comes from adaptation of the cells in the nervous system to the drug's action.

In the case of alcohol, barbiturates, and related hypnotics, a group of depressants of the central nervous system (CNS), chronic use causes the capacity of the enzymes that metabolise the drugs to increase, that is the alcohol can be removed faster. Social drinkers can metabolise ethanol only with the liver's slow-acting enzyme, alcohol dehydrogenase. (Police sometimes use this information to calculate from the level they measure on a breathalyser back to some earlier period of time.) Chronic drinkers are no more efficient with this enzyme, but they induce a new alcohol-destroying enzyme of the P-450 (cytochrome mono-oxygenase) type in the liver. Such persons can perform well on difficult tasks at blood alco-hol levels above 0.2 mg/mL (the legal limit varies: 0.05, 0.08 to 0.1). After several weeks of abstinence, there is a decline in the capacity of this enzyme, so the abstinent alcoholic, and normal individual, metabolise alcohol at the same low rate. Chronic use often means a higher blood concentration is needed to produce the same effects, that is, it produces pharmacodynamic tolerance. This in turn means that to obtain the same effects, the person will consume more of the drug. However, the fatal dose of the drug does not change. The result is often death by overdose.

When the same enzymes are involved, tolerance to one drug can cause cross-tolerance to another. An example is the cross-tolerance of alcohol with benzodiazepines. Thus chronic drinkers will deal not only with alcohol more effectively but also with Valium. If they consume both drugs at the same time, the following happens: the alcohol will monopolise the enzyme, which then is not free to deal with the Valium. The effect of the Valium is enhanced and prolonged (and of course added to that of the alcohol).

There are other aspects, such as linking behaviour to drug level, which are much harder to correlate chemically. The point at which marked intoxication is caused by drinking of alcohol can be monitored by measuring the blood alcohol (or breath) levels. However, there is an interesting difference between the level found to correspond to intoxication 'on the way up', that is, while drinking, compared to 'on the way down', that is, after drinking has ceased. The effect on behaviour (in particular for alcohol) appears to be far less on the way down than on the way up, and this has meant that drivers caught by a breathalyser test the morning after a heavy drinking bout have no idea that their level is still high.

Nicotine and tobacco

Columbus discovered tobacco smoking as well as America, and the name of the plant, *Nicotiana tabacum*, derives from the entrepreneur who promoted its sale, Jean Nicot. The active ingredient, nicotine, was isolated in 1828. New varieties, better methods of curing the leaf, coupled with technology for mass production, allowed the introduction in the mid-nineteenth century of the cigarette — cheaper and neater than the cigar, with a smoke so mild it could be inhaled! About 4000 compounds have been found in cigarette smoke. No other drug of dependence causes cancer, and tobacco is the only environmental cause of cancer that is increasing.

Classification for health purposes has concentrated on the levels of nicotine, tar (which contains the potent carcinogens), and carbon monoxide. The carbon monoxide reacts preferentially with the red corpuscles in the blood. On removal of the source of carbon monoxide, the equilibrium with oxygen is gradually restored.

The nicotine content of tobaccos can vary from 0.2 to 5%, and provides from 0.05 to 2.0 mg (1982 average 1.0 mg) per cigarette to a smoking machine. In cigarettes, the nicotine is nearly always present in a protonated form in which it is less readily absorbed through the mouth (hence the need to breathe the smoke into the lungs) (smoke pH 8.5). Note the analogy to cocaine and 'free base' cocaine. The nicotine is suspended on the minute particles of tar, and absorption from the lung occurs in

The nicotine connection
By impersonating acetylcholine, nicotine alters the way the brain processes information. Like cocaine and amphetamines, nicotine releases dopamine, which makes us feel good — it activates the reward system. It acts on other brain receptors to increase alertness. It is also involved in appetite, mood, and cognition. 'If nicotine didn't kill you, it would be terrific.'

seconds and is almost as fast as intravenous injection. Peak concentrations found in the blood are typically 25 to 50 ng/mL.

A fascinating aspect of smoking is the way smokers titrate their nicotine needs. When heavy smokers are unknowingly given cigarettes with a higher content of nicotine, they subconsciously reduce the number smoked and alter their puffing pattern to maintain about their usual level of nicotine. Conversely when given low-nicotine cigarettes, they increase the number smoked and/or puff more efficiently. Thus another example of counterproductive social engineering is to insist on low-nicotine cigarettes because the smoker then takes in more tar and more carbon monoxide for the same level of nicotine.

This is in stark contrast to marijuana smokers, who do not titrate their desire for the active drug because its absorption and effect are much delayed. It takes seven to 10 minutes for the blood plasma concentration of the active ingredient, Δ^9-THC, to reach its peak, and the physiological and subjective effects do not reach a maximum for 20 to 30 minutes. Titration is thus not possible. (Taking the drug through the mouth delays onset of effects for a half to one hour.) The drug is slowly destroyed in the liver and gradually excreted in the urine and faeces. This long delay makes it difficult for law enforcement authorities to determine whether the drug was in use while, for example, driving, if this is (as it should be) an offence. In fact the legal difficulties of enforcement of non-drug use while performing activities such as driving or operating machinery, would be a major technical obstacle to decriminalisation, if this were otherwise desired.

Drug interactions

The interaction of drugs is an extremely important area of study. One drug can change the pharmacological effect of another — using the term drug in the evident sense to include such substances as alcohol, some foods, and food additives. Some drugs even precipitate in the bottle (e.g. tetracycline and calcium ion).

Drugs that are taken orally have to be absorbed through the gut, and this can be influenced by other material present. By using suitable coatings, a drug can be absorbed either in the acidic stomach or the alkaline duodenum. Once the drug is in the plasma it can become bound to protein and only a small percentage remains free and active. This percentage can be drastically altered by another drug which kicks the first one off its protein site. Often use is made of this method to boost the efficiency of a drug. Also, a drug can affect the efficiency of an enzyme and hence influence the rate at which a second drug is broken down by that enzyme. The monoamine oxidase (MAO) inhibitor drugs, and the consequences of eating cheese while they are being taken, are a classic example.

The way and speed with which drugs are metabolised by the body can also depend on genetic factors, so comparisons between animals and humans, and between individuals, can be misleading. They also depend on physiological factors such as age, diet, hormones (including the effects of pregnancy), and disease states — especially if the liver is involved. The old are particularly liable to be treated with several drugs simultaneously and they, in particular, will have impaired metabolism, which will affect the drugs' effects on them.

Very often a drug is changed in the body to another compound. Sometimes the new compound is inactive, or it may be less active or more active than the original. The original may even be completely inactive, and it is the new compound (metabolite) that is the 'real' drug. Some examples of this are shown in Figure 13.27 (overleaf). (See also Page 180, Fig. 7.7)

Note the tremendous importance of the liver in the metabolism of drugs. You may begin to realise what immense problems this opens up. Not only do we have to have information about the effect of a new drug we may want to introduce, but also about the effect it has on other drugs and the effect they have on it. We have to know what other compounds it forms in the body, and what their properties are. The rat is still one of the most popular animals used in toxicity testing, but it has active microorganisms present in its stomach, so orally administered drugs may be extensively metabolised by bacteria even before absorption, giving a markedly different metabolic pattern from that obtained in humans.

On the other hand, many drugs used in treatment of illness are of high molecular mass (greater than 400) and, as a consequence, are excreted in the bile as well as the urine, so they are frequently subject to bacterial metabolism in the intestines. These products can be reabsorbed and further metabolised by the liver — a cycle of absorption, metabolism by the body, excretion, bacterial metabolism, reabsorption and metabolism by the body.

To top it all, there is also the time factor to be considered. The chemical β-naphthylamine was a very important intermediate in the dyestuff industry. It was found, however, to be a very potent carcinogen — it causes cancer of the bladder. It took a long time to realise this because there is a time lag of about 30 years between contact and cancer. It was only because a large number of ex-employees of a German chemical firm died of the same disease at about the same time that the link with the past could be established. The same story has repeated itself with the chemical vinyl chloride, used in the manufacture of the plastic poly(vinylchloride).

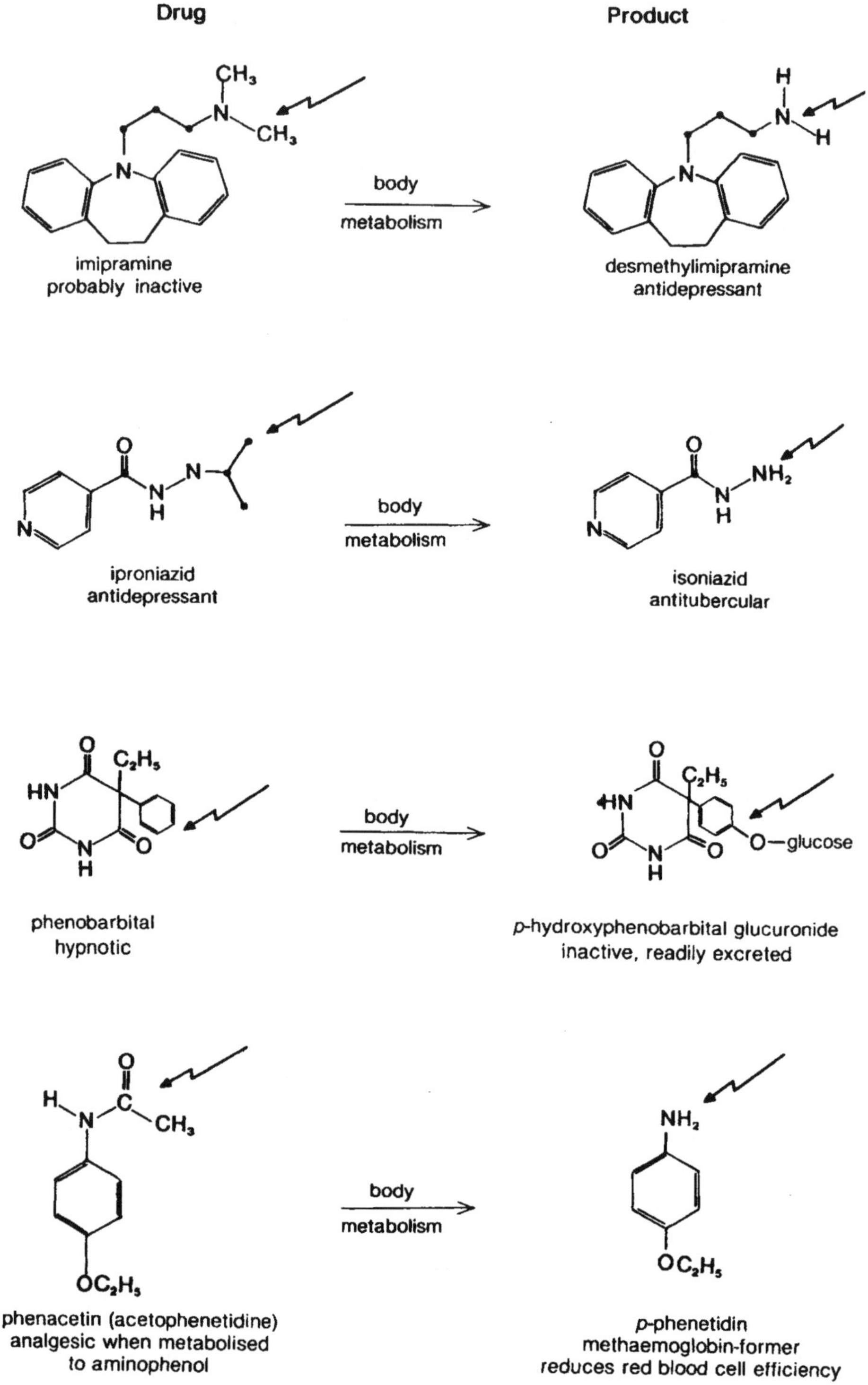

Drug | **Product**

CH_3 N CH_3 N

body metabolism →

H N H N

imipramine
probably inactive

desmethylimipramine
antidepressant

O N N H N

body metabolism →

O NH_2 N H N

iproniazid
antidepressant

isoniazid
antitubercular

O C_2H_5 HN O N H O

body metabolism →

O C_2H_5 HN O N H O O—glucose

phenobarbital
hypnotic

p-hydroxyphenobarbital glucuronide
inactive, readily excreted

O H N C CH_3 OC_2H_5

body metabolism →

NH_2 OC_2H_5

phenacetin (acetophenetidine)
analgesic when metabolised
to aminophenol

p-phenetidin
methaemoglobin-former
reduces red blood cell efficiency

Fig. 13.27
Effects of metabolism on the pharmacological activity of drugs

References

The consumption and production of pharmaceuticals, Bryan Reuben, Chapter 42, pp. 904–938, in *The practice of medicinal chemistry*, Ed. G.C. Wermuth, Academic Press, 1996. Three tables and some discussion comes from this excellent chapter.

Medicos balk over cheaper drugs, Crispin Hull, *Canberra Times*, 4-11-94

Writing about medicines for people: Usability guidelines & glossary for consumer product information, D. Sless & R. Wiseman, 1994, Communications Research Institute, Canberra, Australia.

Drugs in sport, A. George, *Chemistry Review*, Nov. 1994, p. 10.

Bibliography

The Merck index: An encyclopedia of chemicals, drugs, and biologicals, 12th ed. Merck and Co, NJ, 1996, over 10 000 entries, and much else. (First edition 1899.)

Martindale: The extra Pharmacopoeia, Ed. J.E.F. Reynolds, the Pharmaceutical Press, Royal Pharmaceutical Society, London, 31st edn., 1996. (First edition 1883.) Aims to provide unbiased evaluated information of the world's drugs and medicines. Part 3 gives the active ingredient(s) of over 50 000 official and proprietary preparations of the type, Dr Kopfschmidt's corn and callus comforter.

Goodman and Gilman's pharmacological basis of therapeutics, A.G. Gilman, L.S. Goodman, T.W. Ball, F. Murad (Eds.), 7th edn., Macmillan, New York. 1985; classic reference.

Drug interaction facts, 2nd ed. D.S. Tatro et al., J.B. Lippincott, Miss., USA, 1990; 900 pages with simple layout and a significance rating.

MIMS, Intercontinental Medical Statistics (Australasia) Pty Ltd, Sydney. Published bimonthly with an annual.

NHMRC. Standard for the uniform scheduling of drugs and poisons, No. 11. AGPS, Canberra, 1996.

International proprietary names (INN) for pharmaceutical substances, WHO, Geneva, No. 8, 1992.

A *question of balance: Report on the future of drug evaluation in Australia*, Peter Baume, AGPS, Canberra, 1991; smartening up the bureaucrats.

Xenobiosis, A. Albert, Chapman and Hall, London, 1987; a wonderful book developed for a graduate course for chemistry students at the Australian National University.

Murder, magic medicine, J.Mann, Oxford University Press, 1992; lots of fun stories.

Pharmaceutical chemicals in perspective, B.G. Reuben and H.A. Wittcoff, John Wiley, NY, 1989; good solid research and clearly written.

The aspirin wars: Money, medicine and 100 years of rampant competition, C.C. Mann and M.L. Plummer, Alfred A. Knopf, NY, 1991; a history of capitalism gone mad over a small simple molecule — enough to give you a headache.

Chemistry of energy

Jake always prided himself on the fact that he was more energy efficient than the rest of the household fixtures

In this chapter we look at energy in its broadest sense. It is a big topic and a big sector. Particular emphasis is placed on the car as one of our major users of energy and creator of problems.

Introduction

We have become concerned about conserving energy and we ponder the alternatives. We disapprove of the 'wasteful' production of packaging materials and the need to dispose of them. We worry about using large quantities of irreplaceable fossil fuels in our cars and electricity generators. How wasteful it is to produce an aluminium can, fill it with beer, empty it, and throw it away! But is it just as wasteful to eat a breakfast cereal largely provided by fossil fuel (with a small solar contribution) and then proceed to turn it all into waste heat by jogging mindlessly around the block, or by belting a ball against a wall?

We exist as highly complex organisms by ensuring that we waste energy continuously. That's the price of being improbable. What we need to understand is the manner in which energy can be wasted most sensibly.

An energy time scale is given in Figure 14.1.

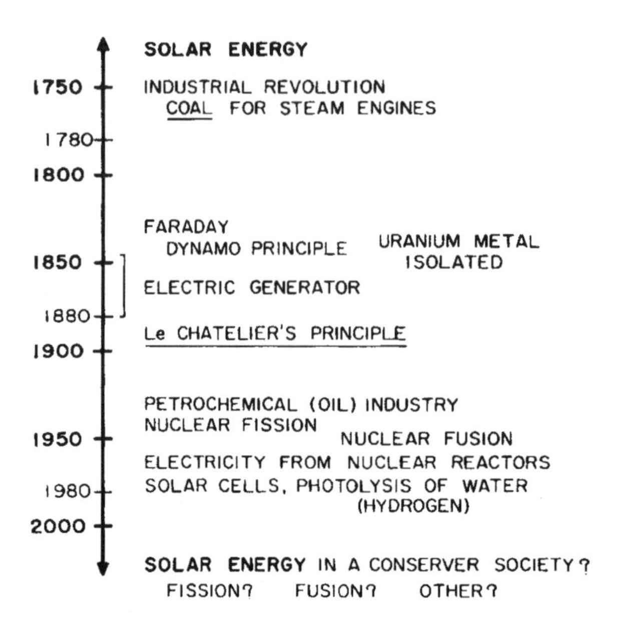

Fig. 14.1 Energy time-scale

What is energy?

What is energy? What do we use? What do we need? What about the sun? What do we pay for it? This section looks at energy from a consumer viewpoint. What things can we do, and ought we to do individually, to optimise our own situations? First we need to understand exactly what energy is. This concept is not as easy as it seems.

Energy equality 'All types of energy are equal — but some are more equal than others' (with apologies to Orwell).

Energy is a means of doing work, or producing heat, and is measured in *joules* and many other common units as well (the calorie, watt-hour, BTU, therm). All the units are interconvertible. Just as there are hard and soft currencies, so there are forms of energy that are desirable (because they are easily converted to other forms), and forms that are less desirable.

Conversion of energy from one form to another *always* involves losses. The size of these losses is set not only by the limits of our technology but also by limits set by the nature of energy itself.

We buy energy to use it, but in fact we do not use it. We *convert* it to work, and (finally) to a less useful form — heat. We extract *usefulness* from energy. Does energy come in a hierarchy of usefulness? It certainly does. If we were offered a certain amount of energy in the form of waste heat from a power station, and at the same price were offered the *same* amount of energy in the form of electricity or oil, we would *not* accept the heat (and it is not even straightforward what 'same' means here). The reason is that the electricity and oil are more readily converted to other forms of energy, whereas heat is not. But using the electricity to drive a heat pump, in say an air-conditioner, we have an efficiency (heat pumped divided by electrical energy used) of much greater than 100%. (We will return to this point.) In fact, we should be offered the waste heat from a power station at a cheap rate because, if the power authority can't sell it, the heat is lost, whereas fossil fuel can be stored.

The idea that the energy we 'use' is never destroyed, but is still around afterwards in the form of heat, is very important from an environmental point of view. All the fossil and nuclear energy we produce and use stays around as heat — to be radiated back out into space, hopefully.

Degradation of the usefulness of energy is measured by a function called *entropy*. The more useless energy becomes, the higher its entropy. It is the entropy that is remorselessly increasing and is a measure of how we run down our energy sources. Consider an analogy. Status in human society is linked to rarity. The least probable situations are the most desirable. You only have to look at 'situation' in a real-estate sense to see that there are fewer more desirable blocks of land than less desirable ones. In fact, this should be stated inversely. Desirability is evaluated by rarity. Perhaps a better analogy comes from a consideration of the distribution of shareholdings in a public company. If a change occurs from a few large holders to a large number of small holders, the effective control of the company by the shareholders is reduced drastically. This follows from the difficulty of obtaining coordinated, cooperative action if the number of shareholders is large. (See Appendix VIII, The entropy game, for further discussion.)

Electricity from ocean gradients
Instead of using electricity to create a temperature difference between two places, we can use an existing temperature difference to generate electricity. The temperature difference between the surface and depth in the ocean can be used to provide electricity by means of an air-conditioner type of device being used in the opposite way, whereby heat going from the hot region to the cold region via the machine produces electricity.

Because heat is so low in status in the energy hierarchy, consumers converting electricity directly into heat in an electric heater are selling themselves short — even though the conversion occurs with 100% efficiency! (Because heat represents about half the end *use* of all our energy, its importance from *this* point of view should not be underestimated. Obtaining heat more efficiently *is* important.)

Instead of *converting* electricity directly into heat, electrical energy can be used more efficiently if it is used to drive a heat *pumping* system. A refrigerator is a good example. Electrical energy is used to pump heat from the inside of the refrigerator to the outside coils. If the unit is set in a wall with a rear heat-exchange coil outside the house (and the door is left open to the room) we have an air-conditioner. Turn it around (or more conveniently, reverse the cycle of inside and outside heat-exchange coils) and we have a reverse-cycle air-conditioner that heats the house and cools the outside. The electricity is converted to mechanical work to *pump* heat. It does this far more efficiently than if the electricity is *converted* directly into heat in a radiator. The efficiencies are greater than the 100% of a direct conversion, and can be quite high for small temperature differences.

Most of the 'higher' forms of energy, such as electrical, mechanical, nuclear, solar, and chemical energy (with reservations), are about equally useful. Heat is the interesting one. Although it is the least useful form of energy, its usefulness is variable. The greater the difference between the temperature at which heat goes into an engine, and that at which it comes out of the engine, the more useful it is. If you think about it, all combustion engines have a hot in, and a colder out (in a car, it is the hot cylinder and colder exhaust; in a steam engine, it is the hot boiler and colder condenser). In Chapter 10, Box, Shape–memory alloys, we saw how a loop of

Hot heat
The 'hotter' heat is, the more useful it is.

Experiment
A bubble in a spirit level can be attracted and pulled along by placing a hot soldering iron near one end. The end of the bubble near the iron heats up and expands and this heat is dissipated through the colder end. If the iron is slowly moved, it follows it. The moving bubble is a simple heat engine.

wire made from shape (memory) alloy, Nitinol, would turn indefinitely around two pulleys if they were kept at two different temperatures.

The *theoretical* (without any losses) efficiency with which heat can be converted into other forms of energy such as mechanical is given by the following formula:

$$\text{Efficiency} = \frac{(t+273)_{\text{hot input}} - (t+273)_{\text{colder exhaust}}}{(t+273)_{\text{hot input}}},$$

where t is the temperature in degrees Celsius. The 273 converts Celsius degrees into absolute degrees of temperature. The ultimate in coldness, or the absolute zero of temperature, is –273°C.

The efficiency is always less than one. The bigger the temperature difference is, the larger the efficiency (Fig. 14.2).

For a refrigerator, the formula is turned upside down and the efficiency is greater than one. Now, the larger the temperature difference is, the *smaller* the efficiency. If the inside of a refrigerator is –18°C (in the freezer) and the outside temperature around the coils is 40°C, the theoretical efficiency is

$$\frac{313}{(313-255)} = 5.4.$$

The ratio is greater than unity — we are *transferring* (not producing) about five times as much heat as the mechanical energy being put in.

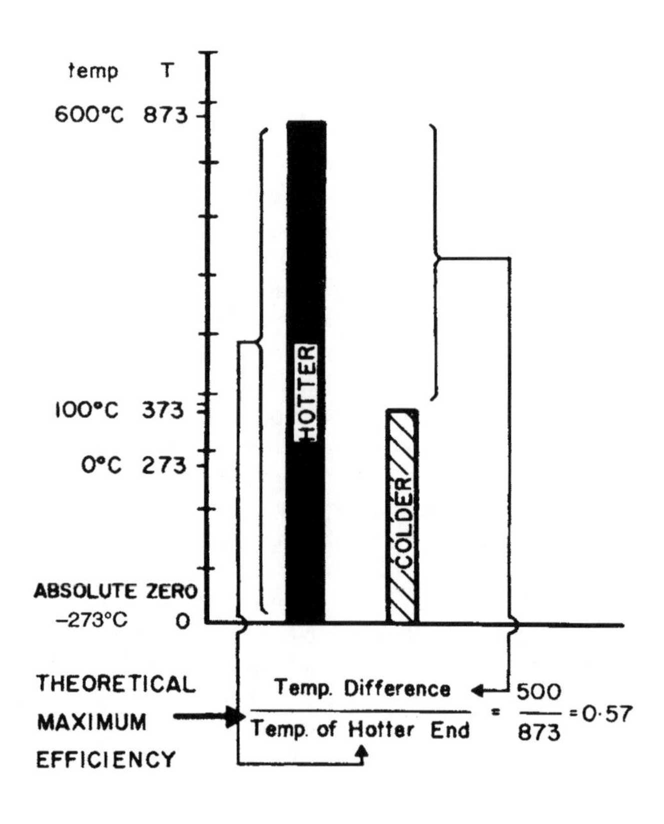

Fig. 14.2
Thermal efficiency

THEORY OF HEAT ENGINES

An ideal heat engine operates by extracting heat (reversibly) from a hot reservoir at temperature T_h and using it to expand a gas in a piston to do mechanical work (reversibly) — step 1. The transfer of the piston to a cold reservoir T_c (step 2) followed by injection of heat (reversibly) to that reservoir allows more work to be done — step 3. Transfer of the piston back to the hot reservoir completes the cycle — step 4 (see Fig. 14.3).

If the engine is regarded as a black box, then it can be regarded as cycling repeatedly, extracting heat q_h from the hot reservoir at temperature T_h and ejecting heat q_c to the cold reservoir at temperature T_c. In the process the engine (ideally) does an amount of work w determined by the first law of thermodynamics (which states that energy can neither be created nor destroyed) given by equation A in Figure 14.3:

$$w = q_h - q_c$$

Entropy is a concept derived from the second law of thermodynamics. The entropy associated with the transfer of heat is determined from the situation where the heat is transferred reversibly. (What 'reversible' means is a bit subtle for our present circumstances. Suffice it to say that we are calculating an unattainable ideal which gives us a measure of relative performance.) This entropy is just the heat transferred, divided by the absolute temperature (°C + 273). The cyclic engine cannot store or lose entropy and this gives us equation B in Figure 14.3:

$$\frac{q_h}{T_h} = \frac{q_c}{T_c}$$

a

HEAT

The greater the temperature the lower the entropy associated with it.
As $T_h > T_c$; q_c can be less than q_h.

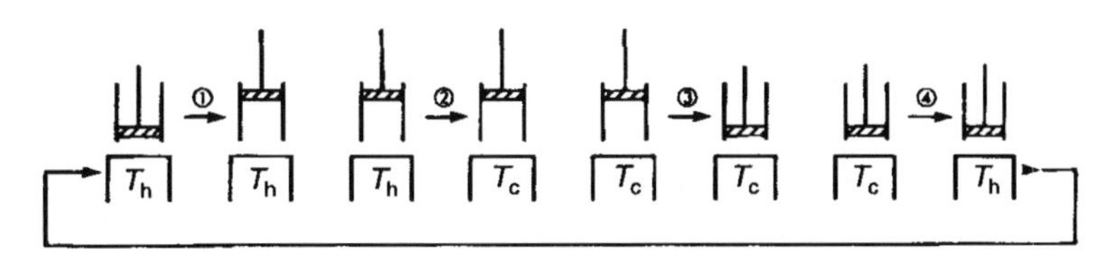

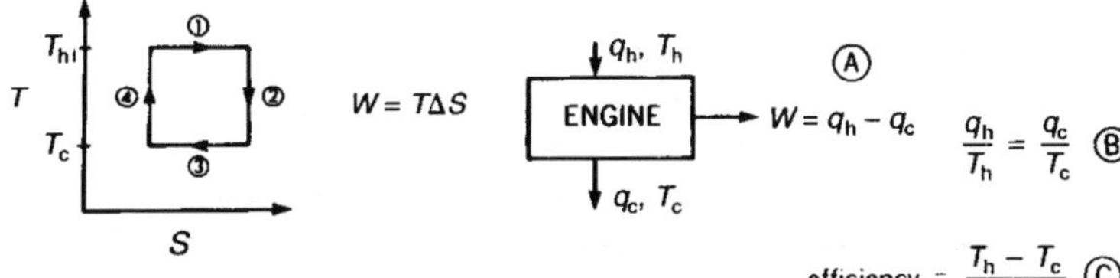

Fig. 14.3
Heat engine

Exercise
Assume that the total cost of operating an air-conditioner as a heat pump (pumping heat from the cold outside into a house) is four times the theoretical power cost for perfect efficiency, whereas the cost of direct electrical heating is just the power cost. If the room temperature is 27°C, at what outdoor temperature would the two systems yield equal cost?

A little algebraic rearrangement gives us the efficiency, which is the work obtained, divided by the heat into the engine or w/q_h in terms of the temperature of the hot and cold reservoir (equation C in Fig. 14.3).

$$\frac{W}{q_h} = \text{efficiency} = \frac{T_h - T_c}{T_h}.$$

So it is all as simple as the ABC!

For enthusiasts, the cycle of the heat engine can be shown on an indicator diagram, which plots temperature T versus entropy S. Step 1, the gas expansion, gives an entropy increase at temperature T_h, while step 3 gives the same entropy decrease at temperature T_c. In each case the (reversible) work done is the product of the temperature and change in entropy, $T\Delta S$, and the work output per (ideal) cycle is the area of the rectangle that the cycle encloses (see Fig. 14.3).

When generating electricity from a temperature difference such as an ocean temperature gradient, the heat arrows in Figure 14.4 are reversed, and heat is taken in at a higher temperature and ejected at a lower temperature. The motor is driven by this and becomes a generator. We are back to a heat engine.

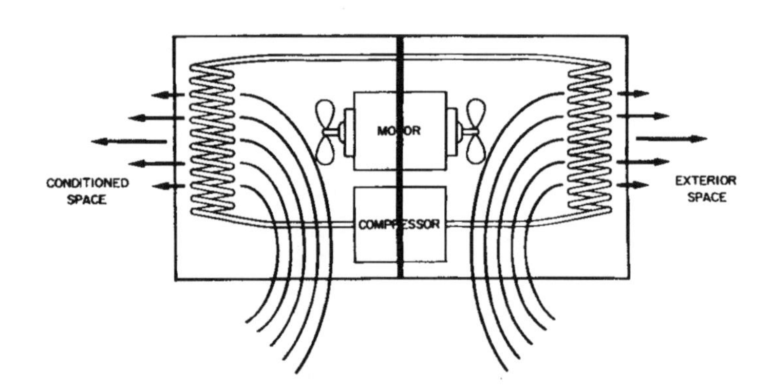

Fig. 14.4
Reverse-cycle air-conditioner

The absolute levels of energy used by consumers in their homes vary with climate. For the year 1980–81 in Australia, it varied from 36 300 MJ in Queensland to 96 000 MJ in Tasmania. In Australia, 40% of the electricity generated is used in the home and electricity represents about 40% of the total home use of energy.

About one-third of home energy is used for space heating, one-third for water heating, and one-third for all other uses. Domestic appliances are the main component of the other uses. The efficiency of *similar* appliances using the *same* fuel and performing the *same* tasks can vary markedly.

What energy do we use?

How much energy is used in obtaining energy and how do we use it? The points to note as domestic consumers are:

1. The energy needed to *obtain* the primary energy (wood, coal, oil, natural gas) is very low. In fact we have been lulled by an historical progression in which each successive primary source has been obtained more easily; that is, with *less* energy. The new sources — uranium, oil shale and coal liquefaction — on the other hand, need more energy for their provision.
2. We *use* our natural gas and oil (after refining) directly, but we use our coal via *electricity* generators, which results in an enormous conversion (and distribution) loss for coal. The reason for this enormous loss is that the production of electricity from coal proceeds via the production of heat (that lowly energy form). The hotter the heat (with respect to the temperature of the cooling water supplied to the condensers) the higher is its status, and the thermodynamic efficiency of conversion. That is why coal-burning power stations are noticeably more efficient in winter than in summer. We also export considerable quantities of coal, both directly and indirectly (in the form of, say, aluminium, produced with electricity generated from coal). Aluminium can be considered a solid stored form of electricity.
3. For transport we use oil. Energy content comparisons are therefore not very helpful because having a scoop of coal in a car is like having an Australian dollar in the back of Brazil — the value is there but not the conversion facility. Hence the interest in coal to oil conversion. In fact, so critical are liquid fuels for transport that they will set the 'parity' price for any other use (e.g. space heating). Some oil will continue to be a feedstock for the chemical industry, but even this will be replaced by coal. Even for transport, oil will become too valuable. Engines powered by hydrogen gas are being developed by Boeing — based on providing the hydrogen from electricity (electrolysis of water).

How much energy do we consume? The total consumption of primary energy in Australia is about a continuous 7 kW per person (about 20 times the use of Third World countries). A comparison of energy from various sources is given in Table 14.1 (overleaf), including those sources humans use as food.

WHAT ENERGY DO WE NEED?

The basal metabolic power used by humans (the rate at which we use energy to keep alive) is about 100 W (a light globe's worth). This can be supplied by 20 g of sugar (or 10 g of fat) per hour — a very modest amount indeed. The recommended daily allowance of usable energy for people is 9 MJ (females) and 12 MJ (males).

An average woman weighing 60 kg and 160 cm tall uses energy as follows:

Basic metabolism	6 MJ/day
Normal activities	3 MJ/day

Increasing activity by 10% uses up 0.3 M J, supplied to her by one slice of bread.

Rowing	4 MJ/hour for a 15 minute race	1 MJ
Gardening	0.8 MJ/hour for 3 hours	2.4 MJ

Even when we do work, like riding a bicycle, the performance is about 700 W. (The longer-term average is about one-third of this.) We are not going to run too many household appliances on these puny efforts. In our Western society, human energy is negligible. See Plate XV 'Peak performance'. See also Sports chemistry and biochemistry in Chapter 4.

TABLE 14.1

The cost of energy to the Australian consumer

Type of energy storage	Inherent energy value (MJ)	Cost per kilo ($)	Cost other ($)	Cost per hundred MJ ($)
1 kg of petrol (1.2 L)	42	0.90		2.1
1 kg of heating oil	42	0.80		2.0
1 kg of butter (1 L of fat, 200 mL of water)	31[a]	4 40		14.2
1 kg of sugar	17[a]	1.10		6.5
Car battery 12V, 50 Ah	2.16		6.0 (charging fee)	277
1 kW radiator	3.6 (per hour)		0.02 per MJ	2.0

[a] This is the energy released on burning the material, not its availability to humans on eating.

APPROXIMATE CONVERSIONS BETWEEN ENERGY UNITS USED IN DIFFERENT DISCIPLINES

In the following table, to convert 1 MHz to metres (wavelength) find the entry at the intersection of the row headed 'MHz' (top row) with the column headed 'metre' (far right), that is, the entry is 300. Note (*) means that the relationship is an inverse one; the bigger one unit, the smaller the other. So for a microwave oven operating at 2450 MHz (see Chapter 4) the wavelength of radiation in metres is 300 divided by 2450, and thus equals 0.12 m or 12 cm.

	MHz	cm^{-1}	eV	kJ mol^{-1}	wavelength nanometre (nm)	wavelength metre (m)
MHz	1	3.3×10^{-5}	4×10^{-9}	0.4	3×10^{11}*	300*
cm^{-1}	3×10^{4}	1	1.24×10^{-4}	0.012	10^{7}*	10^{-2}*
eV	2.4×10^{8}	8000	1	100	1.2×10^{3}*	—
kJ mol^{-1}	2.5×10^{6}	84	0.01	1	1.2×10^{5}*	—
nm	—		1/1240*		1	
m	1/300*	100*	—	—	10^{9}	1

From mine to meal
A perfectly cooked meal on the dinner table marks the end of a long chain of energy-devouring processes. These begin in mining. Mechanised agriculture requires machines. Their production takes a great deal of energy, and so do fertilisers and pesticides, and the provision of irrigation. After passing the farm gate, most food still requires more energy in processing and packaging before it reaches the supermarket. From there it is usually driven home. In the kitchen, it may spend some time in a refrigerator before, finally, it is cooked and becomes a meal.

What about the sun?

Energy down on the farm

In energy terms, our modern food system is expensive. Processing and distribution use up a lot of energy — far more than is actually contained in the food itself. At present much of that energy comes from oil — a resource that is becoming scarce and expensive.

Food and fuel are interchangeable. Farm produce can be chemically converted into fuel, and microbes can convert petroleum into food. However, a point that arose in our introductory comparison of costs of consumer energy must now be clarified. It is important to distinguish between the fuel value of food when it is burnt, and the nutritional energy value when it is eaten. Only about three-quarters of the fuel energy in our food is absorbed when we eat. The figures to be used here refer to *fuel* energy. Of course the indigestible cellulose can be converted to methane.

Two Canberra scientists at CSIRO (Dr Roger Gifford and Dr Richard Millington) studied this energy usage. The energy budget was a very broad one and dealt only with the aggregate of all Australian agricultural production over the years 1965–69. Individual products were not isolated because the web of energy inputs for each product cannot be clearly traced. Plant material harvested by humans and their livestock represents a mere 0.01% of the sunshine (falling on rural land) that could be usefully absorbed by plants. Nevertheless, this plant matter has a fuel value about 1.5 times greater than all the fuel energy burnt by people in Australia. Only about 15% of this massive amount of plant tissue is harvested directly — the rest is eaten by livestock and converted into wool and animal products.

The energy inputs to farm products before they leave the farm are set out in Table 14.2 (overleaf).

Many of the less tangible inputs to farming have been omitted from the budget, such as agricultural research and extension. The energy inputs to farm products after they leave the farm are set out in Table 14.3 (overleaf).

Food processing and distribution use much more energy than food production. Omitted from the table are such indirect items as the energy

TABLE 14.2
Energy input on the farm

	× 10^{15} J per annum	
Direct farm use		
fuel	46.2	
electricity	8.4	54.6
Fertiliser		
mining	3.5	
shipping	9.8	
manufacture	5.5	18.8
Farm machinery (mining and manufacture)		6.8
Agricultural chemicals		4.4
Road transport not included elsewhere		1.0
Farm labour		1.2
		86.8

TABLE 14.3
Energy input from farm to dinner table

	× 10^{15} J per annum	
Transport from farm		
rail	2.1	
road	5.0	
grain handling	0.3	7.4
Factory processing		
bagasse as fuel	9.7	
other fuel	45.6	55.3
Food and drink packaging		
steel cans	10.8	
paper	8.5	
fuel value of paper packaging	4.3	
glass	5.5	29.1
Road transport from factory		7.7
Subtotal to retail store		99.5
Transport from store to home	33.0	
Domestic refrigeration	46.0	
Domestic cooking	42.0	
Subtotal store to dinner table		121
Grand total		220

used for building food factories, refrigerators, and stoves. About 14% of the energy in food produce leaving the farm comes from animals and it costs 3 J of grain (at least) to produce 1 J of meat. Losses during processing, retailing, kitchen preparation, and digestion are difficult to estimate but it seems that we absorb into our bodies only about half the fuel value that leaves the farm (Fig. 14.5).

Fig. 14.5

In summary, what the final balance showed was that for each joule of *digestible* food energy eaten in Australia at least 5 J is expended in making it available, of which

- 0.6 J is used in getting it to the farm gate — 11%;
- 2 J is used in taking it from the farm to the retail store — 38%;
- 2.8 J is used in getting it from the store to the dinner table — 51%.

The major use of energy is past the farm gate. In the food chain, the internal combustion engine guzzles at least 40% of the total energy consumed in the total process. While shopping uses only 2% of the fuel going into petrol tanks of private cars, it represents 10% of the total energy cost of the food. Any improvement in transport would be a substantial saving.

Our method of agriculture has been compared to that of the Tsembaga tribe living in the New Guinea highlands in 8.3 km^2 of tropical rainforest. The comparison is difficult because of different cultural values but a very important difference is that their energy comes from renewable resources (trees) whereas ours comes almost exclusively from non-renewable fossil fuels.

We need shelter — and if we want to cost insulation, we need to understand what is meant by an energy analysis. Before insulation may be said to save energy it must first pay back the *energy* used in its manufacture and insulation. This *energy* payback time can be less than a single heating season.

Along with insulation come all those handy hints for preventing heat loss by convection; for example, close up all the cracks and crevices around windows and doors. In the USA, some houses are now so well sealed that they are also trapping moisture and air pollutants generated inside the house. Among these are fumes from natural gas, household sprays, radon released from radioactivity in building materials, formaldehyde, and other vapours from particle board, plywood, and the insulation itself (see Radon in Chapter 16).

Uranium energy analysis

A similar energy analysis can be done on uranium.

If you assume that there is no inherent energy content in uranium, then nuclear technology can convert about 1 MJ of oil into 1 MJ of electricity (where the oil is used to mine, extract, transport, and upgrade the uranium). This shows an energy advantage over a conventional oil-fuel generating station, which requires 4 MJ of oil (in this case burnt directly) to produce 1 MJ of electricity. Thus uranium can be seen as a component in a technology that allows our oil reserves to last four times as long. If the uranium is counted as a fuel input, then nuclear reactors are poor energy converters (1% conversion).

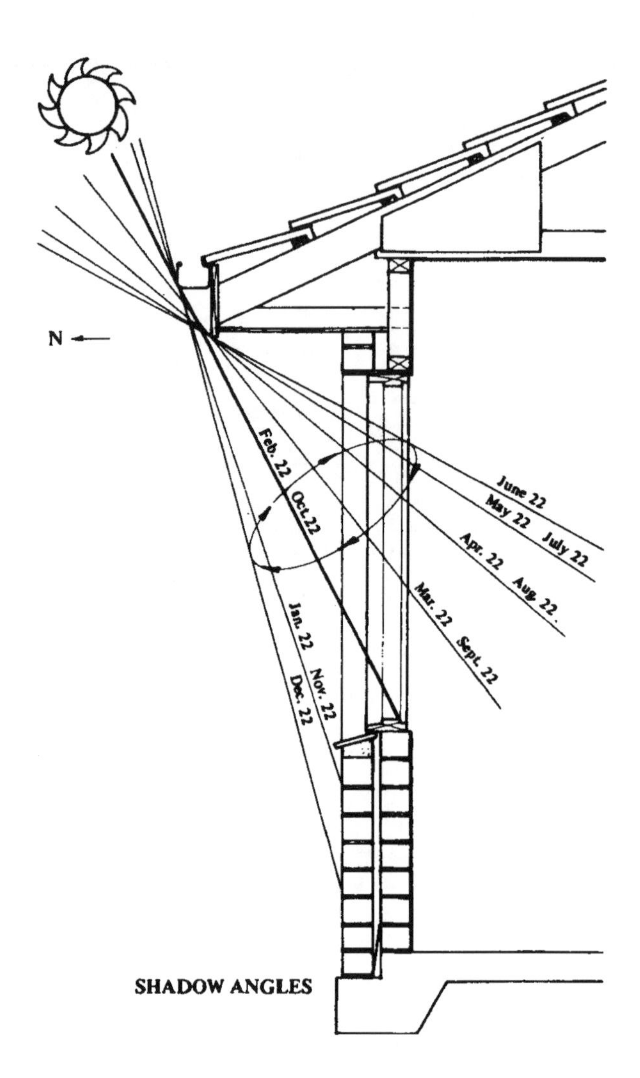

Fig. 14.6 Eaves designed for the Southern Hemisphere at the latitude of Sydney exclude the summer sun but allow winter sunlight into the house

From Rebuild, CSIRO, October 1978.

There are of course many other shelter questions — better design for passive use of solar energy, layout, materials, and other heating and cooling sources — which are dealt with in an expanding popular literature. Simple care with the positioning of windows and eaves can make an enormous difference (Fig. 14.6, above).

Energy efficient houses

If you are building a house, it generally costs no more to design and site the house to be warm in winter and cool in summer. If you own an older house and plan to remodel, there may be quite a lot you can do to make it more energy efficient. Figure 14.7 shows some of the design features to use.

Ideally, the long axis of the house should run east–west, and few windows should be sited in the short east and west walls that get so much

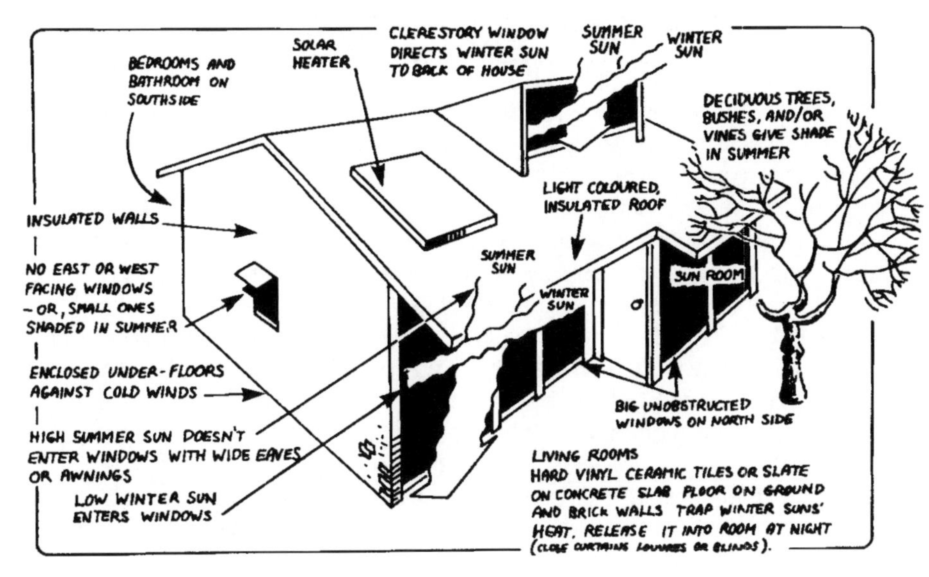

Fig. 14.7
Energy-efficient house design and siting for the Southern Hemisphere

From Choice, May 1983

morning and afternoon sun during summer. Most windows should face north (south in the northern hemisphere) or within about 20° of it, and the window area should be about 20% of the floor area of the space you want to heat. With eaves of the right width, summer sun can be excluded but maximum sunlight can be obtained during winter. (The exact width of eave or awning needed for your latitude, window height, and slope of roof can be worked out.) This winter sunlight will provide plenty of daytime warmth, whatever type of floor you have, and if your house is of concrete slab construction and you choose the right flooring, the floor will store daytime heat and release it slowly into the house overnight.

Remember to check the surroundings, too. Is the house protected from prevailing winter winds? Will it be free of shadow in winter? In southern latitudes (e.g. Melbourne, Victoria) in mid-winter, something north of your house can cast a shadow twice as long as its height. What about your garden? Strategically placed deciduous trees and vines will shade walls and windows in summer and obligingly drop their leaves in winter.

There has been a resurgence of interest in solar energy in recent years because of the realisation of the limitations of our present energy resources, and an immense amount of information on solar energy has been published. Solar hot-water systems consist of panels of collectors, generally copper pipes, installed to face north (south in the northern hemisphere). The water heats up and circulates through collectors into a storage tank which generally has an electric booster for overcast days. The lifetime of commercial plastic collectors is believed to be about ten years. The plastic used in pipes is a black acrylonitrile–butadiene–styrene (ABS) co-polymer (see Chapter 9). It has much lower thermal conductivity (heat transferring properties) than metals.

THE FREIBURG HOUSE

The aim was to answer a technical question: Can an energetically self-sufficient solar house be built in Freiburg, West Germany in the cold of northern Europe? There was little consideration of the capital cost, and the project ended up costing DM1.8 million. (See Plate XVI.)

The house obtains its entire energy demand for heating, domestic hot water, electricity, and cooking from solar energy. It has no external grid connection, nor does it consume fossil fuels.

The large seasonal variation in solar energy supply and heating demand in central Europe, means that the solar energy must be stored for the winter. This is accomplished by converting the excess summer solar energy to electricity with a photovoltaic converter (solar cells). This electricity is used in an electrolysis unit that splits water into hydrogen (and oxygen) gas directly under pressure (i.e. without the use of a compressor). (Unit developed with Siemens — cost about DM 200 000.) A small amount of electricity is stored in lead–acid batteries. More can be produced on demand from a fuel cell that converts hydrogen and oxygen back to water, while converting half the energy into electricity and the other half into heat. Hydrogen is also burnt directly in a catalytic combustor in modified gas cookers and in heaters for extra space heating.

Space heating was minimised through all possible conventional methods, and in particular, the use of excellent thermal insulation of the building, including transparent insulation of windows. Double glazing is standard in Germany and now triple glazing with reflective metal coating and noble gas cavity filling is available. This house had two double glazing systems.

In winter the house is sealed and ventilated by piping in fresh air through underground pipes and prewarming it through a heat exchanger with the exhausted air. Solar collector panel heaters supplied 90% of domestic hot water and the other 10% came from the waste heat of the fuel cells.

Blinds were used for sunlight entry control because louvres obstruct diffuse radiation which is the major energy input in winter. Electrochromic and thermochromic films are quite advanced and will become commercially available soon; these will provide automatic energy inflow control.

Transparent insulation making use of a honeycomb plastic construction removes virtually all convection heat loss and is a major research interest. The greatest heat loss was conduction through the walls.

The average German home uses 3270 kWh of energy per annum. The Freiburg house achieved 894, compared to a theoretical minimum of 700 with the best equipment. Electricity use was minimised by removing the stand-by option universally supplied with electronic equipment such as TVs. Cooking used hydrogen gas. A large research effort went into reducing losses by efficient insulation of walls and windows, and by thermally isolating the foundations from the house.

A comparison of standard and Freiberg energy consumption follows:

Standard energy (kWh/m²/p.a.) consumption, and best achieved at Freiburg

	Appliances	Ventilation	Hot water	Space	Heating total
Standard	20	0	20	100	140
Freiberg	8	0.6	0	0.2	9

No wind and water power
Wind and water power for additional charging in winter were not available.

Details:

- House 140 m^2; solar cells monocrystalline, 30 m^2, power 4.2 kW (14% efficiency, see also solar cells below).
- Lead–acid batteries, 24 in series giving 48 V, 19.2 kWh capacity (one week's supply).
- Electrolyser power 2 kW; fuel cell 1.0 kW.
- Hydrogen storage volume 15 m^3; max pressure 27 bar (1 bar = 100 kPa, $\approx$ 1 atm); oxygen 7.5 m^3.
- The hydrogen storage tank is of modest size and the exercise showed that compressed hydrogen can be used safely in a domestic environment, which is of major significance to a future hydrogen economy. If the storage had been developed for a conventional house with conventional energy usage, the tanks would have been the size of the house!
- Hydrogen is not burnt in the conventional way for two reasons: First, the flame is almost invisible and this creates safety risks. Second, the high combustion temperature of a hydrogen flame produces oxides of nitrogen. Consequently a catalytic porous cylinder burner is used which operates at about 600–700°C and is physically similar to a conventional gas burner (with little hot cylinders instead of the flame burner)
- The fuel cell is 45% efficient in converting hydrogen into electricity, 14% of the energy goes into control mechanisms etc., and 44% comes out as waste heat. The cell operates better with pure oxygen than with air, and this is why oxygen as well as hydrogen is stored. The combined hydrogen electrolysis, plus fuel cell electricity generation efficiency is 26%.
- The energy efficiency of the battery is 92%, but the lifetime is short because of the long discharge cycle in winter.
- The materials are plastics, timber, glass, ceramics, metals, cement. lime, and gypsum. Metals make the major energy contribution, particularly aluminium. Bricks are better than cement. Timber is low (mainly solar).

Market distortion
Conventional economics builds houses by optimising costs, not energy, and the irresponsible cheapness of non-renewable fossil fuel distorts the market completely.

Liveability

The internal temperature of the house is closely controlled. Because the heat comes from the walls, a minimum of 18°C is considered comfortable, whereas with conventional space heating, 20°C is needed. That extra 2°C can double the energy requirements of space heating. In such an energy-efficient house, every bit counts, and the 100 watts or so of heat contributed by each person is missed when the house is empty over a period (plus the lack of use of household appliances and TV, etc.)! Absence can allow the internal temperature to drop significantly.

Except in poorly designed buildings, air conditioning for summer cooling is not necessary in Germany. In hot climates like Australia, air conditioning is needed in summer, and summer is also the time when solar energy is most available. There is thus a good match between energy supply and demand. However, in cold climates the energy needs are for heating, and this occurs in periods of minimum solar supply.

Tenants adapted quickly to the lifestyle and became much more aware of the weather conditions. Thus, they tried to time their use of the appliances with heavy demand for electricity (washing machines, dishwashers) to periods of sunshine when electricity was directly available, rather than place demands on the batteries or fuel cells.

Breakdown
On the odd occasion when things broke down, they quietly ran an extension cord into the neighbour's house!

The project has implications for energy supply to a new shared Australian base on the high plateau in Antarctica, but otherwise a hot climate has other priorities. The projects in the two countries are complementary.

Source: adapted from a talk to the von Humbolt Foundation, Canberra, November 23, 1996, by Prof. A. Götzberger, Fraunhofer Institute for Solar Energy Systems, Freiburg, Germany.

The energy from the sun is spread across a range of energies and the earth responds to it in a variety of ways (Fig. 14.8).

We next explore the part of the energy that resides in the winds.

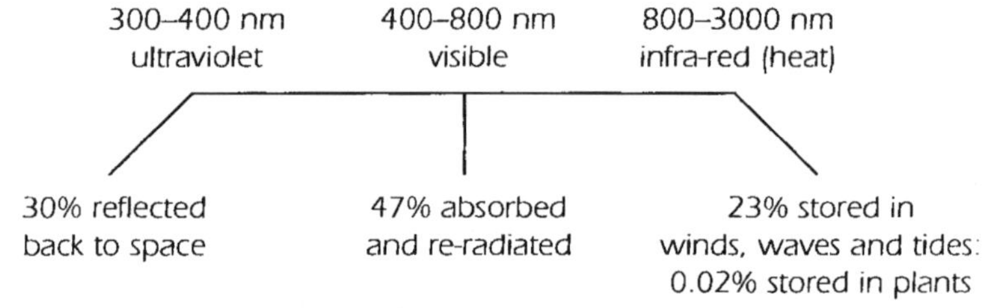

Fig. 14.8
The fate of the solar energy that reaches the earth

Wind energy and turbines

What makes the winds blow? What is the motive force that sweeps masses of air across the surface of the globe? The answer is sunshine, so we are still on solar energy. About two per cent of the radiant energy reaching the earth is converted by this process into the mechanical energy of moving air masses.

The bottom of the 'atmospheric ocean' is heated by the ground, and what happens reproduces, on a global scale, the kind of commotion seen in a pot of boiling water. Basically, the air in contact with relatively warm ground is heated and tends to float upward, and when new air flows in from above chillier regions to replace the warmed air, we call its motion wind. (See Demonstration — inversion layer — later in this chapter).

A mass of moving air can act with awesome force. If the total magnitude of the power blowing through the skies could be converted, without loss, into electricity, it would exceed 10 000 times the combined output of all the generating stations in the world. But in practice, only a small fraction of the raw energy in the winds can be captured and put to use. The behaviour of moving air, the laws of physics, and economics, all limit the wind's potential as an energy resource.

Both the ancient Chinese and the Egyptians probably had some form of wind power. The earliest reliable records describe the windmills that were used in Persia thirteen centuries ago — and which are very similar to windmills used today in eastern Iran. The Persian windmills consisted of sails arranged so that when caught by the wind they turned a vertical shaft. Corn was ground by a grindstone turned by the shaft. From the Islamic world, this invention was brought to Europe — possibly by returning Crusaders. There it was considerably modified.

The primitive Persian windmills evolved into the whirling-armed giants that Don Quixote charged. Four centuries ago there were more than 20 000 windmills at work in the Netherlands, draining the marshes and lakes of the Rhine river delta to create arable land, grinding grains into flour, running sawmills, and crushing seed to make oil. In their day the picturesque Dutch windmills were what oil wells and refineries are today, a prime source of power for a nation's economy.

The use of wind power began to decline with the coming of the steam engine, which unlocked the energy stored in coal, and sparked the industrial revolution. However, windmills never became completely obsolete. Small windmills were common tools on farms scattered over North America and Australia in the early decades of this century. They pumped water for crops, for cattle, and for use in farmhouses. Wind-driven electrical generators were first marketed as an accessory to battery-powered radios, and soon became a common sight, producing modest amounts of power for many thousands of isolated homes. As the electrical distribution grids spread into the countryside, making cheap energy available at the flick of a switch, though subsidised by urban electricity users, both these kinds of small windmills fell into disuse.

In Denmark in the 1890s, the first attempt was made to use a large windmill to supply electrical power to a community. This experiment was followed by similar ones in England, France, Russia, Germany, and the United States. By 2005, Denmark hopes to provide 10% of its electricity from wind power, while California already has thousands of megawatts of windpower, almost equalling that of Victoria's coal-burning capacity.

We know from everyday experience that some places are windier than others and that the faster the winds blow over a spot on the earth's surface, the more power there is to be captured.

Consider a moderate 25 km/h breeze which can set leaves rustling and raise dust or scraps of paper. A stream of air blowing at this speed through a window frame of one square metre in area has a raw power of about 200 watts. If its speed drops slightly to 20 km/h, little difference will be seen — leaves and twigs will still be in constant motion — but the power of the breeze will have dropped sharply to 100 watts, half its former value. In practical terms, where annual average speeds exceed 20 km/h the winds carry useful amounts of power. Where average speeds fall much below this value the winds are too weak to warrant

Wind speed and power

The relationship between wind speed and power follows a cube law, which states that the power of a wind is proportional to the cube of its speed. Thus most of the power in the winds is carried by the faster winds. The cube of two is eight, and so a doubling of wind speed means an eightfold increase in power, and a halving of speed means an eightfold reduction in power.

The reason for the cube law is that a wind of speed v hitting an area of blade A, in a time Δt, will provide a volume of air, $Av\Delta t$. The mass of this air is the volume times the density ρ, and so the kinetic energy, $\frac{1}{2}mv^2$, is $\frac{1}{2}\rho Av^3\Delta t$. The power is the kinetic energy per unit time Δt, and so is proportional to v^3. QED!

harvesting. The cube law narrows the range for optimum energy extraction.

Essentially, a wind turbine works like a water wheel driven by moving water. As air is hundreds of times less dense than water, wind turbines are much larger than water turbines for the same power output. There are two main traditional kinds of wind turbine: those that turn around a horizontal axis, as does a Ferris wheel, and those that turn on a vertical axis, as does a merry-go-round. A wind turbine can never capture more than about 50% of the raw energy of the moving air it intercepts, and the efficiency of practical wind turbines is considered high if it reaches 40 (see Fig. 14.9).

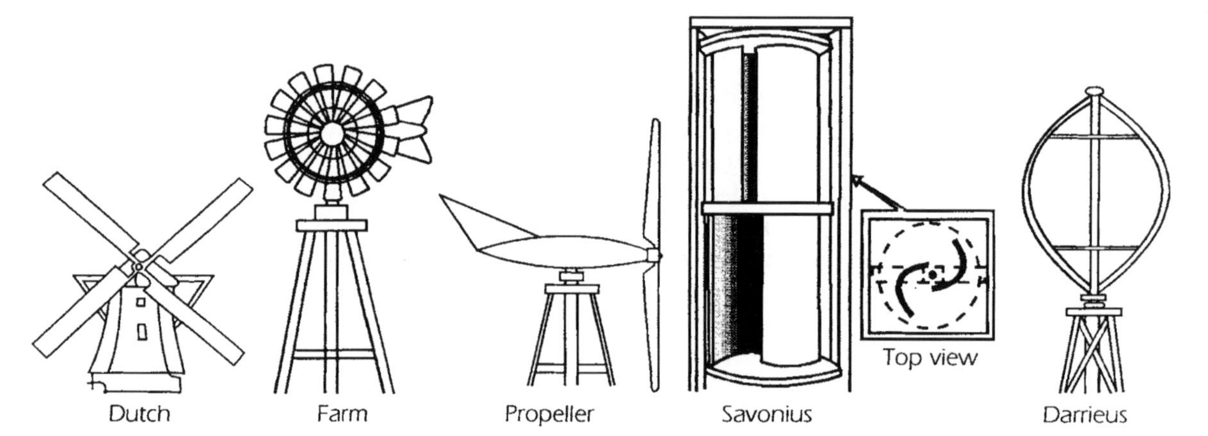

Fig.14.9
Common types of wind turbines

The ponderous Dutch windmills with four wooden sails, and the water pumping farm windmills, each with a fan of many metal blades, turned fairly slowly. Though they did not extract very much of the total energy of the wind, they could start turning in winds of quite low speeds, generating useful amounts of mechanical power in the form of a turning force or torque, and they were thus well suited for mechanical work such as grinding and pumping. Modern propeller wind turbines have a few, light, streamlined blades. They are designed to rotate rapidly, with the outer tips of the blades travelling five or six times the speed of the wind driving them around. Such a design is intended to be used to drive electrical generators — most of which run at high rotation speeds.

Winds can blow at speeds and strengths that exceed the power handling capability of the turbine. If excess power is not spilt, turbines and towers can be toppled, or generators 'burnt' out. There are many methods of controlling the speed of a turbine. Some of the Dutch windmills, for example, had shutters in their sails that could be opened, rather like venetian blinds, to let the high winds pass through. Modern windmills are frequently designed with blades that can be 'feathered', that is, the angle of the blades can be adjusted so that less surface and more edge is presented to the wind as its speed increases.

The Persian windmills of thirteen centuries ago turned on vertical axes, a time-honoured arrangement whose use has been revived in some modern wind turbine designs. What makes this attractive is its mechanical simplicity: unlike horizontal-axis machines, vertical-axis turbines do not need to be turned as the wind changes direction. The simplest of this type is the Savonius turbine, named after the Finnish inventor who patented it in 1931. It can be made by splitting an empty oil drum lengthwise, and then welding the two half-cylinders so that they turn around a central, vertical axis. When seen from above, the Savonius turbine consists of two air scoops forming an S shape.

A Darrieus wind turbine looks somewhat like a large kitchen egg beater. Each of its blades is attached at both ends to a central shaft, and follows a curve close to that of a semicircle. In cross-section, each of the blades is shaped like an aeroplane's wing: it is the lifting force developed by air flowing around them that spins the turbine and generates power.

There are now many ingenious variations that allow a greater range of wind speeds to be harvested, and protect the unit from high-speed wind.

Photovoltaic cells
The French physicist Antoine Henri Becquerel described the photovoltaic effect in 1839 when he found that certain materials would produce small electric currents when exposed to light. (The becquerel is now the unit for radioactive disintegrations, see Chapter 16.) In 1875, W. Siemens extended the research to further materials such as solid selenium, and soon photovoltaic cells were developed that provided 1 to 2% conversion efficiency. The author still has a very reliable 45 year-old photographic exposure meter based on selenium (no batteries needed). It was silicon that made photocells a serious proposition, and it was the space program that launched them commercially.

Solar cells

In contrast to windmills, which become much more economically efficient when they are bigger, solar cell technology is fairly independent of size.

Silicon solar cells

The daily solar radiation reaching the top of the earth's atmosphere varies with latitude and season. At the equator it is about 35 MJ/m^2 all year round. At latitude 30° it varies from 20 to 40 MJ/m^2 between winter and summer. On the earth's surface the variation is much greater. Energy of 35 MJ/m^2 over 24 hours is equivalent to a (day and night) average incident power of 400 watts/m^2.

Large-scale solar power stations provide about 1 MW (e.g. in Toledo, Spain), and can range up to 10 MW. As each cell provides typically two amps at around half a volt (i.e., one watt) these arrays contain about one million solar cells.

Silicon is produced from sand (silica) which is impure silicon dioxide, and some of the purest sands are found in the beaches of Australia's Northern Territory. The production of silicon is well established in steel-smelting processes where it is used for certain steel alloys. For solar cells, the metallurgical-grade silicon must be further purified.

Pure silicon (valency 4) used for solar cells is doped with low levels of boron (valency 3) to cause the occasional absence of an electron (called a positive hole) and this gives p-type (positive) silicon. When later fabricated into silicon wafers, these are then diffused with some phosphorus (valency 5) on one face to provide an n-type (negative) silicon there. These two types of silicon, layered in a single cell, meet to form a p–n junction. Diffusion of electrons across the junction sets up a voltage difference that then opposes further diffusion.

Surface etching
To trap more of the light and stop it reflecting back, a hilly surface is formed on the silicon (with micropyramids of about 10 micrometres in height — solar cell surface feels velvety). This surface is obtained by etching with caustic (KOH plus isopropyl alcohol) which preferentially attacks the crystal face of silicon with less atoms in it (the 100 face) and leaves the face with more atoms (the 111) face less affected (see also the effect of additives on free flowing salt).

Swing now to a mental image of energy states available for electrons arising from the atoms congregated in the solid. Photons of visible sunlight are energetic enough to 'promote' an electron on atoms in the p–n junction to a higher level where it can move freely (in the conduction band) at the same time leaving behind a hole which is now also free to move (more slowly in the valence band; see Plate XVII). This movement of charge, if picked up by wires, constitutes the flow of an electric current.

The p–n junction in silicon has a gap of 1.1 eV (wavelength ≈ 1000 nm), which means it absorbs energy from sunlight that is more energetic than this; all the UV and visible light but not most of the infra-red (see Plate XVIII solar spectrum). In practice, a silicon solar cell generates less than 1.1 volts; actually 0.5 to 0.7 V. The current depends on the surface area exposed to light, and its intensity (see Plate XIX). By connecting several cells in series or parallel, the voltage and current respectively can be increased.

In competition with the collection of current from the p–n junction is the spontaneous recombination of the separated electron and hole, converting their energy into heat, and producing no electricity. This competition is the basic limitation on the theoretical efficiency of a solar cell, and the engineering trick is to collect as many of the charge carriers as possible before they have a chance to recombine. For silicon, this efficiency limit is just under 30% of the solar energy absorbed, and in practice 20–25% has been attained. (See Plate XX.)

Cost of production of solar cells

The cost of production splits roughly into one-third for the raw materials, one-third for the cell itself, and one-third for fabrication of the module, and cells typically cost $4–8 for multicrystal types, and $7–12 for single-crystal types. The single-crystal cells are more efficient (lab. 25%, commercial 15%) but more expensive than the multicrystal types (lab. 18%, commercial 13%). The production process is highly energy intensive but the energy payback time is less than five years, while the lifetime of a cell is of the order of 20 years. (Use of solar concentrators and multiple cells will improve this.)

Many different physical forms of silicon are used. These include single-crystal forms and semicrystalline forms (mixtures of crystals each about a centimetre square in size). The join between two crystals helps charge carriers to recombine, and reduces efficiency, unless other modifications are made. Yet the most widely used commercial thin-film cells are made from amorphous silicon (without well-defined crystal structure, like a glass). Amorphous silicon cells are made ultra-thin to compensate for poorer electrical properties, have an extra layer of undoped silicon within the p–n junction, and are treated with hydrogen to mop up dangling bonds. These cells change their efficiency with time. At first it increases, and then with time it decreases. But they are cheap to produce, both in cost and energy. Other semiconductors are also in the race for solar cells. These include gallium arsenide (GaAs), indium arsenide (InAs), cadmium selenide (CdSe) and telluride (CdTe). As all cells will eventually be discarded, silicon is the most benign (see Heavy metals in Chapter 2 and Chapter 10). (This section has been written with advice from Dr Andres Cuevas, ANU, and more details can be found in the references.)

Alternative energy market

In Australia the off-grid electricity market is about 1% of the total (50 GW), so about 500 MW is immediately open to alternative energy. This makes Australia uniquely placed to exploit solar technology. The future looks sunny!

See Plate XXI showing the Ginninderra College solar car in the Darwin to Adelaide solar race of 1997.

Accumulators (rechargeable batteries)

Normally car engines rely on electricity stored in the accumulator to start them. During running, the electricity drawn from the accumulator is returned by the alternator and the accumulator fails only through mechanical breakdown. Accumulators of this type are called secondary or storage cells; electricity is stored by driving a set of chemical reactions in one direction by connecting the accumulator to a charger. Current may then be drawn from the accumulator when the reactions proceed in the reverse direction.

The lead–acid accumulator

The familiar 12-volt car accumulator contains six cells, each providing 2 volts. Each cell contains two series of plates — positive and negative — immersed in sulfuric acid. The positive plates are made by forcing a paste of lead dioxide (PbO_2) into a grid made of lead alloy, while the negative plates contain lead in a highly active spongy form. The overall reactions are:
at the positive plates:

$$\underset{\text{lead dioxide}}{PbO_2} + 4H^+ + SO_4^{2-} + 2e^- \underset{\text{charge}}{\overset{\text{discharge}}{\rightleftharpoons}} 2H_2O + \underset{\text{insoluble}}{PbSO_4\downarrow}$$

and at the negative plates:

$$\underset{\text{spongy lead}}{Pb} + SO_4^{2-} \underset{\text{charge}}{\overset{\text{discharge}}{\rightleftharpoons}} PbSO_4\downarrow + 2e^-$$

Accumulators and batteries
Rechargeable systems are called accumulators. Non-rechargeable systems are called batteries or cells.

Ideally these reactions can be recycled indefinitely, but with time lead dioxide particles lose electrical contact with the plate. The volume of the positive plate changes by 60% in its transition back and forth between lead and lead dioxide, and a rolled spring-like compression arrangement is used to help compensate for this. The state of charge of a non-sealed lead accumulator can be estimated with a hydrometer, which measures the density of the sulfuric acid. As can be seen from the above reactions, sulfuric acid is consumed by both reactions during discharge and thus the density of the electrolyte falls from about 1.28 (about 37% sulfuric acid) in the fully-charged cell to about 1.15 (about 21% sulfuric acid) in the flat accumulator. The chemical reactions produce a lower voltage at a lower temperature. Not only does the accumulator not function as well, but the engine oil is more viscous, and so the engine turns over with greater difficulty. The starter motor draws over 160 A for a short period; a very heavy demand. A fully-charged accumulator should be able to maintain a voltage of at least 8.4 V for the first 5 to 7 seconds and not drop below 6 V (half the nominal voltage) in under three minutes.

Another use of the accumulator is to supply a small current for a long time to the car lights. The storage capacity of the accumulator under these undemanding conditions is measured in electrical energy units (ampere hours, A h). Accumulators are tested by drawing for 20 hours a uniform current of one-twentieth of the 20-hour capacity. That is, after 20 hours the battery has reached the end of its rated capacity and its voltage should at that stage not have dropped below 10.5 V. A rating of 53 A h for an accumulator should thus allow it to supply 2.65 A for 20 hours.

Another situation occurs when the alternator on a car fails. How long can it be driven before the accumulator goes flat? This requirement is measured by the reserve capacity. An accumulator should be capable of delivering a moderately high current (25 A, enough to run the ignition and headlights) at 25°C before the voltage drops to 10.5 V for 70 minutes.

The accumulator must also *accept* an adequate charging rate from the alternator, otherwise lots of short trips will soon flatten the accumulator.

Maintenance-free accumulators use calcium–lead grids, rather than antimony–lead alloys. This delays self-discharge and, together with immobilised electrolytes and other measures, removes the need for water replacement.

Accumulators for electric cars

The first electric car appeared in the USA in 1847, built by Moses Farmer. It ran on a non-rechargeable battery. Gaston Plante developed the rechargeable lead–acid accumulator in 1859, and Thomas Edison the nickel–iron accumulator in 1901. By 1912 there were 34 000 electric cars in the USA. Two developments killed the electric car. One was the price drop of Henry Ford's Model T from $850 in 1904 to $265 in 1925. The second was the introduction of the electric starter motor in 1911 (but not in the Model T). This removed the major objection of women drivers to hand crank-starting the petrol-driven cars.

In 1899, an electric car held the world land speed record at 105 km/h, but the low energy density of conventional accumulators meant that this was never repeated. A small sedan might require 10 kWh of energy to travel 100 km. This would be supplied by, say, 10 L (8.6 kg) of petrol. The energy density of current lead–acid accumulators is about 35 W h/kg, and so a 280 kg accumulator occupying 180 L would be needed to drive the car the same distance.

It has been the new air pollution laws of the US state of California that has caused a sudden surge in research and development in accumulator research for electric cars. The California Air Resources Board (CARB) publishes regular reports on progress.

Other accumulators

The lead accumulator has a number of advantages. It is cheap, easy to manufacture, uses a cheap electrolyte whose density serves to indicate the state of charge, and maintains a relatively constant voltage over most of the discharge range. However, it is heavy (the energy density in watt-hours per kilogram is consequently low) and the electrolyte is corrosive.

The first of the alternative accumulators was the nickel–cadmium type.

$$2NiOOH + Cd + 2H_2O \underset{\text{charge}}{\overset{\text{discharge}}{\rightleftharpoons}} 2Ni(OH)_2 + Cd(OH)_2$$

Nickel–cadmium accumulators have what is called a memory effect. The chemical reactions may be reversible, but physical changes caused in the electrodes may not be. On both electrodes, changes in crystal size and form can change the behaviour of the surface. If users plug in these accumulators for a recharge when only a portion of the energy stored has been used, the next discharge cycle may stop at that point, and not provide further energy.

The specific energy of a source, the amount of energy it holds per kilogram mass, depends to some extent on the rate at which the energy is being drawn. This is illustrated in Figure 14.10 (overleaf).

The lead accumulator falls off most markedly at higher power. An interesting comparison is made with a cyclist. The racing cyclist is quite competitive at low power.

The lithium-ion accumulator

Lithium has an atomic mass of 7, compared to nickel at 59, and lead at 207. The standard electrode potential for lithium is –3.05 compared to nickel at –0.25 and lead at –0.13. Sony has developed a lithium-ion accumulator. These accumulators have a high energy density of 100 W h/kg, and high operating voltage, a high power density >300 W/kg, long life (1200 cycles), and absence of memory effects found with nickel–cadmium accumulators. A model suitable for electric cars has been developed, with eight cells in series (the Nissan FEV II). Its dimensions are 29 cm width, 15 cm height, it can yield 28.8 volts, 100 A h and it weighs 29 kg. The cathode material is lithium cobaltite, $LiCoO_2$, and the anode is carbon. The energy density is three times that of a lead accumulator. It can operate between –20°C and +80°C. Full charge can be obtained in 2.5 hours, 80% in one hour, and 50% in 30 minutes. When charged over a period of six hours, and discharged over a period of three hours, the trip efficiency of the cycle is 95%, which is remarkably good.

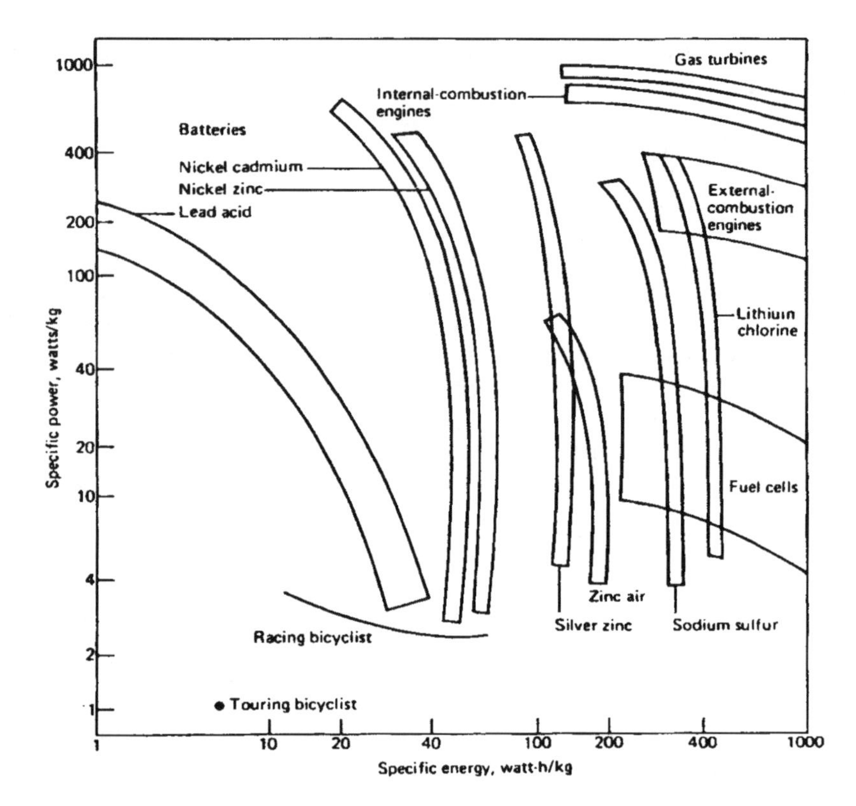

Fig. 14.10
The relationship between power and energy from various sources

(From F. R. White and D. C. Wilson, Bicycling science: Ergonomics and mechanics, 2nd edn, MIT Press, Cambridge, Mass., 1982)

Incorporation of the lithium in a pyrolysed polyacrylonitrile polymer provides a porous carbon support that allows the lithium to 'intercalate'. It is as if the ions can enter a building, rather than congregating on top of each other on the outside wall. The trick is to maximise the amount of lithium that can be housed so as not to have excess carbon weight hanging around, and also not to impede the movement of ions back and forth.

Lithium is likely to be one of the more strategic materials of the future. It is currently extracted from seawater and geothermal brines. Sea farms are found in New Zealand and Chile.

The sodium–sulfur accumulator

The sodium–sulfur accumulator has been used for about 80 years, and a breakthrough has been expected for decades. The cell reaction consists of the reversible formation of sodium sulfide, with the ions travelling through the solid electrolyte. It is about one-fifth the weight of an equivalent nickel–cadmium accumulator, and at least five times the specific energy of a lead-acid accumulator (see Fig. 14.10). However, with a sodium anode, alumina electrolyte, and sulfur cathode, this cell needs to operate between 270°C and 410°C. If it is fully charged, and in operation for at least 80% of the time, it maintains the temperature range by heat released in its internal resistance (the battery is thermally insulated). The high-temperature operation and its inherent flammability limit its options.

Primary cells or batteries

Primary cells are non-rechargeable cells or batteries. The *National Geographic* in 1993 estimated that 2.5 billion batteries (90% non-rechargeable) are bought each year in the USA, costing over a dollar each. About five dry-cell batteries (the zinc and manganese dioxide Leclanché cell) are used every year for every man, woman and child in the Western world. The cell is cheap and easy to manufacture. In spite of developments such as the alkaline cell and more active manganese dioxide (see Chapter 6), its energy density is low, and its discharge characteristics are unsuitable for many applications. The voltage of the cell drops almost from the start of use (see Fig. 14.11).

The mercury cell developed by Ruben–Mallory in the 1930s has the advantage of maintaining a constant voltage of 1.3 V up to 95% of its total capacity. Although its energy per weight is only average, its energy per *volume* of 0.53 W h/cm^3 is as high as that of almost any other system. For consumer applications in watches and calculators, weight is hardly a problem. The disposal of mercury, on the other hand is an environmental problem that is causing concern.

Lithium batteries are now available in the standard sizes. As lithium reacts with water, solvents such as propylene carbonate are used to hold the electrolyte. When used with copper oxide, the cell voltage drops to about 1.5 V, and the cell can be a replacement for conventional batteries. Other systems give much higher voltages (see Fig. 14.11). The military are exploiting lithium–sulfur dioxide cells, which have energy densities of up to 330 W h/kg, can operate in extreme temperature ranges, and have an excellent shelf-life. Because abuse causes explosions, their use in consumer applications at present is prevented. The most promising area of development is in the all-solid cell, which has resulted from the discovery in 1967 of solid electrolyte phases that allow ionic conductance at normal temperatures. The lithium–poly(vinyl pyridine)/iodine battery that generates solid lithium iodide as electrolyte has a high internal resistance, which makes it suitable only for low-current applications. The cell is used, for example, in automatic cameras, small computers, and cardiac pacemakers, and is gradually replacing other batteries.

Battery storage
For a Leclanché dry cell stored at room temperature, the capacity will fall to about 90% after about a year, 70% after two years, and 40% after three. At 45°C it falls to 20 per cent after one year. The loss of capacity on open circuit storage is caused mainly by chemical side reactions such as zinc corrosion. Storing batteries in the freezer at –20°C can hold the capacity to 80% after ten years, but there can be damage if there is a substantial difference in the co-efficient of expansion in its materials. The refrigerator is fine. Lithium forms a protective coating which 'passifies' its surface, so lithium batteries have a 90% capacity after ten years.

Source: New Scientist, 28 Oct., 1995

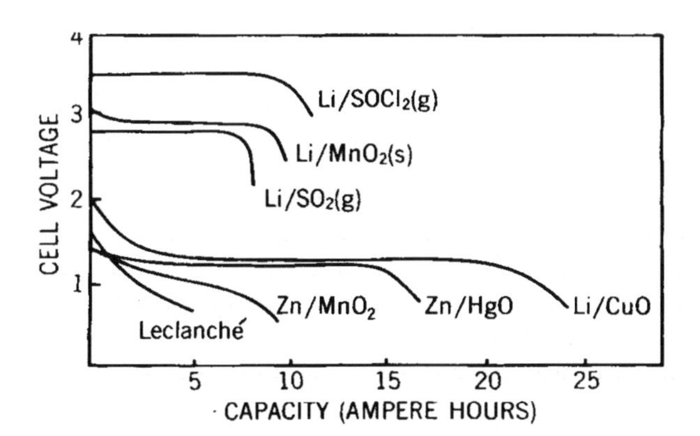

Fig. 14.11
Primary-cell discharge characteristics

Mercury whiskers
A piece of aluminium foil, if rubbed on the edges with mercury or a mercury salt such as mercuric chloride, will grow small whiskers, which can be projected (magnified) with an overhead projector. **Care:** mercury salts are poisonous!

Metal–air batteries

If you can throw away one of the reacting electrodes of a battery and not have to carry it around with you, you can save a lot of weight. Why not use the oxygen in air as the oxidising material, and just blow in air as you go? For a travelling lecture series, some years ago the author decided to make air–aluminium batteries out of beer cans. The first onerous task is to empty out the beer and replace it with the less drinkable caustic potash. Then a graphite electrode from a discarded dry cell is made to act as an anode. When the circuit is closed and air is bubbled over the anode, it is reduced to hydroxide, while the aluminium metal is oxidised to aluminium ion, producing electricity:

$$Al(s) \mid K^+ \, OH^-(aq) \mid O_2(g) \mid C(s),$$

which has a cell reaction:

$$2Al + \tfrac{3}{2}O_2 + 3H_2O \longrightarrow 2Al(OH)_3\downarrow.$$

The aluminium hydroxide does dissolve to a certain extent in the caustic. Salt water can also be used.

This type of battery has been available for a long time. In the US alone, if 1% of cars used aluminium batteries, this would consume 600 000 t of metal per annum. On the basis of the masses of chemically active material (i.e. excluding packing) the battery is said to have a power density of 175 W/kg, and a specific energy density of 400 W h/kg. Zinc–air batteries have a longer history and some of the problems are similar. Both kinds of batteries form oxide layers, which prevent further reaction. For zinc this was overcome (in 1801) by amalgamating (alloying) the surface with mercury. However aluminium becomes too reactive.

A practical solution is provided by alloying with metals such as gallium ($\approx$ 0.04%), indium, or tin, plus the addition of sodium stannate to the electrolyte to reduce corrosion (by almost three orders of magnitude). Examination of the cell under an electron microscope reveals that the current comes from only very small areas of the anode (diameter $\approx$ 1 mm), and with enormous current density (hundreds of amperes per cm^2). This is enough to melt the metal added to the aluminium, and allow the aluminium to form an amalgam with it.

Another problem concerns the formation of aluminium hydroxide gel at the anode, which requires agitation (by the bubbles or air) to cause it to precipitate. A demonstration unit with the required alloys can be obtained commercially.

Various postal services around the world are testing zinc–air batteries, including Sweden, Germany, South Africa, and Israel, with test partners Mercedes Benz and GM–Opel. Spent zinc–air cartridges are being recharged. A plant in Bremen, Germany, reprocesses 100 kg per day. Zinc–air batteries, where air replaces the mercury (or silver oxide), give twice the capacity of a mercury cell of equal volume, and are used in hearing aids.

Fuel cells

The first fuel cell was built in 1839 by Sir William Grove, a Welsh judge and gentleman scientist.

In the hydrogen fuel cell, gaseous hydrogen and oxygen (air) are bubbled over electrodes immersed in potassium hydroxide solution. The cell reaction is:

$$\begin{array}{rcl} 2H_2 + 4OH^- & \longrightarrow & 4H_2O + 4e^- \\ O_2 + 2H_2O + 4e^- & \longrightarrow & 4OH^- \\ \hline 2H_2 + O_2 & \longrightarrow & 2H_2O \end{array}$$

The maximum theoretical voltage is 1.23 V. Because the fuel cell, like a battery, generates electricity directly from chemical reactions, without going through the intermediate phase of a (heat) engine, the theoretical limitations on efficiency of conversion do not apply.

Like batteries, fuel cells produce electricity directly from controlled oxidation without going through a heat producing stage, and they are not rechargeable. They can use a variety of fuels — hydrogen, methanol, ethanol, natural gas, and liquefied petroleum gas. Many different types have been developed for special purposes, such as military use and space flight.

The early US space program chose fuel cells over riskier nuclear power, and more expensive solar energy. Fuel cells furnished power for the Gemini and Apollo spacecraft, and still provide electricity and water for the space shuttle. They are used in the later generation of breathalysers where you do not have to blow but just talk to the device. The alcohol in your breath provides the fuel.

A hydrogen fuel cell-driven car is being developed by Chrysler using petrol to produce hydrogen as it is needed, rather than having to store it. This process basically mimics in your car, on a small scale, what used to be done in earlier times to produce town gas from coal in gasworks. Hydrocarbon fuel is vaporised at high temperature (700–1000°C). The vapour is then fed into a thermal cracker where it is decomposed into hydrogen and carbon monoxide. This mixture is then reacted with steam to convert carbon monoxide to carbon dioxide, and some water to more hydrogen, (the shift, or water gas reaction).

$$H_2O + CO \rightleftharpoons H_2 + CO_2$$

The new aspects here are the need to reduce the level of carbon monoxide to below 10 ppm to prevent poisoning of the platinum catalyst, and to separate out the hydrogen by means of a membrane. The hydrogen is fed into a stack of cells to produce electricity.

Because of the variety of fuel that these units will be able to use, government taxing of fuel may become more difficult.

(For a picture and updated specification of fuel cell vehicles, surf the internet to: http://www.gate.net/~h2_ep/vehicles.htm).

Catalytic lighter
In 1962 the author had a cigarette lighter, in the shape of a lipstick holder, that used alcohol in a cotton-wool soaked pad in the bottom under a platinum gauze catalyst. On removing the top and exposing the fuel and catalyst to air, it burst into flame. It was an impressive conversation opener for girls who smoked. These devices have probably been banned as unsafe.

Gas producers
During World War Two and the post-war period of petrol rationing, some vehicles were fitted with a coal gas producer in the boot (trunk) and a balloon above to hold the gas that then fuelled the conventional (but most unhappy) engine.

Mond's nickel process
Ludwig Mond was one of the first people to attempt to remove carbon monoxide from the hydrogen produced in the water-gas reaction, so as to use the product in a fuel cell, and thus overcome the inefficiency of the early Victorian steam engines. In the 1890s he used nickel oxide as a catalyst. He observed a nickel mirror forming, and discovered the volatile nickel carbonyls. He forgot about fuel cells, and developed instead a unique process for making nickel that is still used today.

Storing energy for later burning

Characteristics of fuels

The most important parameter in heating calculations is the heat value of the fuel — the amount of heat produced when a quantity of fuel is completely burnt, or when any energy source is completely converted to heat. Different heat values for different fuels result from the difference in the individual values for carbon and hydrogen (H = 121 MJ/kg, C = 32.8 MJ/kg). Other factors are the density of the material when costs are given per litre, and non-burning impurities. The heat values for all hydrocarbons lie between the values for hydrogen and carbon, but are closer to the carbon value because carbon is heavier than hydrogen.

Tables of thermodynamic properties of chemicals list values such as heats of combustion as a value per mole. For example, for methane, ΔH°_c is 889 kJ/mol, which divided by the molecular mass, 16, gives 55.56 MJ/kg; for butane, ΔH°_c is 3509 kJ/mol, which divided by 72 (molecular mass) gives 48.74 MJ/kg. Note that the value used for hydrogen is 285.2 kJ/mol (142.6 MJ/kg), which differs from the value given above because all the tabulated values assume combustion to carbon dioxide and water as a *liquid* (i.e. giving up its latent heat of vaporisation) and are thus *gross* rather than *net* calorific values. In practice, water formed during combustion is vented as a gas, taking the 21.6 MJ/kg (of water) latent heat with it to the atmosphere. This means a loss of about 8% of the heat.

The heats of combustion of compounds that are not listed in thermochemical tables can be calculated from a much wider tabulation of heats of *formation* because heats of chemical reactions are additive (Hess's Law).

A second important parameter is the efficiency of the heating device — that is, the heat delivered to the room as a percentage of the total heat available from the heat source. To make the calculation we also need to know the relationships between the various energy units still in current use (see Table 14.4). As well as the calorie, older units of energy include the British Thermal Unit (BTU) and the therm; the unit of electrical energy is the kilowatt-hour, kWh. (The prices used in Table 14.4 are used only for illustrative purposes.)

TABLE 14.4
Energy conversion factors

1 therm	= 100 000 BTU	1 kJ	= 0.948 BTU
	= 106 MJ	1 MJ	= 948 BTU
1 kW h	= 3412 BTU		
	= 3.6 MJ	1 MJ	= 0.28 kWh
1 calorie	= 4.187 J	1 J	= 0.239 calories
1 horsepower hour	= 2.69 MJ	1 MJ	= 0.37 horsepower hours

Oil The heating value of liquid fuels varies little (kerosene 36.7 MJ/L, heating oil 37.7 MJ/L, distillate 38 MJ/L, diesel fuel 38 MJ/L). The efficiency of an unflued oil heater is about 95%, whereas the efficiency of flued oil heaters can vary from 60 to 75%. For comparison we use 75% as the optimum efficiency for a properly serviced heater.

Solid fuel The heating value of coke and coal is 27 MJ/kg and for briquettes it is 24.75 MJ/kg. For split wood it is 12.4 MJ/kg, but it drops to 6.7 MJ/kg for green wood. Comparative heating costs do not take into account any fixed costs (supply or hire of gas cylinders, installation of oil tanks, etc.). The householder must consider capital costs and maintenance.

All currently available commercial sources of energy produce pollution. Coal, briquettes and wood are the worst for air pollution, followed by diesel oil, heating oil, and then liquefied petroleum gas (LPG). Although electricity is non-polluting where it is used, its production causes pollution — air pollution (when coal is used), nuclear waste (from nuclear reactors), or the flooding of river valleys (to generate hydroelectricity).

Gas An unflued gas heater is about 90% efficient. For flued liquid petroleum heaters the maximum and minimum efficiencies are given by the British Standard (BS1250 Part 4) as 78% and 50%. The United Kingdom Ministry of Housing gives an average of 60%. Bottled gas is generally sold by weight, but occasionally also by volume which of course depends on the temperature. If you buy by volume, make sure you buy only on cold days. The heat value of LPG gas is 49 MJ/kg or 25.5 MJ/L at 20°C.

Assuming 50:50 hydrogen (H_2) and methane (CH_4) the heat value of town gas is (121+ 55.6)/2 = 88 MJ/kg, which is over half again of that of natural gas, which is mainly pure methane (55.6 MJ/kg).

Exercise

Why do the burners on gas appliances have to be changed when converting from town to natural gas? (Hint: Write down the combustion equations of the two gases H_2 and CH_4, and work out the amount of oxygen (air) each needs. Remember that equal **volumes** of gas under the same conditions contain equal numbers of **molecules**. If hydrogen gas was H, rather than H_2, would it make any difference?

Wood

About 70% of the mass of dry wood is carbohydrate, of which 40 to 50% is cellulose. The calorific value of carbohydrate is about half that of oil (both as a fuel and as a food). The remainder of the wood is lignin, a complex substance whose major claim to fame is as a source of natural(?) vanillin. Softwoods are higher in lignin. One kilogram of firewood has an energy content of about 16 MJ (equivalent to 0.5 L of oil or 4 kWh of electricity).

When wood burns, a number of processes occur at the same time. The wood is being dried, volatile material is driven off and ignited, and charcoal is being formed and burnt. As a rough estimate, 70% of the energy released by the combustion of wood comes from burning gases, while the rest comes from the glowing coals, the burning of charcoal. The temperature needed to ignite the gases differs for different gases, with some being quite high (carbon monoxide, 600°C; methane, 650°C; acetic acid, 540°C; and hydrogen, 540°C) (see Ignition temperature below). These gases must therefore be kept in the high-temperature zone with sufficient air for long enough to burn completely. Wood stoves with catalytic burners are available that lower the ignition temperature. The *more* creosote the wood produces, the better it burns. The more water in the wood, the

Wood burning
Every time I discuss energy, the emphasis has changed. My first lecture notes in Canberra in 1972 discussed the replacement of wood heaters and stoves with oil — the new wonder heating fuel. With the oil crisis of 1973, everyone was converting to electricity (based on coal). Next time 'round, natural gas was moving into the neighbourhood. Increases in electricity tariffs (one of the 'benefits' of privatisation) caused a shift back to wood and solid fuel again. But pollution from wood is trapped in inversion layers and this was the reason for replacing wood in the first place!

more efficiently creosote will be 'steam distilled' out of the wood. As this occurs at just below the boiling point of water, the creosote will distil away before being reached by the burning edge. The creosote will deposit from the distillate in the cooler parts of the flue. (For more on steam distillation, see Appendix III.) Catalysts can be 'poisoned' by impurities (such as sulfur) which are found in wood.

In order to achieve high efficiencies, a normal heater must burn the wood at high temperatures (to make sure all the gases burn) and with just the right amount of air. (Insufficient air means some of the gas will not burn, and too much means hot air is lost up the chimney.) The air and the wood gases have to be well mixed and the heat must be transferred from the heater to the room.

Open fires are very pleasant, but can funnel a whole room (40 m^3) of air through the chimney every five minutes, and well as losing hot air, the fire is cooled, which means gases are not burnt efficiently. An efficiency of about 10% is reported.

Non-airtight stoves include the Franklin, pot-belly, parlour and the fancy glass-enclosed heaters. They have an efficiency of about 30 to 40%. They are not easy to control and rely mainly on radiation from hot surfaces.

Controlled combustion (airtight) heaters are the most efficient, if not most attractive, heaters, with efficiencies ranging from 40 to 50% at high heat output. When there is a slow-smouldering fire, the efficiency drops off and more creosote reaches the flue. If the area around the fire is insulated the wood gases will stay hot enough to burn, but will need extra preheated air added *above* the fire so that the gases will burn without increasing the burning rate of the wood. The efficiencies given above are somewhat less than those quoted in the sales brochures, but are the results obtained by research workers in Tasmania.

Australian standards for wood burning (domestic) stoves (AS5012, 4013, 4014 (I–IV), Standards Australia, May 1992) set the maximum particulate (smoke) emission limits for all (slow combustion) solid fuel/wood-heating appliances to be sold in Australia. Fuel heaters are certified to operate below 5.5 g/kg fuel for particulate emissions for an average of all burn cycles with the type of fuel to be burnt. Non-complying appliances cannot be sold. With catalytic converters, the maximum emission is set at 3.0 g/kg; open fires are not regulated. A typical stove might run at anything from 2 to 5 kg of fuel per hour. (An oil heater with a common type of burner produces emissions of about 0.1 nanogram per (normal) cubic metre, while a natural gas heater produces 0.05.) Now this 3 gram emission is largely relatively harmless carbon and ash particles, plus an unknown fraction of creosols and related compounds, and certainly some dioxins.

For a ballpark understanding of the magnitudes, it is sobering to remember that Canberra (population 300 000) burns 67 000 tonnes of hardwood fuel per winter season in home heaters. (Heaters sold before 1994 are not subject to the standard.) This use of fuel in Canberra alone is equivalent to six 10 000 tonnes per annum (33 tonnes per day) high temperature incinerators (HTIs) operating all year round. These would, of course, be highly regulated and the flue gas cleaned before disacharge.

FIREWORKS

In the UK, fireworks are associated with Guy Fawkes who plotted to blow up James I and the Houses of Parliament in 1605. In China, fireworks have been used for many centuries at religious festivals and funerals to scare away evil spirits. Simple fireworks are made by packing gunpowder — a mixture of sulfur, charcoal, and potassium nitrate, into bamboo cases or rolled paper tubes.

There are four major types of firework — rockets, roman candles, shells, and fountains. The process is the oxidation of a fuel. Fuels used include charcoal, shellac (a resin from the secretions of an insect) and red gum, all rich in carbon. Elements such as Al, Fe or Mg are also used. Oxidants commonly used are potassium nitrate and potassium perchlorate; potassium chlorate was once used but is now considered too dangerous. The burning characteristics of fireworks are strongly affected by particle size and the degree of mixing, compression, and confinement of the chemicals. A mixture of finely divided aluminium and an oxidant will burn rapidly in the open, but will explode violently when confined (as the bright noisy bangers inside some rockets and shells). If larger pieces of aluminium are used, sparks are produced as the bits of burning metal are thrown out of the firework. With rockets and fountains, the mouth of the firework is constricted to increase the pressure in the firework, so escaping particles come out with greater force. Some mixtures contain a binding agent so that they can be formed into pellets that, when ignited and ejected from the firework, burn brightly, throwing off colours or sparks.

Colour is produced by metals, from either heating the metal or breaking up one of its compounds in a flame. (The process is identical to what happens in the 'flame test' for metals when compounds are inserted into a Bunsen flame.) Because the metal chlorides are generally more volatile, a chlorine donor like PVC is often added to intensify the colour. The best emitters are CuCl for blue, SrCl for red, BaCl for green, and Na atoms for yellow. The species BaCl and SrCl exist only at very high temperatures and have intense colours. Mixtures can be used to provide a wide variety of colour.

Stars are produced by mixing Mg powder and gunpowder to give a high temperature, plus other metals to give the required colour. White sparks are produced by adding Sb; charcoal or Fe can be used for yellow or golden effects, and Al for silver.

Originally, whistling was produced by potassium picrate, which is highly explosive, but now potassium perchlorate and potassium benzoate are used. As the mixture starts to explode, a liquid layer forms that momentarily damps the reaction and then allows it to start again at high frequency, producing a whine.

For a rocket, maximum thrust is needed at the beginning when the rocket is heavy and needs to accelerate from rest. So a conical hole is con-

Cigarette smoke pollution

There are 18 million Australians of whom 25% are smokers. Statistics are only for legal (over 18) smokers, so the population over 12 years of age, say 14 million, might be realistic. Smokers average about 20 cigarettes per day and 20 cigarettes contain about 2 oz of tobacco, that is 56 g. So $0.25 \times 14 \times 10^6 \times 56 \times 10^{-6}$ = 200 tonnes of tobacco is burnt and the noxious fumes exhaled every day in Australia. Thus Australia's cigarette smokers incinerate to the equivalent of about six HTIs. However, because there is no gas cleaning system, cigarette emissions are orders of magnitude more toxic than the emission of an incinerator. But the cigarette incinerators are not all localised in someone's backyard.

structed at the base to expose the maximum surface area of explosive to ignite a lot of material quickly and provide the gas for maximum boost in a few seconds. The conical hole is smaller at the top and this means that the amount of explosive lit decreases with time. However, the weight of the rocket has also decreased, and also the need for further acceleration. As the rocket reaches the top of its flight, a fuse burns through to a separate cavity at the top of the rocket, and this blows out the stars at the top of the firework.

Note: Fireworks require great experience to make them effective and safe. This is no job for amateurs and is illegal in most jurisdictions.

Source: Ed Chem (centrefold) for use in schools.

What about the car — putting chemical-stored energy to work

Combustion engines

All combustion engines convert heat energy into mechanical energy, and the heat is produced from fossil fuels such as petrol, diesel fuel, coal, etc. During the conversion there are further energy losses from friction in the mechanical parts of the engine and, for cars, additional frictional losses on the road and against the air. Quite apart from these 'mechanical' losses there is the much more fundamental loss of all conversions of heat into mechanical work that we discussed earlier.

If the gas in the cylinder of a car engine has an *average* temperature of about 600°C (the peak temperature is much higher) and the exhaust temperature is 100°C, the maximum possible efficiency is $(873 - 373)/873 = 57\%$. Transmission losses cut this back to 20 to 25% in real life. A brief description of some combustion engines follows.

The steam engine

The steam engine is an external combustion engine: the fuel is burnt separately from the motor. Hence a wide variety of fuels can be used and can be burnt efficiently without additives. However, the power-to-weight ratio is unfavourable, compared with the later internal combustion engines, and acceleration is lower. It also requires a driving fluid (water) that has to be transported as well as the fuel.

Efficiency — a reminder

$$\text{Efficiency} = \frac{T_h - T_c}{T_h}.$$

The diesel engine

The diesel engine is an internal combustion engine that explodes its charge by the heating caused by the compression. Both four-stroke and two-stroke engines are used. The four-stroke diesel cycle is:

- Down..........charge with air.
- Up................compression with heating (compression ratio varies from 16:1 to 23:1).
- Down..........fuel injection: burning and expansion with a smaller rise in cylinder pressure than for petrol engines.
- Up................exhaust.

The major problem with diesel is smoke and odour. The smoke can be reduced by about 50% by adding about 0.25% barium additives, which probably work by inhibiting the hydrogenation of hydrocarbons to carbon particles or by promoting their oxidation, or both. The sulfur in the fuel helps to convert about 75% of the barium to insoluble barium sulfate.

Diesels, with their higher compression ratios and higher temperatures, are more efficient than petrol engines, running up to about 35% thermal efficiency.

COCONUT DIESEL

In the South Pacific, coconut oil has been touted as a local substitute for diesel fuel. What are the issues?

Vegetable oils were used during the two world wars during fuel shortages, so they work. Whereas diesel oil contains hydrocarbons ranging from C_{10} to C_{24}, vegetable oils are the glycerol esters of fatty acids (see Oils and fats in Chapter 4). Fats contain one glycerol + 3 fatty acids, about 57 carbon atoms in all, and thus 95–96% of the mass is fatty acids. Coconut oil is one of the most saturated vegetable oils and the fatty acids range from C_8 to C_{18}, with mostly C_{12}. Coconut oil starts freezing at about 24°C which is bad for a cold start but acceptable in the tropics. However, coconut oil has a viscosity 11–18 times that of diesel, so there is a need to modify the injection system.

As 145°C is the optimum injection temperature, one idea is to wrap the inlet around the exhaust pipe! Coconut oil requires an earlier injection setting compared to the late injection setting for diesel oil. Coconut oil is partially unsaturated and so it burns with a sootier exhaust than diesel and this can cause coking on the injectors. This 'gunk' can polymerise on the hot cylinder walls and 'glaze' them. The polymer is swept with unburnt fuel into the oil and raises its viscosity, which reduces lubrication efficiency. Studies at the Department of Public Works in Papua New Guinea show that adding ethanol (10%) to coconut oil eliminates coking of the injectors, possibly because of the formation of ethyl esters of the fatty acids.

Then there is the question of energy value. Diesel oil is just H and C. The theoretical energy value is 48.8 MJ/kg. Coconut oil has O as well; the value is 42.0 MJ/kg, not that much lower, and an energy ratio of 86% by mass, and 95.4% by volume.

Now back at the plantation

Coconut oil is pressed out of ground coconut flesh, and contains fibres and dirt from the husk. Good filtering is essential. Coconut oil pressed from fresh

Methane
Sewage gas is 95% methane, 4% CO_2. The production from four London sewage works back in 1939 amounted to over 28 000 m^3 per day (equivalent to about 5000 litres of petrol). Methane has a high heating value and it does not dilute or destroy the oil film on the cylinder walls.

coconut flesh (coconut cream) tends to form emulsions with water, rather than a separate layer. Proteins and carbohydrates in the coconut flesh act as emulsifiers.

Warm, wet coconut oil will support microbiological growth that will clog injectors and filters, especially if it is stored for long periods. It should be kept dry. It may help to periodically get it very hot (>90°C) to kill bacteria and fungi. In the Western Province of the Solomon Islands, an engineer (Fr Paul Purcell) at the Catholic Training Centre at Loga near Gizo ran a Lister diesel engine on coconut oil for several months (but started it on diesel).

Economic advantages

Use of coconut oil as a fuel gives a floor price to copra (a major South Pacific export) and a ceiling to diesel fuel (a major South Pacific import). It would act as a buffer when there are major declines in copra prices, or increases in petroleum prices. It would also provide import replacement and local employment, particularly if a low pressure (person-driven) extraction is used in each village.

Stationary diesel engines are found primarily at high schools, hospitals, local government facilities and some plantations where fuel costs are the largest cost item after salaries and food.

Further reading

Coconut Power, Bill Morton, Pacific Islands Monthly, Dec. 1993, p. 37.
Vegetable Oil Methyl Ester as a Diesel Substitute, F. Staat & E. Vallet, Chemistry & Industry, **7**, Nov. 1994, 863–865.
Industrial Products from Soya Beans, B. Y. Tao, Chemistry & Industry, **21**, Nov. 1994, 906–909.

The petrol engine

The petrol engine is the most common internal combustion engine for family cars. Like the diesel, both two-stroke and four-stroke cycles are possible, with the latter being the more common and more efficient (but more complex mechanically). The differences between petrol and diesel engines are: petrol engines work on lower compression (usually 7–10:1) and thus need a spark to ignite the charge; and a more volatile, more highly flammable fuel (petrol) is used, and this normally contains a range of additives (see later in this chapter) to 'improve' its performance. The basic four-stroke petrol cycle is:

- Down ....air and *fuel* intake.
- Up...............compression and *spark ignition* near completion of compression.
- Down ....power stroke: expansion caused by increased volume of gas and rise in temperature.
- Up................exhaust stroke.

As increasing the compression ratio increases the thermal efficiency and performance of the motor, this has been a standard 'advance' in automotive design. Increased compression, however, requires higher octane fuel (see later in this chapter). Also, more nitrogen oxides are formed at the higher temperatures of the high-compression engines.

SOME SAFETY RELATED PHYSICAL PROPERTIES OF FUELS AND SOLVENTS

Knowledge of the physical properties of a chemical can indicate the nature or extent of potential health effects, given certain environmental conditions, and enable decisions to be made on correct usage and storage procedures.

- Melting point: The temperature at which a solid turns to liquid (e.g. 0°C for ice). The freezing point is the same for pure single chemicals, but not for mixtures that freeze (and melt) over a range of temperatures (see Appendix III — Phase diagrams).
- Evaporation rate: This measures the rate at which liquid turns into vapour. It depends on the temperature, the surface of liquid exposed, and agitation, etc. Under identical conditions, liquids that require less heat to form vapour do so more quickly and are said to be more volatile.
- Vapour density (relative): This is a measure of how heavy the vapour is compared to air. The more this exceeds 1, the greater is the tendency of the vapour to build up from the ground upwards.
- Solubility: This is the degree to which a substance will dissolve in various solvents (such as water, alcohol, oil, etc.).
- Flash point: This is the temperature at which a liquid produces enough vapour (i.e. reaches the percentage given as the lower flammable range — see Table 14.7, overleaf) to be ignited with a spark (e.g. petrol has a flash point of −38°C). A liquid is said to be flammable if it has a flash point of less than +61°C (Table 14.5). Ignition of solids depends on their fineness of subdivision, for example, coke is non-flammable, but suspended Xerox carbon toner can explode.

TABLE 14.5
Flash points of some common liquids

Liquid	Temp. (°C)	Liquid	Temp. (°C)
Methanol	+11	n-hexane	−21
Ethanol	+13	benzene	−11
Cyclohexanol	+68	n-heptane	−4
Ethylene glycol	+111	n-octane	+13
		n-decane	+46
Diethyl ether	−45	nitromethane	+35
1,4-dioxan	+12	carbon tetrachloride	nf
Cyclohexanone	+44	chloroform	nf
Acetone	−20	dichloromethane	nf

nf = non-flammable

- Ignition temperature: This is the temperature at which gases (liquids and solids) burst into flames without a spark (i.e. spontaneously), see Table 14.6.

TABLE 14.6
Ignition temperatures of gases

Gas	Temp. (°C)	Gas	Temp. (°C)
Hydrogen	580	n-pentane vapour	309
Petrol	550	Carbon disulfide	100
Town gas	600–650	Acetylene	335
Methanol	464	Acetone	538

- The flammability range: This is the range in composition of vapour with air between which explosions can occur. In fuel-rich mixtures, there is insufficient oxygen to sustain combustion. In air-rich mixtures there is insufficient fuel. Table 14.7 shows the flammability limits of various gases.

Infra-red detective work
A new student experiment involved placing some acetylene gas from a cylinder into an infra-red cell to measure and then explain its spectrum. It took some time to realise where extra inexplicable bands in the spectrum came from — acetone.

Note:

- the very broad range for acetylene: 2.5–80%. This makes acetylene a very flammable gas. Acetylene will also explode spontaneously, just from pressure alone. Acetylene is always stored with a solvent (acetone) present in the bottom of the cylinder;
- the wide range for hydrogen sulfide, carbon monoxide, and hydrogen;
- that acetone has a smaller range than methanol, so is safer to use in a demonstration involving flame, (see experiments with acetone, Catalysis, this chapter). Even though too much vapour will not burn, on adding air you can come back into range. The range for petrol is 1.4–7.6%. Thus an 'empty' can of petrol (full of vapour) can be much more dangerous than a full one (see The big bang, p. 405).

TABLE 14.7
Flammability limits in air at ambient temperature (% by volume)

Methane	5 to 15
Propane	2.2 to 9.5
Acetylene	2.5 to 81
Methanol	6 to 36
Diethyl ether	2.0 to 48
Dimethyl ether	3.4 to 18
Hydrogen sulfide	4.3 to 45
Ethanol	3 to 19
Carbon monoxide	12.5 to 74
Hydrogen	4 to 75

Note: Flames or explosions can occur when the gas concentration is between these limits.

THE BIG BANG

This demonstration is for use only by a very competent chemist (never by an unskilled individual).

The combustion of hydrogen and oxygen, an icon for chemical reactions, can be done in many ways.

$$2H_2 + O_2 \longrightarrow 2H_2O$$

One way is to allow the composition of the gas mixture to change with time while the hydrogen is burning. Do the following demonstration only with great care. It is included here only because so much can be learnt from it. While tin cans with detachable lids and gas-filled balloons are variously suggested for this experiment, the results can be erratic. The author has found the following custom-built unit to be reasonably safe and reliable.

The combustion chamber consists of two inverted cones (0.5 mm copper) soldered together around their rims to give a chamber of length about 20 cm and diameter about 15 cm (Plate XXII). At the top is a small hole (about 2 to 3 mm) and at the bottom is a short neck of copper tubing about 2 cm long, with an internal diameter of about 15 mm.

Seal the top hole temporarily with a piece of tape. Flush the chamber thoroughly with hydrogen gas from top to bottom, and long enough to ensure that **all** air is removed. Then close the bottom hole with a rubber stopper and place the chamber, large hole down, in a ring or metal tripod, with a piece of scrap wood underneath. It can be stored this way for a while.

Preferably do the demonstration outdoors, or otherwise in a very large lecture theatre, with the audience at least 7 metres away.

When ready to start, remove the tape from the small hole at the top and light the hydrogen gas at the top. Then remove the bung at the bottom.

Stand clear! The hydrogen gas burns with an almost invisible flame and may appear to go out. Do not be deceived. After a while you will hear a soft whistle, which starts at a high frequency (your dog will hear it earlier while it is still out of your hearing range) and rapidly lowers in pitch, ending in an explosion with flames shooting out of the bottom. Why does this happen?

The speed of sound depends on the average molecular mass of the gas through which it is propagating. As the hydrogen burns at the top, and air moves in at the bottom, the average molecular mass of the gas in the container goes up, and the speed of sound goes down. The chamber acts as a resonant cavity like an organ pipe (or your own voice box) so the wavelength of the resonating sound waves is fixed. If the speed of sound drops, so must the pitch. (The same effect occurs when deep sea divers replace air for breathing by a mixture of helium and oxygen (to avoid the bends from nitrogen) and this gives the diver a squeaky voice.)

While the reaction stoichiometry is simple (two hydrogen molecules and one oxygen molecule to give two water molecules) the mechanism through which it occurs is a complex chain process, similar to the combustion of other fuels. There are certain ranges for the ratios of hydrogen to oxygen that give quiet burning, and others that cause explosions. These are defined by flammability limits (see Table 14.7 above).

The same type of chain reaction occurs in atmospheric processes (see UV radiation and the ozone layer later in this chapter) and in the polymerisation of monomers to form polymers (see Chapter 9).

What about storing energy?

Flywheels

A flywheel is a very compact way of storing energy, coming after nuclear and chemical fuel, and the best on par with accumulators. When a flywheel rotates, it stores kinetic energy, and this can be retrieved by using the flywheel to drive an electric generator. Conversely, the generator acting as an electric motor can speed up the flywheel. Because kinetic energy is $\frac{1}{2}mv^2$, doubling the mass doubles the energy stored, but doubling the speed of rotation increases the storage by a factor of four. A high-tensile steel flywheel spun up to its bursting speed can store the same amount of energy as an amount of water of the same volume as the flywheel, raised to a height of 30 km.

Whereas a petrol engine will not pour petrol back into the tank on coasting downhill, the flywheel will take energy back, and moreover, store enough to drive a bus for half a day. Flywheel buses are used in Switzerland where the routes are up and down. A single-decker Leyland bus in the UK has a glass fibre flywheel that is lighter and safer than a steel one. The bulk of the flywheel is in the rim, which is made of glass and aramid (Kevlar) fibres, bound by epoxy resin. It spins at a maximum of 16 000 rev/min with the rim moving at 1300 km/h, faster than the speed of sound. At this speed the energy is equivalent to taking a 16 tonne bus from start to 48 km/h, while braking returns most of this energy to the flywheel.

A flywheel can repeatedly discharge most of its energy, something that would quickly destroy an accumulator. Flywheels designed for spacecraft (including all the attendant electronics) currently have an energy density of 44 W h/kg and this is increasing. Accumulators needed in space are used in series to provide 110 V, and that means electronic regulation resulting in a final energy density of less than 10 W h/kg.

However, 'when flywheels fail, they explode spectacularly into a cloud of talcum powder-like dust' — (Ben Iannotta, *New Scientist*, 11 Jan. 1997, p. 30) — anyone for space travel?

Petrol and other petroleum-based fuels

With the demise of the steam engine, Australia, following the path of most Western, so-called developed, nations, has switched almost exclusively to *petroleum* as the fuel for its vital transportation industry. World usage is more than 10 million litres of crude petroleum oil, and 3000 million cubic metres of petroleum gas a day.

Petroleum is the fossilised organic remains of minute marine plants and animals that settled to the sea floor millions of years ago. Crude petroleum consists of a complex mixture of compounds, mainly hydrocarbons, but also of smaller amounts of organic molecules containing oxygen, sulfur, nitrogen, and even metals. The precise mixture of compounds present varies widely between oil fields, ranging from almost solid petroleum pitch in Lake Trinidad to crude oil, light enough to be used directly as a diesel tractor fuel in some parts of Kalimantan.

Australian crude oils are lighter than world average so they provide good yields of petrol and diesel, but little bitumen for roads or lubricants. Heavier crudes have to be imported to meet these needs. Table 14.8 shows the approximate yields of intermediates and finished products from (Australian) Bass Strait crude oil, and the percentage volume composition of West Texas crude oil.

The usual first step in refining petroleum is to separate the crude oil into fractions on the basis of their boiling points. Fractions of a typical Middle Eastern crude, arranged in order of increasing boiling point, are given in Table 14.9 (overleaf). The petrol obtained in this separation is known as *straight-run gasoline* and is of too low a quality to be used directly in today's automobiles; it is further refined to finished petrol and solvents. The kerosenes are used in jet engines and as other fuels. Straight-run gas oils are burnt in oil heaters and diesel fuel (see next section on cracking). Imported crudes from the Middle East are used to make heavier fractions such as lubricants or bitumen (asphalt).

Gallons and barrels

We say petrol; in the US they say gasoline. But that's not all, folks! The US gallon of 231 cubic inches was made official by Queen Elizabeth I of England (and was used for wine). It later became the Queen Anne gallon and was used in the US and Canada (until Canada adopted the imperial gallon).

An imperial gallon (based roughly on a gallon of water weighing 10 lb) is 4.546 litres (which tells you that the weight of 1 L of water, that is, 1 kilo, is 1/0.4546 = 2.2 lb). Thus an imperial gallon (as suits its name) is bigger than a US gallon by 20%. The standard oil 'barrel' is 42 US gallons, or 160 L, (but a US liquid barrel is 31.5 US gallons).

The French gave us the metric system, for which we should all be eternally grateful.

Petroleum oil fractionation

Before the advent of electricity as a utility in about 1900, the most useful fraction was kerosene, which was used for home lighting. The advent of electricity and the automobile then made gasoline the most important fraction. Not until the jet engine took over was kerosene in demand again.

TABLE 14.8
Composition of some crude oils

Bass Strait (Australia)	%	West Texas (USA)	%
Liquefied petroleum gases (LPG)	2	Butanes	2
Petrol (straight-run motor gasoline)	38	Gasoline	11
Kerosenes	10	Naphthas (diesel, kerosene)	14
		Furnace oil	17
Diesel (straight-run gas oils)	25	Gas oil (heating oil)	25
Residue for cracking to petrol and diesel	25	Residue (lubricating oil, asphalt)	17

Source: Esso Australia, Sept. 1990.

TABLE 14.9
Petroleum oil fractionation

Fraction	Composition	Boiling range, °C	Principal use
Gas	C_1–C_4	Below 20	Heating fuel
Petroleum ether	C_5–C_6	20–70	Solvents, petrol additive for cold weather
Petrol (straight-run)	C_6–C_{10}	70–200	Motor fuel, solvent
Kerosene	C_{10}–C_{18}	175–320	Jet and diesel fuel
Gas oil	C_{12}–C_{18}	Above 275	Diesel fuel, heating fuel oil, cracking stock
Lubricating oils	Above C_{18}	Distil under vacuum	Lubrication
Asphalt	Above C_{18}	Non-volatile liquid	Roofing and road materials

Straight-run gasoline (octane rating about 70) consists mainly of straight-chain hydrocarbons which detonate causing *pinging* or *knocking*. To get a smooth, constant push on the piston you need a *slow* explosion. Alkanes vary greatly in their ability to burn in an engine without knocking. Thus iso-octane (2,2,4-trimethylpentane) caused little knocking, whereas n-heptane caused a great deal.

The measure that later became the octane rating was devised by Sir Henry Tizard (1885–1959) a wartime science administrator who advocated the development of radar during World War Two. The octane rating of any hydrocarbon is defined as being equal to the proportion of iso-octane in a mixture of iso-octane and n-heptane (Fig. 14.12) that knocked under the same conditions. Methylcyclohexane, for example, knocks under the same conditions as a mixture of 75 per cent iso-octane and 25 per cent n-heptane, and hence has an octane rating of 75.

Fig. 14.12 Iso-octane and n-heptane

A similar system is used for diesel fuel with an index called the *cetane* number, which is set by comparing a fuel with a mixture of cetane (100) and *a*-methylnaphthalene (0). A high cetane number indicates a greater ability of the molecules to continue a burning process.

MIDGLEY — TEL AND CFCs

'To invent one environmental hazard', as Oscar Wilde's Lady Bracknell might have said, 'may be regarded as a misfortune; to invent two looks like carelessness'.

Thomas Midgley was born on 18 May, 1889, graduated in engineering at Cornell in 1911 and joined Dayton Engineering Labs (Delco — founder Charles F. Kettering). Soon after he began to study 'knocking' in internal combustion engines, he showed that, contrary to popular belief, it was not caused by pre-ignition of fuel, but by an abrupt rise in pressure following ignition. Midgley and Kettering felt that the combustion characteristics of the fuel could be improved if the fuel was coloured red, so that it could absorb heat more readily. 'This theory came to us because we happened to know that the leaves of the trailing arbutus are red on the back and that they grow and bloom under snow', wrote Kettering. Midgley could find no oil-soluble red dye in the store, and so used iodine that gives a red solution in organic solvents. Iodine worked, but for the wrong reason. It was not the colour.

During World War Two he worked on aviation fuels. After the war he returned to test a large number of materials for anti-knock problems and he discovered the effectiveness of tetra-ethyl lead (TEL). However, when the fuel burnt it left conductive lead salts deposited on the plugs that shorted them. This required addition of a bromine compound to produce volatile lead bromide that would exit with the exhaust gases (and into our lungs instead) so he then developed a process for extracting bromine from sea water.

At its peak in the 1960s, US consumption of TEL was greater than 300 000 tonnes per year.

Midgley then turned his mind to replacing the gases used in refrigerators, which were dangerous, corrosive, and smelly gases such as sulfur dioxide and ammonia. He studied the periodic table and within three days decided that fluorine compounds were the answer. Conventional wisdom said that fluorine compounds would be toxic. Midgley bought all the antimony fluoride he could find (five one-ounce samples) and proceeded to produce dichlorodifluoromethane. He exposed the gas from one sample to rats who were unharmed. However, gases made from the other four samples were all toxic! (It later turned out that there was an impurity in the SbF_3 that produced phosgene.) He said that if he had studied the samples in another order, then 'I believe we would have given up what would then have seemed like a 'bum hunch'.

At a meeting in 1930, Midgley breathed some CFC and then exhaled it gently around a burning candle, putting it out. He failed to see its use in aerosols. In a paper in 1935 (at a time when there was severe agricultural overproduction) he did make one suggestion that now reads with some irony. 'If...crop curtailment continues as a permanent necessity...the chemist will eliminate the bureaucracy necessary by increasing the amount of ozone in the earth's atmosphere, thereby limiting the ultraviolet radiation available to the amount required (by crop demand)'. He was not to know that CFCs would **reduce** ozone, **increase** UV, but that this would **reduce** crop production — right result, wrong reason — again!

Source: Canberra Times 22 May, 1989, ex Martin Sherwood, The Guardian.

The twin problems of low quality and low quantity of petrol from direct distillation of crude petroleum were solved by a variety of processes known as cracking, alkylation, re-forming, and isomerisation, and by using additives.

Cracking (thermal, catalytic, hydro-)

By heating the high-boiling fractions (C_{10} and above) with a catalyst to 400°–500°C, the larger molecules 'crack' into the smaller hydrocarbons, and at the same time tend to rearrange into the branched-chain hydrocarbons. (The reason for this transformation lies in the fact that the branched-chain products are more 'disordered'— that is, they have a higher entropy than the straight-chain ones.) This not only increases the amount of petrol that can be obtained from crude oil but also improves the octane rating, typically into the 85–95 range. For example:

$$CH_3(CH_2)_8CH_3 \xrightarrow{\text{cracking}} CH_2{=}CH_2 + CH_3{-}\underset{}{\overset{\displaystyle CH_3}{\overset{|}{C}}}H{-}(CH_2)_4{-}CH_3, \text{ etc.}$$

Alkylation

Alkylation involves heating isobutane with the low-boiling alkenes (C_3–C_6) under acid conditions. This causes addition of the isobutane to the alkene — leading to a larger-branched alkane. This process converts some of the lower-boiling gas fractions into high-octane fractions (typically 90–95). For example:

$$\underset{\text{an alkene}}{CH_2{=}CH{-}CH_3} + \underset{\text{isobutane}}{(CH_3)_3C{-}H} \xrightarrow{H^+} \underset{\text{branched chain alkane}}{(CH_3)_3C{-}CH_2{-}CH_2{-}CH_3}$$

Isomerisation

Heating at relatively low temperatures, with special catalysts, causes rearrangement or isomerisation of straight-chain hydrocarbons into their branched-chain isomers. For example:

$$CH_3(CH_2)_6CH_3 \underset{\text{heat}}{\xrightarrow{\text{catalyst}}} CH_3{-}\overset{\displaystyle CH_3}{\overset{|}{C}}H{-}\underset{\displaystyle CH_2CH_3}{\underset{|}{C}}H{-}CH_2{-}CH_3$$

This improves the octane rating to the mid-80s and is used primarily on the straight-run gasoline fraction.

Catalysis

A catalyst is a material that offers an alternative, easier pathway for a chemical reaction. If one describes a chemical reaction as having to go from a valley of starting materials over a mountain range to a valley of product materials, then a catalyst offers a new, lower, mountain pass through the range. Of course the path makes it easier to come back from the product valley to the starting material valley, and so the catalyst does not change the position of equilibrium for the reaction. Catalysts certainly take part in the reaction, but are not used up, and are chemically, if not physically, the same at the end of the reaction as they were at the beginning.

Catalysts are extremely important in making a particular reaction go faster by offering that reaction, rather than competing ones, an easier way. Catalysts in biological systems consist of proteins called enzymes. For chemical engineering purposes, metals (particularly transition metals) and their oxides are more often used.

CATALYSIS EXPERIMENT

A very simple but instructive experiment involves the catalysis of the oxidation of acetone vapour. A catalyst for this reaction is a mixture of the two oxides of copper.

Materials

A 100–250 mL beaker; copper or alloy (e.g. 10 cent) coins suspended by a paper clip, or copper wire wound in a coil around a pencil as a mould; enough acetone to just cover the bottom of the beaker; and a gas burner.
CAUTION Acetone is very flammable, b.p.: 56.2°C, flash point: −20°C ignition temperature: 484–538°C, flammability limits in air: 2.55–12.8% by vol.

Procedure

The copper is heated to red heat in the flame and then supported from a wire to hang just above the acetone in the beaker (see Plate XXIII). The shimmering colours of the copper surface are caused by the heat being maintained in the copper coin from the heat of the chemical reaction taking place on its surface, as well as the colours of copper(II) oxide (black), copper(I) oxide (red), and copper(0) metal (salmon pink). It may be necessary to warm the beaker in warm water to produce sufficient acetone vapour.

Do not sniff the oxidation products — they contain ketene that (like many products of burning) is suspect, as regards health.

The red heat is maintained as long as some acetone remains.

The flammability limits as a percentage of air can be converted to vapour pressure as follows:

$$\text{Lower: } \frac{2.55}{100} \times 1.013 \times 10^5 = 2.6 \text{ kPa (22 torr)}$$

$$\text{Upper: } \frac{12.8}{100} \times 1.013 \times 10^5 = 13 \text{ kPa (97 torr)}$$

The actual equilibrium vapour pressure of acetone at 25°C is 26 kPa, well above the upper flammability limit. Starting with warm acetone, which then keeps warm through the heat of the reaction, sets up this safe situation, except at the top of the beaker where air dilutes the vapour. (Will the same be true for methanol? See Table 14.7.) A flame at the top of the beaker can be avoided by moving the hot copper out quickly. Should a flame occur, it is easily extinguished.

With care, drop the red hot coin or wire into a beaker that has a few millimetres of water or acetone at the bottom. Note that the sizzling sound caused by the cooling of the copper maintains a steady volume until a crescendo heralds the end of the cooling. The same effect occurs when you pour liquid nitrogen into a Dewar flask. Why? (See Bubbles and noise in Chapter 6.)

Catalytic reactions are extremely important industrially: for example, the Haber process for producing ammonia from nitrogen and hydrogen, the decomposition of nitrogen oxides by the catalysts used in car anti-pollution mufflers, and the catalytic converters (cats) used in producing petrol from crude oil. The exothermic (heat given out) nature of the reaction shown is exemplified by the fact that, once heated, the copper is kept hot by the reaction.

Re-formation

Re-formation is by far the most important change process in refining, and involves heating with special catalysts similar to those used in cracking and isomerisation. However, by careful choice of conditions, C_6 and above hydrocarbons are converted into aromatic compounds, mainly benzene, toluene, and xylenes. The improvement in octane rating is marked — into the mid-to-high nineties. However, the aromatic compounds on incomplete combustion in an engine can produce larger molecules which form known carcinogens in the exhaust.

With all this refining of crude oil, the average petrol produced today has the broad composition:

- butane 10%
- paraffins 60–65%
- aromatics 25% to 40%
- alkenes (olefins) small.

The average octane rating is in the high eighties for the 'pure' hydrocarbon mixture. Since the modern car usually has a compression ratio of 8.5–9.5 : 1, and this requires 93–96 octane fuel, lead compounds were added to upgrade the octane rating. Non-lead petrol sometimes contains more branched-chain and aromatic components.

Best ignition occurs in the range air/fuel (A/F) = 10 to 14.5 and the overall ratio must allow for variations within the cylinder, and from cylinder to cylinder. Uncontrolled engines are thus usually tuned to an A/F of about 13. The stoichiometric ratio (where in theory the exact amount of air is present for the fuel) is 15.5. (See Fig. 14.13.)

In rough terms, fast burn refers to an air: fuel ratio of around 16: 1 and lean burn can be anywhere between 18:1 and 22:1. Operation with these

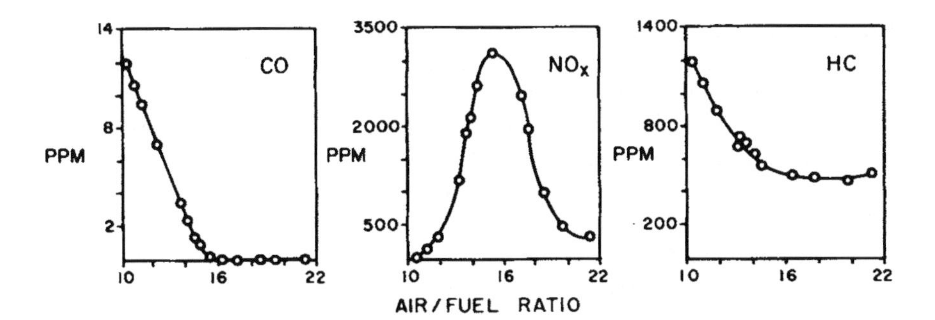

Fig. 14.13
Pollution of exhaust materials as a function of fuel mixture setting

lean mixtures requires exquisite engine design to provide super-efficient mixing and individual timing for each cylinder, which is set to provide the best economy and power under the driving conditions demanded by the driver. Fast combustion of a lean mixture is the key to low emissions, while catalysts try to correct the emission pollutants after the event.

Exhaust catalysts

The catalyst in an exhaust is coated onto a honeycomb filter of sintered aluminium alloy and has two sections, the first to deal with oxides of nitrogen, and the second with carbon monoxide. (Both catalysts use platinum and rhodium, but in different proportions.) To keep the NO_x catalyst hot it must have NO_x on which to operate, and the stoichiometric fuel ratio is used to provide maximum NO_x (maximum power but low efficiency and high emissions). To activate the second catalyst for the oxidation of carbon monoxide, extra air must be fed in at the junction of the two honeycombs to provide oxygen in correct proportion to the unburnt gases. This system requires oxygen sensors and feedback control.

Alcohol as fuel

Alcohol and other oxygen-containing compounds provide less fuel energy than hydrocarbons, but reduce exhaust emissions as well, providing a boost to octane rating.

Alcohol ex biomass
It has often been argued that when produced from biomass, alcohol represents a renewable resource..

Fermentation alcohol is used as a fuel for cars alone, or mixed with the usual hydrocarbons (petrol). In the USA the alcohol is fermented from starch isolated from corn, while in Brazil, cane sugar juice is used. While the primary fermentation is efficient, using the bagasse waste from the cane, the constant boiling (azeotrope) alcohol (95%) must be distilled again at a central location (with benzene) to produce alcohol absolute that will then mix with the hydrocarbons.

The production is not energy efficient overall. In Brazil the brave experiment for attempting to use a (partially) renewable source coincided with a massive drop in oil prices. The production of alcohol as a petrol substitute then meant that there was a shortage of diesel fuel. This meant importing crude oil again, cracking and distilling, and exporting the now excess petrol! There are renewable sources for diesel, of course (see

Coconut diesel in the earlier box) but there are no coconuts in Brazil. Transport costs fuel. Further, if all the corn, wheat, and sugar crops grown by all the world's farmers were converted to ethanol, it would give only 6–7% of the energy equivalent of present world oil production, and demand is still increasing steadily.

Methanol is available cheaply from surplus natural gas, and has been used neat and mixed with petrol, and as a diesel fuel. However, methanol is hygroscopic, absorbing water vapour from the atmosphere, and must be used with a cosolvent alcohol.

Methyl tert-butyl ether (MTBE)

MTBE is added to petrol at a level ranging from 2–15% to improve its combustion performance, in particular to lower ozone and carbon monoxide levels in urban areas. It also acts as an octane booster, replacing aromatic hydrocarbons, like benzene, in non-leaded petrol. It is considered of low toxicity. Production has increased 25% annually since 1984, and in 1994 reached 6.2 billion kg in the US, thereby removing the surplus of methanol.

When spilt petrol leaks into groundwater, the more water-soluble components are preferentially leached, and the mobile components can migrate rapidly into watercourses and aquifers. The monoaromatic petrol components such as benzene, toluene, ethylbenzene, and the xylenes, are the most water soluble, mobile, and toxic hydrocarbon components, and drinking water standards are set in the range 5–100 μg/L. MBTE is even more water soluble and very mobile and is typically found in the 1–400 μg/L range near areas impacted by petrol.

While the monoaromatics are readily biodegraded, MBTE is not. While not very toxic, MBTE has a strong odour and taste, evident if present in water in the 10–100 μg/L range. Natural biodegradation will be very slow. It provides an air pollution bonus, but a groundwater environmental liability. The US Geological Survey tested for MTBE in 210 urban wells and springs, 27% of which had detected levels, none at a health-concern level, but a few at levels causing nuisance. Source: *Preliminary assessment of the occurrence and possible sources of MTBE in groundwater of the United States*, 1993–94 US Geological Survey, Open File Report 95–456.)

Chemical assistance for cars

Lubricating oils

There are at least four tasks for an engine oil to perform. The first is to provide a seal between the rings and the cylinder wall, otherwise there would be no compression and hence no power output at all. The second task is to carry away excess heat, otherwise the piston would melt. The third task is to clean out combustion soot and other particles from the cylinder into the sump. Engine oil has to be changed because much of the debris of burning accumulates, as well as being worn out chemically.

With a car engine running at 5000 revolutions per minute (rpm), each piston is slamming up and down inside the cylinder 83 times every second. Because the piston has to go from being stopped at the top to being stopped at the bottom, the acceleration can be as great as 1500 times gravity.

Against these high stresses, and the temperatures generated, the engine's only protection is a thin layer of oil between the moving metal surfaces. If this oil coverage fails and the metals do come into contact, the high points on both surfaces are welded together and then the weaker side is torn off as one surface slides past the other. This creates a higher peak that is more readily in contact with the other side, and the process escalates. Finally the drag of one metal surface upon another becomes greater than the power produced by the engine, and the engine locks solid. This is known as seizing. The pistons and the bearings that hold the spinning crankshaft are the most likely places for seizing to occur. The engine coming to a grinding halt is the most extreme example, but any metal-to-metal contact will cause wear.

A much more common wear situation arises when a car has been parked for some time, the engine is cold, and the oil has had time to drain back into the sump. When the car is started, there may be some metal-to-metal contact in the top of the engine before the oil is pumped back around it. This is not critical, provided the engine is not revved up. One should always wait at least a few seconds before driving off after a cold start, but waiting until the engine has warmed up (so that the parts have heated to the optimum operating clearances) is not really necessary, and just adds more pollution to the atmosphere.

Modern engine oil also acts as a coolant and a washing agent. As the oil circulates, it spends some time in the hot operating areas, and then some time in the sump where some of the collected heat can be radiated out through the metal walls of the sump, which are cooled by the outside air flow.

The combustion process is not a particularly clean one and some of the by-products are forced past worn or badly fitting piston rings, and accumulate in the engine oil.

Re-refined Oil: One of the problems about recycling oil is the build-up of polycyclic aromatic hydrocarbons (PAHs) formed during the combustion process, and scavenged by the oil. Animal tests involving painting mice skins over their lifetime have suggested that used oil is probably an animal carcinogen. Some used oil is recycled by the oil companies in their production cycle. The basic hydrocarbon oil is mostly intact, but oxidation products must be removed, and additives replenished.

Warning
Oil companies caution users to avoid skin contact with used oil.

SAE viscosity grading: The numbers you see on all car oil containers are the United States SAE (Society of Automotive Engineers) ratings. The numbers indicate viscosity (see Fig. .4.14, overleaf). Viscosity is a measure of a substance's resistance to flow. Honey, for instance, is more viscous than water. With engine oils, the thicker (more viscous) the oil is,

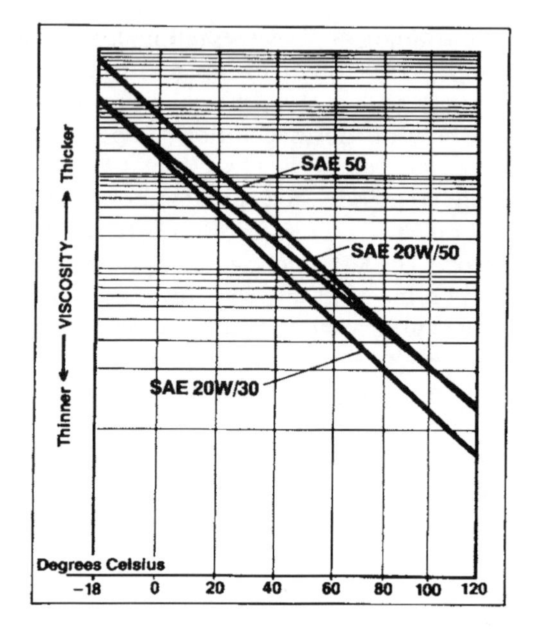

Fig. 14.14 Temperature dependence of oil viscosity. The graph illustrates how although all crankcase oils get thinner as the temperature rises, a multigrade with a large SAE range can be thinner when cold and thicker when hot than a comparable monograde

(Choice, July 1981)

the higher the SAE number. Oils with an SAE rating over 80 are the thicker gearbox and transmission oils.

Lighter solid states
Ice is less dense than water and floats on it. There are very few materials with the solid state less dense than the liquid.

Ethylene glycol
Ethylene glycol is sweet (see 'More on additives — Artificial sweeteners', Chapter 15) — and poisonous! Death on swallowing antifreeze probably occurs, however, from nitrite poisoning. This is unfortunate because the antidote for straight ethylene glycol is alcohol! The reason? The enzymes that would otherwise convert the glycol into the real poison, oxalic acid, are kept busy dealing with the alcohol.

Antifreeze

Water freezes at 0°C and in so doing expands. The expansion can damage the chambers in which radiator water flows. When water has other materials dissolved in it the freezing point is lowered, and the boiling point is raised. The amount by which this happens depends only (to a first approximation) on the *number* of molecules of the added material — not on its mass, or the type of molecule. Thus, for a given *mass* of material, light molecules are more efficient than heavy molecules. For this reason methanol (CH_3OH; MM 32) was commonly used, but it tends to boil away, so ethylene glycol ($HO–CH_2–CH_2–OH$; MM 62) is generally used. While the lowering of freezing point by a solute is great in theory, in practice it is the prevention of solid ice crystals forming that is of major importance (see Alloys, eutectics in Appendix III). The corrosion inhibitors used to protect the metals are generally sodium nitrite with sodium benzoate.

Airbags

An airbag in a car has to protect the occupants very quickly when there is a sudden deceleration. The bag must therefore inflate in a matter of milliseconds, with the actual time depending on the nature of the crash. The pressure in the bag has to start decreasing by the time contact is made because otherwise the bag would appear as hard as stone on impact. A sensor and microprocessor are used to compute the impulse function of the speed, acceleration, braking patterns, and shocks, and to compare these data with crash pattern data stored in memory. The computing is becoming ever smarter. The chemistry is fairly standard.

The gas generator contains a mixture of NaN_3 (sodium azide), KNO_3 (potassium nitrate), and SiO_2 (silica). It is ignited electrically and provides a slow detonation or deflagration that releases a precalculated amount of nitrogen gas to fill the bag.

The reactions are:

$2NaN_3 \longrightarrow 2Na + 3N_2$, at 300°C. ....(1)

$10Na + 2KNO_3 \longrightarrow K_2O + 5Na_2O + N_2$. ....(2)

$K_2O + Na_2O + SiO_2 \longrightarrow$ alkaline silicate ('glass'). ....(3)

Exercise

How much azide is needed to fill a 70 L airbag? One mole of nitrogen (at NTP) occupies 22.4 litres. For a 70 L airbag you need approx 3.2 mol of nitrogen. This nitrogen value comes from Equation 1 which shows that 2 mol of NaN_3 gives 3 mol of N_2. Equation 2 shows that for the additional 0.2 mol of nitrogen required, we need 2 mol of Na. This value has also come from Equation 1. Thus 2 mol of NaN_3 will provide the required amount of gas, and 2 mol has a mass of $2 \times 65 = 130$ g. The 150 g actually found in the bags allows for impurities and wastage.

Source: A Madlung, Journal of Chemical Education, **73**(4), 1996, p. 347–348.

OHS AND ENVIRONMENTAL ASPECTS OF AIRBAGS

There are some important OHS and environmental aspects of airbags. Sodium azide (NaN_3) is a derivative of the explosive hydrazoic acid, a weak acid ($pK_a = 4.92$). While the mercury and other heavy metal azides are used as detonators, the sodium salt needs to be heated or ignited first. It reacts with water, forming hydrazoic acid and caustic.

Sodium azide is very toxic with an allowable workplace concentration, of 0.2 mg/m^3 air, which is half that of potassium cyanide (0.5 mg/m^3 air). An airbag contains 150 g of sodium azide and it is stored in a steel or aluminium canister about 6 cm long and 2 cm across, hidden in the steering column or dashboard. If scrapped, the azide may, however, react with water and form hydrazoic acid, b.p. 37°C, which is extremely toxic. Airbags should be triggered before disposal. Sodium azide pellets should be provided in a reusable form.

What about problems from burning chemical fuels?

Infra-red radiation and the greenhouse effect

The world is warming up, according to most experts who assiduously watch their thermometers like opinion pollsters watch the electorate. Those who live too close to the sea will be flooded, and the weather will be changed unpredictably. The term 'greenhouse effect' is here to stay, describing a very real phenomenon. If there were no greenhouse gases in the atmosphere at all, the earth's temperature would be a bleak –18°C (below zero). Natural levels keep it 33°C higher, at a comfortable 15°C.

A little history

The French mathematician Fourier described in 1827 how carbon dioxide in the atmosphere caused it to warm up, and he also warned that man could affect climate. When Arrhenius coined the phrase 'greenhouse effect' for this in 1896 he believed that the glass in a greenhouse acted like the carbon dioxide in the global atmosphere.

However, Robert W. Wood, at Johns Hopkins University, established in 1909 that two model greenhouses, one covered with glass and one with rock salt (which is transparent to infra-red as well as visible radiation) reached very nearly the same high temperatures. Today plastics (which are fairly transparent to infra-red) are often used instead of glass, and work just as well.

The 'greenhouse effect' in actual greenhouses results mainly from the suppression of ventilation; to cool a greenhouse you open the vents. It was a good attempt an analogy but it just happens to be mostly wrong. (At least Arrhenius was right in his work on how fast chemical reactions go when the temperature is increased, and why all acids taste sour.)

Source: R.G. Fleagle and J.A. Businger, *Science*, Vol. 190, 12 Dec. 1975, pp. 1042–3.

So how does it all work?

Kitchen analogy
Molecules need dipoles to absorb radiation. In Chapter 4 we noted that in a microwave oven, those non-metallic materials with the biggest electric dipoles absorb the microwave radiation most strongly, and heat up most. Melamine becomes much hotter than polythene or glass. The rules for absorbing microwaves are the same as for absorbing heat radiation. Water absorbs microwaves very strongly and the whole exercise of cooking with microwaves depends on this.

The radiation from the 6 000 degree hot sun consists of light and heat. Both are absorbed by earth. The warmed earth radiates heat back into very cold space through the atmosphere.

Our atmosphere on earth consists of approximately 20% oxygen and 80% nitrogen and virtually nothing else, and the formulae for these gases are O_2 and N_2. In each case two identical atoms are joined together and the molecule is symmetrical. This means that the charge due to the electrons in the molecule is uniformly spread. These gases have no plus and minus electrical end.

We are aware that the earth, or a magnet, has a north and south magnetic pole, and in a similar way, most molecules that are not symmetrical have plus and minus electric poles. These are called dipoles (meaning two poles). A molecule needs a dipole for it to be able to absorb heat radiation. That is why the main gases in the atmosphere, O_2 and N_2, do not take part in the greenhouse effect, but some of the minor gases do.

Greenhouse gas: water

Water vapour is a minor component in the atmosphere, but it contributes most to the greenhouse effect of warming. That is why deserts with dry atmospheres are cold at night. There are two reasons for the strong heat-absorbing power of water vapour. One is that in H_2O, hydrogen is joined to oxygen, and these atoms are very different in their power to attract electrons. Their bond has a large electric dipole because the oxygen gets the lion's share of the electrons. The second reason is much more important. It is simply that the centre of gravity of a water molecule is close to the oxygen (because oxygen is sixteen times heavier than hydrogen). As anyone who has played with a science centre exhibit on moment of inertia will remember, when you pull your arms in while spinning around on a chair you will spin faster. It is the small moment of inertia of the water molecule that spreads its absorption of heat over a wide range of the infra-red spectrum.

In Figure 14.15, the two fairly narrow absorption bands around 700 and 2400 cm^{-1} are from carbon dioxide, and the two sets of spiky, spread-out bands around 3800 and 1600 cm^{-1} are from water vapour in the air. The positions of the bands provide the frequencies with which the atoms in these two molecules vibrate and the strength of their bonds, while the internal structure within each band tells us about how the molecules rotate, their moments of inertia, and the length of their bonds. This is only part of the infra-red region, and water in particular is found with discrete absorption bands everywhere. The influence of other greenhouse gases depends in part on whether they absorb in the intervening spaces, in which case they will be important, or on top of these gases, in which case they will not add much to the effect.

Water vapour also forms clouds, which reflect heat radiation from the sun, and keep the earth cooler during the day. The clouds also reflect heat radiation from the earth and keep it warmer at night. Thus there are effects pushing in both directions. Most real problems are complicated by this sort of balance.

Infra-red and plastics

The second edition of this book had an experiment measuring the infra-red spectra of a series of transparent plastics such as polythene, polystyrene (envelope windows), PVC, oven bags (PET), cellophane, and so on. If you have access to a suitable instrument, it is a simple way of identifying plastics and their additives.

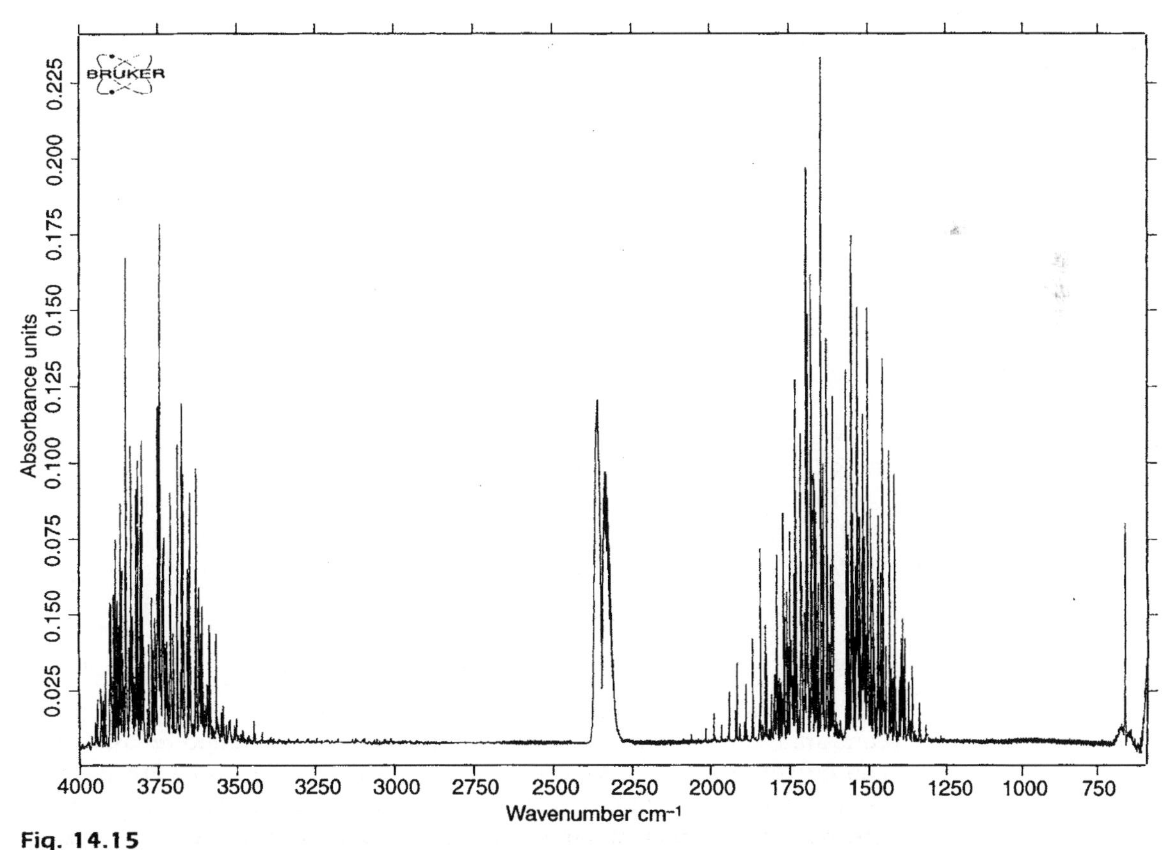

Fig. 14.15
The infra-red spectrum of a few cm of ordinary laboratory air

Greenhouse gas: carbon dioxide

When we burn oil and coal, we are of course just recycling plant material that has become stuck in the natural cycle as fossil fuel. We regenerate the carbon dioxide from which the plants originally came. All the fuss about burning fossil fuels comes about because we do not really want to return to our pristine primeval atmosphere devoid of oxygen! The carbon dioxide, CO_2, has carbon–oxygen bonds that also constitute electric dipoles, but this is a heavier molecule and, with its large moment of inertia, absorbs heat much more selectively than water. Whereas water contributes a rise of temperature of the earth of about 33°C, carbon dioxide contributes only 0.5°C with a possibility of another 1 to 2°C if we continue the way we are going.

Buffering of carbon dioxide

In limestone caves we see an interesting balancing act. Carbon dioxide dissolves in water in streams and then this lightly carbonated water percolates and dissolves limestone rock to form caves. In the caves, the carbon dioxide is later released to form limestone again as stalactites and stalagmites ('the mites run up and the tights come down' was the speleologist's mnemonic we used as regular holiday guides at the Jenolan limestone tourist caves). The same balancing (or buffering) occurs for carbon dioxide and carbonate in the oceans. Extra carbon dioxide allows an increase in shell production.

Minor greenhouse gas: ozone

Ozone (O_3), has three atoms in a row, and is bent in the middle, so it has a small dipole. The removal of ozone would cool us down, a little.

Minor greenhouse gas: CFCs

The CFCs are gases with very large electric dipoles, made by joining carbon to fluorine and chlorine to make chlorofluorocarbons. These gases are very stable, but degrade when they reach the upper atmosphere and are exposed to the ultraviolet radiation. Unfortunately the degradation products, that would quickly disappear if they were near the ground, react instead with the ozone in the upper atmosphere. So, while their greenhouse effect is unimportant, the effect on the ozone layer is critical; see below.

Models of the atmosphere

The general circulation models of the global atmosphere used in computer simulations study the effects on the atmosphere of changing chosen parameters, such as the level of CO_2. This change perturbs the equilibrium, which then shifts to a new position with new consequences for crops and people. What the models do not currently examine is how the atmosphere changes from one steady state to another. There is no information on the dynamics of the process. The change may not be smooth; it could build up to a point of instability and then suddenly flip over to different conditions (like a flip-flop (bistable) circuit in electronics). What such a turbulent change might mean to our lives is hard to imagine.

THE GREEN PAPER EFFECT

As chemists, we learn to deal with complexity slowly and painfully, extracting sense by studying separately the various parts of a system and exploring how they might fit and work together. However, complex systems have emergent properties due to their complexity that are not obtained by summing the parts. The result can very often be counter-intuitive and apparent logical argument can be wildly wrong.

Individual specialist research has to be collated and assessed by skilled generalists before any valid conclusions can emerge, particularly if the discussion is to make sense to a lay audience. Just how do we produce skilled generalists in a society that rewards specialists in proportion to the narrowness of their specialisation?

It is interesting to select the general science topics that need to be covered to understand the greenhouse effect. These include stoichiometry (the balancing of chemical equations), infra-red spectroscopy (moment of inertia and dipole moment of water, etc.), kinetics (the mechanism of those chain reactions in the upper atmosphere), thermodynamics (what compounds will be less reactive with ozone and therefore will be safe to use), and so on, all part of a good (i.e. pre-economic fundamentalism-driven) university chemistry course.

Very high infra-red The infra-red transmitters for consumer electronics like TV remote controls use a very high infra-red frequency which most clear materials allow through.

The rhetoric about boosting science teaching is called the 'Green Paper effect' and paralyses political action. It is much more difficult to deal with than any greenhouse effects!

Ultraviolet radiation and the ozone layer

Depletion of the ozone layer is another important atmospheric effect. It is, however, completely distinct from the greenhouse effect. Even though some molecules may be involved in both (like CFCs, marginally) it is for quite different reasons.

While all the other gases in the atmosphere are all mixed up together, most of the ozone in the atmosphere is found in a layer at altitudes between 20 and 50 km. Why?

Production of ozone in the atmosphere uses oxygen and requires ultraviolet light. As you go higher in the atmosphere, you find less oxygen, but more UV radiation. There is thus an optimum height for ozone production, and this occurs at these intermediate altitudes, 20–50 km. (The ozone in that layer, if it were unmixed with air and compressed to one atmosphere pressure, would only be 3 mm thick.)

The ozone, once formed, absorbs ultraviolet (and some infra-red) radiation and this causes local heating. Hotter, less dense air, thus sits on top of colder, and more dense air, and this is a stable structure (an inversion layer) which leads to the ozone-containing region being called the (peaceful) stratosphere. The troposphere below follows the expected pattern, that is, the temperature decreases the further you go up away from the warm earth. Warming of the lower layers at the earth's surface leads to convection currents and turbulence, that is, weather.

DEMONSTRATION: INVERSION LAYER

Ozone and the stratosphere
Saying that the ozone layer exists in the stratosphere is a tautology. The stratosphere exists only because UV-absorbing ozone heats up to **create** the inversion layer.

You need a transparent (preferable square) tank with an immersion heater at the bottom (or a Bunsen heater). Fill the tank half-full of cold water. Now carefully pour some hot water on top to form a separate layer. Shine a strong torch through the water onto a screen. Because of the difference in refractive index of the water at two different temperatures, a sharp boundary can be seen. It is apparent that there is almost no mixing between the two layers.

You can then turn on the heater at the bottom of the tank. The warm water produced can be clearly seen, rising through the cold layer as a turbulent swirling shadow on the screen, but when the rising water reaches the hot water layer, it rises no further and is trapped.

The turbulence of the (lower) troposphere gives us our weather, but there is little transfer of material to and from the stratosphere. Material that does get into the stratosphere layer stays there for years and can take part in chemical reactions, such as reacting with ozone and oxygen atoms.

The inversion layer prevents mixing between the troposphere and the higher atmosphere, which means that water does not move through, and there are no clouds beyond the stratosphere. If water molecules did move through they would be broken up by ultraviolet light, the hydrogen formed would escape the earth's gravitational field, and we would eventually lose all our water.

Radiation at around 265 nm is most dangerous to living things, including the plants on which we ultimately depend. Ozone prevents radiation below 290 nm from reaching the ground. Ozone also stops a great deal of radiation in the 290 nm to 320 nm range, which causes skin cancer (see Sunscreens in Chapter 5). Concentrations of ozone in the stratosphere fluctuate with natural changes in rate of production and destruction; in any one year, the maximum concentrations in the spring can be half as high again as the minimum in the autumn. While the rates of ozone production appear to be out of our control, the compounds we add to the atmosphere will effect the destruction. The oxides of nitrogen, both natural and from car pollution, account for perhaps two-thirds of the destruction.

One could not wish for a better example of the complexity of chemical feedback loops in the environment than the reactions of the atmospheric gases and the unpredicted effects of quite small perturbations.

Loving ozone
We love our ozone up in the stratosphere because it absorbs parts of the ultraviolet light that would otherwise destroy our crops, and give us skin cancer. We don't love it down on the ground when it is produced (along with oxides of nitrogen) by reactions of car exhaust gases with unburnt fuel and sunlight to produce photochemical smog. On the other hand, we do love it again when we go out after a thunderstorm and breathe in deeply that refreshing smell, which is (again) ozone and nitrogen oxides.

Chemical reactions of ozone

Ozone is produced when ultraviolet (UV) light splits an oxygen molecule, (O_2) into two atoms of oxygen, O.

$$O_2 + UV \longrightarrow O + O.$$

The oxygen atoms can react with other oxygen molecules to produce ozone,

$$O + O_2 \longrightarrow O_3,$$

or react with ozone to produce oxygen molecules again:

$$O + O_3 \longrightarrow O_2 + O_2.$$

When ozone absorbs UV light it breaks up into an oxygen molecule and an oxygen atom again.

$$O_3 + UV \longrightarrow O_2 + O.$$

A delicate balance is set up that regulates the concentration of ozone at about 10 ppm.

Fluorocarbons, chlorofluorocarbons and chlorocarbons

Refrigerators used to be driven with ammonia gas, and aerosols with butane. Ammonia stinks and is poisonous, and butane burns. Two fluorocarbons, CFC-11, CCl_3F, and CFC-12, CCl_2F_2 (see Glossary) were introduced in the 1930s as refrigerants (Freons™) see Box on Midgley, earlier in this chapter. After World War Two they became popular as aerosol propellants. In 1975, about 3000 million aerosol cans ejected more than 500 000 tonnes of fluorocarbons into the atmosphere. Non-flammable, odourless, non-toxic, except at very high concentrations, and chemically inert, so that they do not react with the can contents — what more could we ask for? Because of their different boiling points and vapour pressures, CFC-12 gave the type of high-pressure spray needed for insecticides and paints, while CFC-11 gave a more gentle spray for spraying hair, and even more delicate areas. In 1951 another use was found for these fluorocarbons: a non-flammable replacement for pentane as the foaming agent in the production of the foam plastics polyurethane and polystyrene.

The major use of another fluorocarbon, CFC-13 (CCl_2FCClF_2), was in the electronics industry, dry-cleaning, and aircraft maintenance. It is one of the few solvents that is non-toxic and does not attack electronic components. In Australia the defence force was the greatest user.

(Courtesy of the Canberra Times)

Chemical reactions of CFCs

In the upper atmosphere, the unfiltered UV radiation breaks up these compounds to produce chlorine atoms with varying efficiency. A chlorine atom, Cl, from a chlorofluorocarbon can combine with ozone, O_3, to form ClO and an oxygen molecule, O_2. Then ClO and an oxygen atom, O, combine to produce another O_2 and a free chlorine atom, Cl, again.

$$Cl + O_3 \longrightarrow ClO + O_2$$

$$ClO + O \longrightarrow O_2 + Cl$$

The initial ozone is lost, and since the chlorine atom is regenerated, it can go on and repeat the process in a so-called chain reaction (see Catalysis — p. 410). The chlorine atom is eventually removed from the chain by reacting with some other atmospheric impurity such as methane, forming hydrogen chloride, which in turn contributes to acid rain.

The relative effect that these compounds have on the ozone layer is thus a function of their reactivity as well as their persistence. A table of relative ozone depletion potential, RODP, based on CFC-11 as unity, can be devised for the other compounds assumed to be at the same concentration, and is given in Table 14.10.

Einstein and refrigerators
AB Electrolux in Sweden has files on two patents on refrigerators from Albert Einstein and Leo Szilard from 1920. Their collaboration began when Einstein (then working in the Swiss patent office) read a newspaper report of a family killed by poisonous gases leaking from a refrigerator — the coolants were either ammonia or sulfur dioxide. He convinced Szilard, then a student, that there had to be a better way. The most successful patent was an electromagnetic motor pump to compress refrigerants.

TABLE 14.10
Relative ozone depletion potential (RODP) and lifetime in atmosphere

Compound	RODP	Half-life in atmosphere
CFC-11	1.00	75 years
CFC-12	0.86	112 years
CFC-13		90 years
CFC-22	0.05	20 years
CFC-113	0.80	
CFC-114	0.60	
1,1,1-Trichloroethane	0.15	6.5 years
Carbon tetrachloride	1.11	50 years
Halon-1211	10.00	
Halon-1301	10.00	

Natural organochlorines

Methyl bromide: Methyl bromide is used worldwide because it sterilises the soil, killing virtually all biological organisms, including nematodes, the microscopic worms that love to eat things like the delicate roots of strawberry plants. It also kills rodents and plant pathogens such as fungi, bacteria and weed seeds. There is no alternative with such one-shot effectiveness. As well as agriculture, it is also used in the home pest-control industry.

However methyl bromide is a potent ozone depleter in the stratosphere, 50 times that of the CFCs. There is pressure to remove it, but read on:

Organohalides from marine organisms: There are many natural sources of organohalogens. Marine organisms are a source of enormous variety and amounts of organochlorine and organobromine compounds, many of which are very volatile, and very probably of ozone depletion significance. Of these, the red alga species *Asparagopsis teaxiformis* has one of the highest concentrations and a large distribution. Dry weights of halocarbons exceed 1.3% (13 000 ppm). About 130 halogenated hydrocarbons have been identified in this species. These range from simple bromoform (the bromine equivalent of chloroform), to haloketones (e.g. 3-chloro-1,1,3-tribromoacetone), and so on to more complex structures, many of which are biologically active. *Asparagopsis teaxiformis* is the favourite edible seaweed of most native Hawaiians and has always been known for its strong flavour and so-called 'iodine' odour that it develops on standing. The 'smell' of the ocean is due to these compounds (and not, as commonly believed), ozone! There have been suggestions that these seaweeds may be responsible for excess incidence of stomach cancers in people where this represents a substantial part of the diet.

Although evolution of methyl bromide from marine organisms has not been directly measured, it is the most abundant organobromine species in the lower atmosphere and about 85% is of natural origin. It is also one of the most persistent bromomethanes in the troposphere. Macroalgae may produce up to 10 000 tonnes Br/year of volatile organo-bromines (excluding bromoform). Bromoform emission is estimated at ten times this figure.

Near the Antarctic Peninsula, bromoform is found at levels of 6.3 pptv (parts per trillion per volume) in air and 6.2 ng/L in surface waters. Levels are higher near the coastline, consistent with greater biological activity, approaching 50 ng/L near Elephant and King Islands.

In the stratosphere bromine radicals can act as far more efficient catalysts than chlorine in destroying ozone.

References

Wind energy: Exploring energy 1, NRC and Department of Energy, Mines and Resources, 1981, Canada.

Photovoltaics fundamentals, G. Cook, L. Billman and R. Adcock, Solar Energy Research Institute, US Dept of Energy, Oak Ridge, TN, 1995, 63 pp.

Bibliography

Energy: An introduction to physics, R.H. Romer, W.H. Freeman San Francisco, 1976.

Electric vehicles — an answer to an environmental challenge, K.R. Williams, Proceedings of the Royal Institution of Great Britain, **63**, 1991, p. 137–152.

Fireworks: Principles and practice, R. Lancaster with T. Shimizu, R.E.A. Butler, R.G. Hall, Chemical Publishing Co., NY 1972. The Reverend Lancaster and his colleagues have written an all-time classic recipe book for fireworks.

Punching a hole in the stratosphere, R.P. Wayne, Proceedings of the Royal Institution of Great Britain, **61**, 1989, p. 13–49.

Emissions and performance of wood heaters when burning softwoods, J. Todd, Energy Research and Development Corporation, Commonwealth Government, publ. No. 56, March 1991.

The earth garden book of alternative energy, A.T. Gray, Lothian, Melbourne, 1996.

BP statistical review of world energy. Yearly, British Petroleum plc, London. Also available on http://www.bp.com/bpstats.

Chemistry in the dining room

Introduction

The person who first smoked a herring was putting an additive in food. Thousands of years ago the Chinese are said to have used gases from the incomplete combustion of oil shale to ripen bananas and peas, but the earliest published literature on the use of oil shale would appear to be British Patent 330, issued in 1694 to 'make great quantityes of pitch, tarr and oyle out of a sort of stone'.

Pliny the Elder (AD 23–79) records that wines from Gaul were artificially coloured and flavoured. Pickling in salt, and fermentation processes resulting in the production of lactic acid, alcohol, or acetic acid are methods of food preservation that date from ancient times. In salt caves near the Dead Sea there is evidence of even prehistoric food preservation processes. Human and animal droppings containing protein, and hence nitrogen, on the salt floor produced both sodium nitrate and sodium nitrite, which preserve meat. This accidental (but perhaps socially undesirable) piece of chemistry was apparently put to good use. Curing of meat with brines containing nitrites and nitrates became a long-established process. The nitrites apparently have an essential role in the curing process; they certainly give ham a pleasing colour, which has been attributed to the conversion of myoglobin and haemoglobin into their nitroso-derivatives. The nitrite also prevents the development of the bacterium *Clostridium botulinum*, which causes botulism, the most deadly form of food poisoning. However, over the years there has been some concern about the possibility of a reaction in processed food, or in the digestive system, between nitrite and secondary amines to form nitrosamines, which are highly carcinogenic (i.e. cancer inducing). Small concentrations of nitrosamines (parts per billion) have been found in raw, and particularly smoked, fish. (See Box, nitrate/nitrite.)

Flavouring and seasoning were arts in many ancient civilisations with the result that spices and condiments were important items in commerce. The spices were originally added as preservatives because refrigeration was not available. They also contain antioxidants, and these helped to preserve food for a longer time. No doubt many of us would prefer to eat food straight from the farm, orchard, or sea, but in societies with large heavily populated urban areas that import food produced, perhaps, on the other side of the world, some form of processing, if only for preservation, is necessary. However, people do feel uneasy about 'chemicals' in their food because they are worried about the possible effects of eating substances that we do not eat 'naturally', or simply because some additives make some foods taste different.

Fig. 15.1
Junk food

(©1975 United Feature Syndicate Inc.)

All food, of course, consists of chemicals. The chemicals about which people are concerned are traces of substances that are not themselves food, and that may not be present in foods in their natural or traditional state. The presence of these substances may be either accidental or deliberate. Before allowing ourselves to become carried away by this concern, let us consider a few 'natural' culinary disasters.

In 400 BC an army of Greek mercenaries 10 000 strong became intoxicated and finally unconscious after eating honey in some villages on the shore of the Black Sea. In 1596 the members of an expedition to Novaya Zembla in the polar wastes became ill after eating bear's liver, and three of them lost their skin. Much the same thing happened again in 1913 to members of an Antarctic expedition who ate dog's liver. In 1816 Abraham Lincoln's mother died after drinking milk from a cow that had fed on snakeweed.

What happened in the first case was that the bees had fed on the nectar of rhododendrons, which contained a poison that had been deposited in the honey. Mrs Lincoln's cow had acquired a poison from the pasture. The case of the bears demonstrates that a necessary component in our diet, essential to our health, can be disastrous when eaten in excess. In this case it was vitamin A, stored in the bear's liver. These instances of normal foods being toxic are rare; that of Lincoln's mother is the only one recorded involving snakeweed, but outbreaks of honey poisoning do occur from time to time.

However, there are foods that appear to lead to disease in the long term. In Nigeria the starchy root *cassava* is widely eaten and is believed to be responsible for a nervous disease resulting in, among other things, deafness and difficulty in walking. The cassava root contains compounds that produce *cyanide* when the root is prepared as food. Although it is customary to wash the food well, and the toxic substances pass out of the root on soaking, it appears likely that enough cyanide remains to cause disease over a period of years. It is relevant to note that we have evolved culturally to avoid the dangers from natural poisons (e.g. the Nigerians wash their cassava without ever having done chemistry, or having heard of cyanide!). On the other hand, it is hardly possible to pick up a popular magazine or weekly journal without finding some horrific story of the additives we are all consuming.

Food technology and the law

The first modern processes for preservation were developed empirically half a century before the true cause of food spoilage was known. In 1795 the embryonic revolutionary republic of France was beset by enemies on all sides, and the government was desperate to seek ways of preserving food for its troops. A prize of 12 000 francs was offered, and it was won, after many years of patient experimentation, by a Parisian confectioner, Nicolas Appert. In 1810 he published *The book for all households on the art of preserving animal substances for many years*. That's how it all started.

The State of Victoria enacted the first general food legislation in 1905, followed by New South Wales in 1908. The Australian states adopted a system whereby all substances that are not expressly authorised are prohibited in food — in contrast to allowing what it did not expressly prohibit. Some food products are defined by 'specifications of identity' known as food standards. The Australian and New Zealand Food Authority (ANZFA) develops and maintains standards which are enforced by the individual health departments of Australian states (and territories) and New Zealand.

Adulterated milk?
In the early 1900s in New South Wales, Australia, people were rather pleased with their rich, creamy milk, which seemed to keep so well. The colour, it appears, was by courtesy of the milkman rather than the cow, and formalin acted as a preservative.

Food additives

Types of food additives

In this section some different types of food additives are defined, and then described in the major groups in greater detail.

NATURAL ADDITIVES OFTEN AIN'T SO GREAT

Many legumes contain canavanine, a toxic analogue of the essential amino-acid arginine. Alfalfa (called lucerne in the UK) contains about 1.5% dry weight and feeding quantities of alfalfa sprouts to monkeys causes lupus erythematosus-like symptoms. Lupus in humans is an autoimmune disease which causes a variety of damage to organs and skin. Coumestrol is a heat-stable weak eostrogen (a sterol), and was first found in alfalfa and later in soya beans and sprouts. It can accumulate in the body fat of animals grazing on forage containing the compound and the concentration increases if alfalfa is infected by certain fungi.

Source: Toxicants occurring naturally in food, National Academy of Science 2nd ed. 1973, p. 552).

GRAS: In the USA, a group of additives regarded by the US Food and Drug Administration (FDA) as 'generally regarded as safe' (GRAS), includes a large group of natural flavours and oils that are not specified. To be on this list, an additive must have been in use before 1958 and have met certain specifications of safety. Additives brought into use since 1958 must be approved individually. Occasionally substances are removed from the list by the FDA in the light of new evidence, an example being the cyclamate sweeteners.

Preservatives: Substances used to prevent spoilage caused by bacterial activity, fungus and mould, and which thus prolong the keeping quality of foods.

Antioxidants: Substances used to inhibit the oxidation of fats during storage (i.e. stop them becoming rancid).

Colouring matter: Substances used to colour foods.

Flavouring agents: Aromatic substances, both natural and synthetic, used as components of food flavours, or directly in foods, and artificial sweetening agents. (Note that the use of *aromatic* in chemistry is restricted to chemicals with benzene rings.)

Sweeteners: Substances to make food taste sweet.

Sequesterants: Substances that react with traces of metal ions, tying them up in a manner that prevents their normal reactions, such as catalysing the decomposition of food. Sequesterants such as phosphates are used in detergent formulations to tie up metal ions in water.

Gelling agents, stabilisers, and emulsifiers: Substances used to produce or maintain a certain consistency in foods.

Acids and bases: Acids are used to impart a certain tartness to foods, or to alter the acidity of the medium (i.e. to lower the pH in canned products, or to prevent the crystallisation of jams and jellies). Bases are used as ingredients of baking powders used in pastry production, and in powders for effervescent beverages.

Improving agents: This group includes chemical compounds that enhance one or more of the quality criteria of foods (flavour, consistency) and substances used for polishing and glazing confectionery products.

UNCOMMON SALT

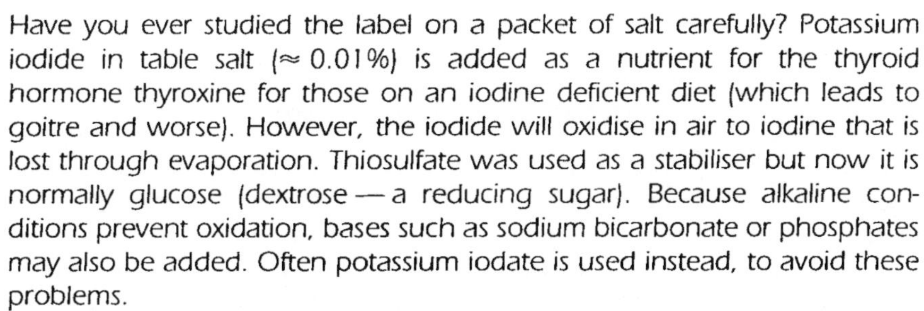

Have you ever studied the label on a packet of salt carefully? Potassium iodide in table salt ($\approx$ 0.01%) is added as a nutrient for the thyroid hormone thyroxine for those on an iodine deficient diet (which leads to goitre and worse). However, the iodide will oxidise in air to iodine that is lost through evaporation. Thiosulfate was used as a stabiliser but now it is normally glucose (dextrose — a reducing sugar). Because alkaline conditions prevent oxidation, bases such as sodium bicarbonate or phosphates may also be added. Often potassium iodate is used instead, to avoid these problems.

A major problem with salt is that with a trace of moisture, the cubic crystals cake together and the salt does not flow. This can be countered with drying agents ($\approx$ 0.5%), such as a range of carbonates (E501–4), and sodium aluminium silicate (E554). See Appendix IX for a full list of additives. The smarter way is to change the shape (habit) of the cubic salt crystals to a form that doesn't provide large flat surfaces to pack together. Salt normally crystallises as cubes because the octahedral faces of the crystal (consisting of either all Na^+ or all Cl^-) grow faster than the cubic faces (with alternating Na^+ and Cl^-). It has been known for a long time that if an impurity adsorbs onto the surface of the fast growing octahedral faces (e.g. urea and certain dyes) the reverse happens, and octahedral crystals are formed instead of cubes.

Cyanide by another name
To hide the 'cyanide', in the US they call it by the old-fashioned, 'yellow prussiate of potash', while in the UK they use the IUPAC–modern 'hexacyanoferrate'.

Adding potassium ferrocyanide $K_4Fe(CN)_6.3H_2O$) at a level above 13 ppm appears to do a similar trick (E536). Cyanide in salt doesn't sound a great selling point (even though this compound is quite safe). If there was enough, you could try to detect it in solution with a ferric salt, with which it will form Prussian/Turnbull's blue.

Sprinkling salt on water causes the surface to contract momentarily towards the crystals, while with pepper, the opposite tends to happen. Why?

Regulation of food components

Setting limits for food components in processed food is a long-standing health regulatory activity. For chemicals deemed to be essential to human health, such as vitamins and minerals, minimum levels are set that prevent corresponding deficiency diseases. However, if vitamins are taken directly, or excess of particular foods is contemplated in unusual diets, there is also a need for maximum limits (particular foods or supplements) to ensure safety. These levels are critical for minerals, and the 'fat-soluble' vitamins such as vitamins A and D which become very toxic at doses much above that recommended. The levels are less critical for the water-soluble ones such as vitamin C and the B group, although megadosing has caused problems with withdrawal symptoms (see Vitamins in Chapter 4).

Checking for compliance with the system

The price of freedom (from health and trade problems) is eternal vigilance. Anyone can write tough laws; it is their enforcement that is crucial. For food this surveillance is provided in two ways:

1. from a survey of food as eaten (to watch the ADI, see Chapter 2) and
2. from a survey of food as produced at the farm gate (to watch the MRL, also see Chapter 2).

There is an administrative distinction between intentional additives and residues on the one hand (e.g. pesticide residues/processing aids), and contaminants on the other (e.g. cadmium). From the consumer standpoint, only the toxicology is of significance, and contaminants often arise from agricultural practice anyway (e.g. use of superphosphate with high cadmium levels, or in the wrong type of soil).

For the ADI, the Australian and New Zealand Food Authority (ANZFA) undertakes Australian Market Basket Surveys (AMBS). These involve more than 50 types of food around the country. They are prepared to 'table ready' state and then analysed for pesticide residues and heavy metal contaminants such as arsenic, cadmium, and lead. Once the concentration of residues in 'table ready' food has been determined, the dietary intakes of each residue are computer simulated through 'typical' diets devised from the results of dietary surveys for different ethnic, lifestyle, and age groups, as well geographic location. Samples are also taken of foods which are known to accumulate residues, so as to give an even more sensitive indicator; for example human milk and lamb liver. The range and size of samples, their consistency from year to year, their relevance to current pesticide use, and, not least, the extent of funding available for the AMBS, has been criticised.

For the second check, the National Residue Survey (NRS) is done by the Bureau of Resource Sciences (BRS). The NRS tests more than 40 000 samples of animal and plant products each year for various residues and environmental contaminants.

AVOID BRUISED OR GREEN POTATOES

Glycoalkaloids are potentially toxic compounds found especially in plants from the Solanaceae family. There are many different glycoalkaloids produced by the various members of this family, for example potatoes, tomatoes, capsicum, and tobacco. However, the only glycoalkaloids recorded as actually causing human death are those produced by potatoes (α-solanine and α-chaconine). Up to 30 deaths and over 2000 documented cases of glycoalkaloid poisoning involving potato have been recorded.

Levels of 380-480 mg of solanine per kg of potato have been fatal. However, the normal levels in properly stored potatoes is only 30–60 mg/kg of potato. Solanine is not destroyed by cooking, but may be washed out to some extent.

The toxicity of potato glucoalkaloids like solanine is far greater in humans than in other animals studied (note the danger of using animal tests!). The lethal level is between 3 and 6 mg per kg body weight (of the person eating it), levels comparable to that of strychnine (5mg/kg body weight). See Box on LD50 on page 21 for more explanation. Take care when reading toxicity numbers not to confuse kg of food with kilograms of person. There appears to be a considerable variation in people's response to high levels of glycoalkaloids. Toxicity is due both to the anticholinesterase activity on the central nervous system, and to membrane disruption activity which affects the digestive system, and also general body metabolism.

Potatoes affected by late blight fungus were suggested . . . as being a major cause of the greatly disabling conditions of anencephaly and spina bifida through teratogenic effects on the early stages of the foetus. Such a hypothesis appears incorrect; however, laboratory studies in which potatoes with high glycoalkaloid content have been fed to animals have indicated that other less incapacitating abnormalities can be produced. Furthermore, studies have repeatedly shown that high foetal mortality and reabsorption occur even with only one high dose of glycoalkaloids given to pregnant animals. Because of this danger it is suggested that any slightly green or damaged potatoes be avoided, especially by women who are pregnant, or likely to become pregnant.

Source: S.C. Morris and T. H. Lee, Food Tech. Aust., **36**(3), 1984, p. 118.

Livestock poison
A large number of livestock have been killed, both by potato glycoalkaloids and by other glycoalkaloids produced by the various members of the Solanaceae family.

Inhibition of sprouting
Potatoes can be irradiated to delay sprouting and prevent greening (see Food irradiation in Chapter 16). While the green chlorophyll is stopped, the production of solanine is not, and so we are left without a warning sign.

International trade

We do not trade in diets, we trade in food. Therefore it is the MRL, not the ADI, that need to be internationally harmonised if we are to sell our produce to other countries, and to have price competition and choice at home through imports. An international body, FAO/WHO Codex Alimentarius, has recommended 3019 MRLs for 187 pesticides as of 1992.

The Codex has difficulty in setting a single MRL because good agricultural practice (GAP) differs so much in the different regions of the world where food is produced. However, the World Trade Organization (WTO) will use Codex standards to resolve disputes, which will mean

EEC additive code
As the first consumer representative, the author put the EEC system on the agenda of the Food Standards Committee of the NHMRC in 1974 and argued strongly for it. In 1983 the NHMRC finally recommended this system for Australia (virtually the same numbers, without the E), after years of further consumer agitation.

that there will be an increasing alignment of national with international levels.

We can outline what this means for Australia by an analogy. In 1983 the Government felt that our increasing international trade meant that we needed to move from a locally controlled currency to one that floated in international markets. This brought trading benefits, but also costs. We are now much more at the mercy of international currency speculation. This often acts against the adoption of sensible local financial policies. The same is true for residues in produce.

Additives — labelling

On the question of additives, the EEC developed a simple rational system in the early 1970s. Each additive is given a code number: E100 to E199 are colours; E200 to E299 are preservatives; E300 to E399 are antioxidants; and E400 to E499 are texture modification agents. Flavours have not been dealt with. The colours are further subdivided: E100 to E109, yellow; E110 to E119, orange; E120 to E129, red; E130 to E139, blue; E140 to E149, green; E150 to E159, brown and black; E160 to E170, unclassified; and E170 to E189, unique surface colorants. They are included in the description of additives discussed below, and a recent list is given in Appendix IX. This system makes it possible to search for specific food additives and to select a diet that avoids them. Technological improvements resulting in the use of fewer additives can then be followed by the consumer who studies the labels.

More on additives

Preservatives or antimicrobials

Benzoic acid (E210–213): Benzoic acid (Fig. 15.2) and its sodium salts are among the bacteriostatic or germicidal agents most widely used in foods. Many berries (e.g. raspberries) contain appreciable amounts (≈ 0.05%) of benzoic acid. Benzoic acid is included on the permitted lists of at least 30 countries for a great variety of foods, particularly soft drinks. Benzoic acid preserves food by inhibiting the growth of bacteria. As it is only the free acid that is effective, it can be used only in foods of pH less than 4.5. (See also HOCl/OCl in swimming pools, Question 6, in Chapter 8; and ionisation of aspirin in Aspirin, Chapter 13.)

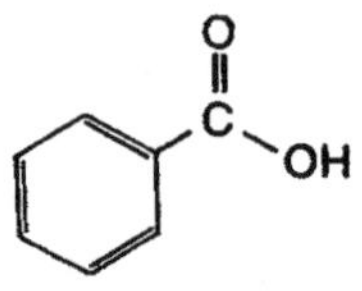

Fig. 15.2
Benzoic acid

The body excretes benzoic acid as hippuric acid within 9 to 15 hours of eating food containing it.

Sulfur dioxide and sulfites (E220–224): Sulfur dioxide and sulfites were used by the ancient Egyptians and the Romans. Sulfur dioxide is unique in being the most effective inhibitor of the deterioration of dried fruits and fruit juices. It is used widely in the fermentation industry to prevent spoilage by microbes and as a selective inhibitor. It is also used as an antioxidant and antibrowning agent (for *casse brune*) in wine making. Sulfur dioxide destroys thiamine (a vitamin); therefore its use is restricted to foods that are not important sources of thiamine. In Australia it is forbidden in meat, except for cooked manufactured meat (such as some salamis, processed meat products, and sausages) where it is allowed because it protects against the danger of bacterial contamination during processing at elevated temperatures. Some people are allergic to sulfur dioxide and there are products (including some wines) free of it.

Propionates, CH_3CH_2COOH (E280–283): Flour contains the spores of the bacterium *Bacillus subtilis* which are not likely to be killed by the baking temperature. Under summer conditions, these bacteria become active and produce a condition called *rope,* which renders bread inedible. The calcium and sodium salts of propionic acid are used in bread (0.2%) to inhibit the growth of micro-organisms.

Sorbic acid (E200–203): Sorbic acid (2,4-hexadienoic acid, $CH_3–CH=CH–CH=CHCOOH$) is naturally present in some fruits. It is a selective growth inhibitor for certain moulds, yeasts, and bacteria. It is used in cheese, pickles, fish products, cordials, and carbonated drinks.

Nitrates and nitrites (E249–253): As indicated in the introduction to this chapter, nitrates and nitrites occur naturally in many foods, particularly vegetables. The human infant is extremely sensitive to nitrites because of low ability to deal with a modification of blood haemoglobin caused by nitrites. Additional nitrite is derived from nitrate by bacterial activity in the gut, which is particularly efficient in the very young infant because of the inadequacy of acid production in the stomach. Nitrate/nitrite (Fig. 15.3) is one of many substances for which there are dipstick indicators.

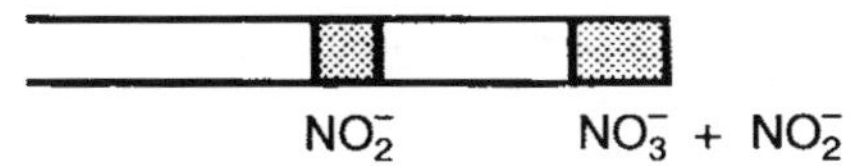

Fig. 15.3
Nitrate/nitrite dipstick

SOME EXPERIMENTS YOU CAN DO WITH NITRATE/NITRITE DIPSTICKS

Sensitivity: for nitrate 10–500 mg/L ; for nitrite 1–50 mg/L

1. Test for oxides of nitrogen in air. Sensitivity 1 mL of NO_2/m^3 of air.
2. Test for nitrite in saliva, expected average 7 mg/L, except after eating foods with high nitrate level (celery, beets, etc.), where you can obtain elevated levels for the following 24 hours.

3. Test fermented raw meat (salami type). Legal limit 500 mg/kg, nitrate. Cured meat (corned beef). Legal limit 125 mg/kg, nitrite. Canned ham. Legal limit 50 mg/kg, nitrite.
4. Nitrite in vegetables.
   Conventional carrots, expect 40–100 mg/kg.
   Some organically grown carrots 200–400 mg/kg (from all that excess natural manure used).
   Fresh spinach ≈ 5 mg/kg (if refrigerated for two weeks it can rise to ≈ 300 mg/kg).
5. Test for denitrification in waterlogged soils
   (Soil + nitrate + glucose ⟶ N_2O).

References

Dipstick chemistry, B. Selinger, in Consumer Science, Senior Science Seminar 9. Mitchell College of Advanced Education, NSW, Australia, July 1980, pp. 28–32.

Semi-quantitative dipsticks are available from chemical supply houses such as Merck, and their variety and sensitivity are improving all the time. The range and fascination of what you can do with them is limited only by your imagination.

On the basis of cost–benefit, preservatives score well on the benefit side. But no case can be made for the indiscriminate use of preservatives. It is often better to encourage the use of refrigeration and good handling techniques.

Antioxidants

Preservatives for fatty products and oils are called antioxidants. They prevent the occurrence of oxidation, which is the cause of *rancidity.* Vitamin C (ascorbic acid E300–301) is commonly used for water-soluble products, but the most common antioxidants are the fat-soluble BHA (butylated hydroxyanisole, E320) and BHT (butylated hydroxy-toluene E321). They have similar properties to the 'natural' oxidant, vitamin E (α-tocopherol). The word *butylated* is not widely used in chemical nomenclature. It is applied here because the usual names for BHA and BHT include the words *cresol* or *phenol,* which generally have connotations of toxicity. To avoid consumer rejection of these 'safe' compounds, the names were made to sound safe as well. Various esters (propyl, octyl and dodecyl) of gallic acid (3,4,5-trihydroxybenzoic acid, E310–312) are allowed in Australia in edible oils, margarine, table spread, salad oils, lard, and dripping.

Antioxidants are additives that we will probably have to accept if we want convenience and reasonable shelf-life. On the other hand we should expect producers to use the maximum technological skill to reduce the required amount to a minimum.

Colouring matter

Colours are put in food mainly for aesthetic reasons. The way food looks has an effect on its palatability. Both natural and synthetic colours are used. The synthetic colours are mainly coal-tar dyes; many of the earlier ones had been found to be carcinogenic. The list of permitted red dyes has halved over 30 years. No two countries seemed to agree on which colours are safe. The old USSR put its faith in natural colours and allowed only three synthetic ones (although the distinction is questionable when a natural dye is synthesised). When the British had to conform to the EEC norm (1 January 1977), the 'gold' put in the kippers, and 'pink' in the sausage, had to be changed.

A typical example of the type of compound used as a food dye is allura red which was added to allowed food colours in Australia in 1977 (Fig. 15.4).

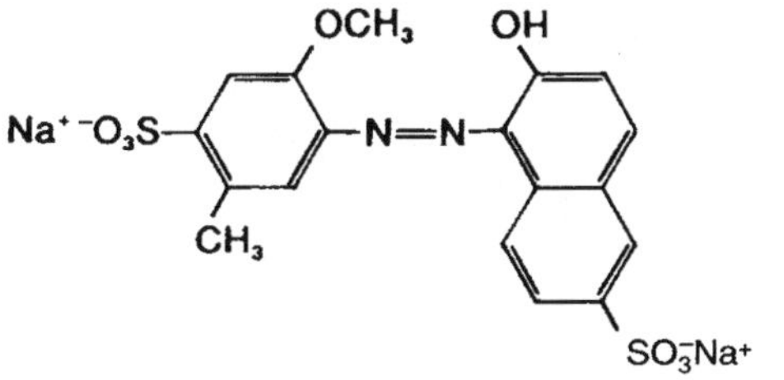

Fig. 15.4 Allura red (E129)

The coal-tar designation comes from the presence of aromatic rings, mainly benzene and naphthalene (Fig. 15.5). The colour is introduced by one or more *diazo* (dinitrogen) groups: –N=N–. To make the dyes soluble in water, one or more sulfonic acid (SO_3^{2-}) groups are attached, with Na^+ or NH_4^+ being the other ion. The dyes are generally made from two halves which are brought together by joining the nitrogens in a diazo coupling reaction (E100s).

Fig. 15.5 Aromatic hydrocarbons

A case has been reported of a young boy who was passing red stools. Unfortunately he was subjected to extensive tests in hospital before it was found that the source of the red was *erythrosine* which was used to colour the cereal he liked. Apparently very little of the colour is absorbed: most is excreted in the faeces, which explains its lack of toxicity.

This metabolic inertness is not found with the food azo dyes described above. The azo linkage is split by the bacteria in the gut of humans and animals, and it is probable that the products are the problem. The process has been known since the discovery in 1935 of the first antibacterial sulfa drug, Prontosil, which was later found to form the active drug sulfanilamide in the bowel (Fig. 15.6). (See also Antibacterial drugs in Chapter 13.)

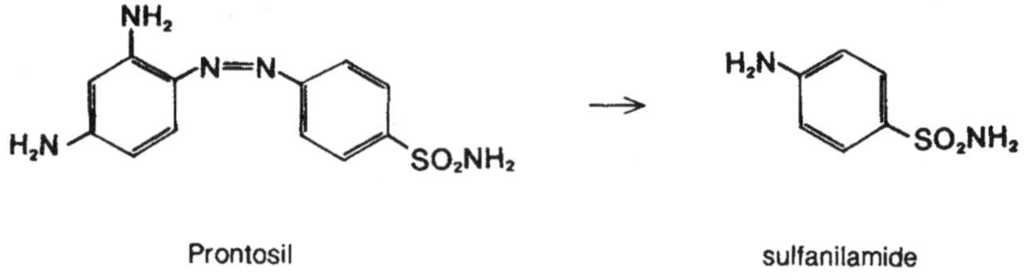

H_2N — NH_2 — $N{=}N$ — SO_2NH_2 → H_2N — SO_2NH_2

Prontosil   sulfanilamide

Fig. 15.6
The action of Prontosil

In 1960 the USA had 14 permitted food dyes and the UK had 30. But nine of the permitted US dyes were banned in the UK. Because the chemical structures of all azo dyes are very similar, one might imagine that there is not much to choose between them from the point of view of danger to health. However, two of the basic materials in the preparation of some of these dyes are the naphthylamines (Fig. 15.7) . One of these, 2-naphthylamine, is a very potent carcinogen, and causes cancer of the bladder with an induction period between intake and disease of about 20 to 30 years. On the other hand, 1-naphthylamine is less potent (i.e. compared to 2-naphthylamine) and its greatest danger is that it may be contaminated by the 2-compound.

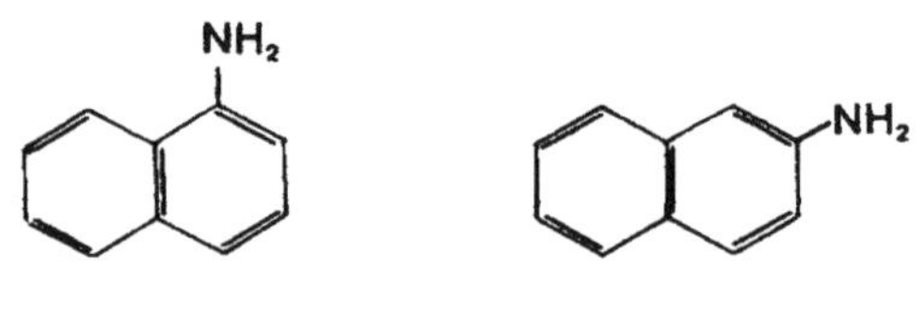

1-naphthylamine   2-naphthylamine

Fig. 15.7
Naphthylamines

Although it can be argued that preservatives are essential under modern conditions, the case for adding colours is not strong. In 1976 Australians drank 70 L of soft drinks per head of population, about one-third of which were cola-type drinks (and beer consumption was about 140 L a head). The permitted concentration of dyestuff in soft drinks as consumed was 70 mg/kg, in cordials 130 mg/L, and where allowed in solid foods, 290 mg/L. Because it may take decades for these substances to be cancer producing, protection of our children should be a minimum objective. In addition, drinks of the cola variety can contain caffeine (a stimulant) and phosphoric acid, whose use is questionable.

Colour as flavour

The confusion between colour and flavour is not entirely fanciful. A carefully designed survey was carried out by CSIRO in an ice-cream parlour in a Sydney beach suburb for four weeks in summer. The aim of the survey was to determine the influence of artificial colour in what consumers choose. Four flavours were surveyed as follows in a 'cross-over' mode. (The flavours are normally sold with and without colour.)

Week 1. Uncoloured passionfruit and coloured butterscotch, plus controls.
Week 2. Coloured passionfruit and uncoloured butterscotch, plus controls.

Week 3. Uncoloured rockmelon (cantaloupe) and coloured peppermint, plus controls.

Week 4. Coloured rockmelon and uncoloured peppermint, plus controls.

Controls. Ten other flavours of ice-cream; vanilla (white), chocolate (brown), vanilla chocolate chip (white), orange chocolate chip (orange), peach mango (white), boysenberry (purple), honeycomb (white), caramel toffee (brown), rum raisin (white) and coffee walnut (brown); that is, five coloured and five uncoloured flavours were used as an unchanging 'background' for the full four weeks.

For the surveyed samples, coloured and uncoloured ice-creams of the same flavour were never presented together, but for the controls they were. The colours used were orange/yellow (beta carotene), brown (caramel), and green (chlorophyll). For each of the surveyed flavours the time taken for the purchase of about 12 litres (121 ice-creams) was recorded as an inverse measure of popularity (Table 15.1).

Exercise
Note how carefully the experiment was designed to avoid answering of the wrong question or biasing the result. Think of other interesting topics, such as artificial versus natural colourings. Devise your own survey to test out a marketing hypothesis.

TABLE 15.1
Time taken (hours) for the purchase of 121 ice-creams
(the longer the time, the less popular)

Flavour	Uncoloured	Coloured
Passionfruit	44	15
Rockmelon	40	16
Butterscotch	27	9
Peppermint chip	36	9

All flavours were clearly labelled but no reference was made to colour (i.e. both coloured and uncoloured passionfruit were labelled 'passion-fruit'). The coloured outsold the uncoloured by approximately 3:1! While there is clearly a difference in popularity of flavours, it is not as large as the effect of colour.

Flavours

Flavours constitute the largest class of food additives. There are about 1300 flavouring substances in the USA that are generally regarded as safe (GRAS) because of their long time use. This sets a tremendous task of checking for dangerous effects, which, needless to say, has hardly been attempted. Some countries publish lists of permitted and prohibited flavours; others have a short list of prohibited flavours, many of which are natural; and others allow flavourings that are found only in the aromatic oils of edible plants.

The International Organisation of the Flavour Industry divides flavourings into five groups:

- Aromatic raw materials of vegetable and animal origin, such as pepper or meat extract.
- Natural flavours such as concentrates prepared from aromatic raw materials by extraction and concentration. For example, the flavour in tarragon vinegar is extracted from tarragon.

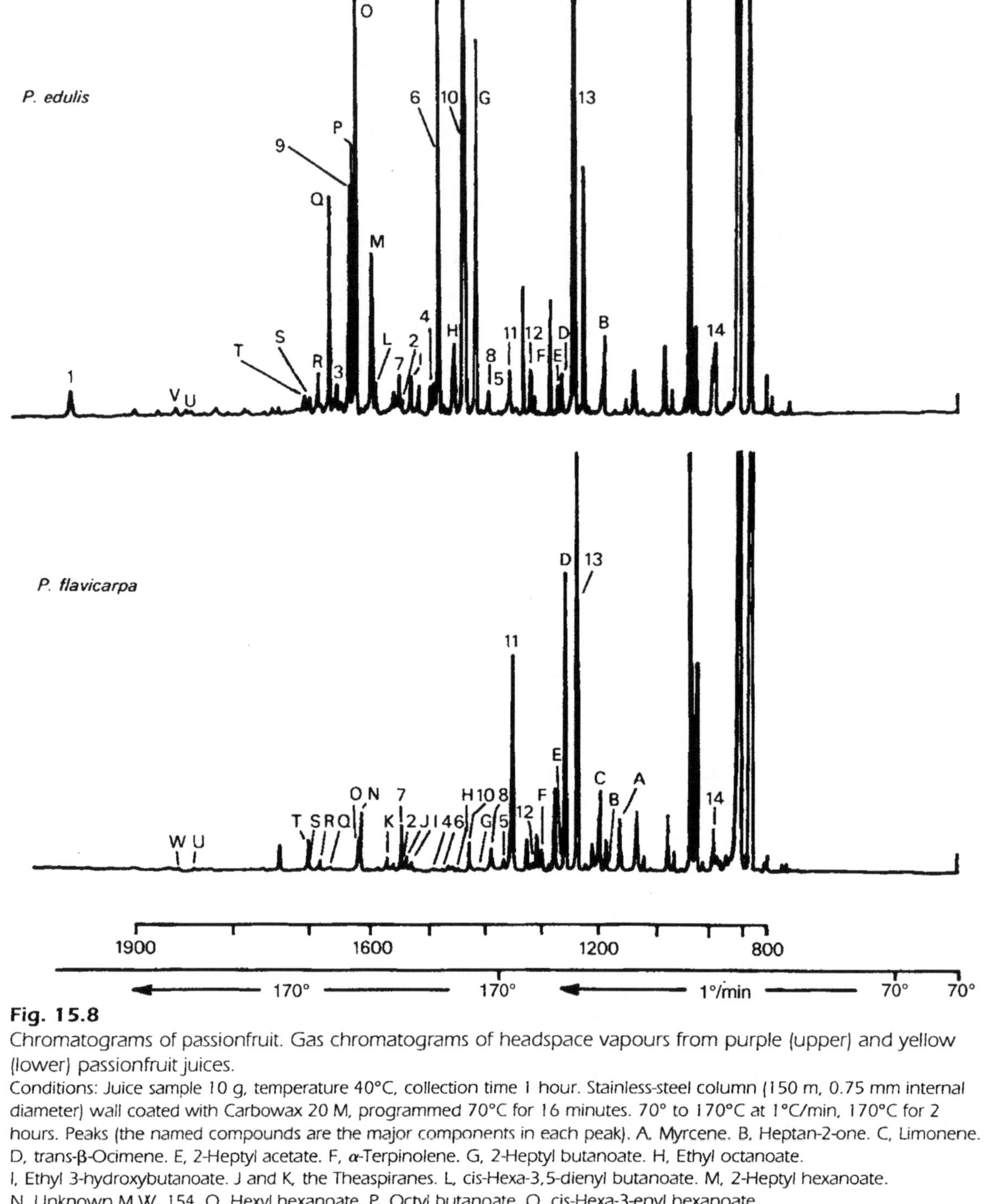

Fig. 15.8

Chromatograms of passionfruit. Gas chromatograms of headspace vapours from purple (upper) and yellow (lower) passionfruit juices.

Conditions: Juice sample 10 g, temperature 40°C, collection time 1 hour. Stainless-steel column (150 m, 0.75 mm internal diameter) wall coated with Carbowax 20 M, programmed 70°C for 16 minutes. 70° to 170°C at 1°C/min, 170°C for 2 hours. Peaks (the named compounds are the major components in each peak). A, Myrcene. B, Heptan-2-one. C, Limonene. D, trans-β-Ocimene. E, 2-Heptyl acetate. F, α-Terpinolene. G, 2-Heptyl butanoate. H, Ethyl octanoate. I, Ethyl 3-hydroxybutanoate. J and K, the Theaspiranes. L, cis-Hexa-3,5-dienyl butanoate. M, 2-Heptyl hexanoate. N, Unknown M.W. 154. O, Hexyl hexanoate. P, Octyl butanoate. Q, cis-Hexa-3-enyl hexanoate. R and S, the 2-(3-Hydroxybutyl)-butanoates. T, α-Terpineol, U, Dihydro-β-ionone. V, Dihydro-ionone. W, Geranyl acetone. 1, β-Ionone. 2, Ethyl cis-octa-4,7-dienoate. 3, Megastigma-4,6,8-triene. 4, Ethyl cis-oct-4-enoate. 5, Rose oxide. 6, cis-Hexa-3-enyl butanoate. 7, Linalool. 8, cis-Hexa-3enol. 9, Edulan 1. 10, Hexyl butanoate. 11, Hexanol. 12, Heptan-2-ol. 13, Ethyl hexanoate. 14, Methyl butanoate (F.B. Whitfield, 'The chemistry of food acceptance', Food Res. Q. **42**, 1982, 52.)

- Natural flavouring substances that are isolated from aromatic raw materials by physical means. Lemon oil bought in a bottle, for example, contains the same chemicals as the zest of lemon, which can be collected by rubbing a sugar cube over a lemon.
- Natural substances that are isolated by chemical processes but are identical with the natural substance. For example, monosodium glutamate prepared synthetically is the same as monosodium glutamate found naturally in tomatoes, mushrooms, parmesan cheese, and sweet corn.
- Artificial flavouring substances that have not been identified in natural products, but that simulate natural flavours.

Most flavours fall into the first four categories and are therefore natural products normally found in food. However, not all are. Many synthetic flavours are identical to the natural ones, but is vanillin produced from wood lignin as natural as vanillin produced from a vanilla bean?

Gas chromatography is a very sensitive physical chemical technique used to separate components in a chemical mixture, particularly if they are fairly volatile. Chromatograms of two varieties of passionfruit, *Passiflora edulis* (Sims) and its mutant *P. edulis f. flavicarpa* (Degener) are shown in Figure 15.8.

It is possible to detect compounds present at concentrations as low as 10 μg/kg (1 in 10^8), The differences in composition shown in Figure 15.8 lead to quite a different taste. During the 1950s, Australian plant breeders set out to combine the flavour of *P. edulis* with the hardiness and juice yield of *P. flavicarpa.* Today *P. edulis* is no longer grown commercially; it has been replaced by four purple-skinned hybrids, but some of these lack the ionone-related compounds that gave *P. edulis* the flavour we remember in earlier days. (So it is not your taste buds that have dulled!)

Among the important flavours are the 2-methoxypyrazines, one of which, the 3-isopropyl derivative, is responsible for the characteristic aroma and flavour of green capsicums. It also occurs in pea *pods* but far less in the pea *seed.* The beautiful aroma from the kitchen when peas are shelled comes from crushed pea pods. Hence the practice of adding a few pods during cooking to enhance the pea flavour — all lost when you buy them shelled (dried or frozen). (See also Perfumes in Chapter 5.)

ALKYL METHOXYPYRAZINES

Alkyl methoxypyrazines are potent vegetable odour chemicals with one of the lowest odour thresholds known (Fig 15.9).

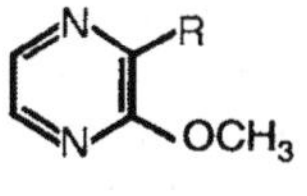

Fig. 15.9 Structure of alkyl methoxypyrazines

For the isobutyl and isopropyl compounds found in some wines, the taste can be detected at 2 ng/L or 2 parts in 10^{12} (equivalent to detecting a bottle of wine in a cubic kilometre of water, say in Sydney Harbour). These compounds are destroyed on ripening at a rate that depends on ambient temperature. In warm climates the levels can be so low as to cause loss of character of the wine, while in cool climates the level can be so high as to give a decidedly unpleasant capsicum taste. In the 1985 and 1986 vintage New Zealand Sauvignon Blanc wines, analyses ranged from 34.5 down to 20.9 ng/L, whereas Australian wines of the same vintage ranged from 12.0 to 0.6 ng/L. While the Cabernet Sauvignon grapes are similar, growing and processing ensure that the levels in these wines are very low.

Source: Proc. Second Intern. Cool Climate Viticulture and Oenology Symposium, Auckland, NZ, Jan. 1988, pp. 344–5, Roger Harris et al., Hall, ACT, Australia.

Avocados must be eaten at just the right time to obtain the best flavour. Left too long, the bland, slightly nutty, flavour becomes fatty-tallow and putty-like. The flavour chemicals have not changed, but the gas chromatograph reveals that the *relative* amounts are quite different. The avocado is very rich in unsaturated fatty acids (see Chapter 4), and these oxidise during ripening to carbonyl compounds. Too much oxidation, and the taste becomes rancid; too little, and there is no taste at all.

There can be little that is more nauseating than a rotting potato. The cause is a bacterium that breaks down two amino acids, tryptophan and tyrosine, to produce compounds found in faeces and horse manure (however, see also Perfumes in Chapter 5). These compounds can be absorbed by other potatoes nearby, and render the lot unacceptable. We are incredibly sensitive to some compounds used in fungicides and other protective chemicals. Chlorophenols and their precursors are now environmentally ubiquitous and the related chloroanisoles can be detected in food by their musty–mouldy flavours at concentrations well below 1 $\mu g/kg$ (1 in 10^9). These compounds are formed when micro-organisms detoxify chlorophenol-based antimould preparations. One of these compounds can be detected in water at a concentration of 3×10^{-7} $\mu g/kg$ (3 in 10^{16}) and another occurs when cork comes in contact with chlorine cleaning agents. Chloraniline is formed, which spoils the taste of wine at very low levels. At such low concentrations we are concerned with aesthetics and not safety.

Monosodium glutamate (E620–621)

Related to flavours are the additives known as *flavour enhancers*. The commonest of them is monosodium glutamate, MSG, which is the monosodium salt of glutamic acid, one of the natural amino acids (Fig. 15.10). In excess, this compound is alleged to cause the unpleasant complaint known as *Kwok's disease* or, more commonly, *Chinese restaurant syndrome*, because of the generous application of MSG in the dishes of many of these establishments. However, this is still disputed. It is no longer permitted in baby foods, where its purpose was to make food taste nice for the purchasing mother.

common to all common α-amino acids

Fig. 15.10 Monosodium glutamate E621

Cooks throughout South-East Asia firmly believe that a pinch of monosodium glutamate can bring out the flavour of every single ingredient in any complex dish they are frying up. Cans, jars or plastic bags of this seasoning powder are found in every kitchen in the subcontinent, and in many Chinese restaurants throughout the world.

Test for borax
Ground turmeric is extracted from tubers into ethanol (or methylated spirit), and filter paper (or newspaper edges) are soaked in the solution and dried. The suspected crystals can be scattered on the paper, and then drops of 1 M hydrochloric acid (HCl) added. Alternatively the paper can be added to the borax acidified with HCl. In either case a pink coloration develops. When then placed in ammonia vapour, turmeric paper turns from yellow to pink, and the pink borax stain turns blue. Turmeric is also used as a colouring in mustard.

The most efficient form of production is the fermentation of glucose, or sucrose, produced by the acid hydrolysis of any cheaply available carbohydrate such as (in Thailand) molasses and tapioca, in a suitable nutrient medium containing nitrogen (e.g. urea). The organism involved is *Corynebacterium glutanicum*. The product comes as short, needle-like crystals that look dull. They smell like sauerkraut and taste sweet and a bit salty. The reason for this detailed description is that adulteration by shopkeepers in the region was widespread during a visit by the author in 1973. In spite of its relative cheapness (about $1 per kilo in Thailand in 1973), similar-looking crystals of borax or sodium metaphosphate (and sugar and salt) were often added, with disastrous results for the consumer in the case of borax. Borax was such a common illegal additive in the region (in meatballs, fish, etc.) that methods for its detection suitable for local use were publicised.

Sugar

Sugar was known in India 5000 years ago, and one of the earliest forms of Indian sweet, *khandi*, has its name preserved in a modern American equivalent. Sweets join jewellery and perfume as one of the earliest forms of gift. However, sugar was not known to the Romans, who had to make do with honey for confectionery. They did, however, invent the first artificial sweetener *sapa* (hence sapor).

Sir Edward Barry, an historian of the ancient art of making wine, wrote a book in 1775 called *Observations historical, critical and medicinal on the wines of the ancients*. He commented that the Romans boiled down grape juice in lead pans to give a concentrated sweet syrup. (When other fruits were used it was called *defrutum*.) The syrup contained lead salts such as lead acetate (called sugar of lead) which had about the same sweetness as sucrose. This both sweetened and preserved the wine (lead kills microbes). As lead affects the brain (see also Chapter 10), it is alleged that this sweetener (along with many other lead products used), finally led to the decline of the Roman Empire.

Sugar cane (*Saccharum officinarum*) was cultivated in southern Europe only around AD 800, and sugar beet (*Beta vulgaris*) much later in around

AD 1800. Sugar infiltrated European consciousness because it was sweet and expensive and thus prompted early entrepreneurs to establish slave-based empires in the tropics. Sugar helps preserve meat and fruit. It is used in soft drinks because it makes water feel more 'refreshing', and in tomato sauce and peanut paste, because of its 'go-away' properties — the ability to degrease the palate after eating fat. The average Western person's daily consumption of 170 g provides about 1.8 MJ of energy.

The percentage of sugar deliberately added to food in the home has declined, but this has been more than compensated by the increase in the amount of sugar now added directly to manufactured foods; hence the interest in low-joule sweeteners.

Nevertheless, we must never forget that sugar is a very important food and it is only our overindulgence that makes artificial sweeteners attractive. Raw sugar is a cheap source of minerals, and some protein as well as carbohydrate, and keeps indefinitely and without refrigeration. Australia exports almost 80% of its sugar as raw sugar. However, sugar is also exported in other forms, as shown in Table 15.2.

Sugar in soft drinks

Try the following as a simple demonstration. Into a bucket of water, toss a can of 7-Up and a can of diet 7-Up. Generally, the first one sinks and the second one floats (it can depend on how much air is left in the can). There is enough sugar dissolved in normal soft drinks to increase the density of the whole can above 1.

TABLE 15.2

Export of sugar in other forms, by tonnage, 1985-86

Category	Tonnage
Preserved foods	11 155
Confectionery	5 373
Dairy foods	1 899
Non-alcoholic beverages	1 583
Alcoholic beverages	1 057
Bakery products	937
Other groceries	1 086
Total	23 090

Over 12 million children a year die from the effects of diarrhoea, mainly in the Third World. A simple treatment consists in feeding them with water containing an 8:1 ratio of sugar and salt.

The joint FAO/WHO report *Carbohydrates in human nutrition* found that 'there is no conclusive evidence that the consumption of simple sugars is of aetiological significance in diabetes mellitus'. This common form of diabetes occurs in obese adults, and control of total kilojoule intake is required.

Sugar *is* implicated in dental caries because it acts as food for the bacterium *Spectrococcus mutans*, which converts it into acid. Sticky forms of sugar (and starch, etc.) are more effective because they stay on the teeth.

Sugar is a most natural food, consumed with relish in the form of honey by our hunter-gatherer forebears and is produced in all green plants. The nutritional problem with sugar is that it replaces other sources of energy which provide additional nutrients not present in sugar. It is for this reason that controlled intake is recommended.

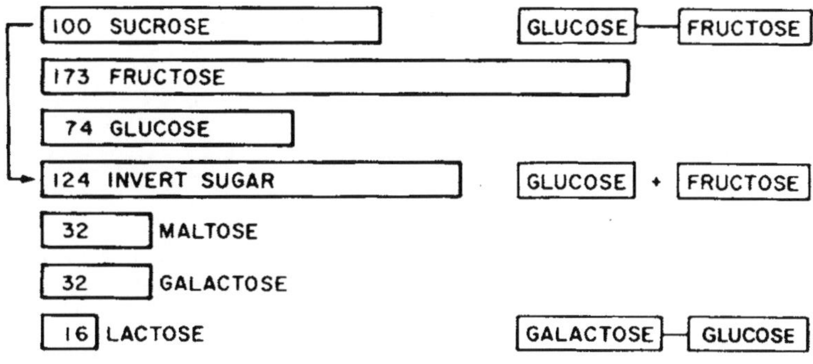

Fig. 15.11
Relative sweetness of sugars. Sucrose is a chemical combination of glucose and fructose, and lactose is a chemical combination of galactose and maltose. When sucrose is 'inverted' it breaks up into glucose plus fructose, which is thus sweeter than the sucrose it came from.

It is interesting to compare the relative sweetness of sugars. The 'inversion' (splitting) of sucrose to form glucose and fructose increases the sweetness by 24% (see Fig. 15.11, above).

Artificial sweeteners

As a chemical educator, the author never ceased to emphasise the need for care and cleanliness in a chemistry laboratory. To his shame, the story of the discovery of modern artificial sweeteners is one story after another of unhygienic serendipity (see Table 15.3, overleaf).

The sweet-tasting property of an *ortho*-toluenesulfonamide was discovered during a routine research project at a time when the taste of a new chemical was considered one its noteworthy properties. Saccharin's main attraction was that it is excreted unchanged, and hence not metabolised. It may be a very weak carcinogen in rats (which concentrate urine very highly in their bladders). Cyclamate (sodium cyclohexylsulfamate) dominated the US market in the mid-1960s until it was banned. Unlike saccharin, cyclamate survives cooking.

Aspartame (NutraSweet™ — US) is something quite different chemically. It is just two amino acids joined together to give a dipeptide, aspartyl phenylalanine (methyl ester) (Fig. 15.12, overleaf). Phenylalanine is an essential amino acid, which, however, must be avoided by the one person in 15 000 who has the genetic condition called phenylketonuria (PKU). Aspartame decomposes at about 10% per month at ambient temperature, so it can be used only in foods with a fast turnover such as soft drinks and fruit yoghurts. At higher temperatures the rate increases, so it cannot be used in cooked foods. There are some anecdotal reports that aspartame has an effect on mental behaviour, allegedly through the formation of methanol, although the quantities seem insufficient for this to be the cause (it would produce less methanol than found naturally in pineapple juice, for instance).

Aspartame (NutraSweet™ and Equal™) is permitted in a number of foods in Australia and must be included in the ingredient list. Up to

TABLE 15.3
The discovery of artificial sweeteners

Sweetener	Date	Discoverer and laboratory	Comments
Saccharin	1878	I. Remsen and C. Fahlberg (Johns Hopkins University, US)	Tasting of new compounds was accepted practice at the time.
Dulcin	1884	J. Berlinerblau (Bern University, Switzerland)	Tasting of new compounds was accepted practice at the time. Introduced 1890s; withdrawn in US, 1951.
Cyclamate	1937	M. Sveda (E. I. DuPont de Nemours, US)	The chemical contaminated a cigarette he was smoking in the lab. Banned in US and UK, 1970; approved in Europe (production and sales, 2000 tonnes p.a.); approved Canada and Australia.
Aspartame	1965	J. Schlatter (G. D. Searle, US)	Accidentally discovered sweet taste while working on anti-ulcer drugs. Approved in US, 1974; banned, 1975; ban lifted, 1981. Monsanto took over Searle on the basis of this one patent; annual sales now over $US1 billion. Patent expired 1987.
Acesufame K	1967	K. Claus (Hoechst, Germany)	While in lab., licked fingers to pick up piece of paper.
Sucralose	1970s	S. Phadnis (Queen Elizabeth College, UK)	During the planning of a project, a foreign research student misinterpreted a request (from a sugar company), for **testing** of a compound as a request for **tasting**, which he promptly did. The lab. immediately applied for the patent.
Alitame	1981		A dipeptide amide patented by Pfizer.

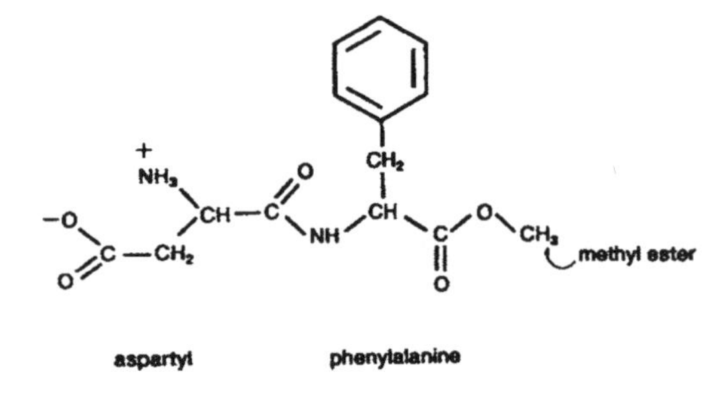

Fig. 15.12 Aspartame (N-L-α-aspartyl-L-phenylalanine 1-methyl ester) — sweeteners from a pair of amino acids

10 g/kg it may be present in low-joule chewing gum, other low-joule foods may contain up to 5 g/kg, and brewed soft drinks may have up to 1 g/kg. Products containing aspartame must carry a warning of risk to PKU sufferers. One Equal™ tablet (equivalent in sweetness to a level teaspoon of sugar) contains 18 mg of aspartame; one Equal™ sachet (equivalent to two level teaspoons of sugar) contains 38 mg. A cup of tea (250 mL) with a sachet contains more than twice the dose recommended for pregnant women.

A set of five related polypeptides (with 207 amino acids) provides another sweetener, thaumatin (Talin™) which comes from a west African plant, ketemfe (*Thaumatococcus danielli*). Intensely sweet, with a long, lingering licórice after-taste, about 750–1600 times sweeter than sugar on a weight basis (30 000–100 000 on a molar basis). The molecule loses sweetness on heating and at low pH (< 2.5) which points to the importance of the tertiary protein structure to the effect. Acesulfame K has a structure similar to saccharin and cyclamate, and a sweetness comparable to aspartame. Unlike aspartame it is stable in water and heat. It is excreted and not metabolised. However, it has not had the same commercial success.

The relative sweetness of some artificial sweeteners is seen in Table 15.4.

TABLE 15.4

Relative sweetness of some artificial sweeteners

Sweetener	E code*	Relative sweetness[a] (mass for mass)
Sucrose (cane sugar)	food	1.0
Lead acetate (sugar of lead)	toxic	1.0
Glycerol (glycerine)	422	0.6
Ethylene glycol (anti-freeze)	toxic	1.3
D-tryptophan (single amino acid)		35
Cyclamate	952	30–80
Acesulfame-K	950	150
Aspartame (dipeptide ester)	951	160
Dulcin (sucrol) ((4-Ethoxyphenyl)urea)	now banned	250
Saccharin (o-toluenesulfonamide)	954	300–500
Sucralose (trichlorogalactosucrose)	955	650
Alitame (dipeptide amide)	956	2000
Thaumatin (five forms)	957	750–1600

* Those indicating an E-number are approved in Australia.

[a] Sweetness is measured by comparing the taste in water to a 4% solution of cane sugar. The results vary with the people in the taste panel.

As chemists have been able to pin down more closely the section of molecules that are responsible for taste, molecular design has become more logical (Fig. 15.13, overleaf).

Another approach has been to modify the sugar molecule, although this can have the opposite effect. One sugar derivative, sucrose octaacetate, is so bitter it is used as a harmless denaturant for alcohol to make the 'denatured' spirit undrinkable. (In this it replaces methanol, which causes blindness in determined drinkers.) Adding chlorine atoms to sucrose has a variety of effects, but a tetrachloro substitution in the correct place gives an increase in sweetness of 2200. With a substitution in the wrong place the sugar is as bitter as quinine. Sucralose™ is 4,1',6'-trichloro-4,1',6'-trideoxygalactosucrose, a trichloroderivative 650 times as sweet as sucrose.

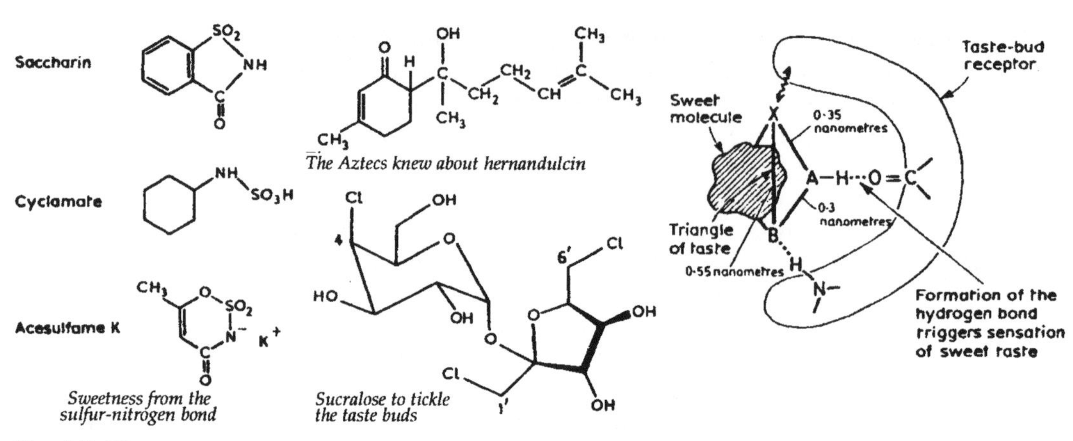

Fig. 15.13
Triangle of sweetness

(From L. Hough and J. Emsley, 'The shape of sweeteners to come', New Scientist, 19 June 1986, p. 50.)

Sequesterants

Metals such as copper, iron and nickel get into food from processing machinery or because of chemical reactions with the container. A sequesterant such as citric acid (chief acid in citrus fruit, 6–7% in lemon juice, Fig. 15.14) thus acts as a synergist (a helping agent) for antioxidants. Sequesterants are used in shortenings, mayonnaise, lard, soup, margarine, cheese, etc.

Fig. 15.14
Citric acid
(E330–333)

```
        H  OH       H
        |  |        |
HOOC—C—C———C—COOH
        |  |        |
        H  COOH     H
```

Stabilisers and thickeners

Stabilisers and thickeners are added to improve the texture and blends of foods. An example is *carrageenan* (E407, a polymer from edible seaweed) which belongs to a group of chemicals called polysaccharides (carbohydrates of high molecular mass, including sugars, cellulose, and starch). They are particularly effective in icings, frozen desserts, salad dressing, whipped cream, confectionery, and cheeses.

Emulsifying (surface-active) agents

The food 'soaps' (E433–444) are used to stabilise emulsions of oil and water components in foods.

Polyhydric alcohols

Polyhydric alcohols are additives used as humectants (to prevent foods from drying out). Tobacco is also kept moist by them. An added feature of these compounds is their sweetness. Some particularly effective alcohols added to sweeten sugarless chewing gum are mannitol (E421), sorbitol (E420), and glycerol (E422). These polyhydric alcohols have the same energy (calorific) value as cane sugar — 16.5 kJ/g.

Water-retention agents

Polyphosphates (E450–452), are increasingly used in the processing of poultry and mammalian meats to bind water and to minimise 'drip', and as an aid to further processing. They are also used to a large extent in processed fish. Phosphates are also used in soft drinks and in the production of modified starches. Because of these many uses, phosphate, although an essential mineral, is likely to be consumed in larger amounts than would be the case if the diet consisted of unprocessed foods. There are indications that an excessive daily intake can lead to the premature cessation of bone growth in children, with a consequent significant reduction in adult height.

Microbial contamination

Contaminants of microbiological origin are critical. *Aflatoxin* is one of a number of naturally occurring toxic products (poisons) found in various moulds. There is usually no visual evidence of its presence. It has been found in greatest amounts in peanuts and other nuts, in corn, and products manufactured from these commodities. It has been shown to cause liver cancer in some test animals and has been a suspected, but unproven, cause of liver cancer in certain African and Asian countries, where high amounts of these toxins have been detected in foods normally consumed. Not all moulds produce dangerous toxins, and some are useful in food processing, such as the moulds used to produce Roquefort cheese. Other moulds are used to produce antibiotics.

Precautions to help control aflatoxins involve the prevention of mould formation by proper drying and storage of crops, removal of damaged material before storage or processing, and provision of adequate moisture and humidity control of stored foodstuffs. A limit of 5 μg/kg has been set for all foods except peanut paste (peanut butter), where, for 'practical reasons', the limit is 15 μg/kg.

FOOD BORNE ILLNESS

In the USA, estimates of food borne illness range from 6 million to 80 million cases per year, with about 9000 deaths. The economic impact is around $5billion. Several factors are promoting the increase of food borne illness.

- The proportion of Salmonella infections that are resistant to antimicrobials has increased from 17 to 31%.
- Increased global trade has placed people in contact with previously unfamiliar food borne pathogens. In 1992, an outbreak of cholera in

Alcohol clearance
Ethanol is metabolised mainly in the liver's cytosol where it is slowly oxidised to acetaldehyde, which is rapidly converted to acetyl-CoA, in turn, providing energy for the citric acid (Krebs) cycle (see also ATP in 'Smart-card' energy transmissions in Chapter 4). While we can use alcohol as a food, supplying calories, it can only be oxidised slowly, and not on demand like glucose. Therefore its clearance rate from the blood cannot be increased by muscular activity. If the energy it provides is not used immediately, then it can **only** be stored as fat (and not as glycogen) in the characteristic rotund shape of the heavy imbiber.
See also Tolerance in Chapter 13.

Maryland was caused by coconut milk from Asia. In 1994, outbreaks of Shigella sonnei in the US, Norway and Sweden were asociated with lettuce from southern Europe.

- Food industry demographics. Increased market size and greater geographic distribution means that problems from centralised food processing extend wider. In 1985, an outbreak of salmonella in contaminated milk in midwest USA caused 200 000 illnesses. One brand of ice-cream caused 250 000 cases in at least 30 states.
- International travel. There were 25 million international tourist arrivals worldwide in 1950. By 2010, 937 million are expected. Traveller's diarrhoea is a catch-all for food disease often caused by faecal contamination of food or water.
- Social demographics. In 1995, about one million people in the US were infected with HIV, which compromises their immune system, and these people are more frequently infected. An aging population and longer survival with disease increases the overall population susceptibility. These include undiagnosed diabetes, and Hodgkin's lymphoma (the five year survival rate in the 1970s was 50%; in 1985 it was 80%).
- Eating habits. Fresh fruit and vegetables can have surface contamination. Raw oysters are a major problem. In the US, teaching of food preparation skills has given way to drug and sex education. One-third of adults did not wash their hands or wash cutting boards after handling raw meat or poultry .

Alcoholic products

If you buy a tin of soup, the ingredients are listed on the label. If you buy a bottle of beer there's no way of knowing what's in it. This is not simply a matter of curiosity. Not so long ago cobalt sulfate was used in brewing in the United States: it gave the beer a lovely head but it gave consumers lousy hearts, and after more than 40 people had died, its use was discontinued.

In the traditional method of making beer, hops (or hops extract) was added to malted grain and then fermented. The yeast converts glucose to alcohol, but it also destroys some of the bitter flavouring substances. By extracting the hops with *liquid carbon dioxide* the flavour can be obtained and added to the beer *after* fermentation. Solvent extraction of foods to remove certain components has always had the problem of residual solvent. By using a material that is normally a harmless gas this problem is solved. Carbon dioxide is normally a gas or a solid (dry ice) but, under pressure in a cylinder, it can be a liquid, and it is then a marvellous solvent. Above a critical temperature, the distinction between a liquid and a gas disappears. This is the critical point, and beyond this the CO_2 is supercritical (see Fig. 15.15).

For carbon dioxide this occurs at 31°C and 73 atmospheres pressure.

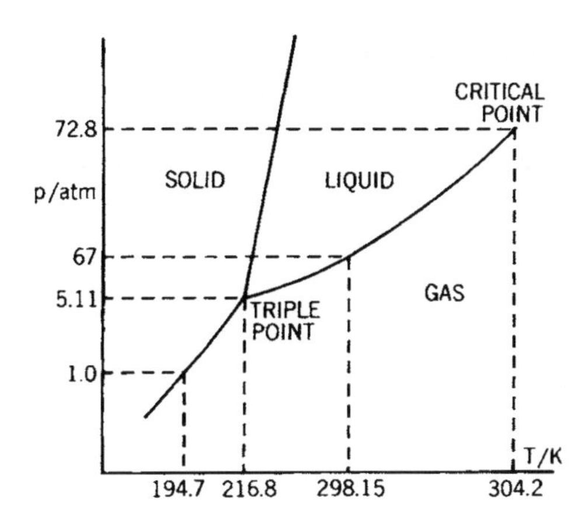

Fig. 15.15
Phase diagram for carbon dioxide

SUPERCRITICAL SOLVENTS

Supercritical CO_2 is a very selective solvent and is used, for example, in extracting caffeine from coffee beans while leaving their flavour and aroma virtually unchanged. This gives decaffeinated coffee, and surplus caffeine which goes into analgesics and cola drinks.

Nicotine can be extracted from tobacco to give a cigarette that still causes cancer but has not the kick — not much demand for this! Nicotine in chewing gum has been introduced as an alternative product — to give the kick without the cancer. Nicotine affects you very quickly because it reaches the brain from the lungs in just eight seconds. (See Nicotine and tobacco, Chapter 13, for further details.)

Oilseeds (such as soya and sunflower) can be extracted with supercritical CO_2. This results in less gum being extracted than with conventional solvents. The oil separates as the CO_2 pressure drops, and so a very neat cyclic process is possible. The natural insecticide pyrethrin (see Chapter 7) is now extracted from plants this way and essential oils for perfumes are being studied.

For water, the critical constants are 374°C and 218 atmospheres pressure. Supercritical water is used in modern steam-turbine electricity generation because it simplifies the design calculations for heat transfer (there is only one phase, not two) and the higher temperature provides higher efficiency (see Fig. 14.2).

Wine

And much as Wine has play'd the Infidel
And robb'd me of my hale of honour — well
I often wonder what the Vintners buy
One half so precious as the Goods they sell.

Good wines are grown in the Canberra district

White or red wine? The process of manufacture differs from country to country. In Australia the shifting popularity from red to white can possibly be attributed to the cleaner chemistry in the production of white wines (see box).

RED OR WHITE?

Red wine	White wine
Skin contact	No skin contact
Phenol/colour extraction from skins 700–2000 mg/kg.	Juice phenol only present at levels of 200–500 mg/kg.
Extraction of acid salt from skin higher at pH 3.5–4.0.	Limited extraction of acid salt at low pH 3.0–3.5.
Unavoidable presence of wild micro-organisms, yeast and bacteria, leading to off-odour formation.	Juice can be clarified prior to fermentation; microbiological control high; no off-odours.
Extraction of growth activators from skins and stimulation of fermentation both primary and malolactic.	Controlled fermentation because of absence of solids and growth stimulators.
Believed by wine makers to require high temperatures — fermentation conditions conducive to high temperature loss of fruit volatiles.	Low-temperature fermentation: retention of fruit volatiles.
Addition of SO_2 is limited because of its effect on colour and it is less effective because of high pH and binding by phenol. Malolactic fermentation all but inevitable; causes loss of acid and fruit.	Addition of SO_2 at low pH leads to free levels of SO_2 effective in controlling malolactic fermentation and oxidation.
Promotion of belief in requirement for oxidation and extraction of wood flavour to complete wine complexity.	Promotion of simple, identifiable fruit flavours.
Promotion of belief in requirement for bottle aging to complete complexity.	Fresh fruit flavours promoted; product consumed early.
Taste incompatibility of high tannin and high extraction with presence of sugar.	

Soda hangover
A scientific connoisseur of spirits had a hangover every time he drank scotch and soda. When he switched to gin and soda without improvement, he realised immediately it was the soda.

The effects of wine

There are headaches and headaches, and careful neurological study quite clearly distinguishes migraine, cluster headaches, and tension headaches, and the mechanism of the effect of wine on each is quite distinct. What does seem to be clear is that it is highly unlikely that any of the amines (e.g. histamine) in alcoholic beverages can be directly connected with the onset of a migraine attack. In Australia, red wines are believed to be more potent in this respect. In France, the opposite is true. Clarets and port are often associated with the attack, and yet are very low in amines.

This brings us back to the scotch and soda story — it now appears that it is the *alcohol* that is important. Alcohol is known to be able to release histamine from the liver into the bloodstream in dogs, and so possibly also in humans. Histamine release is also known to be the cause of headaches similar to migraine. Any carbohydrate meal rapidly increases serotonin levels in the brain, which probably accounts for the attraction of sugar. The Bedouin, forbidden alcohol, sucked sherbets. The level of serotonin is known to fall during a migraine attack. This is interesting in that it lends some credence to the belief that sugar acts as a prophylactic in forestalling a hangover.

The waiter test
Dipstick test strips (see earlier this chapter) for tartaric acid are basically used to test wine. The author tried it out in a very expensive restaurant in Canberra. An elegant wine waiter arrived with the ordered vintage and poured some in a glass for taste. Out came the test-kit and the wine was right on the borderline of acceptability. I slowly shook my head, more in sadness than in anger. The wine waiter's face had to be seen to be believed. How good the test strip is for wine I am not sure, but for testing the sang-froid of a wine waiter, it is absolutely superb. The fact that I dare not return to that restaurant can only be a saving on the family budget — so I have no regrets.

Labelling of wine

Even the question of accurate labelling of wine raises highly technical discussion. It apparently wasn't until we had a visit from a French ampelographer that we began to know something about the origin and nature of our vines in Australia. The nature versus nurture argument in regard to human intelligence is one with which we are all familiar, but there is a parallel argument applied to 'wines from vines'.

It is often assumed that any variety will give a specific flavour in its wine, which can be recognised by those who have developed enough skill. Often it is the *process* that is characteristic, such as low-temperature fermentation, giving very flowery-scented aromas. A characteristic flavour is produced by initially fermenting the grapes *before* crushing, whereby the first alcohol is produced in the grape without added yeasts.

Of course, all this has legal consequences in labelling. Wines that are subsequently blended are distinct from wines made from grapes that are initially blended. The sugar content at harvest of a Rutherglen muscat may be much more important than the grape variety.

There is general agreement that the French agricultural trade-mark system of 'Appellation Controle' has little relevance in Australia, and even in France it might appear to be a method of protecting 'inferior' technology. It depends whether you regard the 'malolactic fermentation', which often gives the 'unique character' appellation of wines, as secondary fermentation, or bacterial spoilage.

Fig. 15.16
'If this was made from grapes, why no tartaric acid?'

What is the place of new technology in European winemaking? The German consumers are interested in what is done to their grapes, and reserve judgment on what is good for themselves. French consumers are more concerned with the pedigree, and adjust their taste accordingly. On the other hand if you want to improve your Latin, then Chaucer's Summoner had the answer:

> And whan that he wel dronken hadde the wyn,
> Thanne wolde he speke no word but Latyn.

Caffeine

Coffee is consumed in one form or another by about one-third of the world's population. The two most important commercial varieties are *Coffea robusta* and *Coffea arabia*. *C. robusta*, grown in West Africa and Indonesia, has a higher caffeine content. *C. arabia*, grown in East Africa, Central and South America, the Caribbean and New Guinea, yields a stronger flavour. Blending and roasting change the character of the crop considerably.

Instant coffee was popularised through the US armed forces in 1938 and is now the most widely used form of coffee in Australia. Australian food regulations set a minimum level of 3% *robusta* beans.

Decaffeinated coffee is even older, having been introduced by the German firm Kaffee HAG in the early 1900s, but has only now attracted a loyal and growing public. There was initially some concern with the safety of the solvent, trichlorethylene, used to extract the caffeine from coffee. However, the safer trichlorethane, and liquid carbon dioxide are now used (see box Supercritical solvents above).

Worldwide consumption of caffeine has been estimated at 120 000 tonnes per annum, which works out at 70 mg/per person per day. Approximately 54% of this is from coffee, 43% from tea and 3% from other sources. In spite of the image of the USA as a heavy coffee-consuming country, Scandinavians consume almost three times as much caffeine (340 mg/day) from coffee as Americans, while the British match the Scandinavians in their caffeine intake from tea (320 mg/day). Only the USA and Canada show significant caffeine intake from soft drinks (35 and 16 mg/day respectively). Caffeine is allowed in non-cola drinks in these countries (but not in Australia). The total caffeine intakes of USA and Canada (211 and 238 mg/day respectively) are well below those of Sweden (425 mg/day) and the UK (444 mg/day). The estimated level of consumption of caffeine in Australia is 240 mg/day from all sources.

The approximate caffeine content of beverages is given in Table 15.5.

Respiratory stimulant
Modern medicine makes use of caffeine as a respiratory stimulant, and the related theophylline for bronchial asthma (doses for adults are in the 250 mg range). But note that caffeine and the benzodiazepam drugs (e.g. Valium) are directly antagonistic.

In addition to caffeine, tea contains about 1 mg per cup of the much more active alkaloid theophylline, while cocoa contains about 250 mg per cup of the much less active alkaloid theobromine (which does not contain bromine; see Chocolate in Chapter 4). Plants store these chemicals in their leaves as a natural insecticide. Tea and coffee grounds are therefore also effective as insecticides or repellents.

The relation between the chemical structures of xanthine, caffeine, theophylline and theobromine can be seen in Figure 15.17.

Conversely, there have been many studies attempting to link coffee with disease. Unfortunately, the many symptoms attributable to excess caffeine are those for which a cup of coffee is often self-prescribed. These are insomnia, irritability, headache, palpitations, diuresis, and diarrhoea.

Folklore has it that people vary considerably in their response to caffeine. Some claim to sleep like a log after several cups at night, while others find a single cup causes a violent reaction. Closely controlled experiments do not bear this out, however, and effects such as tolerance

TABLE 15.5
Approximate caffeine content of beverages

Beverage	Container	Volume (mL)	Content (mg)
Percolated coffee	cup	150	100
Drip coffee	cup	150	80–200
Instant coffee	cup	150	60–70
Cocoa	cup	150	5
Chocolate bars	bar	100 g	20
Cola drink	can	375	35–55
Tea (av.)	cup	150	50
Tea (weak)	cup	150	30
Tea (instant)	cup	150	30
Tea (Chinese)	cup	75	15
Decaffeinated tea	cup	150	2
Coffee beans (green)		0.8–1.8%	
Tea leaves (undried)		0.68–2.1%	

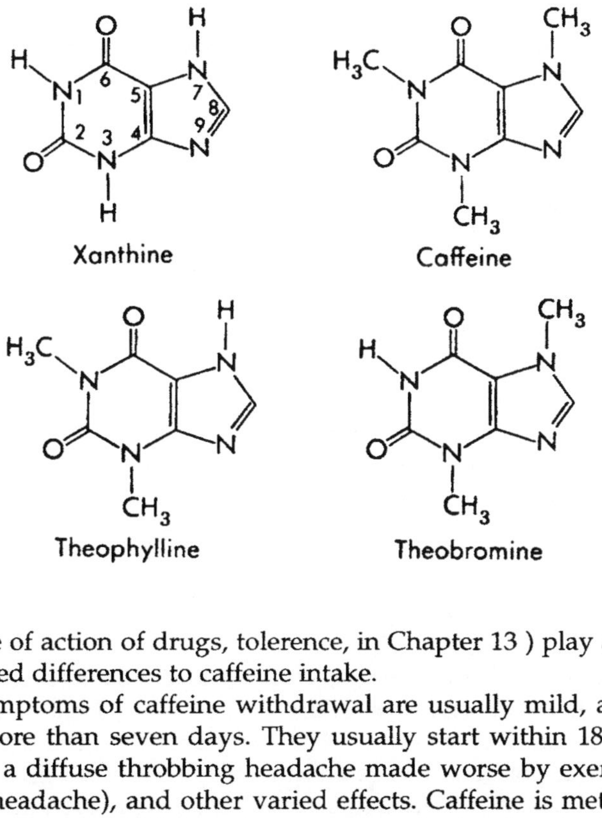

Xanthine

Caffeine

Theophylline

Theobromine

Fig. 15.17
Chemical structure of some xanthines

(see Mode of action of drugs, tolerence, in Chapter 13) play a large part in perceived differences to caffeine intake.

The symptoms of caffeine withdrawal are usually mild, and seldom last for more than seven days. They usually start within 18 hours and consist of a diffuse throbbing headache made worse by exercise (like a 'tension' headache), and other varied effects. Caffeine is metabolised in the liver and the half-life (the time to reduce the level in the body to half its initial value) is three to four hours. The rate is increased by smoking and other causes of increased activity of 'mixed function oxidases', and is greatly slowed in the later stages of pregnancy, an effect related to endocrine function. It is well established that women automatically reduce their caffeine intake during pregnancy. In fact, the consumption of caffeine, unlike many other drugs, is self-regulating; you stop drinking when you've had enough.

KOSHER COKE

While Coca-Cola has been on the market since 1886, only since 1935 has it been certified kosher. Rabbi Tobias Geffen, an Orthodox rabbi who served a congregation in Atlanta USA from 1910 until his death in 1970, was asked to rule on the acceptability of Coke by many of his colleagues who were under considerable pressure from the children of their congregants. To do that, the rabbi had to contact the Coca-Cola company for a list of ingredients. This list is one of the most closely guarded corporate secrets in American capitalist history, but a decision was made that under the utmost confidence, access would be allowed. The list revealed that glycerine, made from non-kosher beef, was included. The concentration was only one part per thousand, and to be kosher, up to one part per 60 is acceptable. However, this only applies if the contamination is accidental, not if it is a deliberate addition. Thus its inclusion could not knowingly be tolerated by observant Jews.

Back at the lab in the Coca-Cola company, chemists contacted Procter and Gamble who had routinely prepared glycerine from cottonseed and coconut oils. Coke chemists followed the same course, the seal of rabbinical approval was then forthcoming, and a whole community was allowed to sample the real thing.

But there was a second problem. The formula for Coke included traces of alcohol as a by-product of the fermentation of grain kernels. Since anything derived from grains is not-kosher for Passover (where stricter rules apply — no beer, but wine is fine), this was a problem. So for the weeks before the Passover, Coke contains sweeteners derived from beet and cane sugar, rather than corn and 'a taste-uncompromised-kosher-Coke' became available for Passover.

Original source: American Jewish Historical Society, 7 Feb, 1997.

The question of caffeine intake by children is a vexed one. An 18 kg child consuming a 375 mL can of soft drink with a legal caffeine level of 145 mg/L consumes 55 mg of caffeine, which represents 3 mg/kg body mass. (Actual levels found in soft drinks in the USA are given in Table 15.6.) As a level of 1.5 mg/kg is already effective in postponing sleep, one wonders about the 'technological need' for such a high legal level.

TABLE 15.6
Caffeine content of various soft drinks (US)

Soft drinks containing caffeine[a]	Caffeine content (mg/375 mL)	Soft drinks containing caffeine[a]	Caffeine content (mg/375 mL)
Jolt	76.1	Sugar-Free Dr Pepper	41.8
Sugar-Free Mr PIBB	62.1	Big Red	40.6
Mountain Dew	57.1	Sugar-Free Big Red	40.6
Mello Yello	55.8	Pepsi-Cola	40.6
TAB	49.5	Aspen	38.0
Coca-Cola	48.2	Diet Pepsi	38.0
Diet Coke	48.2	Pepsi Light	38.0
Shasta Cola	46.9	RC Cola	38.0
Shasta Cherry Cola	46.9	Diet Rite	38.0
Shasta Diet Cola	46.0	Kick	33.0
Shasta Diet Cherry Cola	46.4	Canada Dry Jamaica Cola	31.7
Mr PIBB	43.1	Canada Dry Diet Cola	1.3
Dr Pepper	41.8		

[a] There are many types of soft drinks, manufactured by the leading bottlers, that contain no caffeine.
Note: Australian limit for caffeine content is 55 mg/375 mL.

Source: Food Tech. Aust. **40**(3) 1988, 106

CAFFEINE AND OTHER MEDICAL CONTROVERSIES

When considering the more serious accusations made against coffee, it is interesting to see the problem from the point of view of a publisher of scientific research. Dr Arnold Relman, editor of the prestigious New England Journal of Medicine was interviewed on his approach by David Dale in the Sydney Morning Herald, 30 August, 1986 (p. 41).

'About five years ago we got a paper which said that there was a statistical association between drinking more than five cups of coffee a day and cancer of the pancreas. Now cancer of the pancreas is an increasingly common cancer, not as common as cancer of the bowel or breast, but it is getting there, and there's no treatment for it. It's usually rapidly fatal.

'So this correlation with coffee drinking was a very interesting finding. It didn't prove that drinking coffee **causes** cancer of the pancreas; there are many possible explanations — but it's a piece of evidence; it may provoke further research.

'We debated a long time whether to publish. We could just see the headlines. We sent the paper to our statistical referees; they said the methodology was OK; the conclusions were justified. It was a close call, but we went ahead.

'Oh boy, did that hit the fan. Headlines all over the world. Reuters called me up and said that the bottom had dropped out of the coffee market. It was terrible.

'There was a lot of criticism that we published it. And now it turns out that it's probably not going to hold up. People have done studies since — we have published them — which suggest that the initial association is probably not true (the effect may have been due to the solvent used for decaffeinated coffee, see above). 'The problem is in conveying to journalists and to the public that science works in small steps.

'We publish on biodegradable material, not stone. (Emphasis added.)

'We publish in the full knowledge that tomorrow we may get a paper in the mail which gives a completely opposite explanation of the facts . . . All we can promise is that in the judgment of competent, hardworking people with no axe to grind, the report you are reading is interesting, and original, and important.'

Eighty-five per cent of papers rejected by the journal are published elsewhere. The prestige given by a journal's rigorous refereeing of submitted papers itself gives credence to what is published (see also Why should we trust the experts? in Chapter 2).

These then are the issues that must be balanced. Total avoidance of caffeine seems unwarranted and an intake of less than 300 mg/day seems acceptable by all the evidence. What is disturbing is the attitude that for children, coffee is undesirable, but cola drinks are acceptable.

TABLE 15.7
Required levels of vitamin C in fruit juices

Juice	Vitamin C (mg/L)
Orange	400
Lemon	350
Grapefruit	300
Guava	1000
Pawpaw	400
Mango	250
Blackcurrant	700
Pineapple	100

Fruit juices

The minimum levels of vitamin C required in fruit juices are shown in Table 15.7 (at left). As apples have little vitamin C, apple juice is not considered a source. However, vitamin C is added to all the juices on the list because it can be destroyed during processing. Preservatives are permitted in some fruit juices and these are sulfur dioxide (maximum 115 mg/kg) or benzoic acid, or sorbic acid (maximum 400 mg/kg), along with citric acid and carbon dioxide.

Flour and bread

At one time the demand for white bread was met by treating flour with *nitrogen trichloride* (agene), but this process was abandoned some years later after dogs fed large amounts of agenised bread developed running fits, or canine hysteria. The toxic compound in the treated flour was later shown to be methionine sulfoximine, which is formed by the action of nitrogen trichloride on methionine, an essential amino acid of proteins. Today the situation has improved.

Bread represents nearly half the Australian domestic use of wheat. Cereals are currently our most important source of complex carbohydrates, thiamine, iron and energy; the second most important source of protein; and the third most important source of calcium. Australia produces enough wheat to provide each citizen every day with 10 plates piled with about 300 g of grain. One plate is eaten here, and the other nine are exported.

Gluten

Gluten is the most important protein component in wheat and gives flour its unique physical properties for bread making. Gluten can be isolated by kneading dough under running water, which removes most of the starch and other water-soluble components. The dry matter of gluten consists of two main groups of proteins, *glutenin* and *gliadin*, plus some bound fats and carbohydrates.

The addition of water to flour hydrates the protein. Kneading, or high-speed mixing in bakeries, 'works' the dough. This breaks the disulfide bonds between protein chains, allowing the chains to move. Then the bonds reform in this new position. The protein then changes shape to accommodate the new bonding positions (Fig. 15.18). The process is thus analogous to what happens in setting hair. (See also Hair in Chapter 5; Wool, Silk and Leather in Chapter 11.)

Bakeries sometimes use a gluten-softening agent, such as L-cysteine or sodium metabisulfite to help break bonds and aid mixing, and gluten-strengthening agents such as ascorbic acid, to help set the new bonds.

During 'proofing', starch breakdown and fermentation occur. The carbon dioxide produced opens up the gluten network, giving it a cellular structure. Baking denatures the protein, gelatinises the starch granules with water from the gluten, and sets the bread.

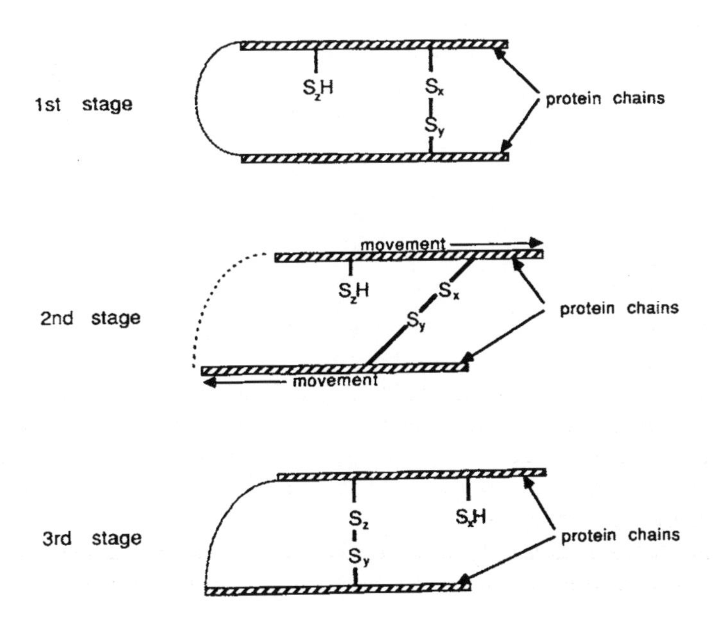

Fig. 15.18
Making and breaking of disulfide linkages during kneading of dough

(Courtesy of Bread Research Institute of Australia)

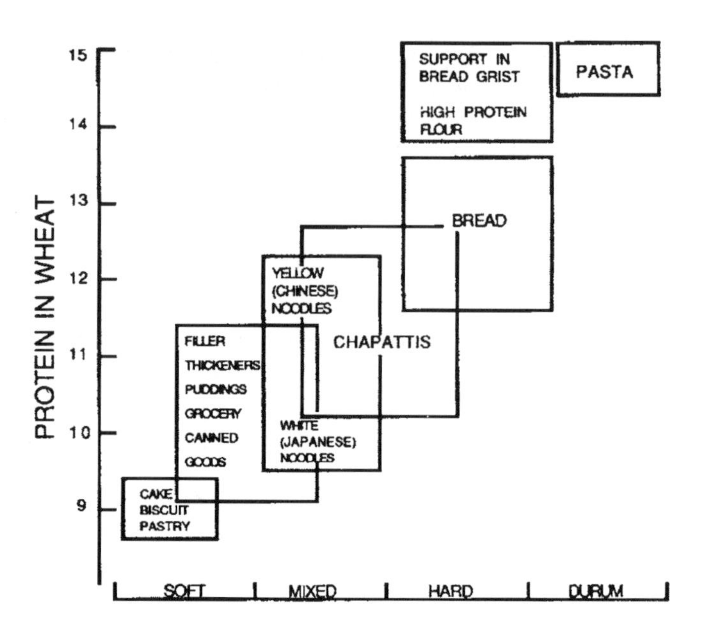

Fig. 15.19 Requirements of certain wheat products in terms of protein percentage and hardness

(From H. J. Moss, J. Aust. Inst. Ag. Sci. 1973, 109; courtesy of Bread Research Institute of Australia)

Dough

'You pays your money and you gets your dough'. Protein content and hardness are two of the most important physical characteristics of dough. Hard wheats with strong doughs are suitable for bread flours. Soft wheats are cultivated to give weaker doughs and low protein. As such, they are ideal for biscuits. The different requirements of wheats for different products are shown in the two-dimensional graph in Figure 15.19 (above).

In practice, a somewhat more complex set of tests and parameters is used.

Pasta is made from durum wheat (protein content 14% to 15%) milled to produce semolina. A minimum of 12% protein is needed to make pasta that remains firm on cooking (some hard bread wheat is sometimes

Fig. 15.20 From wheat to bread (and money!). The old Australian $2 note shows William Farrer, who pioneered the development of wheat in Australia.

mixed in). The semolina is blended with water, extruded through various shaped nozzles and dried to 12 to 13% moisture level.

Noodles require flours of between 9.5 and 12% protein level. Higher levels giving stronger gluten are liable to cause tearing during processing. With lower levels, the noodles are liable to fall apart during processing. The flour is also required to be very free from bran specks, as these are responsible for discoloration.

Flat breads (such as chapatti, pita, mafrood, Lebanese) rely for their quality on steam produced when the dough is heated. Flour with a high degree of mechanical damage to the starch grains absorbs more water. Such flour is best made from hard bread wheats. Relatively low strength is required so that the dough will not distort during preparation.

Biscuit doughs are expected to be weak, as tough texture is undesirable. Low-protein flours from soft wheats are used.

The number of parts by mass of flour extracted per 100 parts of wheat is called the extraction rate. Milling to produce white flour results in the removal of varying amounts of bran, pollard, and germ, which together constitute 12 to 15% of the mass of the grain. The limit of white flour extraction is 75 to 79%. Figure 15.21 shows the influence of extraction rate on nutrient retention. Australian white flour has increased in extraction

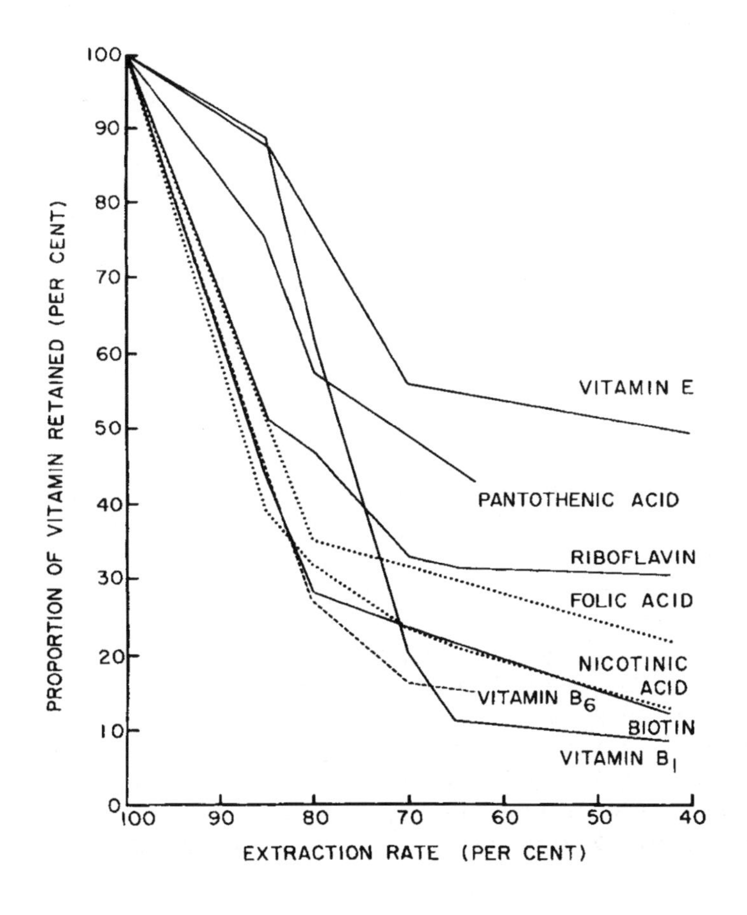

Fig. 15.21
Milling extraction rate and proportion of total vitamins retained in the grain in flour

(From Magnus Pyke, Man and food, World University Library, London, 1970)

rate from 71% in 1940 to 77 per cent in 1985, and is expected to plateau at about 79 per cent.

White or brown bread?

The small difference in nutritional value of white (70% extraction) versus wholemeal (100% extraction) flour in the context of a Western diet today is unimportant. However, the selling point of wholemeal bread is strong. The advantages of white flour are as follows:

- Bread made from it is lighter, finer, and is white.
- It contains less fat. It is thus less prone to rancidity, and keeps longer.
- It contains less phytic acid and fibre, which tie up minerals such as calcium, iron, magnesium, and zinc. Thus minerals in white bread are more available.

The disadvantages of white flour are:

- It contains less of the B-group vitamins.
- It contains less of certain minerals and trace elements.
- It contains less fibre.

The limiting amino acid in wheat flour is lysine, which in the normal diet is compensated for by other sources of protein.

Digestion

RAISING THE WIND

'Full of sound and fury signifying nothing' said Macbeth after consuming Jerusalem artichoke soup. (Well not really! In fact the Jerusalem artichoke is a form of sunflower, Helianthus tuberosus. In Italian, the characteristic 'turning towards the sun' is 'girasole', which was turned by the fine English ear for language into 'Jerusalem'.)

The scientist-cum-statesman Benjamin Franklin proposed the following project to the Royal Academy of Brussels in the 1770s as a problem for their proposed prize for 'useful science':

> It is universally well known, that in digesting our common food, there is created or produced in the Bowels of human creatures, a great quantity of Wind. That the permitting of this Air to escape and mix with the Atmosphere, is usually offensive to the Company, from the fetid smell that accompanies it. That all well bred People therefore, to avoid giving offence, forcibly restrain the Efforts of Nature to discharge that Wind. That so retained contrary to Nature, it not only gives frequently great present pain, but occasions future Diseases such as habitual Cholics, Ruptures, Tympanies, &c., often destructive of the Constitution and sometimes Life itself.
>
> Were it not for the odiously offensive smell accompanying such escapes, polite People would probably be under no more restraint in discharging such Wind in Company, than they are in spitting or in blowing their Noses.

My Prize Question therefore should be: To discover some Drug, wholesome and not disagreeable, to be mixed with our common Food, or Sauces, that shall render the natural discharges of Wind from our Bodies not only inoffensive, but agreeable as Perfumes'.

The answer is not to try to disguise the output, but to modify the input. The human gut cannot digest certain medium-size sugar molecules such as raffinose and stachyose, found particularly in beans. Not so the bacteria in the colon, which attack them with vigour, producing a variety of gases and vapours, both inoffensive (carbon dioxide, hydrogen, and methane) and offensive (possibly hydrogen sulfide or rotten egg gas, indole, skatole, and ammonia). Bacteria are responsible for the production of about one-third of the gas, while the other two-thirds comes from swallowing.

Both hydrogen and methane are flammable, and have caused explosions during operations involving electrocautery on the colon. (The explosion limits in air for H_2 are 4 to 74%, and for CH_4 are 5 to 15%. See Table 14.7, Flammability limits.)

Normal people pass about half a litre of gas by the anus per day, although amounts as high as 168 mL per hour have been recorded in volunteers when half their diet was baked beans. Gastric gas causing a bloated feeling is also released upwards and it is interesting that certain substances relax the sphincter muscle at the *top* of the stomach, releasing the gas in that direction. These include nutmeg, ginger, caraway, cinnamon and peppermint. Taking peppermints after meals (as a sweet or liqueur) is obviously a tradition with a purpose.

It is only today, using genetic engineering, that this problem is being solved. Because soya beans are used in so many processed foods, research in the US has concentrated on locating the genes responsible for producing the indigestible sugars, and replacing them with genes that will produce more acceptable sugars.

Allergies

While all people react in identical ways to some poisonous substances, there are other substances to which individuals react quite differently. These are called *idiosyncratic* reactions. Idiosyncratic reactions to food were known to Hippocrates, but systematic studies have really been documented only since the early 1940s.

There are believed to be three mechanisms of adverse reaction to food. The first of these reactions results from the body's *metabolism* and relates to a genetic fault that affects the way an individual metabolises certain food components. Examples of such genetic faults are diabetes, phenylketonuria (PKU), lactase deficiency, and favism. The second and most common mechanism is *pharmacological* (see Table 15.8, overleaf) and is really an adverse 'drug' reaction in which the 'drug' is a natural or

TABLE 15.8
Clinical features of adverse reaction to food

Pharmacological	Allergic
Common	Uncommon
All ages	Mostly children
Non-atopic	Atopic[a]
Many foods	Specific food
Delayed reaction	Immediate reaction
Difficult diagnosis	Easy diagnosis

[a] Atopic is a description of the type of person who has the constitutional make-up that responds with allergic reactions to certain stimuli; that is, the sort of person who gets dermatitis, hay fever, hives, etc. People who have other allergic reactions are the sort who get allergic reactions to food.

synthetic food component. The component is a relatively small molecule that is biologically active, just like a drug. For example, in Chapter 13 we saw that the *drug* sulfonamide is modelled on the *natural* food component *para*-aminobenzoic acid, and the molecule is the same size. Far less common than drug reactions is the third mechanism, which is a true *allergic* reaction to food. Allergic reactions involve the immunoglobulins, the carriers of specific antibodies. Some of these, the gammaglobulins, carry well-identified antibodies such as tetanus antitoxin, and Rh antibodies, whereas IgE (immunoglobulin E) was found responsible for forming antibodies against food proteins. Note that the chemical causing the production of the antibody in this case — namely a protein — is very large. The *allergic* reaction to a bee-sting is a reaction to a foreign *protein*. The response in the victim in both cases may be very similar.

Allergic reactions in any one individual generally involve only a couple of foods (e.g. eggs, nuts, milk, seafood) and, in particular, their protein components. Local swelling, itching and burning around the mouth, are common and may be followed by nausea, vomiting, abdominal cramps, and diarrhoea.

Combined reactions
Sometimes pharmacological and allergic reactions are combined in an individual to cause severe symptoms, with the former increasing the severity of the latter.

Pharmacological food reaction is much more prevalent. The reaction is to a chemical that may be common to a number of foods (and drugs). The time between exposure and reaction may be delayed for periods varying from hours to days, which makes linkage between cause and effect quite difficult. Detection is complicated by the fact that the amount of chemical needed to cause any adverse reaction does not stay the same but can depend on how much of it has recently been eaten. Just as antihistamines can make you very drowsy initially, and continuous use reduces this effect, the same masking can occur with pharmacological reactions to food.

Skin tests detecting specific IgE may provide useful information on true food allergy, although high levels of IgE occur in normal blood donors, and are of no significance in the absence of symptoms.

Natural and synthetic food chemicals to which there are common reactions are salicylates (including aspirin), benzoates, metals, synthetic food colours such as tartrazine, and sulfur dioxide.

Consumer reaction

Consumers who find this chapter interesting should extend their reading and keep it, and their eating habits, as broad as possible. Eating a narrow selection of foods in excess, or reading only articles with which you agree, are both undesirable.

Food additives are an emotive issue even if the risks are minute compared to those from natural food components and microbial contamination. Proper disclosure of food ingredients is not only necessary for reasons of idiosyncratic reaction, but also to give people choice on other grounds, such as religious beliefs.

Concern with *process* and *motive*, rather than the 'objective' properties of the end-product, is an area where industrial scientists and community activists beg to differ. Recent examples include protests against irradiated and bioengineered foods.

References

People prefer coloured flavours, G.G. Severino, R.L. McBride and B.M. Cox, *Food Research Quarterly* 47, CSIRO, 1987, 64–65.

Chinese-restaurant syndrome, R.H.M. Kwok, *New England Journal of Medicine*, 278, 1968, 796 (letter to the editor).

The shape of sweeteners to come, L. Hough and J. Emsley, *New Scientist*, 15 June 1986.

Colin Wrigley, Officer-in-Charge, CSIRO Wheat Research Unit, Sydney, and John Moss, Bread Research Institute of Australia, Epping Road, North Ryde, NSW 2113.

A Christmas digest, S. Young, *New Scientist*, 19/20 December 1985.

Begone with the wind, B. Dixon, *New Scientist*, 24 April 1986.

Bibliography

Food Chemicals Codex, 2nd edn, Committee on Specifications, Food Chemicals Codex of the Committee on Food Protection, National Research Council, National Academy of Science, Washington DC, 2nd edn., 1972.

The Australian Model Food Act, F.H. Reuter, Supplement, *Food Australia*, Aus. Inst. Food Sc. & Tech., March 1997, 31 pp. A personal history of the development of uniform food standards in Australia, by Professor Reuter (at age 91). Awesome.

On food and cooking: The science and lore in the kitchen, H. McGee, Unwin Hyman, London, 1987. A delightful book.

Toxic hazards in food, D.M. Conning and A.B.G. Lansdown, Croom Helm, Canberra, 1983. A little gem published in a forgotten place.

Toxicants occurring naturally in foods, National Research Council, National Academy of Science, Washington DC, 2nd edn, 1973. Still the classic reference.

The causes of cancer, R. Doll and R. Peto, Oxford Medical Publications, Oxford, 1981. The classic on cancer epidemiology.

Human ecology and susceptibility to the chemical environment, T.G. Randolph, C.C. Thomas, 1962. This book predates the fashionable interest in the environment and ecology.

Chemistry of ionising radiation

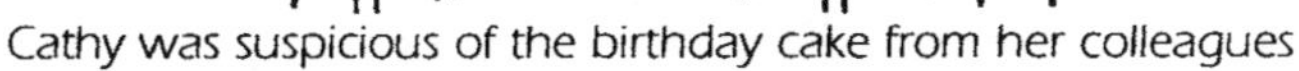
Cathy was suspicious of the birthday cake from her colleagues

Yellowcake
Gentle reader, colour this cake yellow.

Chant of the radioactive workers

We're not afraid of the alpha ray,
A sheet of paper will keep it away!
A beta ray needs much more care,
Place sheets of metal here and there.
And as for the powerful gamma ray
(Pay careful heed to what we say)
Unless you wish to spend weeks in bed,
Take cover behind thick slabs of lead!
Fast neutrons pass through everything.
Wax slabs remove their nasty sting.
These slow them down, and even a moron
Knows they can be absorbed by boron.
Remember, remember all that we've said,
Because it's no use remembering when you're dead.

Radiation

Radiation has two forms. First there is electromagnetic radiation, such as light, but with a range of frequencies from radio waves to gamma rays.

All this radiation travels at the speed of light (3×10^8 m/s) and can also be regarded as a stream of massless particles called photons. When photons hit you, the *effect* depends on their *energy*, which increases — the shorter the wavelength (higher the frequency) of the radiation. The *amount* of radiation depends on the *intensity* (i.e. the number of photons). The small percentage of ultraviolet radiation in sunlight can cause burning in 20 minutes, whereas you can spend weeks behind glass (which cuts out the ultraviolet, but lets through the visible light) without getting a tan.

The second type of radiation consists of a stream of particles *with* rest mass travelling at various speeds, depending on their kinetic energy. Examples are alpha-particles (helium nuclei), beta-particles (electrons), and neutrons (discovered after the wave–particle duality was understood, and so there is no special name for their radiation).

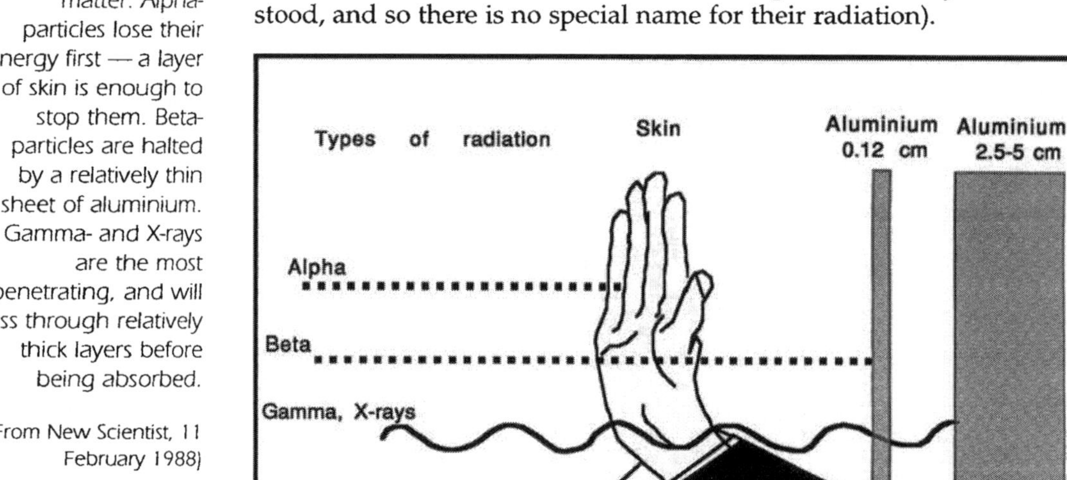

Fig. 16.1 How far radiation travels through matter. Alpha-particles lose their energy first — a layer of skin is enough to stop them. Beta-particles are halted by a relatively thin sheet of aluminium. Gamma- and X-rays are the most penetrating, and will pass through relatively thick layers before being absorbed.

(From New Scientist, 11 February 1988)

Radioactive disintegration

Radioactive decay is a random process. We do not know which atom in a sample will decay, nor do we know when. In fact we can use the model presented in the Entropy Game, (Appendix VIII) to describe radioactive decay. In Figure 16.2 the horizontal axis is time, and the vertical axis is the average number of atoms emitting at that time.

The general application of this simulation game to quite diverse areas takes a while to be appreciated. The fixed half-life of a radioactive species, which is the average time for half the atoms to emit radiation (i.e. to decay), corresponds to the fixed number of counters per square in a particular run of the game.

The rate at which a large collection of radioactive nuclei emits particles is called the *activity* of the source. One becquerel (Bq) is one radioactive disintegration per second. (The older unit, the curie, is the activity of 1.02 g of radium, and thus 3.7×10^{10} Bq.)

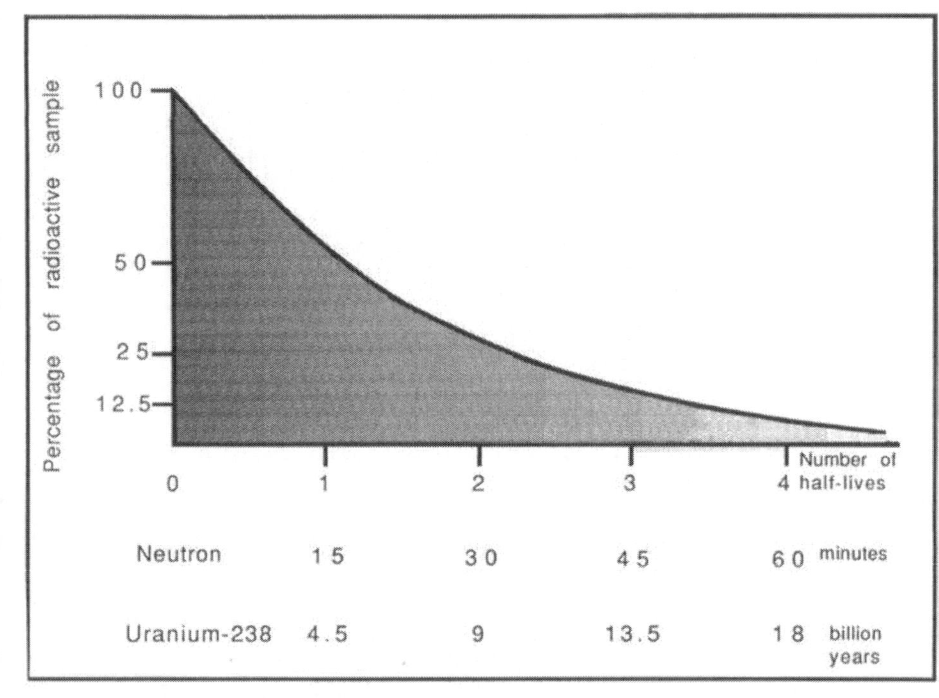

Fig. 16.2
Exponential decay. The decay pattern for all radioactive nuclei is the same, regardless of the half-life.

(From New Scientist, 11 February 1988)

Biological effect of radiation

The biological effect of radiation depends on the amount of energy that the radiation deposits in the body. This is the average energy of the ray as emitted, measured in joules, deposited per kilogram of body mass. The unit used is the gray (Gy) corresponding to 100 rads (*r*adiation *a*bsorbed *d*ose) in older units. When beta-particles are emitted they are accompanied by an anti-neutrino (which is not absorbed) and hence the energies of beta-particles are spread over a wide range. The average energy of the beta-particle is about one-third of the maximum value quoted. Alpha emission has a sharply defined energy, and the energy value quoted is used directly. Note that the gray is a strict physical conversion of radiation energy to energy per kilogram of body mass.

We now have a measure of our source in terms of a rate of production of particles, and a measure of biological damage in terms of energy deposited. The difficulty is to connect these two.

For the same amount of energy deposited, alpha-particles do approximately 20 times as much damage as beta-particles or gamma-rays, mainly because they are more effective at forming ions in the body. The designation *ionising* radiation comes from this property of these three emissions. Don't think from the poem that alpha-rays are harmless because they are easily stopped. If you ingest or breathe in alpha emitters, the tissues stopping the alpha rays are severely damaged in the process.

The proper unit with which to assess the effects of radiation on health is the *dose equivalent*, measured in a unit called the sievert, corresponding to 100 rems (*r*öntgen *e*quivalent *m*an (rem) in older units). Its units are also joules/kg. The amount of energy absorbed (Gy) and the dose

Energy deposit
Consider the following. If I punch your nose, I am transmitting energy from my fist to your face. A hard punch can draw blood. The amount depends on how often I repeat the process. A soft blow has little effect, no matter how often it is repeated. In fact, a whole series of soft blows may far exceed in energy the energy of a single hard blow, but it is not absorbed in a manner that causes damage. As we saw, the same argument holds for the UV versus visible radiation in sunlight.

equivalent (Sv) are connected by a quality factor, which, for example, is 20 when comparing alpha- with beta- and gamma-rays.

The average energy deposited by the radiation in joules/kg or grays is multiplied by the quality factor (see Table 16.1) to give the dose in sieverts.

TABLE 16.1
The quality factor in radiation dose

Type of radiation	Quality factor
X- and gamma-rays and beta-particles	1.0
Thermal neutrons	3.0
Fast neutrons or protons	10
Alpha-particles or ions heavier than $^{4}He^{2+}$	20

In summary, disintegrations/s (Bq) and average energy absorbed per disintegration, measured in millions of electronvolts (MeV), give energy deposited per kg of body mass (Gy) which when modified by a quality factor gives dose equivalent (Sv).

Later we will see that because different organs have a different sensitivity to radiation, another weighting factor, W_t, is used to accommodate this and give an *effective* dose equivalent, also measured in Sv. No wonder it is hard to see what facts mean in this field!

Internal exposure to radiation

Body potassium In the body, natural potassium undergoes about 4000 radioactive disintegrations per second.

A naturally occurring radioactive element in the body, potassium-40 (^{40}K) has a half-life of over one billion (10^9) years. The average human (70 g) contains about 140 g of potassium. Since the isotopic abundance of potassium-40 is 0.0117%, there is 16.4 mg, or 2.47×10^{20} atoms of potassium-40 in the average body. From the number of atoms N, and the half-life $t_{½}$ of 1.28×10^9 years, we can calculate the number of disintegrations per second:

$$\frac{dN}{dt} = kN = \left(\frac{0.6931}{t_{½}}\right) N.$$

This yields 4250 Bq, that is, 4250 radioactive disintegrations per second in the whole body (i.e. all 70 kg of it).

The average energy of the beta radiation from potassium-40 is 0.548 MeV (max. 1.35) and coupling this to the activity per kg body mass, we obtain a dose of 170 microsieverts (μSv) per year. The isotope also emits high-energy gamma-rays which contribute further to the dose.

It is interesting to compare potassium-40 to another naturally occurring radioisotope in the body, carbon-14. Assuming 1.6×10^4 g of carbon in the average 70 kg body, an isotopic abundance for carbon-14 of 1 in 10^{12}, and a half-life of 5730 years, we obtain an activity of 3000 Bq, which is roughly the same as for potassium-40. However, because of the energy difference of their beta rays (average 0.0441 MeV, max. 0.156 for carbon-14), the dose is 10 μSv/yr, almost 20 times less. Thus potassium-40 is the dominant contributor to the 200 μSv/yr whole-body exposure of beta and gamma radiation from internal sources.

The beta-particle (electron) emitted by the potassium-40 has a large energy. It travels at less than the speed of light in a vacuum, but its speed in a medium such as water would be greater than the speed of light in water. If KCl is dissolved in water, the electron is slowed down instantly and this leads to a shock wave (like that from a supersonic plane) and the emission of bluish-white light, called Cerenkov radiation. The light is far too weak even in a saturated solution of KCl to see with the naked eye, but it can be measured by very sensitive light detectors. (The author once used a saturated KCl solution to provide an absolute calibration of sensitivity for counting single photons in a new instrument developed as part of a research project in the late 1970s). Cerenkov radiation is seen as the eerie glow that surrounds water-cooled nuclear reactor fuel elements.

Cerenkov radiation
With a scintillation counter, this effect can be used to do an experiment in radioactivity without using special radioactive materials, such as to determine the percentage of potassium in dietary salt substitute, where KCl replaces some or all of the NaCl.

External exposure to radiation

Now let us consider some external sources of radioactivity, and study the radioactive decay series, namely that from uranium-238 (^{238}U). (The other external sources are from thorium-232 and actinium-235.) A simplified radioactive decay series for uranium-233 is shown above Figure 16.3 (overleaf).

The half-lives are in yr (years), d (days), m (minutes), and s (seconds). Eight elements in the series decay with the loss of an alpha-particle (a helium nucleus) and their atomic number goes down by two, while their atomic mass goes down by four. The other mode of decay is emission of a beta-particle (an electron from the nucleus), which converts a neutron into a proton. This does not alter the mass appreciably but raises the atomic number by one. This series can be shown on a graph of atomic number versus atomic mass (Fig. 16.3, overleaf).

External exposure from rocks and soil comes from potassium-40 (clay), uranium, thorium, and their descendants (called daughter products).

Cosmic rays contribute about 300 μSv per year at sea level. Living at 3000 m trebles the dose because there is less atmosphere to absorb the rays. Short periods at high elevation, such as 10 trans-Atlantic plane trips, give an exposure equivalent to that in a whole year at sea level.

Rocks contribute radon gas, and its descendants, which are usually considered as internal exposure sources.

Radioactivity in food

Radioactivity in food comes mainly from the alpha emitter polonium-210. Other contributions are from the beta emission from lead-210 and bismuth-210. Relatively high concentrations of polonium are found in seafoods, such as in the muscles of fish and in molluscs.

Because reindeer and caribou eat lichen, and these accumulate lead-210 and polonium-210, the peoples of the sub-Arctic that live off these animals take in 140 Bq and 1400 Bq per year respectively from these

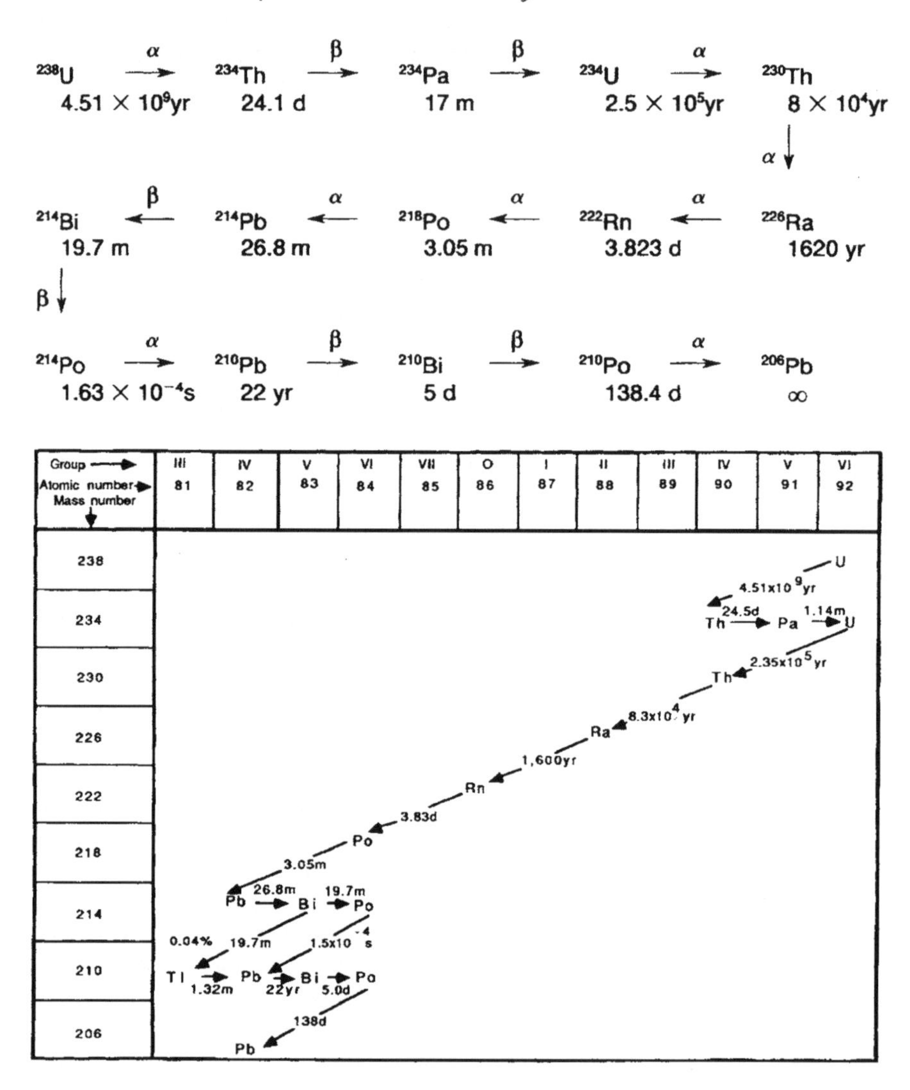

Fig. 16.3 Uranium-radium series (mass numbers 4n + 2)

isotopes, compared to a normal 40. Whereas most alpha emitters such as lead accumulate in the bone minerals, polonium is the exception, and distributes in the soft tissues after intake, settling in the bones after decay to lead. For the average citizen, the yearly radiation intake from food amounts to about 370 μSv per year.

Let us carry out a typical calculation on polonium levels for non-smokers (smokers receive extra doses in their lungs) in areas of normal dietary intake. Because different organs in the body have different sensitivities and masses, the dose they receive is weighted with a factor W_t to compute an *effective dose equivalent* for each organ. This weighting factor is the first entry in Table 16.2. The next entry gives the average activity of polonium-210 found in each organ. Then comes the absorbed dose, which is the energy deposited by the radiation. The final column is the effective dose equivalent. This is the absorbed dose in grays, converted to the effective dose in sieverts by a factor of $Q = 20$ because polonium is an alpha emitter. Then this is multiplied by the weighting factor W_t for each organ given in column 1, to give the *effective dose equivalent*, also in sieverts in the last column. That is the figure that is usually quoted.

TABLE 16.2
Polonium levels in non-smokers

Tissue	W_t	^{210}Po Bq/kg	Absorbed dose μGy/yr	Effective dose equiv. μSv
Gonads	0.25	0.2	5.4	27
Breast	0.15	0.2	5.4	16
Red bone marrow	0.12	0.11	5.1	12
Lung	0.12	0.1	2.7	6
Thyroid	0.03	0.2	5.4	3
Bone lining	0.06	—	36	43
Remainder	0.24	0.2	5.4	26
Σ (sum)	1.00			133 μSv

Thus the dose from polonium-210, (130 μSv/yr), is about the same as we calculated for potassium-40 (170 μSv/yr).

When exposures are given for medical X-rays of a particular part of the body, it is the *effective dose equivalent* to the whole body that is quoted. The actual dose of the X-ray is higher by the inverse of the weighting factors. A single chest X-ray deposits an *effective dose equivalent* of about 100 μSv. Among the higher exposures, X-rays of the small intestine, for example, vary from 3000 μSv for radiography to 8000 μSv for fluoroscopy per X-ray.

For medical reasons, X-ray exposure is thus characterised by high dose rates and uneven distribution of dose in the population. They are by far the largest contribution from non-natural sources. However, when averaged over the whole world population, medical exposure contributes an effective dose equivalent of 400 μSv/yr.

This highlights by how much individual exposure can vary. It can be very high for a citizen in the developed world having easy access to diagnostic X-rays and very low for a citizen in the Third World far from medical services. The low figure for the average 'dilutes' the high figure for the individual.

The same argument applies to an accident such as at Chernobyl in the Ukraine, which for the following 12 months contributed only 30 μSv per person to the average world exposure, about the same as one person would obtain from a trans-Atlantic flight. However, for people downwind of the accident, or in the Third World eating cheap (contaminated) European food exported at the time, the situation is dangerous.

TABLE 16.3
Personal annual summary of radiation dosage (US average data)

Source of radiation	Quantity per year (μSv)
1. In the body from ^{40}K (abundance 0.01%)	200
2. Location of your town or city	
(a) Cosmic radiation at sea level	300
(b) Additional mrem based on elevation above sea-level	
1000 m	100
2000 m	300
3000 m	900
3. House construction	
brick	750
wood	400
concrete	850
4. ground radiation from rocks	150
5. Food, water and air	250
6. Fallout from nuclear weapons testing	40
7. Medical X-rays	
(a) Chest X-ray	(100 per visit)
(b) gastrointestinal tract	(2000 per visit)
(c) Dental	(100 per visit)
8. Jet travel five hours at 9000 m,	30 per flight

(After Chemistry in the Community (ChemComm), Am. Chem. Soc. 1988) with permission.

Nuclear explosions in the atmosphere diminished in intensity from the mid-1950s and ceased altogether in the 1960s, but the fall-out effects continue to expose the world population. It is estimated that the exposure resulting from all these tests until 1980 is equivalent (for the present world population) to being exposed to the natural background for an additional four years. Underground testing continued, but contributed far less to world exposure and this has also ceased, with a final imperial fling in the Pacific Ocean from France in 1996. The most significant exposure is through carbon-14, strontium-90 and caesium-137. The total committed dose varies from 3100 μSv in the southern hemisphere to 4500 μSv in the northern hemisphere.

See also Plate XXIV showing sources of radiation in the UK.

Risk from occupational exposure

For radiation workers, the whole-body exposure limit is set at 50 000 μSv/yr, while for the citizen it is set so that the exposure over a 70-year lifetime does not exceed, on average, 1000 μSv/yr. Thus for the citizen, this additional exposure from industrial sources is equal to about one-half of what is received from natural sources (1000 μSv from radon descendants, and 1000 μSv, other) not including medical exposure.

It should be noted that the International Commission on Radiological Protection (ICRP) has insisted since 1977 that these levels should no longer be regarded as the boundary of acceptable practice, but rather the point at which practices become distinctly unacceptable. In 1977 the emphasis shifted to the so-called 'alara' principle, (*a*s *l*ow *a*s *r*easonably *a*chievable, economic and social factors being taken into account).

Radiation effects are called *somatic* if they become manifest in the exposed individuals themselves, and *hereditary* (or genetic) if they affect the descendants of the exposed people. These effects are now categorised in two ways.

1. *Non-stochastic:* where the severity of the effect increases with the dose — for example, skin reddening, cataract of the lenses of the eyes, cell depletion in bone marrow, and impaired fertility of the gonad cells.
2. *Stochastic:* where the *probability* of the effect, rather than its severity, is proportional to the dose. This generally refers to the induction of later cancer or genetic damage.

Crudely, the difference is determined by whether cells are killed and therefore have no long-term influence, or whether they are only damaged, in which case they have a latent ability to cause harm later on. Thus, ironically, the very heavily exposed population of Hiroshima and Nagasaki often showed less than expected cancers because of the high death rate in their cells.

The stochastic effect is of the same type as, say, between cigarette smoking and lung cancer, or the various risk factors associated with heart disease. You do not see a direct cause and effect relation in each individual, but a statistical correlation in an exposed population.

The international occupational level is based on an assumed acceptable level of fatalities in comparable industries, namely *100 fatalities per year per million workers.* This 'magical' figure is heavily quoted by occupational health and safety authorities but it is hard to find a really quantitative basis for it (see also Chapter 2, The anthropology of regulatory culture — risk assessment).

Radon

On the basis of existing measurements, and an average of one change of air per hour in buildings, the UK National Radiological Protection Board has estimated that the average annual dose of radon received in Britain is equivalent to a dose of nearly 1000 μSv/yr, about half the annual exposure of members of the public from natural background sources.

Radon has a half-life of 3.8 days and decays to polonium-218, another alpha emitter with a half-life of three minutes, to give lead-214, with a half-life of 27 minutes. Two beta emissions and we are at polonium-214, another short-lived alpha emitter, which produces lead-210, a beta emitter of 22 years half-life. (Incidentally, this is one of the few isotopes that cannot be made artificially.) These solids attach to aerosols that can

Radon in UK limestone caves
Limestone caves can also have very high radon levels; not a risk for the occasional visitor but a problem for those involved in sport or work in caves. In the UK, the mean levels in 47 caves ranged from 454 to 8868 Bq/m^3. In one system in the Peak district of Yorkshire a level of 155 000 Bq/m^3 was recorded.

Radon deaths
In 1992 the US EPA estimated that the number of deaths from radon in the US was somewhere between 7000 and 30 000.

be trapped in the lungs and deliver their dose there. Many alpha emitters also concentrate in bone. The uranium-238 series contributes about 1000 μSv (including 800 μSv from ^{222}Ra and its descendants), and the thorium 232 series, 330 μSv (including 170 μSv from ^{220}Ra and its descendants) per year.

Problems occur if a granite used in home building has been crushed to a gravel because the fracturing allows an easier escape for radon. In the case of housing built on rock fill, the deciding factor is the depth and permeability of the soil. Normally a metre or so of soil is sufficient to slow down the gas diffusion so that the radon has decayed to solid descendants (which are trapped) before escaping. Radon in the house wall-construction materials is generally effectively trapped.

The usual problem, particularly in cold climates, is that the urge to conserve heating inevitably involves restricting ventilation. This leads to a build-up of radon and its descendants. Removal of uranium ore in mining increases the radioactivity in the local environment because of the break-up of the radioactive ores and release of radon. Diffusion of radon from the ground decreases when the ground is wet as the surface becomes sealed. Radon is quite soluble in water.

The major entry of radon into buildings is through the underground basements. It has been suggested in the USA that the levels in homes be recorded with other information on the title deeds. The risk level (7000–30 000) compares to 23 000 fatalities from drink driving, 4400 from fires and 1000 from aeroplane crashes.

Sweden has a big problem because of the high percentage of uranium in rocks and soils. A 1994 survey concluded that 'residential exposure to radon is an important cause of lung cancer in the general population'. A survey of 300 000 homes found that 25 000 needed radon proofing and the government provides half the cost of remediation if the level is over 400 Bq/m^3.

Hong Kong has a major radon problem in its high-rise buildings (most unusual because generally only low floors are affected) with mean levels of about 178 Bq/m^3. The reason is that the building materials have some of the highest levels of radium-226 in the world. It turns out that the best sealant is not paint or plaster but wallpaper.

By 1996, 250 000 homes in the UK had been tested. Of these 100 000 had levels above 200 Bq/m^3 and were regarded as 'affected'. In regions where the number of affected houses exceeds 10%, radon-proof barriers and extraction units are required in new houses. For regions with greater than 1% affected, extraction only is required. There are no regulations for established houses. The British Geological Survey has published maps which include radon release potential as well as actual levels, including levels in the soils and underlying geology, such as uranium ore bodies and the porosity of the rock.

Risk factors for radon are derived from epidemiological studies of uranium miners who were heavily exposed and who contracted lung cancer, as well as modelling studies. Very high urban regions, such as the village of Umhausen in the Austrian Tyrol, have provided statistically significant correlations with lung cancer incidence, despite the confounding effect of smoking. Houses on a particular granite rock slide have an annual mean level on the ground floor of 1868 Bq/m^3 (max. 274 000 Bq/m^3).

Consensus appears to be emerging that for radon, 20 Bq/m^3 = 1000 μSv dose for adults; 1200 μSv for 10-year-olds; and 2000 μSv for the newborn. At 200 Bq/m^3 = 10 000 μSv, householders in most countries are advised to contemplate corrective action (such as installing good ventilation, venting of basements to the outside air, etc.) whereas in workplaces the action level is 400 Bq/m^3.

Australia has completed a year-long survey of 3864 homes volunteered from 12 000 chosen randomly from the electoral roll in which radon monitors have been set up.

The apparatus to collect the radon samples over 12 months used a CR-39 polyester resin in a solid state track detector (SSTD) that is later etched, and the tracks caused by the radiation were measured under an optical microscope. (See Fig. 16.4, overleaf.)

The preliminary results reported in the 4th edition of this book, showed that Jabiru, a town near the Ranger uranium mine in the Kakadu National Park in the Northern Territory, and Armidale, an inland NSW town in an area of granitic rock containing uranium and thorium at higher than average levels, had high readings. The highest level measured in a home was 395 Bq/m^3.

The final results were shown to follow a log-normal distribution (see Appendix II) and are shown in Table 16.4.

The average radon concentration in Australian homes over a year is 11 Bq/m^3. This leads to an estimated annual average effective dose equivalent of 500 μSv. Another 900 μSv comes from exposure to gamma-rays. The survey suggests that about 2000 to 3000 homes in Australia would be expected to exceed the action level of 200 Bq/m^3.

Beach sands
Concentrates of beach sands set a bigger risk from radioactivity than uranium mining.

Radiation levels are higher in solid brick than brick veneer construction, and in homes with poor ventilation or that rely on air-conditioning.

While one might have thought that mining and processing of uranium ores constituted the major industrial radiation hazard in Australia, it appears that beneficiation of beach sand for the separation of rare earths

TABLE 16.4
Average radon concentration for Australian homes

State	Sample size	Radon concentration Bq/m^3 Geom. mean	Geom. s.d
ACT	248	13.3	1.7
NSW	907	6.6	2.1
NT	196	10.2	1.9
QLD	404	5.1	1.9
SA	331	10.9	2.0
Tas	285	7.1	2.1
VIC	731	7.9	2.0
WA	311	10.1	1.8

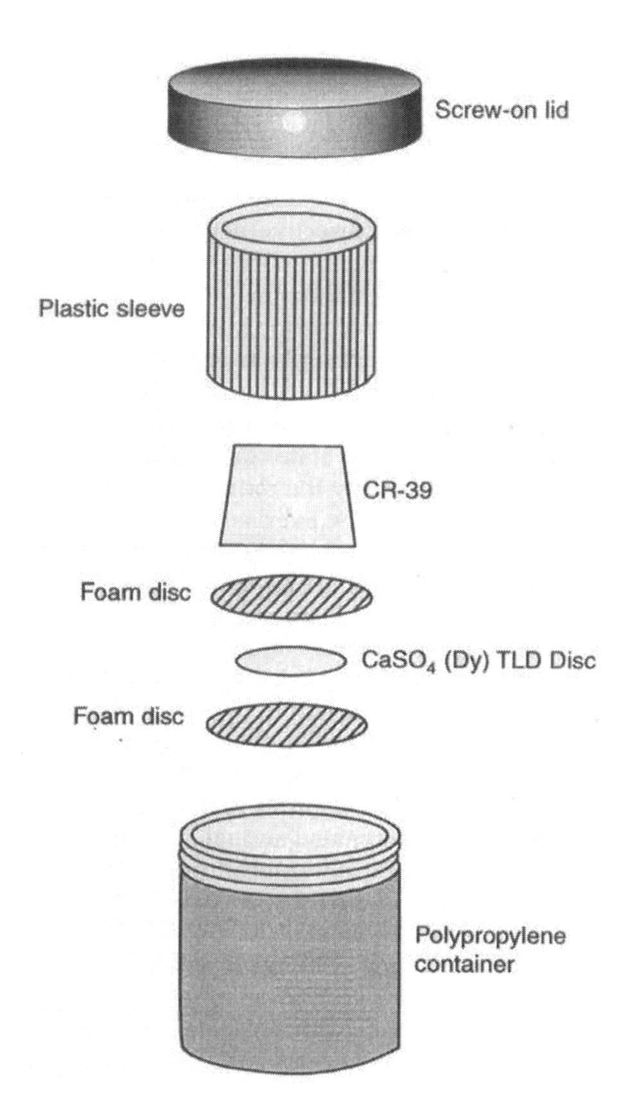

Fig. 16.4
Radon dosimeter

(From A Nationwide survey of radon-222 and gamma-radiation levels in Australian homes, M. K. Langroo, K.N. Wise, J.C. Duggleby and L.H. Kotler, Health Physics **61** (6), 1991, pp. 753–761.)

is the real problem area. The decay of thorium-232 provides a series of hard-gamma-ray emitters with high energies. As these operations tend to be dusty, quite high levels of exposure are probable. The thorium decay series is given below in simplified form and in Figure 16.5.

Radiation-emitting consumer products

The energy emitted during the radioactive decay of radium-226, promethium-147, and tritium (^3H) can be converted into light by use of a scintillator. The application is to luminous instrument dials. Cancer of the lips was common among women painting dials for military instruments and watches during World War Two, because of their habit of sharpening the paint brush between their lips. There are also long-term dangers

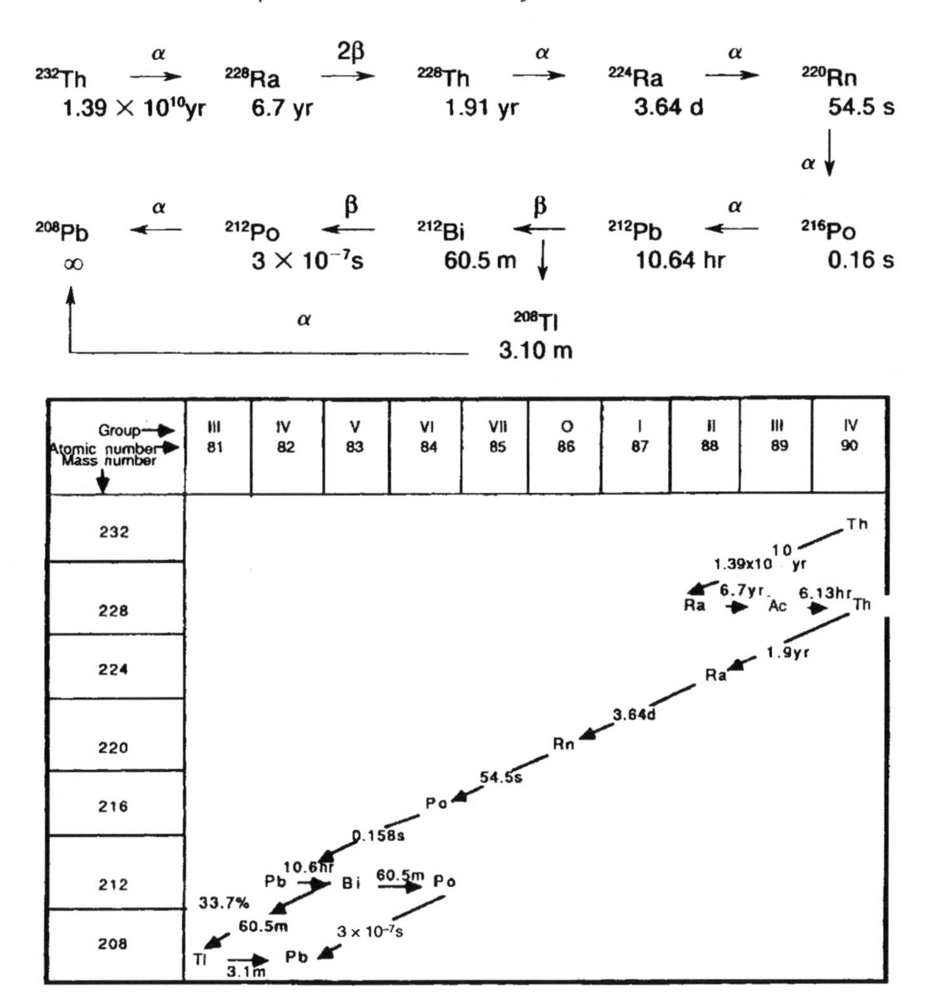

Fig. 16.5 Thorium series (mass numbers 4n)

to users. Today tritium is used, almost exclusively, such as in gaseous tritium light sources for liquid crystal displays.

Television screens once contributed about 5 μSv from soft X-rays, but modern colour sets under normal operation and servicing have negligible emission.

Starters for fluorescent lamps, and trigger tubes in electrical appliances, incorporate radioisotopes for faster and more reliable operation. These are krypton-85, promethium-147, and thorium-232. The level of activity is appreciable, but not a hazard, except in the case of breakages, or careless disposal.

In some countries, anti-static devices use polonium-210 to ionise the air. The only significant hazard is from the very small gamma component emitted. Smoke detectors usually contain americium-241 with a half-life of about 450 years. These devices have a useful life of about 10 years, and their safe disposal is important.

Uranium is used as a pigment in ceramics and glassware, while thorium is used in the incandescent mantles of most kerosene lamps and in some optical products. The beta emission from the decay products can be hazardous under special circumstances, for example from optical lenses containing high levels of thorium.

Uranium added to the porcelain used in dentistry to simulate the natural fluorescence of teeth could give an undesirably high dose to the gums.

Art forgery detection

In 1937 a Dr Abraham Bredius, a noted art expert, announced that he had discovered *the* masterpiece of Johannes Vermeer (1632–75), *Christ and His disciples at Emmaus.* Vermeer was perhaps one of the finest of the seventeenth century Dutch genre painters, whose work was wrongly attributed until 1866. Because there were never more than 36 authentic Vermeers catalogued, they are highly prized. The owner of the painting, a certain Hans van Meegeren, netted $280 520 from Boyman's Museum in the Netherlands for a painting he had produced himself. It was so well done that the museum even had to restore it before the official unveiling.

A gifted artist in his own right, but with little recognition because of his old fashioned style, van Meegeren accepted that consumers bought on brand name, and that art connoisseurs were no exception. He continued to produce what the public wanted by forgery and amassed the tidy sum of three million dollars. The buck(s) stopped in 1945 when the art collection of Hermann Goering was discovered in a salt mine at Alt-Ausee in Austria. Among the treasures was a Vermeer, *Woman taken in adultery.*

The Dutch rapidly traced the sale to van Meegeren, who was tried as a Nazi collaborator. This carried the death penalty at the time. He changed his tune rapidly and pleaded that his Vermeers were forged, and that therefore he had thwarted the Nazis' art acquisition program by selling them his own works. However, the experts disputed this (there were after all considerable investments threatened, not to mention reputations) and the art world closed ranks and rejected van Meegeren's claims as a reckless attempt to save his neck.

The chemical examinations at the time were ambiguous and a desperate van Meegeren set out to prove his skills by painting, under police supervision, another 'Vermeer', *Jesus amongst the doctors.* This allowed him to escape execution, but many did not accept that the *Emmaus,* among other works, was forged. The controversy continued after his death and was not finally resolved until 1968 when radiochemical analysis of the lead paint was undertaken.

The manufacture of lead pigment involves the roasting of galena, a lead sulfide ore. The ore always contains ^{238}U together with its descendants, all in equilibrium. (It is not strictly an equilibrium situation, although that term is used. It is a steady state.) In this situation, as fast as isotopes in the chain are emitting particles and converting to daughter products, their concentrations are being replenished from further up the chain. A good analogy is to consider a series of dams of different sizes (amounts of isotopes). The time taken to half-empty an isolated dam is constant and fixed. This is its characteristic half-life. It can vary from a trickle to a flood, depending on the nature of the dam (milliseconds to millennia). However, the amount of water leaving an isolated dam at any time is not constant, but is proportional to the amount of water currently in that dam. As the amount in the dam drops, so does the outflow, giving the characteristic falling exponential decay.

Because our dams are linked, the outflowing water from one dam flows into the dam(s) below, giving a connected stream of water flow. Some dams will slowly fill, and others empty, until a steady state is reached.

Now look at the radioactive series for ^{238}U again, at the point at which ^{226}Ra occurs. It has a half-life of decay of 1600 years, and its daughter ^{210}Pb has a half-life of only 22 years. Radium and lead behave quite differently in the refining of lead ore. While ^{210}Pb is carried through with the normal lead, most of the ^{226}Ra stays in the waste slag products. The ^{210}Pb, now without most of its parent, decays with the 22-year half-life, and reaches a low level when it re-equilibrates with the remaining ^{226}Ra after about 200 years. The amount of ^{226}Ra is measured by its alpha emission (4.78 MeV) while the amount of ^{210}Pb is measured through the alpha emission of its descendant (^{210}Po) because the beta emission of ^{210}Pb is too weak. The count rates for ^{210}Po and ^{226}Ra in the *Emmaus* painting were found to be 142 and 13.3 Bq/kg lead respectively. The radium level is low, indicating proper purification of the lead had been carried out. However to explain the current polonium level, the level back in the mid-seventeenth century (about 13 half-lives back) would have had to have been $142 \times 2^{13} = 1.2 \times 10^6$ Bq, which is unrealistically high. Using our analogy, if the upstream dam (radium) had really been emptied in the seventeenth century, then there is just too much water today in the (lead/polonium) dam below.

There are many other fascinating chemical aspects of art examination and you should consult some of the references for further reading.

Food irradiation

A method for killing parasites in meat with X-rays appeared in a patent more than 60 years ago. In the 1950s, Eisenhower's 'Atoms for Peace' initiative experimented with irradiated food for combat and space rations, but could not overcome the problems of changing taste. About 20 countries have approved irradiation for about 30 foods. The core of an irradiation unit is either a radioactive source, usually cobalt-60 (gamma-rays of energy 1.17 and 1.33 MeV) or caesium-137 (gamma-rays of energy 0.66 MeV), or an electron beam.

Radiation breaks up molecules, and big molecules such as the DNA heredity material are more susceptible. Micro-organisms are destroyed in this way. Plant development and ripening can also be inhibited. The physical process involves the gamma-ray hitting a molecule, which ejects an electron and ionises. Each ejected electron may ionise further molecules. As electrons are slowed down, they emit their excess kinetic energy as weak X-rays called *Bremsstrahlung*, which can cause nuclear transformations. The maximum energy levels of irradiation are set to prevent food becoming radioactive. The susceptible atoms are the common carbon-12 and oxygen-16.

In the UK, limits have been set to ensure that these atoms are not made radioactive and these levels are as follows

Max. overall dose	10 kilograys
Max. energy (gamma- or X-rays)	5 MeV
Max. energy (electrons)	10 MeV

The *dose* is measured in a unit (now) called the gray and this gives the amount of energy that the radiation deposits in the irradiated material. The other limits set how energetic (million electronvolts) the different types of radiation are allowed to be. The threshold for making carbon radioactive is 18.6 MeV, and for oxygen it is 15 MeV.

While a dose of 5 grays can kill a person, a dose of 1 kilogray is needed to reduce the population of most microbes to 10% of their initial value, the D_{10} value. *Clostridium botulinum* (an organism responsible for botulism) is not very susceptible and it can withstand up to 50 kGy. Irradiation can thus kill its competitors and leave it better off to multiply.

Irradiation does affect food components; a change in taste in (boiled) rice irradiated even at a level of 1 kGy can be detected by Australian consumers, while the more discerning Japanese can detect 0.5 kGy. There are some rather subtle aspects of the effects of irradiation. The greening of potatoes on exposure to light is a significant problem; losses of up to half a million dollars per annum are reported in New South Wales alone. The greening due to chlorophyll formation is in itself quite harmless, but the greening is accompanied by the simultaneous synthesis of the poisonous substance, solanine, which may be dangerous for pregnant women (see box 'Avoid bruised or green potatoes' in Chapter 15). Solanine is not inhibited by irradiation.

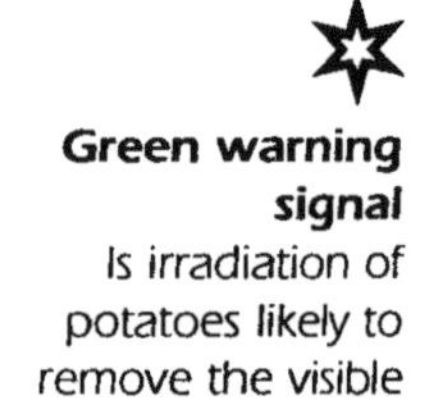

Green warning signal
Is irradiation of potatoes likely to remove the visible green warning signal of a toxic presence?

The large capital cost of irradiation facilities limits their application to a few fixed localities.

Irradiation is successful for such things as sterilising pharmaceuticals and for bulk sterilisation of manure for mushroom media. Its future in the food area is problematical.

References

Radioactivity: Inside science, C. Sutton, *New Scientist,* 11 February 1988.
The food irridation section is based on S. Sonsino, *Radiation meets the public's taste, New Scientist,* 19 February 1987; and S.C. Morris, *The practical and*

economic benefits of ionising radiation for the post-harvest treatment of fruit and vegetables: An evaluation, Food Tech. Aust. **39**(7), 1987, p. 336.

Natural sources of ionising radiation, G.M. Bodner, and T.A.J. Rhea, *Chem. Ed.* **61**(8), 1984, p. 687.

Silent but deadly, P. Phillips, T. Denman and S. Barker, *Chem. Brit.*, Jan 1997, pp. 35–38.

ICRP — history and developments, K.H. Lokan, Australian Radiation Laboratory, Melbourne, Victoria, 1988.

A nationwide survey of ^{222}Rn and γ-radiation levels in Australian homes, M.K. Langroo, K.N. Wise, J.C. Duggleby and L.H. Kotler, *Health Physics* **61**(6), 1991, pp. 753–761.

Bibliography

Food irradiation: Who wants it?, T. Webb, T. Lang and K Tucker, Thorsons Publ. Rochester, 1987.

Authenticity in art: The scientific detection of forgery. S.J. Flemming, The Institute of Physics, London 1975.

Appendixes

Nomenclature in chemistry

Simple organic chemistry

As its name implies, organic chemistry originally dealt with chemical substances produced by living organisms. The key element in all organic molecules is carbon, which forms many more compounds than any other element (with the possible exception of hydrogen). In its stable compounds, carbon always has a valency of four, and when it is bonded to four other atoms, they are arranged tetrahedrally about the carbon atom (see Plate XXV). Carbon also has the property of *catenation* (chain forming); that is, many carbon atoms can bond together to form chains or rings. The simplest group of these chain molecules are the *paraffins,* or *alkanes,* which are found in natural gas and petroleum. The first members of the series are methane, CH_4; ethane, C_2H_6; propane, C_3H_8; and butane, C_4H_{10} (Fig. A1.1). Propane and butane are the main constituents of bottled fuel gas, whereas petrol contains principally the members with seven (heptane) or eight (octane) carbon atoms. The lubricating oils have much longer chains.

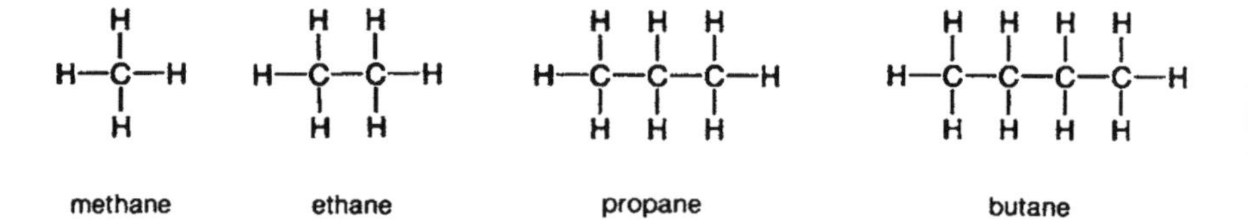

Fig. A1.1 Formulae for some paraffins

In all of the paraffin compounds, the carbon atom shows its constant valency of four and the atoms bonded to it are arranged tetrahedrally. However, it is possible for two or three of the four valencies to be used up in forming a *double* or a *triple* covalent bond with another atom. Carbon dioxide, O=C=O, and hydrogen cyanide, HC≡N, are familiar examples of double and triple bonds.

There is a series of compounds, called the *olefins or alkenes,* which are closely related to the alkane series, but each member has two less hydrogen atoms than the corresponding alkane, which means that each has one double bond. The olefins are the simplest unsaturated compounds.

Figure A1.2 gives the structural formulae for ethene (ethylene), C_2H_4 (compare ethane), the basic molecule from which polythene (polyethylene) is made, and for propene (propylene), C_3H_6, which is used in the plastic polypropylene. Note that the ending *-ene* is indicative of the alk*enes*, as *-ane* is of the alk*anes*.

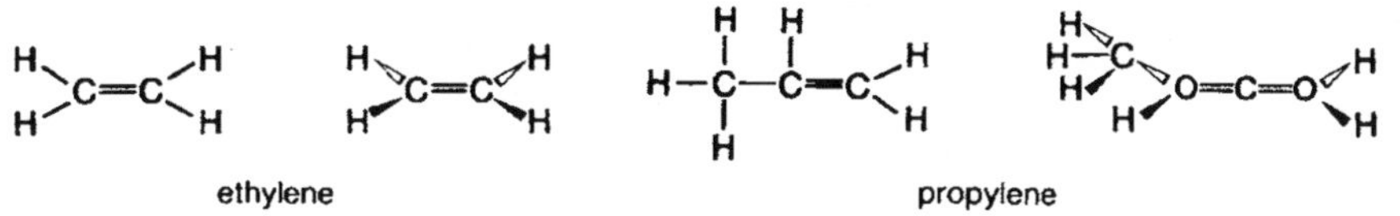

ethylene propylene

Fig. A1.2 Structural formulae for ethylene and propylene

Another type of unsaturated hydrocarbon is butadiene, C_4H_6, which contains two double bonds (Fig. A1.3). It is the basic constituent of the earliest synthetic rubbers.

Fig. A1.3 Butadiene

The simplest member of the triple-bonded series, the *alkynes*, is acetylene, HC≡CH, the gas used in welding. Such carbon compounds containing *multiple* bonds are said to be *unsaturated* because it is possible to add more atoms to their molecules, thereby *saturating* them. For example, ethylene can be converted simply to ethane by the addition of two hydrogen atoms (Fig. A1.4). We discuss polyunsaturated margarine in Chapter 4.

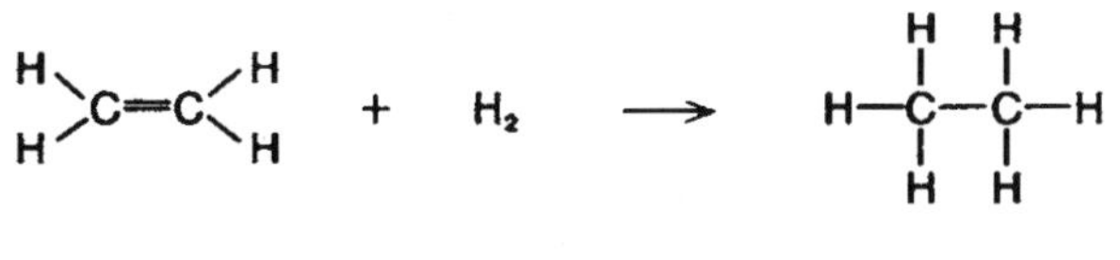

Fig. A1.4 Hydrogenation of ethylene

Another important group of carbon compounds are those containing ring molecules, such as benzene, toluene, and naphthalene (Fig. A1.5).

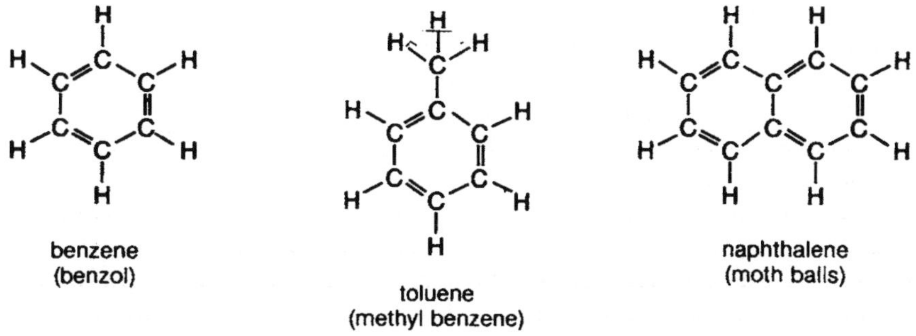

benzene (benzol) toluene (methyl benzene) naphthalene (moth balls)

A1.5 Some aromatic hydrocarbons

One small step
One small step for a chemical; a big effect on humankind.

The alternating double and single bonds are not a true description, because the electrons are smeared around the ring and this means these molecules do not behave as if they had real double bonds.

So far we have considered only organic molecules containing carbon and hydrogen — the hydrocarbons — but it is possible, by replacing some of the hydrogen atoms with other atoms, to make whole new series of compounds, with different properties, depending upon the nature of the *substituting* atoms. It is important to note at this stage that apparently minor changes to a molecule can produce major changes in physical, chemical and physiological properties. Thus, from methane, CH_4, we can prepare CH_3Cl (methyl chloride), CH_2Cl_2 (dichloromethane), used in paint strippers, $CHCl_3$ (chloroform), the first anaesthetic other than alcohol, and CCl_4 (carbon tetrachloride), once widely used in dry cleaning until its toxic properties were admitted (see Table A1.1).

TABLE A1.1
Chlorinated hydrocarbons

Methane	Methyl chloride Chloromethane	Methylene chloride Dichloromethane	Chloroform Trichloromethane	Carbon tetrachloride Tetrachloromethane
CH_4 gas, b.p. –262°C Fuel (natural gas)	CH_3Cl gas, b.p. –24°C	CH_2Cl_2 liq., b.p. 40°C Paint stripper	$CHCl_3$ liq., b.p. 61°C First anaesthetic	CCl_4 liq., b.p. 77°C Former dry-cleaning fluid; former (electrical) fire extinguisher
non-toxic	toxic	relatively non-toxic	damages liver	more toxic than $CHCl_3$

Some other examples of property changes resulting from small changes in a molecule can be seen in the different structures of ethane, C_2H_6, ethanol (ethyl alcohol) C_2H_5OH, and ethanoic acid (acetic acid), CH_3COOH; and of benzene, C_6H_6, benzoic acid, C_6H_5COOH, and acetylsalicylic acid, $CH_3COOC_6H_4COOH$ (Fig. A1.6).

Yet another example is given by the successive replacement of the hydrogen atoms of acetic acid with fluorine (Fig. A1.7).

Acetic acid, in vinegar, is relatively non-toxic, whereas fluoroacetic acid is used as the sodium salt in '1080' rabbit poison and is highly toxic to humans. It is also the poisonous constituent in some poisonous plants (see box, Murderous organoflourides in Chapter 7). Difluoroacetic acid is not primarily toxic, whereas trifluoroacetic acid forms the basis of some weed killers, such as Dalapon.

Many of the changes in physical and chemical properties accompanying substitutions of this sort can be predicted, but changes in physiological properties are often unexpected. By the same token, certain geometric groupings of atoms within a molecule are known to produce certain physiological effects, and much drug research is devoted to modifying these basic structures in the hope of enhancing their pharmacological activity and, at the same time, reducing undesirable side effects. Organic compounds may be classified according to the nature of the

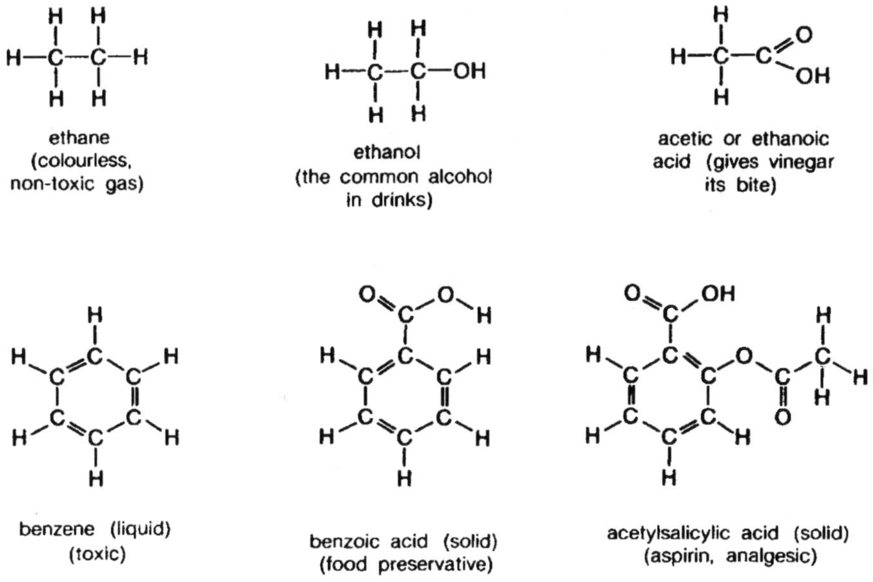

ethane (colourless, non-toxic gas)

ethanol (the common alcohol in drinks)

acetic or ethanoic acid (gives vinegar its bite)

benzene (liquid) (toxic)

benzoic acid (solid) (food preservative)

acetylsalicylic acid (solid) (aspirin, analgesic)

Fig. A1.6
Property changes due to molecular changes

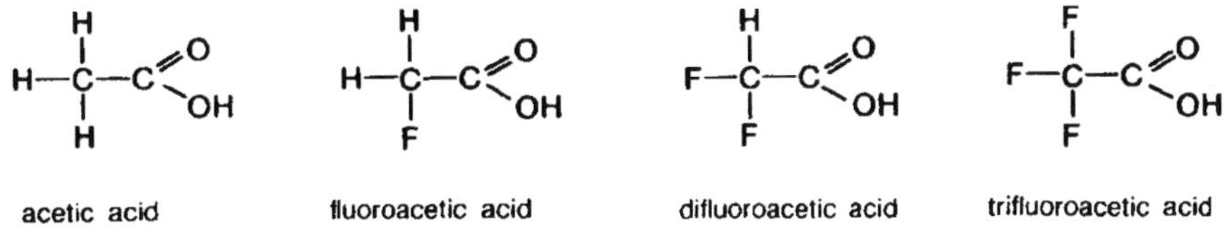

acetic acid

fluoroacetic acid

difluoroacetic acid

trifluoroacetic acid

Fig. A1.7
Replacement of hydrogen atoms of acetic acid with fluorine

functional groups that have replaced hydrogen in the basic molecule. Three important functional groups are alcohols, organic acids and esters.

Alcohols have the general formula R-OH, where R stands for the rest of the hydrocarbon skeleton. The simplest alcohol, derived from methane, CH_4, is methanol, CH_3-OH. This substance, often called 'wood alcohol' because it is obtained by heating wood, is more toxic than the familiar ethanol, CH_3CH_2-OH.

Most *organic acids* contain the characteristic carboxyl group –COOH. Any molecule containing this group is called a *carboxylic acid.* Some examples are illustrated in Figure A1.8 (overleaf).

Esters are produced by the combination of alcohols and acids in an *esterification* reaction (Fig. A1.9, overleaf). Ethyl acetate is an important solvent in the paint and adhesives industries. Many of the natural flavours and smells result from the presence of volatile esters in flowers and fruits, whereas meat fat consists of solid esters, and oils are liquid esters.

Acids and bases

Common acids are sulfuric acid and hydrochloric acid. Caustic soda is a member of a group of compounds called bases.

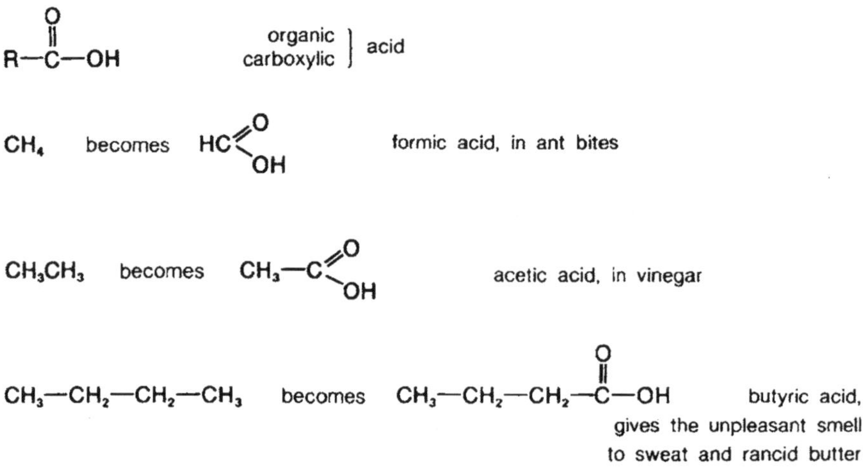

$R{-}\overset{O}{\overset{\|}{C}}{-}OH$ organic carboxylic } acid

CH_4 becomes $HC(=O)OH$ formic acid, in ant bites

CH_3CH_3 becomes $CH_3{-}C(=O)OH$ acetic acid, in vinegar

$CH_3{-}CH_2{-}CH_2{-}CH_3$ becomes $CH_3{-}CH_2{-}CH_2{-}\overset{O}{\overset{\|}{C}}{-}OH$ butyric acid, gives the unpleasant smell to sweat and rancid butter

Fig. A1.8
Organic acids

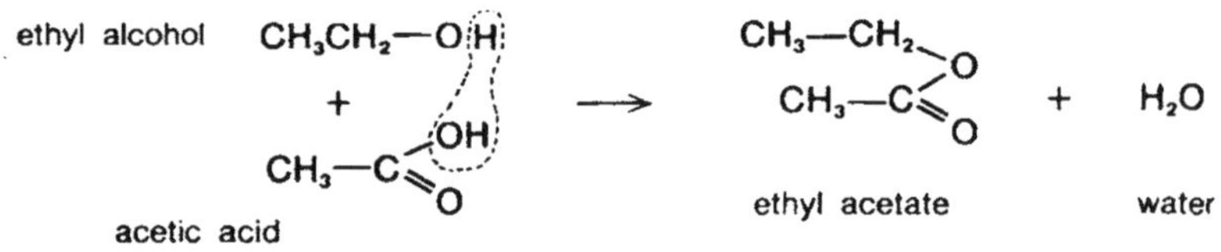

ethyl alcohol $CH_3CH_2{-}O{-}H$ + $CH_3{-}C(=O){-}OH$ acetic acid $\longrightarrow$ $CH_3{-}CH_2{-}O{-}C(=O){-}CH_3$ ethyl acetate + H_2O water

Fig. A1.9
Esterification reaction

Fig. A1.10
The chemical goanna (made by Terry Sedgwick)

If caustic soda is mixed with hydrochloric acid in just the right proportions, the corrosive properties of both solutions are overcome, or *neutralised*. The chemical basis for the concepts of acidity and basicity is rather complex and we must begin with a simplified approach. For our purposes, all substances that we will consider as acids, such as hydrochloric acid, HCl, sulfuric acid, H_2SO_4, nitric acid, HNO_3, and acetic acid, CH_3COOH, are characterised by the presence, in the molecule, of one or more reactive hydrogen atoms that can be displaced by a base in the process called neutralisation. The reactive or *acidic* hydrogen atoms in the molecules shown in Figure A1.11 are marked by an asterisk.

$H^*{-}Cl$	$(H^*O)_2SO_2$	$H^*O{-}NO_2$	$CH_3{-}C(=O){-}OH^*$
hydrochloric	sulfuric	nitric	acetic

Fig. A1.11
Reactive or acidic hydrogen atoms in acid

Note that the methyl hydrogens of acetic acid, the ones attached directly to the carbon atom, are non-acidic. Among the common bases are sodium hydroxide (caustic soda), NaOH; calcium hydroxide (slaked lime), $Ca(OH)_2$; and ammonium hydroxide ('cloudy' ammonia), NH_4OH.

The neutralisation reaction between an acid and a base follows the general pattern:

BI-CARBONATE OF SODA
100g NET
JOHN BULL
CITRIC ACID
ACETIC ACID
HARPER'S
TARTARIC ACID
GUARANTEED PURE

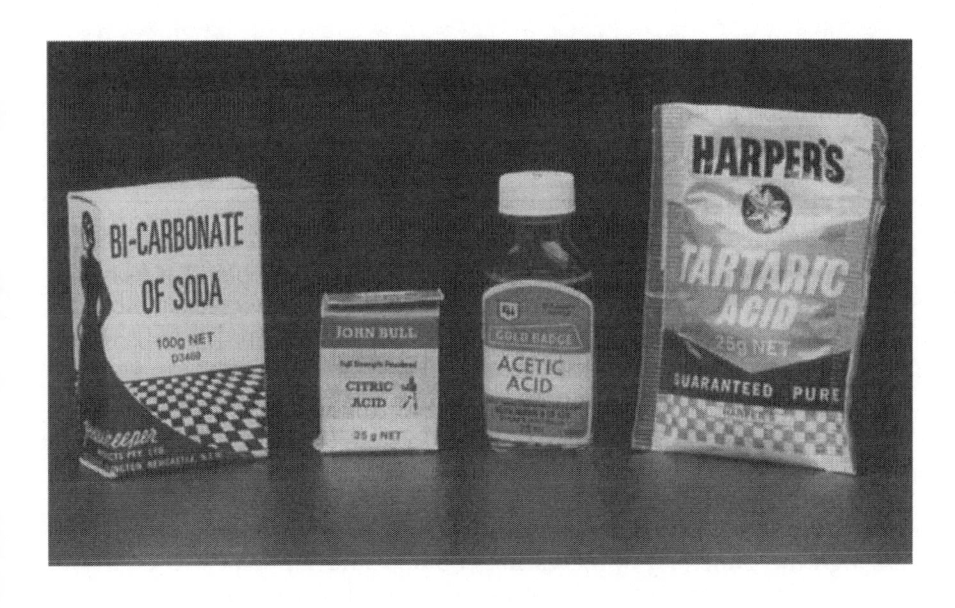

Fig. A1.12
Common acids and bases

$$\text{an acid} + \text{a base} \rightleftharpoons \text{a salt} + \text{water}$$

$$\text{e.g.} \quad \overset{*}{H}\text{–Cl} + Na^{+}OH^{-} \longrightarrow Na^{+}Cl^{-} + H\text{–}OH$$

Fig. A1.13
The neutralisation reaction

which is clearly very closely related to the esterification reaction between organic acids and alcohols already discussed.

Simple acids have one other important property: when dissolved in water, to a greater or lesser extent the reactive hydrogen atom can transfer to the water molecule and thus form an ion. Thus:

$$\overset{*}{H}\text{–Cl} + H_2O \longrightarrow H_3O^{+} + Cl^{-}$$

$$CH_3COO\overset{*}{H} + H_2O \longrightarrow H_3O^{+} + CH_3COO^{-}$$

Fig. A1.14
Ionisation in water

On solution in water then, acids produce the H_3O^+ ion. Strong acids such as HCl transfer the active hydrogen completely; some weak acids, such as CH_3COOH, transfer their hydrogen only partially.

The amount, or rather the concentration, of the species H_3O^+ produced when a given acid is dissolved in water may be expressed on a scale known as the *pH scale* (see Fig. A1.15). Pure water has a pH of 7 (neutral

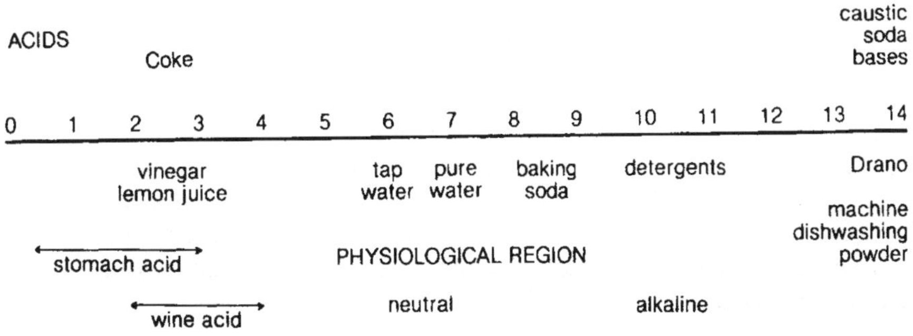

ACIDS
Coke
caustic soda
bases
0 1 2 3 4 5 6 7 8 9 10 11 12 13 14
vinegar
lemon juice
tap water
pure water
baking soda
detergents
Drano
machine dishwashing powder
stomach acid
PHYSIOLOGICAL REGION
wine acid
neutral
alkaline

Fig. A1.15
pH scale

pH). A solution of a strong acid in water at unit concentration (1 mole/litre) has a pH of 0, whereas a solution of a strong base at unit concentration has a pH of 14.

Nomenclature

At school you may have studied some inorganic chemistry, for example the reactivity of halogens, including fluorine. In the world outside the classroom you are more likely to hear about fluoride in drinking water. How can a violently reactive element be good for teeth? And as for hydrofluoric acid, which can etch through glass like no other acid . . . Fluorine, *fluoride:* chemists are finicky about naming (nomenclature). A one-letter misprint can be the difference between life and death! You will have heard less about the naming of organic compounds, so we explain it in more detail. In what follows, you will be led as gently as possible through the minefield of the various nomenclature practices used in the world today. Even this will not be sufficient for you to read chemical names, let alone write them, but the underlying logic will be exposed.

The aim of any systematic nomenclature is to describe a chemical structure in words. Verbal building blocks can be moved across from one substance to another, thereby preserving a particular piece of chemical information. There are a number of naming conventions, each with its own role. The International Union of Pure and Applied Chemistry (IUPAC) has established a systematic naming procedure (which fills several books with its rules).

DDT

The mosquito was heard to complain
That chemists had addled its brain.
The cause of the sorrow
Was para-dichloro diphenyl-trichloro-ethane.

DDT (see Organochlorine compounds in Chapter 7) is shorthand for dichlorodiphenyltrichloroethane:

$$(C_6H_4Cl)_2CH\text{–}CCl_3$$

However, this name is ambiguous and is no longer acceptable. The IUPAC name is:

1,1,1-trichloro-2,2-bis (4-chlorophenyl)ethane

and we need a new poem!

On the right of the name, the root word *ethane* establishes a chain of two carbon atoms $H_3C\text{–}CH_3$, on which the other structural parts are built. On the left, *1,1,1-trichloro* says there are three chlorine atoms (for atoms the Latin *tri* is used to mean thrice) replacing the three hydrogen atoms on carbon atom 1 of the ethane, that is $Cl_3C\text{–}CH_3$. Then, *2,2-bis* says there are two identical groups (for groups the Greek *bis* is used to mean twice) replacing two hydrogen atoms on the carbon atom 2. Then, *(4-chlorophenyl)* says what the two groups are. Each group is a benzene ring with a bond to the carbon on the ethane (an attached benzene ring is called a phenyl group), and with one chlorine atom replacing one hydrogen in position 4, that is, diametrically opposite the point of attachment. The third hydrogen at C2 on the ethane is not replaced.

IUPAC names are not always unique because you can chunk some molecules in different ways, with different rules. For example, the most common method involves a *substitution* sequence, which was used for DDT. Let us try another example, the drug barbital (Fig. A1.16).

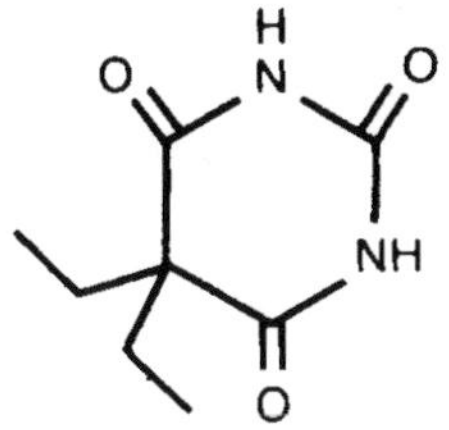

H
O N O
NH
O

IUPAC 5,5–diethylbarbituric acid
IUPAC 5,5–diethyl(1H,3H,5H)-pyrimidine-2,4,6-trione

Fig. A1.16
Barbital (Barbitone)

As soon as a compound appears in the chemical literature it needs to be classified. This work is done by the Chemical Abstracts Service (CAS). CAS abstracts the chemical literature of the whole world and insists on a unique name for every one of the eight million or so chemicals it has currently indexed. Whereas IUPAC names begin at the beginning and proceed to the end, CAS needs a system that lends itself to indexing, and so a name usually appears in the form of a parent compound, followed by bracketed phrases denoting the nature of a particular derivative. For barbital:

CAS (index form) 2,4,6(1*H*,3*H*,5*H*)-Pyrimidinetrione, 5,5-diethyl–

The CAS reg. number is [57-44-3]. (This number is unique to this compound, but a compound may have more than one number.)

We can now list derivatives other than the 5, 5-diethyl. This inverted presentation is in keeping with the concept of an index. The alphabetical listing of IUPAC names does not collect together compounds of the same chemical class. For example, bromobenzene would be found under 'b', while dibromobenzene would be under 'd'. Because barbitone is a drug, it will appear in the *British Pharmacopoeia* with a British approved name (BAN). The World Health Organization (WHO) agrees on a list of international non-proprietary names (INNs), for example, WHO barbital (INN).

These names consist of a few syllables, devised to be distinct from names already in use. At the same time, the name should tell the doctor or nurse something about the drug. Thus the various barbiturates will be 'generically' related in name, like phenobarbital, pentobarbital, etc. Australia follows the *British Pharmacopoeia* and uses BANs in the first instance in preference to the INNs where these differ. Thus our barbiturates are barbitones rather than barbitals. These names are given in a publication called *Australian approved names and other names for therapeutic substances* (AGPS. 4th edn, 1986). From the third edition onwards, this book has adopted the international nomenclature wherever possible.

Sometimes a name comes from a condensation of a systematic name:
for example, 2-p**hthalimido**glutar**imide**.

INNs are published in Latin, English, French, Spanish, and Russian, along with the names used in some national pharmacopoeias. While INNs may have arisen from lapsed trade names, they must be distinct from, and not used as, current trade marks (see Fig. A1.17).

The International Standards Organization (ISO) sets the common names recommended/approved for international use, and Standards Australia does the same for local use. The IUPAC name is given as an alternative.

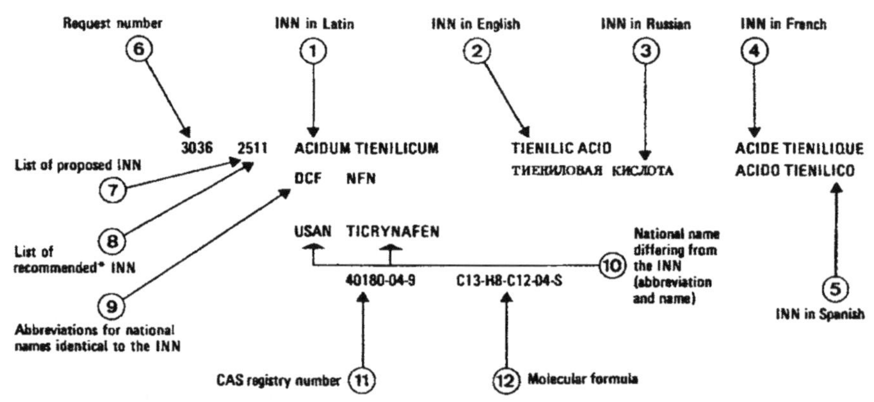

Fig. A1.17 Layout of information for INNs

• An asterisk in place of a recommended list number signifies that an objection has been raised to the proposed name.

Color or colour!

For dyestuffs and colourants, the internationally recognised source is the *Colour Index (CI)*, a joint publication of the UK Society of Dyers and Colourists and the American Association of Textile Chemists and Colorists (with the UK spelling winning out in the name for the index). The CI numbers are used for food colours. The index contains the CI common names, commercial name(s) and a CI number. The *Colour Index* gives the structure, where available, or otherwise a description of the method of manufacture. Food colours have additional names given to them by national regulating bodies which are so indicated (e.g. the US uses the FD & C — Food, Drug and Cosmetics Act and the European Economic Community uses E.)

For food chemicals and additives, the *US Food Chemical Codex* of the US National Academy of Sciences gives systematic names (and/or structures). *McCutcheon's Detergents and Emulsifiers* (Allured Publishing Corporation) is an index of trade names, with the name of manufacturer, class and formula, a type classification, and further remarks.

In common use and in industry, names are used that relate to properties (e.g. *caustic* soda, *slaked* lime, *killed* spirits), use (e.g. *bleaching* powder, *heating* oil, *battery* acid), or source (e.g. muriatic acid — how is your

Latin?). It becomes more complicated when only initials are used. TCP can mean trichlorophenol, trichloracetophenone, tricresyl phosphate, etc. BHC stands for benzene hexachloride, which it is not. In fact, it is hexachlorocyclohexane, a completely different compound (see Fig. A1.18).

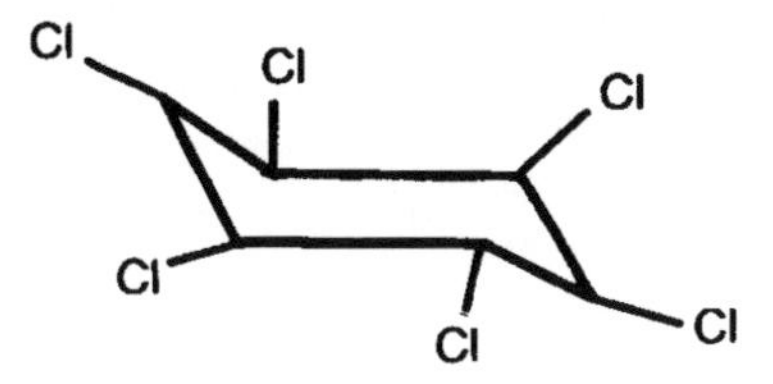

Fig. A1.18
Hexachlorocyclohexane (BHC)

Puzzle
Guess the following:
DDT
MEK
RDX
BHT
KIOP
TDI
DNA
PVC
QUAT
ABS
SAN
and PCB.

Once you know the logic in some code names they can be very useful. For example, in discussing the depletion of the ozone layer by 'freon' fluorocarbons note the following:

freon 11 is	methane, trichlorofluoro-	CCl_3F	[75-69-4]
freon 12 is	methane, dichlorodifluoro-	CCl_2F_2	[75-71-81
freon 22 is	methane, chlorodifluoro-	$CHClF_2$	[75-45-6]
freon 115 is	ethane, chloropentafluoro-	$CClF_2CF_3$	[76-15-3]
freon 132 is	ethane, 1, 2-dichloro-1,2-difluoro-	CHClFCHClF	[413-06-1]

The last digit is the number of fluorine atoms. The next to last digit is one plus the number of hydrogen atoms. The next digit forward is the number of carbon atoms minus one. The balance is the number of chlorine atoms. The designation freon 11, etc. is used to avoid brand names.

Some rules for IUPAC naming

We have tried a few simple examples of interpreting chemical names; now we will see how they are assembled. Naming of compounds fills several books, so all we can do here is give a few of the simpler rules, first for linear (non-cyclic) compounds, and then for cyclic compounds.

Linear compounds

Step 1: Identify all the functional groups present. Select the *principal functional group* (PFG) according to the order of priority given in Table A1.2.

Step 2: Check for unsaturation of the carbon skeleton, that is, the presence of double (C=C) or triple (C≡C) bonds.

TABLE A1.2
Order of priority of functional groups

Order of priority	Functional group	Formula	Order of priority	Functional group	Formula
1	Carboxylic acid	–COOH	8	Ketone	–CO
2	Sulfonic acid	$-SO_3H$	9	Alcohol	–OH
3	Ester	–COOR	10	Phenol	–OH
4	Acid chloride	–COCl	11	Thiol	–SH
5	Amide	$-CONH_2$	12	Amine	$-NH_2$
6	Nitrile	–CN	13	Ether[a]	–OR
7	Aldehyde	–CHO	14	Sulfide[a]	–SR

[a] Simple ethers and sulfides are often named using the radico-functional system, ethyl methyl ether, instead of methoxymethane; but more complex ones, for example 2-ethoxypropane, are not.

Step 3: Select the longest continuous carbon chain that contains the principal functional group and the maximum number of unsaturated bonds. This is called the *principal chain.* This supplies the stem name (see Table A1.3).

Step 4: Number the carbon chain from one end to the other so that the PFG has the lowest number possible.

Step 5: The PFG now provides the suffix for the chain name. Any other functional groups present are cited as prefixes.

The list in Table A1.4 is in order of priority for principal functional groups. The alternative radico-functional names for some of these are given later.

Step 6: Insert all the lesser priority functional groups and branching alkyl groups (radicals) as prefixes, using the names from Table A1.4. These prefix names are listed alphabetically, ignoring, for ordering purposes, any multiplicity modifiers. (Multiplicity terms are *di* or *bis* (×2), *tri or tris* (×3),

TABLE A1.3
Simple stem names

	Hydrocarbon chain	Stem name	Alkane/radical	Alkane/radical name
1	carbon atom	meth-	CH_4/CH_3^-	methane/methyl
2	carbon atoms	eth-	$C_2H_6/C_2H_5^-$	ethane/ethyl
3	carbon atoms	prop-	$C_3H_8/C_3H_7^-$	propane/propyl
4	carbon atoms	but-	$C_4H_{10}/C_4H_9^-$	butane/butyl
5	carbon atoms	pent-	$C_5H_{12}/C_5H_{11}^-$	pentane/pentyl
6	carbon atoms	hex-	$C_6H_{14}/C_6H_{13}^-$	hexane/hexyl
7	carbon atoms	hept-	$C_7H_{16}/C_7H_{15}^-$	heptane/heptyl
8	carbon atoms	oct-	$C_8H_{18}/C_8H_{17}^-$	octane/octyl

TABLE A1.4
Suffixes and prefixes

		Group structure	Prefix	Suffix
1.	Carboxylic acid	X–COOH	carboxy	-oic acid
2.	Sulfonic acid	X–SO_2–OH	sulfo-	-sulfonic acid
3.	Ester	X–CO–OR	R-oxycarbonyl-	-alkyl-oate
4.	Acid chloride	X–CO–Cl	halo-formyl-	-oyl halide
5.	Amide	X–CONRR	carbamoyl-	-carboxamide
		X–NR–COR	acylamido-	
6.	Nitrile	X–CN	cyano-	-nitrile
7.	Aldehyde	X–CH=O	formyl-	-al
8.	Ketone	X–COR	oxo-	-one
9.	Alcohol	X–OH	hydroxy-	-ol
10.	Phenol	X–C_6H_4–OH	hydroxy-	-ol
11.	Thiol	X–SH	mercapto-	-thiol
12.	Amine	X–NRR	amino-	-amine
13.	Alkene[a]	X–C=CRR	alkenyl-	-ene
14.	Alkyne[a]	X–C≡CR	alkylnyl-	-yne
Groups cited only as prefixes				
15.	Fluoride	X–F	fluoro-	
16.	Chloride	X–Cl	chloro-	
17.	Bromide	X–Br	bromo-	
18.	Iodide	X–I	iodo-	
19.	Azide	X–N_3	azido-	
20.	Nitroso	X–NO	nitroso-	
21.	Nitro	X–NO_2	nitro-	
22.	Ether	X–OR	R-oxy-	
23.	Sulfide	X–SR	R-thio-	

[a] 13,14 as suffix only if no PFG present.

tetra (×4), *penta* (×5), *hexa* (×6), etc.; for example, bis(dimethylamino)). The numbers, indicating the location along the chain, are placed in front of the functional or branching alkyl groups they locate.

Step 7: Assemble the complete name as a single word (there are some exceptions), using commas to separate numbers from numbers and hyphens to separate numbers from words:

2,2-dichlorohept-3-en-l-ol

Breakdown of molecule for nomenclature purposes

The following steps are followed when naming an organic compound. Table A1.2 gives the order of priority for choosing a suffix.

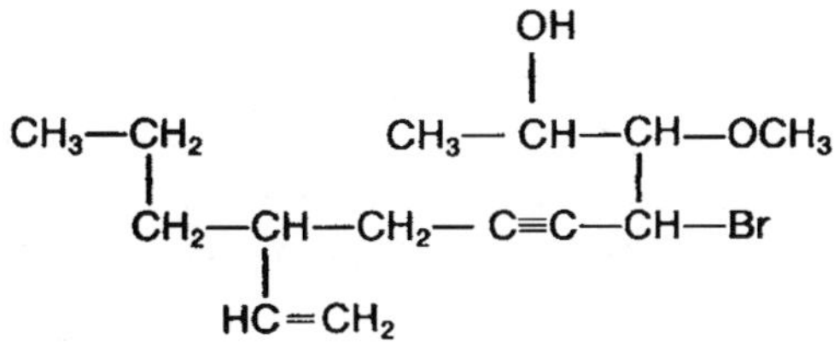

```
                               OH
                               |
CH3—CH2              CH3— CH—CH—OCH3
    |                          |
    CH2—CH—CH2— C≡C—CH—Br
        |
       HC=CH2
```

Fig. A1.19

Step 1: Functional groups are determined
OH — hydroxyl will determine the suffix (see Table A1.2).
OCH_3 — methoxy; Br — bromo.

Step 2: Unsaturated? Yes, it contains both C=C and C≡C.

Step 3: Selecting principal chain
The stem is chosen to include the PFG and maximum number of double and triple bonds.

```
                OH
                |
             C—C—C
                   |
     C—C—C≡C—C
     |
     C=C
```

Fig. A1.20

Principal chain (stem) name: dec-en-yn. This order is maintained.

Steps 4–5: Suffix for PFG and numbers added

```
                    OH
                    |
             C1—C2—C3
                        |
     C8—C7—C6≡C5—C4
     |
     C9=C10
```

Fig. A1.21

Basic name: dec-9-en-5-yn-2-ol.

If a suffix begins with a vowel (or y), when adding the suffix to a hydrocarbon name the final *e* is dropped. Thus from pentane comes pentanol, pentanoic acid, pentanal, etc. When assembling the stem name, if both a double and a triple bond are present, the *e* is elided between -ene and -yne: thus pentenyne.

Step 6: Inclusion of lower priority functional groups and branching chains as substituents

$CH_3CH_2CH_2$–	propyl
CH_3O–	methoxy
Br–	bromo

These substituents are listed alphabetically as prefixes.

Step 7:
IUPAC name: 4-bromo-3-methoxy-8-propyldec-9-en-5-yn-2-ol

The rules for Chemical Abstracts are not quite the same. The CAS name for the same compound is 4-bromo-7-ethenyl-3-methoxyundec-5-yn-2-ol. The CAS sequence of application of principles (neglecting heteroatoms and rings) is as follows:

1. Greatest number of principal chemical functional group (PFG).
2. Largest index heading parent (i.e. longest chain).
3. Greatest number of multiple bonds (but double bonds preferred over triple bonds — inferred from example).
4. Lowest numbers for PFGs.
5. Lowest number for all multiple bonds.
6. Lowest number for double bonds.

In the IUPAC system, rule 3 takes precedence over rule 2.

Cyclic compounds

Aromatic compounds are related to benzene either by replacing a hydrogen atom with a substituent, or by fusing the rings together to give naphthalene or anthracene, etc. as shown in Figure A1.22.

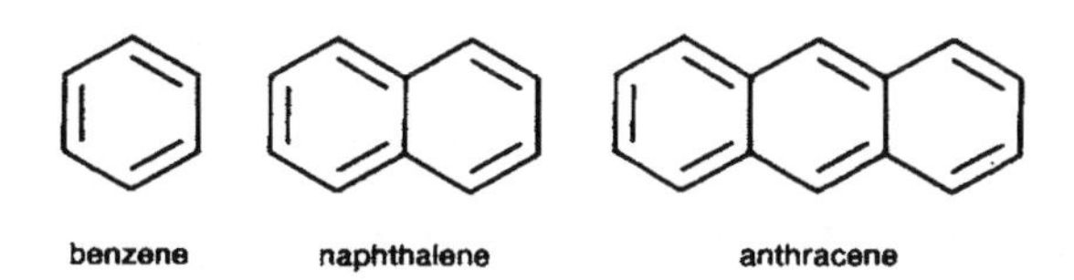

Fig. A1.22 Aromatic hydrocarbons

Note: the radical derived from benzene, C_6H_5–, is called *phenyl* (not benzenyl); other related radicals are $C_6H_5CH_2$–, called *benzyl*, and C_6H_5CO–, called *benzoyl*. Most simple aromatic compounds are known by trivial names: for example C_6H_5OH, called phenol; $C_6H_5NH_2$, aniline; $C_6H_5CH_3$, toluene; $C_6H_5CH{=}CH_2$, styrene; $C_6H_5(CH_3)_2$, xylene; $(C_6H_5)_2CO$ benzophenone.

Semi-systematic names are formed as shown in Figure A1.23.

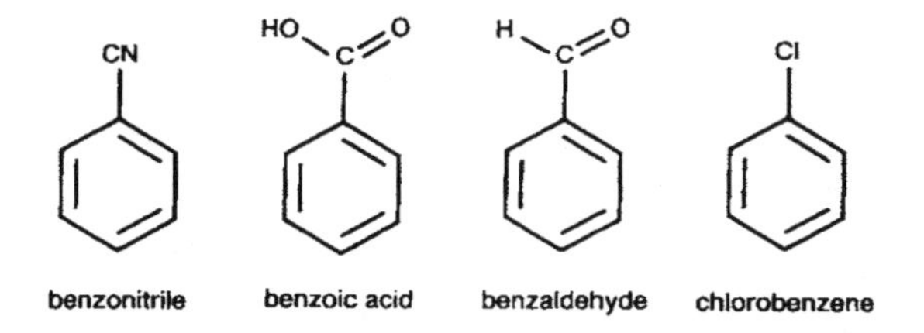

Fig. A1.23 Aromatic compounds

Two substituents in the benzene ring on adjacent positions 1,2 are labelled ortho or *o*-; on positions 1,3 are labelled meta or *m*-; on opposite positions 1,4 are labelled para or *p*-. IUPAC prefers the use of the numbers rather than these names, which apply only to benzene.

Again the PFG attracts the lowest number, as shown in Figure A1.24.

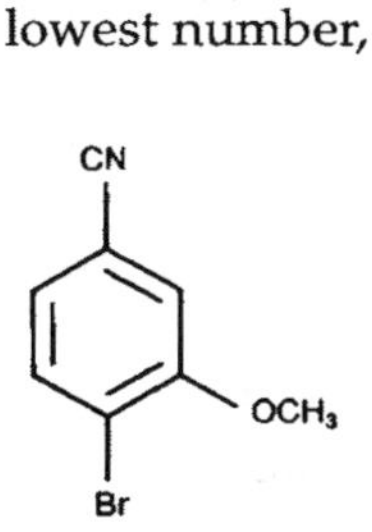

CN

OCH_3

Br

Fig. A1.24 4-bromo-3-methoxy-benzonitrile

Non-aromatic cyclic systems follow similar rules. Thus cyclohexane is a C_6 ring and the numbering is the same as for benzene.

A good learning resource, and the reference used for this section, is *Chemical nomenclature* by J.D. Coyle and E.W. Godly. The standard reference, 'The Blue Book', is *Nomenclature of organic chemistry,* in particular 'Characteristic groups containing carbon, hydrogen, oxygen, nitrogen, halogens, sulfur, selenium and/or tellurium', issued by the IUPAC Commission on Nomenclature in Organic Chemistry. As a reference for particular compounds, be they trade, trivial, or systematic, you cannot better the *Chemical index guide* which comes with each volume of *Chemical Abstracts*, the major chemistry reference series.

Bibliography

Australian approval names and other names for therapeutic substances, AGPS, 4th edn, 1986.

Nomenclature of organic chemistry, International Union of Pure and Applied Chemistry (IUPAC). Butterworths, London, 1969.

Chemical nomenclature. J.D. Coyle, and E.W. Godly, Open University and Laboratory of the Government Chemist, Milton Keynes, UK, 1984.

APPENDIX

The prevalence of logarithmic scales

Unlike the other appendixes, and indeed the rest of this book, this one is the author's indulgence for one of his pet ideas: that our perception of risk, like our perception of just about everything else, is logarithmic. It gives one reason why we tend to overestimate the very small risks (e.g. food additives) and underestimate the very large ones (car accidents). There are many other factors, including voluntary versus involuntary, but still, the author hopes you find the argument interesting.

Intensity
Our observation of many waves, for example light, quantum mechanical wave functions, and X-ray diffraction patterns, is generally via their **intensity**. The intensity of a wave is the square of its amplitude. With electric current we can measure both amplitude (voltage) or power (watts); it depends on the detector used.
Our ears always sense sound as pressure differences in the amplitude of the sound wave, not an intensity difference. It is thus possible to 'cancel' sound by feeding in a signal of opposite phase (negative pressure, that is, less than atmospheric at that point and time).

Sound science — dB or not dB?

Our senses, such as hearing (sound) and sight (light), cover huge dynamic ranges. We can hear down to levels just above the random movement of molecules, and yet cope with aircraft and jackhammer

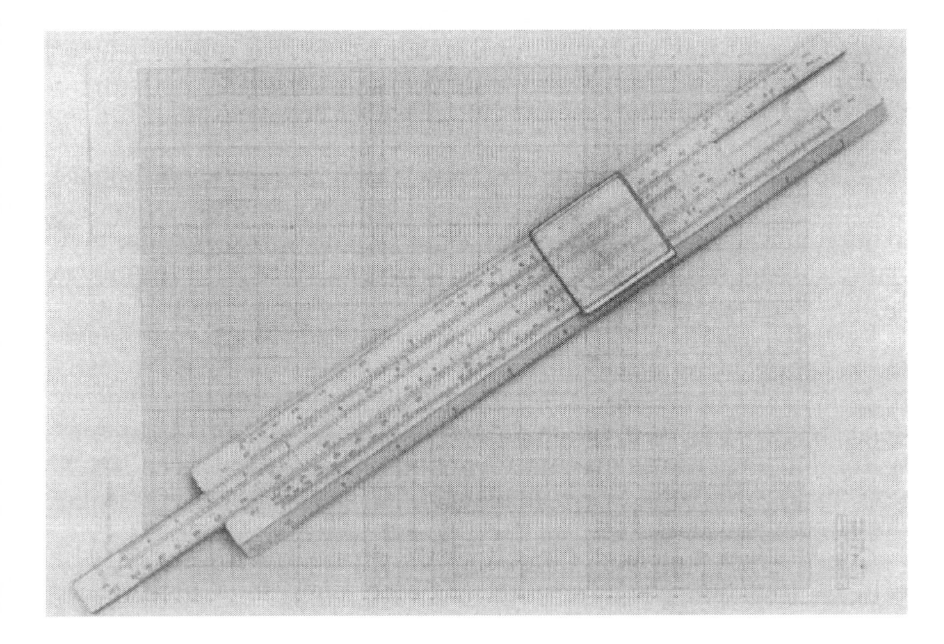

Fig. A2.1
Logarithmic scale

noise, at least for a short exposure. We can see in the dimmest of moonlight and in the intensity of the bright sun. These dynamic ranges (depending on the parameter used to measure them) cover ranges of tens or thousands of millions. Touch and taste also cover a large dynamic range.

Our senses cope with large dynamic ranges on a single scale by using a logarithmic conversion.

Perception of sound

A psychological observation given as the Weber–Fechner law suggests that the perceived increase in loudness of a sound changes logarithmically with the physical increase. A single car outside might cause a particular level of noise. With ten cars, the level of loudness we perceive doubles. An increase from ten to one hundred trebles the perceived loudness. This is the basis of using a logarithmic scale for measuring sound. There are other factors . . .

Weber's law
dx/x = constant.
The smallest change we can detect is proportional to whatever is already there.

Measurement of sound

There are two ways of measuring sound. One is to follow our ears, and measure differences in sound pressure (which is after all what pushes the ear's cilia to sense sound). This is measured in units of pressure which are:

Newton.m^{-2} (or Pa or kg.m^{-1}.s^{-2} or J.m^{-3}).

The second way is to measure sound intensity (as for light intensity) and this is a measure of the energy of sound:

Watts m^{-2} (or kg.s^{-3} or J.s^{-1}.m^{-2}).

Consistent with other waves, intensity (energy) is set proportional to the square of the pressure (amplitude).

A bel is defined as a ratio of an 'after' to a 'before', using a logarithmic scale to base 10. It has no units. A bel equals ten decibels (dB).
So if dB = 10 log (intensity measured)/(reference intensity),
then dB = 20 log (pressure measured)/(reference pressure).
(Note that $\log x^2 = 2 \log x$.)
We'll stick to using sound pressure (rather than intensity) and describe common sounds in that dB scale with an arbitrary zero corresponding to minimum audibility, which is just above the kinetic energy of gas molecules at room temperature (10^{-16} J/s), hitting an average eardrum (area 1 cm^2). See table A2.1.

We find in practice that a doubling of perceived loudness corresponds to an approximate increase of 20 dB (i.e. a tenfold increase in sound pressure) right down Table A2.1. (Forget about correcting for frequency dependent effects, 'A' scales, and phons and sones, and so on.) In other words, loudness relates in a linear way to dB, but pressure goes up exponentially with both; hence the danger of damage at high levels that don't sound much louder.

Conversely, a doubling of sound pressure corresponds to an increase of 6 dB. (But those bloody sound engineers jump from pressures to

TABLE A2.1
dB and sound pressure
Sound pressure range of 0 to 140 decibels.
('A' scale decibel meter)

dB (pressure scale)	Ear's response	Sound pressure units (Pa)
0 dB:		2×10^{-5}
10 dB:	just audible	
20 dB:	empty broadcasting studio	2×10^{-4}
30 dB:	soft whisper at 5 m	
35 dB:	quiet library	
40 dB	bedroom, no conversation	2×10^{-3}
50 dB:	'very quiet'	
55 dB:	light traffic at 15 m	
60 dB:	air conditioning at 6 m.	2×10^{-2}
65 dB:	normal conversation	
70 dB:	light freeway traffic	
75 dB:	conversation noticeably difficult	
80 dB:	'annoying' sound level	2×10^{-1}
85 dB:	pneumatic drill at 15 m	
90 dB:	heavy truck at 15 m	
95 dB:	'very annoying'	
100 dB:	loud shout at 15 m	2
105 dB:	jet plane take-off at 600 m	
110 dB:	riveting gun close by	
115 dB:	max. vocal voice without amplification	
117 dB:	discotheque at full blast	
120 dB:	jet take-off at 60m	2×10
130 dB:	limit of amplified speech	
135 dB:	'painfully' loud	
140 dB:	on aircraft carrier deck	2×10^{2}

Source: Noise, R.Taylor, Penguin, 2nd edn, 1975

Relationships

- Pressure is measured in pascals (Nm^{-2}).
- Every 20 dB doubles the loudness.
- Every doubling of pressure increases loudness by 6 dB.

intensities for their ghetto blasters using a doubling of sound *intensity*, which therefore corresponds to 3 dB.)

As this is not a lesson in sound science but only an example of the fact that our senses respond logarithmically to stimulae in order to cover huge dynamic ranges, we won't labour the point further.

There are also very many examples for the processing of information as logarithms of quantity in the physical and biological sciences.

If decibels had been in use earlier, Newton might have said that doubling the distance between objects reduces gravitational pull by 6 dB and Coulomb might have said the same about electrical forces. Richter insisted on a logarithmic measure for earthquake intensity.

Brønsted would not have invented a separate scale for measuring acidity. Each doubling of the hydrogen ion concentration in water means

Saying
We would now be saying: 'Pure water at 7 bels!'

a pH decrease of 0.3 pH units. A tenfold increase in $[H^+]$ means a pH decrease of 1 pH unit.

However, more interesting still is the example of an intellectual 'sense', quite removed from our physical senses. I refer to the way economists, against all mathematical logic, have also had to adopt a logarithmic measure.

Economic rationalisation

If you save a dollar on a $4.50 packet of coffee, you feel the 'special' has been worth while. On the other hand if you save a dollar on a $600 suit, you think it is a joke. To take off 100% *extra*, namely $2, would just add insult to injury. Why? Logically, from the point of view of the family budget, there is a dollar (or two) saved in each case, but this is not our (economic) perception.

ST PETERSBURG PARADOX

Peter tosses a fair coin repeatedly until it lands heads. He agrees to pay Paul $1.00 if it lands heads on the first toss, $2.00 if it lands heads on the second, $4.00 if it lands heads on the third toss, and so on; each time the coin lands tails the payment is doubled. The question arises, what is a fair price for Paul to pay for participating in the game?

The maths

Peter pays Paul 2^{n-1} dollars if the first head occurs on the nth toss, and the chance of this is $(1/2)^n$; so Paul's (statistical) expectation (summed from n = 1 to n = infinity) is:

$$\sum_{n=1}^{n=\infty}(2)^{n-1}(\tfrac{1}{2})^n = \sum_{n=1}^{n=\infty}\tfrac{1}{2} = \text{infinity.}$$

Thus Paul needs to pay an infinite amount of dollars for a finite chance of probably quite a small gain from the game with Peter. This seems intuitively unreasonable because, for example, there is a probability of 7/8 that Paul will receive at most $4.00.

Bernoulli's application
Bernoulli applied Weber's law for sound and sight to perception of money

The history of the development of the marginal utility concept started in 1738 with Daniel Bernoulli (1700–1782). He argued that the value or utility to a person of a small amount of money was inversely proportional to the amount of money he already had. Thus it is the ratio dx/x that is considered the measure of utility. We are led inexorably to the (modern) economic theory of decreasing marginal utility, and an intellectual logarithmic response to add to our well accepted physical responses.

See for example The St Petersburg paradox, in Encyclopedia of statistical sciences, **8**, Ed. S. Kotz & N.L. Johnson, 1988, Wiley NY.

I would argue that there is just a small perceptual jump from the economic measure of utility to what I believe is a logical measure of risk. Our perception of risk *must* in some way be proportional to the background level of (that) risk to which we are already exposed.

All our physical senses respond roughly logarithmically. One major intellectual response (economic utility) behaves likewise. Why not another, such as risk perception? What I am suggesting is a table of risk (or safety); let us call it sels (dB) that mirrors the table on sound loudness (dB) with the same approach to typical examples. Interestingly, a dB scale (being logarithmic) has no real zero. For sound, an arbitrary zero dB is set at the point below which a sound cannot be heard (just above the random motion of molecules hitting the eardrum). There is also no maximum, but above about 140 dB your eardrums pierce and that seems a useful place to stop. Also a chance of one in a million of death or injury over a lifetime makes a similarly sensible zero for 0 dB sels. A maximum occurs at certain death or injury, and it comes at 120 sels dB; above that and it's overkill! Instead of quoting odds of risky events, quoting the logarithmic sels would compress the risk scale in a manner more attuned to our perceptions, and allow a more logical spread of resources on risk reduction.

There are many confounding factors in risk perception. For even the simple argument of dx/x = constant to work, people have to be aware of the magnitude of x. If you are not aware of the high risk of some factor, for example food poisoning by bacteria, then you will be over-alarmed by a minute risk of food additives dx, and misplace your vigilance.

A final note on Acute toxicity in Chapter 2. A logarithmic-normal distribution is assumed for the dose response to toxic chemicals when producing LD_{50} plots. This distribution is assumed whenever the variability of a parameter increases with the size of the parameter.

Response to change
Given a group of ten people in a room, an average observer can estimate at a glance if the group has dropped to say, 8 or increased to 12. With a group of 100, ± 2 cannot be observed at a glance; on average ± 20 probably can. With a group of 1000, no to ± 2, and no to ± 20, but probably on average ± 200 can.
The ability to respond on average to a change roughly proportional to the magnitude is very common in biological systems.

Use of phase diagrams

The following four examples relate directly to matters discussed in the book. They are all closely related concepts but usually taught separately and not connected. We start with a eutectic mixture, then discuss steam distillation, follow this by a Dean and Starke distillation, and end with an environmental example of Henry's law.

Part 1: Mixing tin and lead — have a dialectic with a eutectic

To understand the behaviour of mixtures of solids (and liquids) we must understand equilibrium phase diagrams. A phase, in a chemical sense, is a uniform, physically distinct form of matter that can be present in a system. Ice, water, and steam are different phases of a single component, water. Each is physically uniform. In this case they are also chemically uniform (just water) but this need not be.

We will examine the equilibrium phase diagram for mixtures of the metals tin and lead as a function of temperature (Fig. A3.1a,b), as an examplar. You need to put yourself on a spot in the diagram. A 'spot' corresponds to a particular temperature (read off the vertical scale) and a particular overall composition for the total material in the sample (read off the horizontal scale(s)). The lines in the diagram tell you how the overall composition is shared among the different phases (liquid (gas) solids). It takes some getting used to but is very informative. Phase diagrams are used a great deal in fields such as corrosion studies, distribution of elements among different species in soils and rivers, and so on.

In Figure A3.1a the bottom axis shows increasing per cent of tin by mass. (Along the top axis it is given in terms of per cent of atoms of tin.) The vertical axis is temperature, and you will note that pure lead melts at 327°C (left-hand side) and pure tin melts at 232°C (right-hand side).

This phase diagram is obtained by making up a series of mixtures ranging from 0–100% tin, (which is the same as from 100–0% lead), heating each of them until they completely melt, and then cooling them very slowly. The plot follows the temperature where (for each composition) two phases are present together, and microscopic examination indicates the nature of the phases.

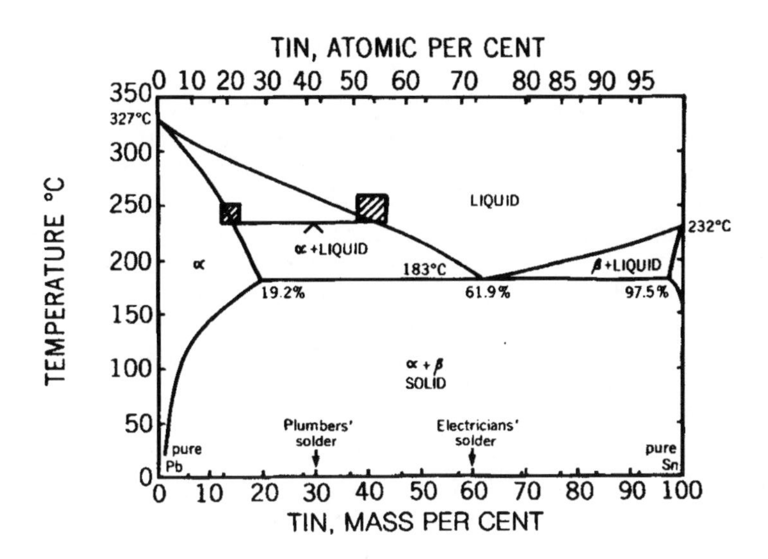

Fig. 3.1a
Phase diagram for the lead–tin system

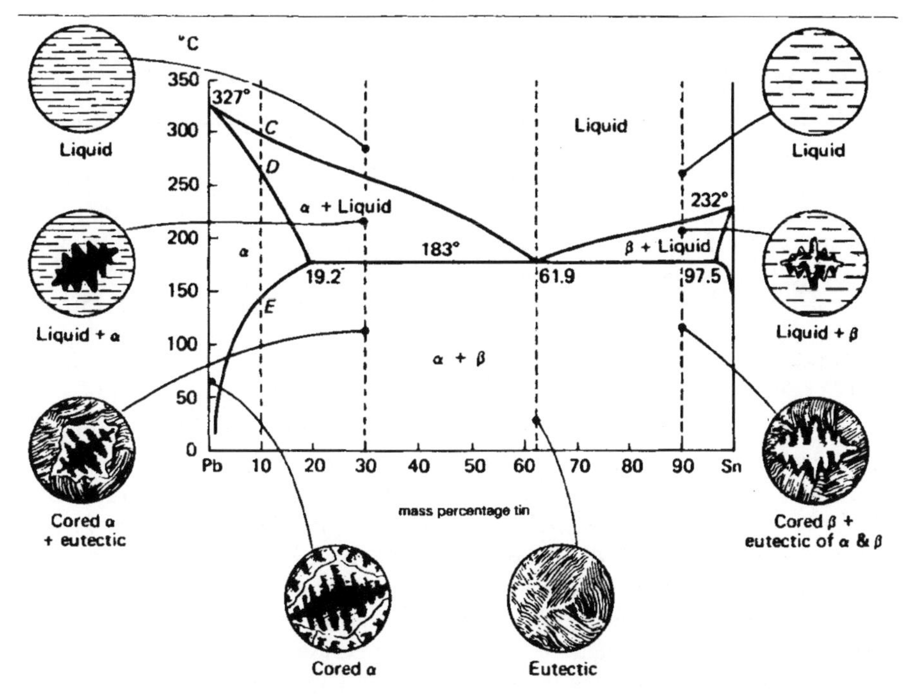

Fig. 3.1b
The tin–lead equilibrium diagram. The cored alpha and beta grains result from industrial cooling.

(Courtesy of R.A. Higgins, from Engineering Metallurgy, Vol. 1, Hodder and Stoughton, London, 1972)

The first thing to note is that adding tin to lead lowers the melting point of the lead, and adding lead to tin lowers the melting point of tin. The second observation is that these two falling lines meet at the spot 183°C and composition of 61.9% tin. This is a minimum temperature and is called a *eutectic* temperature. The corresponding composition is called the *eutectic* mixture for tin and lead (Greek *eu* good; *tektos* melted).

To the immediate left and right are regions where solid and liquid are both present. To the left, there is a solid called phase α, plus liquid. To the right, there is a solid called phase β, plus liquid.

To the far left (against the left axis) is an all solid region with solid phase α, and to the far right (against the right axis) is a (small) all solid region with solid phase β. Below the eutectic point is an all solid region with two solid phases, $\alpha + \beta$. The solid phases α and β are each solid *solutions*. Phase α is solid lead with some tin dissolved in it, and phase β is solid tin with some lead dissolved in it.

In the region $\alpha + \beta$ these two distinct homogeneous phases are *mixed*. Note the distinction between a true *solution*, where atoms are mixed giving a single phase, and a *mixture*, where different solid phases (both containing both tin and lead atoms, but in different proportions) are mixed — a mixture of crystals. Figure A3.1b shows the nature of the phases formed. So far so good.

What is the composition of phase α; how much tin is dissolved in the lead in this phase? Well this depends on the temperature at which solid phase α forms. Looking at the left of the diagram, at 25°C the amount of tin in phase α is only a couple of per cent, increasing to a maximum of 19.2% at the eutectic temperature of 183°C. Going to higher temperatures, the amount of tin drops again, reaching zero at 327°C. (A similar argument applies to phase β crunched up on the far right.) Tin is overall more soluble in solid lead than lead is in solid tin.

Solder
Electricians' solder freezes sharply, but plumbers' solder does not.

Now mark the spot for a 60:40 mixture of tin and lead at 250°C. Slowly drop the temperature. The liquid cools and it *all* freezes at 183°C; it freezes *sharply*. That is, below 183°C it is a solid, and above 183°C it is a liquid. A mixture that has a sharp melting point (as if it were a pure substance) is called a *eutectic* mixture. The eutectic mixture of tin and lead is at the ratio 60:40. It is this composition of tin and lead that constitutes electricians' solder. Electricians need a low melting metal that conducts electricity, 'wets' copper wire, and sets quickly, so that the electronic components do not overheat. Repeating, although the eutectic mixture behaves like a pure substance (in that it has a sharp melting point) it is in fact a solid mixture of two phases ($\alpha + \beta$), the different crystals of which can be seen in a cut and polished sample under a microscope.

Now mark the spot for a 30:70 mixture of tin and lead at 250°C. As you cool this liquid, it *begins* to solidify at 250°C, but now the freezing point is not 'sharp' because both liquid and solid (phase α) are present and stay present for another 70°C cooling. It is not until the temperature drops to 183°C that the mixture freezes completely and no liquid remains. This solid is a mixture of phase α (which has been crystallising out during cooling) and feathery crystals of the eutectic (which solidifies suddenly when the temperature hits 183°C). When viewed under a microscope, the 'suddenly solidified solid' can be seen packed in between the phase α crystals. The feathery eutectic crystals are a solid mixture of phase α and phase β.

Now the 30:70, tin:lead mixture is *plumbers' solder* (Latin *plumbum* lead). When the plumber solders a joint with molten solder at 250°C the solder takes a long time to cool to 183°C, at which point the mush of liquid and solid finally sets completely to solid. This gives the plumber time to set the joint. The mechanical properties of the sludge give time for the joint to be moved by the plumber. (The electrician has no time.) The cooling sludge is interesting because it is not a liquid from which more and more of the *same* crystals grow.

If you look at the line enclosing the α phase at the left from the point 19.2% tin, 183°C to 0% tin (pure lead), 327°C, you will see how the composition of the solid phase α changes with temperature. You can see that the higher the temperature, the less tin the phase α contains. As our liquid plumbers' solder begins to form crystals below 250°C, the composition of the crystals changes as the solder cools. Each new batch of crystals forming has more tin in it, and the liquid remaining behind is also richer in tin. In fact the changing composition of the *liquid* phase can be obtained from looking at the line from the eutectic (61.9% tin, 183°C) to melting lead (0% tin, 327°C).

Thus molten plumbers' solder at 250°C has 30% tin, but, as it cools, the composition of the crystals forming is obtained by taking a horizontal line across to the left to where it hits the phase α curve, and the composition of the liquid remaining is obtained by taking a horizontal line across to the right to where it hits the liquid curve. The complete horizontal line is called a *tie line* and it ties together the composition of the solid formed and the liquid remaining. As the mixture cools, the tie line drops lower and the composition of the crystals (phase α) and the liquid are *both* richer in tin! (Good grief, the alchemist's dream come true! We should start working on a *gold*–lead alloy!) Alas, we have considered only the *composition* of each phase, not the *amount* that is there. We start off with liquid, and on cooling, we have more and more solid and less and less liquid. So, even though both phases become richer in tin, the phase with less tin overall (solid α) is increasing in amount, while the phase with more tin overall (liquid) is decreasing. The change in amount of each phase is described by placing a hypothetical fulcrum at the point you are at (temperature, overall composition) and balancing the amount of liquid and solid on the tie line where it touches the respective curves (See Fig. A3.2).

Alchemy
Turning lead into gold — the alchemist's dream.

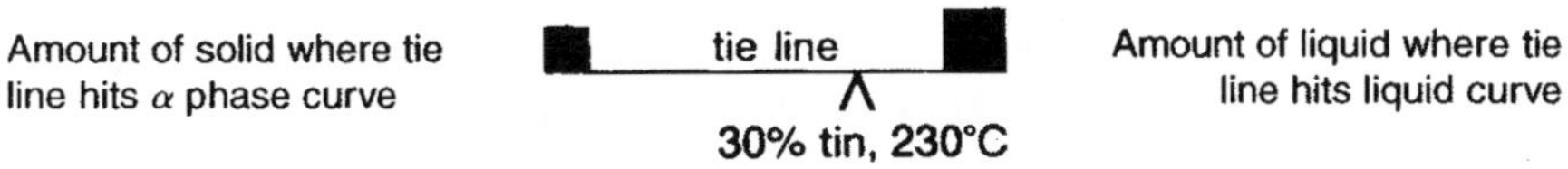

Fig. A3.2
Tie lines

The lever shifts horizontally as the temperature changes. This is called (surprise!) the *lever rule.* The changing amount of each phase and the changing composition of each phase combine to keep the *overall* composition the same (30% tin in this example), and so we have not yet found the philosopher's stone.

We could also go on and look at the right-hand side of the diagram, where we have lead dissolved in tin, lowering its melting point by up to 40°C to produce pewter.

Well if you are not yet 'phased', there are plenty of even more interesting diagrams to study— such as, for example, the phase diagrams of copper and nickel and of iron and carbon to form steel and the effect on steel of adding additional metals. Phase diagrams give the results of experiments carried out at *equilibrium*. What happens when materials are heated and then cooled quickly is even more complicated, especially if the materials are metals that can take on a variety of crystal structures.

Part 2: Oil and water don't mix

EXPERIMENT A3.1
Steam distillation of eucalyptus leaves

Equipment

- Steam distillation apparatus
- Eucalyptus leaves

Exercise
You might also like to produce oil of cloves (eugenol, b.p. 164°C) from cloves.

Method

The oil produced from eucalyptus leaves contains eucalyptol (1,8-cineole) in many cases as the main component in an oil yield of about 1–3%. The oil is useful in cough drops, mouthwashes, gargles, dental preparations, inhalants, room sprays, and medicated soaps. It is an effective disinfectant. (See Fig. A3.3.)

eucalyptol

eugenol

Figs. A3.3
Essential oils

Because the oil decomposes when boiled (b.p. 176°C) the method used to extract it from the leaves and purify it is to distil the leaves with steam. The amount of oil carried over with the steam depends on how volatile the oil is. Using this method of steam distillation, the temperature of the oil never exceeds 100°C and the oil is not destroyed. In fact this technique is identical to the one used in the Dean and Starke experiment (see Part 3) except that in this case we are interested in the oil phase rather than the water phase (see Fig. A3.5).

The frying of wet foods involves the evolution of steam, which helps to distil out these (phenolic) antioxidants from vegetable oils. (See Fig. 4.12.) It is interesting to note that the natural antioxidant α-tocopherol, because of its long chain, is less volatile and thus longer acting

Hot coals
Aromatic oils thrown on sauna 'hot coals' would char. Mixing with water first prevents this.

Fig. A3.4
A bush eucalyptus still, Braidwood, NSW, c. 1900. Cooling pipes run into a creek (just visible).

(Courtesy of Professor A. J. Birch)

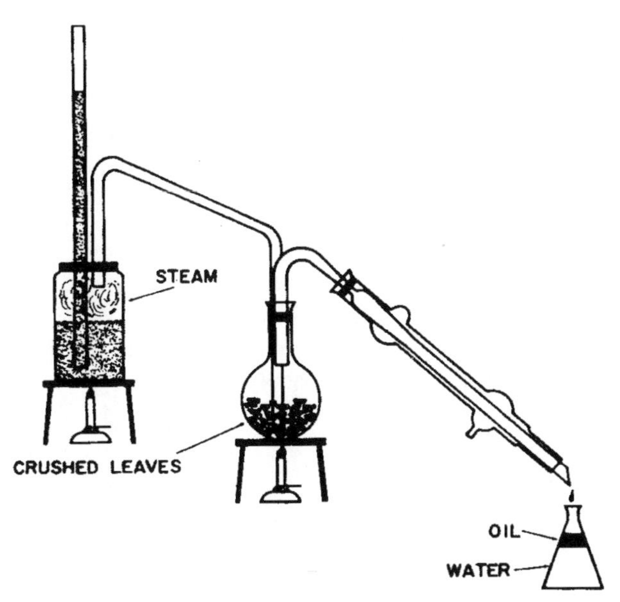

Fig. A3.5
Steam distillation of eucalyptus leaves in the laboratory: the oil settles out as a separate phase (layer) on top of the water (compare with Dean and Starke distillation, Part 3, overleaf).

Part 3: Water and oil don't mix either

Steam distillation, see part 2, has the double objective of separating the volatile from non-volatile components, with a reduction in boiling point to reduce thermal decomposition. In this experiment the emphasis is on the inverse; a volatile organic carrier is used to distil over water.

The Dean and Starke apparatus is shown in Figure A3.6 (overleaf). It is used to determine the amount of water present in such materials as

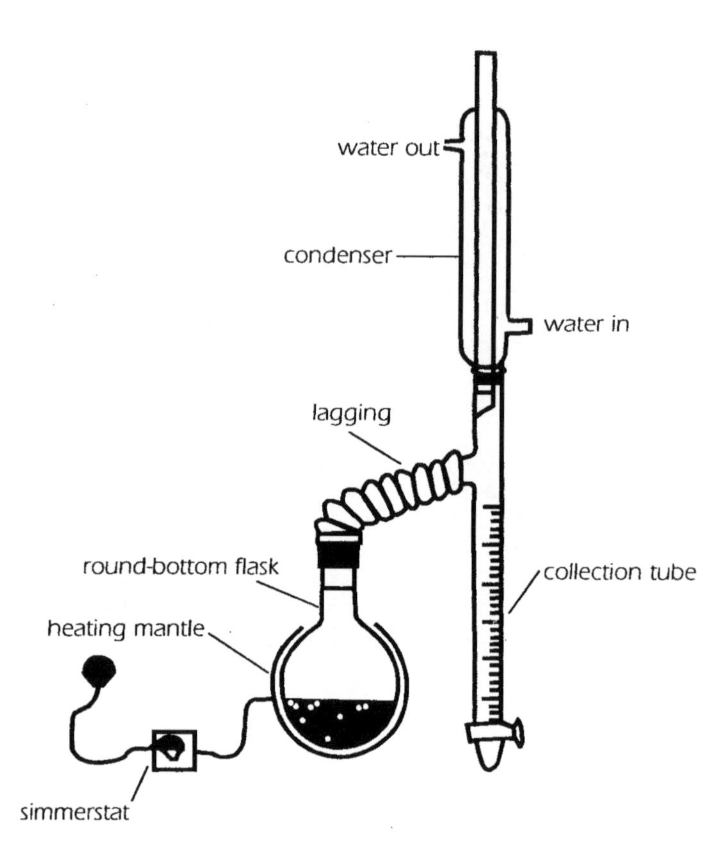

Fig. A3.6 Dean and Starke apparatus

detergents, oils, meat, cheese, etc. It can also be used to determine the amount of fat or fat-like material present in a product.

The principle of the experiment is that, in a mixture of toluene and water, boiling occurs at 85°C, and both water and toluene are given off in the vapour. On condensing, two layers of liquid are formed: a bottom layer of water (with 0.06% toluene dissolved) and a top layer of toluene (with 0.05% water dissolved). The relative volumes are 18% water and 82% toluene. The excess toluene flows back into the flask and distils back over with more water. If powdered laundry detergents are being examined, then there will be water of hydration tied up in some of the components (e.g. tripolyphosphate). A higher boiling solvent is needed to ensure that the hydrate is broken down (xylene or petroleum spirit, b.p. 140°C).

Figure A3.7 shows a phase diagram for the toluene–water system (with the mutual miscibility exaggerated). At a temperature below 85°C the two liquid phases coexist. At 85°C the sum of the individual vapour pressures (of water and toluene) is equal to the external pressure (of one atmosphere, $\approx$ 100 kPa) and boiling takes place.

Since three phases and two components are present and the pressure is fixed, the phase rule ($P + F = C + 2$) tells us that the system is invariant. The composition of the vapour phase is fixed, and the mole fraction of each component is proportional to its vapour pressure at that

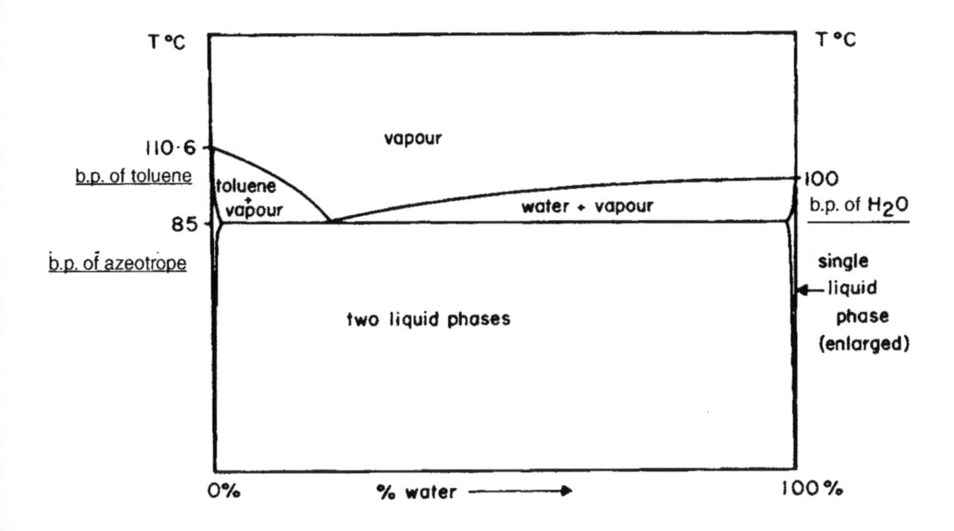

Fig. A3.7
Phase diagram of the toluene-water system

(From B. K. Selinger, 'Water, water, everywhere', Education in Chemistry. **16**, 1979, 125)

temperature. At 85°C the vapour pressure of water is 57.7 kPa and the vapour pressure of toluene is about 50.6 kPa, so the composition (by mass) of water will be (MM H_2O/MM toluene) × (v.p. H_2O/v.p. toluene) = 0.22 or 22%, which is slightly higher than found in practice (entry 1 in Table A3.1, overleaf).

Normally toluene is used as the carrier because of this favourable transfer rate for water. The azeotrope vapour consists of 20.2% water, and the composition of the lower layer of the distillate is 99.94% water. The toluene upper layer contains 0.06% water; this represents a small error, which is progressively reduced as the volume of water collected increases and the toluene in the upper layer is reduced by running back into the still. The distillation is finished when all the water has been transferred. The temperature of distillation rises as the system moves into the liquid toluene + vapour region in Figure A3.7.

The alkyl ethers appear to be ideal carriers for ethanol plus water mixtures (entry 3 in Table A3.1 overleaf), giving a lower layer virtually free of carrier. But these are notorious peroxide formers, and so are undesirable in a student laboratory.

Part IV: Between water and air — a use of Henry's Law

Transfer of organic compounds between bodies of water and the atmosphere is a key factor in environmental dispersion. The Henry's law constant (H) is effectively an air/water partition coefficient, and describes the equilibrium between solution in water and the vapour pressure P (in air) for neutral organic compounds at low concentration (x) in water.

Henry's law: $P = Hx$, a linear plot.

TABLE A3.1
Properties of azeotropes with water

Components		Azeotrope					
			Per cent composition				
				Distillate			
Compounds	b.p. (°C)	b.p. (°C)	In azeotrope vapour	Upper	Lower	Relative volume of layers at 20°C	Relative density of distillate layers
1. toluene	110.6	85.0	79.8	99.95	0.06	U 82.0	U 0.868
water	100.0		20.2	0.05	99.94	L 18.0	L 1.000
2. ethanol	78.5		37.0	15.6	54.8	U 46.5	U 0.849
toluene	110.6	74.4	51.0	81.3	24.5	L 53.5	O 0.855
water	100 0		12.0	3.1	20.7		
3. ethanol	78.5		21.2	20.6	32.0	U 95.5	U 0.722
vinylpropyl ether	65.1	57.0	73.7	77.8	0.2	L 4.5	L 0.923
water	100.0		5.1	1.6	67.8		
4. ethanol	78.5		12.0	3.0	75.0	U 90.0	U 0.672
hexane	69.0	56.0	85.0	96.5	6.0	L 10.0	L 0.833
water	100 0		3.0	0.5	19.0		
5. hexane	69.0	61.6	94.4			U 96.2	U 0.660
water	100.0		5.6			L 3.8	L 1.000

Source: After R. C. Weast, Handbook of chemistry and physics, 58th edn, CRC, Cleveland, Ohio, 1977. D180, D198.

Henry's law constants vary over 6 orders of magnitude. A high value of H indicates a great tendency to leave water bodies (large positive deviation from ideal behaviour).

For organic materials with only slight solubility in water, the linear plot of vapour pressure versus mole fraction has meaning only at the tiny bits at each end of the mole fraction scale, with a very large gap in the middle. Rather than try and measure a slope, it is simpler to proceed as follows:

Pure benzene will dissolve only a tiny amount of water, and and is saturated with water at 0.0023 mole fraction water. Not surprisingly, the vapour pressure of benzene saturated with this small amount of water is practically the same as that of pure benzene, 1.27×10^4 Pa. Conversely, pure water will dissolve only a tiny amount of benzene, and is saturated with benzene at 0.000 41 mole fraction. The vapour pressure of benzene from this water (that contains so little benzene) is the same as that of pure benzene. That is surprising.

The Henry's law constant H can be calculated from data on the organic component alone; one needs the maximum solubility in water and vapour pressure. For benzene the calculation is:

$$1.27 \times 10^4 / 0.000\,41 \text{ Pa},$$

agreeing well with the direct measurement of H for benzene of around 3×10^7 Pa.

Lead pottery glazes*

Introduction

For centuries, lead compounds have been included as constituents in pottery glazes because of the many advantages that accrue to both the potter and the consumer from their use. Lead compounds readily dissolve other essential ingredients such as alumina and silica to form a glaze with a high gloss and brilliance. Lead glazes have a relatively low melting point and a wide softening range. The low surface tension and viscosity of the melt allow minor imperfections on the surface of the clay body to be covered, and a ready release of trapped air and good healing of the surface. Generally, there is sufficient reaction between the molten glaze and the underlying clay to form an intermediate layer which relieves stresses and offers a high resistance to crazing and devitrification.

The affinity of lead glazes for colouring agents permits the development of a wide range of colours, many of which are difficult to attain by other means.

Not replaceable?
It is often claimed that lead cannot be replaced by any other material that will provide comparable aesthetic effects with equivalent ease and effort.

Lead is a toxic metal. When lead glazes are applied to surfaces used in contact with food and beverages, lead may be leached out by acids present in the food and beverage at levels which are toxicologically unsafe for human consumption. This hazard, primarily due to insufficiencies in glaze formulation, application and firing, has been recognised for a long time. Research into formulation, application and firing procedures has enabled safe lead glazes to be produced.

Ceramic glaze

A ceramic glaze is a thin glossy coating fused onto the clayware body. An analogy exists between a glaze and a mixture of rock components fused together at a high temperature, in much the same way as nature produces

* Source: reprint from *Some facts about lead glazes for workshop and studio potters*, a pamphlet issued by the National Health and Medical Research Council in 1975. Approved by the NHMRC, 77th session, November 1973.

molten lava in an erupting volcano, which then coats the earth with a glossy substance. Volcanic rocks can be pulverised and used as glazes. However, the major difference between a lava and an artificial glaze is that, in the latter, a flux is added.

Glazes are formulated from basic compounds such as alumina and silica with derivatives of barium, boron, calcium, lead, lithium, magnesium, potassium, sodium, strontium and zinc. In addition the formulations may include colouring agents containing antimony, cadmium, chromium, cobalt, copper, iron, manganese, nickel or selenium, as well as opacifiers such as tin, titanium and zirconium. Consequently, the chemistry of glazes is extremely complex, being further complicated by the reaction between the glaze and the clay body during the firing operation.

Flux

A flux is an additive which permits the basic components of a glaze to fuse together at a lower temperature to form a homogeneous mass. The more commonly used fluxes are compounds of boron, calcium, lead, potassium, and sodium.

Frit

A frit is a pre-formulated glaze. Selected raw materials are carefully proportioned, mixed, and fused in a high-temperature furnace to form a glass. The glass mass is then milled to a fine powder. The frit is evenly dispersed in water and applied to the surface of the shaped piece. During firing the particles re-melt to form a thin layer of finished glaze. Frits are generally designed to be either lead-free or contain a high percentage of lead. In the latter case, lead is chemically bound with the other constituents so that the level of leachable lead from the finished surface is negligible, even under the most stringent conditions. Lead leachability depends on different formulation parameters and is not necessarily related to the actual content of lead in the glaze.

Frits
By the use of frits it is possible for the potter to exercise rigid control over the formulation of the glaze.

Frits are designed to be safe when used by themselves or with compatible on-glaze decorations and other accessories. Their indiscriminate use with incompatible materials can lead to increased lead leachability. Such practices are to be strongly discouraged. Not all manufactured lead frits are safe. In some cases lead frits are merely fluxes. Many commercial glazes are designed exclusively for the decoration of artware such as tiles, sculptures and architectural ceramics. The normal use of such wares does not present a health hazard. Care should, however, be exercised to ensure that such commercial glazes are not used on wares that could contain food and beverages.

Formulation

Instead of purchasing a commercial frit, potters may elect to prepare glazes from their own basic constituents. A multitude of published recipes are available for this purpose. In the selection of a glaze recipe, only those proved reliable by laboratory examination of the finished article should be considered.

It is frequent practice to reduce the melting point of a glaze by the addition of a greater amount of flux. In so doing, the potter should realise that serious health hazards may be presented both personally and to the consumer. Some potters also, for economic reasons, pool excess glazes, and then use the haphazard mixture. This practice is to be strongly discouraged because of the uncontrolled imbalance of formulation that inevitably results.

With a full understanding of the physical chemistry of glazes, and giving proper care to formulation, safe glazes can be obtained. A very wide freedom of choice is permissible when glazes are intended for ornamental ware alone.

Colouring agents

Unsafe frits
Colours can make lead frits unsafe.

Problems sometimes occur when colouring agents are introduced. Although a lead frit or a balanced formulation may be safe, the incorporation of oxides or carbonates of copper or cobalt, and to a lesser extent those of nickel and of other metals, can cause a release of lead from the silicate matrix. Regardless of the care taken in firing, research has shown that the association of copper or cobalt with a lead glaze triggers off chemical interactions which predispose to a marked increase in lead leachability. In general, any glaze that finishes with a blue or green colour should not be applied to a surface of a utensil intended to contain food or beverage. In some instances black surfaces can also be unacceptable.

Leadless glazes

A wide range of leadless glazes with well-balanced formulations are available. When doubt remains on the reliability of a lead glaze intended for domestic ware, preference should be given to the use of a leadless glaze from a reliable source. Both leadless frits and many leadless formulations have been developed that provide excellent results with an attractive and serviceable finish. Leadless glazes that mature at temperatures above 1200°C and below 1200°C respectively for stoneware and ceramicware are available. In some instances where a high percentage of alkali is present, advanced knowledge and skill are required to avoid subsequent deterioration of the surfaces. Some formulations are intended entirely for art glazes on ornamental ware.

Kiln operation

Adequate glaze maturation requires due care to firing conditions, particularly kiln temperature, firing time, and the kiln atmosphere. At temperatures below 1080°C, glaze constituents may not react sufficiently with each other or with the clay body to provide adequate maturation. Sufficient time is also required to allow the products of reaction between the glaze and clay body to migrate through the glaze network, thereby strengthening the glaze and making it more impervious. For adequate maturation, a heavy glaze application (2 mm or more) requires considerably longer firing time than that of a thin layer. The correct firing schedule for temperature and time can be conveniently achieved by the use of pyrometric cones.

Oxidising conditions in the kiln atmosphere are essential. Without proper controls, products of combustion in gas, oil and wood-fired kilns may result in reducing conditions. These conditions favour the reduction of metallic compounds to a form in which they can no longer be securely bound to silica, alumina, and other adjuncts.

No alternative
An improperly fired glaze cannot be made safe by refiring, washing with acids, or baking in an oven.

Occupational hygiene

Unless handled with due care and with the use of proper equipment, lead glazes are hazardous to the health of the potter. Good housekeeping is important. The workshop should be vacuumed and mopped regularly. Any spill of material should be immediately damp sponged and, if dust appears, the workshop vacuumed thoroughly. Dust of any kind is to be avoided. All operations that disperse dust and fumes should be controlled by forced exhaust ventilation. A dust respirator complying with the specification of Australian standard AS Z18 should be used when dry glazes are mixed or ground. The spraying of glazes should only be done in a well-ventilated booth exhausted to the exterior.

As lead volatilises during the firing procedure, the atmosphere around the kiln may, under certain circumstances, constitute a health hazard. The kiln should therefore be located in an area where children or adults are not unwittingly exposed to lead fumes. The kiln should also be carefully fitted with a hood exhausting to the exterior.

Personal hygiene

Extreme care should be taken to avoid transferring lead from the hands to the mouth. If gloves are not worn when handling glazes, hands and fingernails should be thoroughly scrubbed upon finishing work. Food, drink and tobacco should not be brought into the workshop area. Changes of protective clothing should be provided for use in the workshop; but never worn elsewhere. Children should not be allowed to enter a workshop unless supervised by a responsible person.

Teaching of pottery crafts in schools

Council recommended that in the teaching of pottery crafts in primary and secondary schools, lead compounds should not be used in the making of utensils that could be used as containers for food and beverages.

Child poisoning

The problem of accidental poisoning (in children) is a serious one. The figures in the following tables for 1996 are from the Royal Adelaide Women and Children's Hospital. Earlier (and more detailed) statistics are given in previous editions of *Chemistry in the marketplace.*

TABLE A5.1
Poisoning cases treated

Category	Number of cases	% of total
Internal medicines	258	33.8
Foreign bodies	218	28.6
Animal hazards	152	19.9
Household, fertiliser, DIY	72	9.4
Plants	28	3.7
Chemicals and solvents	11	1.4
External medicines	9	1.2
Pesticides, herbicides and fungicides	8	1.0
Cosmetics	5	0.65
Noxious foods	1	0.1
Unspecified	1	0.1
Total	763	100

TABLE A5.2
Reason for poisoning

Reason	Number of cases	% of total
Accidental	441	57.8
Environmental (incl. bites and stings)	156	20.5
Intentional	119	15.6
Adverse eaction	47	6.2
Unknown	0	0
Total	763	100

TABLE A5.2
Age groups

Age group (years)	Number of cases	% of total
0–1	65	8.5
1–2	155	20.3
2–3	129	16.9
3–4	69	9.0
4–5	45	5.9
5–12	141	18.5
13–19	159	20.8
Adult (>20)	0	0
Unknown	0	0
Total	763	100

During 1996 the hospital received 14 628 calls related to poisoning but statistics of their nature are no longer kept. Cases treated in the hospital are recorded and the summary appears in the above tables.

Source: Courtesy of Di Milne, Pharmacy Department.

Drug-taking data — Australian Bureau of Statistics 1995

The 1995 National Health Survey gathered data on the (legal) drug taking habits of Australians. The sample was based on 1/3% of Australia's population and this equates to approximately 16 400 dwellings. This is considered reliable for broad categories and relatively large subpopulations spread fairly evenly around the country.

Only the more popular of the many drugs recorded in the survey have been reproduced in the table, and they are ranked in each group according to frequency of persons using them.

The relevant questions were:

- *In the last two weeks, have you taken any of these (prompt card) medications?*
- *What are the names and brands of all the medications you took in the last two weeks?*
- *Which of the categories best describes the medication?* (15 categories).

A number of subsidiary questions were asked relating to reasons, source of advice, prescription, length of use, and the like.

TABLE A6.1

Persons ('000) in Australia taking drugs

Type of generic medication	Males	Females	Persons
ANALGESICS			
Aspirin	548.8	519.4	1068.2
Paracetamol	1306.4	1841.5	3147.9
Paracetamol combinations	283.8	415.5	699.3
ARTHRITIS			
Anti-inflammatory, antirheumatic — non-steroidal			
Naproxen	79.6	112.4	192.0
Diclofenac	76.8	111.6	188.4
Piroxicam	43.8	60.8	104.5
Ketoprofen	40.9	48.8	89.7
Indomethacin	43.9	22.1	66.0

TABLE A6.1 (Continued)

Type of generic medication	Males	Females	Persons
ALLERGY			
Nasal decongestants			
Oxymetazoline	27.1	33.1	60.1
Tramazoline	11.7	9.4	21.1
Antihistamines (by mouth)			
Terfenadine	36.8	43.3	80.1
Promethazine	16.9	31.3	48.2
Dexchlorpheniramine	21.6	25.6	47.2
Loratadine	15.8	25.1	40.9
ASTHMA			
Adrenaline-type inhalants			
Salbutamol	478.3	495.5	973.8
Terbutaline	88.1	107.3	195.5
Other-type inhalants			
Beclamethasone dipropionate	191.7	244.9	436.6
Budesonide	139.5	159.0	298.4
Sodium cromoglycate	54.6	41.7	96.3
Non-inhalants (systemic)			
Theophylline	54.0	55.5	109.5
CHOLESTEROL LIPID LOWERING			
Simvastin	121.9	122.3	244.3
Gemifibrozil	33.5	15.4	49.0
Pravastatin	12.0	10.5	22.5
Nicotinic acid (niacin)	2.1	1.9	4.0
DIABETES			
Insulin	42.2	35.0	77.2
Oral blood glucose-reducing			
Metformin	45.8	51.0	96.9
Gliclazide	30.1	33.2	63.3
Glibenclamide	31.0	27.4	58.4
HEART AND BLOOD PRESSURE			
Digoxin	74.3	87.6	161.9
Isosorbide	58.0	56.1	114.1
Glyceryl trinitrate	53.0	36.0	89.0
Methyldopa	12.0	36.9	48.9
Enalapril	115.0	136.2	251.2
Beta-blockers			
Atenolol	121.0	118.8	239.8
Metropolol	77.9	86.1	164.0

TABLE A6.1 (Continued)

Type of generic medication	Males	Females	Persons
Calcium channel blockers			
Verapamil	78.2	122.7	201.0
Felodipine	66.9	74.3	141.2
Nifedipine	61.9	59.2	121.1
ANTI-ANXIETY			
Diazepam	43.9	64.1	108.0
Temazepam	40.2	64.5	104.7
Oxazepam	26.6	42.3	68.9
ANTI-DEPRESSANTS			
Dothiepin	26.8	37.7	64.5
Amitryptyline	25.1	34.3	59.5
Fluoxetine	23.2	29.4	52.6
SKIN OINTMENTS, CREAMS, ANTISEPTICS	788.9	947.6	1736.5
COUGH AND COLD	477.4	509.6	987.0
STOMACH MEDICATIONS	372.9	372.0	744.9
VITAMINS AND MINERALS	73.1	193.0	266.1
LAXATIVES	43.8	77.9	121.7
OTHER including ANTIBIOTICS	1323.7	2123.1	3446.8
TOTAL	8993.7	9067.3	18 061.0

Source: 1995 National Health Survey; Unpublished Data

Top 50 Pharmaceutical Benefits Prescriptions 1995–1996

Pharmaceutical Benefits Scheme: most prescribed items — 12 months to 30 June 1996

Rank #	Item	Form	Example of proprietary name	Example of drug or treatment	Script volume #	% of total script vol.	Total cost Gov. cost $	Total cost (incl. co-pay) $	Ave price $
1	Paracetamol	Tablet 500 mg	Panamax	Analgesic	3,754,416	3.01%	21,252,566	28,914,247	7.70
2	Codeine Phosphate with Paracetamol	Tablet 30 mg–500 mg	Dymadon & Panadeine	Analgesic	2,704,936	2.17%	15,304,754	21,104,801	7.80
3	Ranitidine Hydrochloride	Tablet 150 mg (base)	Zantac	Antacid (ulcers)	2,475,063	1.98%	68,640,700	81,665,468	33.00
4	Temazepam	Capsule 10 mg	Normison	Psycholeptic	2,361,686	1.89%	9,931,238	14,778,429	6.26
5	Salbutamol	Oral inhalation 100 mcg, 200 doses	Ventolin	Anti-asthmatic	2,078,215	1.66%	14,400,974	18,973,699	9.13
6	Atenolol	Tablet 50 mg	Tenormin	Beta-blocking agent	1,929,549	1.55%	16,108,714	20,652,123	10.70
7	Simvastatin	Tablet 10 mg	Zocor	Serumlipidreducing agents	1,720,489	1.38%	61,931,820	72,241,865	41.99
8	Influenza vaccine	Injection (trivalent) 0.5 mL	Fluvax	Vaccine	1,714,985	1.37%	19,653,898	29,938,305	17.46
9	Oxazepam	Tablet 30 mg	Serepax	Psycholeptic	1,422,935	1.14%	5,461,581	8,315,479	5.84
10	Enalapril Maleate	Tablet 20 mg	Renitec	ACE Inhibitor	1,375,609	1.10%	43,473,231	52,387,725	38.08
11	Diazepam	Tablet 5 mg	Valium	Psycholeptic	1,175,071	0.94%	5,342,396	7,784,462	6.62
12	Simvastatin	Tablet 20 mg	Zocor	Serumlipidreducing agents	1,151,640	0.92%	62,463,268	69,762,944	60.58
13	Enalapril Maleate	Tablet 10 mg	Renitec	ACE inhibitor	1,143,663	0.92%	26,316,789	33,616,425	29.39
14	Frusemide	Tablet 40 mg	Lasix	Diuretic	1,102,498	0.88%	6,089,755	8,278,001	7.51
15	Amoxycillin	Capsule 500 mg	Amoxil	Antibacterials for systemic use	1,086,867	0.87%	9,456,897	12,065,585	11.10

(Continued overleaf)

Appendix VII (Continued)

Rank	Item	Form	Example of proprietary name	Example of drug or treatment	Script volume	% of total script vol.	Total Cost Gov cost	Total Cost (incl. co-pay)	Ave price
#					#		$	$	$
16	Roxithromycin	Tablet 150 mg	Rulide	Antibacterials for systemic use	987,447	0.79%	9,903,745	12,232,932	12.39
17	Isosorbide Mononitrate	Tablet 60 mg (sustained release)	Imdur Durule	Cardiac therapy	979,717	0.78%	19,765,398	22,906,019	23.38
18	Verapamil Hydrochloride	Tablet 240 mg (sustained release)	Cordilox	Calcium channel blockers	978,830	0.78%	11,078,764	17,080,474	17.45
19	Amoxycillin with Clavulanic Acid	Tablet 500 mg–125 mg	Augmentin Forte	Antibacterials for systemic use	976,416	0.78%	11,511,758	16,541,982	16.94
20	Diclofenac Sodium	Tablet 50 mg (enteric coated)	Voltaren	Anti-inflammatory and antirheumatic	956,939	0.77%	7 982 577	10 165 267	10.62
21	Amlodipine Besylate	Tablet 5 mg (base)	Norvasc	Calcium channel blockers	905,072	0.72%	16,641,414	22,101,536	24.42
22	Cephalexin	Capsule 500 mg	Keflex	Antibacterials for systemic use	888,175	0.71%	7,906,935	10,018,647	11.28
23	Felodipine	Tablet 5 mg (extended release)	Agon / Plendil	Calcium channel blockers	853,017	0.68%	12,432,192	17,531,058	20.55
24	Nitrazepam	Tablet 5 mg	Mogadon/Alodorm	Psycholeptics	829,637	0.66%	3,581,976	5,192,638	6.26
25	Metformin Hydrochloride	Tablet 500 mg	Glucophage/ Diaformin	Drugs used in diabetes	823,731	0.66%	10,465,115	12,268,417	14.89
26	Omeprazole	Capsule 20 mg	Losec	Antacids,drugs to treat peptic ulcers	814,244	0.65%	75,425,122	80,792,921	99.22
27	Beclomethasone Dipropionate	Oral inhalation 250 mcg, 200 doses	Becloforte	Anti-asthmatics	805,806	0.65%	20,331,471	25,384,244	31.50
28	Famotidine	Tablet 20 mg	Pepcidine/Amfamox	Antacids,drugs to treat peptic ulcers	803,502	0.64%	21,255,790	25,820,405	32.13
29	Felodipine	Tablet 10 mg (extended release)	Agon/Plendil	Calcium channel blockers	787,691	0.63%	21,784,494	26,872,137	34.12
30	Oestradiol	Transdermal patches 4 mg/24 hours, 8	Estraderm 50	Sex hormones and modulators	770,904	0.62%	9,021,472	16,733,605	21.71
31	Prochlorperazine	Tablet 5 mg	Stemetil	Antiemetics and antinauseants	766,057	0.61%	3,653,120	5,258,797	6.86
32	Salbutamol Sulfate	Nebuliser sol. 5 mg in 2.5 mL, 30	Ventolin nebules	Anti-asthmatics	754,566	0.60%	19,866,985	22,669,066	30.04

(Continued)

33	Beclomethasone Dipropionate	Oral inhalation 100 mcg, 200 doses	Becotide 100	Anti-asthmatics	698,098	0.56%	6,629,835	12,000,857	17.19
34	Budesonide	Powder for inhal. 400 mcg, 200 doses	Pulmicort turbuhaler	Anti-asthmatics	680,971	0.55%	25,182,582	29,999,343	44.05
35	Enalapril Maleate	Tablet 5 mg	Renitec M/ Amprace 5	Agents acting on Renin-angiotensin	648,374	0.52%	10,277,267	13,716,451	21.16
36	Timolol Maleate	Eye drops 5 mg (base)	Timoptol/Tenopt	Ophthalmologicals	630,295	0.50%	6,666,280	8,042,910	12.76
37	Fluoxetine Hydrochloride	Capsule 20 mg (base)	Prozac 20/Lovan	Psychoanaleptics	617,394	0.49%	38,632,781	43,795,170	70.94
38	Aspirin	Tablet 300 mg (dispersible)	Solprin	Analgesics	611,591	0.49%	2,388,517	3,668,556	6.00
39	Budesonide	Aqueous nasal spray 100 mcg, 200 doses	Rhinocort aqueous	Nasal preparations	607,066	0.49%	5,936,908	11,545,680	19.02
40	Perindopril Erbumine	Tablet 4 mg	Coversyl	Agents acting on Renin-angiotensin	601,392	0.48%	15,174,533	19,503,414	32.43
41	Ketoprofen	Capsule 200 mg (sustained release)	Orudis SR 200	Anti-inflammatory & antirheumatic	600,283	0.48%	7,540,861	8,890,000	14.81
42	Potassium Chloride	Tablet 600 mg (sustained release)	KSR/Slow-K/Span-K	Mineral supplements	569,336	0.46%	4,506,074	5,639,805	9.91
43	Famotidine	Tablet 40 mg	Pepcidine	Antacids, drugs to treat peptic ulcers	568,443	0.46%	15,501,679	18,738,411	32.96
44	Allopurinol	Tablet 300 mg	Zyloprim/Progout 300	Antigout preparations	557,630	0.45%	4,358,231	5,554,595	9.96
45	Moclobemide	Tablet 150 mg	Aurorix/Arima	Psychoanaleptics	556,753	0.45%	20,703,563	24,841,949	44.62
46	Cefaclor	Tablet 375 mg (sustained release)	Ceclor CD	Antibacterials for systemic use	552,317	0.44%	6,640,980	7,955,111	14.40
47	Amoxycillin	Capsule 250 mg	Amoxil	Antibacterials for systemic use	550,856	0.44%	2,897,092	4,237,044	7.69
48	Indapamide	Tablet 2.5 mg	Natrilix/Dapa-Tabs	Diuretics	545,646	0.44%	7,322,153	10,448,327	19.15
49	Flucloxacillin	Capsule 500 mg	Floxapen/Flopen	Antibacterials for systemic use	536,741	0.43%	5,250,025	10,547,721	19.65
50	Ranitidine Hydrochloride	Tablet 300 mg (base)	Zantac	Antacids, drugs to treat peptic ulcers	531,440	0.43%	14,902,317	17,956,166	33.79
					52,543,999	42.07%	868,948,586	1,087,131,211	
Total PBS scripts 95-96 financial year					124,888,280		2,207,446,099	2,685,548,163	

The entropy game

The following game was developed to illustrate the statistical approach to entropy, discussed at the beginning of Chapter 14. Perhaps not quite the chemist's answer to Dungeons and Dragons, but Counters and Grids™ not only introduces you to the Einstein solid and statistical thermodynamics, but also to social theory, organisational structure, political prioritisation and epidemiology. It displays in one *connected* swoop, the main probability distributions of interest to chemists.

You proceed as follows: set up a grid, say 6 × 6, and on it place any number of counters, say 108 (to give a ratio of three counters per grid). You can place them on randomly or evenly, or all on one square, or choose your own unique scheme. You then plot out a histogram of the number of squares with 0, 1, 2, etc. counters versus the number of counters 0, 1, 2, etc. For random placement, you place a counter on each square as specified by the throw of a pair of dice (6 sided for a 6 × 6 grid, or 4, 8, 10 sided, etc. for other grids), to specify a particular square, like in a city map reference. One die defines the horizontal coordinate and the other the vertical coordinate. The result of one attempt is seen in Figure A8.1.

Now throw the dice again but use the result in the following manner. On the first throw of the dice, you pick up a counter, if there is one, from

2	2	1	3	4	6
1	4	3	4	2	2
1	3	2	4	1	1
3	6	6	2	4	1
5	6	1	4	2	5
2	4	4	1	1	5

Fig. A8.1 Number of counters on each square (random placement)

the square selected by the throw. If you hit a blank square you throw again. On the next throw you place the counter down again, on the square selected by the throw. You repeat this hundreds or thousands of times and note the changing shape of the histogram along the way. What happens is very interesting. No matter how you start you end up with the same result, a fluctuating, approximately exponentially falling distribution. Let us start with 108 counters placed randomly on a 6 × 6 grid. You usually obtain a histogram with a hump in the middle, as in Figure A8.2 (approaches normal for a high ratio of counters to squares).

The *random* experiment of dice throwing (say 1000 times) causes the histogram to change shape (see Fig. A8.3) to an exponential distribution.

This becomes more obvious if you smooth out the fluctuations and display the *average* of a number of distributions (say 500), after the 1000 iterations (see Fig. A8.4).

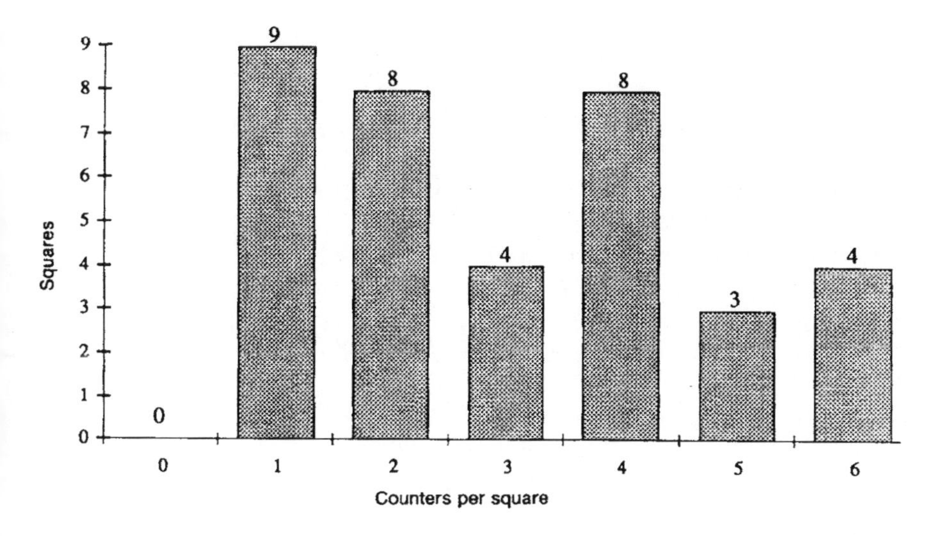

Fig. A8.2
Number of squares with 0,1,2, etc. counters. 108 counters on 36 squares. $W = 1.825 \times 10^{23}$; variance = 2.857

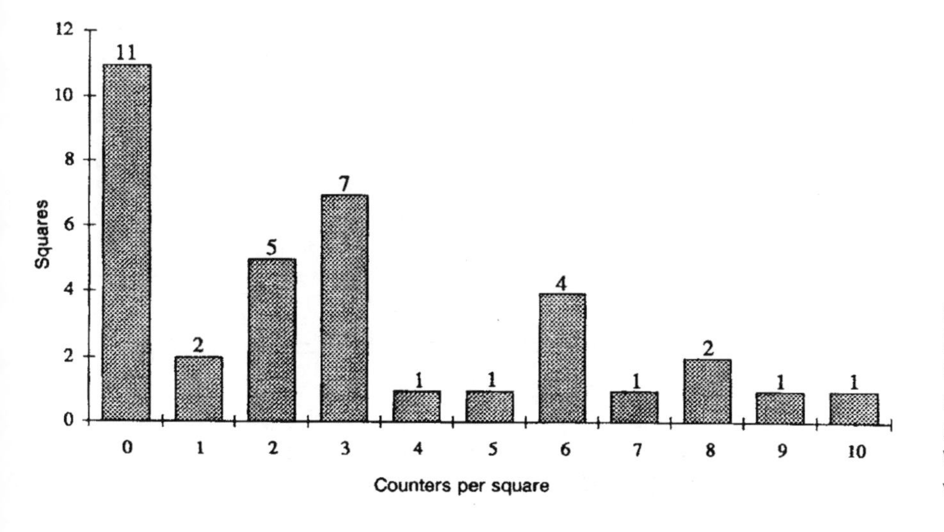

Fig. A8.3
Histogram after 1000 iterations. 108 counters on 36 squares.
$W = 1.6 \times 10^{26}$; variance = 8.686

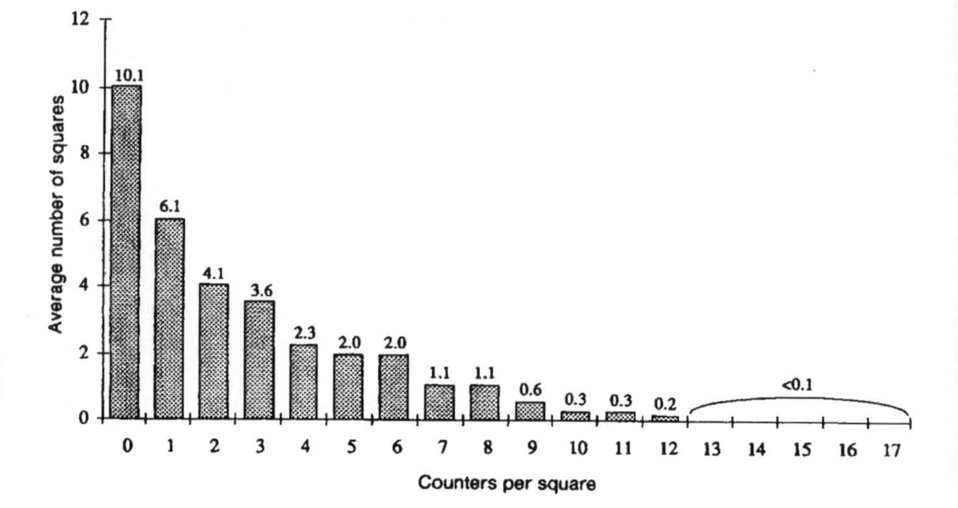

Fig. A8.4
Histogram averaged for 500 iterations, after 1000 iterations, variance = 10.747

The explanation for the changes can be seen in Table A8.1 if we calculate the number of possibilities for each arrangement.[1] Consider a starting position that is uniform and calculate the possibilities of each consequent position. N is the number of squares while the n_0, n_1 etc. are the number of squares with 0, 1, etc. counters. W is the number of possibilities of distributing counters amongst squares without changing the histogram. The sample variance is given by:

$$S^2 = \frac{\Sigma(y_i - y)}{(N-1)} = \frac{[\Sigma y_i - (\Sigma y_i)^2/N]}{(N-1)}$$

where y_1 is the sample value and y is the mean.

TABLE A8.1
Number of squares with 0,1,2 . . . counters q

Number of throws	n_0	n_1	n_2	n_3	n_4	n_5	n_6	$W = \frac{N!}{n_0!\,n_1!\,n_2!\ldots}$	Variance
uniform	0	36	0	0	0	0	0	1	0
5	5	26	5	0	0	0	0	6×10^{10}	0.29
10	13	12	9	2	0	0	0	2×10^{17}	0.57
20	18	9	4	2	2	1	0	1.67×10^{18}	1.83
40	16	9	7	3	1	0	0	1.62×10^{18}	1.26
60	15	11	6	3	1	0	0	1.65×10^{18}	1.17
80	15	10	8	2	1	0	0	9.72×10^{17}	1.14
100	19	9	3	2	1	1	1	7×10^{17}	2.29
All on one	0	0	0	0 . . . n_{36} = 1				36	36

Thus for

n_0	n_1	n_2	n_3	n_4
16	9	7	3	1

$$S^2 = [\{(16 \times 0^2) + (9 \times 1^2) + (7 \times 2^2) + (3 \times 3^2) + (1 \times 4^2)\} - 36]/35 = 1.26$$

The maximum number of possibilities means keeping the product of the numbers on the bottom line of the expression for W_0 as small as possible, that is keeping the numbers themselves as small as possible, subject to two constraints:

1. $(n_1 \times 1) + (n_2 \times 2) + (n_3 \times 3) + \ldots$ = no. of counters
2. $n_0 + n_1 + n_2 + n_3 + \ldots$ = no. of squares

Note the following:

1. Very few changes in the original distribution cause a rapid increase in W.
2. W_{max} is the value of W for the most probable configuration and is of the order of 10^{18}.
3. The distribution can fluctuate at equilibrium provided the value of W stays near to W_{max}.

W_{total} is the sum of all the possibilities for all the configurations and is given by[2]:

$$W_{total} = \frac{(N+q-1)!}{(N-1)!\,q!} = \text{sum of } W \text{ over all configurations.}$$

For $N = 36$, $q = 36$, $W_{max} = 1 \times 10^{18}$, $W_{total} = 2 \times 10^{20}$.

Some idea of the magnitude of the number of possibilities can be obtained when you consider the number of possibilities for setting Rubik's Cube. Hofstadter[3] estimates this as 4.3×10^{19} (43 252 003 274 489 856 000, to be precise). You can imagine that throwing a dice to select a move will never bring a random cube configuration back to the single START.

This, then, is the rationale of the Second Law; systems drift towards configurations that are made up from the greatest number of possibilities or indistinguishable microstates.

There are some other interesting observations to make about the distributions. When you start placing counters at random one at a time on the squares, you produce a *binomial* distribution. For 36 squares and 36 counters (using the usual statistical notation), $n = 36$ and $p = 1/36$. The mean is $np = 1$, and the expected variance is $np(1-p)$. Under these conditions, the binomial is approximately *Poisson* where the mean is μ and equals the expected variance, because $np(1-p) \longrightarrow np$. When, in addition, μ is small, the Poisson distribution becomes a geometric distribution. On the other hand, keeping p the same but increasing n to give a *large* $np(\geq 20)$, we find that the binomial $\longrightarrow$ Poisson, tends to a *normal* or bell-shaped error distribution (still with equal mean and variance). With 108 counters on 36 squares we are on the way towards this limit.

When we start the game with 108 on 36, we quickly see that the random starting distribution does not have the largest number of possibilities W, and thus the distribution changes in shape with play. Seeing that we started with a *random* placement, to say that the drive is now dominated by a *tendency to randomness* is unhelpful, unless we explain how the constraints have changed when we changed the rule. We started by *depositing* the counters binomially. This fixed the expected variance at $np(1-p)$ (= 3). However, as the game proceeds, the variance is no longer constrained, but increases, and another distribution takes over.

For *aficionados* you can actually derive the equilibrium to which this distribution tends. For i counters per square the probability distribution $\Pi(i)$ is given by:

$$\Pi(i) = \frac{(N+q-i-2)C(N-1)}{(N+q-1)C(N)}$$

where **C** is the combination symbol.

In the limit as $N \longrightarrow \infty$, $q \longrightarrow \infty$ (or taking an average of a large number of runs), the distribution becomes a *geometric* distribution. The expected equilibrium probability $\Pi(i)$, for this limit is now given[4] as:

$$\Pi(i) = \{1/(1+\theta)\} \times \{\theta/(1+\theta)\}^i$$

$$\text{where } \theta = q/N$$

$$\text{The variance} = \{\theta/(1+\theta)\} \div \{1/(1+\theta)^2\}$$

Thus for 36 counters on 36 squares the equilibrium distribution $\Pi(i)$ is given by $(½) \times (½)^i$, while for 108 counters it is $¼ \times (¾)^i$ (The expected equilibrium variances are 2 and 12 respectively. Try some experiments to show that this is so.) These geometric distributions are discrete *exponential* in the limit of large (or averaged) samples.

This game illustrates the fundamental concept of the Second Law of Thermodynamics, the understanding of which the author, C. P. Snow, declared in his Rede lecture[5] as *the* test of scientific literacy, the equivalent to being able to appreciate a play by Shakespeare. Interestingly enough, he later drew back from this test. 'This law', he later said, 'is one of the greatest depth and generality; it has its own sombre beauty; like all major scientific laws it evokes reverence. There is of course no value in the non-scientist knowing it by the rubric in an encyclopaedia. It needs understanding, which cannot be attained unless one has learned some of the language of physics. That understanding ought to be part of the common twentieth century culture. Nevertheless, I wish I had chosen another example.' He goes on to say that he would now have chosen molecular biology, but adds, 'the ideas in this branch of science are not as physically deep or of such universal significance as those of the Second Law. The Second Law is a generalisation which covers the cosmos'.

Counters and grids is a game that provides a small-scale simulation of the probability interpretation of entropy, the Einstein solid, Bose–Einstein statistics with the Boltzmann distribution as limit. In fact the Boltzmann limit is found for any large-scale distribution for which the only constraint, or sole information, is the constant value of the mean.

Examples include the intensity of light passing through an absorbing solution (constant absorbance); first order kinetics (constant rate constant); radioactive or fluorescence decay (constant lifetime); radial electron density in an s orbital (fixed Bohr radius) etc.

However what is incredibly interesting is that the thinking behind the Second Law is by no means limited to these physical examples.

Consumer examples — naturally experimental

Just call the counters *money* and the squares *people*. You then discover a truism. In a laissez-faire economy in which money is exchanged freely between individuals for goods and services without any restrictions (in particular redistribution via taxation and social benefits; the model fits authoritarian economies better than democratic ones), most people end up with few dollars and a few end up with most of the dollars. Even if you increase the mean number of counters per square from less than one to greater than one, the final distribution does not change significantly in shape, so people's relativities hardly change, even though in absolute terms they are better off.

A closer model is where hundreds of people enter a casino with, say, an equal modest stake and at the end of the evening leave with their pockets emptied or filled. Their group winnings follow very closely the exponential distribution. It is extremely important to note that we cannot predict who will be rich and who will be poor. The more possibilities for distributing the money *between* people without affecting the *overall* distribution, the more probable is that distribution.

Even more interesting is that even where we ourselves control the result, as for example in the way the organisers deliberately distribute prizes in lotteries, there are few large prizes, moderate numbers of moderate prizes and many small prizes. With a fixed amount of prize money and number of prizes, the exponential distribution is the most *natural* way to do things.

Lottery odds
Would you buy a ticket in a lottery if all the prizes were equal and only modest?

You say that you're not into gambling. Well you are, you know! As a policy holder in any insurance scheme (car, health, house, etc.), you contribute a small premium to protect yourself against the *certainty* that natural events will dictate that a few people will need large payouts and more will need smaller ones. The small premium monies will redistribute themselves as claims, roughly along the lines of the exponential distribution.

From the probability of occurrence of individual letters in this text to the size of cities in a country, or oil wells in the world, or stars in the galaxy, the distribution as determined by the Second Law of Thermodynamics (with minor reservations) gives the gist of the answer. Further examples can be seen in the loss of control of a private company when it distributes shares to a large public (look at the share distribution of shareholders in a large company).

If we move slightly laterally to the random exchange of power and influence in an organisation, then its distribution follows the same pattern. The number of people at any level decreases with the height of the level. Hierarchies behave just like the real atmosphere and for the same reason; they become rarer the higher you go.

The thinking behind the Second Law has other everyday analogues. For example, what is the fundamental reason for it being so much harder to park a car in a tight spot than to drive it out again afterwards?

(Consider the number of possibilities for being parked compared to being 'unparked'.)

One could stretch the argument and say that the reason for lack of support for teaching innovation is a consequence of the Second Law. It is much more natural to distribute funds to provide a small spectrum of more spectacular items to fill annual reports and for parliamentarians to make speeches about, than to uniformly improve the whole field. Laying foundation stones for high tech hospitals beats small and unnoticed public health improvements across the board. For the same reason, it is very difficult to replace the concept of a large centralised power generation system (be it coal-fired or nuclear) with many small, widely distributed solar units. Finally, no lottery would attract custom if, for the same total amount of prize money, it offered only equal, relatively small prizes and no outstanding ones.

In regard to quantum theory, Einstein once said 'God does not play dice'. Einstein was probably wrong. Anyway, in thermodynamics he definitely does play dice. And so do we, all the time.

Epidemiology (See also Chapter 2.)

For our last example, we return to the random distribution with which we set up the game in Figures A8.1 and A8.2 and study it in more detail. We use the distribution to study epidemiology — incidence of disease in the community.[6] Leukaemia (which is a combination of related diseases) is responsible for about one in 40 deaths in the West, so it is rather rare. However it is the main cancer found in children. While there are several postulated causes of leukaemia, there have been suggestions that childhood leukaemia, in particular, may also be caused by low levels of radioactivity, such as those emitted by nuclear installations. A UK TV program in 1983 revealed an apparent cluster of four children with leukaemia close to the nuclear reprocessing plant at Sellafield, formerly Windscale. The statistical occurrence of clusters when the expected value is quite low can be illustrated with our random distribution.

Choose a 12 × 12 grid and enter 72 counters at random. The average number of counters per square is 0.50. However notice the occupancy of the grids. In the particular throw seen in Figure A8.5, there are 94 squares with no counters, 35 with one counter, 9 with two counters, 5 with three counters and none with four counters. Although the average number of counters per grid is 0.5, one grid contains 4 counters, eight times the average. The expected number is given by the Poisson distribution:

$$N(x) = \frac{\mu^x \exp[-\mu]}{x!}$$

where the mean $\mu = 0.5$, and x = number of counters per square. Thus $N(x)$ is the number of squares with x counters.

The Poisson distribution predicts, on average, 87 squares with zero counters, 44 with one counter, 11 with two counters, 2 with three counters and 0.02 with four counters. Our result is in good agreement with the expected (as can be shown by a statistical test of goodness of fit). We note that four counters on one grid would be expected on average every 1/0.02 = five throws.

Although on average two-thirds of the grids are empty, this is clearly not the probability of obtaining empty squares, since it is impossible to cover 144 squares with 72 counters. The probability of finding one or more empty grids is 100%. The frequency with which multiple occupancy occurs is [9 + 5 + 1]/144 or 10%, and yet the probability that at least one square has multiple occupancy is:

$$1 - (144/144) \times 143/144) \times (132/144) \times \ldots (73/144).$$

which is over 99.9%.

Exercise
Fill a large jar with jellybeans or Smarties with a light and a dark colour (ratio 8:1). Shake the jar. Note the clusters.

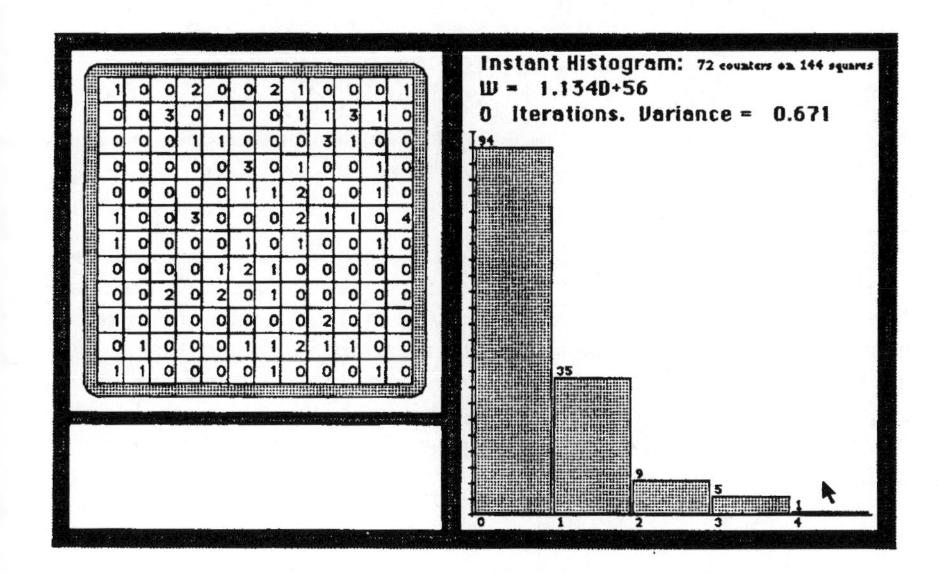

Fig. A8.5 Random distribution of 72 counters on 144 squares

For this section you need to have access to a computer, because you need to throw the dice 72 times (or your choice) each time. However you can play a similar game by packing the bottom of a shallow dish (such as a petri dish) with, say, 400 white beads and 50 (i.e. one-eighth) coloured beads. Pour the balls from one dish to another and then let them settle flat. If the balls are fairly close packed, then the distribution will look fairly even and each ball will be surrounded by about six others (see Fig. A8.6 (overleaf) and also the section and plates on models for atom packing in the Introduction in Chapter 10). However the random distribution is far from even in colour, and will on average reveal several colour clusters. A random distribution is not uniform.

If you divide the space up into smaller regions, you can produce a greater or smaller number of regions with clusters, depending on how you gerrymander the boundaries. In fairness, you should mark out the divisions before you shake the balls, not afterwards!

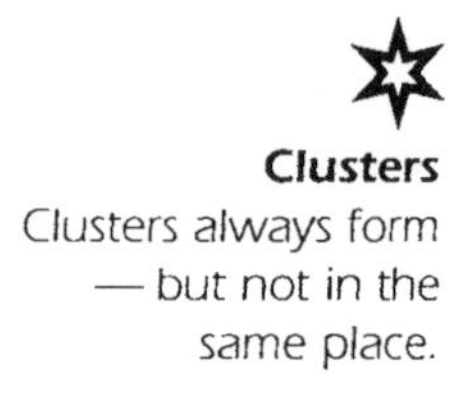

Clusters
Clusters always form — but not in the same place.

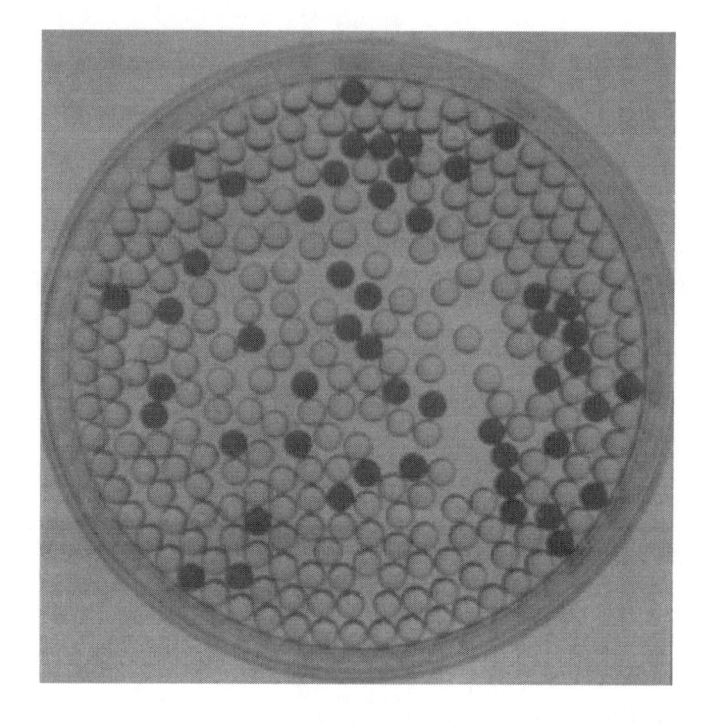

Fig. A8.6

Crossing the road
Why you can always cross a busy road — at some time.

Producing 'honest electorates' is a central problem in epidemiology.

The game can also be used to illustrate another interesting occurrence in everyday life. When you are waiting to cross at a busy intersection, eventually you get a break. Assume that the traffic is random, that is there are no obvious traffic lights or other obstacles that are bunching up the traffic. The 10-second interval you need to cross the intersection can be made to be equivalent to say four adjacent empty squares. In our example there are, on average, half as many counters as squares, so on average every second square should be occupied. On this basis you should never be able to cross the intersection! However, moving along a row at a time in Figure A8.5 (above) shows that several clusters of four empty squares occur, and so you can cross the intersection.

References

1. The teaching of thermodynamics: a teaching problem and an opportunity', *School Science Review* **57**(210), 1976, 654; 'A picture of shuffling quanta', in *Nuffield Advanced Science Chemistry, Teachers' Guide I*, Topics 1-11, pp. 75–79, Longman, Harlow, Essex, 1984.
2. H. A. Bent, *The Second Law*, Oxford University Press, New York, 1965, ch. 21.
3. D. Hofstadter, *Metamagical themas: questing for the essence of mind and pattern*, Penguin, 1985, p. 305.
4. B. Selinger, and R. Sutherland 'Counters and grids', *J. Chem. Ed.*, 66, (1989) 506-508.
5. C. P. Snow, *The two cultures and a second look*, Cambridge University Press, Cambridge, 1969.
6. D. Taylor and D. Wilke, 'Drawing the line with leukemia', *New Scientist*, 21 July 1988, 53.

Approved additive numbers

The internationally established food additive numbers, approved by the Australian and New Zealand Food Authority (ANZFA) are listed below in numerical order. Vitamins and minerals that may be added to food to supplement the diet are not classed as additives, and are therefore not assigned a number, unless they perform an additional function, for example as an antioxidant.

Symbols used in this list are as follows:

α = alpha, β = beta, = δ = delta, γ = gamma

* may be contained in salt substitutes

\# Erythrosine is limited to use in maraschino cherries from March 1997.

No.	Substance	Use
100	Curcumin or turmeric	Colouring
101	Riboflavin	Colouring
102	Tartrazine	Colouring
103	Alkanet	Colouring
104	Quinoline yellow	Colouring
110	Sunset yellow FCF	Colouring
120	Carmines or cochineal	Colouring
122	Azorubine	Colouring
123	Amaranth	Colouring
124	Ponceau 4R	Colouring
127	Erythrosine	Colouring
129	Allura red AC	Colouring
132	Indigotine	Colouring
133	Brilliant blue FCF	Colouring
140	Chlorophyll	Colouring
141	Chlorophyll-copper complex, chlorophyllin copper complex, sodium and potassium salts	Colouring
142	Food green S	Colouring
150	Caramel	Colouring
151	Brilliant black BN	Colouring
153	Carbon black	Colouring
155	Brown HT	Colouring
160	Carotene, others	Colouring
160(a)	β-carotene	Colouring
160(b)	Annatto extract	Colouring
160(e)	β-apo-8′ carotenal	Colouring
160(f)	β-apo-8′ carotenoic acid methyl or ethyl ester	Colouring
162	Beet red	Colouring
163	Anthocyanins	Colouring
170	Calcium carbonate	Mineral salt, colouring
171	Titanium dioxide	Colouring
172	Iron oxide (black, red, yellow)	Colouring
181	Tannic acid	Colouring
200	Sorbic acid	Preservative
201	Sodium sorbate	Preservative

No.	Substance	Use
202	Potassium sorbate	Preservative
203	Calcium sorbate	Preservative
210	Benzoic acid	Preservative
211	Sodium benzoate	Preservative
212	Potassium benzoate	Preservative
213	Calcium benzoate	Preservative
216	Propylparaben	Preservative
218	Methylparaben	Preservative
220	Sulfur dioxide	Preservative
221	Sodium sulfite	Preservative
222	Sodium bisulfite	Preservative
223	Sodium metabisulfite	Preservative, flour treatment agent
225	Potassium metabisulfite	Preservative
228	Potassium sulfite, potassium bisulfite	Preservative
234	Nisin	Preservative
235	Natamycin	Mould inhibitor
249	Potassium nitrite	Preservative, colour fixative
250	Sodium nitrite	Preservative, colour fixative
251	Sodium nitrate	Preservative, colour fixative
260	Acetic acid, glacial	Food acid
261	Potassium acetate	Food acid
262	Sodium diacetate	Food acid
263	Calcium acetate	Food acid
264	Ammonium acetate	Food acid
270	Lactic acid	Food acid
280	Propionic acid	Preservative
281	Sodium propionate	Preservative
282	Calcium propionate	Preservative
283	Potassium propionate	Food acid
290	Carbon dioxide	Propellant
296	Malic acid	Food acid
297	Fumaric acid	Food acid
300	Ascorbic acid	Antioxidant, flour treatment agent
301	Sodium ascorbate	Antioxidant

No.	Substance	Use
302	Calcium ascorbate	Antioxidant
303	Potassium ascorbate	Antioxidant
304	Ascorbyl palmitate	Antioxidant
306	Tocopherols, concentrate, mix	Antioxidant
307	dl-α-tocopherol†	Antioxidant
308	γ-tocopherol	Antioxidant
309	δ-tocopherol	Antioxidant
310	Propyl gallate	Antioxidant
311	Octyl gallate	Antioxidant
312	Dodecyl gallate	Antioxidant
315	Erythorbic acid	Antioxidant
316	Sodium erythorbate	Antioxidant
319	tert-butylhydroquinone	Antioxidant
320	Butylated hydroxyanisole	Antioxidant
321	Butylated hydroxytoluene	Antioxidant
322	Lecithin	Antioxidant, emulsifier
325	Sodium lactate	Food acid
326	Potassium lactate	Food acid
327	Calcium lactate	Food acid
328	Ammonium lactate	Food acid
329	Magnesium lactate	Food acid
330	Citric acid	Food acid
331	Sodium citrates	Food acid
332	Potassium citrates	Food acid
333	Calcium citrate	Food acid
334	Tartaric acid	Food acid
335	Sodium tartrates	Food acid
336	Potassium tartrates	Food acid
337	Potassium sodium tartrate	Food acid
338	Phosphoric acid	Food acid
339	Sodium phosphates	Food acid
340	Potassium phosphates*	Food acid
341	Calcium phosphates*	Food acid
342	Ammonium phosphates*	Food acid
343	Magnesium phosphates*	Food acid
349	Ammonium malate	Food acid
350	DL-sodium malates	Food acid
351	Potassium malates	Food acid
352	DL-calcium malates	Food acid
353	Metatartaric acid	Food acid
354	Calcium tartrate	Food acid
355	Adipic acid	Food acid
357	Potassium adipate	Food acid
365	Sodium fumarate	Food acid
366	Potassium fumarate	Food acid
367	Calcium fumarate	Food acid
368	Ammonium fumarate	Food acid

† dl (and DL in 350 and 352) are optical isomer terms commonly used in medicine

No.	Substance	Use
375	Niacin	Colour retention agent
380	Ammonium citrates	Food acid
381	Ferric ammonium citrate	Food acid
385	Calcium disodium ethylenediaminetetraacetate	Sequestrant, preservative
400	Alginic acid	Thickener, vegetable gum
401	Sodium alginate	Thickener, vegetable gum
402	Potassium alginate	Thickener, vegetable gum
403	Ammonium alginate	Thickener, vegetable gum
404	Calcium alginate	Thickener, vegetable gum
405	Propylene glycol alginate	Thickener, vegetable gum
406	Agar	Thickener, vegetable gum
407	Carrageenan	Thickener, vegetable gum
409	Arabinogalactan	Thickener, vegetable gum
410	Locust bean gum	Thickener, vegetable gum
412	Guar gum	Thickener, vegetable gum
413	Tragacanth	Thickener, vegetable gum
414	Acacia	Thickener, vegetable gum
415	Xanthan gum	Thickener, vegetable gum

No.	Substance	Use
416	Karaya gum	Thickener, vegetable gum
420	Sorbitol	Bulking agent, humectant
421	Mannitol	Thickener, vegetable gum
422	Glycerin	Humectant
433	Polysorbate 80	Emulsifier
435	Polysorbate 60	Emulsifier
436	Polysorbate 65	Emulsifier
440	Pectin	Vegetable gum
442	Ammonium salts of phosphatidic acid	Emulsifier
444	Sucrose acetate isobutyrate	Emulsifier, stabiliser
450	Sodium and potassium pyrophosphates*	Mineral salt
451	Sodium and potassium tripolyphosphates*	Mineral salt
452	Sodium and potassium metaphosphates, polymetaphosphates and polyphosphates*	Mineral salt
460	Cellulose, microcrystalline and powdered	Anti-caking agent
461	Methylcellulose	Thickener, vegetable gum
464	Hydroxypropyl methylcellulose	Thickener, vegetable gum
465	Methyl ethyl cellulose	Thickener, vegetable gum
466	Sodium carboxymethylcellulose	Thickener, vegetable gum
470	Magnesium stearate	Emulsifier, stabiliser
471	Mono- and di-glycerides of fatty acids	Emulsifier
472a	Acetic and fatty acid esters of glycerol	Emulsifier
472b	Lactic and fatty acid esters of glycerol	Emulsifier

No.	Substance	Use
472c	Citric and fatty acid esters of glycerol	Emulsifier
472d	Tartaric and fatty acid esters of glycerol	Emulsifier
472e	Diacetyltartaric and fatty acid esters of glycerol	Emulsifier
473	Sucrose esters of fatty acids	Emulsifier
475	Polyglycerol esters of fatty acids	Emulsifier
476	Polyglycerol esters of interesterified ricinoleic acid	Emulsifier
477	Propylene glyco mono- and di-esters	Emulsifier
480	Dioctyl sodium sulfosuccinate	Emulsifier
481	Sodium oleyl or stearoyl lactylate	Emulsifier
482	Calcium oleyl or stearoyl lactylate	Emulsifier
491	Sorbitan monostearate	Emulsifier
492	Sorbitan tristearate	Emulsifier
500	Sodium carbonates	Mineral salt
501	Potassium carbonates*	Mineral salt
503	Ammonium carbonates*	Mineral salt
504	Magnesium carbonate*	Anti-caking agent, mineral salt
507	Hydrochloric acid	Acidity regulator
508	Potassium chloride*	Mineral salt
509	Calcium chloride*	Mineral salt
510	Ammonium chloride	Mineral salt
511	Magnesium chloride	Mineral salt
512	Stannous chloride	Colour retention agent
514	Sodium sulfate	Mineral salt
515	Potassium sulfate	Mineral salt
516	Calcium sulfate	Flour treatment agent, mineral salt
518	Magnesium sulfate	Mineral salt
519	Cupric sulfate	Mineral salt
526	Calcium hydroxide	Mineral salt
529	Calcium oxide	Mineral salt
535	Sodium ferrocyanide	Anti-caking agent
536	Potassium ferrocyanide	Anti-caking agent
541	Sodium aluminium phosphate, acidic	Acidity regulator, emulsifier
542	Bone phosphate	Anti-caking agent
551	Silicon dioxide	Anti-caking agent
552	Calcium silicate	Anti-caking agent
553	Talc	Anti-caking agent
554	Sodium aluminosilicate	Anti-caking agent
556	Calcium aluminium silicate	Anti-caking agent
558	Bentonite	Anti-caking agent
559	Kaolin	Anti-caking agent
570	Stearic acid	Anti-caking agent
575	Glucono δ-lactone	Acidity regulator, raising agent
577	Potassium gluconate	Stabiliser
578	Calcium gluconate	Acidity regulator, firming agent
579	Ferrous gluconate	Colour retention agent
520	L-glutamic acid	Flavour enhancer
621	Monosodium L-glutamate	Flavour enhancer
622	Monopotassium L-glutamate	Flavour enhancer
623	Calcium di-L-glutamate	Flavour enhancer
624	Monoammonium L-glutamate	Flavour enhancer
625	Magnesium di-L-glutamate	Flavour enhancer
627	Disodium guanylate	Flavour enhancer
631	Disodium inosinate	Flavour enhancer
635	Disodium 5'-ribonucleotides	Flavour enhancer
636	Maltol	Flavour enhancer

No.	Substance	Use
637	Ethyl maltol	Flavour enhancer
900	Dimethylpolysiloxane	Emulsifier, anti-foaming agent, anti-caking agent
901	Beeswax, white and yellow	Glazing agent, release agent
903	Carnauba wax	Glazing agent
904	Shellac, bleached	Glazing agent
905a	Mineral oil white	Glazing agent
905b	Petrolatum	Glazing agent
920	L-cysteine monohydrochloride	Flour treatment agent
925	Chlorine	Bleaching agent
926	Chlorine dioxide	Bleaching agent
928	Benzoyl peroxide	Bleaching agent
941	Nitrogen	Propellant
942	Nitrous oxide	Propellant
950	Acesulfame potassium	Artificial sweetening substance
951	Aspartame	Artificial sweetening substance
952	Cyclamates	Artificial sweetening substance
953	Isomalt	Humectant
954	Saccharins	Artificial sweetening substance
955	Sucralose	Artificial sweetening substance
956	Alitame	Artificial sweetening substance
957	Thaumatin	Flavour enhancer, artificial sweetening substance
965	Maltitol and maltitol syrup	Humectant
966	Lactitol	Humectant
967	Xylitol	Humectant
1100	Amylases	Enzyme — flour treatment agent
1101	Proteases (papain, bromelain, ficin)	Enzymes — flour treatment agent, stabiliser, tenderiser, flavour enhancer
1102	Glucose oxidase	Enzyme, antioxidant
1104	Lipases	Enzyme, flavour enhancer
1105	Lysozyme	Enzyme, preservative
1200	Polydextrose	Humectant
1201	Polyvinylpyrrolidone	Stabiliser, clarifying agent, dispersing agent
1202	Polyvinylpolypyrrolidone	Colour stabiliser, colloidal stabiliser
1400	Dextrin roasted starch	Thickener, vegetable gum
1401	Acid treated starch	Thickener, vegetable gum
1402	Alkaline treated starch	Thickener, vegetable gum
1403	Bleached starch	Thickener, vegetable gum

No.	Substance	Use
1404	Oxidised starch	Thickener, vegetable gum
1405	Enzyme-treated starches	Thickener, vegetable gum
1410	Monostarch phosphate	Thickener, vegetable gum
1412	Distarch phosphate	Thickener, vegetable gum
1413	Phosphated distarch phosphate	Thickener, vegetable gum
1414	Acetylated distarch phosphate	Thickener, vegetable gum
1420	Starch acetate esterified with acetic anhydride	Thickener, vegetable, gum
1421	Starch acetate esterified with vinyl acetate	Thickener, vegetable gum
1422	Acetylated distarch adipate	Thickener, vegetable gum1440
1440	Hydroxypropyl starch	Thickener, vegetable gum
1442	Hydroxypropyl distarch phosphate	Thickener, vegetable gum
1450	Starch sodium octenyl succinate	Thickener, vegetable gum
1505	Triethyl citrate	Thickener, vegetable gum
1518	Triacetin	Thickener, vegetable gum
1520	Propylene glycol	Humecant, wetting agent, dispersing agent

Chemicals available in the marketplace

It should be noted that some of the chemicals listed in Table A10.1 as readily available are very poisonous (e.g. cadmium salts) and many are very dangerous in mixtures (e.g. strong oxidising agents, potassium dichromate, calcium hypochlorite 'chlorine', etc. with oxidisable material: see Chapter 8).

TABLE A10.1
Chemicals readily available in the marketplace

Chemical name	Common name	Availability
Acetic acid (dilute)	Vinegar	Supermarket
Acetone	Nail polish remover	Hardware
Aluminium	Aluminium (foil, cans)	Supermarket
Aluminium oxide (hydrated)	Alumina hydrate	Craft
Ammonia solution	Cloudy ammonia	Supermarket
Ammonium chloride	Sal ammoniac	Hardware
Ammonium nitrate	Ammonium nitrate	Garden supplies
Ammonium sulfate	Sulfate of ammonia	Garden supplies
Ascorbic acid	Vitamin C	Pharmacy, supermarket
Beeswax	Beeswax	Craft
Barium carbonate	Barium oxide flux	Craft
Barium chromate	Barium chromate	Craft
Benzoic acid	Preservative	Health food
Boric acid	Boracic acid	Pharmacy
Cadmium selenide	Red overglaze, cadmium red, pigment red 108	Craft
Cadmium sulfide	Yellow overglaze, cadmium yellow, pigment yellow 35, 36	Craft
Calcium carbide	Carbide	Speleologist's supplies
Calcium carbonate	Marble chips, eggshells, snail shells, sea shells, white wash, calcamine	Garden supplies, craft
Calcium fluoride	Fluorspar	Craft
Calcium hexametaphosphate	Calgon	Supermarket
Calcium hydroxide	Hydrated or slaked lime	Garden supplies
Calcium oxychloride	'Solid chlorine' 70%	Swimming pool supplies

TABLE A10.1 (Continued)

Chemical name	Common name	Availability
Calcium phosphate	Bone ash	Craft
Calcium sulfate	Gypsum, modelling powder	Craft
Camphor	Some mothballs	Supermarket
Carbon	Coke, charcoal	'Lead' pencil, craft
Citric acid	Citric acid	Supermarket, health food
Chromic oxide	Chromium green oxide	Craft
Cobalt carbonate	Blue glaze	Craft
Cobalt oxide	Blue glaze	Craft
Collagen	Sausage casings	Butcher
Copper	Copper wire, foil	Electrician, hardware, craft
Cryolite (sodium aluminium fluoride)	Cryolite	Craft
Cupric carbonate	Green glaze	Craft
Cupric oxide	Black copper oxide	Craft
Cupric sulfate	Bluestone	Garden supplies, hardware
Cuprous oxide	Red copper oxide	Craft
Cyanuric acid	Cyanuric acid	Swimming pool water-purifying tablets
Dolomite (calcium magnesium carbonate)	Dolomite	Craft
Ethanol	'Metho', methylated spirits (denatured)	Supermarket, hardware
Ethylene glycol	Antifreeze (contains additives)	Garage
Feldspar (aluminium silicates)	Feldspar	Craft
Fluorspar (calcium fluoride)	Fluorspar	Craft
Gelatin	Gelatin	Supermarket, pharmacy
Glucose	Glucose, dextrose	Supermarket
Glycerol	Glycerine	Pharmacy, hardware
Gum Arabic	Gum Arabic	Health food, craft
Hydrochloric acid	Muriatic acid	Pharmacy, hardware, pool supplies
Hydrogen peroxide	Peroxide	Supermarket, pharmacy
Iron	Nails, steel wool	Hardware, supermarket
Iron chromite	Chromite	Craft
Iron (ferric) oxide	Rust, rouge, red iron oxide	Craft
Iron (ferrous) sulfate	Iron sulfate	Garden supplies
Lead	Lead 'sinkers'	Hardware, sports stores
Lead antimonate	Antimonate of lead, Naples yellow	Craft
Lead carbonate, basic	White lead flux	Craft
Lead dioxide	Car battery plates	Old batteries
Lead oxide, mono	Litharge, flux	Craft
Lead oxide, red	Red lead, flux	Craft
Lead sulfide	Galena	Mineral, rock shop
Lecithin	Lecithin	Health food
Linseed oil	Linseed oil	Hardware, craft
Lithium carbonate	Flux	Craft

TABLE A10.1 (Continued)

Chemical name	Common name	Availability
Magnesium metal	Fire starters	Camping supplies
Magnesium carbonate	Carbonate of magnesia, magnesite	Pharmacy, craft
Magnesium sulfate	Epsom salts	Pharmacy
Manganese carbonate	Manganese carbonate	Craft
Manganese dioxide	Manganese oxide	Used dry cells, craft
Methyl red indicator	pH test solution	Swimming pool supplies
Naphthalene	Some mothballs	Pharmacy, supermarket
Nickel (II) oxide	Green glaze, nickel oxide	Craft
Nickel (III) oxide	Black glaze, nickel oxide	Craft
Para-amino-benzoic acid	PABA	Health food
Phosphoric acid	Phosphoric acid	Hardware
Phthalocyanines	Phthalocyanine green, blue	Craft
Polyester resin	Fibreglass resin	Craft, hobby, speciality suppliers
Potassium aluminium sulfate	Alum	Pharmacy, pool chemicals
Potassium chloride	Salt substitute 'No Salt', 50% KCl	Supermarket, health foodl
Potassium hydrogen tartrate	Cream of tartar	Supermarket
Potassium ferricyanide/ferrocyanide	Case hardening agents	Speciality hardware
Potassium permanganate	Condy's crystals	Pharmacy
Silica gel	Drying agent in packaging (blue/red)	Parcels
Silicon carbide	Carborundum	Hardware, craft
Silicon oxide	Silica, sand	Craft
Silver chloride	Silver chloride glaze	Craft
Sodium bicarbonate	Baking soda, carb. soda	Supermarket, health
Sodium bisulfite	Sodium metabisulfite	Beer/wine kits
Sodium carbonate	Washing soda, soda ash	Supermarket, craft
Sodium chlorate	Sodium chlorate	Hardware (weed-killer)
Sodium chloride	Table salt	Supermarket
Sodium hydrogen glutamate	Monosodium glutamate	Health food, supermarket, Chinese restaurants
Sodium hydrogen sulfate	Pool acid	Swimming pool supplies
Sodium hydroxide	Caustic soda	Hardware
Sodium hypochlorite	Bleach, chlorine water	Supermarket
Sodium perborate	Non-chlorine bleach	In detergent formulations
Sodium silicate	Water glass (egg preservative)	Pharmacy, craft
Sodium sulfate	Glauber's salt	Pharmacy
Sodium tetraborate	Borax	Supermarket, hardware, craft
Sodium thiosulfate	Photographic 'hypo'	Pharmacy, photo supplies
Stannic oxide	Tin oxide	Craft
Starch	Starch	Supermarket
Sulfur	Sulfur	Pharmacy
Sulfuric acid	Battery acid	Garage, pharmacy
Tartaric acid	Tartaric acid	Health food, supermarket
Tin	Modern pewter, can plating	Pewterware
Titanium dioxide	White paint and 'white out' pigments, titanium white	Stationers, hardware, craft.

TABLE A10.1 (Continued)

Chemical name	Common name	Availability
Tungsten	Light bulb filaments	Home, supermarket
Tunsten carbide	Tungsten carbide	Hardware
Urea	Urea	Garden supplies
Zinc	Zinc	Metal casings of used dry cells
Zinc chloride (32% in HCl)	Solder flux	Hardware, craft
Zinc oxide	Zinc oxide	Craft
Zirconium	Zirconium	Photoflash bulbs
Zirconium oxide	Zirconium oxide	Craft

Glossary

Note: *This glossary contains brief definitions of some technical terms and abbreviations. The index will also guide readers to meanings.*

ABC — Australian Broadcasting Corporation (a statutory authority responsible for the national broadcaster).
ACA — Australian Consumers' Association.
acetylcholine — Chemical that transmits a signal from one muscle nerve cell to the next.
acid — Originally a sour material [L *acidus,* sour]; then a substance that releases hydrogen ions in solution and reduces the pH. More generalised definitions are now in use.
acid demand — A measurement of the amount of acid that needs to be added to, say, swimming pool water, in order to lower the pH and total alkalinity to acceptable levels.
acidic — Describes water with a pH below 7.
ACS — American Chemical Society, the largest professional association in the world.
ADI — Acceptable daily intake; see detailed discussion in Chapter 2.
aerobic — Term coined by Pasteur in 1863 to denote bacterial processes occurring only in the presence of oxygen; opposite is anaerobic. [Gk *aero-* (*aer* air), + *bios* life.]
aflatoxins — A very toxic group of substances formed by the mould, *Aspergillus flavus.*
AGAL — Australian Government Analytical Laboratories.
agar (agar-agar) — A gelatinous substance obtained from red seaweed (Rhodophyceae); used in bacterial cultures and in food. [From Malay.]
AGPS — Australian Government Publishing Service. The equivalent in the UK is Her Majesty's Stationery Office (HMSO); in the USA, Superintendent of Documents, US Government Printing Office.
alchemy — Pre-chemistry stage where the chief aim was to transmute base metals into gold.
alcohols — Organic compounds with the functional group –OH (but not when attached to an aromatic ring; they are then phenols).
aldehyde — A compound containing the group –CHO. Coined by Liebig in 1837.
algacide — A chemical that is capable of killing algae.
aliphatic — Term introduced by Hjelt, c. 1860, to distinguish carbon compounds found in fats from those found in aromatic substances; i.e. open-chain hydrocarbons and derivatives, as distinct from derivatives of benzene. [Gk *aleiphar,* oil, fat.]
aliquot — A portion taken for analysis and which is a known fraction of the whole sample. [From L *aliquot,* some, so many.]
alkali — A water-soluble base yielding a caustic solution, i.e. pH > 7. [Arabic *alqili,* the roasted — product of roasting marine plants] hence any substance having properties similar to those of roasted calcium carbonate.
alkalinity, total — A measure of the total amount of dissolved alkaline compounds in, say, swimming pool water. Total alkalinity is a measurement of the resistance of the pool water to a change in pH. Better called *buffer* capacity.
alkaloid — Like an alkali; basic organic nitrogen compounds occurring in plants and having powerful action on animals; e.g. nicotine, morphine, quinine, strychnine, cocaine, curare; but also mild compounds such as caffeine and piperine (in black pepper). [From Gk, used by Dumas in 1835.]
alkanes — Saturated hydrocarbons with the general formula C_nH_{2n-2}, like methane CH_4, ethane, C_2H_6, C_3H_8 and so on. The higher members are solid, called paraffin wax and used for candles. In the UK 'paraffin' means 'kerosene', a middle ranking boiling fraction of petroleum used for heating and jet fuel.
alkenes — unsaturated hydrocarbons that contain one or more double bonds, such as ethene (ethylene), propene, etc. The position of the double bond(s) is indicated by a chain number, i.e. but-1-ene and but-2-ene.

alkyd resin — A group of adhesive and coating resins made from glycerol and unsaturated organic acids.

alkylation — Replacing a hydrogen on a cyclic compound with an alkyl (e.g. CH_3 or longer chain) group.

alkylolamide — An alkylolamine is an alkylol (fatty alcohol) in which one of the hydrogens (on carbon) has been replaced by an amine group. The product from the reaction of this amine part with an organic acid gives an alkylolamide.

alkynes — Acetylenes; unsaturated hydrocarbons with one or more triple bonds. The most common is acetylene (ethyne) C_2H_2.

allergen — A substance, often a protein, that elicits an allergic reaction.

alloy — An intimate association (which may be a compound, solution or mixture) of two or more metals (also used in reference to plastics). [L *ad,* to, + *ligare,* bind.]

alpha particle — Helium nucleus (2 protons plus two neutrons) emitted from heavy atoms during one of the radioactive decay processes.

alum — Name given to certain double salts that crystallise readily as octahedra; e.g. $KAl(SO_4)_2.12H_2O$, common alum. [L *alumen,* bitter salt.]

AMBS — Australian market basket survey; sampling the supermarket for foods corresponding to a range of diets occurring in a multicultural Australia to ensure that levels of additives and contaminants (including pesticide residues) do not exceed the acceptable daily intake (*ADI*).

amide — The group —$CONH_2$; the amine (from ammonia NH_3) replaces the –OH of the acid –COOH (See *alkylolamide).*

amine — The group –NH_2 made by replacement of one or more hydrogens in ammonia by alkyl or other hydrocarbon radicals.

amino acid — An organic acid with an amino group attached. The alpha amino acids are components of proteins.

amorphous — Without crystalline, or regular, structure. [Gk *amorphos,* without shape.]

amphibole — A group of rock-forming minerals of variable composition; Haüy (1743–1822). [L *amphibolos,* ambiguous.]

amphoteric — Capable of acting as an acid or a base. [Gk *amphoteros* — comparative of *ampho,* both.]

amyl — The group C_5H_{11}. [Gk *amylon,* L *amylum,* starch.] An alcohol obtained originally from starch (amyl alcohol); coined by Balard, 1844.

anaerobe — A microbe that thrives only in an oxygen-deficient environment.

angstrom (Å) — The unit of measurement for wavelength, equal to 10^{-10} metres. Convenient for molecular dimensions but being phased out for SI units. [Named after A.J. Ångstrom, 1814–74, the Swedish physicist.]

antibody — Protein essential to the immune system, produced in the plasma specifically in response to bacteria, viruses and other substances. The agent causing this response is called an antigen. Each antibody has a name relating to its specific activity.

antimetabolite — Chemical that mimics (replaces) a natural substance produced (and needed) by an organism but which then causes damage.

antioxidant — A substance that inhibits oxidation (loosely, the reaction with oxygen) of another material to which it has been added.

ANZAAS — Australian and New Zealand Association for the Advancement of Science. Compare BA (British Association), AAA (American Association for the Advancement of Science).

ANZECC — Australian and New Zealand Environment and Conservation Council (the state and commonwealth ministers for the environment).

ANZFA — Australian and New Zealand Food Authority; sets uniform standards for food in both countries including levels of contaminants and additives.

ANU — Australian National University, Canberra, ACT 0200.

arachidic — Arachidic acid, $C_{19}H_{39}COOH$, found in peanuts (ground nuts, genus *Arachis*).

aramid — aromatic polyamide; e.g. Kevlar, Nomex; contrast aliphatic polyamide, nylon.

argon — A colourless, odourless, chemically inactive, monatomic gaseous element found at a level of 1% in the air, used in incandescent light bulbs, lasers and for welding. [Gk *argos,* inactive.]

aromatic — Having planar ring-type groups usually composed of carbon atoms (e.g. benzene and naphthalene) which have alternating double and single bonds (these in fact *don't* alternate but are smeared around the ring uniformly as an average of one and one-half bonds); in contrast to *aliphatic*.

aryl — The term used to refer to a derivative of an aromatic group such as benzene or naphthalene [ar(omatic) + yl].

asbestos — A group of fibrous silicate minerals. [Gk, unquenchable.]

aspirin — Synthetic acetylsalicylic acid ['A' for acetyl and L *spiraea*, plant source of salicylate]; coined by Dreser in 1899 for the synthetic but nature-identical substance found in *Spiraea ulmaria* (willow tree).

atactic — *Isotactic, syndiotactic* or *atactic* refer to the packing of polymer chains and describe situations of maximum, intermediate and random regularity. This in turn affects the physical properties of the material.

atoms — All matter is composed of atoms, of which there are about 90 distinctly different kinds occurring naturally on the earth. In very simple terms, each atom consists of a *positively-charged nucleus* containing protons and neutrons surrounded by a cloud of *negatively-charged electrons*, so that, overall, the atom has no net electrical charge.

atopic — Hereditary tendency to develop an immediate allergic reaction, such as asthma, dermatitis, hay fever, because of the presence of an antibody in the skin.

ATP — Adenosine triphosphate is a molecule that acts as the 'currency' of chemical energy processes in the body. It allows the extraction of energy from food in a manner that the body can utilise, by cycling between a second form called ADP, adenosine diphosphate.

atropine — Poisonous alkaloid found in deadly nightshade (*belladonna*). Used medicinally; e.g. to widen pupils for eye examination.

austenite — Iron in a face-centred cubic (*fcc*) crystalline form. Above around 700°C it can dissolve up to 1.7% carbon to form a non-magnetic steel.

azeotrope — A mixture of liquids that boils at constant temperature. [Gk *a* without, *zein* boil + *tropos* turning, change.] See also *eutectic*.

bactericide — A chemical that is capable of killing bacteria.

becquerel — Unit of radioactivity, symbol Bq, corresponding to one nuclear transformation per second.

bel — See *decibel*.

Bessemer process — Process for converting pig iron from a blast furnace into steel by blowing air (or pure oxygen) into the molten impure metal, which converts impurities into a separating slag, and also adding controlled amounts of carbon.

beta-particle — Emission of an electron (or positron) by a nucleus as one of the processes of radioactive decay.

bifunctional — Describing a molecule with two reactive groups.

billion — The value 10^9, one thousand million, is used in this book.

biocide — Pesticide.

biodegradable — The property of a complex chemical compound able to be broken down into simpler components under naturally occurring biological processes, such as those which form part of the normal life cycle in a river or soil.

biological half-life — The mean time required for half of a quantity of specified material in a living organism to be biologically eliminated.

biological priming — Transformation of an inactive chemical in an organism to an active form.

birefringent — Having a different refractive index for light in different directions.

bonding — Molecules are composed of atoms held together and arranged to satisfy the rules of valency. The forces holding the atoms together in that particular way are the *chemical bonds* and they arise from the sharing of electrons between the atoms. Thus, in the molecule of hydrogen, H_2, the two atoms are bonded together by the sharing of two electrons, one from each atom. We write it H : H, where the two dots represent the electrons, or, more commonly, H–H, where the dash represents the electron pair or the chemical bond between the atoms. A bond of this sharing type is called a *covalent bond*, which is by far the most common kind.

bonding, polar — For hydrogen chloride, we can write H: Cl or H–Cl, but because the two atoms of the molecule are different, the bonding pair is not shared equally; rather, the chlorine has more than its fair share. Because the chlorine not only has its own electron but a share of the electron from the hydrogen as well, it has gained a slight additional negative charge:

$$\overset{\delta^+}{H}–\overset{\delta^-}{Cl}$$

Because this molecule has positive and negative ends, or *poles,* it is said to be a polar molecule, the bond is called a *polar bond,* and the molecule has a dipole. This uneven distribution of electrons can have most important consequences for the behaviour (or *properties)* of the molecule.

bonding, covalent — The bond in hydrogen chloride is a covalent (or sharing) bond, albeit a polar one, but when the molecule dissolves in water, the chlorine atom takes the bonding electron completely from the hydrogen atom, assuming in the process a full unit negative charge (the charge of one electron) and becoming what is called a *negative ion* or *anion.* Similarly, the hydrogen atom has become a *positive ion* or *cation.*

bonding, ionic — Compounds composed of ions are called *ionic compounds* and are said to be held together by *ionic bonds.* As a general rule, ionic compounds are water soluble and fat insoluble, while the reverse is true of covalent compounds. Where a molecule has both types of bonds, intermediate behaviour is to be expected. Sodium chloride, common salt, is another example of an ionic compound. We show this as Na^+Cl^-, even in the solid state.

borax — Sodium tetraborate decahydrate. [Arabic, Persian.]

borosilicate glass — Addition of borate allows the formation of a glass that melts at a lower temperate than silica, and expands less on heating than soda (window) glass, as well as more plastic over a wider temperature range. Examples are Pyrex and glass wool.

BP — See *pharmacopoeia*; British Pharmacopoeia; official listing of drugs, their purity and directions for use; British was one of the first.

bremsstrahlung — X-rays are emitted when a charged particle is rapidly slowed down near an atomic nucleus. Thus electrons are shot at metal targets to produce a broad frequency spread of X-rays.

broad spectrum — Non-selective.

bromine — A halogen, intermediate between chlorine (gas) and iodine (solid); a dense, fuming liquid at room temperature. [Gk *bromos,* stink.]

buffer — A mixture of substances that tend to hinder large changes in acid or basic properties of a solution. Used in a more general sense outside chemistry. See also *alkalinity.*

bumping — Sudden formation of a large amount of vapour from the bottom of a (super) heated vessel of liquid, rather than the usual controlled boiling.

burette — A graduated glass tube open at one end and fitted with a tap at the other.

1,3-butadiene — $CH_2{=}CH{-}CH{=}CH_2$, butane with two double bonds, bivinyl.

butyric acid — C_3H_7COOH [L *butyrum,* butter.]. The acid from rancid butter.

C & EN — *Chemical and Engineering News,* published by the American Chemical Society.

calorimeter — Device for measuring heat output.

carbamate — Ester of carbamic acid; from carb(onic) and am(ide).

carboxylic — Organic acids having the functional group –COOH.

carcinogen — An agent capable of inducing cancer.

CAS — Chemical Abstract Service; produces the unique CA number for each chemical.

casein — The phosphoprotein of milk and cheese. [L *caseus,* cheese.]

catalyst — An agent that speeds up a chemical reaction without itself being used up in the process. Enzymes are catalysts for biological reactions, whereas the transition metals (Co, Ni, Pt, etc.) are common chemical reaction catalysts. [Gk *cata,* down, + *lysis,* a loosening, setting free; coined by Berzelius, 1835.]

catecholamines (biogenic amines) — A series of biologically active amines; e.g. dopamine (3-hydroxytyramine), noradrenaline, adrenaline.

catenation — Formation of chains of atoms.

caustic — Very alkaline — capable of dissolving skin, fat, etc. to form soap.

Cerenkov radiation — Bluish light emitted when a high-energy charged particle passes through a transparent medium at a speed greater than the speed of light *in that medium* (but not faster than the speed of light in a vacuum which is not possible). (Pavel Cerenkov 1934). Analogous to the sonic boom shock wave produced by a supersonic jet.

chalking — Loose pigment powder on the surface of a weathered paint film left by erosion of the outer layer of binder under the action of ultraviolet light. Some chalking is desirable to give a self-cleaning surface.

chamois leather — Leather that can absorb and desorb large volumes of water. It is an oil-tanned sheepskin.

checking — Slight fine breaks in the surface of a paint film visible directly or under × 10 magnification.

Chem. Aust. — *Chemistry in Australia*, published by the Royal Australian Chemical Institute (RACI), replacing the Proceedings of the RACI in July 1977.

chemical active — Active ingredient in a formulation which may contain other chemicals (misleadingly often called inerts) such as solvent (adjuvant), spreading agent, granulated adsorbent, preservative, etc.

chemical equations — Equations are chemical sentences, composed of words (molecules) and conveying information about the transformation of chemical substances; that is, they describe *chemical reactions*. Thus $2H_2 + O_2 \longrightarrow H_2O$ tells us that two molecules of hydrogen combine with one molecule of oxygen to give two molecules of water. These chemical equations are statements of the overall change and make no attempt to indicate *how* the change occurs.

Chem 13 — A publication of the chemistry department of the University of Waterloo in Ontario, Canada.

chemistry — From *alchemy*, the art of transmutation of elements. [Arabic *al-kimiya*.]

Chemistry and Industry — Journal published by the Society of Chemical Industry, UK.

Chemistry in Britain — Journal published by the UK Royal Society of Chemistry and sent to members.

Chem.Tech. — Journal published by the American Chemical Society.

chiral — Any geometrical figure which like a hand, cannot be brought into coincidence with its mirror image, is said to be chiral or exhibit chirality [Gk *chei*, hand.] Coined by Lord Kelvin, 1894.

chloramphenicol — A broad-acting powerful antibiotic, natural and synthetic [named from chlor-amide-phenyl-nitro-glycol]. One of the very few natural products containing a nitro group. It can have severe side effects.

chlorine — A yellowish green gas. Pool chlorine is a compound of chlorine and calcium hydroxide. It is approximately 70% calcium oxychloride. [Gk *chloros*, pale green.]

chlorine, combined — Chlorine that has combined with ammonium compounds or organic matter containing nitrogen to form chloramines. Bactericidal properties of combined chlorine (i.e. chloramines) are only approximately one hundredth that of a similar level of free chlorine in water.

chlorine, free — Chlorine that is not combined with ammonia but is free to kill bacteria and algae in a pool and to destroy organic contamination introduced into the pool water. Free chlorine is also known as 'free available chlorine' and 'free residual chlorine'.

chlorine, total — The sum of combined chlorine and free chlorine.

chlorosis — Plant abnormality in which chlorophyll production is stopped, resulting in a pale yellow coloration.

Choice — Monthly magazine of the Australian Consumers' Association.

cholesterol — A fat-like molecule (chemically not a fat, but an alcohol) with a structure on which all the steroids (e.g. sex hormones, bile acids, vitamin D and cortisone) are based. Produced by the body. [Gk *chole*, gall, bile, + *stereos*, solid.]

chromatography — Literally, the separation of colours. Now a general technique of separation of chemicals based on the difference in the strength of adsorption onto another phase (solid or liquid). [From Gk *chroma*, colour, + *graphe*, writing.]

chrome tanning — Conversion of skin to leather through the application of chromium salts.

chrysotile — Chief asbestos mineral, a magnesium silicate. [Gk *chrysos*, gold; iron impurities give it a yellowish colour.]

cis — On this side of [L]; cf. *trans*, on opposite sides.

citric acid cycle — Also called the Krebs cycle after its principal discoverer. A cyclic set of biochemical reactions in which ADP is recharged to ATP as part of the energy conversion processes in the body.

clostridium botulinum — Spore-forming bacterium that grows anaerobically (in the absence of air) and causes the rare, but often fatal food poisoning called botulism. The spores are heat stable at 100°C and several minutes at higher temperature are required to destroy them.
CNS — Central nervous system; includes the brain and the spinal cord which produces the rapid responses to stimuli.
cobalt — A diluent of silver ores. The ore could not be made to yield a useful metal [Gk *kobalt,* an evil sprite, a goblin.]
Codex Alimentarius Commission — (Codex). United Nations body which establishes international codes of produce and food quality standards.
collagen — A protein in animal connective tissue that gives structural support to the skin. Yields gelatin on boiling. [Gk *colla,* glue, + *gennan,* to generate.]
colligative — Describing properties of a solution which depend on the number, but not nature, of dissolved particles. [L *colligare,* to bind together.]
compounding — A polymer is formed from monomer units and is sometimes called a resin. In order to make a useful plastic material, the resin must be mixed with other materials or *compounded.*
compounds — Substances formed by specific combinations of different elements. The basic combination of atoms characteristic of each compound is called a *molecule,* for which a formula can be written using the symbols of the elements making up the compound. The formula also includes a number to indicate the number of each type of atom present (but not '1', if only one atom is present). For example, carbon, C, forms two compounds with oxygen, O: carbon (mon)oxide, CO, and carbon (di)oxide, CO_2. Water is a compound of hydrogen and oxygen, H_2O.
conjugated — Alternating double and single bonds. Note that polyunsaturated chains have 'cis-methylene interrupted' or 'skipped' double bonds which are not conjugated. In *aromatic* compounds (like benzene) the bonds are not alternating but averaged.
continuous phase/outer phase — The continuous 'outside' liquid that surrounds a second liquid (its droplets being discontinuous) in an emulsion.
co-polymer — A polymer formed from linking two (or more) different monomer types.
copper — The copper ores of Cyprus were the Romans' main source of the metal. [L *cuprum,* from the island of Cyprus.]
corrosion — An example of slow oxidation caused by oxygen or other oxidising agents on a metal.
cosmic rays — High energy particles that enter the earth's atmosphere from space.
crocidolite — Blue asbestos. [From Gk *crocydos,* a nap of woollen cloth, + *lithos,* stone.]
CSIRO — Commonwealth Scientific and Industrial Research Organisation, engaged in research for Australian industry.
CT — *Canberra Times* (newspaper).
curie — The former unit of radioactivity, based on one gram of radium. Symbol Ci.

daughter products — Resulting from the decay or fission of another atom (parent).
decibel (dB) — Logarithmic unit used for human audibility measurements ranging from 1 (just audible) to 120 (just causing pain). [The corresponding linear scale ranges from 1 to 10^{12} change in sound pressure — a doubling of sound *pressure* (anywhere in the range) corresponds to 6 dB.] A doubling of sound *loudness* (anywhere in the range) corresponds to a tenfold increase in sound pressure [20 dB]. A closely related but different decibel scale is used for measuring the output of audio-amplifiers in terms of *intensity* (power). See also *Weber-Fechner law,* and Appendix II.
denature — Destroy structure. The tertiary structure of a protein collapses or is 'denatured' by heating, acid, or agitation in air.
dermal toxicity — Poisonous nature when applied to the skin.
detergent — Synthetic surfactants, not including soaps (which are the sodium salts of natural fatty acids). [L *detergere,* to cleanse, wipe away.]
diastereoisomer — Stereoisomers that are not identical and yet not mirror images, e.g. the *d* form of tartaric acid and the *meso* form. [From Gk *dia,* through; *stereos,* solid; *isos,* equal; *meros,* part.] Not an easy concept!
dilatant — See *Newtonian fluid.*

dioxan — The cyclic ether 1,4-dioxan is a solvent miscible with water. Not to be confused with *dioxin*.

dioxins — Generally refers to a series of chlorinated dioxins such as 2,3,7,8-tetrachlorodibenzo-*p*-dioxin (2,3,7,8-TCDD).

dipstick — A test strip for fast convenient chemical analysis, such as for glucose in the urine.

Ed.Chem. — *Education in Chemistry*, published by the Royal Society of Chemistry (UK).

effective half-life — The time required for the activity of a radionuclide (in a living organism) to fall to half its original value as a result of both biological elimination and radioactive decay.

efflorescence — Crystals containing water of crystallisation lose some of that water by evaporation and begin to powder. Literally 'burst into flower'. Also used to describe the dissolved salts rising to the surface and crystallising (of the ground or a wall).

elastomer — A polymer material with elastic properties, namely the ability to snap back to the original dimensions after distortion.

electrodeposition — Depositing paint by attracting charged paint particles to an oppositely charged (metal) surface to be painted, thus avoiding the use of solvents. Used for cars and whitegoods.

electrolysis — Electricity is carried through a solution by ions. At the electrodes, these (or other) ions are neutralised and discharged as neutral atoms or molecules. [From Gk *electron*, amber, + *lysis*, setting free.]

electromagnetic radiation — The spectrum of visible, ultraviolet and infra-red light, gamma rays, X-rays, microwave radiation, radio waves, etc., all represented by photons (of different energy) travelling at the speed of light. These different regions interact energetically with molecules in different ways and this interaction is one of the most powerful methods of studying those molecules.

element — Elements are substances that contain only atoms with the same number of protons. Consequently, there are only about 90 natural elements. Familiar examples are copper, tin, iron, aluminium, oxygen, nitrogen, hydrogen and carbon. Each element is given a symbol — O for oxygen, H for hydrogen, C for carbon, and so on. These symbols may be regarded as the letters of the chemical alphabet.

emollient — A substance that softens and soothes the skin.

empirical formula — A formula that lists only the types and proportions of atoms present in a molecule. For example, the empirical formula for acetic acid is CH_2O, whereas the *group* formula is CH_3COOH.

emulsions — The suspension of one liquid as fine droplets in another with which it does not mix. Hence also *emulsifiers*. See also *glycerides*.

enantiomer — One of a pair of molecules which are optical isomers of each other (see *chiral)*.

endoscope — Tubular device for looking or sampling inside the body.

entropy — Measure of the change from more to less useful forms of energy. Measure of randomness only in the sense of a ratio of more probable to less probable arrangements (of, say, polymer chains). Measure of a lack of information about a microscopic situation.

enzyme — A biological molecule that can promote or catalyse a particular reaction (to the exclusion of others). [From Gk *en*, in, + *zyme*, yeast. Coined by Kuhne, 1878.]

eosin — Tetrabromofluorescein. [Gk *eos*, dawn.]

epidemiology — The study of disease in populations, with the ultimate aim of establishing cause and effect.

epoxy — Oxygen directly linked to two adjacent bonded carbon atoms forming a triangle. [Gk *epi*, beside.]

equilibrium — In chemistry, this has a very specific meaning and refers to reactions in which the forward and reverse rates are matched so that the composition of the mixture appears unchanging in time. [L *aequus*, equal, + *libra*, balance.]

ergot — A disease of rye caused by a fungus that causes bread made from diseased rye to become poisonous. The word also refers to a toxic substance, which is also used medicinally.

erythema — Surface inflammation of the skin. [Gk *eruthema*, be red.]

essential oils — Natural oils with pleasant distinctive scent secreted from the glands of certain aromatic plants — often containing terpenes (unsaturated hydrocarbons).

esterification — Forming an ester; reaction of an (generally organic) acid(s) with an alcohol. The reverse process is ester hydrolysis, or saponification (the making of soap from fat because fat is the ester of glycerol and three fatty acids).

esters — Combination of (organic) acids and alcohols. The carboxylic esters with short chains are often pleasant smelling. Fats and oils belong within the classification of esters. [G *essig*, vinegar, + *ather*, ether, coined by Gmelin, 1848.]

ether — Compound in which two hydrocarbon groups are linked by one oxygen. [L *aether*, Gk *aither*, pure upper air.]

eutectic — A mixture of two or more substances at the composition yielding a lowest (local) melting point. [Gk *eu*, easily, + *tekein*, to melt.] Contrast *dystectic*, a composition of maximum melting point. See also *azeotrope*.

Faraday's laws — Relate the amount of chemical reaction required to produce a certain amount of electricity and vice versa.

fat bloom — White deposit of fat formed on the surface of chocolate that has not been stored properly.

fatty — Having a long chain of carbon atoms, usually 10–18 members. These chains are the backbone of the fatty acids in fats.

FAO — Food and Agricultural Organisation (of the UN).

FAO/WHO/JMPR — Joint meeting on pesticide residues of the two UN bodies.

FDA — Food and Drug Administration (USA).

ferrite — (Pure) iron without carbon (or iron carbide). *Ferritic* means having the same structure as ferrite but not necessarily pure iron or iron at all.

flammable — Easily set on fire. 'Inflammable' (derived from inflame is philologically more correct but is no longer used.) The opposite is non-flammable.

flammability (explosion) limits — outer limits for the ratio of fuel to air within which the mixture will burn.

flash point — Temperature at which a chemical produces enough vapour to catch fire in the presence of a flame.

flocculate — To coagulate in fluffy lumps. [L *flocculus*, a little flock of wool.] Hence *flocculant*.

fluorescence — The rapid emission of light at longer wavelengths than that which is absorbed; e.g. adsorption of ultraviolet light can yield blue fluorescence.

flux — A substance added to lower the melting temperature in metallurgy (and soldering). [L *fluere*, to flow.] Hence, also, *flux* as a measure of particles flowing per second through a surface of 1 m^2.

Food Technology in Australia — Published by the Council of Australian Food Technologists Association, Inc. (CAFTA), now called *Food Australia*.

free radical — See *radical*.

froth flotation — Use of adsorption of chemicals on solid particles along with a foam to preferentially float off certain minerals and leave others behind.

fuel cell — Device with a cathode and anode, which converts a fuel directly into electricity without burning. The simplest case is hydrogen bubbled over a porous sintered nickel anode in alkali solution, while oxygen is bubbled over a similar cathode separated by a porous membrane. An electric current is produced in an external circuit. Like a battery, except that fuels (methanol) rather than metals are consumed, and the reaction is not reversible.

gas chromatograph — Analytical instrument based on *chromatography* in the gas phase.

galvanise — To cover metal by electrodeposition of zinc. Incorrectly used when referring to covering steel with zinc by other means. [After Luigi Galvani (1737–1798) who investigated electric effects on frogs' legs — jerking their muscles — hence galvanise into action.]

gelatin — See *collagen*, protein from animal tissues; used as a glue. [L *gelare*, to freeze, set solid; hence Italian *gelato*.]

GAP — Good agricultural practice (in the approved use of chemicals) that ensures proper performance of the product while at the same time ensuring that no unwanted side effects occur, such as excess residues in the produce.

GLP — Good laboratory practice is an agreed set of codes of practice for doing things so that the results are believable by others. It is the 'reasonable' behaviour that one professional in the field expects of another. Also GMP — good manufacturing practice.

glycerides — Esters of the tri-alcohol glycerol; sometimes called triglycerides (fats). Monoglycerides are made synthetically and used as emulsifiers. [Gk *glyceros*, sweet.]

glyco- — Sugar attached.

glycosides — Natural compounds linked to glucose, which are easily broken off.

GRAS — Generally recognised as safe; the approach in the US of accepting without further toxicology, materials that have a long history of safe use.

gray — Unit of absorbed dose of radiation corresponding to one joule per kilogram of matter. Symbol Gy.

group formula — Places atoms together in groups that correspond to the grouping in the actual molecule, and uses prefix symbols to give some indication of how the groups fit together. For example, aspirin, $CH_3CO.O.C_6H_4.COOH$.

gutta-percha — Latex of the family Sapotaceae, obtained mostly from the *Palaquium* species. Same chemical composition as natural rubber (polyisoprene) but with a *trans* rather than *cis* geometry of bonding.

haemoglobin — One of a group of globular proteins occurring widely in animals as oxygen carriers.

half-life — The length of time for a substance to drop in concentration to half its original level. Strictly useful only for exponential decay (which never reaches zero) but used more generally.

halogens — The chemical family: fluorine, chlorine, bromine, iodine (and astatine).

hardness, calcium — A measure of the amount of dissolved calcium (and magnesium, iron) compounds in water.

HCB — Hexachlorobenzene. Once used as a pesticide and also formed as a residue from the manufacture of chlorinated solvents. Now a waste classified as a *POP*.

heavy metal — The metals with higher atomic mass tend to form compounds which are more poisonous (e.g. Hg, Cd, Pb, etc.). Some light metals are included in common or legal usage, (e.g. Zn).

Hess's law — Energy content of molecules is additive and so unknown values can be obtained from known values by simple arithmetic.

hiding power — The ability of a pigment to hide a painted surface depends on its refractive index and on the particle size. The measure of hiding power is the area of a black-and-white check design obscured completely by a kilogram of pigment.

high temperature incinerator — An incinerator that operates at temperatures in excess of 1100°C and with residence times in excess of one second. HTI are of various types including rotary kilns and multichambers, and have a variety of gas cleaning add-ons. They are used to destroy hazardous wastes including medical waste from hospitals. They are subject to community controversy because they can produce low levels of dioxins.

histamine — An amine found in human tissue and released upon injury. Histamine is the cause of some of the symptoms of an allergic reaction.

HLB — Hydrophilic–lipophilic balance (hydrophilic = water loving; lipophilic = oil loving). The relative affinity of an emulsifier for the oil phase is expressed as a number ranging from 1 to 20.

HMSO — Her Majesty's Stationery Office (UK).

hood — A fume hood is a roomy enclosed reinforced cupboard with facilities for chemical reactions (water, gas, power) and is used under negative air pressure. It has a vertical sliding front window and often has gloved access for hands.

Hooke's law — Stress is proportional to strain — the amount of stretching depends directly on the amount of pulling.

humectants — Additives for keeping a product moist, or a product for keeping something else (e.g. skin) moist.

hydrogen [Gk *hydros*, water, + *genes*, forming. Compare G *Wasserstoff*, stuff of water.] A chemical element.

hydrogen bond — A special bond in which hydrogen already bonded to an electronegative atom, such as oxygen, nitrogen, or sulfur, forms an additional 'extra' attraction for a second such atom close by. Hydrogen bonds account for the unique structure of water and determine the overall secondary structure of proteins (enzymes) and nucleic acids. The double helix of DNA is held together by hydrogen bonds. [Gk *hydro*, water, + L *geinomai*, produce.]

hydrogenation — Addition of hydrogen to a molecule; converting unsaturated to saturated, reducing double bonds to single bonds.

hydrolysis — Splitting a molecule using a reaction with water.
hydrophilic — Water loving, polar materials that mix with water.
hydrophobic — Water hating, non-polar (often oily) materials that do not mix with water.
hygroscopic — Materials which absorb water from the air. This property depends on how much moisture is in the air. *Hydroscopic* has a different meaning.
hyper- — Prefix meaning high, generally restricted to medical terms.
hypo- — Prefix meaning low, generally restricted to medical terms. Photographic 'hypo' (sodium thiosulfate) has one sulfur atom with a low valency of two.

ICRP — International Commission on Radiological Protection. Sets agreed standards for exposure to radioactivity emissions.
-ide — Suffix used for inorganic compounds containing two elements.
imine — As $-NH_2$ is an amine, =NH became an imine. A compound of $-NH_2$ is amino (amide), so a compound of =NH became an imino (imide).
indole — A low-melting solid with a faecal smell but used in low concentrations in perfumes. [L *ind(igo)* + *ol(eum)*, oil from indigo.]
-ine — As a suffix it can indicate (a) organic base; (b) amino acid; or (c) halogen element. Very versatile!
inert ingredient — Material with no pesticide action, but it may have other effects; see also *chemical active*.
infra-red — The region of the sun's radiation which lies beyond the red colour as seen in a rainbow. It is not visible but is sensed as heat. Just as materials absorb different colours from white light, so they also absorb different sections of infra-red radiation.
inhibit — Slow down a chemical reaction by blocking a part of the mechanism. [L *inhibitus*, curb, restrain.]
initiator — A substance used to start a polymerisation reaction; often a *free radical*.
INN — International non-proprietary names; agreed generic names for drug active components.
inorganic — Chemistry of elements other than those with carbon as the most important constituent.
interstice — Space between packed atoms or ions in a crystal. [L *inter* between, *store* to stand.]
inversion layer — In a normal atmosphere the air is colder, the higher up you are, and less dense because you are further away from the earth's gravity. During the day the ground is warmed much faster than the air by radiation from the sun. The ground then warms the air in direct contact with it. Warm air rises. As the ground-warmed packets of air rise, they expand so as to match the lower density of the air around them (like a hot air balloon). Expansion of an air packet causes it to cool (at about 10°C per km height). However, this rising air may still remain warmer than the air surrounding it, and it then continues to rise, causing unstable turbulent conditions in which the air mixes.
On dry cloudless nights, the ground is cooled much faster than the air by radiating heat out to space. The ground now cools (rather than heats) the air in contact with it, which then does not rise. In contrast, the air is now *warmer* the higher up you are and a stable (non-mixing) situation called a temperature inversion occurs. Pollutants (such as smoke from fires, exhausts from cars, as well as pesticide spray) are trapped in this lower layer. The boundary where the switch of temperature change occurs can be clearly seen from a vantage point above the inversion layer, like a hill.
In the morning, the sun heats the ground and as the day progresses, the ground heats up more quickly, and the unstable conditions begin, any inversion layer is broken, and the usual cycle starts again.
iodophors — A group of compounds that release iodine slowly in solution to act as local disinfectants (in milking).
ion/ionic — Chemical entity carrying electric charge (positive or negative). Ions occur in equal number of oppositely-charged members, either close together, as in solid salt ($Na^+ Cl^-$), or free to wander, as in a salt solution in water.
ionise — When hydrochloric acid gas is dissolved in water, it breaks up into ions; i.e. it ionises. Ions can come together and form an un-ionised compound (not union). [Gk *ion (ienai)*, to go, coined by Faraday, 1834.]
ionising radiation — Radiation with sufficiently energetic photons to cause the atoms/molecules in the medium through which it is travelling to form ions or free radicals. Includes radioactive particle emission (alpha, and beta) and short-wavelength e.m radiation, UV, gamma rays and X-rays. *Excludes* lower energy e.m. radiation such as microwave, radio waves, IR and light waves.

IPM — Integrated pest management; use of a variety of non-chemical methods along with reduced chemical use.
irradiation — Exposure to radiation; usually refers to UV or radioactive radiation.
ISO — International Standards Organisation; sets standards internationally which are taken up more or less by national standards associations such as Standards Australia. We thus have ISO standards, SA standards, BS (British), DIN (German) standards, and so on.
isomerisation — Rearrangement of the geometry of a molecule without changing its overall formula.
isotopes — Atoms with the same atomic number and hence almost identical chemistry, but with different atomic mass, hence often having different radioactive properties.
-ium — Suffix normally used for metals or groups supposedly having properties of metals; e.g. *ammonium*.
IUPAC — International Union of Pure and Applied Chemistry. Sets international chemical standards, and has a naming system for substances.

JCE — *Journal of Chemical Education*, published by American Chemical Society.
joule — Symbol J. SI unit of energy or work. One calorie equals 4.187 joules.

keratin — Fibrous proteins occurring in hair, feathers, hooves and horns, imbedded in a matrix that makes them strong and elastic. The proteins contain sulfur and are held together by disulfide bonds.
ketone — Organic compound containing the group –C=O. [L *acetum*, vinegar, with Gk suffix *-one*, a female patronymic used in chemistry to denote a weaker derivative.] Acetic acid CH_3COOH leads to acetone $(CH_3)_2C=O$.
Kevlar — See *aramid*.
kilowatt — Symbol kW, a unit of power equal to 1000 watts.
kilowatt-hour — Symbol kWh, a unit of energy obtained when 1000 watts of power runs for one hour; equivalent to 3.6 MJ.
Krebs cycle — See *citric acid cycle*.

labile — Unstable, liable to change to another form or to move away.
lanolin — Wool fat (the palmitate and stearate esters of cholesterol). [L *lan(a)*, wool, + *ol(eum)*, oil, fat.]
latex — An emulsion of rubber globules (in water), extended to include globules of synthetic materials such as in paints.
lattice — Regular three-dimensional array of atoms or ions in a crystal. [Compare *lath* (OE *laett*), piece of sawn or split timber in the form of a thin strip used for support for slates, plaster, or trellis, etc.]
Law of Definite Proportions — Atoms react to form molecules in simple ratios reflecting the valencies of the atoms involved (mostly).
LC_{50} — The median concentration in a food (water/air) of a chemical that kills, on average, 50% of the sample of test organisms.
LD_{50} — The median lethal dose (MLD) of a substance. The dose that, on average, produces death in 50% of the sample tested animals, expressed as mg of chemical per kg of body weight.
Le Chatelier's principle — Systems at equilibrium respond to external changes (temperature, pressure, added material) by adjusting so as to annul or tend to annul those changes. First stated in 1888 by Henri Louis Le Chatelier (1850–1936).
leather — Skin that has been chemically treated to make it resistant to bacterial attack.
lecithin — A biological fat and cell wall component with phosphate polar head group.
Leclanché cell — Most common consumer battery with a zinc outer cathode and carbon rod inner anode.
lectin — Proteins that can clump red blood cells in a specific manner like antibodies. Word derived from the L *legere*, to choose. They are found mainly in plants, and a typical example is ricin from castor (oil) beans. Because they are proteins, their activity is destroyed on cooking.
light emitting diode — LED, a semiconductor rectifier that emits light when a voltage is applied.
linoleic acid — An unsaturated fatty acid, $C_{17}H_{31}COOH$. [L *linum*, flax, + *ol(eum)*, oil; hence also *linoleum*.]
lipid — General term for biological fats covering both simple glycerides of fatty acids and more complex forms.
lipoprotein — Combination of a fat with protein found in cell membranes.

litmus — Blue colouring matter from certain lichens used as an acid/base indicator. [Old Norse *litmose*, lichen for dyeing.]

logarithm — Of a number; the power to which a base must be raised to give the number. [Gk *logos*, word, ratio, + *arithmos*, number; thence mod. L *logarithmus*, coined by J. Napier, 1614.]

lyophilic — Solvent-loving; refers to colloid-size molecules (surfactants, proteins, etc.). [Gk *lysis*, a loosening, dissolution, + *philos*, loving.] Hence also *lipophilic*, fat soluble; *hydrophilic*, water soluble and then opposites (see also *lyophobic*).

lyophobic — Solvent-hating. [Gk *phobia*, fear of.]

-lysis — Breaking down, decomposition. [Gk *lysis*, a loosening, dissolution.] Hence *hydrolysis*, breaking down *by* water (hydro); *pyrolysis*, breaking down *by* heat (pyro); *haemolysis*, splitting *of* blood corpuscles (haem); *electrolysis*, splitting *with* electricity; *photolysis*, splitting with light.

M, mega — A million times; 10^6.

macromolecule — A very large molecule, such as a polymer.

magnetron — A modification of a thermionic valve (as once used for radios and amplifiers before the invention of the transistor) designed for generating microwaves.

malic, maleic — Organic acids. [L *malum*, apple.]

malt — Malt is the grain of barley which has been caused to sprout by being kept moist. [Indo-European *mel*, Gk *mill*, L *molere* to grind.]

margarine — Butter substitute.

martensite — Solid solution of carbon in iron formed on rapid cooling, responsible for the hardness of quenched steel, *martensite hardening*. In shape-memory alloys (like Nitinol), a specimen in a martensite condition may be deformed in what appears to be a plastic manner but is actually deforming as a result of the growth and shrinkage of self-accommodating martensite plates. When the specimen is heated to the temperature of the parent phase, a complete recovery of the deformation takes place.

melanin — Polymers derived from the amino acid tyrosine that provide the pigmentation of eyes, skin and hair.

memory effect — See *shape-memory alloy*.

meta- — [Gk *meta*, changed (in form, etc.), next to, between.] Hence meta-substituted benzene, metabisulfite, *metabolism*, metastable form.

metabolic rate — Rate at which the body uses energy.

metabolism — Chemical reactions occurring in a living organism.

metabolite — Breakdown product of a chemical in a living organism.

methylene — Group $-CH_2-$. [Gk *methy* + wine, *hyle*, wood. Methyl alcohol was originally made by distilling wood.]

microcosmic salt — Sodium ammonium hydrogen phosphate, a crystalline salt obtained from evaporation of urine at a time when humans were regarded as the centre (microcosm) of the universe.

microemulsion — Emulsion of two liquids with suitable surfactants which is thermodynamically stable and hence does not break.

mineral spirits (turpentine) — A petroleum fraction boiling between 150°C and 200°C, containing aliphatic hydrocarbons. Abbreviated 'turps'.

miscible — Two liquids are miscible when they mix completely in all proportions. Often liquids are partially miscible. [L *miscere*, to mix.]

MM — Molecular mass (older term, molecular weight, m.w.) is the mass of one *mole* of that material.

mole/moles/mols/mol — Bakers have their dozen (= 13) and so do chemists. The chemist's one is called a mole and contains 6.023×10^{23} single units (atoms, molecules, electrons, etc.). This number is chosen because the number of atoms and molecules always weighs in grams an amount equal to the atomic (molecular) mass. Thus a mole of carbon weighs 12 g and a mole of oxygen weighs 16 g. A mole of carbon monoxide (CO) weighs 28 g. [Coined by Ostwald 1853–1932.]

monomer — A simple molecule that is joined to others (the same or different) to form a dimer, trimer, or polymer.

morphine — An alkaloid, $C_{17}H_{19}O_3N$. [Gk *Morpheus*, god of dreams, son of sleep.]

MRL — Maximum residue limit of a chemical residue, the maximum concentration (in, say, mg of chemical per kg of produce) that is legally permitted in that produce. It is a quality control measure for produce to check whether good agricultural practice (GAP) has been followed in the use of pesticides. Called a 'tolerance' in the USA.
MSDS — Material safety data sheet; occupational health and safety information on a chemical.
msg — Monosodium glutamate; the sodium salt of a natural amino acid that is used as a flavour enhancer in oriental cooking and is accused of causing a reaction in some consumers.
mutagen — An agent capable of causing mutations in the genetic material, which can affect either the organism or its offspring, depending on which cells are affected.
mycotoxin — A poison produced by a mould.

nanometre, nm — One billionth of a metre, 10^{-9}.
napalm — Petrol gelled (originally) with the aluminium salts of naphthalenic and palmitic acids and used in war in flame-throwers and bombs.
neoprene — A chlorinated synthetic rubber (made from 2–chloro–1,3–butadiene) which when vulcanised is very resistant to oils, chemicals, sunlight, ozone, and heat.
neurotoxin — A substance that causes defects in nerve tissue.
neutron — A subatomic particle. Neutrons, along with protons, form the nucleus of atoms. While a proton has a unit of positive charge, a neutron (of almost identical mass) is neutral. Emitted from heavy atoms during one of the radioactive decay processes. Part of the nuclear fission chain reaction.
New Scientist — UK science weekly.
Newtonian fluid — Fluid in which the velocity gradient is directly proportional to the shear stress — the analogue for gooey materials of Hooke's Law for springs. *Non-Newtonian* fluids are complex; two types are described as follows:
(1) *dilatant:* the faster the liquid moves, the *more* viscous it becomes, [dilate, to expand].
(2) *thixotropic*: the faster the liquid moves, the *less* viscous it becomes. [Gk *thixis,* a touch, + *tropos,* a turning, change.]
NHMRC — National Health and Medical Research Council (Australia).
nicotine — Jean Nicot introduced tobacco into France in 1560. [Mod. L *(herba) Nicotiana,* herb of Nicot = tobacco.]
nitrosamines — Compounds produced by the reaction of nitrous acid on (secondary) amines. Suspected of being formed in the gut when nitrites react with amino acids from protein, and possibly initiating cancer.
NOEL — No observable effect level. The highest amount of substance found experimentally to cause no detectable (usually adverse) change of morphology (i.e. in the nature of an organ), functional capacity, growth, development, or life span of the most sensitive test organism. (See also *threshold.*)
NOHSC — National Occupation Health and Safety Commission.
non-miscible — Property of two liquids which do not mix (e.g. oil and water).
non-Newtonian fluid — See *Newtonian fluid.*
non-volatile — Not very volatile. The material has a very low, but not negligible volatility, (in contrast to involatile materials such as granite, which have no vapour at all).
NRA — National Registration Authority for Agricultural and Veterinary Chemicals (a statutory authority responsible for regulating chemicals other than for industrial use or human therapeutics).
NRS — National Residue Survey; samples produce 'at the farm gate' to monitor for good agricultural practice (*GAP*) by checking that the maximum residue limits (*MRL*) of pesticide residues and contaminants in produce (before processing) are not exceeded, which might jeopardise exports and purchase by large domestic retailers (overseas importers and local food chains also carry out their own analyses).
NTP — Normal temperature and pressure (now standard temperature and pressure, (STP)), set as a benchmark so that comparisons can be made for materials under different conditions. It is 0°C and 1 atm. pressure.

OECD — Organisation for Economic Co-operation and Development.
OHS — Occupational health and safety.

oil tanning — Method of tanning sheepskin to produce chamois leather.
olefines; olefins — See *Alkenes.*
oleic acid — A mono-unsaturated fatty acid. [L *oleum,* oil; Gk *elaion,* olive oil.]
organic — Originally referred to chemicals produced by living organisms. Today it simply refers to the chemistry of compounds containing carbon. [Gk *organor,* instrument, bodily organ.]
organochlorine — A compound which is generally composed only of carbon and hydrogen, to which chlorine has been added. Such compounds are often biologically very active but not easy to break down; hence they are persistent.
ortho- — Regular form. [Gk *ortho.*]
osmosis — Selective passage of a solvent molecule, but not solute molecule, through a semipermeable membrane. [Mod. L.]
oxalic acid — The simplest organic acid with two carboxylic acid groups — poisonous. Found in some plants such as rhubarb.
oxidation — Originally reaction with oxygen as in burning. Now generalised and can include reaction with chlorine or other oxidation agents. [Gk *oxys eidos,* species; coined by de Morveau, 1790.] Contrast *reduction.*
oxidation number — A method of bookkeeping in balancing reactions in which oxidation and reduction are simultaneously occurring.

para- — Alongside, beyond, near, contrary to. [Gk *para.*] Very flexible prefix in chemistry.
paraffins — See *alkanes.*
parenteral — (Of drugs) absorbed into the body by a route other than by the intestinal tract.
Pa s — Pascal second, N s m^{-2}, a measure of viscosity replacing the c.g.s. unit, poise (= 0.1 Pa s).
peptide — Molecule formed when two amino acids are joined; hence peptide bond, polypeptide. A protein is a large polypeptide.
periodic table of the elements — A table reflecting the way in which the chemical elements with increasing atomic number show periodic similarity in properties. For example, lithium, sodium and potassium.
persistent organic pollutants — POPs; organic materials that do not biodegrade satisfactorily and can disperse through air and water far from their source. For example DDT.
Perspex — (Plexiglas, Lucite). A transparent polymer. [L *perspicere,* to look through.]
pesticide — A chemical used to kill pests; *insecticide* — kills insects, *herbicide* — kills weeds, *rodenticide* — kills rats and other rodents; *acaricide* — kills spiders, mites, ticks, etc.
petroleum fraction — A fraction of oil selected in a refinery on the basis of boiling point.
pH — A scale (ranging from 0 to 14) that indicates the amount of acid or alkali present in water. Water with a pH of 7 is neutral. Since pH is logarithmic, a solution with a pH of 5 is *ten* times more acidic than a solution with a pH of 6, while a solution with a pH of 4 is *one hundred* times more acidic than a solution with a pH of 6.

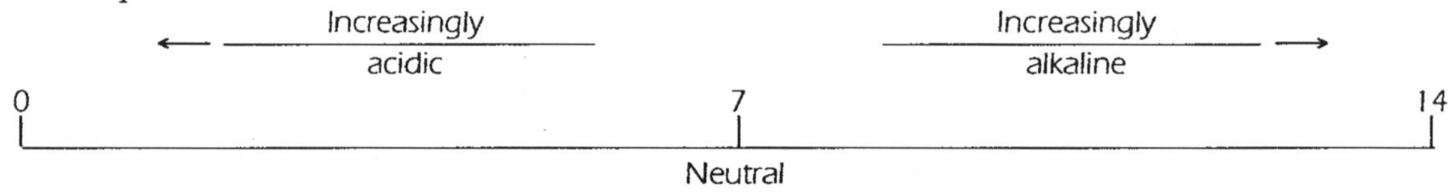

pharmacodynamics — Chemotherapy; attempting to kill target cells in the body (e.g. cancer cells) while sparing normal cells. *Chemotherapy* was actually coined by Paul Ehrlich, (1854–1915, founder of the modern drug use) as 'the use of drugs to injure an *invading* organism without injury to the host'. Common usage has changed this meaning.
pharmacokinetics — The study of the rate at which drugs are absorbed, distributed, metabolised, and excreted in the body.
pharmacology — The study of the action of drugs.
pharmacopoeia — Official listing of drugs, their purity and directions for use; British (BP), was one of the first.

phase — A term introduced by Willard Gibbs in his treatise on thermodynamics (1876–78). [Gk *phasis*, mod. L *plasis*, appearance (of a star).] Thus water can occur in three phases, solid, liquid, and gas. See Appendix III.

phase transfer catalysis — A technique whereby water-soluble and oil-soluble components, each in its own immiscible solvent (one on top of the other), can be reacted by use of a third chemical which 'escorts' the water-soluble component into the oil solvent by formation of an ion pair.

phenols — Aromatic groups such as benzene and naphthalene with the functional group –OH attached; contrast *alcohols*.

pheromones — Chemicals released as specific signals, usually to other members of the same species, to regulate social behaviour such as attracting mates, marking trails, and promoting social cohesion.

phosphorylation — Adding a phosphate group to a molecule.

photodynamic — Caused by light.

photolysis — A chemical reaction brought about by light including ultra-violet light. Radiolysis is the equivalent when radioactive emissions are involved.

photon — A packet of energy of size $h\gamma$, where γ (Gr. nu) is the frequency of the corresponding wave (and h is Planck's constant). A zero rest mass particle (photon) is required for the description of e.m. radiation as both a particle and a wave.

PIC — Prior informed consent; given by an importing country before the export to it of hazardous material can take place.

pK_a — A measure of the degree to which an acid or base will dissociate in water. The (negative) logarithm of the acid dissociation constant K_a. When the pH of solution is at the value of pK_a for a dissolved acid, that acid will be 50% dissociated.

PKU — Phenylketonuria; the inability (through a genetic defect) to metabolise the amino acid phenylalanine; hence causing its build-up in the body and damage to the brain. This compound is present in some foods and artificial sweeteners (e.g aspartame).

plasticiser — This is an additive that makes a polymer material more flexible or less rigid.

poise — See Pa s.

poison — A substance that, when introduced into or absorbed by a living organism, destroys life or impairs health. Also used anthropomorphically by chemists when referring to catalysts, which when 'poisoned' can no longer function.

polyamide — A polymer based on a condensation reaction forming an amide link in an *aliphatic* chain; e.g. nylon.

polyester — A polymer based on a condensation reaction forming an ester link; e.g. terylene, PET.

polymerisation — The process by which single units (monomers) are joined, like linked paper-clips, to form a giant molecule called a *polymer*. If the links are in only one dimension forming a chain, the product is a *linear polymer*. If there are in addition, some cross-links between chains, the product is a *cross-linked polymer*.

polymorphism — Occurrence of a substance in a number of distinct solid forms.

polythene — Polyethylene.

POPs — See *persistent organic pollutants*.

potentiation — Where one substance (with no effect of its own) enhances the effect of one or more other substances (contrast *synergism*).

ppm — Parts per million (by mass), equivalent to a grain of sugar in a cup of tea (very approximate). Now called milligrams per kilogram, or given as 0.0001%.

precipitate/precipitation — To fall out of solution as a sediment. [L *praecipitatus*, thrown down headlong.]

prostaglandins — Biological chemicals based on fatty acids isolated most readily from sperm and prostate glands of sheep, but very widespread in animals. These compounds are hormones with widely differing functions.

protein — Essential constituent of life. Proteins consist of one or more long chains (polypeptides) of amino acids linked in a specific characteristic sequence. [Gk *proteion*, the first place, the chief rank; coined by G.J. Mulder, 1838.]

proton — A hydrogen atom without its electron; that is, a hydrogen ion. Hence protonation, which is adding a proton.

Prussic acid — Hydrogen cyanide (hydrocyanic acid), hence the old term yellow prussiate of soda for the more recent potassium ferrocyanide, and the IUPAC correct potassium hexacyanoferrate (II), (a flow additive for salt).

QA — Quality assurance is a set of requirements that guarantee that procedures have been followed and records are kept, so that each repetition of a practice is the same and gives the same outcome. That outcome may not necessarily be correct or desirable, only reproducible.

quantum mechanics — The theory that describes the behaviour of very small things such as electrons and atoms in terms of waves as well as discrete particles. It becomes identical to the mechanics of Newton for larger objects. [L *quantus*, how much, how many; coined by Planck, 1900.]

racemic — A one-to-one mixture of left-handed and right-handed (*chiral*) forms of the same molecule. Most chemical reactions produce products as *racemic* mixtures, whereas biological reactions generally produce one or the other form only. [L *racemus*, cluster of grapes.]

radical — A group of atoms that behaves like a single atom in a chemical reaction; such as the ammonium radical, NH_4^+. A *free radical* has an unpaired single electron, such as the methyl radical, $CH_3^{\bullet}$.

radioactive decay — Radioactivity is the spontaneous disintegration of certain heavy nuclei through the emission of alpha-particles, beta-particles and gamma rays.

radiology — The study and application of X-rays, gamma rays, etc., in the diagnosis and treatment of disease.

radon — Rn; A colourless radioactive water-soluble gas formed from the decay of radium, which in turn has come from heavier atoms such as uranium and thorium.

RDA — Recommended daily allowance (USA) of vitamins and minerals.

redox — Reduction coupled with oxidation reactions.

reduction — Reduction and oxidation go together. When a material is oxidised by oxygen for example, the oxygen is reduced to water. Such processes are known as *redox* reactions.

refractive index — A measure of the ability of a material to bend a ray of light. The larger the value, the greater is the refraction (e.g. water, $n = 1.33$; crown glass, $n = 1.5$; diamond, $n = 2.42$). It is related to the density of the material.

RI — Royal Institution of Great Britain. Founded in 1799 by Benjamin Thompson, Count Rumford, and has occupied 21 Albemarle St, London since then. Michael Faraday was one of its most important Directors. It is devoted to research and public understanding of science.

R, R′ — The R designates an undefined organic group, in our case generally a hydrocarbon chain. The R′ just says it is not necessarily the same as R.

RSC — Royal Society of Chemistry, London, publishers of *Chemistry and Britain*, *Education in Chemistry*, and many excellent series of books.

SA — Standards Australia, formerly Standards Association of Australia (SAA).

salicylic acid — *o*-hydroxybenzoic acid. Acetylsalicylic acid = aspirin (analgesic). Methyl salicylate = oil of wintergreen (used as a liniment for sore muscles). Phenyl salicylates (salol) = used in sunscreens and as a stabiliser in plastics. Salicylanilide = a compound of salicylic acid and aniline derivatives, used as an antiseptic in soap.

saponify — To make soap from fat. [L *sapo*, soap, + *facere*, to make, do.]

semiconductor — Metals such as copper conduct electricity; non-metals such as sulfur do not. In between are a number of materials such as silicon and germanium which have a low conductivity that (in contrast to metals) increases with temperature. These materials can be 'doped' with certain impurities to increase their conductivity. Doped semiconductors are the basis of the transistor and the photovoltaic solar cell.

sequester — To take out of circulation, to tie up metal ions so that they don't interfere (by precipitating soaps, etc.). [L *sequestare*, to commit for safe keeping.]

sequestering agent — A chemical that ties up metallic ions in solution.

serotonin — 5-hydroxytryptamine transmits signals between nerves in the brain and affects moods. Drugs such as dopamine and LSD affect the level of serotonin in the brain.

shape-memory alloy — An SMA is a metal alloy (nickel–titanium is most common) that below a particular temperature (depending on composition) can be twisted into any shape, but on heating above that temperature reverts immediately to a 'remembered' shape.

shapes of molecules — The arrangement of the atoms in the water molecules, something of the form H–O–H, where one oxygen atom with a valency of two forms two covalent bonds with two hydrogen atoms. These bonds are polar, and we could show this as

$$\overset{\delta^+}{\text{H}}\text{–}\overset{\delta^-}{\text{O}}\text{–}\overset{\delta^+}{\text{H}}$$

But do the three atoms of the water molecule lie in a straight line? If not, what is the shape of the water molecule? (Consideration of molecular shapes and their consequences is known as *stereochemistry*, a most important aspect of the properties of the molecules.)

Consider the three compounds: methane CH_4; ammonia, NH_3; and water, H_2O. Carbon has a valency of four, and in methane the four hydrogen atoms lie at the corners of a tetrahedron with the carbon atom at the centre. Ammonia, containing the central atom of nitrogen with a valency of three, has a similar shape, except that one of the corners of the tetrahedron does not have a hydrogen atom; the ammonia molecule is pyramidal. The water molecule, with only two hydrogen atoms, is bent. [See Plates I and XXV.]

sievert — The unit of radiation dose equivalent corresponding to the absorption of one joule in one kilogram of biological matter, taking into account the quality factor and other modifying factors. Symbol Sv.

SI units — Système international d'unités. The international system of units based on a restricted metric system. It replaces older c.g.s. (centigrade gram second) and m.k.s. (metre kilogram second) schemes, and British units.

solute — Dissolved material.

solvent — Dissolving material (usually liquid).

spectroscopy — The separation and analysis of light into its component parts and study of the interaction of materials with the different regions of light. Now applied over the whole range of electromagnetic radiation from radio waves to gamma and X-rays. The most powerful and most common source of information on the structure of molecules.

spontaneous — In chemistry this has the specialised meaning of referring to a process that has the potential to occur on its own without further input. However, it may occur so slowly as to be unmeasurable. A mixture of LP gas and air will burn *spontaneously*, but to do so requires a catalyst or a match. Otherwise the mixture sits there 'for ever'. [Latin L *spontaneus*, of one's own free will; coined by Hobbes, 1656.]

stereospecific — As applied to polymers, it is the ability to cause a polymer to be formed in a single geometry rather than as a mixture of structures. This often means the polymer can pack together more efficiently to give a material of greater order and higher density. See also *atactic*.

sterilisation (commercial) — Reducing the number of micro-organisms to below a very low, but finite, agreed level.

stick and space-filling model — Used to indicate the actual geometry of a molecule. This may be essential for explaining its biological activity. The stick and space-filling model for aspirin is shown in Plate I.

stoich(e)iometry — Ratio of atoms in a molecule or a reaction. [Gk *stoicheion*, element, + *metria*, process of measuring; coined by Richter, 1792.]

St Petersburg paradox — A change in the perception of the value of money depends (mainly) on the amount of money already there. Birth of the concept of marginal utility in economic thinking. See also *Weber–Fechner law*, *decibel* and Appendix II.

substrate — A basis on which something else is placed; a starting material.

suede — Leather that has been buffed on an emery wheel so that some of the fibres are loosened at one end.

sulfonation — Addition of the function group $-SO_3H$ to a molecule.

sulfonic acid/sulfonate — An organic compound with the functional group $-SO_3H$ or $-SO_3^-$.

supercritical — Above the critical temperature and pressure, there is no longer any distinction between a liquid and its vapour.

superheating — Heating a liquid above the temperature at which it would normally boil and form vapour. See also *bumping*.

surface energy (tension) — The force of attraction for itself that gives a liquid such as water an apparent skin, which contracts so as to form drops rather than sheets (on surfaces it does not wet). On some surfaces (e.g. clean glass) the attraction of the water is greater for the glass than for itself, so the water wets the glass.

surfactant — A molecule attracted to the surface of water and capable of changing the properties of the surface, generally by lowering the surface tension.

synergism — Where two or more substances together produce an effect that is greater than the sum of the individual separate effects. (See also *potentiation*.)

systematic name — For any substance, this name is built up according to strict rules. Each compound has only one correct systematic name, and that name conveys the complete information about the detailed structure of the compound; it is a stylised written description of the chemical structure. For example, Aspro, is a trade name; aspirin, or acetylsalicylic acid, is the generic term; and 2-acetyloxybenzoic acid, is the systematic name.

tannins — Chemicals (polyphenols) extracted from plants that can tan skin to leather. Also found in red wine and tea.

tautomer — When an atom (often hydrogen) moves backwards and forwards between (two) different places on a molecule, the new and original molecules form a tautomeric pair.

tempering — Time–temperature treatment for modifying the mechanical properties of complex materials such as steel and chocolate.

teratogen — An agent capable of inducing foetal abnormalities (monstrosities). [Gk *teras*, monster.]

theobromine — A stimulant related to caffeine, found in chocolate (does not contain bromine).

therapeutic index — A ratio of the toxic dose to the effective dose; the bigger this is, the safer the drug.

thermodynamics — (In chemistry) the study of whether chemical reactions are feasible and what the energy effects will be. No information is obtained on how fast, and hence how practical, any process may be.

thermoplastic — Property of a polymer that softens or melts on heating.

thermosetting — Property of a polymer that once formed does not melt on heating, charring instead. Its structure has cross-linking from one chain to the next.

thio- — [Gk *theion*, brimstone.] A prefix used when one atom is replaced by a sulfur atom. For example, sodium sulfate Na_2SO_4 becomes sodium thiosulfate $Na_2S_2O_3$ by replacing one oxygen by one sulfur.

thixotropy — See *Newtonian fluid*.

threshold — (Of poisons) the maximum level of intake that produces no clinically detectable effect — no response dose. This is not necessarily the level below which no damage is done.

thyroxine — An iodine-containing hormone produced by the thyroid gland that controls the rate of all metabolic processes in the body.

titrate — Determine the strength (concentration) of a solution by finding the volume of a standard solution with which it reacts. [F *titre*, to give title to, to determine strength.]

TLV — Threshold limit value, is a maximum concentration set for a material in air, to protect workers.

tolerance — In describing pesticide residues, this term is used in the US for the maximum residue limit, *MRL*. Tolerance in pharmacology refers to the long-term use of a substance (drug) whereby more is needed to obtain the same effect (but the amount causing toxic effects is often unchanged).

toxicity — The property of poisons.

toxicology — The study of poisons.

toxin — The term originally used for poisons secreted by microbes.

trade mark — A device or word legally registered to distinguish a manufacturer's or trader's goods. It is spelt with a capital letter to distinguish it from a common name; e.g. Coke™ produced by the Coca Cola Company. Both *coca* and *cola* are common names.

trans — On opposite sides. ([L] See also *cis*.).

trans fatty acid — An unusual fatty acid formed as a result of hydrogenation.

UNEP — United Nations Environment Protection Program.

valency — Just as there are spelling rules for putting letters together to form words, so there are the rules of *valency* (or *combining power)* for constructing the formulae of molecules from the atomic symbols. In water, H_2O, for example, hydrogen has a combining power of one, while oxygen has a combining power of two. Thus one oxygen atom may combine with two hydrogen atoms. Elements often show different valencies in different compounds. For example, lead, Pb, forms two compounds with chlorine, $PbCl_2$ and $PbCl_4$, showing valencies of two and four, while chlorine in both compounds has a valency of one. [L *valentia,* strength, capacity.]

vapour — Gas at a temperature below its critical temperature, where it can be liquefied by pressure alone.

varnish — A solution of a natural or synthetic resin in a solvent, sometimes with the addition of a drying oil.

vermiculite — A group of mica minerals, so called because when slowly heated they open into long worm-like structures. [L *vermiculus,* diminutive of *vermis,* worm; hence vermin.]

viscosity — Resistance to change in shape or form of a material — 'internal friction'.

vitamin — One of a group of unrelated chemicals that cannot be produced by the body and must be taken in externally. [L *vita,* life, + G *amin,* amine; coined by C. Funk, 1913.]

volatile — Readily forming a vapour.

v/v — Volume/volume.

waterglass — Colloidal solution of sodium silicate in water; used in chemical gardens, egg preserving, paper sizing, and much else.

'wetting' — The covering of a solid by a liquid with a thin film. The contact angle the liquid makes on the solid is small. See also *surface energy (tension).*

WHO — World Health Organisation (United Nations).

WTO — World Trade Organisation; the organisation that replaced the UN committee, GATT (General agreement on tariffs and trade); regulates disputes about allegations of unfair trade between nations.

w/v — Weight/volume.

w/w — Weight/weight (or more correctly, mass/mass).

xylene — Dimethyl benzenes [Gk *xylon* wood], from destructive distillation of wood.

Young's modulus — Elastic modulus is the ratio of the stress applied to a material to the strain produced. The stiffer a material, the larger is the modulus. See also *Hooke's law.*

Weber–Fechner law — Perceived intensity (S) of a stimulus (I) varies as the logarithm of the physical intensity [$S = K \log I$ + constant] (Fechner). The magnitude of a reference stimulus (I) and the amount of change necessary to produce *a just-noticeable difference.* (ΔI), are related [$\Delta I/I$ = constant] (Weber). These laws are in fact equivalent and so are often combined, as here.

zeolite — A group of hydrated aluminosilicates (natural or synthetic minerals) [Gk *zein* to boil and *lithos* stone; coined by A.F. Cronstedt, 1756, for a mineral species which appeared to boil when heated in a blowpipe.] They retain pores or channels in their crystal structure which can trap a wide variety of ions and molecules. Used in detergents as water softeners, and as catalysts for reforming petroleum products.

Index

A

B

C

D

E

F

I

J

K

L

M

N

O

P

Q

R

S

T

U

V